Proceedings of the 55th Conference of the ISTVS (ISTVS 2025)

Lebanon, New Hampshire, USA
6-9 October 2025

ISBN: 979-8-3313-3088-0

Printed from e-media with permission by:

Curran Associates, Inc.
57 Morehouse Lane
Red Hook, NY 12571

Some format issues inherent in the e-media version may also appear in this print version.

Printed with permission by Curran Associates, Inc. (2026)

For permission requests, please contact the International Society for Terrain Vehicle Systems at the address below.

International Society for Terrain-Vehicle Systems
c/o Cold Regions Research and Engineering Laboratory
72 Lyme Road Hanover, New Hampshire 03755-1290
USA

http://www.istvs.org

Additional copies of this publication are available from:

Curran Associates, Inc.
57 Morehouse Lane
Red Hook, NY 12571 USA
Phone: 845-758-0400
Fax: 845-758-2634
Email: curran@proceedings.com
Web: www.proceedings.com

Proceedings of the 55th Conference of the ISTVS

ISTVS 2025
October 06-09, 2025
Lebanon, NH, USA
https://2025.istvs.org

Proceedings of the 55th Conference of the ISTVS

Additional copies of this publication are available from
https://www.istvs.org/proceedings

Editors

Susan Frankenstein
ERDC – Cold Regions Research and Engineering Laboratory, USA

Massimo Martelli
National Research Council, Italy

Assistant Editors

Maria Stevens
ERDC – Geotechnical and Structures Laboratory, USA

Sally Shoop
ERDC – Cold Regions Research and Engineering Laboratory, USA

Sarah Piedmont
ERDC – Cold Regions Research and Engineering Laboratory, USA

Contents

Conference overview

The 55th Conference of the International Society for Terrain-Vehicle Systems (ISTVS) is an opportunity to present and disseminate the latest scientific and technological achievements in the field of vehicle-terrain systems.

Representatives of science in the fields of mechanical engineering, vehicles and machines, robotics and automation, soil science, civil engineering, agricultural engineering, environmental engineering, as well as professionals from industry, academia and government agencies are invited to participate. The organizers particularly welcome participation of students from under- and postgraduate courses.

The intention of the organizers is to create a friendly atmosphere and comfortable conditions for holding plenary sessions. The conference includes the following thematic tracks:

- Innovative concepts in tires, wheels, tracks, and other ground contact elements
- Propulsion systems and vehicle electrification
- Creative terrain mechanical characterization and deformation modeling
- Advances in off-road mobility modeling
- Off-road autonomy, robotics and navigation
- Planetary terramechanics
- New instrumentation and analysis methods in terramechanics testing
- Climate impacts on terrain, terramechanics, and off-road vehicles
- International collaborations

Conference Chair

Michael Parker | ERDC – Cold Regions Research and Engineering Laboratory, USA

Conference Co-Chair

Sally Shoop | ERDC – Cold Regions Research and Engineering Laboratory, USA

Local Organizing Committee

Michael Parker | Conference Chair | ERDC – Cold Regions Research and Engineering Laboratory, USA

Sally Shoop | Conference Co-Chair | ERDC – Cold Regions Research and Engineering Laboratory, USA

Heidi Howard | ISTVS Treasurer for the Americas | ERDC – Construction Engineering Research Laboratory, USA

Renee Melendy | ISTVS Administration for the Americas | ERDC – Cold Regions Research and Engineering Laboratory, USA

Lindsey Wetzel | Program Committee | ERDC – Cold Regions Research and Engineering Laboratory, USA

Clifford Witte | Program Committee | ERDC – Cold Regions Research and Engineering Laboratory, USA

Conference Scientific Committee

Michael Parker | Conference Chair | ERDC – Cold Regions Research and Engineering Laboratory, USA

Sally Shoop | Conference Co-Chair | ERDC – Cold Regions Research and Engineering Laboratory, USA

Schalk Els | University of Pretoria, South Africa

Susan Frankenstein | ERDC – Cold Regions Research and Engineering Laboratory, USA

David Gorsich | Ground Vehicle Systems Center, USA

Brad Hansen | ERDC – Geotechnical and Structures Laboratory, USA

Henry Hodges | Nevada Automotive Test Center, USA

Heidi Howard | ERDC – Construction Engineering Research Laboratory, USA

Mike Letherwood | Ground Vehicle Systems Center, USA

Massimo Martelli | National Research Council, Italy

Sarah Piedmont | ERDC – Cold Regions Research and Engineering Laboratory, USA

Lutz Richter | SoftServe, Inc., Germany

John Rushing | ERDC – Geotechnical and Structures Laboratory, USA

Maria Stevens | ERDC – Geotechnical and Structures Laboratory, USA

Mehari Tekeste | Iowa State University, USA

Junya Yamakawa | National Defense Academy, Japan

Conference venue

HILTON GARDEN INN
35 Labombard Road
Lebanon, NH 03766, USA

Sponsors

TEAM O'NEIL | Founded in 1997 in Dalton, New Hampshire, operates a 583-acre purpose-built driver training facility specializing in loose-surface vehicle control. As a global leader in rally, off-road, winter, and tarmac driving instruction, Team O'Neil trains individuals, corporations, and government agencies using a progressive, hands-on curriculum. The school's full-time instructors deliver high-quality training across front-, rear-, and all-wheel-drive platforms on varied gravel and paved courses. Team O'Neil also offers security and tactical driving programs, drift training, and corporate or special event hosting. With decades of experience, the team is dedicated to improving driver capability, safety, and confidence in all conditions.
https://teamoneil.com/

NEVADA AUTOMOTIVE TEST CENTER (NATC) | Founded in 1957, is a privately owned, independent test and engineering facility serving global transportation industries. With over 50 million test miles and experience across 1,000+ vehicle systems, NATC advances state-of-the-art test procedures, data acquisition, and analysis. The company specializes in development and certification testing for automotive, commercial, construction, military, and utility applications. Leveraging an extensive database of worldwide duty cycles and operating conditions, NATC delivers accurate, service-representative test programs. Its integrated capabilities – from prototype fabrication to engineering analysis – provide a complete, one-stop solution from design through product verification.
https://natc-ht.com/

Special thanks

The 55th Conference of the ISTVS held in Lebanon, NH, USA in October 2025 was a huge success. In part, this was due to the nearly 125 participants in attendance as well as to organizations who assisted in the background. Specifically, the organizing committee would like to acknowledge:

1. Nevada Automotive Test Center, NATC, contributed considerable financial support as well as staff and expertise in conference hosting and planning, setting up live demonstration events, and off-road autonomy knowledge. The conference would not have been possible without them.
2. Team O'Neil Rally School generously provided the facility, skills, and expertise for live demonstrations as well as discussions on off-road vehicle mechanics and dynamics from expert rally drivers during the technical tour. They also hosted the optional post-conference Rally Experience at their Rally School in Dalton, NH.
3. Algoryx Simulations taught an excellent pre-conference workshop on off-road vehicle and construction simulation.
4. The University of Pretoria Vehicle Dynamics Group generously provided considerable IT help for every technical session as well as technical tour chaperones.
5. Keweenaw Research Center of Michigan Technological University and Mississippi State University stepped in at the last minute to assist where needed. We are extremely grateful.

The shutdown of the US Government on 1 October 2025, along with its precursors, created additional challenges to hosting this conference. The ISTVS community was exceptionally helpful and supportive in overcoming these hurdles. Finally, we also wish to thank our numerous ISTVS colleagues who presented, served as session chairs and tour chaperones, and helped wherever needed to ensure that everyone had an excellent conference experience.

To all of you, our deepest gratitude.

The ISTVS 2025 Organizing Committee

Keynote Speech

MINING & CONSTRUCTION ON THE MOON - TESTING & DESIGN ON EARTH

Dr. Paul van Susante

Assistant Professor, holder of the Lou & Herbert Wacker professorship in Mechanical Engineering, department of Mechanical and Aerospace Engineering; Affiliate Assistant Professor in the Civil, Environmental and Geospatial Engineering department, Michigan Technological University, USA

Dr. van Susante is an endowed assistant professor at Michigan Technological University in the Mechanical and Aerospace Engineering Department and the founder and PI of the Planetary Surface Technology Development Lab (PSTDL), also known as HuskyWorks. Dr. van Susante's research focuses on robotic systems for In-Situ Resource Utilization (ISRU) and construction and in particular, robotic mechanical systems under lunar and mars conditions interacting with the lunar and Mars surface.

He is actively researching, developing and testing technologies under lunar and mars environmental conditions for detecting volatiles in lunar regolith, determining geotechnical properties of lunar regolith, excavating, conveying and compacting (icy, cemented) lunar regolith, excavating (cemented) lunar regolith, excavating gypsum and extracting water from excavated gypsum and buried glaciers for Mars applications as well as developing power transfer via (superconducting) tether into the lunar Permanently Shade Regions.

Dr. van Susante is PI on multiple NASA and industry funded ISRU projects with a combined value of over $5.6M.

More information can be found at https://www.huskyworks.space.

Keynote Speech

A REAL-TIME, WEATHER-INFORMED, CROSS-COUNTRY GROUND VEHICLE MOBILITY ANALYSIS AND FORECASTING CAPABILITY

Dr. John B. Eylander

Research Physical Scientist, Hydrologic Systems Branch, Coastal and Hydraulic Laboratory, U.S. Army Engineer Research and Development Center, USA

Dr. Eylander is a research physical scientist with expertise in developing and managing interdisciplinary research and development programs that integrate weather and terrain data into decision support systems. Dr. Eylander's technical background is focused on weather, surface hydrology, land surface modeling, land-atmosphere interactions, high-performance computing, and satellite remote sensing. Dr. Eylander has led numerous multi-organization research initiatives, supports end-users across the U.S. Government, and works with military and non-military groups supporting weather effects on trafficability and international diplomacy on transboundary hydrology. The interagency and interdisciplinary collaborations have resulted in the development and integration of award-winning numerical modeling and decision support software, such as the NASA-USAF-ERDC Land Information System hydrometeorological modeling and data assimilation framework, the Geospatial Weather Affected Terrain Conditions and Hazards application.

Dr. Eylander collaborates with researchers across multiple US Government organizations, including NASA, NOAA, NSF, and USGS, multiple academic institutions, private sector companies, the national intelligence community, and across military services. He is a member of several national and international committees and collaborates internationally on technical aspects of land data assimilation and hydrologic modeling.

Prior to joining ERDC, Dr. Eylander was the Chief Technology Officer for the formerly named Air Force Weather Agency (AFWA; now 557th Weather Wing) 16th Weather Squadron, serving as the lead technical expert for AFWA's numerical weather prediction system, and leading strategic development and integration of NASA's Land Information System into AFWA operations and AFWA's $200M cloud analysis and prediction system, the Cloud Depiction and Forecast System-Version II. Prior to working for AFWA, Dr. Eylander supported the NASA Goddard Space Flight Center Sea Ice climate research team.

Conference program

Monday, October 06, 2025

08:00 – 12:00 | **ISTVS Board of Directors meeting**

08:00 – 12:00 | **Algoryx Simulation Workshop**

13:00 – 16:00 | **Optional CRREL Tour** (CANCELED due to US Government Shutdown)

17:00 – 19:00 | **Ice Braker Reception**

Algoryx Simulation Workshop

Algoryx

About

This hands-on workshop introduces AGX Dynamics, a powerful multibody dynamics simulation toolkit for virtual prototyping and designing heavy machines and earthmoving equipment.

AGX Dynamics enables the modeling and simulation of complex mechanical systems in motion, with a strong focus on vehicle dynamics and terrain interaction. It supports a range of applications, from engineering analysis to interactive 3D training and AI/ML-based virtual testing environments.

Instructor

Tomas Berglund, MSc Engineering Physics, is a distinguished R&D engineer at Algoryx, renowned for his pivotal role in developing AGX Terrain, a cutting-edge simulation tool for real-time modeling of soil and heavy equipment interactions.

With a strong background in physics and engineering, Tomas has significantly advanced computational modeling techniques. His work includes presenting "Impact force analysis with the nonsmooth discrete element method" at the ECCOMAS Particles 2017 conference, showcasing his expertise in granular dynamics, Algoryx at ECCOMAS 2017.

Tomas's contributions have been instrumental in enhancing the realism and precision of terrain-vehicle interaction simulations, benefiting industries such as construction, mining, and robotics. His dedication to advancing simulation technologies continues to drive innovation in the field.

About Algoryx

Founded in 2007 as a spin-off from Umeå University, Algoryx is a global leader in real-time physics simulation for complex machinery, vehicles, and robotics. Our physics-informed digital twins are essential for innovation in off-highway autonomy, vehicle-terrain interaction, and AI-driven control systems. These simulations support the entire development cycle – from conceptual design and systems engineering to validation, verification, and machine learning-based autonomy.

Algoryx's technology is widely used in construction, mining, forestry, agriculture, and defense, where vehicles must navigate extreme terrains with high precision. By enabling realistic, high-fidelity simulation of soft soil, granular materials, and complex vehicle dynamics, we help industry leaders reduce development time, optimize designs, and train AI-driven control systems in virtual environments before real-world deployment.

Our solutions also extend beyond Earth: the space industry leverages Algoryx's technology for developing and testing robotic systems and autonomous vehicles for planetary exploration, where Earth-based trials are impractical or impossible.

Tuesday, October 07, 2025

08:00 Session 1: General Session
Chair: Sally Shoop

08:00 Conference Welcome
Michael Parker and Dr. Sally Shoop

ERDC-CRREL welcome
Dr. Ivan Beckman, CRREL Director

ISTVS President Welcome
Dr. Lutz Richter

Keynote speech: Mining & construction on the Moon - Testing & design on Earth
Dr. Paul van Susante

10:00 Break

10:40 Session 2A: Planetary Terramechanics 1
Chair: Lutz Richter

10:40 1171 - Effect of stress distribution model in wheel traveling analysis using the extended terramechanics model
Mai Shimizu and Shingo Ozaki

11:00 9165 - Simulation of the relationship between the decreased bearing capacity while imparting vibration to ground and vibration frequency in the discrete element method
Tomohiro Watanabe, Dai Watanabe, and Kojiro Iizuka

11:20 7222 - A simulation framework for autonomous lunar construction work
Mattias Linde, Daniel Lindmark, Sandra Ålstig, and Martin Servin

11:40 1960 - Efficient modelling of the effect of tire deformation during contact with obstacles
Mahdi Maleki and Jozsef Kovecses

11:40 Session 2B: Terrain 1
Chair: Russ Alger and Adrian Doran

10:40 6905 - Investigating data collection and processing strategies for UAV-based terrain models in vehicle dynamics simulation
Herman Hamersma, Christian van Aswegen, Glenn Guthrie, Schalk Els, and Carl Becker

11:00 6965 - Correlation between optical reflectance and load-bearing capacity of soils
György Pillinger and Peter Kiss

11:20 5658 - Winter route planning algorithm
Mark Bodie and Michael Parker

11:40 7265 - Evaluation of a new friction measurement device on Arctic terrain
Sally Shoop, Elder Bruce, Stephen Meurer, Owen Parker, Nathan Kamprath, and Don Halliday

10:40 Session 2C: Modeling 1
Chair: Daniel Carruth

10:40 0375 - A general model to approximate the slippage curve for applications from analysis of the tire and soil contributions to longitudinal slippage
Vladimir Vantsevich, Jesse Paldan, Jordan Whitson, David Gorsich, and Lee Moradi

11:00 1536 - Automated system for determining military road and terrain passability based on terrain, meteorological and hydrological conditions
Krzysztof Pokonieczny, Marek Wyszyński, and Wojciech Dawid

11:20 0954 - Enhancing combined slip manoeuvres in high centre-of-mass off-road vehicles using a semi-active suspension system
Christos Kapareliotis, Schalk Els, and Andries Peenze

11:40 6366 - Simulation-based optimization of a tracked vehicle suspension system
Markus Pogulis, Martin Servin, and Joel Nilsson

12:00 Lunch

13:00 Session 3A: Planetary Terramechanics 2
Chair: Paul van Susante

13:00 4122 - DEM analysis of effect of interparticle forces on sand-outflow behavior from hopper under low-gravity environment
Shunsuke Muryokoji and Shingo Ozaki

13:20 7125 - A data-driven parameter identification method of characterization of wheel-regolith interaction for lunar surface vehicles
Hiroaki Kawamura, Tadashi Matsuura, Shinichiro Noda, Kohei Shintani, Nobutaka Tanishima, and Natsu Fujioka

13:40 6650 - Reliability of gravity offload for lunar rover wheel performance
James Hurrell, Keisuke Takehana, Kentaro Uno, and Kazuya Yoshida

14:00 4368 - Improving the expanding/contracting speed of a push-rolling rover on loose surfaces with steep slopes
Daisuke Fujiwara, Hiroto Fujisawa, Tasuku Hoshino, and Kojiro Iizuka

13:00 Session 3B: Terrain 2
Chair: Schalk Els

13:00 2908 - Drawbar pull experiment and analysis of grousered-track unit on sandy terrain
Ryota Azuma, Tatsuya Maeda, Kosuke Horie, Shinichiro Miyai, Shunsuke Sugimura, and Genya Ishigami

13:20 0270 - Performance assessment of an off-road vehicle across different terrains
Luca Zerbato, Carl M. Becker, Andries J. Peenze, Enrico Galvagno, and P. Schalk Els

13:40 2264 - Mobility testing on frost susceptible soils at field capacity
Michael Parker, Clifford Witte, Allan Wheeler, and Susan Frankenstein

14:00 7475 - Validation of experimental terramechanics modelling approaches through physical testing
Eric Karpman, Jozsef Kovecses, and Marek Teichmann

14:20 0583 - Dynamic off-road route planning: integrating terrain passability and weather for optimal mobility
Wojciech Dawid, Krzysztof Pokonieczny, and Marek Wyszyński

13:00 **Session 3C: Modeling 2**
Chair: Herman Hamersma

13:00 0938 - AI-based terrain passability modeling
Marek Wyszyński, Krzysztof Pokonieczny, and Wojciech Dawid

13:20 4335 - Extended brush model for use with gravelly terrain
Devon Cox, and P. Schalk Els

13:40 3375 - Real-time simulation of flexible tracks for accurate vehicle-terrain interaction
Mahdi Maleki, Wing Hang Ho, Joe Hewlett, Marek Teichmann, and Jozsef Kovecses

14:00 4558 - In-situ soil-property estimation and bayesian mapping with a simulated compact track loader
W. Jacob Wagner, Ahmet Soylemezoglu, and Katherine Driggs-Campbell

14:20 4416 - Prediction of soil rut depth and tire traction forces using Bekker and DEM-FEM techniques
Mehari Tekeste, Pius Jjagwe, Thomas Way, Kumaran Periannan, Raghuram Thiagarajan, Mohamed Abdeldayem and John Sheriff

14:40 Break

15:00 Session 4: Poster Session
Chair: Clifford Witte

8964 - Application of LSTM networks to prediction of a wheel traveling phenomena
Haruki Endo, Shingo Ozaki and Kai Xing

1056 - Optimizing vehicle navigation over terrain
Anthony Farkas, Henry Hodges, and Cooper Madrazo

3332 - Calibration and sensitivity analysis of DEM models using triaxial test of 6 lunar regolith simulants
Isabel Casasbuenas

7256 - Data assimilation analysis of footpad impact behavior of lander on soft ground
Ryuta Kaneda, Masatsugu Otsuki, and Shingo Ozaki

7714 - The role of the evanescent slow wave in non-contact moisture measurement of field trafficability
Zhihe Jin, Joseph Dvorak, and Michael Peterson

4970 - L-Band microwave radiometry for non-contact moisture measurement of field trafficability
Maik Wolleben, Joseph Dvorak, Kaleb Dempsey, and Michael Peterson

16:00 **Session 5A: Autonomy Practitioners**
Chair: Travis Novak and Henry Hodges

16:00 6752 - Foundational software architectures: the cornerstone of off-road autonomy and multi-mission UGV performance
Matthew Foglesong – Parry Labs

16:20 0442 - Lessons learned from global UGV and autonomous systems
Patrick Shepherd – Milrem Robotics

16:00 Session 5B: Terrain 3
Chair: Orian Welling

16:00 0051 - Ground vehicle detection metrics and terramechanical characterization derived from a distributed acoustic sensing system deployed on an Arctic ice road
Aaron Meyer, Adrian Doran, Abhishek Bhagat, Meghan Quinn, Keith Wilson, Matthew Kimball, Jennifer Picucci, Wei Huang, and Russ Alger

16:20 8342 - Effect of snow subduction on vehicle mobility
Mark Bodie and Michael Parker

16:40 9563 - Physical testing and digital twin simulations for prediction of ride quality of light utility vehicle
Eric Karpman, Wei Huang, Nikita Yovchev, and Patrick Kehoe

17:00 0280 - Real-time ground segmentation and terrain topography mapping using point clouds
Jyotirmoy Mukherjee, Giri M Kumar, Andrea L'afflitto, Hannah White, and Corina Sandu

16:00 Session 5C: Modeling 3
Chair: Susan Frankenstein

16:00 4342 - Cornering performance of rigid wheel in granular media using coarse-scale DEM models
Aidan Dickerson, Bohumir Jelinek, George Mason, Thomas Skorupa, Michael Cole, and Jody Priddy

16:20 6010 - Validation of mechanistic DEM calibration
Zamir Syed

16:40 6198 - Vehicle impact analysis using remote datasets
Heidi Howard, Jaime Miranda Ramirez, and Israel Lopez Toledo

17:00 2154 - Simulation of dynamic behavior in armored vehicles using modified off-road vehicles for civil protection applications
Jorge Lopera, and Juan Pablo Angulo

Wednesday, October 08, 2025

Technical Tour

07:30 | **Departure from Hilton hotel**

08:00 – 17:30 | **Technical activities**

Team O'Neil Rally Demonstration

Autonomy Demonstration

The Technical tour takes up the whole day.

The first part consists of three events at the Team O'Neil Rally School asphalt facility in Canan, NH: hands on presentations of Rally Car mechanical enhancements; discussion of on- and off-road vehicle dynamics from expert rally driver and five-time North American Rally Champion, Tim O'Neil, and Team O'Neil CEO Chris Cyr; and a ride-along rally experience putting vehicle dynamics principles into practice!

The second part of the tour is a live Off-Road Autonomy Demonstration of capabilities by Milrem Robotics and a presentation including remote links to live field tests by Field AI.

Team O'Neil Rally Demonstration

Experience professional rally driving techniques

Autonomy Demonstration

Observe autonomous off-road navigation in action

Thursday, October 09, 2025

08:00 Session 6: General Session
Chair: Michael Parker

08:00 Sponsors and Technical tour recap
Michael Parker

Journal of Terramechanics update
Dr. Vladimir Vantsevich

9668 - Innovating mobility: a student competition in wheel and track design
Devin Chen, Chaitanya Shekhar Sonalkar, Riku Kikuta, Andries Peenze, Varsha S Swamy, John Ethan Salmon, Bohumir Jelinek, George Mason, P. Schalk Els, and Corina Sandu

Keynote speech: A real-time, weather-informed, cross-country ground vehicle mobility analysis and forecasting capability
Dr. John Eylander

10:00 Break

10:40 Session 7A: Autonomy 1
Chair: Lutz Richter and Jacob Wagner

10:40 9360 - Optical velocity and slip estimation for off-road UGV control
Glenn Guthrie, Andries Peenze and Schalk Els

11:00 4111 - Off-road autonomy using Koopman theory
Umesh Vaidya

11:20 2824 - Local path planning on rough terrain for unmanned ground vehicle by reinforcement learning considering search reward
Ryosuke Eto, Fushiki Taguchi and Junya Yamakawa

11:40 0733 - Impact of operational environment on mobility in unmanned mode of wheeled UGVs
Marek Nowakowski

10:40 Session 7B: Terrain 4
Chair: Carl Becker

10:40 8352 - Assessing the strength of river sediments for vehicle fording
Sally Shoop, John Eylander, Theodore Letcher and Kathleen Staebell

11:00 4417 - The status of soil physical, mechanical, and chemical properties two years after deforestation in a military training area
Kersti Vennik and Tõnu Tõnutare

11:20 4755 - Relationship between seasonal terrain properties and vehicle rut depths
Susan Frankenstein, Allan Wheeler and Bruce Elder

11:40 6540 - KRC bevameter development and mobility data collection
Russell Alger and Peter Alger

10:40 Session 7C: Modeling 4
Chair: Travis Novak

10:40 6775 - Modeling and simulation of autonomous vehicular systems at ERDC - I Software integration laboratory: development and applications
Peilin Song, Burhman Gates, Garrett Holden, Brent Towne, Joshua Fairley

11:00 1868 - Modeling and simulation of autonomous vehicular systems at ERDC - III Verification and validation of m&s for autonomous vehicles
Peilin Song, Burhman Gates, Garrett Holden, and Joshua Fairley

11:20 6001 - Modeling and simulation of autonomous vehicular systems at ERDC - II Soft soil model Support in vehicle dynamics
Peilin Song, Burhman Gates, Garrett Holden, Brent Towne, Joshua Fairley

11:40 3559 - Modeling light detection and ranging (LiDAR) response in snowy environments
Brian Brady, Aaron Meyer, Sergey Vecherin, Mike Parker, and Han Nguyen

12:00 Lunch

13:00 Session 8A: Autonomy 2
Chair: Dan Carruth

13:00 9257 - Acoustic winter terrain classification for offroad autonomous vehicles
Anthony Fragoso

13:20 3985 - One AI driver for multi-mission ground Operations
Bobby Bonello and Rob Brown

13:40 2469 - TERRA-V toolchain: virtual validation and optimization of autonomous vehicles in unpaved terrain
Gerhard Skoff, Daniel Hassler and Gernot Hasenbichler

14:00 5493 - Effects of electronic control systems on autonomous vehicle performance in granular terrain
Bohumir Jelinek, Aidan Dickerson, Angela Card, Greg Henley, George Mason, John Ethan Salmon, Ch. Michael Gibson, Tyler Hannis, Tom Skorupa, Michael Cole, Jody Priddy, Miriam Figueroa-Santos, Sara Boyle and Jeremy Mange

14:20 8969 - Survey of LiDAR effects for autonomous driving simulations
John Kaniarz, Brian Brady, Laura Walizer, Phillip Price and Mike Parker

13:00 Session 8B: Tires 1
Chair: Travis Novak and Clifford Witte

13:00 2458 - Obtaining first order tyre parameters with insitu testing
Carl M. Becker and P. Schalk Els

13:20 4579 - Analysis of passenger tire-salted pavement interaction: a traction perspective
Erin Fenton and Zeinab El-Sayegh

13:40 7559 - Dynamic contact patch estimation from 3D contact patch measurements inside a moving tyre under combined slip conditions
Amal Patel and Schalk Els

13:00 Session 8C: Deformation 1
Chair: Corina Sandu

13:00 0398 - A contact dynamics approach to analysing slope collapse triggered by heavy vehicles
Tomas Berglund, Joakim Rydman and Martin Servin

13:20 8140 - Numerical modeling of plate sinkage and in situ bevameter shear testing using FEM and the density-dependent NorSand critical state constitutive soil model
Ray Kruger, Petrus Arnoldus Crous, Yuderka Trinidad Gonzalez and Mehari Tekeste

13:40 7470 - Constraint-based terramechanics simulation for realtime and faster simulation
Björn Ahlman, Tomas Berglund, Mikael Lundbäck, Pontus Nordin, Petter Persson, Joakim Rydman, Viktor Wiberg and Martin Servin

14:00 4217 - A computationally efficient plug-in for highfidelity off-road vehicle dynamics studies on deformable terrains
Amir Khosravian, Sadegh Yarmohammadisatri, Corina Sandu, Yinglong He and Davide Tavernini

14:20 Break

15:00 Session 9: Awards, Closing Ceremony, Leave for Banquet
Chair: Michael Parker

Friday, October 10, 2025

Post-Conference Events

Team O'Neil Rally School Tour

You are invited to enjoy a free Off-Road Rally Experience at the Team O'Neil Rally School in Dalton, NH!

Two time slots available: 10:00 - 12:00 and 14:00 – 16:00.

Autonomy Workshop

07:00 **Check-in**

07:30 **Dr. David Gorsich, GVSC**
Moderator and Keynote Address
Off-road Ground Vehicle Mobility and Assessment

08:10 **Vendor Overviews**

08:30 **Dr. Tulga Ersal, University of Michigan**
Panel Discussion on Classical & Hybrid Approaches

09:30 **Dr. Gaurav Pandey, Texas A&M University**
Panel Discussion on Machine Learning Approaches

10:30 **Break**

11:00 **Breakout Discussion Sessions**
Pros, Cons, and Recommendations on Machine Learning, Classical & Hybrid Approaches

12:00 **Leader-Led Breakout Summary**
Recommendations and next steps

12:45 **Drs. Schalk Els and Lutz Richter**
Publication lead authors discuss and formulate outline of the peer review publication and solicit/assign author and reviewer participation

14:00 **Break**

14:30 **Drs. Schalk Els and Lutz Richter**
Continuation of peer review publication outline and section assignments

The results of the conference autonomy demonstrations, technical presentations, and workshop discussions will be synthesized in a peer-reviewed publication in the *Journal of Terramechanics*.

This publication on the State of Off-Road Vehicle Autonomy will serve as a foundational technology reference, capturing the current performance of off-road autonomous systems, defining capability benchmarks, and identifying science and technology gaps most worthy of continued investment.

List of submissions

Full papers

Abstract-only

ISTVS

Conference papers

0051 | Ground vehicle detection and environment characterization derived from a distributed acoustic sensing system deployed on an Arctic ice road

Aaron Meyer [a,*], Adrian Doran [a], Abhishek Bhagat [a], Meghan Quinn [a], Keith Wilson [a], Matthew Kimball [a], Jennifer Picucci [b], Wei Huang [c], Russ Alger [d]

[a] *Cold Regions Research and Engineering Laboratory, U.S. Army Corps of Engineers Research and Development Center, Hanover, NH*

[b] *Geotechnical and Structures Laboratory, U.S. Army Corps of Engineers Research and Development Center, Vicksburg, MS*

[c] *National Research Council Canada, Ottawa, Canada*

[d] *Michigan Technological University, Houghton, MI*

* *Corresponding author: aaron.c.meyer@usace.army.mil*

ABSTRACT

We present results from vehicle testing conducted on an Arctic ice road in February 2025 and monitored with a fiber optic distributed acoustic sensing (DAS) system. The DAS sensing element, a 1-km telecommunications-grade fiber-optic cable, was deployed in three configurations: surface-laid on snow, surface-laid on ice, and frozen into ice. We utilized an amplitude-only DAS system that recorded strain-rate time series with a channel spacing of 5 m. Detection metrics as a function of deployment configuration were derived against multiple vehicles transiting at variable speeds. Detection statistics are shown to vary with deployment conditions and target characteristics, suggesting that sensor performance can be optimized by accounting for environment and source type. Additionally, we used the fiber-optic sensor and vehicle excitations to interrogate the propagation medium. We present evidence that observed differences in frequency-dependent transmission loss can discriminate DAS cables embedded in ice and snow. This work resulted from an international collaboration enabled by the International Cooperative Engagement Program for Polar Research (ICE-PPR).

Keywords
detection
Arctic
ice
snow
vehicles
mobility
distributed acoustic sensing

1. Introduction

Ground vehicle mobility in the Arctic presents a unique suite of challenges (Shoop, 2024). There is a diverse set of Arctic surface conditions, including wind-swept tundra, marshy peatlands, frozen rivers and lakes, among others. To complicate matters, each environment may experience extreme seasonal changes or change substantially on much shorter time scales (Parker et al., 2021; Parker et al., 2024). Monitoring and forecasting of mobility conditions in the Arctic is therefore highly desirable. The current methods to characterize dynamic frozen soil conditions rely heavily on coarse surface models, that have noted deficiencies (Letcher, 2024).

We present here an exploratory investigation of the use of a fiber optic distributed acoustic sensing (DAS) system to fill the gaps in our monitoring and prediction capabilities for ground vehicle mobility in the Arctic. To understand the viability of DAS for Arctic mobility monitoring, we first investigate if the system can robustly detect the vehicles of interest in these austere conditions. We present results for a vehicle detection study conducted along an Arctic ice road with a DAS system.

A DAS is a modern acoustic sensor that consists of a central processing unit, called an interrogator unit (IU), and a fiber optic (FO) cable (Parker et al., 2014). When a ground vehicle generates vibrations in the air or ground, those vibrations may transmit into the medium where the DAS cable is emplaced, coupled to the cable itself. Those vibrations cause a local strain of the cable, which is measured by the IU. For detecting ground-based sources, a DAS cable is typically buried in shallow ground. However, in recent work, a surface-laid cable is shown to still be an effective sensor (Quinn et al., 2025). A surface-laid cable can be rapidly and easily deployed. This is a notable practical advantage, especially in Arctic conditions. We therefore considered three different cable environments or mediums in our study: surface-laid on snow, surface-laid on ice, and frozen into ice. For the results presented in proceeding sections we refer to these three as simply "Snow" (surface-laid on snow), "Ground" (surface-laid on ice), and "Ice" (in ice), respectively.

The first goal of the study is to investigate the detectability of each ground vehicle for each cable environment type. The detectability is a function of two primary variables: vehicle speed, and vehicle-sensor separation distance. We present receiver operating characteristic (ROC) curves for each vehicle tested, within each medium, and for varying speeds and distances. The secondary goal is a corollary to the first: to understand if the ground vehicle signals can characterize the cable's medium. Regarding the second goal, a similar investigation has been recently published (Yuan et al., 2024), although the environment, methods, and objectives are different than our own.

The next section explains in more detail the DAS system, the geometry of the experiment, and the three ground vehicles that were tested.

2. Experiment

2.1. Distributed Acoustic Sensing System

A DAS system consists of a fiber optic cable connected to a laser interrogator unit. In this study, an OLA-2 interrogator unit manufactured by OptaSense was used. The IU sends pulsed laser light down the length of the FO cable. Local strains of the cable cause Rayleigh backscatter of the laser light, which is interpreted and recorded by the IU. The device works on the principles of

https://doi.org/10.56884/ISF3IT7O

interferometry, meaning the measured strains are calculated using phase delays observed in the backscattered laser light.

The measurement outputs are discretized along the length of the cable into channels. Each channel measures the average strain over a discrete segment, called a gauge length. The time series recorded for each channel are a strain-rate, in units of micro radians per unit time. A 5-meter gauge length and spatial sampling were used in our experiments. The OLA-2 is an amplitude-only DAS system. This means the times series measured on neighboring channels are not coherent or related by the phase of their signals. More sophisticated DAS systems can resolve phase, but the capability is not necessary for a detection study of this kind.

The OLA-2 was connected to an ordinary telecommunications-grade FO cable. The total length of the cable was approximately 1 km. DAS systems can, however, interrogate cables that are 10s of kms long. The long, linear span of a DAS has the potential to provide a useful compromise between the wide-scale, but coarse surface models, and the very limited in-situ surface observations. DAS systems are simultaneously wide aperture and have fine spatial resolution. Furthermore, once installed into the ground, the cables are essentially permanent fixtures, that can remain operable for years (Quinn, 2022, Chapter 4). Finally, the cables themselves require no active power. The only powered component is the interrogator unit, which can be located many miles away from any given cable channel. These are important practical advantageous for the Arctic environment, where power usage and durability of sensors are chief concerns.

2.2. Test Geometry

Figure 1 shows the geometry of the experiment. The vehicle routes are shown as the pink lines, the location of the DAS cable channels is depicted by the grey circles, and the camp that housed the interrogator unit is annotated in the scene as well. The vehicle routes extend at maximum approximately 1 km away from the DAS channels.

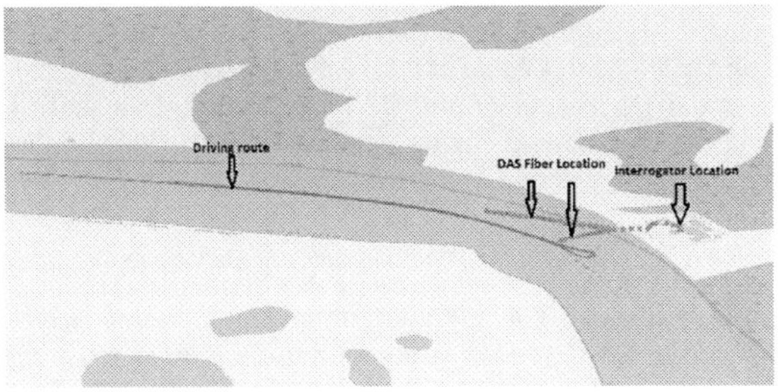

Fig. 1. Experiment geometry. DAS cable is seen as the grey circles, and driving routes are given by the pink lines. The extent of the pink lines is approximately 1 km.

The cable was laid within two snow trenches. One snow trench was short and perpendicular to the road, and the other was longer and orientated at a diagonal to the road. The trenches were approximately 1 meter deep, dug atop the river ice. The trenches are connected, forming a "V" shape. The two arrows labeled "DAS Fiber Location" in Fig. 1 point to the short and long trenches.

Fig. 2. View of the shorter of the two snow trenches. Each snow trench was approximately 1 meter deep. The vehicle tracks can be seen at the bottom of the image.

Figure 2 shows a view of the short snow trench as it was dug. The longer snow trench can be seen in the foreground of the picture. The bottom of the image shows a part of the ice road where the vehicles were driven. The vehicle tracks are visible moving left-to-right.

Along the length of the short snow trench a groove was cut into the river ice. A section of the FO cable was placed into the groove, then water was poured back over and allowed to refreeze. The sections of cable frozen into the river ice are the "Ice" channels in our study. The surface-laid on ice or "Ground" channels were simply laid onto the bare ice floor of each snow trench. The "Snow" channels were laid on top of the snowpack adjacent to and along the length of the long snow trench.

2.3. Vehicles

We tested three different ground vehicles: an MRZR Alpha, a 2025 Toyota Tundra, and a Lynx Brutal Snowmobile. Figure 3 shows each test vehicle. The vehicles weighed approximately 2200 kg, 2700 kg, and 250 kg for the MRZR, Toyota, and Snowmobile, respectively.

Each vehicle performed the same battery of fixed-speed driving tests. The vehicles traversed back-and-forth by the DAS array at 10 kph, 20 kph, 30 kph, 40 kph, and 50 kph. For each speed run, the vehicle continued down the ice road away from the DAS until the apparent signal disappeared. The vehicle then reset and repeated the run. Each speed run was repeated between 10-15 times to acquire good statistical sampling.

Fig. 3. The three test vehicles: An MRZR Alpha (A), 2025 Toyota Tundra (B), and Brutal Snowmobile (C).

The dataset we collected is multi-faceted. Phenomenology related to vehicle type, cable environment type, signal spectra, vehicle-sensor separation distance, and vehicle speed can all be explored. In the next section we provide a succinct set of results highlighting our most important findings. These results, however, only represent a fraction of the analyses and information that could be obtained from the dataset.

3. Results

In this section we present the main results of our analyses, primarily in the form of ROC curves. The ROC curves are calculated for various fixed vehicle speeds, vehicle-sensor separation distances, vehicle type, and environment type.

3.1. Receiver Operating Characteristic Curves

ROC curves are a standard and effective way of summarizing the detection performance of a given sensor, environment, and target. An ROC curve is traditionally plotted with probability of detection (Pd) on the y-axis and probability of false alarm (Pfa) on the x-axis. To calculate an ROC curve, the noise and signal power distributions must be known. These could be derived in theory, or measured empirically, as is the case for this study. Each point on the curve represents a detection threshold, which is a hidden dependent variable. When the detection threshold is varied, the probability of detection and probability of false alarm both simultaneously change. An ROC curve begins at the origin, Pd = Pfa = 0, which corresponds to the limit of high threshold, and ends at Pd = Pfa = 1, which corresponds to the limit of low threshold.

A simple detection algorithm, used in this work to compute the empirical ROC curves, is described by the following pseudo code:

```
(define detection threshold)
For (each time step)
    (calculate signal power)
    (check if calculated power is above threshold: logical 1)
    (check if test vehicle is present: logical 2)
    If (logical 1 = true) & (logical 2 = true)
        (assign event to be true positive)
        (save vehicle-sensor separation data)
        (save vehicle speed data)
    elseif (logical 1 = true) & (logical 2 = false)
        (assign event to be false positive)
    elseif (logical 1 = false) & (logical 2 = true)
        (assign event to be false negative)
        (save vehicle-sensor separation data)
        (save vehicle speed data)
    elseif (logical 1 = false) & (logical 2 = false)
        (assign event to be true negative)
    end
end
```

After executing the above code, probability of detection may be calculated as

$$Pd = \frac{N_{TP}}{N_{TP} + N_{FN}}, \qquad (1)$$

and probability of false alarm calculated as

$$Pfa = \frac{N_{FP}}{N_{FP} + N_{TN}}, \qquad (2)$$

where N_{TP} , N_{FP} , N_{FN} and N_{TN} are the number of true positive, false positive, false negative, and true negative events, respectively.

ROC curves can be obtained by using the above procedure while varying detection thresholds incrementally through the relevant range of signal and noise powers. The functions of Pd and Pfa derived are then filtered by vehicle type, vehicle speed, and vehicle-sensor separation distance.

Often plotted with the ROC curves is a diagonal dashed line with a slope of 1. This line represents a random detector whose performance is equivalent to random 50-50 guessing. The ideal detector would have a value of Pd = 1, Pfa = 0, which graphically is in the upper left corner of an ROC plot. Therefore, sensors with ROC curves that have a higher "elbow" towards the upper-left corner have better detection performance.

3.2. Cable Environment and Vehicle-Sensor Separation Distance

Figure 4 shows three plots containing nine ROC curves. It is the main result of the detection study. The three plots correspond to the three test vehicles, indicated by the title of each plot. The ROC curves are calculated for three separate vehicle-sensor separation distance bins, called "Near" (solid lines), "Mid" (dashed lines), and "Far" (dot dashed lines). The vehicle-sensor separation data, which is continuous, needs to be averaged into these bins to obtain adequate statistical sampling when calculating Pd and Pfa. The distance ranges of each bin are given as an inset in each plot. The curves labeled "Ground" (black lines), "Snow" (red lines), and "Ice" (blue lines) correspond to the surface-laid on ice, surface-laid on snow, and encased in ice channels, respectively. All curves are calculated for the 30 kph speed trials. Similar sets of curves were obtained for the other speed trials, but the main inferences and conclusions can be conveyed from the 30 kph results.

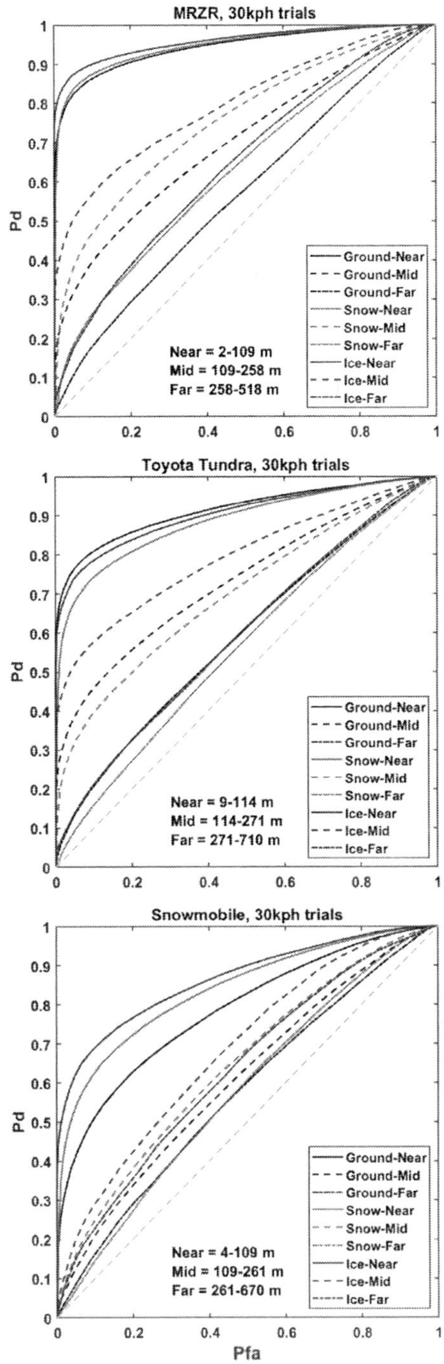

Fig. 4. Sets of fixed-speed ROC curves for the MRZR (top plot), Toyota (middle), and Snowmobile (bottom). Each cable environment is depicted by the color (black = Ground, red = Snow, blue = Ice). Three vehicle-sensor distance bins, called "Near", "Mid", and "Far", are given by the solid, dashed, and dot-dashed lines, respectively. The ranges of each bin are provided in the plot insets.

The ROC curves in Fig. 4 demonstrate that the overall best detection performance was achieved by the ice channels. There are two plausible explanations for this outcome. First, the ice channels experienced the lowest amount of acoustic noise from the surrounding environment. Although care was taken to collect data during quiet moments at the camp, acoustic noise generated from human activity in the area is still present. The ice encasing those DAS channels reflects much of this acoustic energy, effectively lowering the noise floor on those channels, and improving the rate of false alarms. Second, the coupling of the cable to the medium is likely most consistent in the ice, as opposed to the bare cable laid on top of the ice or snow. Better coupling enhances the transmission of seismic vibrations to those cable channels.

Figure 4 also shows that the detectability of the vehicles degrades expectedly with greater distances. The Mid curves lay in between the Near and Far curves, except for the snowmobile, where there is some overlap between Mid and Far. The overlap is evidence that even at the Mid ranges the snowmobile signal is too weak to be easily distinguished from the noise. Evidently in Arctic conditions, at ranges >250m the detection of the vehicles with a surface laid DAS system becomes unreliable. The complete story is, however, more complicated. In the next subsection we demonstrate detectability is also a function of vehicle speed.

3.3. Effect of Vehicle Speed

Figure 5 shows a representative set of variable-speed ROC curves, for the same vehicle, at fixed distance bin and channel environment. For the slowest speed we tested, 10 kph (black line), the Toyota was hardly detectable. The detectability gradually increases with increased vehicle speed, as indicated by the rising elbow of the ROC curves.

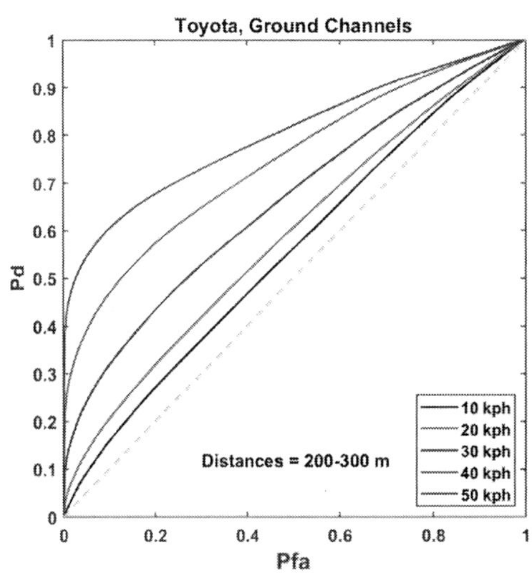

Fig. 5. Fixed-distance ROC curves for different vehicle speeds. Faster vehicles are more detectable, as indicated by the rising ROC curves.

Vehicles moving at faster speeds produce louder engine sound, noise generated by the tire-road interaction, and seismic waves imparted into the ground by frictional forces. All three types of energy caused by the motion of the vehicle contribute to the strain measured by the DAS. Interestingly, we have uncovered evidence that suggests for some distance ranges and cable environment types, one contribution may dominate over another. This is shown in Fig. 6 in the next section.

3.4. Vehicle Comparison

The top plot of Fig. 6 shows a direct comparison of the three vehicles for the same cable environment type (Snow), the same distance bin, and vehicle speed. The MRZR was the most detectable, followed by the Toyota Tundra, and then the snowmobile. The Snow channels were elevated off the ice surface by the 1-meter snowpack. Therefore, we expect very little seismic energy to contribute to the Snow channel signals. The Snow channel signals appear to be dominated by acoustic waves traveling through the air. On this point, when both vehicles are idling, the MRZR is noticeably louder by ear than the Toyota Tundra. Modern commercial vehicles like the Tundra are engineered to be quiet, whereas the MRZR is not. This fact is borne out in the top plot of Fig. 6.

Now contrast the pattern seen in the top plot with the bottom plot of Fig. 6 The three ROC curves again represent the three vehicle types, but the results pertain to the Ice channels, and for a more distant vehicle-sensor distance bin. Note that now the Toyota Tundra is slightly more detectable than the MRZR. The change can be explained by considering the seismic contribution to the received signal. As mentioned earlier, the Ice channels are more shielded from acoustic waves compared to the surface-laid snow and surface-laid ice channels. They are, however, well coupled to the river ice itself, where the seismic waves traveling through the ice can be intercepted. Therefore, we suspect the heavier Toyota Tundra is transmitting a comparable amount, or more energy into the ice than the MRZR.

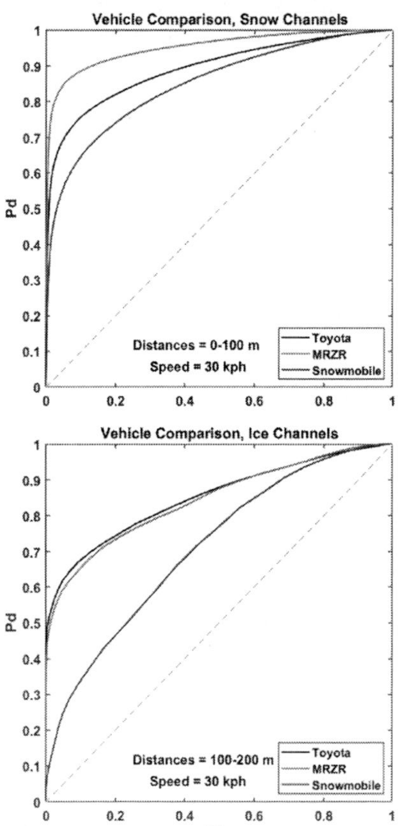

Fig. 6. Vehicle comparison ROC curves. Top plot shows detectability of each vehicle in the DAS Snow channels, for fixed speed and distance. Bottom plot shows the same results for the DAS Ice channels.

Such a result suggests certain cable environments and installations are more suitable for specific source types. For example, a DAS system intended to detect slower, heavier targets may be installed within a different medium, and/or configurations than a system intended to detect light, fast moving targets.

3.5. Environmental Characterization

Figure 7 shows power spectrums of the signal generated by the Toyota Tundra during the 40 kph trials. The signal power is calculated for seven consecutive 1/3 octave bands (solid lines), each plotted as a function of vehicle-sensor distance. The center frequency of each band is given in the plot legends. The curves in the top plot are calculated for the Ice channels, and the curves in the bottom plot are calculated for the Snow channels. Figure 7 shows how much signal energy is contained in each frequency band, and how each attenuates with range. The horizontal dashed lines in each plot show the 75th percentile of the noise, calculated for the same frequency bands with matching colors. They are provided to illustrate where the descending signal curves merge with the noise floor, providing an effective range of detection for each frequency band.

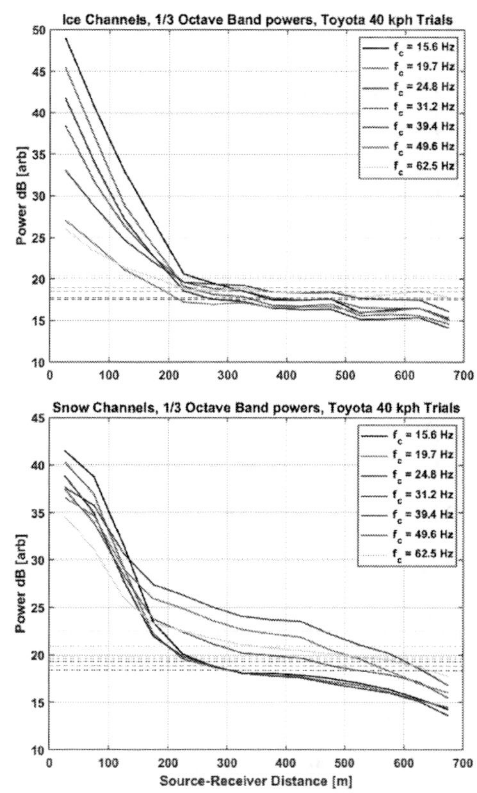

Fig. 7. 1/3 octave band spectra for the Toyota 40 kph trials. Each spectrum is plotted as a function of vehicle-sensor distance. Horizontal dashed lines represent the noise floor of each frequency bin. Top plot shows the spectra computed from the DAS Ice channels, bottom plot shows Snow channels. The center frequency of each band is given in the plot legends.

The purpose of Fig. 7 is to show that an excitation of the DAS cable caused by a moving vehicle can potentially be used to discriminate the environment of the cable. Compare the top and bottom groups of signal curves. Not only has the ordering of the frequencies changed, but the shape of the curves is also different. For the Ice channels, the frequencies are arranged neatly by power from low to high frequency. The attenuation behavior before the noise floor is also nearly linear (in dB units). For the Snow channels, the ordering has changed, and the curves have developed a shoulder. Such features in the measured signal may be exploited to classify the medium of the cable.

4. Conclusions and future work

In this paper we have demonstrated a DAS system installed on an Arctic ice road can robustly detect ground vehicles of varying sizes and driving speeds. Detection performance appears to be enhanced for cables encased in ice compared to those surface-laid on ice or snow. The enhanced performance is likely due to the more consistent cable coupling, and lower noise floor of the ice channels. The MRZR Alpha was generally the most detectable vehicle, followed by the Toyota Tundra, and the Brutal Snowmobile. Finally, we observed the reliability of detections degrades beyond a couple hundred meters but is improved for faster moving vehicles.

We have also provided preliminary evidence that the DAS excitations caused by a source may be used to identify the cable's environment. More research observing the change in the DAS signal would be uniquely useful for monitoring Arctic mobility. The ground in the Arctic routinely experiences strong seasonal fluctuations and freeze-thaw cycles. The same section of cable frozen solid in February may be surrounded by wet mud come April. Analyzing and modeling how the DAS signal features evolve with the changing ground environment may lead to a real-time, wide range, mobility monitoring tool. We believe this is a ripe area for future research.

5. Nomenclature

DAS	Distributed Acoustic Sensing	
Pd	Probability of detection	[unitless]
Pfa	Probability of false alarm	[unitless]
ROC	Receiver Operating Characteristic	

6. Acknowledgements

The authors would like to acknowledge the ERDC FLEX-4 program for funding this work, and the International Cooperative Engagement Program for Polar Research (ICE-PPR) for their cooperation and support during the field experiment.

7. Declaration of competing interest

The authors declare that they have no known competing financial interests or personal relationships that could have influenced the work reported in this paper.

8. References

Letcher, T., Eylander, J., Shoop, S., Frankenstein, S., 2024. Are the Noah and Noah-MP land surface models accurate for frozen soil conditions? Cold Regions Science and Technology 220, 104149. https://doi.org/10.1016/j.coldregions.2024.104149

Parker, M., Stott, A., Bodie, M., Frankenstein, S., Shoop, S., 2021. Vehicle mobility on highly organic soils. Journal of Terramech. 98, 16–24. https://doi.org/10.1016/j.jterra.2021.09.001

Parker, M., Witte, C., Frankenstein, S., Wheeler, A., 2024. Mobility During The Transition Seasons In The Arctic. Proceedings of the 21st International and 12th Asia-Pacific Regional Conference of the ISTVS. Presented at the 21st International and 12th Asia-Pacific Regional Conference of the ISTVS, International Society for Terrain-Vehicle Systems, Yokohama, Japan. https://doi.org/10.56884/VBHYODO9

Parker, T., Shatalin, S., Farhadiroushan, M., 2014. Distributed Acoustic Sensing – a new tool for seismic applications. First Break 32. https://doi.org/10.3997/1365-2397.2013034

Quinn, M.C., Baxter, C.D.P., Winters, K.E., Picucci, J.R., 2022. Geotechnical Effects on Fiber Optic Distributed Acoustic Sensing Performance, in: Geo-Congress 2022. Presented at the Geo-Congress 2022, American Society of Civil Engineers, Charlotte, North Carolina, pp. 53–62. https://doi.org/10.1061/9780784484067.006

Quinn, M.C.L., Wagner, A.M., Doran, A., Coclin, C., Winters, K.E., 2025. Non-Destructive Distributed Fiber Optic Sensing Considerations, in: Geotechnical Frontiers 2025. Presented at the Geotechnical Frontiers 2025, American Society of Civil Engineers, Louisville, Kentucky, pp. 366–376. https://doi.org/10.1061/9780784485989.037

Shoop, S., Parker, M., Olivier, J., Garibay, E.A., 2024. Understanding Mobility in the Arctic. Presented at the 2024 NDIA Michigan Chapter Ground Vehicle Systems Engineering and Technology Symposium, Novi, Michigan, United States, pp. 2024-01-4132. https://doi.org/10.4271/2024-01-4132

Yuan, S., Liu, J., Noh, H.Y., Clapp, R., Biondi, B., 2024. Using Vehicle-Induced DAS Signals for Near-Surface Characterization With High Spatiotemporal Resolution. JGR Solid Earth 129, e2023JB028033. https://doi.org/10.1029/2023JB028033

0270 | Performance assessment of an off-road vehicle across different terrains

Luca Zerbato [a,*], Carl M. Becker [b], Andries J. Peenze [b], Enrico Galvagno [a], P. Schalk Els [b]

[a] *Department of Mechanical and Aeronautical Engineering, Politecnico di Torino, Torino, Italy*

[b] *Vehicle Dynamics Group, University of Pretoria, Pretoria, South Africa*

* *Corresponding author: luca.zerbato@polito.it*

ABSTRACT

The prediction of vehicle performance during off-road missions has historically been a complex task due to the interaction between tires and soft soil. A comprehensive characterisation of vehicle performance across different terrains is essential for a full understanding of tractive and handling behaviour. Moreover, it is crucial to discuss the tools and methods needed to extract relevant information from off-road testing, as standardisation in this area is lacking compared to on-road testing. This paper focuses on characterising vehicle performance using experimental data collected during an off-road testing campaign conducted across various terrains, including Kalahari Desert sand and gravel pavements. For longitudinal dynamics, coast-down, acceleration, and drawbar pull tests are performed to assess rolling resistance and traction capabilities under off-road conditions. For lateral dynamics, constant-radius tests are conducted on gravel terrain to evaluate cornering behaviour. The synthesis of test analysis results serves as the foundation for developing simplified lumped-parameter handling models and estimating contact parameters for tire–terrain interaction models.

Keywords
Off-road performance
Experimental testing
Tire-terrain interaction

1. Introduction

The investigation of vehicle mobility on deformable and hard off-road terrains has always been a crucial point for the evaluation of wheeled and tracked vehicle performances in many research areas such as agriculture, mining, and exploration (Wong, 2009). Differently from what is available for the on-road testing (ISO 4138:2021, 2021; ISO 7401:2011, 2011; ISO 15037-1:2019, 2019), the characterisation of the vehicle performance on off-road terrains requires using appropriate tools and methods (i.e., experimental procedures, equipment, and analysis) to extract meaningful performance indicators. Moreover, the synthesis of the vehicle performance can be used as reference for the validation of semi-empirical Terramechanics models and the estimation of contact parameters (Vella et al., 2024; Zerbato et al., 2024). Off-road mobility and trafficability performance have been investigated by the pioneers of the Terramechanics; these approaches involve using a small set of empirical coefficients and vehicle information (Bekker, 1969; Wong and Reece, 1967a, 1967b). In these approaches the Cone Penetrometer allows the characterisation of the terrain strength, as the Cone Index (CI) and the Vehicle Cone Index (Stevens et al., 2013). However, these indicators have been used to assess only the vehicle mobility (go or no-go pass) rather than extracting the vehicle performance. To do this, the correlation between CI and traction performance has been done by (Brixius, 1987; Brixius and Wismer, 1975) which introduced the wheel mobility number to derive the Gross Traction, Motion Resistance and Net Traction characteristics. Hegazy and Sandu (2013) and Wong et al. (2020) updated the numerical coefficients to improve the correlation with the experimental evidence. Other authors (Kruger et al., 2023) used the Bevameter testing machine to derive the terrain parameters such as the pressure-sinkage coefficients (Wong, 2009), cohesion, and friction angle.

In this context, this paper aims at presenting an experimental methodology for the assessment of vehicle performance moving off-road; this methodology combines testing procedures from off-road and on-road research areas.

The methodology involves the following steps:

- characterising the test surface (i.e., the off-road terrains).
- preparing the test surfaces before vehicle testing.
- defining the test procedure.
- data processing and derivation of off-road key performance indices.

The manuscript is divided as follows: Section 2 is dedicated to the description of the experimental campaign, including the characterisation of the testing terrains, vehicle sensors and setup, and testing procedures. In Section 3, the details of the experimental data processing and the definition of performance indicators are shown. In Section 4, the results of the experimental campaign and the comparison of the vehicle performance on different terrains are presented. Finally, conclusions are drawn.

2. Experimental campaign

The experimental campaign has been conducted between the Sports Field campus of the University of Pretoria and the Gerotek Test Facilities ("https://www.armscor.co.za/gerotek," n.d.). In this section, the experimental environments, comprehensive terrain

https://doi.org/10.56884/4MCVKUXF

characteristics and preparation, vehicle setup, and testing procedure are presented.

2.1. Terrain granulometry characterisation

The vehicle dynamics tests have been carried out on four different surfaces. For each off-road terrain, a granulometry analysis distribution has been carried out by collecting samples of the soil from the different tracks. By using standardised sieves and process described in the South African National Standards (SANS) 1953_2020 edition 2 (SANS1953_2020), the different terrains have been classified (Natural Resources Conservation Service, 2012) as reported in Fig. 1.

- Kalahari sand (T1): terrain constituted by a 96% of homogeneous fine-grained sand and a residual part of silt (black line with square markers).
- Gravel road (T2): terrain composed by 20% of gravel, 75% of sand (fine and coarse-grained), and a 5% of clay and silt (red line with circle markers).
- Compacted gravel soil (T3): terrain composed by 38% of gravel, 47% of sand (fine and coarse-grained), and a 15% of clay and silt (blue line with diamond markers).

To compare the vehicle's off-road performance with on-road behaviour, a rigid surface, specifically, a concrete skid pad, was used as the test track and reference condition. In the manuscript, this concrete road is referred to as T4.

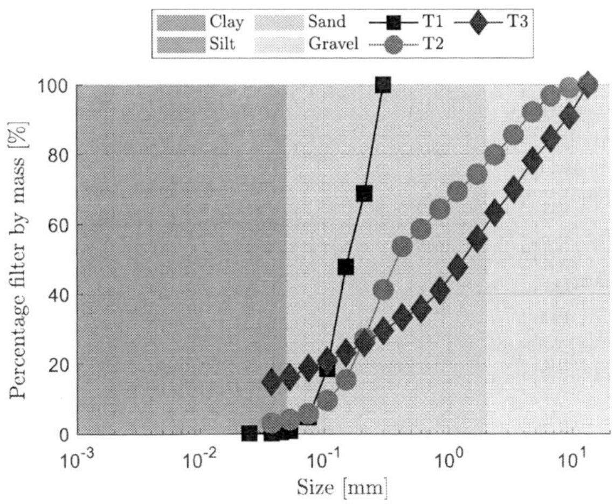

Fig. 1. Granulometry distribution of the terrain tested.

Drawbar pull and acceleration tests have been performed on T1 and T2, coast-down tests on T1 and T4, and handling manoeuvres have been carried out on T3 and T4.

2.2. Soil preparation

The Kalahari Desert sand (T1) is housed in a covered sand trough 62 meters long, 6 meters wide. The location of the Gerotek test facility in South Africa is at a high altitude (in the order of 1430m above sea level) and has, in general, low ambient humidity. This assists with keeping the Kalahari Desert sand as dry as possible and representative of the desert sand. For the tests in question, the sand was levelled out with the use of a scraper towed behind a vehicle to smooth all wheel ruts from previous passes. Figure 2 shows the condition of the sand before (Fig. 2a) and after (*Fig. 2b*) preparation.

Fig. 2. Kalahari Desert sand in a sand trough before (a) and after (b) soil preparation.

A level section on the natural gravel road (T2) was selected as the test area where part of the longitudinal dynamic tests was conducted. This terrain was not specifically prepared, as this section of road is in constant use and was deemed to be a good representation of a typical gravel road. A section of the T2 gravel road section is shown in Fig. 3.

Fig. 3. Natural gravel road (T2) used for acceleration and drawbar pull tests.

The area used for the lateral dynamics testing was a section of compacted gravel (T3), large enough to conduct a constant radius test, with a slightly different grading compared to T2. The test area is depicted in Fig. 4.

Fig. 4. Gravel road section (T3) where handling tests were performed.

2.3. Vehicle and sensors

The vehicle used in this study is a 1997 Land Rover Defender 110 TDi, extensively modified to serve as a vehicle dynamics test platform. The vehicle is fitted with a five-speed manual gearbox with high-range and low-range options, as well as a central locking differential. The vehicle is fitted with Michelin LTX A/T2 tyres with a size of 235/85R16 (Fig. 5). All the tests are conducted using all-wheel drive with a locked centre differential.

Fig. 5. Detail of the equipped tyre on test vehicle.

The original suspension springs and dampers have been removed and replaced with a controllable hydropneumatic suspension system similar to the system developed by Theron and Els (2007). This system features hydropneumatic springs that can switch between four discrete stiffness settings on demand. Damping characteristics are adjusted using a continuously variable bypass valve, allowing smooth variation between the softest and hardest settings. Additionally, the suspension system includes a slow-active ride height function that hydraulically adjusts each strut's ride height independently. The spring and damper settings are controlled in a semi-active manner.

The test vehicle is equipped with a range of sensors to capture test data accurately. For the suspension system, hydraulic pressure and displacement are measured at each strut. Vehicle attitude is recorded using a VBOX 3i GNSS system with an integrated IMU (RACELOGIC, 2024), measuring the position, velocity, acceleration, and angular velocity of the vehicle body. The IMU is mounted near the vehicle's centre of gravity. A ground speed sensor (DICKEY-john ,2025) is also installed to provide speed measurements when GNSS signals are unavailable. The front wheel steering angle is measured using a rope string potentiometer connected to the drag link of the steering system. For the drawbar pull tests considered in this study, a 5-ton ULP-type load cell (Loadcell Services, 2024) is installed in place of the rear tow hitch and used to measure drawbar forces via a suitable strain gauge amplifier.

All measurements are recorded using a dSPACE MicroAutoBox II system at 200 Hz (dSPACE, 2025). The same system also serves as the embedded controller for all controllable subsystems like the suspension. Figure 6 shows the relative location of the sensors installed on the vehicle.

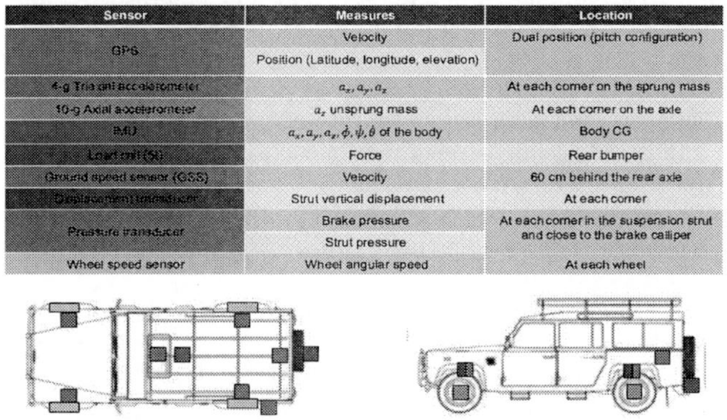

Sensor	Measures	Location
GPS	Velocity	Dual position (pitch configuration)
	Position (Latitude, longitude, elevation)	
4-g Triaxial accelerometer	a_x, a_y, a_z	At each corner on the sprung mass
10-g Axial accelerometer	a_z unsprung mass	At each corner on the axle
IMU	$a_x, a_y, a_z, \dot{\phi}, \dot{\psi}, \dot{\theta}$ of the body	Body CG
Load cell (5t)	Force	Rear bumper
Ground speed sensor (GSS)	Velocity	60 cm behind the rear axle
Displacement transducer	Strut vertical displacement	At each corner
Pressure transducer	Brake pressure	At each corner in the suspension strut and close to the brake calliper
	Strut pressure	
Wheel speed sensor	Wheel angular speed	At each wheel

Fig. 6. Relative sensor installation locations on test vehicle

Key vehicle parameters required in this study are provided in Table 1.

Table 1
Vehicle parameters

Parameter	Symbol	Value	Unit
Gross Vehicle Weight	GVW	22	kN
Vehicle mass	m	2252	kg
Undeformed tyre radius	R_u	0.403	m
Tyre effective rolling radius	R_e	0.3875	m
Wheelbase	L	2.8	m
CG to front axle	a	1.5	m
CG to rear axle	b	1.3	m
Track width	T_f , T_r	1.49	m

2.4. Manoeuvres

For all the tests performed in this study, the suspension ride height adjustment was used to level the vehicle to the same desired static ride height before each experiment, after which it was kept at this setting during the test. Due to some variations in the ground surface slope and inclination when performing this ride height adjustment, some differences are expected in the static pressure in the hydropneumatic suspension struts. These variations are, however, deemed small enough to not significantly affect the results in this study. For all the tests, the semi-active suspension system was set to the hard spring and high damping setting for all four struts to keep the tests consistent, repeatable, and safe.

For the drawbar pull tests, the test vehicle instrumented with the equipment described in Section 2.3, is towing another Land Rover Defender with a tow strap connected to the 5-ton load cell at the lead vehicle tow hitch location. A drawbar pull test is performed, with the towed vehicle's transmission left in neutral while the towing vehicle is driven at a constant speed of 15 km/h in low range, second gear, with the central differential in the locked position. These tests are performed on the T1 and T2 roads. After both vehicles are driving at a constant speed, the towed vehicle gradually increases its brake pressure by progressively applying the hand brake, causing both vehicles to slow down; the towing vehicle continues applying a driving torque to maintain the desired speed. The load cell connected between the two vehicles measures the drawbar force while the wheel speed sensors and ground speed sensor measure the wheel and vehicle speeds respectively.

For the acceleration tests, the instrumented vehicle is accelerated from a standstill to a constant speed with the gearbox in second gear, low range, with a locked central differential. These tests were conducted on the T1 and T2 roads, using the IMU accelerometer measurements, the ground speed sensor, and the wheel speed sensors.

Coast-down tests are conducted on the T1 and T4 road surfaces, where the vehicle is driven to a speed of 15-20 km/h, after which the clutch is disengaged, and the vehicle coasts down to a standstill. Similar to the acceleration tests, the IMU and ground speed sensors are used to measure the vehicle acceleration and speed.

Finally, the steady-state lateral performance of the vehicle is assessed with constant radius tests. These tests were performed on the T3 and T4 roads. On the T3 gravel road, a small 10-meter constant radius test was conducted, with the speed of the vehicle slowly increasing until excessive sliding was experienced. For the T4 concrete road, a larger 40-meter constant radius is used to assess the on-road steady-state handling performance of the vehicle. For both tests, the VBOX GNSS and IMU were used to measure the vehicle speed, accelerations, and estimated side slip angles. Additionally, the front wheel steering angle is measured and used to determine the understeer characteristics of the vehicle on these different road surfaces.

3. Data processing and performance quantification

In this section, the description of the data processing applied to the experimental measures to derive the synthesis of the results of Sections 4 and 5 is presented. A fourth-order digital Butterworth lowpass filter is designed and applied to attenuate the signal noise, as can be observed through Fig. 7, which depicts the time history of the raw (blue line) and filtered (orange line) loadcell force data gathered during the drawbar pull test on T2.

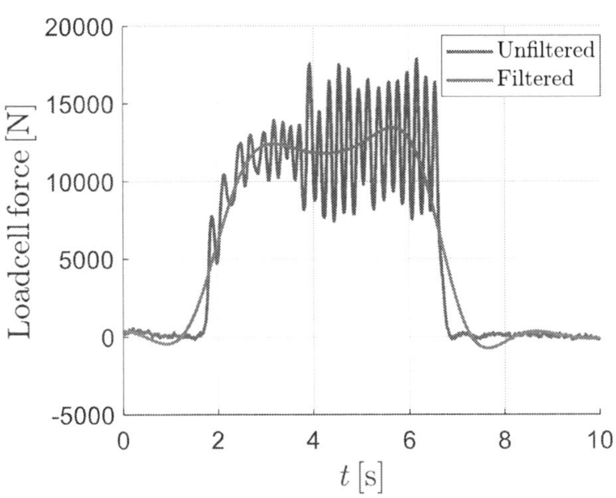

Fig. 7. Experimental signal filtering.

For each manoeuvre, the traction and handling performance assessment is made by using a dedicated routine.

To quantify the drawbar pull performance, the drawbar pull coefficient is defined as the ratio between the Drawbar Pull Force (DBF) and the Gross Vehicle Weight (GVW) Eq. 1:

$$\mu_{DB} = \frac{\text{DBF}}{\text{GVW}} \tag{1}$$

This quantity is used in the drawbar pull-slip curve. The average value of the four-wheel slip ratio $\overline{\kappa}$ is defined Eq. 2:

$$\overline{\kappa} = \frac{1}{4}\sum_{i=1}^{4}\left(1 - \frac{V_i}{\omega_i R_e}\right) \tag{2}$$

where V_i and ω_i are the translational speed and angular speed of the wheel, respectively, and R_e is the effective rolling radius (He et al., 2020). Previous research by Hamersma et al. (2016) found the effective rolling radius of the equipped tyre to be 0.3875 m in on-road condition. Moreover, from the drawbar pull-slip curve, two performance indices are evaluated: the slope of the curve near the origin (k_{DB}) and the maximum value of the drawbar coefficient (namely $\mu_{DB,max}$).

The coast-down test allows the quantification of the vehicle motion resistance; from the experimental measures of the vehicle speed V and the longitudinal acceleration a_x, the vehicle motion resistance F_{res} during the deceleration phase is computed as follows in Eq. 3 (Genta and Morello, 2009):

$$F_{\text{res}} = -ma_x = f_0 + f_2 V^2 \tag{3}$$

where m represents the vehicle mass. Vehicle motion resistance is a quadratic function of the velocity, where f_0, f_2 are the cost-down coefficients. These coefficients are typically determined through a quadratic regression between F_{res} and V, fitting a second-order polynomial function. However, in this paper only the constant coefficient f_0 is determined for terrain T1 and T4, due to the low starting velocity of the coast-down test: the quadratic coefficient, mainly due to the aerodynamic resistances, is neglected. *Figure 8* depicts the trend of the vehicle speed (blue line) and the motion resistance (orange line) during a run of the cost-down test on terrain T1. The grey shaded area identifies the region where the data fitting has been performed: two thresholds (i.e., t_1 and t_2) are identified when the vehicle speed is in the range of $15 \geq V \geq 6$ km/h.

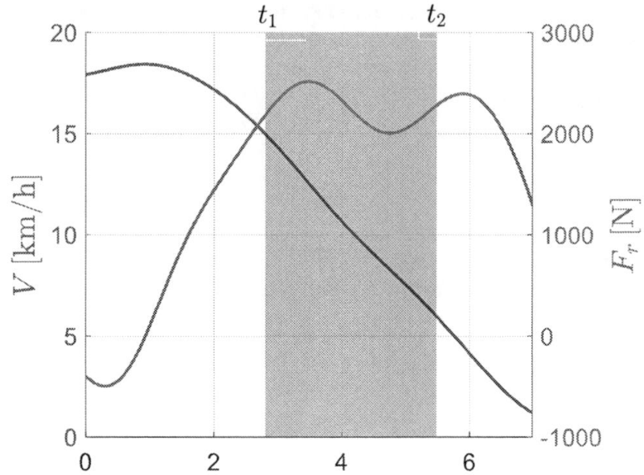

Fig. 8. Experimental results from the coast-down test: vehicle speed and motion resistance.

The handling quasi-static performance is assessed through the computation of the understeer characteristic and the sideslip characteristic. These quantities are defined through Eqs. 4-5 (Galvagno et al., 2021; Genta and Morello, 2009; Guiggiani, 2018)

$$\delta_{F,dyn} = \delta_F - \delta_{F,0} = \alpha_F - \alpha_R \tag{4}$$
$$\beta_{dyn} = \beta - \beta_0 = -\alpha_R \tag{5}$$

where δ_F is the average value of the front steer angle (δ_{FL}, δ_{FR}), δ_0 is the kinematic front wheel steering angle, α_F and α_R are the slip angles of the front and rear axle, β is the sideslip angle of the vehicle, and β_0 is the kinematic sideslip angle of the vehicle (Genta and Morello, 2009). During the test, only the quantity δ_{FR} is measured. The value of δ_{FL} is quantified through a nonlinear map of the steering characteristics previously identified (Fig. 9). In the range $[-5\ 5]$ ° the steering mechanism enables parallel steering, while above that range the inner wheel steers more than the outer one, enabling Ackermann steering as highlighted by the comparison with the parallel steering curve (yellow line) and the Ackermann one (orange line) (Milliken et al., 1995).

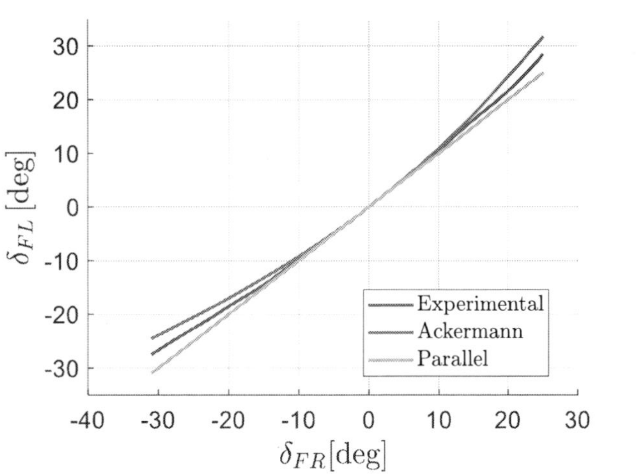

Fig. 9. Map of the experimental left wheel steer angle as a function of the right wheel.

The kinematic steer angle $\delta_{F,0}$ and the kinematic sideslip angle β_0 are evaluated using the geometrical relationship of the single-track model Eq.s 6-7, which involves the quantity L, b and the turning radius R; under the assumption of kinematic conditions (i.e. null slip angles)(Genta and Morello, 2009). The turning radius is computed using the measures of the yaw rate $\dot{\psi}$ and the vehicle speed V, as expressed in Eq. 6.

$$\delta_0 = \frac{L\dot{\psi}_{ss}}{V} \tag{6}$$
$$\beta_0 = \frac{b\dot{\psi}_{ss}}{V} \tag{7}$$

The slip angles α_F and α_R are evaluated through the computation of the tyre slip angles $\alpha_{FL}, \alpha_{FR}, \alpha_{RL}, \alpha_{RR}$ expressed through Eqs. 8-11 (Guiggiani, 2018):

$$\alpha_{FL} = \frac{v + \dot{\psi}a}{u - \dot{\psi}t_F/2} - \delta_{FL} \tag{8}$$
$$\alpha_{FR} = \frac{v + \dot{\psi}a}{u + \dot{\psi}t_F/2} - \delta_{FR} \tag{9}$$
$$\alpha_{RL} = \frac{v - \dot{\psi}b}{u - \dot{\psi}t_R/2} \tag{10}$$
$$\alpha_{RR} = \frac{v - \dot{\psi}b}{u + \dot{\psi}t_R/2} \tag{11}$$

The quantities u and v are the longitudinal and lateral components of the vehicle COG velocity: they are evaluated through the projection of the experimental velocity V along the x and y axes of the vehicle with the estimated sideslip angle β. α_F and α_R are computed with Eqs. 12-13:

$$\alpha_F = \frac{1}{2}(\alpha_{FL} + \alpha_{FR}) \tag{12}$$
$$\alpha_R = \frac{1}{2}(\alpha_{RL} + \alpha_{RR}) \tag{13}$$

Finally, the experimental trends of Eq. 4 and Eq. 5 are fitted through an analytical expression that approximates the characteristics into two parts (De Novellis et al., 2014):

- Linear range: linear fitting of characteristics through the definition of the understeer gradient K_{US} and sideslip gradient K_β up to a threshold of lateral acceleration $a_{y,lin}$.
- Nonlinear range: nonlinear fitting of the characteristics using an exponential function up to the maximum lateral acceleration $a_{y,max,\delta}$ and $a_{y,max,\beta}$.

The values of thresholds $a_{y,lin}$, $a_{y,max,\beta}$, $a_{y,max,\delta}$ are listed in *Table 2*. The performance indices from the handling test are the values of K_{US}, K_β and $a_{y,max}$ derived from the test.

Table 2
Handling analysis: steady-state thresholds.

Parameter	Unit	T3	T4
$a_{y,lin}$	[m/s²]	3.7	4
$a_{y,max,\delta}$	[m/s²]	5	5.5
$a_{y,max,\beta}$	[m/s²]	5.5	5.2

4. Results

4.1. Acceleration and Drawbar Pull tests

The results of the traction performance of the vehicle negotiating off-road conditions are discussed in this paragraph. A comparison between the two terrains (T1 and T2) is made by identifying black colour for terrain T1 and red colour for terrain T2.

Figure 10 depicts the trends of the vehicle velocity (solid line) and the front left wheel peripheral speed (dotted line) during the transient phase of the acceleration test. Only one wheel result is reported for the sake of compactness, since the front right exhibits similar trends, while the rear axle does not show high slip conditions. It is evident that the wheel starts to spin on both the terrains, more on the gravel section (T2); however, in both cases, the front wheels reach 100% slip. The vehicle can accelerate faster on terrain T2 than on terrain T1.

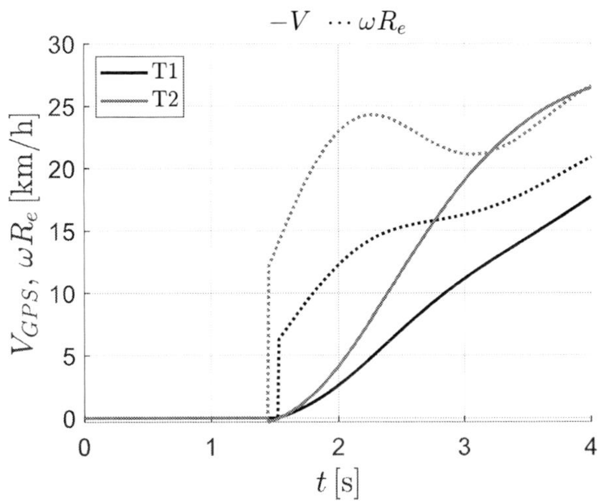

Fig. 10. Comparison between the vehicle speed (solid line) and front left peripheral speed (dotted line) during the acceleration test on T1 and T2.

Figure 11 depicts the trend of the drawbar pull-slip characteristics obtained from the tests on T1 (black dots) and T2 (red dots). The results show that the off-road traction performance on T2 is higher than on T1, as the peak of the drawbar coefficients and the linear slope are almost 42% higher, as highlighted in *Table 3*. This result agrees with the acceleration tests: the off-road performance on T1 is influenced by the wheel sinkage, which leads to an increase in the motion resistance. This aspect is underlined in the next paragraph dedicated to the characterisation of the vehicle motion resistance on T1.

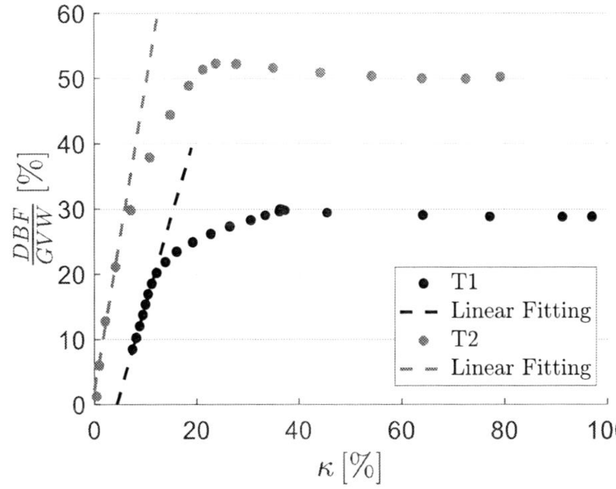

Fig. 11. Drawbar pull-slip characteristic obtained on T1 and T2.

Table 3
Drawbar pull performance.

Parameter	Unit	T1	T2
k_{DB}	[-]	2.71	4.62
$\mu_{DB,max}$	[%]	29	50

4.2. Coast down test

The results of the assessment of the vehicle motion resistance on T1 are summarised in this paragraph. Figure 12 depicts the trend of the motion resistance F_{res} against the vehicle velocity during the coast down test on T1 (black circles) and T4 (magenta circles). As the vehicle is running at low velocity, the trend of both characteristics is constant. In this condition the coefficient f_0 of Eq. 3 is identified with a value of $f_{0,\mathrm{T1}} = 0.095$ on T1 and $f_{0,\mathrm{T4}} = 0.008$ on T4. Therefore, the results show that moving on soft terrain such as terrain T1, the motion resistance is an order of magnitude higher than moving on hard terrain like asphalt.

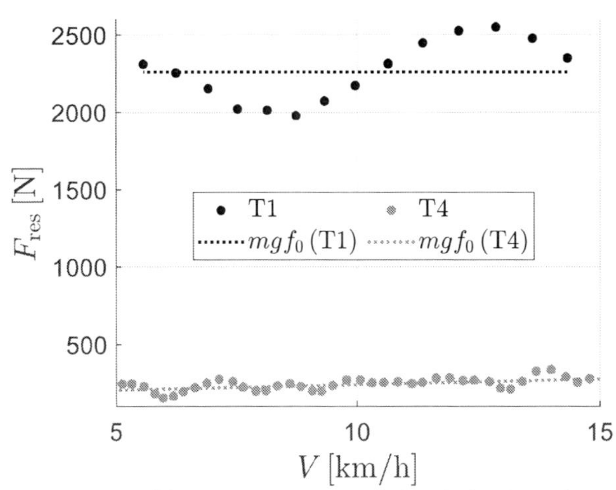

Fig. 12. Fitting of the motion resistance with the vehicle velocity for a vehicle moving on soft (T1) and hard (T4) terrain.

4.3. Constant radius cornering

The results of constant radius cornering of the vehicle negotiating off-road conditions are summarised in this paragraph.

A comparison between the gravel terrain T3 and a hard terrain T4 is made.

Figure 13 depicts the comparison of the trend of the understeering characteristics, Eq. 4, on the two terrains. The on-road T4 result highlights a slight oversteering tendency of the vehicle, whereas the off-road one (T3) shows an understeering one. The results show a maximum value of lateral acceleration of ≈ 5 m/s² on T3 and a value of ≈ 5.2 m/s² on T4; no asymptote is highlighted by the experimental points of the T4; the test is terminated before the vehicle starts to spin. On the contrary, an asymptote, due to the saturation of the front axle, is evident in the trend of T3. Moreover, both the characteristics reach the limit on friction before the limit on roll-over. In both cases the reason is due to the axle saturation (front first for T3 and rear first for T4).

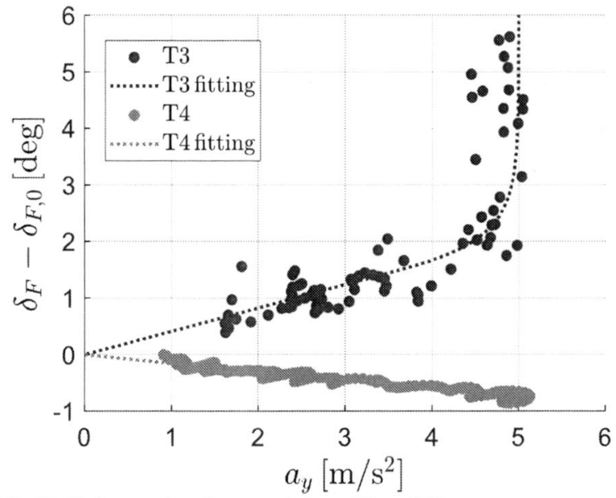

Fig. 13. Understeering characteristic on T3 and T4.

Figure 14 depicts the comparison of the trend of the sideslip characteristics, Eq. 5. Both the sideslip gradient and the limit of the characteristics are higher on T3. The trends suggest that at equal lateral acceleration (i.e., equal total lateral force), both front and rear slip angles are higher during off-road conditions.

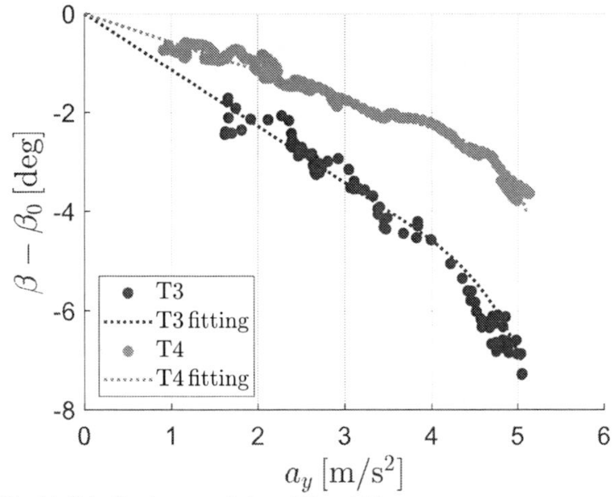

Fig. 14. Sideslip characteristic on T3 and T4.

Finally, the experimental points are fitted with Eq. 14-15; the comparison of K_{US} and K_β is presented in Table 4.

Table 4
Handling performance characteristics and KPIs

Parameter	Unit	T3	T4
K_{US}	[deg/(m/s²)]	0.472	-0.168
K_β	[deg/(m/s²)]	-1.140	-0.581
$a_{y,max}$	[m/s²]	5	5.2

5. Conclusions and future work

This paper presents a methodology to assess vehicle performance negotiating off-road conditions. The methodology involves the definition of the proper tools and methods to extract meaningful traction and handling performance indicators. This has required combining a well-defined experimental methodology (terrains, tools, sensors, and vehicle) with physical equations from the vehicle dynamics research area. The results highlight a degradation of the vehicle traction performance on sand through acceleration, drawbar pull, and coast down tests; this result is linked to the high wheel sinkage experienced and the consequent increase of the rolling resistance (ten times higher than with typical on-road conditions). The gravel terrains show better performance due to the limited amount of sinkage; however, the handling results highlight an increase in the understeer and sideslip of the vehicle. The performance indices, together with the experimental results, will be used to derive the contact parameters of semi-empirical Terramechanics contact models.

6. Acknowledgements

This work was supported in part by the European Union's Horizon 2020 research and innovation program under the Marie Skłodowska-Curie grant 872907.

7. Declaration of competing interest

The authors declare that they have no known competing financial interests or personal relationships that could have appeared to influence the work reported in this paper.

8. References

Bekker, M.G., 1969. Introduction to terrain-vehicle systems: Part 1: The terrain. Part 2: The vehicle. University of Michigan Press.
Brixius, W.W., 1987. Traction prediction equations for bias ply tires.
Brixius, W.W., Wismer, R.D., 1975. Traction prediction equations for wheeled vehicles. John Deere Report No. 109. Project 9008. Technical Center.
De Novellis, L., Sorniotti, A., Gruber, P., 2014. Design and comparison of the handling performance of different electric vehicle layouts. Proceedings of the Institution of Mechanical Engineers, Part D: Journal of Automobile Engineering 228, 218–232. https://doi.org/10.1177/0954407013506569
Galvagno, E., Galfrè, M., Mari, G., Velardocchia, M., Tota, A., 2021. A Methodology for Parameter Estimation of Nonlinear Single Track Models from Multibody Full Vehicle Simulation, in: SAE Technical Papers. SAE International. https://doi.org/10.4271/2021-01-0336
Genta, G., Morello, L., 2009. The automotive chassis: vol. 2: system design. Springer.

Guiggiani, M., 2018. The science of vehicle dynamics: handling, braking, and ride of road and race cars. Springer.

Hamersma, H.A., Botha, T.R., Schalk Els, P., 2016. The dynamic rolling radius of a pneumatic tyre on hard terrains.

He, R., Sandu, C., Mousavi, H., Shenvi, M.N., Braun, K., Kruger, R., Els, P.S., 2020. Updated standards of the international society for terrain-vehicle systems. J Terramech. https://doi.org/10.1016/j.jterra.2020.06.007

Hegazy, S., Sandu, C., 2013. Experimental investigation of vehicle mobility using a novel wheel mobility number. J Terramech 50, 303–310. https://doi.org/10.1016/j.jterra.2013.09.005

https://www.armscor.co.za/gerotek [WWW Document], n.d.

ISO 4138:2021, 2021. Passenger cars — Steady-state circular driving behaviour — Open-loop test methods.

ISO 7401:2011, 2011. Road vehicles — Lateral transient response test methods — Open-loop test methods.

ISO 15037-1:2019, 2019. Road vehicles - Vehicle Dynamics test methods - Part 1: General conditions for passenger cars.

Kruger, R., Els, P.S., Hamersma, H.A., 2023. Experimental investigation of factors affecting the characterisation of soil strength properties using a Bevameter in-situ plate sinkage and shear test apparatus. J Terramech 109, 45–62. https://doi.org/10.1016/j.jterra.2023.06.002

Milliken, W.F., Milliken, D.L., Metz, L.D., 1995. Race car vehicle dynamics. SAE international Warrendale.

Natural Resources Conservation Service, 2012, 2012. National Engineering Handbook: Chapter 3 - Engineering Classification of Earth Materials, in: National Engineering Handbook.

Stevens, M.T., Towne, B.W., Mason, G.L., Priddy, J.D., Osorio, J.E., Barela, C.A., 2013. Procedures for One-Pass Vehicle Cone Index (VCI 1) Determination for Acquisition Support.

Theron, N.J., Els, P.S., 2007. Modelling of a semi-active hydropneumatic spring-damper unit, Int. J. Vehicle Design.

Vella, A.D., Zerbato, L., Galvagno, E., Vigliani, A., Data, S., Sacchi, M.E., 2024. A Model-Based Parameter Estimation Algorithm for Tire-Soft Soil Contact Model from Off-Road Longitudinal Tests, in: The International Conference of IFToMM ITALY. Springer, pp. 214–221.

Wong, J.Y., 2009. Terramechanics and off-road vehicle engineering: terrain behaviour, off-road vehicle performance and design. Butterworth-heinemann.

Wong, J.Y., Jayakumar, P., Toma, E., Preston-Thomas, J., 2020. A review of mobility metrics for next generation vehicle mobility models. J Terramech. https://doi.org/10.1016/j.jterra.2019.10.003

Wong, J.-Y., Reece, A.R., 1967a. Prediction of rigid wheel performance based on the analysis of soil-wheel stresses part I. Performance of driven rigid wheels. J Terramech 4, 81–98.

Wong, J.-Y., Reece, A.R., 1967b. Prediction of rigid wheel performance based on the analysis of soil-wheel stresses: Part II. Performance of towed rigid wheels. J Terramech 4, 7–25.

Zerbato, L., Vella, A.D., Galvagno, E., Vigliani, A., Data, S., Sacchi, M.E., 2024. A Numerical Analysis of Terrain and Vehicle Characteristics in Off-Road Conditions through Semi-Empirical Tire Contact Modelling. SAE Technical Paper.

0375 | A general model to approximate the slippage curve for applications from analysis of the tire and soil contributions to longitudinal slippage

Vladimir Vantsevich [a], Jesse Paldan [a,*], Jordan Whitson [b], David Gorsich [b], Lee Moradi [a]

[a] *Worcester Polytechnic Institute, Worcester, Massachusetts, United States*

[b] *U.S. Army DEVCOM Ground Vehicle Systems Center, Warren, Michigan, United States*

* *Corresponding author: jpaldan@wpi.edu*

ABSTRACT

The traction-longitudinal slippage curve characterizes the interaction of a driving wheel and deformable terrain and has important considerations affecting the performance of the wheel and of the vehicle. Knowledge of the slippage is crucial to off-road vehicle mobility and energy efficiency assessment and wheel control applications. To be able to analyze and compare the efficiency of different tires on the same terrain, or a tire on different terrains, the tire slippage should be split into two components to determine contributions of the tire and soil to the tire traction characteristic. A new exponential formulation of the longitudinal slippage was previously derived for this application.

Many analytically-inspired-empirical based math models exist to describe slippage curves, which are validated, recommended, and being used in software products to approximate tire-terrain performance. To use these models in research with split tire and soil contributions would require rigorous research effort on re-deriving equations for each model because each is based on different mathematical theory and statistical constraints; such effort may not lead to a positive result. As an alternative approach, the analytical split traction-slippage curve is proposed as a general model to approximate other formulations of the slippage curve which can then be split into separate tire and soil slippage curves. Computational simulations of other slippage models are performed and their resulting curves are fit to the new exponential function to analyze the effectiveness of the split slippage model to approximate other models of traction-slippage curves and soil characterization. Using the split traction model, equations are derived to quantify tire and soil contributions the slip power losses.

Keywords
Off-road mobility
Soil characterization
Tires

1. Introduction

The traction-longitudinal slippage curve characterizes the interaction of a driving wheel and deformable terrain and has important considerations affecting the performance of the wheel and of the vehicle. The tire slippage is defined as a loss of linear velocity of the center of the driving wheel due to the tire and terrain longitudinal deformations and relative movement between the tire and terrain.

$$s_\delta = \frac{V_t - V_x}{V_t} = \frac{V_\delta}{V_t} \tag{1}$$

V_δ is the slip velocity, which is the difference between the actual linear velocity V_x and the theoretical velocity V_t at zero slip.

$$V_\delta = V_t - V_x \tag{2}$$

Some slippage is necessary to create traction, and will always exist when torque is applied to the wheel, while high slippage can lead to inefficiency, immobilization, and loss of control. Therefore, knowledge of the slippage is crucial to off-road vehicle mobility and energy efficiency assessment and wheel control applications, and researchers have developed many models of the traction-slippage curve. Studies have also been performed on the applicability of models to different terrain conditions and the ability to convert between parameters of different models. Tiwari et al., (2010) reviewed many traction models to check their applicability to different soil conditions. Lopez-Arreguin and Montenegro (2019) derived equations for wheel forces of rovers intended to require fewer physical parameters to measure. Huang et al. (2020) performed a study to examine the ability to convert cone index data into Bekker-Wong terrain parameters and developed an improved model for cone-terrain interaction. Maclaurin (2014) developed an adaptation of the Magic Formula used for on-road tires for tires in soft cohesive soils. Mason et al. (2022) proposed a unified gross traction equation for wheels in towed, braked, and powered modes on clay soils. Farhadi et al., (2022) developed a finite-element based model of tire-soil interaction to estimate power losses.

The slippage is affected by an interaction of multiple factors of the vehicle/tire and of the ground. Tire factors include load, inflation pressure, size, and tread type. For off-road vehicles, ground factors such as soil strength and moisture content significantly affect the slippage. To be able to analyze and compare different tires on the same terrain, or a tire on different terrains, it is valuable to split the tire slippage into two components to determine contributions of the tire and soil to the tire traction characteristic. A model was developed to split the tire slippage into two components $s_{\delta t}$ (made by the tire compression) and $s_{\delta s}$ (made by soil compaction) in order to quantitatively characterize the contributions of the tire and soil to the tire slippage s_δ (Andreev and Vantsevich, 2017). The model was based on the soil shear stress-strain developed by Janosi and Hanamoto which is widely known and used in terramechanics (Janosi and Hanamoto, 1961). From this model, a version of the traction-slip curve with an exponential form was derived as

$$F_x = R_z \mu_{px} \left(1 - \frac{s_{\delta c}}{2 s_\delta}\left(1 - \exp\left(-\frac{2 s_\delta}{s_{\delta c}}\right)\right)\right) \tag{3}$$

where F_x is the wheel circumferential force (the gross traction developed by the wheel torque), s_δ is the tire slippage, R_z is the normal reaction of the wheel load, μ_{px} is the peak friction coefficient, and $s_{\delta c}$ is the tire characteristic slippage. When the

https://doi.org/10.56884/IIP1JS9W

circumferential force is normalized by R_z, the traction characteristic is expressed as

$$\mu_x = \mu_{px}\left(1 - \frac{s_{\delta c}}{2s_\delta}\left(1 - \exp\left(-\frac{2s_\delta}{s_{\delta c}}\right)\right)\right) \tag{4}$$

where $\mu_x = F_x/R_z$ is the current friction coefficient.

This formulation of the traction curve was used to derive a method to mathematically separate the tire and soil contributions to the slippage, and thus compute separate tire and soil slippage curves and characteristic slippage values to analyse tire traction data at a range of various loads, inflation pressures, and on multiple terrain surfaces (Whitson et al., 2021). The sum of $s_{\delta s}$ and $s_{\delta t}$ gives s_δ in Eq. 4.

$$s_\delta = s_{\delta s} + s_{\delta t} \tag{5}$$

The separation allows estimation and comparison of the efficiency of tires on different terrains and the efficiency of similar tires on the same terrain.

Many math models exist to describe slippage curves, and are validated, recommended, as well as being in use in multiple software products. To use these models in research with split tire and soil contributions would require rigorous research effort on re-deriving equations for each model because each is based on different mathematical theory; such effort may not lead to a positive result. Alternatively, it is proposed in this paper to use Eq. 3 as a general model of the slippage curve for separating the contributions of the tire and soil. As an investigation in the practicality of this approach, the applicability to fit the curve of Eq. 3 to other formulations of the slippage curve is investigated in this paper. Furthermore, the contributions of the tire and soil components of the slippage to the wheel slip power losses is considered. Expressions of the slip power efficiency characterized by the tire and soil contributions are derived to demonstrate the application of the split traction model to analysis of energy efficiency assessment.

2. Split Tire Soil Model

A computational algorithm to study contributions of the tire and soil to the tire slippage was derived in (Andreev and Vantsevich, 2017) for the traction model described by Eq. 4 parameterized by μ_{px} and $s_{\delta c}$. A summary of the algorithm is provided here. First, the characteristic slippage due to the tire longitudinal stiffness, $s_{\delta tc}$ is computed from the tire longitudinal stiffness, K_{xt} using Eq. 6.

$$s_{\delta tc} = \frac{0.568\mu_{px}R_z}{K_{xt}} \tag{6}$$

The characteristic slippage due to the soil compaction in the longitudinal direction $s_{\delta sc}$ is computed from Eq. 7.

$$s_{\delta c} = s_{\delta sc} + s_{\delta tc} \tag{7}$$

The three characteristic slippages $s_{\delta c}$, $s_{\delta sc}$, and $s_{\delta tc}$ are then used to compute the combined and separate traction characteristics from Eq. 4.

3. Comparison of Traction Models

Additional traction models were then computed to evaluate the ability of Eq. 4 to approximate other traction models. A 1500x600-635 tire with a nominal wheel load of 50 kN and inflation pressure of 0.35 MPa was used for the simulations performed in this study. Characteristics of the tire are listed in Table 1.

Table 1
1500x600-635 tire characteristics

Characteristic	Value
Unloaded diameter d_0	1.5 m
Width b_w	0.6 m
Section height h	0.4325 m
Lug height h_r	0.025 m
Lug spacing t_r	0.13 m
Inflation pressure p_w	0.35 MPa
Vertical stiffness K_{zt} @ p_w	817 kN/m
Longitudinal stiffness K_{xt} @ p_w	3627 kN

The load/deflection characteristics of the tire at 0.35 MPa inflation pressure are given in Fig. 1.

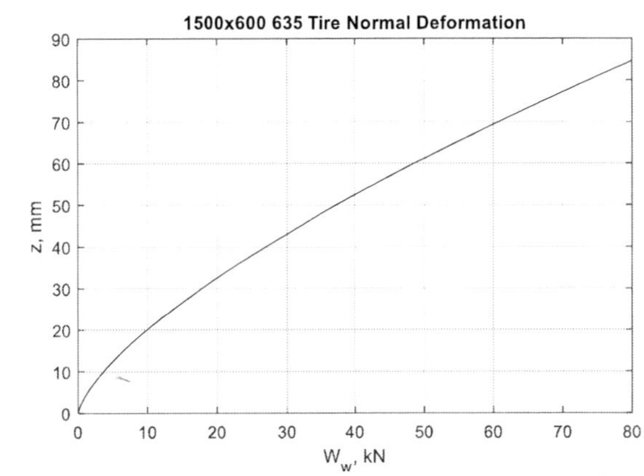

Fig. 1. Load-deflection characteristics of 1500x600-635 tire, p_w = 0.35 MPa

3.1. Bekker-Wong Model

As a first example, a traction model using the Bekker-Wong terrain parameters was used for comparison. Bekker proposed the following equation for the F_x-slippage curve of a tire (Wong, 2024)

$$F_x = (Ac + R_z \tan\varphi)\left(1 - \frac{G}{s_\delta l_t}\left(1 - \exp\left(-\frac{s_\delta l_t}{G}\right)\right)\right) \tag{8}$$

where A is the contact area, l_t is the contact length, c is the soil cohesion, φ is the angle of internal shearing resistance, and G is the shear deformation modulus of the soil. To determine the contact length, an iterative process was used. The vertical force equilibrium is (Wong, 2022)

$$R_z = b_w l_t p_g + W_{cu} \tag{9}$$

where b_w is the tire width, p_g is the ground pressure and the component of the vertical reaction W_{cu} is calculated from Eq. 10 (Wong, 2022).

$$W_{cu} = \left[b_w\left(\frac{k_c}{b_w}+k_\varphi\right)\sqrt{d_0}(z_s+z)^{n-1}\right]\times \frac{1}{3}\left[(3-n)(z_s+z)^{\frac{3}{2}}-(3-n)z^{\frac{3}{2}}-3z_s\sqrt{z}\right] \quad (10)$$

d_0 is the unloaded tire diameter, z is the tire normal deflection, z_s is the sinkage of a tire operating in the elastic mode. k_c, k_φ, and n are parameters of the Bekker-Wong pressure sinkage model (Eq. 11) (Wong, 2022).

$$z_s = \left(\frac{p_g}{k_c/b_w + k_\varphi}\right)^{1/n} \quad (11)$$

The contact length l_t is approximated from (Wong, 2022)

$$l_t = 2\sqrt{d_0 z - z^2} \quad (12)$$

The contact length is determined from Eq.s (9-12) in an iterative solver process and used to compute points on the F_x-s_δ curve from the Bekker-Wong model with Eq. 8. F_x was then converted to μ_x by dividing by R_z.

The terrain condition used for the simulation was a loam stubble field referenced from Dwyer (1974) and listed in Table 2.

Table 2
Silsoe loam stubble field characteristics (Dwyer, 1974)

Characteristic	Value
Cone index CI	960 kPa
Cohesion c	5 kPa
Soil friction angle φ	33.8°
Shear modulus G	11.3 mm
Cohesive modulus k_c	-1380 kN/m$^{(n+1)}$
Frictional modulus k_φ	20000 kN/m$^{(n+2)}$
Sinkage Exponent n	1.2
Moisture content	20.7%

A nonlinear least squares curve fitting tool was then used in MATLAB to find the coefficients of μ_{px} and $s_{\delta c}$ which fit a curve from Eq. 4 to the data points obtained from the Bekker-Wong model. The results are plotted in Fig. 2, where $s_{\delta m}$ is the input model slippage curve computed from the Bekker-Wong model and s_δ is the slippage curve fitted to Eq. 4. Characteristic values for the fitted curve and the quality of the curve fit are given in Table 3. Critical friction coefficient μ_{x-cr} is the value of μ_x which corresponds to the characteristic slippage. The mean absolute error (MAE) between the points of the curves was computed using

$$MAE = \sum_{i=1}^{n}\frac{|\mu_x - \mu_{xm}|}{n} \quad (13)$$

where μ_{xm} is the input model values and μ_x is the curve approximation from Eq. 4. The coefficient of determination r^2 was determined from MATLAB's curve fitting tool based on the sum of squared error.

$$r^2 = 1 - \frac{\sum_{i=1}^{n}(\mu_x - \mu_{xm})^2}{\sum_{i=1}^{n}(\mu_{xm} - \bar{\mu}_{xm})^2} \quad (14)$$

The close agreement between the models is expected by the similar forms of Eq. 3 and Eq. 8.

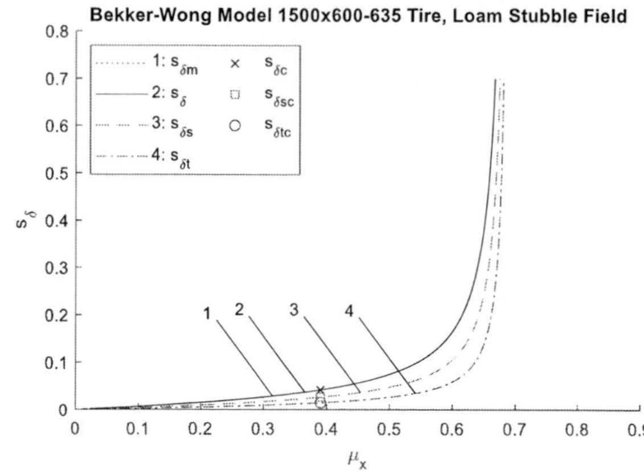

Fig. 2. Bekker-Wong model curve fit and split slippage curves for 1500x600-635 tire on loam stubble field, R_z = 50 kN, p_w = 0.35 MPa

Table 3
Bekker-Wong model characteristic values

Characteristic	Value
Peak friction coefficient μ_{px}	0.6879
Characteristic slippage s_δ	0.0424
Characteristic slippage (soil component) $s_{\delta sc}$	0.0271
Characteristic slippage (tire component) $s_{\delta tc}$	0.0153
Critical friction coefficient μ_{x-cr}	0.3902
Mean absolute error MAE	1.08x10^{-8}
Coefficient of determination r^2	1.000

3.2. Guskov Model

A second comparison was made using Guskov's traction model (Guskov, 1977). In this model, the circumferential force as a function of the slippage is computed using

$$F_x = \frac{\mu_k K_\tau R_z}{s_\delta l_t}\left(\ln\left(\cosh\left(\frac{s_\delta l_t}{K_\tau}\right)\right) - \mu_r\left(\frac{1}{\cosh\left(\frac{s_\delta l_t}{K_\tau}\right)} - 1\right)\right) + 2\tau_{cp}\frac{h_r l_t}{t_r} \quad (15)$$

where h_r is the height of tire lugs, t_r is the distance between lugs, $K_\tau = 0.4t_r$ is a proportional coefficient, τ_{cp} is the longitudinal shear stress, μ_k is the sliding friction coefficient, and μ_r is a reduced friction coefficient computed from

$$\mu_r = 2.55\left(\frac{\mu_s - \mu_k}{\mu_s}\right)^{0.825} \quad (16)$$

where μ_s is the static friction coefficient from Eq. 17.

$$\mu_s = \tan\varphi + \frac{\tau_0}{p_g} \quad (17)$$

p_g is the average ground pressure and τ_0 is the gripping force/internal cohesion of particles. The contact length l_t is determined from

$$l_t = \frac{d_0}{2}\alpha_0 + \sqrt{d_0 z_s} \quad (18)$$

where

$$\alpha_0 = \sqrt[3]{\frac{d_0 z_s + {z_s}^2}{0.5d_0 - z_s}} \quad (19)$$

The sinkage z_s is

$$z_s = \sqrt[3]{\frac{R_z^2}{K_K b_w^2 d_r}} \quad (20)$$

where d_r is the reduced diameter of an elastic tire

$$d_r = d_0\left(1 + \frac{z}{z_s}\right) \quad (21)$$

and K_K is the reduced coefficient of side compression computed from the volumetric compression coefficient of the soil K and the tire dimensions (Katsygin, 1964).

$$K_K = \frac{K}{100\sqrt{b_w d_0}} \quad (22)$$

Parameters for the terrain condition (Table 4) were from a field consisting of light clay, clay & sand after a wheat harvest (Guskov, 1977, Kacigin and Guskov, 1968).

Table 4
Light clay, clay & sand post wheat harvest field characteristics (Guskov, 1977, Kacigin and Guskov, 1968)

Characteristic	Value
Sliding friction coefficient μ_k	0.67
Internal cohesion τ_0	2.5 kPa
Longitudinal shear stress τ_{cp}	1940 kPa
Soil friction angle φ	35.8°
Volumetric compression coefficient K	111000 kN/m³

Slippage curves for the 1500x600-635 tire on the terrain condition from Table 4 are plotted in Fig. 3. Characteristic values for the curve fit are given in Table 5.

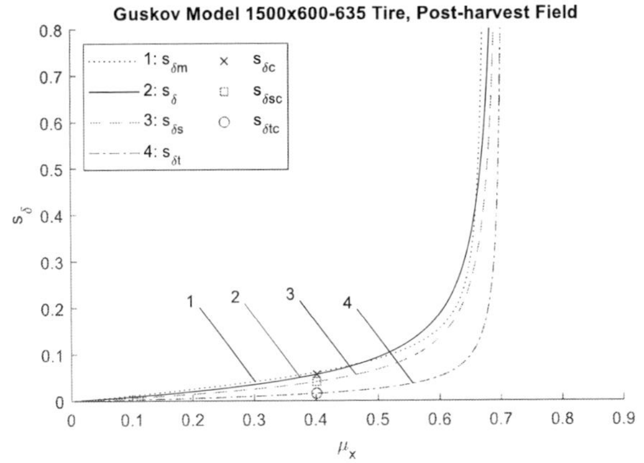

Fig. 3. Guskov model curve fit and split slippage curves for 1500x600-635 tire on light clay, clay & sand field post wheat harvest, R_z = 50 kN, p_w = 0.35 MPa

Table 5
Guskov model characteristic values

Characteristic	Value
Peak friction coefficient μ_{px}	0.7071
Characteristic slippage $s_{\delta c}$	0.0567
Characteristic slippage (soil component) $s_{\delta sc}$	0.0409
Characteristic slippage (tire component) $s_{\delta tc}$	0.0157
Critical friction coefficient μ_{x-cr}	0.4014
Mean absolute error MAE	0.0107
Coefficient of determination r^2	0.9892

The low error of 0.0107 and high coefficient of determination indicates that Eq. 4 approximates the model well.

3.3. Cone Index Models

Multiple traction prediction models were developed based on the cone index, a widely-used measurement of soil strength. Three such models were computed and used to approximate the split tire-soil slippage curve.

Wismer and Luth (1973) used a dimensionless wheel numeric C_n in Eq. 23 to relate the wheel torque T_w to the slippage.

$$\frac{T_w}{r_f W_w} = 0.75(1 - e^{-0.3C_n s_\delta}) \quad (23)$$

C_n is defined as

$$C_n = (CI)\frac{d_0}{W_w} \quad (24)$$

where CI is the cone index and W_w is the wheel load. r_f is the rolling radius defined as the distance traveled per wheel revolution divided by 2π when operating in a self-propelled condition on a hard surface with zero drawbar load (Wismer and Luth, 1973), also known as the rolling radius in the free mode (Andreev et al., 2010). The difference between the rolling radius in the free mode r_f and the rolling radius in the driven mode r_w^0 can be neglected for hard surface roads and used as the reference for zero slip ratio and for relating the wheel torque to the circumferential force (Andreev et al., 2010).

$$F_x = \frac{T_w}{r_w^0} \quad (25)$$

Equation 23 can therefore be rewritten in terms of the current friction coefficient μ_x

$$\mu_x = 0.75(1 - e^{-0.3C_n s_\delta}) \quad (26)$$

Dwyer (1974) and Gee-Clough (1980) investigated empirical relations between the cone index and tractive performance. Their approach utilizes the mobility number M developed by Freitag and Turnage, which was created to combine independent soil and tire parameters into dimensionless forms that correlate closely with dimensionless tire performance coefficients (Turnage, 1972)

$$M = \frac{(CI)b_w d_0}{W_w}\sqrt{\frac{z}{h}}\left(\frac{1}{1 + \frac{b_w}{2d_0}}\right) \quad (27)$$

where h is the tire section height. Empirical equations were developed intended for making choices of tire sizes for off-road agricultural vehicles.

The equation for the coefficient of traction (current friction coefficient μ_x) is

$$\mu_x = (C_T)_{max}(1 - e^{-ks_\delta}) \tag{28}$$

where $(C_T)_{max}$ is the max coefficient of traction and k is a rate constant related to the mobility number through Eq. 29 and Eq. 30.

$$(C_T)_{max} = 0.796 - \frac{0.92}{M} \tag{29}$$

$$k(C_T)_{max} = 4.838 + 0.061M \tag{30}$$

Brixius (1987) developed a traction prediction equation using a modified mobility number, B_n

$$\frac{T_w}{r_f W_w} = 0.88(1 - e^{-0.1B_n})(1 - e^{-0.75s_\delta}) + 0.04 \tag{31}$$

where B_n is defined as

$$B_n = \frac{(CI)d_0 b_w \left(1 + 5\frac{z}{h}\right)}{W_w\left(1 + 3\frac{b_w}{d_0}\right)} \tag{32}$$

Equation (31) can be expressed in terms of μ_x as before:

$$\mu_x = 0.88(1 - e^{-0.1B_n})(1 - e^{-0.75s_\delta}) + 0.04 \tag{33}$$

Results of the three models for the 1500x600-635 tire are plotted in Fig. 4, Fig. 5, and Fig. 6 for the loam stubble field soil parameters from Table 2, which has a cone index of 960 kPa.

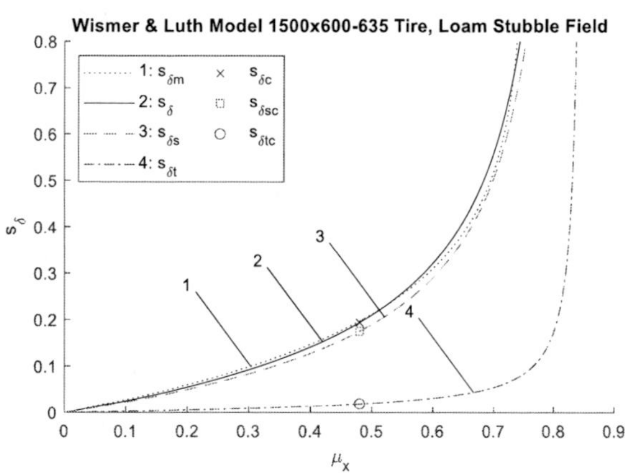

Fig. 4. Wismer and Luth model curve fit and split slippage curves for 1500x600-635 tire on loam stubble field, R_z = 50 kN, p_w = 0.35 MPa, CI = 960 kPa

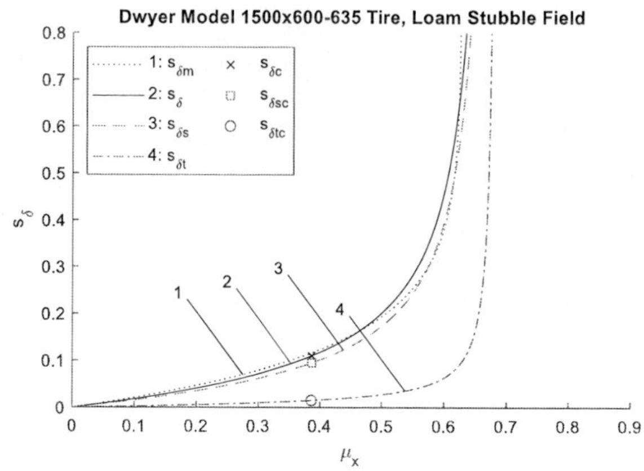

Fig. 5. Dwyer model curve fit and split slippage curves for 1500x600-635 tire on loam stubble field, R_z = 50 kN, p_w = 0.35 MPa, CI = 960 kPa

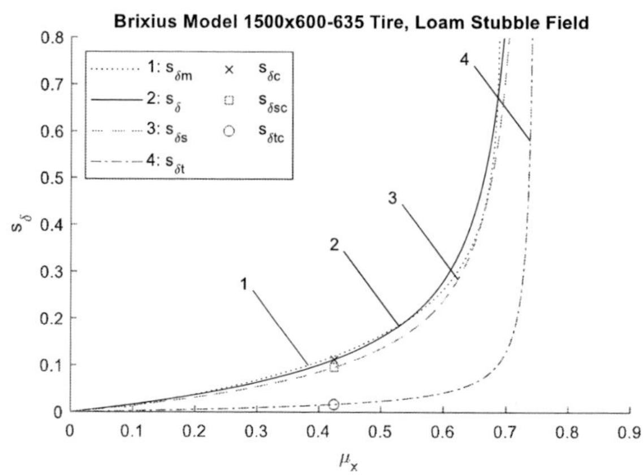

Fig. 6. Brixius model curve fit and split slippage curves for 1500x600-635 tire on loam stubble field, R_z = 50 kN, p_w = 0.35 MPa, CI = 960 kPa

Characteristic values for all three models are given in Table 6.

Table 6
Cone index models characteristic values

Characteristic	Wismer and Luth	Dwyer	Brixius
Peak friction coefficient μ_{px}	0.8454	0.6814	0.7483
Characteristic slippage $s_{\delta c}$	0.1939	0.1101	0.1123
Characteristic slippage (soil) $s_{\delta sc}$	0.1751	0.0950	0.0957
Characteristic slippage (tire) $s_{\delta tc}$	0.0188	0.0151	0.0166
Critical friction coefficient μ_{x-cr}	0.4800	0.3868	0.4247
Mean absolute error MAE	0.0086	0.0109	0.0094
Coefficient of determination r^2	0.9970	0.9909	0.9941

Comparing the results of all five models, there were small differences in the characteristic slippage component of the tire $s_{\delta t}$. The bigger difference between the models is accounted for by the soil component $s_{\delta s}$. The models were computed using the same tire data but differ in how they structure and parameterize the tire-soil interaction. The proposed general exponential function approximated all five models well as indicated by the high r^2 values above 0.98 and mean errors below 0. 011.

4. Slip Power Losses

A model of the tire and soil contributions to the slip power losses is derived as an application of the split tire soil model to vehicle and tire energy efficiency assessment for a straight-line motion of the wheel. The wheel-terrain interaction is described as a power-consuming system with an input power, an output power, and components that account for their difference.

For a wheel moving at a constant speed, the wheel power balance can be expressed as (Andreev et al., 2010)

$$P_w^{in} = P_f + P_\delta + P_w^{out} \tag{34}$$

The input power P_w^{in} comes from the driving torque applied to the wheel.

$$P_w^{in} = T_w \omega_w \tag{35}$$

P_w^{out} is the output power transmitted from the wheel to the vehicle body in the longitudinal direction. Rolling resistance power P_f is the power used to overcome the rolling resistance force R_x.

$$P_f = R_x V_x \tag{36}$$

Slip power P_δ is the power lost in slippage (due to deflection of the tire and the soil in the longitudinal direction and relative motion between the tire and surface). Slip power results in loss of the linear velocity of the wheel given by V_δ from Eq. 2.

$$P_\delta = F_x V_t s_\delta = F_x V_\delta \tag{37}$$

4.1. Split of Power Losses

Based on the above formulation for the slip power, the power losses due to the slippage can be split into the tire and soil contributions to the slip power P_δ. The s_δ term in the slip velocity is replaced with the sum of the tire and soil contributions to slippage

$$V_\delta = V_t s_\delta = V_t(s_{\delta t} + s_{\delta s}) \tag{38}$$

making two components of the slip velocity, $V_{\delta s}$ and $V_{\delta t}$.

$$V_{\delta t} = V_t s_{\delta t} \tag{39}$$

$$V_{\delta s} = V_t s_{\delta s} \tag{40}$$

The slip velocities, theoretical velocity, and actual velocity are plotted in Fig. 7 for the slippage computed from the Guskov model for the 1500x600-635 tire, showing the breakdown of the decrease in V_x from V_t as the slippage rises when the wheel is assigned a theoretical velocity of 5 m/s on an ideally flat surface.

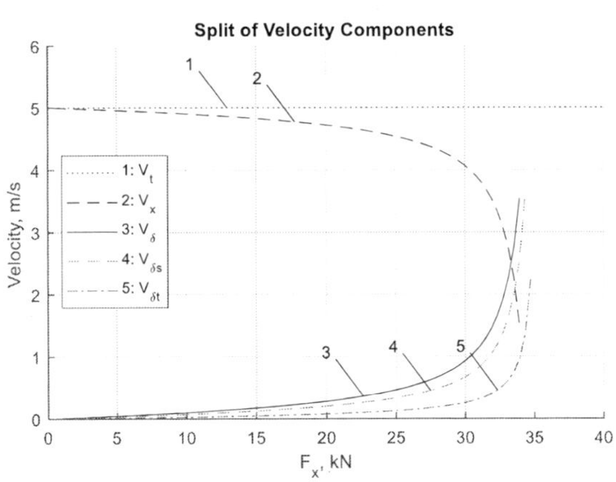

Fig. 7. Split of velocity components: 1500x600-635 tire on light clay, clay & sand field post wheat harvest, R_z = 50 kN, p_w = 0.35 MPa (Guskov model)

The contributions to the slip power losses are determined by splitting P_δ from Eq. 37 into the tire and soil components.

$$P_{\delta s} = F_x V_t s_{\delta s} = F_x V_{\delta s} \tag{41}$$

$$P_{\delta t} = F_x V_t s_{\delta t} = F_x V_{\delta t} \tag{42}$$

Added together, the two terms make up the total slip power loss P_δ.

$$P_\delta = P_{\delta t} + P_{\delta s} \tag{43}$$

The slip power loss components are plotted in Fig. 8.

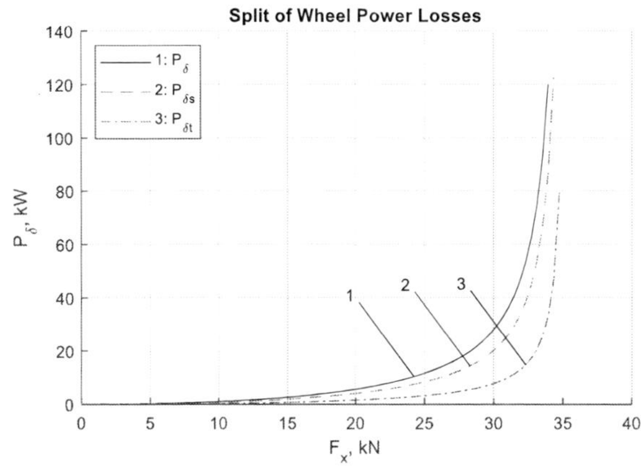

Fig. 8. Split of wheel power loss components: 1500x600-635 tire on light clay, clay & sand field post wheat harvest, R_z = 50 kN, p_w = 0.35 MPa (Guskov model)

Comparing Fig. 3 to Fig. 8, the higher contribution to slippage from the soil results in it making up a greater component of the overall power losses due to slippage.

4.2. Split of the Wheel Power Efficiency

To complete the analysis of the power losses, the contributions of the tire and soil components to the efficiency of the wheel power is derived. The theoretical velocity of wheel represents the maximum velocity that would be possible if there were no power losses. Since some slippage is inevitable when a torque is applied to a wheel, actual velocity V_x is always less than V_t. The reduction in linear velocity caused by the tire slippage can be therefore expressed as an efficiency $\eta_{\delta w}$ using Eq. 44 representing the velocity lost due to slippage.

$$\eta_{\delta w} = \frac{V_x}{V_t} = \frac{V_t - V_\delta}{V_t} \tag{44}$$

By breaking V_δ into the tire and soil contributions $V_{\delta t}$ and $V_{\delta s}$, $\eta_{\delta w}$ becomes

$$\begin{aligned} \eta_{\delta w} &= \frac{V_t - V_{\delta t} - V_{\delta s}}{V_t} = \frac{V_t - V_{\delta t}}{V_t} - \frac{V_{\delta s}}{V_t} + 1 - 1 \\ &= \frac{V_t - V_{\delta t}}{V_t} + \frac{V_t - V_{\delta s}}{V_t} - 1 \end{aligned} \tag{45}$$

resulting in

$$\eta_{\delta w} = \eta_{\delta t} + \eta_{\delta s} - 1 \tag{46}$$

where the split contributions to the efficiency are expressed as

$$\eta_{\delta s} = \frac{V_t - V_{\delta s}}{V_t} \tag{47}$$

and

$$\eta_{\delta t} = \frac{V_t - V_{\delta t}}{V_t} \tag{48}$$

The efficiency components are plotted in Fig. 9.

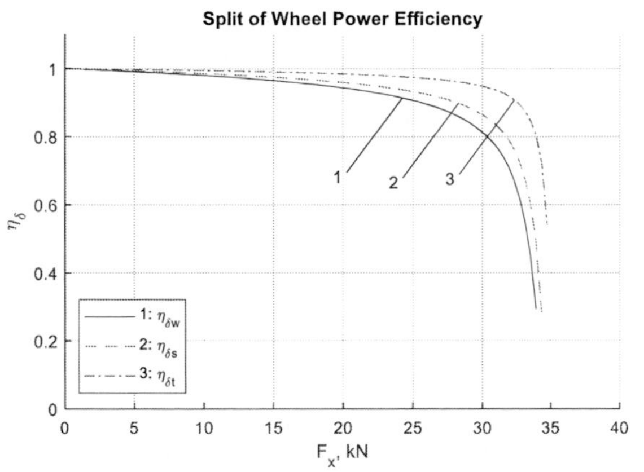

Fig. 9. Split of wheel power efficiency components: 1500x600-635 tire on light clay, clay & sand field post wheat harvest, R_z = 50 kN, p_w = 0.35 MPa (Guskov model)

From Fig. 9, it can be seen that the soil component's effect on the efficiency dominates over most of the range of F_x, with the tire component only increasing at very high values of F_x.

5. Conclusion

In this paper, a method for analyzing the tire and soil contributions to the tire slippage and slip power losses was considered. A model of traction-longitudinal slippage curve was investigated as a general model to approximate the traction-slippage curve of other models with a low degree of error. By approximating the traction curve with this model, the slippage can be split into the tire and soil components without the need to derive new equations for each model. Equations for separating and analyzing the tire and soil contributions to the slip power losses and efficiency were derived for the model. Estimating the slip power loss separately for the tire and soil enables better understanding the tire contribution and therefore also has applications to tire selection and conceptual design for energy efficiency.

6. References

Andreev, A.F., Kabanau, V.I., Vantsevich, V.V., 2010. Driveline systems of ground vehicles: theory and design, CRC Press, Boca Raton.

Andreev, A.F., Vantsevich, V.V., 2017. Tyre and soil contribution to tyre traction characteristic, in: Spiryagin, M., Gordon, T., Cole, C., McSweeney, T. (Eds), Dynamics of Vehicles on Roads and Tracks Vol. 1. CRC Press/Balkema, Leiden.

Brixius, W.W., 1987. Traction prediction equations for bias ply tires. Am. soc. of Agricultural Engineers 87-1622, St. Joseph.

Dwyer, M.J., Comely, D.R., Evernden, D.W., 1974. The Field Performance of Some Tractor Tyres Related to Soil Mechanical Properties. J. Agric. Engng Res. 19, 35-50.

Farhadi, P., Golmohammadi, A., Malvajerdi, A.S., Shahgholi, G., 2020. Tire and soil effects on power loss: Measurement and comparison with finite element model results. J. Terramech. 92, 13–22.

Gee-Clough, D., 1980. Selection of Tyre Sizes for Agricultural Vehicles. J. Agric. Engng Rcs. 25, 261-278.

Guskov, V.V., 1977. Tractors. Volume II, Theory, Vysheishaya shkola Publishing House, Minsk.

Huang, W., Wong, J.Y. Preston-Thomas, J., Jayakumar, P., 2020. Predicting terrain parameters for physics-based vehicle mobility models from cone index data, J. Terramech. 88, 29–40.

Janosi, Z., Hanamoto, B., 1961. The Analytical Determination of Drawbar Pull as a Function of Slip for Tracked Vehicles in Deformable Soils. Proceedings of the ISTVS 1st International Conference, Torino.

Kacigin, V.V., Guskov, V.V., 1968. The Basis of Tractor Performance Theory, J. Terramech. 5(3), 43–66.

Katsygin, V.V., 1964. Fundamentals of the theory of selection of optimal parameters of mobile agricultural machines and implements. In: Questions of agricultural mechanics, Volume 13, 5–165, Minsk.

Lopez-Arreguin, A.J.R., Montenegro, S., 2019. Improving engineering models of terramechanics for planetary exploration. Results in Engineering 3, 100027.

Maclaurin, B., 2014. Using a modified version of the Magic Formula to describe the traction/slip relationships of tyres in soft cohesive soils. J. Terramech. 52, 1–7.

Mason, G.L., Vahedifard, F., Caster, T.J., Priddy, J.D., 2022. A Unified Equation for Predicting Gross Traction for Wheels on Clay Over a Range of Braked, Towed, and Powered Operations, J. Terramech. 104, 1–13.

Tiwari, V.K., Pandey, K.P., Pranav, P.K., 2010. A review on traction prediction equations. J. Terramech. 47, 191–199.

Turnage, G.W., 1972. Performance of Soils Under Tire Loads Report 8: Application of Test Results to Tire Selection for Off-Road Vehicles. US Army Engineer Waterways Experiment Station Technical Report 3-666, Vicksburg.

Wismer, R.D., Luth, H.J., 1973. Off-Road Traction Prediction for Wheeled Vehicles. J. Terramech. 10(2), 49–61.

Whitson, J.A., Paldan, J.R., Vantsevich, V.V., Gorsich, D.J., Letherwood, M.D., 2020. A Split Tire-Soil Traction Model for Analysis of Tire Characteristic Slippage and Tire Database. Proceedings of the ISTVS 20th International Conference, Montréal.

Wong, J.Y., 2022. Theory of Ground Vehicles, fifth ed., Wiley, Hoboken.

Wong, J.Y., 2024. Terramechanics and Off-Road Vehicle Engineering, third ed., Butterworth-Heinemann, London.

0954 | Enhancing combined slip manoeuvres of high centre-of-mass off-road vehicles using a semi-active suspension system

Christos Kapareliotis, P. Schalk Els *, Andries Peenze

Department of Mechanical and Aeronautical Engineering, University of Pretoria, Pretoria, South Africa

** Corresponding author: schalk.els@up.ac.za*

ABSTRACT

Emergency driving scenarios require combined slip, where tyres generate longitudinal and lateral forces to slow the vehicle while following a path. Stability systems e.g. ABS brakes and Electronic Stability Control assist in regaining control after loss of stability. These systems do not actively prevent instability. ABS and ESC systems do not have direct rollover prevention, which is necessary in vehicles with a high centre of mass.

For off-road vehicles, this challenge is exaggerated by uneven terrain, loose soil, and rollover risk. Stability control systems struggle with unpredictable road inputs, often compromising handling. A preventative system is needed that manipulates load transfer to improve tyre contact before traction limits are exceeded, and where the roll angle of the vehicle can be directly controlled.

This study proposes using a semi-active suspension system to improve performance during combined slip manoeuvres. By adjusting suspension parameters, the system manipulates load transfer and enhances overall traction limits, reducing instability and rollover risk while maintaining steering and braking.

A vehicle, equipped with a controllable semi-active suspension system, is used as a test platform. A high-fidelity model is updated and used for parametric studies. Results are validated using experimental tests, with open-loop manoeuvres using brake and steering robots for repeatability and controlled assessment.

Preliminary findings indicate that a semi-active suspension system can manipulate combined slip performance, with potential to reduce instability and rollover. While this research is performed on smooth road surfaces, it provides a critical foundation for off-road applications, where terrain variability further challenges the vehicle's stability.

Keywords
Semi-Active Suspension
Combined Slip
High centre-of-mass vehicle
Vehicle Model
Experimental testing

1. Introduction

Every journey on the road carries an expectation of safety, yet for many, this expectation is not met. In countries like South Africa, where the majority make use of vehicles as their primary form of transportation, the challenge of maintaining road safety is evident. Every two hours, lives are lost to road traffic incidents according to the World Health Organization (WHO), which places the country's road traffic mortality rate at 24.5 deaths per 100,000 population (World Health Organization, 2024).

Many accidents that occur are due to bad weather and/or driving above the speed limit. During such events, the vehicle typically loses control when the brakes are being applied while steering in an attempt to avoid the incident. This requirement by the driver is unfavourable for tyre force generation as the tyres lose all lateral force capabilities when they lock up (Gillespie, 1992). The amount of force that the tyre can supply relies on various factors such as tyre composition, pressure, temperature, wear, tread pattern, contact patch, vertical load, and road surface characteristics (Jazar, 2008). On off-road terrain, the situation becomes even more complex. This is due to factors such as sinkage, lower traction and the bulldozer effect.

Many of these variables are not easily controllable, if at all, during emergency scenarios. At the heart of this issue is the dynamics of the vehicle during such manoeuvres. Vehicle dynamics is a field that governs how vehicles respond to driver inputs and external forces. Enhancing vehicle dynamics, particularly through advanced systems or strategies like Anti-lock Braking Systems (ABS), Electronic Stability Control (ESC), and controllable suspension, holds the promise of improving these statistics by improving vehicle performance and potentially saving lives. Where vehicle safety systems such as ABS and ESC may fall short in preventing collisions is that they do not have the means of manipulating load transfer to the tyres. This means that tyres could lose contact, which makes these systems unable to do what they were intended for.

The objective of this paper is to investigate the feasibility of using a semi-active suspension system to manipulate the load transfer of the vehicle, improving stability, safety, and overall performance during combined braking and steering manoeuvres. The semi-active suspension system allows for adjustment of the way the tyres are loaded during dynamic manoeuvres and thus aids the vehicle's performance with or without existing safety systems such as ABS or ESC. The feasibility study will be done by running simulations with a high-fidelity vehicle model while comparing the results to experimental tests using various suspension settings. While this research assesses smooth road surfaces, it provides a critical foundation for off-road applications, where terrain variability further challenges the vehicle's stability.

This paper presents a comparison between experimental and simulation data of a Land Rover Defender 110 performing a combined braking and cornering manoeuvre. Vehicle setup is described, followed by simulation and experimental setups. The results are compared and discussed, highlighting the key performance indicators. These findings are consolidated, and future work is further discussed.

2. Vehicle and Suspension System

The vehicle used in this investigation is a 1997 Land Rover Defender 110, which is equipped with a hydro-pneumatic semi-active suspension system derived from Els (2006). The Land Rover provides a good test platform since it has a high centre of mass and

https://doi.org/10.56884/GB4ZTZZK

can be easily modified. The 4S4 suspension system (4-state semi-active suspension) by Els (2006), named for its capability to achieve four distinct suspension states, is achieved through two spring settings and two damping settings, which can be combined by controlling solenoid valves. Since then, the 4S4 suspension system has been modified by De Wet (2020) by adding a variable valve allowing for continuous variable damping. A schematic of the modified 4S4 is displayed in Fig. 1.

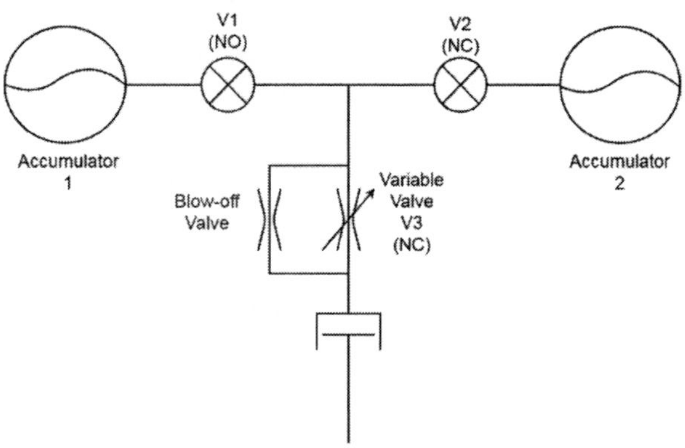

Fig. 1. Modified 4S4 suspension system schematic (De Wet, 2020)

3. Simulation Model

Thoresson et al. (2009) developed a comprehensive model of the Land Rover using MSC ADAMS, with some improvements done by Hamersma and Els (2014). When using the model for combined braking and steering manoeuvres, further modifications had to be made to account for the reaction forces caused by the brakes. These reaction forces significantly affect the dynamics of the vehicle and thus have to be accounted for. The current model being used for simulation is illustrated in Fig. 2. This non-linear model incorporates 15 unconstrained Degrees of Freedom (DOF) representing the Land Rover Defender. This model runs in a co-simulation with MATLAB Simulink where all the parameters and controllers are set.

The simulation model is currently using a Pacejka '89 tyre model on a smooth road. The simulations are done on a smooth road to ensure that the model is capturing the expected physics. The experimental tests to date could only be done on a smooth road due to facility limitations and thus it is appropriate to simulate on a similar surface.

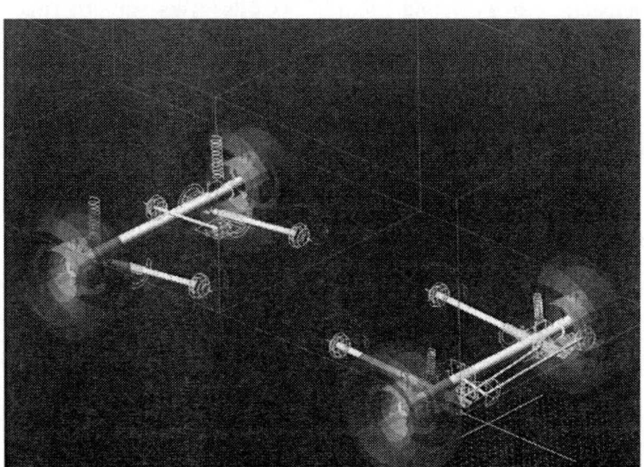

Fig. 2. 15 DOF MSC.ADAMS model of Land Rover Defender

4. Experimental Setup

As mentioned in Section 2, a 1997 Land Rover Defender 110 will be used as the experimental vehicle. The vehicle is equipped with a hydro-pneumatic semi-active suspension system, a steering robot, and a brake robot. The purpose of the steering and brake robots is to ensure controlled and repeatable testing. The steering robot is only used to lock the steering wheel, while the brake robot is controlled to maintain a constant desired brake Pressure. This setup is depicted in Fig. 3. and is explained further in Section 5. For safety purposes, the vehicle is fitted with outriggers on the front and rear to ensure that the vehicle does not roll over during the testing.

Fig. 3. Steering and brake robot on test vehicle

The control for the suspension system, steering robot and brake robot is set up in MATLAB Simulink and deployed to a dSPACE MicroAutoBox II (MABX-2). The MABX-2 additionally serves as the data acquisition device to record all the data. The GNSS and IMU data are obtained from VBOX 3i with dual GNSS antennae, RTK base station and IMU integration.

5. Test Scenarios

Identifying an effective performance metric for combined braking and steering manoeuvres is a complex task. To initiate the analysis of combined braking and steering manoeuvres, ISO 7975 (International Organization for Standardization, 2019) was studied. ISO 7975 (International Organization for Standardization, 2019) is an open-loop test which involves driving at a constant radius as carried out by ISO 4138 (International Organization for Standardization, 2012) followed by braking at a constant deceleration while keeping the hand steering wheel in a fixed position. This emphasises the importance of the braking and steering robot to ensure consistent and repeatable experimental testing.

Many performance metrics are used in the standard, such as yaw rate ratio (displayed by Eq. 1), maximum sideslip angle, and path deviation. Figure 4 depicts the definition of path deviation as used in ISO 7975 (International Organization for Standardization, 2019).

$$\psi_{ratio} = \frac{\psi_{max}}{\psi_{Ref,t_{max}}} \tag{1}$$

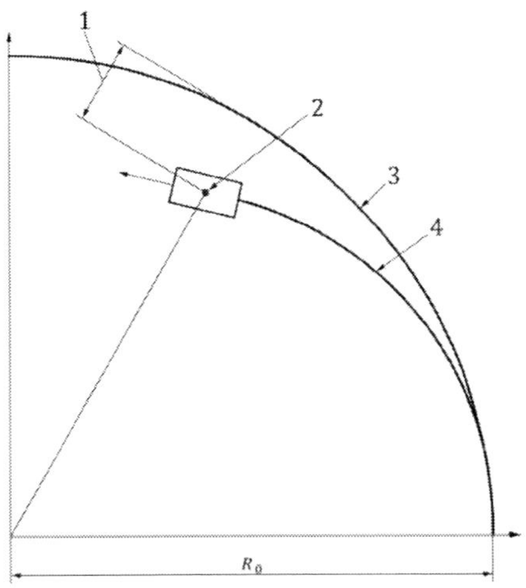

Fig. 4. Definition of path deviation from ISO 7975(International Organization for Standardization, 2019). (1 - Path deviation, 2 - Vehicle referenced point, 3 - Initial circular path, 4 - Path of reference point, Ro – initial radius.)

All tests presented in this paper begin from rest on a 100-meter radius asphalt test track, with the vehicle progressively accelerating until it reaches a speed of 20 metres per second (approximately 70 kilometres per hour). Once the vehicle stabilises at this speed, the steering wheel is locked in position, and the brake robot applies the brakes to achieve a constant desired force. It should be noted that in the simulation, a perfectly smooth road surface was used.

For the parametric studies in the simulation, the standard procedure was modified when determining the desired lateral acceleration. ISO 7975 (International Organization for Standardization, 2019) recommends adjusting the vehicle's speed to achieve the desired lateral acceleration; however, this approach resulted in varying braking dynamics due to differences in vehicle speed when the brakes were applied. Since there are no size limitations in the simulation environment, the longitudinal velocity of the vehicle was kept constant across all tests, while the path radius was adjusted to achieve the desired lateral acceleration. This approach allows for a fair comparison when braking is applied at different lateral accelerations. In contrast, during experimental tests, this method is not feasible, and the approach prescribed in the standard must be followed.

6. Results and discussion

This section compares the experimental and simulation results for two edge cases: the soft and hard suspension modes. The soft suspension mode, typically used for optimal ride comfort, features soft springs and low damping on all four suspension struts. In contrast, the hard suspension mode represents a common, naive approach to improving handling and consists of stiff springs and high damping on all four wheels. Comparing these two extreme configurations provides a clearer indication of the model's general validity for different suspension setups.

The tests were conducted similarly to ISO 7975 (International Organization for Standardization, 2019) and followed the procedure outlined in Section 5. Only the results for a constant velocity of 20 metres per second are presented in this paper.

Due to the unreliability of the roll angle estimation obtained from the VBOX system, the relative roll angle between the body and the axle was used as an approximation for evaluating roll dynamics. This approximation is supported by the availability of reliable suspension travel measurements on the vehicle.

To ensure comparable conditions between the simulation and experimental tests, the simulation model uses the same velocity profile and path followed by the vehicle during testing. If the model is correctly configured, the resulting dynamics should closely match those observed in the experiment.

6.1. Soft suspension mode

Figures 5 to 7 compare the experimental and simulation results for the vehicle while it is in soft suspension mode. Both the lateral acceleration and longitudinal deceleration for this manoeuvre in the soft suspension mode are approximately 3 meters per second squared. The longitudinal deceleration for this test was low due to the vehicle's rollover tendency in this suspension mode.

Figure 5 illustrates the relative roll angle between the axle and the body for the front and rear axles. The simulation results indicate that the effective roll stiffness of the vehicle is stiffer in simulation when compared to the experimental tests. This observation is made due to the lower gradient of the roll angle as the speed of the vehicle increases, thus increasing the lateral acceleration.

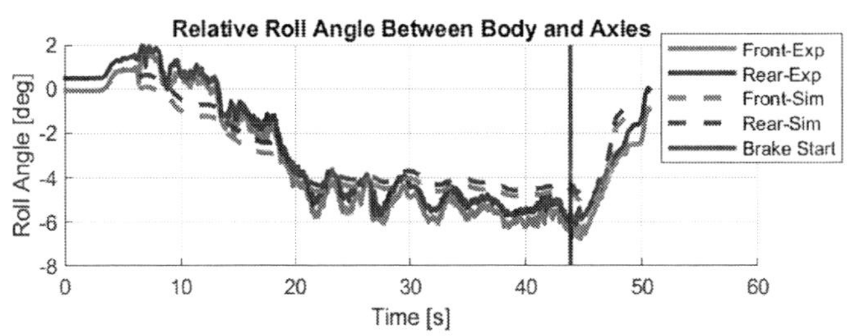

Fig. 5. Relative Roll Angle Between Body and Axles (Front and Rear) in soft suspension mode

Figure 6 illustrates the roll and yaw rates experienced by the vehicle during the manoeuvre. The simulation and experimental roll rate results are generally similar; however, the experimental data exhibits higher fluctuations. These discrepancies may be attributed to factors such as road camber and surface imperfections, which were not included in the simulation and therefore do not appear in the simulated results.

The yaw rate simulation results show a good fit with the experimental data. Road imperfections are not expected to significantly affect the yaw rate, as they typically do not excite the vehicle in the yaw plane as strongly as they do in the roll plane.

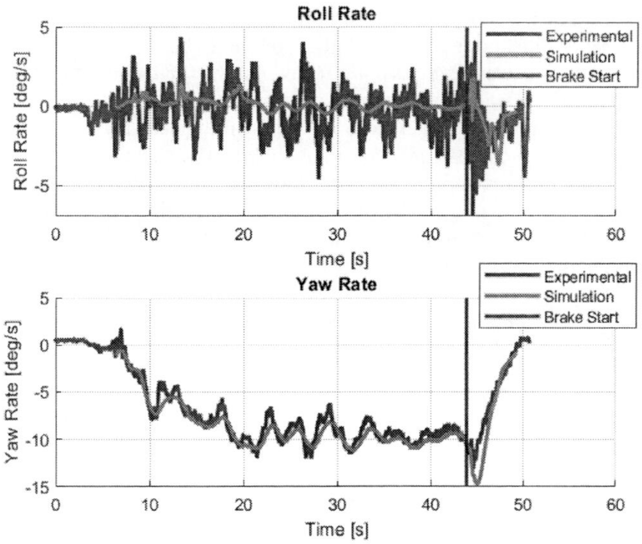

Fig. 6. Roll and Yaw rates in soft suspension mode

Figure 7 illustrates the sideslip angle of the vehicle during the manoeuvre. It is evident that the discrepancy between the experimental and simulation results increases with higher lateral accelerations. One proposed explanation for this error is that the tyre model used in the simulation does not fully represent the real tyre behaviour. As a result, the simulated vehicle is able to follow a similar path as in the experimental test, but with a lower sideslip angle.

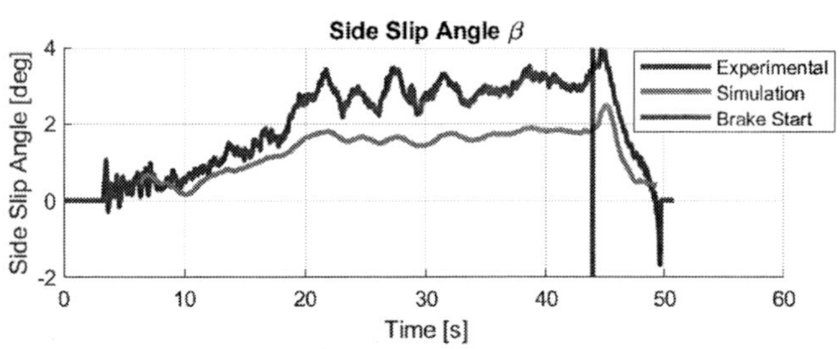

Fig. 7. Side Slip Angle of the vehicle body in soft suspension mode

6.2. Hard suspension mode

Figures 8 to 10 compare the experimental and simulation results for the vehicle while it is in hard suspension mode. Similar results were achieved, although there are higher discrepancies in the roll angle comparison. However, in general, the model is a good representation of the test results as it follows the trends, although the magnitudes are off. Thus, the model is suitable for future analysis. These discrepancies are thought to be tyre model related. However, further investigation is needed before a reason can be concluded.

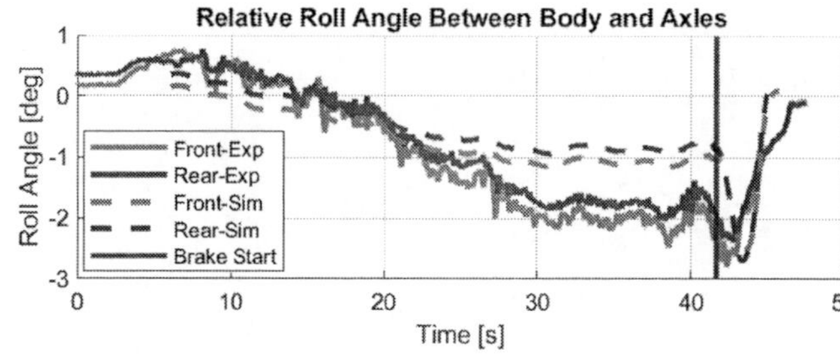

Fig. 8. Relative Roll Angle Between Body and Axles (Front and Rear) in hard suspension mode

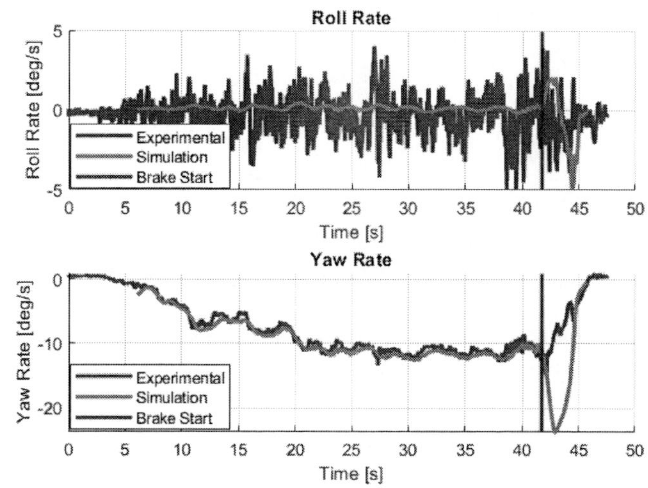

Fig. 9. Roll and Yaw rates in hard suspension mode

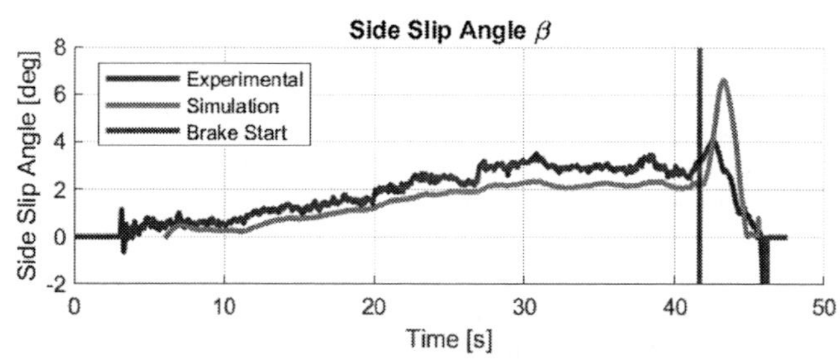

Fig. 10. Side Slip Angle of the vehicle body in hard suspension mode

6.3. Performance metrics

Figures 11 to 14 illustrate selected performance metrics defined in ISO 7975 (International Organization for Standardization, 2019), with the exception of the RMS of the path deviation. While ISO 7975 recommends using the path deviation metric as defined in Fig. 4, it was observed in practice that a vehicle could deviate from the path and subsequently return to it, which may result in a misleadingly low deviation value. To better capture the true path-following behaviour, a metric that considers the vehicle's path history from the moment braking is initiated until the vehicle comes to rest is more appropriate. Therefore, the RMS of the path deviation over this interval is proposed, as it provides a more comprehensive representation of the vehicle's deviation from the desired path.

As discussed in Section 5, these simulation tests were performed at a constant longitudinal velocity, with varying path radii used to achieve the desired lateral accelerations. The longitudinal velocity was set to 20 metres per second, consistent with the experimental tests. The legend in Fig. 11 indicates the lateral accelerations in terms of gravitational acceleration, and this legend applies to Fig.s 11 through 14.

In this study, the suspension settings for the front and rear axles were allowed to differ, while the damper setting was kept the same for all four wheels. To differentiate between the suspension configurations, the following notation is used: the first letter represents the front suspension setting, the second letter the rear suspension setting, and the subscript indicates the damping setting. The letter "h" denotes a hard setting, and "s" denotes a soft setting. For example, "hs_h" indicates a hard setting in the front, a soft setting in the rear and high damping.

Figure 11 presents the maximum sideslip angle for the various suspension configurations. For lateral accelerations greater than 0.1g, configurations with a hard front suspension generally result in larger sideslip angles. The worst-case scenario occurs with a hard front, soft rear, and soft damping configuration. In contrast, the vehicle demonstrates more consistent behaviour when the front suspension is soft. This suggests that, based on the maximum sideslip angle metric, a soft suspension setting in the front of the vehicle generally outperforms a harder configuration. Although this may go against vehicle dynamics intuition, it is true due to the fact that the Land Rover Defender's leading arm to axle bushes act as a stiff antiroll bar, and thus softening the suspension in the front helps minimise the lateral load transfer on the front axle. Allowing for more lateral force generation by the front tyres. Since the rear has a very low roll stiffness, the inside rear wheel will lift off the ground in high lateral manoeuvres when the rebound stops are reached, thus inducing large slip angles due to the low lateral force generation by the rear tyres. This higher lateral force generation of the front tyres will result in less slip experienced by the front wheels, which will improve the stability of the vehicle and achieve a lower slide slip angle. In future work, it would be useful to extract the slip angles of the front and rear axles independently so oversteer and understeer tendencies could be discussed.

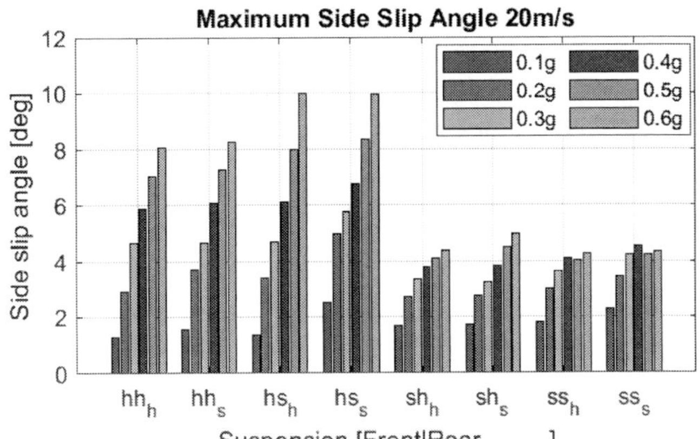

Fig. 11. Maximum Side Slip Angle for various suspension modes

Figure 12 illustrates the maximum roll angle for the various suspension settings. The results indicate that a stiffer front suspension generally results in a lower roll angle compared to configurations with softer front springs, with the exception of the "hs_s" setting. This observation suggests that the vehicle's roll angle is more strongly influenced by the front suspension stiffness than by the rear.

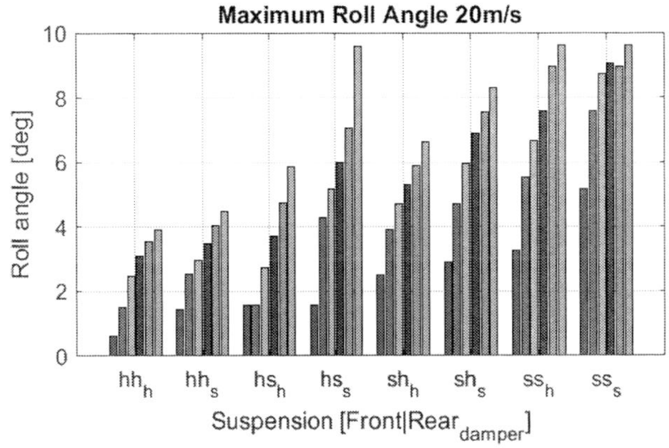

Fig. 12. Maximum Roll Angle for various suspension modes

Figure 13 illustrates the RMS of the path deviation from the moment the brakes are applied until the vehicle comes to rest. No single suspension setting is ideal across all lateral accelerations. The results suggest that for lateral accelerations below 0.3g, a stiff front spring reduces the RMS of the path deviation. In contrast, for accelerations above 0.3g, a softer front spring yields better performance.

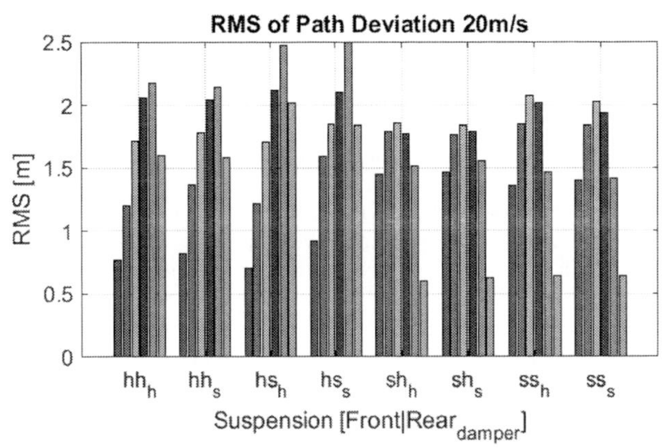

Fig. 13. RMS of Path Deviation for various suspension modes

Figure 14 illustrates the yaw rate ratio for the various suspension settings. For lateral accelerations greater than 0.2g, a softer front spring is more favourable. Conversely, for accelerations below this threshold, a stiffer front spring tends to perform better, with the exception of the "hs_s" setting.

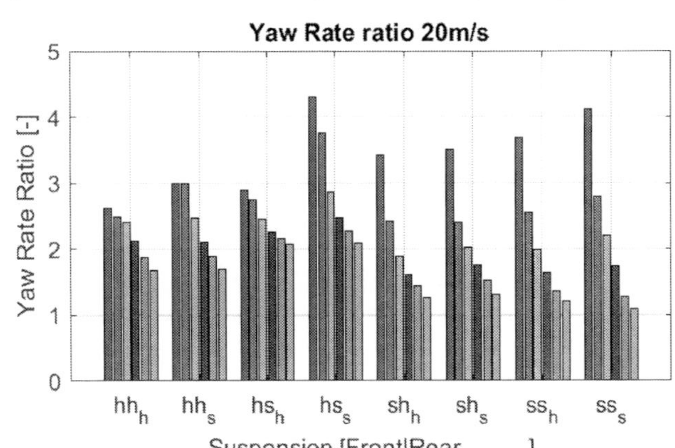

Fig. 14. Yaw Rate Ratio for various suspension modes

Based on the results in Figures 11 to 14, damping has a small effect but is not a significant determining factor when analysing the performance metrics. During a combined braking and steering manoeuvre that begins from steady-state cornering, the vehicle is already close to the posture it will adopt when braking is applied. As a result, strut velocities are relatively low compared to a scenario in which braking occurs without prior steady-state cornering. Therefore, the observed damper sensitivities may not apply to fully transient manoeuvres.

The evidence also suggests that a stiff suspension is not always advantageous in combined braking and steering scenarios, although it does offer certain benefits.

7. Conclusions and future work

Based on these findings, it is evident that a vehicle's dynamics during combined braking and cornering manoeuvres are sensitive to suspension settings, however, discrepancies were noted between simulation and experimental results, particularly in the sideslip and roll angles in the hard suspension setting, suggesting that the tyre model needs further refinement to improve correlation.

Given the current results from the model, the maximum side slip angle performance metric indicates that the SHh setting shows the best performance. This is due to the compromise that allows low lateral load transfer in the front of the vehicle by having the soft springs while stiffening the rear springs to minimise the tendency of wheel liftoff.

The best suspension setting to improve the roll angle of the vehicle is the HHh setting. This is the naive approach to better vehicle performance overall, although this setting does not do well in the other performance metrics when focusing more on the higher lateral accelerations. Thus, HHh is only the best setting to prevent rollover.

The best suspension setting to improve the rms of the path deviation is the SHh and SHs settings. This result complements the best setting for side slip angle, further indicating that this is a more stable suspension setting.

The sensitivity of the yaw rate ratio varies significantly with lateral acceleration. The best setting for low lateral accelerations is the HHh settings, while the best setting for the high lateral accelerations is the SSs setting. This is due to the metric being a normalised yaw rate with respect to the vehicle's steady state, and thus the trends will vary with different lateral accelerations.

Overall damping tends to be relatively insensitive for these results, which is expected since the test manoeuvre is performed close to quasi-static conditions and thus has very few transients.

The simulation model captures the general physical trends; however, further work is required to improve its correlation with experimental results. Secondary effects, such as suspension bushes, will be studied, and a sensitivity analysis of the tyre model is needed to identify the sources of discrepancy. An optimisation technique can be implemented to improve vehicle performance based on the discussed performance metrics.

Once the model has been refined, the study can progress to additional open-loop testing to evaluate the robustness of the improved suspension settings. A controller could then be implemented to adapt the suspension settings in real-time based on the vehicle's dynamic behaviour. Additionally, the vehicle could then be analysed on uneven terrain, off-road, and on soft soils. The damping effects could possibly be more pronounced on rough terrain and thus need to be studied.

8. Nomenclature

ψ_{ratio}	Yaw Rate Ratio	[-]
ψ_{max}	Maximum Yaw Rate	[deg/s]
$\psi_{Ref,t_{max}}$	Yaw Rate during steady state	[deg/s]

9. Acknowledgements

We would like to give thanks of gratitude to the Vehicle Dynamics Group (VDG) at the University of Pretoria for their expertise and research opportunities.

10. Declaration of competing interest

The authors declare that they have no known competing financial interests or personal relationships that could have appeared to influence the work reported in this paper.

11. Declaration of Generative AI and AI-assisted technologies in the writing process

During the preparation of this work, the author(s) used Grammarly and ChatGPT in order to correct grammar and improve readability. After using this tool/service, the author(s) reviewed and edited the content as needed and take(s) full responsibility for the content of the publication.

12. References

De Wet, B., 2020. Achieving Controllable Continuous Variable Damping within a Semi-Active Hydro-Pneumatic Suspension System. University of Pretoria (South Africa).

Els, P.S., 2006. The ride comfort vs. handling compromise for off-road vehicles. PhD, University of Pretoria.

Gillespie, T., 1992. Fundamentals of vehicle dynamics, SAE international.

Hamersma, H.A., Els, P.S., 2014. Improving the braking performance of a vehicle with ABS and a semi-active suspension system on a rough road. Journal of Terramechanics, 56, 91-101.

International Organization For Standardization, 2012. ISO 4138. Passenger cars — Steady-state circular driving behaviour — Open-loop test methods. Geneva, Switzerland.

International Organization For Standardization, 2019. ISO 7975. Passenger cars - Braking in a turn - Open-loop test method. Geneva, Switzerland.

Jazar, R.N. 2008. Vehicle dynamics, Springer.

Thoresson, M.J., Uys, P.E., Els, P.S. & Snyman, J.A., 2009. Efficient optimisation of a vehicle suspension system, using a gradient-based approximation method, Part 1: Mathematical modelling. Mathematical and Computer Modelling, 50, 1421-1436.

World Health Organization, 2024. Road traffic mortality rate (per 100 000 population). https://data.who.int/indicators/i/B9D9E6A/D6176E2 [Accessed 23 June 2024].

1868 | Modeling and simulation of autonomous vehicular systems at ERDC – III Verification and validation of M&S

Peilin Song *, Burhman Gates, Garrett Holden, Joshua Fairley

Mobility System Branch, Engineer Research and Development Center, Vicksburg, MS, USA

** Corresponding author: peilin.song@erdc.dren.mil*

ABSTRACT

Rigorous verification and validation (V&V) processes are essential for modeling and simulation (M&S) of autonomous vehicles in military applications, ensuring system reliability and safety within complex, dynamic environments. Only properly verified and validated Software Integrated Laboratories (SILs) can be effectively employed in broader operational scenarios, including challenging edge cases such as rain, fog, and dust.

This paper provides an overview of the V&V activities undertaken at the U.S. Army Engineer Research and Development Center (ERDC) for autonomous vehicle simulation. Recognizing the hierarchical structure of the SIL, a corresponding hierarchical V&V methodology was developed across various mission types. This methodology mandates component-level V&V prior to system-level assessment, providing a structured approach to ensure simulation accuracy and reliability. The case study presented herein demonstrates the effectiveness and utility of ERDC's V&V process in validating autonomous vehicle systems. Furthermore, the paper examines the frameworks and processes utilized, highlighting the challenges encountered and insights gained from applying V&V practices to autonomous vehicle development.

Keywords
Off-road mobility
Verification
Validation
Hierarchical Approach
Snow
Rain
Fog

1. Introduction

The U.S. Army Engineer Research and Development Center (ERDC) has been a key contributor to the advancement of off-road modeling and simulation (M&S), particularly in support of autonomous ground vehicles and robotics operating in complex, unstructured environments. Unlike on-road systems, off-road autonomy presents unique challenges due to variable terrain types, vegetation, weather conditions, and limited infrastructure.

To ensure the reliability of simulation results, we aim to perform Verification and Validation (V&V) of the M&S in well-understood, normal cases—such as standard terrain profiles, typical environmental conditions, and predictable vehicle behavior. Once the simulation is verified and validated under these controlled scenarios, it can then be applied to edge cases—such as degraded environments, unknown obstacles, or mission-critical conditions—with a high level of confidence in its predictive performance and decision-support value.

In this paper, we present our efforts in V&V on the M&S of autonomous systems, which are inherently complex and hierarchical simulation systems. Our contributions include:

- A methodology for V&V of hierarchical simulations, addressing the layered architecture and multi-resolution interactions within such systems.
- V&V of the M&S of autonomous systems for selected missions and operational contexts, focusing on both mobility and sensing performance.
- Application of the verified and validated simulation system to edge-case scenarios, such as driving in rain, fog, and reduced visibility conditions, to demonstrate the robustness and practical utility of our approach.

This work not only supports the development of resilient autonomous systems for off-road environments but also contributes to broader simulation science by offering a structured V&V framework for complex, mission-driven applications.

2. Verification and Validation of M&S of Autonomous Systems

The ERDC System Integration Laboratory (SIL) supports Modeling and Simulation (M&S) of autonomous systems by providing a credible platform for evaluating the performance of navigation and control capabilities within the Army Test and Evaluation (T&E) efforts for autonomy. A primary purpose of M&S in this context is to simulate Army T&E tasks, which are often costly, time-consuming, and resource-intensive when conducted through physical testing. High-fidelity simulation environments enable early-stage testing, system refinement, and risk reduction before committing to full-scale field trials.

However, performing Verification and Validation (V&V) on M&S for autonomous systems remains a significant challenge. These systems are not only large in scale but also hierarchical in architecture, involving complex interactions among perception, planning, control, and communication subsystems. Despite growing interest, a mature and standardized V&V methodology for such hierarchical simulation systems has yet to be established.

In this paper, we adopt a phase-based approach to address the V&V of hierarchical systems. This method begins with the V&V of individual subsystems and components, then incrementally elevates to the full system level. Such a layered V&V strategy closely mirrors the hierarchical structure of the autonomous system itself, making it both practical and conceptually aligned

https://doi.org/10.56884/VZRN23KK

with the system under test. V&V have been the topic since inception of computer-code based simulation and general discussion of V&V is beyond the scope of this paper. Durst (2017) gives a comprehensive view of V&V.

While Durst (2019) introduced a framework for V&V of autonomous robot simulations, his work focused primarily on validating individual subsystems (e.g., cameras) and did not address system-level integration. In contrast, our V&V approach targets both the subsystem and system levels within the Army's autonomy T&E framework, ensuring that individual components function correctly, and that the system as a whole behaves as expected.

As the Army's autonomy T&E activities continue through phase-based development, this report presents our initial V&V efforts, providing a foundational approach that will be expanded and refined as the simulation systems evolve to support increasingly complex scenarios and edge-case conditions.

2.1. *Hierarchical View of SIL*

The System Integration Laboratory (SIL) is structured using a hierarchical architecture (Fig. 1), reflecting the layered and modular nature of autonomous systems. This hierarchical view organizes the simulation environment into distinct functional layers:

- Top-Level Modules: Core subsystems such as Vehicle Dynamics and Sensor Simulator form the backbone of the SIL.
- Supporting Components: These subsystems are driven by underlying modules, including:
    - Scenes
    - Sensor Models
    - Vehicle Models

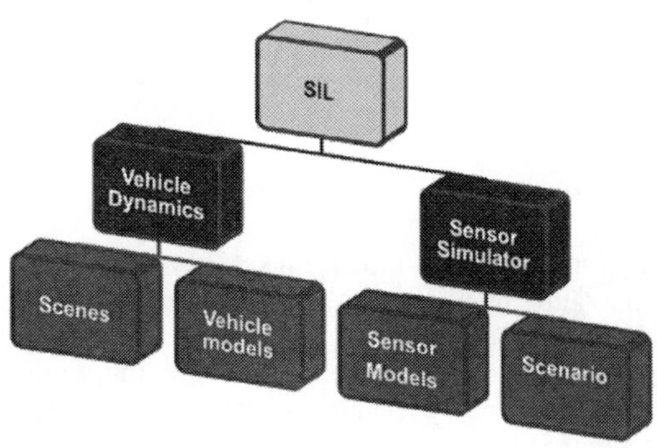

Fig. 1. Hierarchical View of SIL

This structured layering supports a hierarchical V&V approach, where verification and validation are performed progressively from individual components up to the integrated system level. This mirrors the architecture of real-world autonomous systems, ensuring each module is verified in isolation and in context.

2.2. Hierarchical V&V Process

The V&V process starts at the component level and builds upward:

- Vehicle Model in Unreal: Validates the fidelity of the vehicle's behavior within the simulation environment.
- Terrain Model: Ensures the simulated terrain accurately represents real-world operational conditions.
- Sensor Models: Verifies sensor behavior under various scenarios and environmental conditions.

While successful V&V at the component and subsystem levels provides confidence in those modules, it only guarantees partial verification at the system level. This is because the interconnections and interactions between subsystems—such as data flow, timing, and integration behavior—remain unverified until full system-level testing is conducted.

System-level verification builds upon successful lower-level V&V, but goes further by:

- Ensuring that communication protocols and data exchanges between components function correctly.
- Validating the integrated behavior of all subsystems as a cohesive whole.

System-level validation is performed by comparing SIL simulation results with actual test data collected from field tests (e.g., at Grayling, Michigan). This comparison confirms that the SIL accurately reproduces the performance of the real-world system under test conditions relevant to Army autonomy missions.

Importantly, the V&V results are conditional—they are valid only within the specific terrain types and mission profiles used in the verification and validation process. However, the verified and validated M&S system is designed to be extensible, enabling adaptation and application to new scenarios, including diverse environmental conditions such as adverse weather (e.g., rain, fog) and different mission contexts.

Some components within the SIL architecture are exempt from the V&V process:

Scenes: These are directly defined using measured field data and are treated as input data, not models to be verified.

Navigation Stack: This is the subject of evaluation, not part of the evaluation environment. It is a software suite used identically in both the simulation and physical vehicle, ensuring consistency of behavior across domains. Because it is the target of testing, it is not subject to verification and validation within the SIL itself.

3. V&V of Components in SIL

Verification and validation begin at the component level, focused on ensuring that foundational subsystems behave as intended:

Vehicle Dynamics in Unreal Engine 4 – Verified for accurate simulation of physical vehicle behavior, including motion, control response, and terrain interaction.

Sensor Simulation via ESE – Verified for accurate reproduction of environmental sensor outputs under different conditions and scenarios.

3.1. Lidar Model V&V in ESE

The lidar model in ESE is

$$P_R = \frac{C_a \rho \cdot e^{(-2\alpha L)}}{L^2} \quad (1)$$

$-P_R$ = Power Received $[W]$
$-C_a$ = LiDAR system constant $[W \cdot m^2]$

$-\rho =$ target reflectivity [%]
$-\alpha =$ extinction coefficient [m^{-1}]
$-L =$ range to target [m]

Implementation of governing equation in ESE lidar simulator guarantees the theoretical foundation and verifies the lidar model in ESE

3.2. Lidar validation in rain

Simulation has been conducted in the rain for two objects of different reflectivity, A and B.

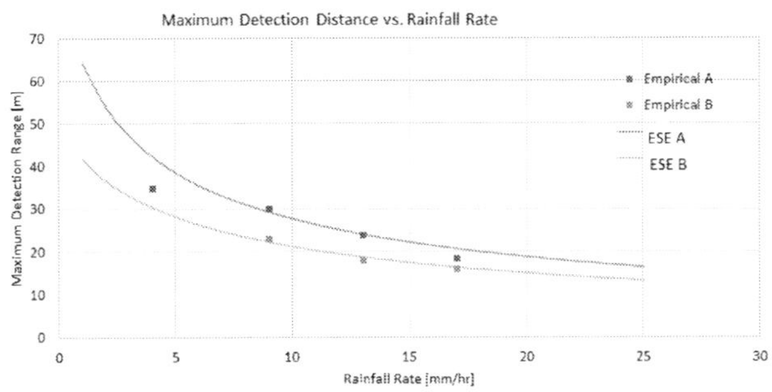

Fig. 2. Lidar Detection Range in Rain: Simulated vs Empirical

Figure 2 show a comparison of detection distance between simulated and empirical data for a type of lidar of interests. good fits are observed for both A and B. The empirical data are from Rasshofer (2011).

3.3. Lidar ESE Model Validation in Fog

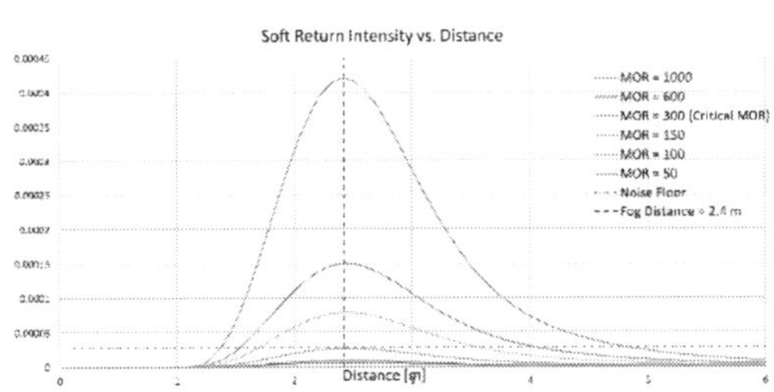

Fig. 3. Mean Optical Range and normalized Intensity

Figure 3 shows the relationship between normalized return intensity with distance for a variety of Mean Optical Range (MOR), which is validated by our test. Simulation results of the interested lidar with ESE are in Fig. 4, which demonstrates the donut shape in dense fog.

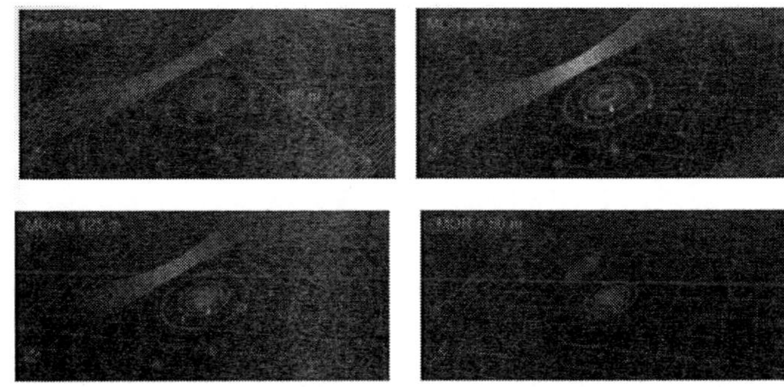

Fig. 4. Simulated Fog Effects with Variable MOR

3.4. Vehicle Dynamics V&V

The vehicle dynamics of the subject vehicle for mobility analysis are modeled and simulated in Unreal Engine using the PhysX real-time physics engine. The ERDC soft-soil model is employed to represent the contact mechanics between the terrain and the vehicle's wheels or tracks. The simulated data are compared with experimental test data in terms of acceleration, braking, lane change, and steering performance.

Tables 1 and 2 show the comparison results for acceleration and braking with a light tactical vehicle. The test and simulation are conducted on the concrete surface.

Table 1
Acceleration Comparison Between Test and Simulation

Goal(mph)	Test Time(s)	Sim Time(s)	Error (%)
10.0	1.1	1.2	8.18
20.0	2.5	2.5	0.0
30.0	4.6	4.6	0.0
40.0	7.8	7.8	0.0
60.0	27.4	37.0	1.46

Table 2
Braking Comparison Between Test and Simulation

Speed (mph)	Test (feet)	Simulation (feet)	Error (%)
20	18.0	18.1	0.6

4. System Level V&V of SIL

4.1. Verification of SIL

At the system level, verification and validation have been conducted to ensure model integrity and integration fidelity. Individual components have been verified and validated for the relevant application scenarios. System integration is tested using a SIL configuration, verifying the complete navigation pipeline. This includes: (1) simulated sensor signals—such as LiDAR point clouds, GPS, and IMU—being correctly fed into the navigation stack; (2) throttle and steering commands generated by the navigation stack being properly delivered to the vehicle dynamics model; and (3) the position and orientation feedback from the vehicle dynamics model being accurately provided to the sensor simulators. These checks ensure consistency and coherence across all subsystems, supporting reliable system-level performance.

4.2. Validation of SIL

ERDC SIL validation is conducted for a test vehicle at a designated test range under normal weather conditions, using two distinct missions—Mission A and Mission B—executed with different motion planners: the Maverick Motion Planner and the Vaquerito Motion Planner. The objective is to validate the ERDC SIL environment for the specific test vehicle operating under the tested autonomy system for specific missions. Only the process of validation of Mission A is presented in this paper.

By validating the SIL simulation against real-world test results in normal weather conditions, confidence is established in the simulation's fidelity. This validated SIL can then be used to predict autonomy performance in edge-case scenarios—such as rain, fog, or low-visibility environments—which are often difficult, costly, or hazardous to reproduce in physical tests.

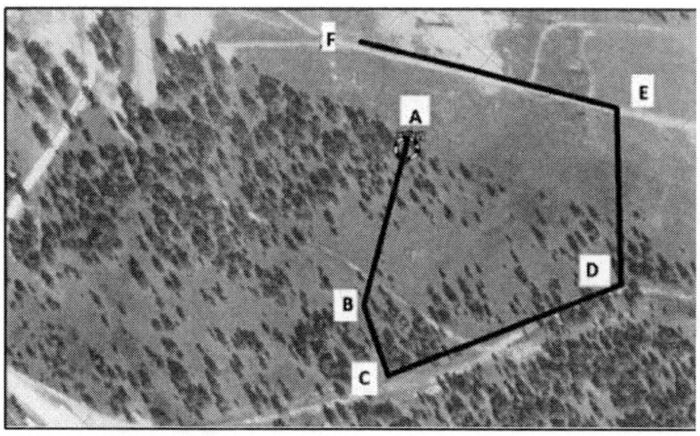

Fig. 5. Mission A Trajectory: A To F

Figure 5 illustrates the mission trajectory, progressing from point A to F. The trajectory can be segmented further with the density of vegetation: light medium and heavy (Fig. 9). To establish a baseline trajectory, three tests were conducted. However, trajectory forking is observed at two locations as a result of three-tree passing. This forking creates two distinct sets of trajectories, consequently leading to two baselines that encompass areas (as depicted in the upper-right image of Fig. 6). This forking phenomenon complicates the comparison between the collected test data and the three sets of simulation data. The comparison is performed in two steps: First, the variation among the simulation trajectories and their average is analyzed (Fig. 7), revealing a maximum Root Mean Square (RMS) difference of 2.5 meters. Second, the average of the simulated trajectories is compared against the two baselines derived from the test data (Fig. 8), yielding a maximum RMS difference of 2.5 meters as well.

Therefore, it can be concluded that the simulation for Mission A is validated, exhibiting a mean deviation of less than 5.0 meters.

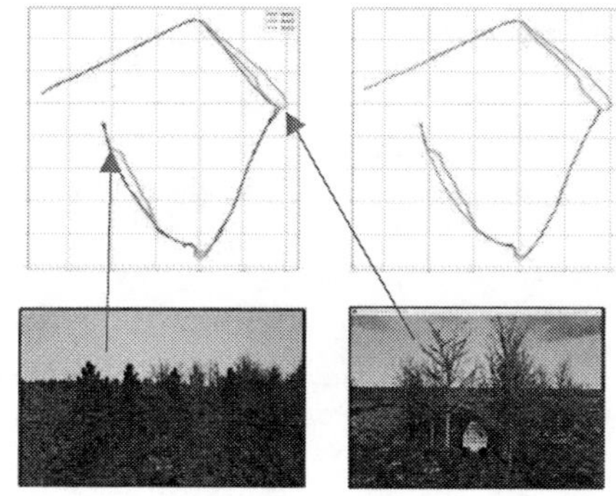

Fig. 6. Test Trajectories and Forking

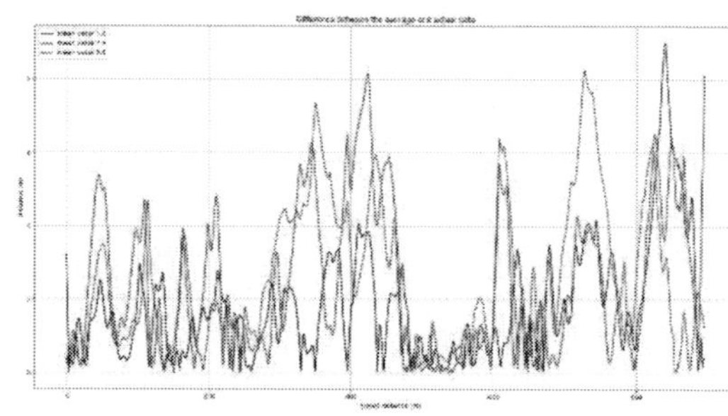

Fig. 7. Variations Between Average and Simulated Data

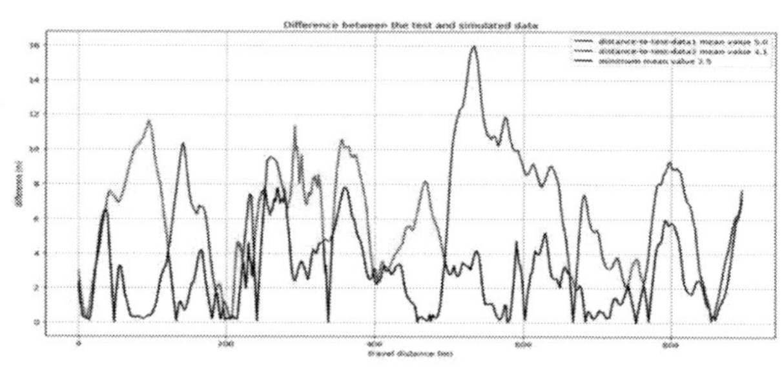

Fig. 8. Mean Distance of Simulated Average to Test Data of Two Baselines.

5. Applications of SIL to Edge Cases

With the M&S for normal weather conditions verified and validated, and with lidar models for both rain and fog already verified and validated, efforts have been made to extend the simulation for edge cases, specifically examining the mission's performance under:

- Rain: Simulating various intensities of rainfall and its impact on lidar measurements, including potential attenuation of the laser beams, increased noise from water droplets, and reduced effective range.
- Fog: Simulating different levels of fog density and its effect on lidar's ability to accurately perceive the environment due to scattering of the laser pulses by water particles. This includes assessing the degradation in point cloud density and accuracy at different distances.

5.1. Experimental Design for Edge Cases

Employing Design of Experiments (DoE) techniques, variations of these edge cases will be simulated across different levels of vegetation. This systematic approach will allow for the efficient exploration of the combined effects of weather conditions (rain and fog) and varying vegetation density on the mission's performance, specifically focusing on the validated lidar sensor model.

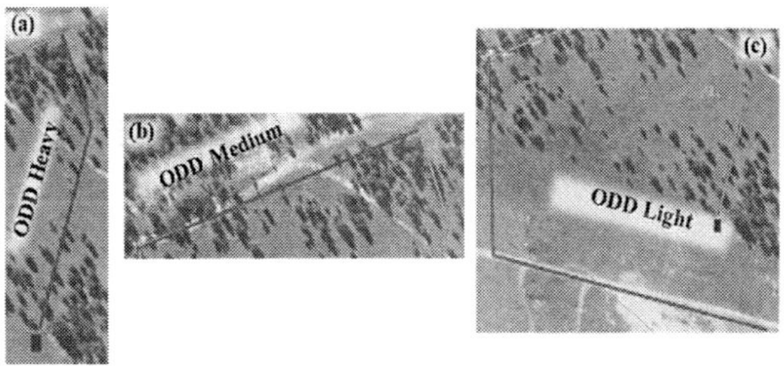

Fig. 9. Segmentation of Mission A Trajectory according to Vegetation Density (Blue Mark is the starting point of Complete Mission)

The Operational Design Domains (ODD) are presented in Fig. 10.

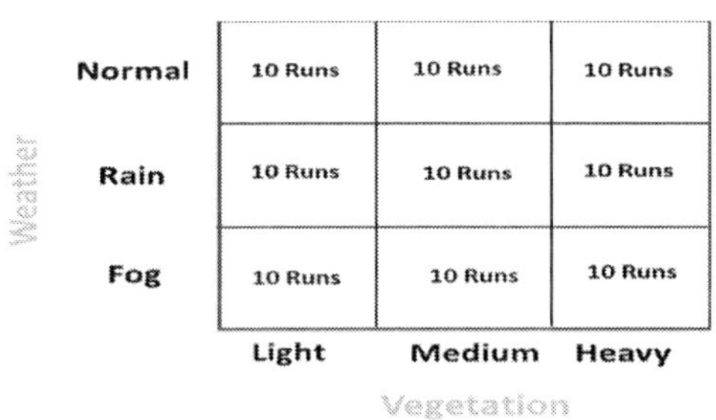

Fig. 10. Operational Design Domains for DOE

5.2. Simulation Result Analysis

Table 3
Perception-Based Vehicle Speed vs. Weather Effects

vegetation	Weather	Max Speed (km/h)	Ave Speed (km/h)	Distance (m)	Duration(s)
heavy	normal	4.86	1.85	250	136
heavy	rain	4.99	2.16	228	106
heavy	fog	4.96	2.10	242	115
medium	normal	5.10	1.83	260	154
medium	rain	5.18	2.31	250	110
medium	fog	5.15	2.24	255	115
light	normal	5.01	2.10	409	193
light	rain	5.15	2.51	386	155
light	fog	5.08	2.35	386	168

Out of 118 total runs, 90 were successful. The success rates for the tested conditions are Caldwell et al. (2024):

- Normal: 66.7% (indicated in red, suggesting a lower success rate compared to the others)
- Fog: 88.2%
- Rain: 100% (indicated in green, suggesting a perfect success rate)

In addition, it is also observed that rain and fog result in increased vehicle speeds (Table 3). This is an interesting and potentially counterintuitive finding, as one might typically expect adverse weather conditions like rain and fog to lead to reduced speeds for safety reasons. Further investigation would be needed to understand the underlying causes of this observed increase in vehicle speeds during simulated rain and fog conditions.

The anomaly exposed by simulation of edge cases leads to the examination of the cost map and lidar point cloud in navigation stack.

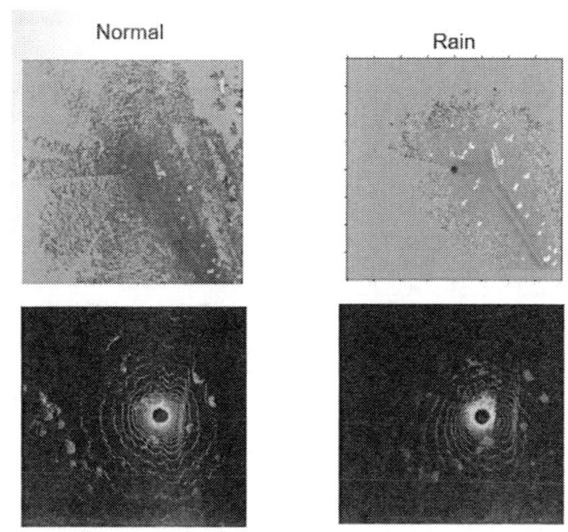

Fig. 11. Comparisons of Lidar Point cloud and Cost map between Normal and Raining Cases

Further investigation finds that rain significantly attenuates and scatters lidar signals, resulting in a noisy and less complete point cloud (Fig. 11). This degraded data leads to a cost map that obscures obstacles and misrepresents the traversable space. Consequently, the navigation stack, receiving an inaccurate perception of a more open environment, plans and executes more aggressive vehicle maneuvers (e.g., higher speeds, sharper turns) due to the underestimation of environmental complexity and potential hazards.

6. Conclusions

In summary, this work has underscored the critical importance of verified and validated (V&V) models for the development and deployment of safe and reliable autonomous systems. We have presented a detailed roadmap for V&V, encompassing methodologies and processes applicable across the development lifecycle, from the component level through integrated system testing. Furthermore, we have demonstrated the practical utility of a verified and validated model in extending simulation capabilities to explore challenging edge cases and generate predictions about system behavior under diverse operational conditions, as exemplified by the analysis of lidar performance in rainy environments. This comprehensive V&V approach is essential for building confidence in the robustness and dependability of autonomous systems before real-world deployment.

The value of verified and validated (V&V) models is further underscored by the observed anomaly of aggressive motion in degraded environments. This phenomenon highlights the critical role of V&V in autonomous system development, particularly within simulation. By rigorously verifying the components and validating the integrated system against expected behaviors across a range of conditions – including degraded environments – such unexpected and potentially unsafe behaviors can be identified, analyzed, and ultimately mitigated before real-world deployment. This emphasizes that a robust V&V process within simulation is not just a quality assurance step, but a fundamental requirement for ensuring the safety and reliability of autonomous systems.

7. References

Durst, P.J., Anderson, D.T., Bethel, C.L, 2017. A historical review of the development of verification and validation theories for simulation models. International Journal of Modeling and Simulation for Scientific Computing, 8(2), 1730001 (24 pages). https://doi.org/10.1142/S1793962317300011

Durst, P. J., 2019. The Reference Autonomous Mobility Model: A framework for predicting autonomous unmanned ground vehicle performance. Doctoral dissertation, Mississippi State University. https://scholorsjunction/msstate.edu/td/4600

Rasshofer, R.H., Spies, M., Spies, H., 2011. Influences of weather phenomena on automotive laser radar systems. Advances in Radio Science, 9, 49-55. https://doi.org/10.5194/ars-9-49-2011

Caldwell, J., Cross, J., Oglesbee, P., Thompson, B., 2024. Autonomous Ground Vehicle Simulation: Usability and Evaluation Study. Proceedings of the Annual General Donald R. Keith Memorial Conference, West Point, NY, May 2. https://www.ieworldconference.org/content/WP2023/Papers/GDRKMCC23_50.pdf

2264 | Mobility testing on frost susceptible soils at field capacity

Michael Parker *, Clifford Witte, Allen Wheeler, Susan Frankenstein

U.S. Army Engineer Research and Development Center, Cold Regions Research and Engineering Laboratory, 72 Lyme Rd, Hanover, New Hampshire, 03755, United States

** Corresponding author: Michael.w.parker@erdc.dren.mil*

ABSTRACT

Frost susceptible soils include a wide variety of soil types from highly organic to fine grained silty soils. These soils can pose significant mobility challenges during each of the four seasons especially during the freeze/thaw transition periods between seasons. These mobility challenges have been on display during several conflicts throughout history with the most recent being the war in Ukraine. Maintaining mobility superiority during conflict is critical to avoiding immobilization and becoming a target. Similarly, maintaining mobility in the Arctic can have similar consequences because immobilization can mean the difference between life or death due to the harsh environmental conditions. This research studies the effects on vehicle mobility across three vehicle classes including both wheeled and tracked platforms operating on peat and silty soils at field capacity. Testing occurred on unfrozen, partially frozen, fully frozen, and partially thawed soil conditions to capture the maximum tractive force during these transitional seasons and better predict cross-country vehicle mobility.

Keywords
Off-road mobility
Soil characterization
Vehicle testing
Arctic
Freeze/thaw
Field capacity

1. Introduction

Spring brings new life and vegetation with budding trees, blooming flowers, and green grass, while fall signifies the end of the growing season where the grass becomes dormant, and the leaves change colors (fall foliage). Despite the drastic differences in the growing season between spring and fall, the two seasons share a common problem for off-road vehicle travel. Both are considered transition periods, with the spring thawing and the fall freezing up respectively. These periods are especially important when considering mobility in the Arctic. Peat and other highly organic soils take a significant amount of time to freeze deep enough to support vehicle movement over these soft terrain conditions. In the spring these soils along with the road ways begin to thaw out causing soupy and slippery surface conditions. These conditions rapidly increase motion resistance while reducing vehicle traction. The reduction in tractive force is partially caused by the soil filling up the tread on the tracks and tires reducing the ability to engage with the soil. The ability to accurately predict the effects on vehicle mobility during these two transitional periods is critical to maintaining mobility superiority for a fight in the Arctic.

Work conducted within this report specifically focuses on light tracked all-terrain vehicles, light tactical wheeled vehicles, and larger over snow tracked vehicles. These vehicles were tested on unfrozen, partially frozen, frozen, and partially thawed frost susceptible peat and silt soils at field capacity.

2. Background

The freeze and thaw seasons have always challenged mobility in warfare. World War II is a prime example when vehicles struggled to move through the soupy conditions found on and off the roadways. Figure 1 shows several images of vehicles stuck in the muddy conditions (Mud on the Eastern Front in WW2).

Fig. 1. Images of vehicles stuck in the thawing road surface conditions of WW2.

Looking ahead eighty years to February 2022 to the Russian-Ukraine war, the same immobilizing conditions are seen to cause catastrophic mission failure. Figure 2 illustrates these challenges as the Russians struggled to move off-road in Ukraine during the thawing season.

https://doi.org/10.56884/9ZVW2E8H

Fig. 2. Spring thaw immobilizing both wheeled and tracked vehicles. Photo Credit AP News.

Continuing to deal with immobilization during the transition seasons despite repeated identification of the issues through military history highlights the complexity of predicting and modeling mobility on these surface conditions. Rushing (2024) and others have studied the use of ground reinforcement mats, logs and other techniques to improve surface conditions, reduce rutting, and improve trafficability. In some instances, the vehicle traveled entirely on top of the matting never coming in contact with the soil surface. This significantly reduces the resistance caused by the soil on the vehicle and keeps the tread clear of soil, both have a positive impact on overall vehicle traction/mobility.

3. Experimental Testing

Testing for this study was conducted in the Frost Effects Research Facility (FERF) at the Cold Regions Research and Engineering Laboratory in Hanover, New Hampshire, United States during the winter of 2025. This is a large temperature-controlled facility with drive through testing capability. Temperatures for the facility can reach as low as 5-10 degrees Fahrenheit depending on outside ambient temperatures.

3.1. Test Vehicles

This study included testing for three separate vehicles. One vehicle was a light all-terrain tracked vehicle weighing 2.0 tonnes, the second vehicle was a light tactical wheeled vehicle weighing 2.9 tonnes and the third vehicle was a medium tracked vehicle weighing 12.6 tonnes. The first and third vehicles are considered light and medium over snow vehicles, LOSV and MOSV respectively. These vehicles were chosen for the study because they are representative of vehicles that would be used in the Arctic environment for both winter and summer operations, which includes the transition periods leading both into winter and out of the winter into the summer.

3.2. Surface conditions

Testing was conducted on a 6.4 meter wide test lane that was 82.3 meters long. The continuous soil test section was constructed to include the ability to control the water table. As seen in Fig 3, water is added using surface sprinklers as well as vertical standpipes which drain into a series of horizontal pipes distributing water along the base of the test section. As the standpipes are filled, the water naturally wicks towards the surface via capillary action. While this phase of testing was conducted at a single target moisture content, the section was designed to allow relatively natural moisture changes between events, enabling future studies on moisture's role in mobility and soil behavior.

Parker (2024) describes the main test section construction with new soil types added to the surface for 2025 to closely represent the silty and highly organic conditions frequently encountered in the Arctic.

One of the primary test surfaces for 2025 testing consisted of a peat-sand blend (designated PT), prepared by mixing peat and SP-SM in a 20:80 volumetric ratio. This high-organic material was installed at a moisture content of 20–40% intending to reflect the softness and rutting behavior of natural peat. The silt layer was installed to the specified depth shown in Fig. 3, with moisture content controlled at 8–9%. Following the main construction phase, the top layers were ripped to 46 cm and tilled to about 23 cm deep to produce a consistent, soft testing surface for each ensuing test.

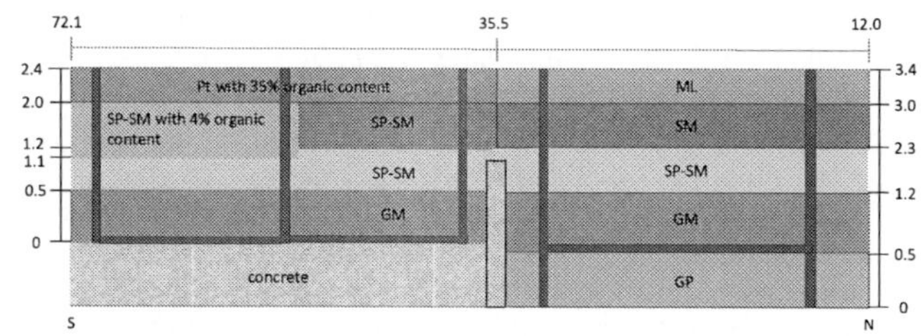

Fig. 3. Test section construction schematic for 2025 testing. (not to scale).

Construction of the added soil layers for 2025 is shown in Fig. 4 and Fig 5.

Fig. 4. Peat (PT) soil construction on the South end of the FERF test section.

Fig. 5. Soil (ML) soil construction on the North end of the FERF test section.

3.3. Soil instrumentation

The soil characterization for this research effort was completed with a range of instrumentation designed to measure the required parameters for mobility testing on soils. Measurements were completed on un-trafficked soil before the vehicle traversed the test section and in the vehicle ruts at the conclusion of the motion resistance test event. Parameters that were measured include vehicle rut depth, frost depth, thaw depth (if present), density, cone index, shear strength, California bearing ratio (CBR), and volumetric water content.

Instruments used for soils characterization for this testing include the Geonor shear vane for soil shear strength, dynamic cone penetrometer for California bearing ratio (ASTM D6951/D6951M-09 2015), Clegg impact hammer for soil strength (Lafayette Instrument Co 2005) and ASTM D5874-16 (2016), Field Scout SC900 (Spectrum Technologies, 2022a) for soil compaction, FieldScout TDR (Spectrum Technologies, 2022b), and the Troxler E-Guage 4590 for soil density. Refer to Parker (2024) and Frankenstein et al. (2025) for more information on the test procedure for each instrument and additional details the soil measurements for these tests.

3.4. Vehicle instrumentation

All vehicles for this test were outfitted with the AIM Evo 5 data logger and IMU, which is appropriate for testing in cold and harsh conditions with its IP65 weather rating and small form factor. The unit measures vehicle speed, wheel speed, throttle position, along with roll, pitch, yaw and the associated accelerations using the internal IMU. Wheel speeds were measured with hall effect sensors at each corner of the vehicles, and true vehicle speed was measured with a 5th wheel located at the right rear of the vehicle. For the MOSV, individual track speeds were measured at the front track sprockets. Throttle position was measured using a string potentiometer which indicated displacement from 0-100 percent.

Since the gross traction is measured by a single load cell behind the vehicle instead of at each wheel it provides a total vehicle traction value. With traction being measured as a total vehicle parameter the ground pressure should be measured in a similar fashion. The ground pressure was measured as an average ground pressure for each of the test vehicles instead of a single wheel/track ground pressure. Tire pressures on the LTV were set to equal pressures front and rear before calculating the average ground pressure. Values for each of the vehicles are 12.4 kPa (LOSV), 83.4 kPa (LTV), and 16.6 kPa (MOSV). Gross tractive effort was calculated for each vehicle on the various test surfaces by measuring the total motion resistance (MR_{Total}) and drawbar pull (DBP) values directly from the load cell in accordance with SAE J2014 and then adding them as indicated in Eq. 1. Motion resistance and drawbar pull were measured at 5 MPH for the LOSV and LTV and at 3 MPH for the MOSV. The intent with the lower ground speed on the MOSV is to provide improved tractive force slip control and generate greater track slip. See Parker et al. (2019) for additional details on MR and DBP test procedures.

$$Gross\ Traction = DBP + MR_{Total} \quad (1)$$

3.5. Surface Characterization

The peat and silt soils were ripped down to a depth of 41-46 cm then tilled to a depth of 18-20 cm (Fig. 6.). This was done to remove rutting from the previous test, returning the test surface to its original condition and as uniform as possible between individual vehicle tests. During this process the soil moisture was measured to a depth of 12.2 cm and increased to a target value of 25-30% by adding water to the stand pipes (increasing the water table) and by adding water to the surface with sprinklers. This brought the surface moisture content to the field capacity for the given test soils. Table 1, Table 2, and Table 3 show the soil parameters that were recorded for each test pre and post test event.

Fig. 6. Soil reconditioning process between vehicle tests. Near end is the silt and the far end is peat.

Table 1
Virgin Soil and Rutting info for Tracked Light All-Terrain Tracked Vehicle (LOSV)

variable	2024-12-18		2025-01-06		2025-02-24		2025-03-27	
	Pt	ML	Pt	ML	Pt	ML	Pt	ML
rut depth (cm)	9.81	7.13	2.94	0	0	0	5.93	3.25
frost depth (cm)	0	0	5.40	6.49	15.24	14.48	15.25	25.38
thaw depth (cm)	0	0	0	0	0	0	1.70	2.55
dry density 0-5cm (g/cm³)	0.65	1.19	0.72	1.00	0.70	1.05	0.88	1.59
wet density 0-5cm (g/cm³)	0.85	1.37	0.78	1.06	0.75	1.10	1.00	1.71
cone index 0-15cm (psi)	9	92	34	227	151	599	622	583
cone index 15-30cm (psi)	9	245	88	339	48	too hard	too hard	too hard
Clegg Impact Hammer CBR (%)	0	1.27	9.18	11.69	6.20	25.80	5.61	13.63
surface shear (kPa)	0.70	2.45	too hard	too hard	112.4	too hard	69.2	56.00
DCP 0 blow penetration depth (cm)	18.25	15.75	1.50	2.75	0.60	0	3.50	3.00

Table 2
Virgin Soil and Rutting info for Light Tactical Wheeled Vehicle (LTV)

variable	2024-12-17		2025-01-22		2025-02-24		2025-03-28	
	Pt	ML	Pt	ML	Pt	ML	Pt	ML
rut depth (cm)	16.80	11.50	9.28	5.65	2.53	0	1.10 MR 6.62 DBP	0.60 MR 5.00 DBP
frost depth (cm)	0	0	6.25	6.25	15.24	14.48	15.62	25.88
thaw depth (cm)	0	0	0	0	0	0	3.05	2.10
dry density 0-5cm (g/cm³)	0.64	1.27	0.64	1.13	0.70	1.05	0.93	1.68
wet density 0-5cm (g/cm³)	0.82	1.48	0.73	1.20	0.75	1.10	1.04	1.77
cone index 0-15cm (psi)	8	19	37	178	151	599	560	603
cone index 15-30cm (psi)	78	231	102	252	48	too hard	too hard	too hard
Clegg Impact Hammer CBR (%)	0	3.33	7.89	20.50	6.20	25.80	7.84	9.09
surface shear (kPa)	0.95	2.25	too hard	too hard	112.4	too hard	5.60	17.00
DCP 0 blow penetration depth (cm)	14.00	17.25	2.50	0.25	0.60	0	4.75	4.00

Table 3
Virgin Soil and Rutting info for Medium Tracked Vehicle (MOSV)

variable	2024-12-20		2025-02-04		2025-02-25		2025-03-28	
	Pt	ML	Pt	ML	Pt	ML	Pt	ML
rut depth (cm)	8.43	8.19	3.43	1.71	2.50	0	4.40 MR 5.81 DBP	4.95 MR 5.20 DBP
frost depth (cm)	0	0	4.27	4.02	15.24	14.48	15.62	25.88
thaw depth (cm)	0	0	0	0	0	0	3.05	2.10
dry density 0-5cm (g/cm³)	0.65	1.19	0.64	1.12	0.70	1.05	0.93	1.68
wet density 0-5cm (g/cm³)	0.85	1.42	0.74	1.22	0.75	1.10	1.04	1.77
cone index 0-15cm (psi)	6	35	13	153	151	599	560	603
cone index 15-30cm (psi)	75	221	82	168	48	too hard	too hard	too hard
Clegg Impact Hammer CBR (%)	0.02	1.86	2.13	11.50	6.20	25.80	7.84	9.09
surface shear (kPa)	0.70	4.25	41.60	too hard	112.4	too hard	5.60	17.00
DCP 0 blow penetration depth (cm)	23.05	26.18	1.50	2.00	0.60	0	4.75	4.00

The December testing in each of the table is for unfrozen soil, the late January or early February date is for the partially frozen soil condition (5 cm), Late February date is for completely frozen soil (13-15 cm), and the late March test date is the partially thawed (5 cm) soil condition. Each table shows that the largest rut depths occur in the unfrozen soil configuration and the shallowest rut depths are for the fully frozen soil condition. The fully frozen soil condition was targeted to ensure minimal if not zero sinkage. The target partial thaw depth was targeted to provide a shallow slippery tawed soil surface (5 cm) over a competent 13-15 cm frozen layer below.

Figure 7 shows soils measurements taken on the frozen peat soil.

Fig. 7. CRREL researchers, Bruce Elder and Susan Frankenstein, taking Clegg measurements on the partially frozen Peat soil.

4. Experimental Results

Results in the following graphs show the performance of each of the test vehicles on the four different test conditions (unfrozen, partially frozen, fully frozen, and partially thawed). Figure 8 shows the motion resistance values for all three vehicles on both soil types along with a hard asphalt surface for comparison. Hard surface motion resistance was measured for each test because colder temperatures cause the added stiffness of the rubber in the tires and tracks and increased viscosity of the greases to increase the hard surface motion resistance values. Parker et al. (2019) details this effect for both wheeled and tracked vehicles. This phenomenon is also noticed in the gray bars in Fig. 8 because the warmer temperatures were experienced for the unfrozen and partially frozen tests and the coldest air temperature was experienced during the heart of the winter for the fully frozen test. Motion resistance caused by the soil/terrain can be calculated by subtracting the hard surface value from the individual soil values for each individual test. The unfrozen soil condition had the greatest resistance for each vehicle respectively. This is expected since this is soil condition where the highest sinkage was recorded for each vehicle. With the exception of the fully frozen LTV test the lowest motion resistance for each was the thawing soils condition for all vehicles. Figure 8 shows the LTV having similar motion resistance values for the fully frozen silt and hard surface on the same day. This is expected because there is zero sinkage, but an increased surface roughness since the silt was ripped and tilled leaving a semi-rough surface. For reference, if the surface had been rolled smooth then the values would likely have been almost identical for this fully frozen test because the surface would have behaved more like a hard surface. This is most noticeable on the wheeled LTV results because this vehicle experienced the highest sinkage values on each of the tests, which was expected because the LTV had the highest average ground pressure of the three vehicles tested.

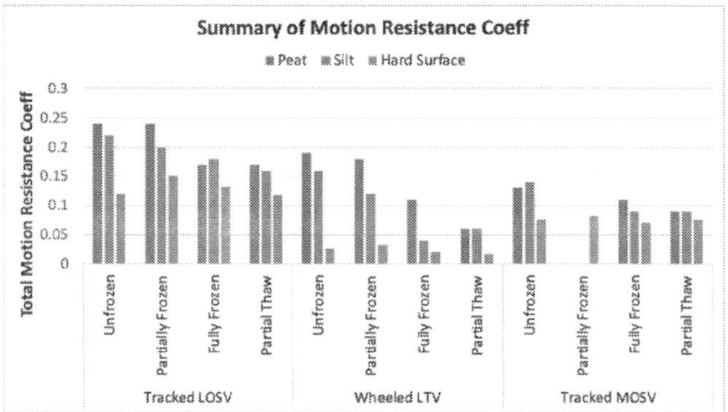

Fig. 8. Total Motion resistance coefficient for each vehicle on each soil condition.

Figure 9 shows the gross traction for each vehicle on the different soil conditions. The tracked vehicles generally exhibit higher gross traction coefficients than the wheeled vehicle on similar conditions, with the LOSV having greater than twice the gross traction coefficient for partially thawed soil conditions. The unfrozen tracked vehicles both experienced significantly higher unfrozen gross traction values than the wheeled vehicle, approaching 0.8, partially because they experience a much lower terrain resistance (less sinkage) and because both tracks had good mechanical grouser engagement. In both vehicles where the partially frozen test was run it produced the highest absolute peak traction coefficient. The gross traction differences between the partially frozen and fully frozen Peat and Silt for the LOSV are attributed to the mechanical grouser engagement with the partially frozen soil surface, still allowing sinkage, while the fully frozen Peat did not.

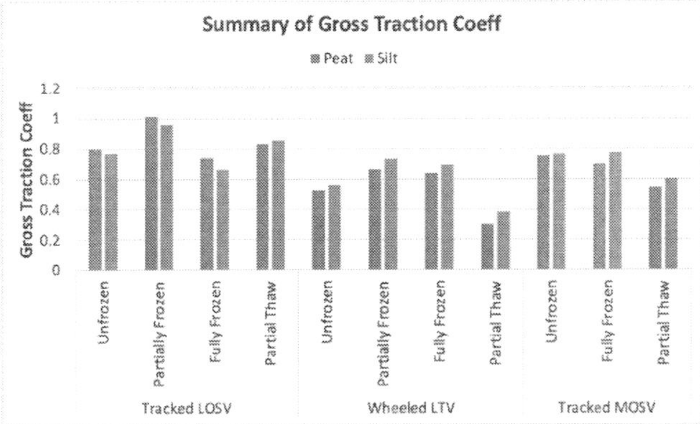

Fig. 9. Gross Traction coefficient for all vehicles on each soil condition.

5. Conclusions and future work

This testing provides valuable results critical to understanding the impact of soil state on Arctic soils at or near field capacity. This work focused on three vehicles (LOSV, LTV, and MOSV) typically used in Arctic conditions. Results were limited to a single soil moisture condition where each soil type was at field capacity placing them around 25-30% soil moisture. This moisture level or higher is typical for soils found in the high moisture environments in the Arctic. Each vehicle, with the exception of the LTV on frozen silt, experienced the lowest motion resistance under the partially thawed condition. The partially thawed condition softened up the top surface of the soil creating a shallow wet thawed layer that reduced surface roughness of the frozen tilled surface for the fully frozen test while still experiencing minimal sinkage values. The high moisture partial thaw condition impacted the traction coefficient for all vehicles because the wet soil acted as a lubricant allowing the vehicles to spin on the frozen surface below.

Future work should focus on the last data point critical to this test matrix. A deep-frozen soil layer (60-80 cm) with a deep-wet thaw condition (25-30 cm). This should significantly increase the motion resistance on the vehicle while decreasing the gross traction coefficient and in extreme cases may immobilize the vehicle.

6. Nomenclature

ASTM	American Society for Testin and Materials	
CBR	California Bearing Ratio	
DBP	Drawbar Pull	[lbs]
FERF	Frost Effects Research Facility	
LOSV	Light Over Snow Vehicle	
MOSV	Medium Over Snow Vehicle	
MR_{Total}	Total Motion Resistance	[lbs]
SAE	Society of Automotive Engineers	
WW2	World War 2	

7. Acknowledgements

All work described and data presented in this report are based upon testing conducted by the US Army Engineer Research and Development Center Cold Regions Research and Engineering Laboratory supported under congressional program increase lines of effort.

8. Declaration of competing interest

The authors declare that they have no known competing financial interests or personal relationships that could have appeared to influence the work reported in this paper.

9. References

ASTM D5874-16, 2016. Standard Test Methods for Determination of the Impact Value (IV) of a Soil. ASTM International. West Conshohocken, PA.

ASTM D6951/D6951M-09, 2015. Standard Test Method for use of the Dynamic Cone Penetrometer in Shallow Pavement Applications. ASTM International. West Conshohocken, PA.

Frankenstein, S., Wheeler, A., Elder, B., 2025. Relationship between seasonal terrain properties and vehicle rut depths. Proceedings of the International Society for Terrain-Vehicle Systems (ISTVS), Lebanon, New Hampshire, USA, October 6-9.

Lafayette Instrument Company, 2005. Clegg Manual. Lafayette, IN.
Spectrum Technologies, Inc. 2022. Field Scout TDR 350 Soil Moisture Meter Product Manual.

Mud on the eastern front during WW2, n.b. https://www.youtube.com/watch?v=rhstOECIm_0.

Parker, M.W., Stott, A.R., Bodie, M.O., Frankenstein, S., Shoop, S.A., 2019. Vehicle Mobility on Highly Organic Soils. Proceedings of the International Society for Terrain-Vehicle Systems (ISTVS), Prague, Czech Republic, September 9-11.

Parker, M.W., Witte, C.F., Frankenstein, S., Wheeler, A.C., 2024. Mobility During the Transition Seasons in the Arctic. Proceedings of the 21st International and 12th Asia-Pacific Regional Conference of the International Society for Terrain-Vehicle Systems (ISTVS), Yokohama, Japan, October 28-31.

Rushing, J.F., Hansen, B.S., Parker, M.W., 2024. Improved trafficability over soft soils using ground matting. Proceedings of the 21st International and 12th Asia-Pacific Regional Conference of the International Society for Terrain-Vehicle Systems (ISTVS), Yokohama, Japan, October 28-31.

Society of Automotive Engineers (SAE), 2021. Pneumatic Tire/Wheel/Runflat Assembly Qualifications for Military Tactical Wheeled Vehicles. SAE J2014, SAE Int. Surface Vehicle Standards.

Spectrum Technologies, Inc., 2022a. FieldScout SC 900 Soil Compaction Meter Product Manual.

Spectrum Technologies, Inc., 2022b. FieldScout TDR 350 Soil Moisture Meter Product Manual.

Troxler Electronic Laboratories, Inc. 2016. Manual of Operation and Instruction, E-Gauge™ Model 4590 Surface Density Gauge. Research Triangle Park, NC.

2469 | TERRA-V toolchain: virtual validation and optimization of autonomous vehicles in unpaved terrain

Gerhard Skoff *, Daniel Hassler, Gernot Hasenbichler

AVL List GmbH, Graz, Austria

** Corresponding author: gerhard.skoff@avl.com*

ABSTRACT

To ensure safe and robust autonomous offroad operation, specific challenges must be managed, including robust terrain-specific obstacle detection. Virtual approaches have the potential to significantly shorten development time and effort. Adequate methods need to consider changing environmental conditions, such as e.g., changes of the vegetation during the seasons of the year as well as different weather conditions. To manage all those requirements, a modular simulation platform has been developed.

The tool chain consists of six independent and flexible modules, which are all linked together to conduct simulations seamlessly and with best correlation to the real world:

1. Vehicle-specific multibody dynamic simulation module of the vehicle, which can be derived from different multibody dynamic simulation tools, whereby a mathematical soft-soil model applies,
2. A terrain model, which utilizes a comprehensive set of specifically designed terrain elements, and which undergo AI-supported modifications, e.g., darkness, fog, or rain during the validation process,
3. A sensor module, which can operate with different kinds of active and non-emitting sensors. Sensor selection and arrangement on the vehicle are optimized in the virtual sensor lab,
4. An autonomous algorithm, which is the system-under-test (SuT), and which gets optimized and validated,
5. Set of terrain-specific KPI's, which define application-specific boundaries and qualification criteria, and
6. A terrain-specific scenario database, where ontology-based automated offroad-scenarios are generated.

The sensor model and SuT get integrated into the Vehicle model, and the vehicle is executing the automatically generated, ontology-based testcase-combinations. Application-specific KPI's are calculated and compared with target values.

Keywords
Autonomous Systems
Validation
Offroad-Mobility
Simulation

1. Introduction

The specific challenges for the validation of automated vehicles on unpaved terrain include robust obstacle detection, intelligent path planning and safe maneuver execution. Virtual validation methods are needed to effectively develop these functions and test them in a variety of scenarios and environmental conditions. AVL's modular simulation platform TERRA-V enables the time- and cost-effective development of such autonomous systems.

2. Challenges for autonomous offroad driving

The technological advances in the automotive industry in recent years regarding automated driving are not only highly relevant in civilian road traffic, but also in off-road applications, especially in agriculture and for military applications. Intensive work is being done on the implementation of autonomous driving systems, for example for unmanned supply convoys on unsurfaced roads. The qualifications of these unmanned driving systems in the agricultural and military sectors follow different rules than those that have now become standard in on-road operation.

The particular challenges lie in the undefined terrain structure, where no easily classifiable traffic infrastructure can be used. Off-road obstacles come in many forms and sometimes change over time, for example due to vegetation over the course of the seasons. Due to vegetation, some obstacles can also be hidden or not clearly recognized, for example in high grass.

Furthermore, especially in military applications, only limited sensor technology is available. GPS cannot be used and emitting sensors such as lidar or radar are not desirable in many cases, so that only inert navigation and optical sensors like cameras, which are additionally exposed to demanding environmental conditions, can be used.

Due to these demanding constraints and the lack of standardized approaches to scenario description, it has not yet been possible to establish standardized qualification rules. In many cases, limited practical testing is still used, but this is neither reproducible nor able to evaluate the extensive critical use cases.

With the virtual validation toolchain TERRA-V, AVL has created a product that closes this technological gap and provides vehicle developers with a tool for optimizing their driving algorithms on unpaved roads and off-road terrain, as well as for validating these algorithms against key performance indicators developed specifically for this application.

3. Simulation Platform TERRA-V

The TERRA-V simulation platform provides a state-of-the-art environment for developing and evaluating autonomous vehicle systems. As shown in Fig. 1, this simulation framework consists of several essential components that work together to enable realistic and precise simulations. The environmental model replicates the real world in detail and ensures that simulations take place under realistic conditions.

An accurate vehicle model replicates all relevant physical properties and behaviors of the vehicle. This model can be correlated in the TERRA-V simulation platform using measurements from the real vehicle and then being integrated into the simulation platform. It includes the dynamics of the vehicle and the actuators.

https://doi.org/10.56884/JCVHRORX

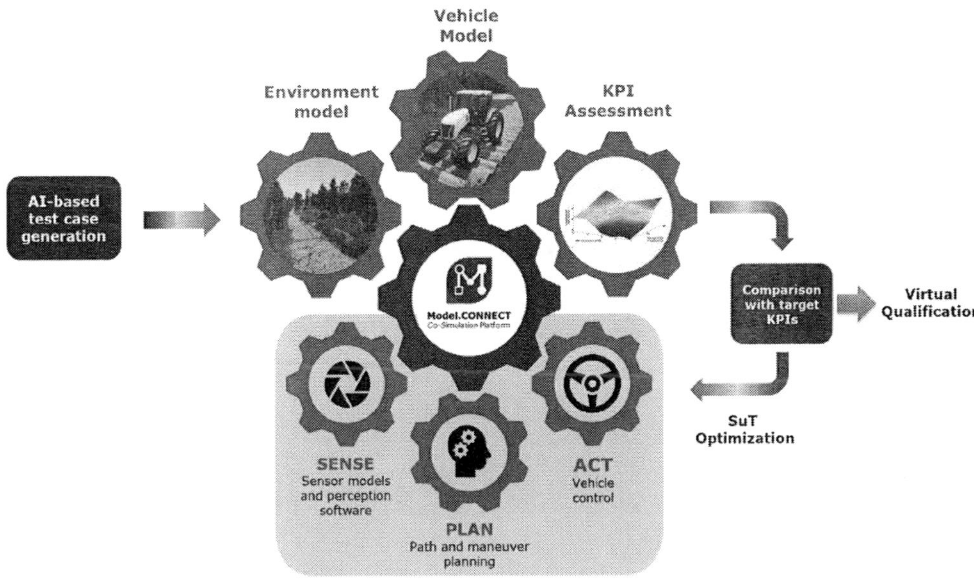

Fig. 1. Elements of the AVL Simulation Platform TERRA-V

The autonomous modules of the platform, consisting of SENSE, PLAN and ACT, play a leading role. The SENSE module is responsible for perceiving the environment and processes data from various sensors such as cameras, lidar and radar to create a comprehensive picture of the surroundings. Virtual sensors can be selected from a catalog and adapted to specific requirements, or correlated sensors can be integrated based on measurements.

The planning module PLAN develops strategies and routes based on the information provided by the SENSE module. It considers obstacles and other objects that may obstruct the originally planned path.

The ACT module implements the planned actions and controls the vehicle movements to navigate safely and efficiently.

In the TERRA-V platform, AI-generated test cases are used to test autonomous systems under various scenarios. These test cases can cover a wide range of situations, from everyday driving situations to rare driving events in extreme weather conditions. In particular, the AI algorithms are used to generate scenarios to simulate unknown and critical situations.

After each simulation, the performance of the autonomous system is analyzed using a KPI evaluation tool. This tool assesses performance against predefined key performance indicators (KPIs), which are selected from an extensive database for the respective use case. This database is divided into individual autonomous modules and covers important aspects such as safety, effectivity, and comfort. By comparing the achieved KPIs with the target values, targeted improvements can be made to the autonomous functions. This enables continuous optimization and ensures that the autonomous systems meet the highest standards.

The TERRA-V platform is characterized by its high flexibility in integrating the systems under test. It provides a highly detailed virtual environment that can also be used as a digital twin if required.

4. Generation of Critical Scenarios

The legally required assessment of the safety and robustness of a highly automated vehicle in on-road applications cannot be conducted without virtual validation in practice. Depending on the Operational Design Domain (ODD) and the system's range of functions, up to several hundred million kilometers must be tested to take all eventualities into account. Additional critical factors arise on unpaved terrain, which expose the autonomous system to challenges. The complexity lies in the fact that, in addition to the driving maneuver of the unmanned ground vehicle in the terrain itself, other objects in the immediate vicinity – moving or stationary – must also be appropriately modelled and their interactions, taking into account highly dynamic environmental conditions, must be considered in terms of their effect on dynamic driving behavior (Zhou et al., 2021; Almanee et al., 2021; Young et al., 2022).

Fig. 2. AI-supported generation of test scenarios

There are numerous methods available for generating scenarios, such as full factorial selection, pseudo-random sampling or "corner case" identification using optimization methods. In recent years, standardized approaches to scenario description have been established, e.g., ASAM OpenSCENARIO (ASAM, 2025). For civil on-road use cases, scenario databases can be derived from various data sources such as accident records, fleet data, traffic observations, NCAP, etc. However, for military off-road use cases, such scenario databases and/or open formats are not yet available.

AVL has developed a new method and implemented it in TERRA-V to generate critical scenarios using game-based testing,

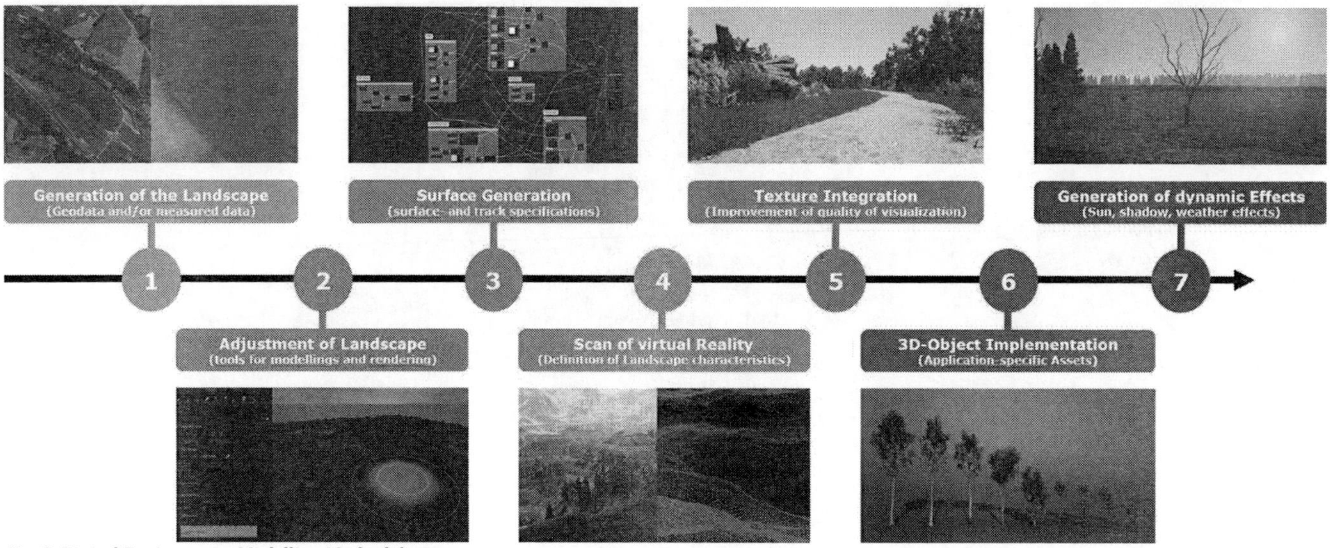

Fig. 3. Digital Environment Modelling Methodology

which incorporates artificial intelligence techniques to autonomously explore and refine scenarios, enabling efficient identification of challenging and critical test cases. (Fig. 2) This method uses an initial scene with static elements and background agents that can perform various actions. A behavioral and dynamics model realistically simulates the interactions, which are optimized over several iterations to finally generate a critical scenario (Almanee et al., 2021; Deniz et al., 2024; Young et al., 2022; Zhou et al., 2021).

5. Digital Environment Creation

The process (Fig. 3) of digital environment creation for off-road proving grounds is a comprehensive, multi-stage workflow that ensures both visual fidelity and physical accuracy for simulation purposes, relying on advanced data assimilation techniques such as those described by Chen and Lu (2017), which utilize high-resolution point clouds to accurately reconstruct road surfaces and terrain features essential for realistic simulation environments.

It starts with the generation of the landscape using geodata and/or measurement data, such as high-resolution point clouds or satellite imagery, to capture the real-world terrain's geometry and features. This process often relies on advanced LiDAR surveying techniques that provide dense and precise 3D point clouds by emitting laser pulses to measure distances, which enables the creation of highly accurate digital terrain models essential for realistic simulation environments, as described by Finelli et al. (2025). This raw landscape is then refined by environment modelling experts, who use sculpting and smoothing tools within platforms like Unreal Engine to adapt the terrain to specific requirements or to introduce features that may not exist in the real world. Material creation follows, where detailed surface properties are defined for different soil and track types, enabling realistic tire-ground interactions and supporting advanced vehicle dynamics modeling.

A key innovation in this workflow is the virtual environment scan (Fig. 4), which analyzes the digital landscape and generates a high-density point cloud. This process benefits greatly from advanced interpolation methods, such as midpoint interpolation and spatial refinement, which densify the sparse point cloud data

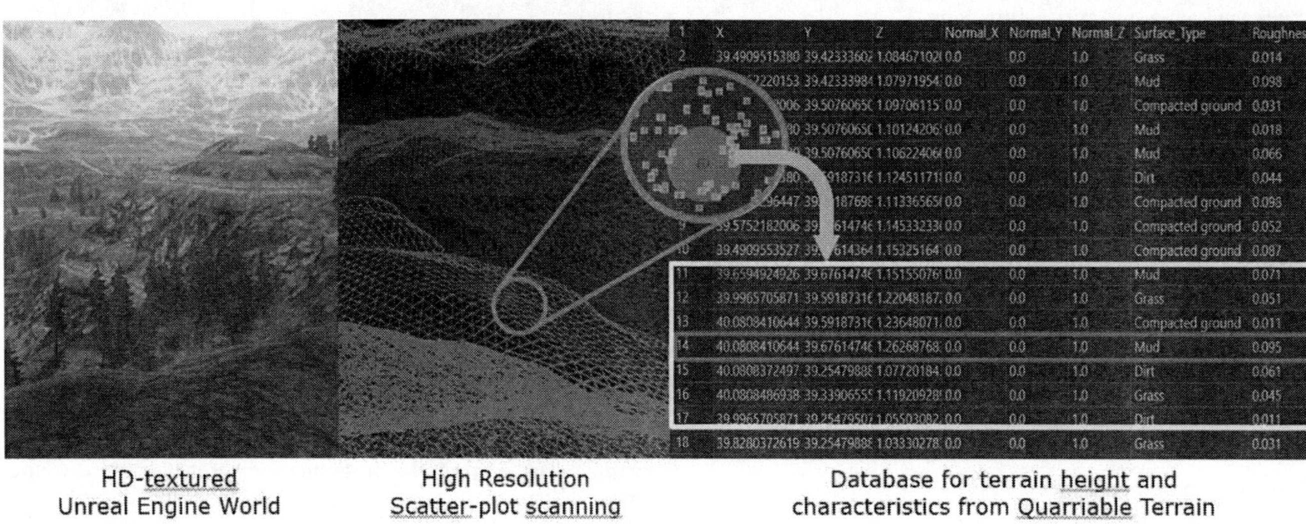

Fig. 4. Virtual Environment Scan

while preserving geometric fidelity and surface details, thereby enhancing the accuracy and visual quality of the reconstructed environment (Li et al., 2021).

Each point in this cloud is automatically assigned parameters such as elevation, surface type, and soil characteristics, creating a queryable database of terrain properties. This enables precise simulation of vehicle-environment interactions, especially under varying conditions. High-quality textures are then added to enhance visual realism, supporting both human-in-the-loop and camera-based validation. The environment is further populated with 3D objects-such as vegetation, rocks, or man-made structures-tailored to the use case. Finally, dynamic effects like sun position, shadows, and weather phenomena (e.g., rain, fog) are integrated, allowing the simulation of seasonal changes and challenging environmental scenarios. The modular nature of this toolchain also allows for the automated creation of custom environments, enabling the testing of edge cases and rare scenarios that are not present in the physical proving ground, thus significantly broadening the scope and robustness of virtual validation.

6. Digital Vehicle Twin Creation

The creation of a digital vehicle twin is an equally detailed and iterative process, designed to ensure that the virtual representation of the vehicle matches real-world behavior with the model within the AVL VSM™ driving dynamics tool. AVL's patented correlation methodology enables rapid and accurate adjustment of the virtual vehicle's parameters, iteratively refining them until the simulated dynamic vehicle data correlates more than 95% with the real vehicle's dynamic responses across lateral, longitudinal, and vertical domains (Fig. 5).

The workflow begins with the execution of measurement runs using the actual vehicle, during which a variety of driving maneuvers are performed and recorded with high-precision sensors. These real-world data sets form the foundation for the parameterization and correlation of the vehicle.

Once the dynamic model achieves the required level of correlation, it is complemented by a detailed 3D geometric model of the vehicle, which is integrated into the TERRA-V toolchain. This allows the digital twin to be tested in a wide range of virtual environments and scenarios, including those that are difficult or impossible to replicate in reality-such as extreme terrain, rare obstacle configurations, or adverse weather conditions. The digital twin can also be substituted by a benchmark vehicle from a catalog for early-phase studies or when real vehicle data is not yet available. This approach not only accelerates the development process but also enables comprehensive validation and optimization of autonomous driving functions, sensor configurations, and control algorithms before deployment in the field. The combination of highly realistic environment models and precisely correlated vehicle twins forms the backbone of the TERRA-V simulation platform, supporting robust, efficient, and repeatable virtual validation for autonomous off-road vehicles.

7. Virtual Validation by using TERRA-V

The TERRA-V simulation platform enables realistic virtual validation by supporting various test environments, such as Model-in-the-Loop (MiL), Software-in-the-Loop (SiL) and Hardware-in-the-Loop (HiL). This flexibility allows autonomous vehicle systems to be tested at various stages of development. The core of the simulation framework is the AVL Model.CONNECT™ co-simulation platform. Model.CONNECT™ is used to integrate and link different models and tools. This platform offers interfaces to more than thirty modeling tools and enables precise and high-performance co-simulation through advanced synchronization technologies.

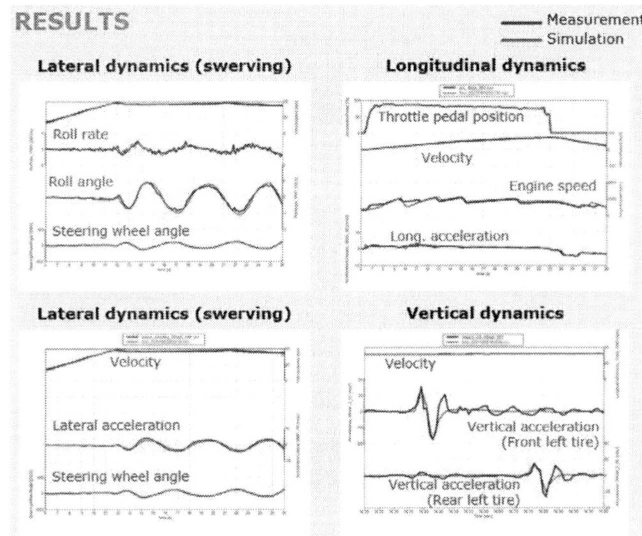

Fig. 5. Example Correlation Results

The integration of the vehicle dynamics models is done using AVL VSM™ and is based on correlations with real test drives. MATLAB/Simulink or similar software can also be used to simulate driving dynamics if required. The environmental model is based on the open-source software CARLA or the Unreal Engine, which enables detailed and realistic reproduction of the environment. The virtual sensor system is also modeled using CARLA and replaces physical sensor signals with simulated signals, which are then used for algorithms for perceiving the environment. The "system under test" consisting of the autonomous functions SENSE, PLAN and ACT, is integrated into the simulation platform by means of AVL Model.CONNECT™ using suitable software interfaces. Fig. 6 shows an example simulation with the TERRA-V simulation platform for an agricultural application. The figure shows the Graphical User Interface (GUI) of Model.CONNECT™. The left part of the figure shows the visualization of the 3D environment in CARLA or Unreal Engine, the right part of the figure shows the visualization of the Scenario Engine Esmini; and in the lower right part of the figure examples of simulation data are shown, which illustrate the results and parameters of the simulations performed and are used for the KPI assessment.

8. Autonomous Algorithm optimization using key performance indicators

The assessment and optimization of the autonomous functions SENSE, PLAN and ACT in the TERRA-V simulation platform is done by means of key performance indicators (KPIs). After each simulation, the autonomous functions are evaluated using predefined KPIs that are selected from a comprehensive database for the respective application. (Fig. 7).

Fig. 6. Example simulation with the TERRA-V simulation platform

These KPIs cover important aspects such as safety, efficiency, and comfort. For example, the SENSE module is tested for its ability to perceive the environment quickly and accurately, while the PLAN module is evaluated in terms of the effectiveness and efficiency of its route planning. The ACT module is analyzed for the accuracy and reliability of the vehicle control. By comparing the achieved KPIs with the target values, targeted improvements can be made to the autonomous functions.

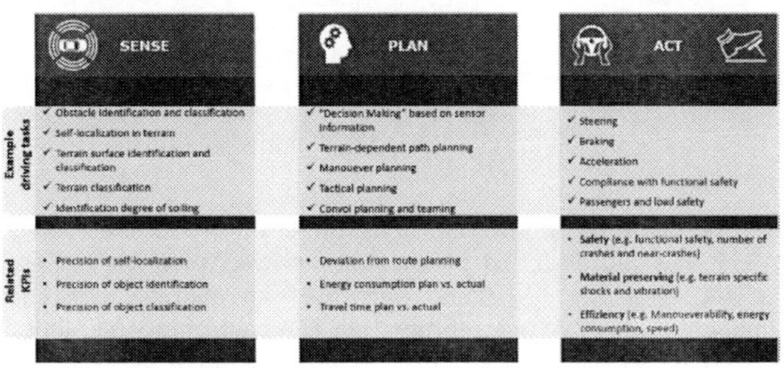

Fig. 7. Examples of KPI selection

9. Application Potential

Due to the flexibility and modular design of the simulation platform, TERRA-V can be used in many different application areas for the development and validation of highly automated systems. The focus is on the various off-road applications on unsurfaced or off-road terrain, as used primarily in agricultural and military environments, but also in mining operations (Skoff et al., 2025).

10. Conclusions and future work

Only through advanced simulation approaches, in addition to real-world testing, can highly automated vehicles be sufficiently validated for off-road scenarios. The TERRA-V simulation platform provides realistic models for vehicle dynamics, sensor technology, and environmental conditions, as well as modular integration of software for sensor data perception, path planning, and vehicle control. The innovative scenario generation approach, combined with off-road-relevant key performance indicators (KPIs), enables comprehensive, quantitative assessment and subsequent optimization of autonomous systems. This ensures that the autonomous functions SENSE, PLAN, and ACT can be continuously improved by comparing achieved KPIs with target values and iteratively optimizing algorithms until all requirements are met in critical maneuvers.

Building on these foundations, TERRA-V aligns closely with the objectives of current defense research. Military UGVs face particularly demanding requirements, such as robust obstacle detection, identification of road characteristics, and reliable navigation using only passive, non-emitting sensors-especially under GNSS-denied conditions. Following up, a methodology will be developed to model and analyze 3D environments and to create an off-road ontology for military-relevant scenarios. This involves defining and characterizing extensive road surfaces, obstacles, and environments based on real data, resulting in a scenario database that enables the derivation of necessary driving maneuvers for specific military deployment scenarios. These maneuvers can be used as a basis for safe path planning in any real UGV, significantly hardening automated driving algorithms for military off-road operations.

The advancement of unmanned ground vehicles (UGVs) also requires specialized test environments in addition to simulations. The latest version of Regulation UNR No. 152 mandates that at least 30% of the required tests must be conducted on real road sections to demonstrate the robustness of autonomous systems (UNECE, 2024). This is particularly important in European defense research, as there is currently a lack of sufficient empirical data at the national and EU levels to safely integrate autonomous systems into combat situations. Such tests are necessary to verify the performance and reliability of the systems under real conditions and ensure they meet the requirements in military operational scenarios.

The main objective is to precisely allocate real test environments and link these with digital simulations. Flexible, modular virtual test environments for autonomous systems in military environments, which correlate with real test environments, will be created. Methodologies will be developed to plan, conduct, and correlate virtual test drives in various terrains with simulations that can be compared to real practical tests in the identical environment.

11. Acknowledgements

Significant parts of TERRA-V have been developed with funds from the Austrian defense research program FORTE. The authors would like to thank the BMF, the BMLV and the FFG for their support.

12. Declaration of competing interest

The authors declare that they have no known competing financial interests or personal relationships that could have appeared to influence the work reported in this paper.

13. References

Almanee E., Balachandran W., Hossain M., 2021. scenoRITA: Generating Less-Redundant, Safety-Critical and Motion Sickness-Inducing Scenarios for Autonomous Vehicles. arXiv preprint arXiv:2112.09725. https://arxiv.org/abs/2112.09725

ASAM, 2025. ASAM OpenSCENARIO User Guide Version 2.0. ASAM e.V., Association for Standardization of Automation and Measuring Systems. https://www.asam.net/standards/detail/openscenario/v200/

Chen D.-l., Lu Y.-Y., 2017. 3S road surface reconstruction based on point clouds data assimilation algorithm. MATEC Web of Conferences 139, DOI:10.1051/matecconf/201713900005

Deniz O., Murphy K., Zhao Y., 2024. Reinforcement Learning-Based Autonomous Landing of AirSim Simulated Quadcopter in Unreal Engine. AIAA SciTech Forum. https://arc.aiaa.org/doi/10.2514/6.2024-2485

Finelli R., Sena, P., Lorusso, A., Cecere, L., Villecco, F., 2025. Reproduction of Road Scenarios for Simulated Driving Using LiDar Surveying Technique. https://doi.org/10.3390/machines13010004

Li, R., Li, X., Heng, P.-A., Fu, C.-W., 2021. Point Cloud Upsampling via Disentangled Refinement. https://arxiv.org/abs/2106.04779

Skoff G., Hassler D., Nitsche P., Hassler, 2025. Virtual Validation and Optimization of Autonomous Vehicles in Unpaved Terrain. ATZ electronics 03-04 2025, 8-13.

UNECE, 2024. UN Regulation No. 152 Revision 2: Uniform provisions concerning the approval of motor vehicles with regard to the Advanced Emergency Braking System (AEBS) for M1 and N1 vehicles. United Nations Economic Commission for Europe. https://unece.org/sites/default/files/2024-04/R152r2E.pdf

Young L., Jung J., Park H., Lee B., 2022. Enabling Artificial Intelligence Studies in Off-Road Mobility Through Physics-Based Simulation of Multiagent Scenarios. ASME Journal of Computational Nonlinear Dynamics, 17(5):051001. https://doi.org/10.1115/1.4054437

Zhou J., Yu F., Yang R., Song L., 2021. SMARTS: Scalable Multi-Agent Reinforcement Learning Training School for Autonomous Driving. Proceedings of the 38th International Conference on Machine Learning, PMLR 139, 11474-11486. https://proceedings.mlr.press/v139/zhou21a.html

2824 | Local path planning on rough terrain for unmanned ground vehicle by reinforcement learning considering search reward

Ryosuke Eto *, Fushiki Taguchi, Junya Yamakawa

Department of Mechanical Engineering, National Defense Academy of Japan, Yokosuka, Japan

** Corresponding author: eto@nda.ac.jp*

ABSTRACT

Autonomous unmanned ground vehicles are expected to be utilized at disaster sites. The previously obtained map information for the site is often unreliable because disaster areas can change rapidly and unpredictably. Therefore, the vehicles must be able to recognize the surrounding rough terrain environment in real time and plan optimal routes to approach the destination safely. However, local path planning may fall into the local minimum solution and get stuck. Therefore, in this study, a method of avoiding a standstill by utilizing information entropy and route history to encourage exploratory behavior was investigated. In the proposed method, routes are selected by determining the vehicle behavior using reinforcement learning. For observation information, a Digital Elevation Model, an information entropy map of elevation, and a map that records the number of times the vehicle has traveled on the grid, in addition to the vehicle's state were used. The aim is to acquire the behavior of going to the target with selecting a safe route to avoid getting stuck by rewarding the vehicle for going to the target, avoiding rough terrain, reducing information entropy, and not repeatedly driving in the same place. In this study, we tested the proposed method on each terrain with a single large obstacle and a dead-end by simulations. The simulation results showed that the vehicle can reach the target efficiently by introducing rewards for exploratory behavior.

Keywords
local path planning
autonomous unmanned ground vehicle
reinforcement learning
rough terrain

1. Introduction

The shortage of workers is an urgent problem in Japan. Therefore, robots are expected to replace humans in the workforce and for safety reasons. Autonomous unmanned ground vehicles are being introduced for work and security in agriculture, construction, and disaster recovery. However, the road surface conditions outdoors and the arrangement of supplies and materials at these sites often change and differ from the map information provided in advance. Therefore, local path planning is necessary to recognize the surrounding environment in real time and select an appropriate route for the vehicle to move safely. Local path planning methods include the Dynamic Window Approach (Dieter et al., 1997) and reinforcement learning (Weerakoon et al., 2021). However, since local path planning uses only the surrounding information around the vehicle, it sometimes results in a local solution for large obstacles or dead-end, causing the vehicle to get stuck. In this study, we investigate a method to generate avoidance actions against these obstacles by reinforcement learning.

In reinforcement learning, the agent learns behaviors to maximize a set reward. In path planning on rough terrain, rewards are set for moving toward the target and for avoiding roughness (Bo et al., 2022). In the case of a large obstacle, the agent needs to move away from the goal to avoid the roughness, resulting in a local solution. In this study, we introduced a reward that encourages the vehicle to go to a place it has never been in order to avoid staying in the same place. It was tested that the vehicle can explore unknown terrain, escape from complex terrain, and select safer paths to reach the goal.

2. Simulation Environment

In this study, MATLAB was used to implement reinforcement learning. Gazebo, a physical engine simulator, was used as the virtual environment for training and testing, and the Robot Operating System was used for data communication with MATLAB.

2.1. Robot model

Figure 1 shows the model of the robot used in the simulator. ROBOTIS TURTLEBOT3 Waffle Pi was modified to have four wheels, and the wheel diameters were increased from 7 to 10 cm to improve its traveling performance on rough terrain. The size of the robot is approximately 30×30 cm^2. The robot is equipped with a 16-layer 3D-LiDAR at 20 cm from the ground to measure the surrounding terrain. The robot's behavior was dictated by continuous values of body velocity and yaw angular velocity in the ranges of [0, 0.3] m/s and [-1, 1] rad/s. The maximum measurable distance of the 3D-LiDAR was set to 4 m. The measurement horizontal angle range was set from -90 to 90 degrees and vertical one was from -15 to 15 degrees.

https://doi.org/10.56884/HVHA4780

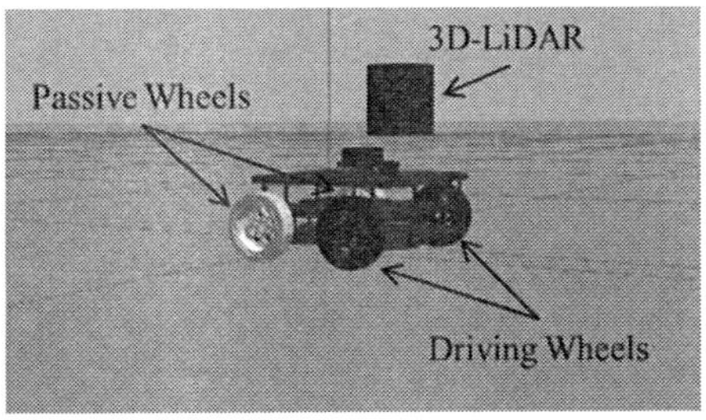

Fig. 1. Robot model

2.2. Terrain

In this study, Digital Elevation Model data was loaded into a simulator to construct a terrain with obstacles. Learning was conducted by changing the difficulty level of the obstacles to acquire behaviors corresponding to the obstacles, such as stepping over or avoiding the obstacles. Two types of obstacles were tested: a large straight line-shaped obstacle and a U-shaped dead-end. Fig. 2 shows examples of the terrain used. The linear obstacles were set longer than the measurement range of 3D-LiDAR, and the difficulty level increased as the obstacle height was raised. When the obstacle height is low, it can be traversed, but when the obstacle height is high, a search is required to pass through a place where there is no obstacle. The difficulty level is increased by setting the height of the obstacle in the dead-end shape to 25 cm and by lengthening the lateral projection parts. To avoid the obstacle and reach the target, the robot needs to acquire more complex behaviors than in a straight-line obstacle.

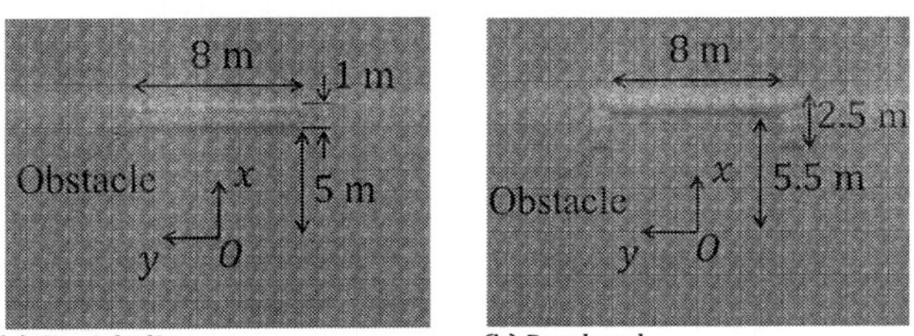

(a) Straight line *(b) Dead-end*
Fig. 2. Terrain model

3. Map

In this study, an elevation map with a grid size of 0.3 m×0.3 m, consisting of a100 × 100 grid as a whole, a grid pass count map, and entropy map of elevation probability distribution were created.

Elevation maps were created based on Bayesian estimation after converting point cloud data obtained from 3D-LiDAR into a map coordinate system. The elevation probability distribution of grid (i,j) was represented by a normal distribution N ($\mu_{\text{pos}}^{(i,j)}$, $\{\sigma_{\text{pos}}^{(i,j)}\}^2$). The initial distribution is N (0, 10000) and is updated by the following equation.

$$\{\sigma_{\text{pos}}^{(i,j)}\}^2 = \frac{\{\sigma_{\text{pri}}^{(i,j)}\}^2\{\sigma_{\text{obs}}^{(i,j)}\}^2}{\{\sigma_{\text{pri}}^{(i,j)}\}^2 + \{\sigma_{\text{obs}}^{(i,j)}\}^2} \tag{1}$$

$$\mu_{\text{pos}}^{(i,j)} = \{\sigma_{\text{pos}}^{(i,j)}\}^2\left(\frac{\mu_{\text{pri}}^{(i,j)}}{\{\sigma_{\text{pri}}^{(i,j)}\}^2} + \frac{h_{\text{obs}}^{(i,j)}}{\{\sigma_{\text{obs}}^{(i,j)}\}^2}\right) \tag{2}$$

where $\{\sigma_{\text{pos}}^{(i,j)}\}^2$ is the variance of the posterior elevation probability distribution, $\{\sigma_{\text{pri}}^{(i,j)}\}^2$ is the variance of the prior elevation probability distribution, $\{\sigma_{\text{obs}}^{(i,j)}\}^2$ is the variance of the normal distribution indicating the measurement error of LiDAR. The LiDAR assumed in this study is the Velodyne VLP16, and the $\sigma_{\text{obs}}^{(i,j)}$ was set to 0.01 m based on the accuracy of this sensor. $\mu_{\text{pos}}^{(i,j)}$ is the mean of the posterior elevation probability distribution, $\mu_{\text{pri}}^{(i,j)}$ is the mean of the prior elevation probability distribution, and $h_{\text{obs}}^{(i,j)}$ is the highest point cloud z value in the grid obtained by LiDAR. The robot's position and posture were obtained using output values from the simulator instead of self-position estimation by LiDAR. A 19 ×19 local map centered on the robot's current position was used for observations of reinforcement learning agent in the linear obstacle condition, and a 47×47 map was used in the dead-end condition.

To facilitate robot movement to grids that have not been passed, a map was created to record the number of times the robot has passed each grid in the environment (visit count map). Each grid has a value indicating how many times the robot has passed through it, and each time the robot passes through a grid, the value is incremented by one. A 47×47 local map centered on the robot's current position was used for observation.

We also created an entropy map to investigate the use of entropy in rewards as another method of facilitating robot movement. Entropy is a measure in information theory that quantifies the uncertainty of a random variable (Shannon, 1948). The entropy of a normal distribution is expressed as:

$$H_{(i,j)} = \frac{1}{2}\log[2\pi e\{\sigma_{pos}^{(i,j)}\}^2] \tag{3}$$

A 47×47 local map centered on the robot's current position was used for observation. Figure 3 shows the size of the local map. The local elevation map is input to the neural network as a square in order for the robot to see the information behind it and recognize where to exit the dead-end. Since less observation data is better, we determined the size of the map through trial and error, setting it to the minimum size necessary for successful learning. Although similar results can be expected with a rectangular map, this study used a square shape, which can be ignored in terms of aspect ratio, in order to reduce the trial time required to determine the minimum size.

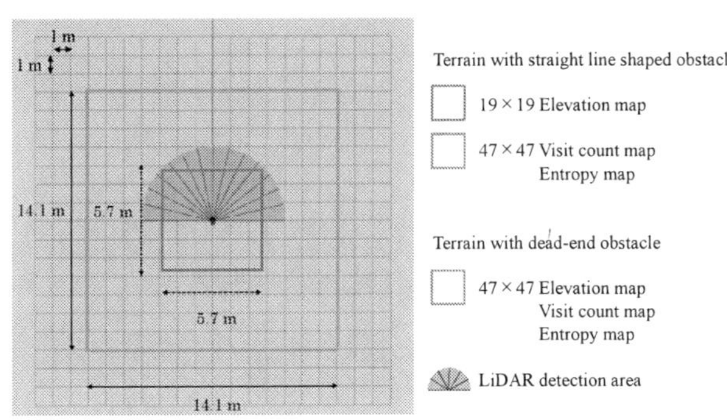

Fig. 3. Size of local map

4. Reinforcement learning

4.1. Model

Reinforcement learning is a method in which an agent learns optimal strategies through trial and error while receiving rewards from the environment. In this study, we employed Proximal Policy Optimization (Schulman, J., et al., 2017), which has an Actor-Critic structure (Konda and Tsitsiklis, 2020).

Figure 4 shows the structure of the Actor network, in which the local map of visit count or entropy (47×47) and the local map of elevation (19×19 or 47×47) are input to extract map features. The robot's body speed, yaw angular velocity, distance and angle between the robot and the target, and the robot's roll and pitch angles (6×1 vector) were also input to extract features. From each feature, through an intermediate layer, the mean probability of action was finally derived using the Tanh function, and the standard deviation of the probability of action was derived using the Softplus function.

Curriculum learning is a method that allows agents to efficiently learn complex tasks by increasing the difficulty of learning in stages (Yoshua, et al., 2009). For the straight obstacles, the height was increased as the learning progressed, and the number of episodes for trials was 30 for 10, 15, and 20 cm obstacles, respectively, and 310 or more for 25 cm. The number of episodes was determined by trial and error. The number of episodes was set low because it was judged that traversing low obstacles was a simple task and thus easy to learn. The number of episodes for 25 cm obstacle increased because the obstacle height was considered difficult to learn because it was a relatively difficult task that required avoidance. In the dead-end, the learning task was conducted with 100 episodes on a flat terrain with no obstacles, 200 episodes on a dead-end with a 1.5 m side protrusion, and 200 episodes on a 2.5 m dead-end shown in Fig. 2(b), so that the goal would be gradually more difficult to reach the goal.

In the dead-end, many actions are required to reach the goal, and learning takes a large amount of time. Therefore, we employed imitation learning that uses supervised actions to improve learning efficiency. The supervised behavior was designed so that the robot normally moves straight toward the target, and when it detects an obstacle, it follows waypoints for avoidance to reach the goal. Waypoints were set by the authors along the obstacle as shown in Fig. 5. The number of episodes for performing the supervised behavior was 1 to 5 and multiples of 10, and the agent was encouraged to observe the supervised behavior periodically while learning to improve it in other episodes.

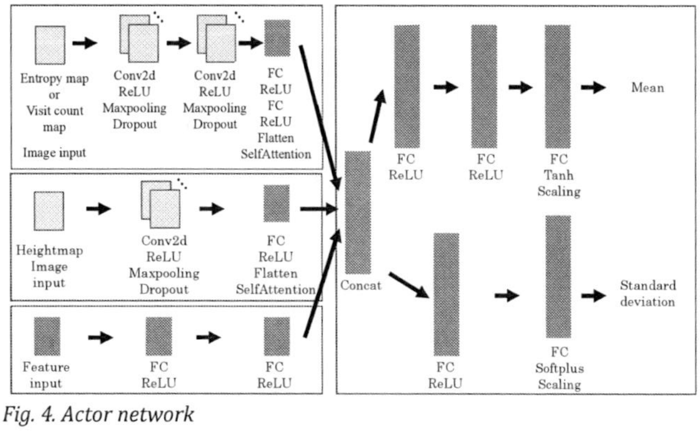

Fig. 4. Actor network

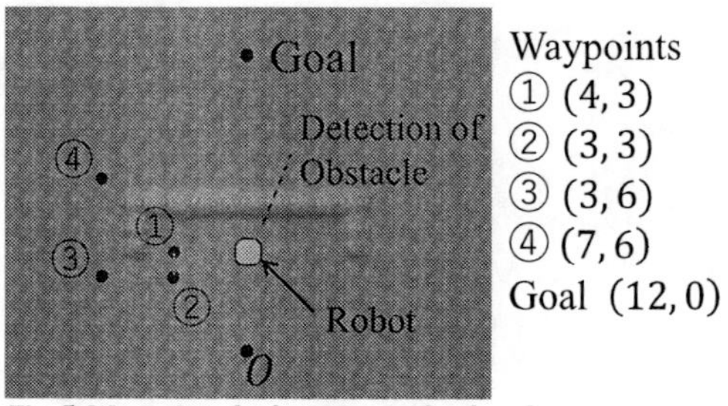

Fig. 5. Waypoints for learning in dead-end

4.2. Rewards

The reward design is an important setting that determines the progress of reinforcement learning. The reward r_v per step for the method using the visit count map (visit count method) and the reward r_e for the method using entropy map (entropy method) are calculated by the following equation, respectively.

$$r_v = r_{reached} + r_{notreached} + r_{distance} + r_{angularvel} + r_{step} + r_{unvisited} \tag{4}$$

$$r_e = r_{reached} + r_{notreached} + r_{distance} + r_{angularvel} + r_{step} + r_{emtropy} \tag{5}$$

The success reward $r_{reached}$ was a large positive reward when the robot reached the target, as shown in Eq. 6. The reward was set to a larger value to encourage success because it is difficult to learn in a dead-end. In this case, we set the threshold of the distance at which the target is judged to have been reached to 1 m.

$$r_{reached} = \begin{cases} 200 & \text{Straight line} \\ 400 & \text{Dead-end} \end{cases} \tag{6}$$

The failure reward $r_{notreached}$ was negative when the robot did not reach the target or when the robot's pitch angle exceeded 60°.

$$r_{notreached} = -5 \tag{7}$$

The distance reward $r_{distance}$ was designed so that the closer the distance between the robot and the target, the larger the positive reward, and the further away the distance, the larger the negative reward. d_p^n is the distance between the target and the robot at step n, and d_p^{n-1} is the distance between the target and the robot at step $n-1$. We chose small values because a large reward toward the target in a dead-end makes it difficult for the robot to learn escape behaviors.

$$r_{distance} = \begin{cases} 10 * (d_p^{n-1} - d_p^n) & \text{Straight line} \\ d_p^{n-1} - d_p^n & \text{Dead-end} \end{cases} \tag{8}$$

The angular velocity reward $r_{angularvel}$ is defined as follows in order to control sudden or excessive turning of the robot. ω is the yaw angular velocity of the robot.

$$r_{angularvel} = \begin{cases} 0 & \text{if} |\omega| \le 0.1 \\ -0.01 * |\omega| & \text{otherwise} \end{cases} \tag{9}$$

The step reward r_{step} was designed as follows in order for the robot to travel on as flat a terrain as possible.

$$r_{step} = -\sum_{(k,l)\in N} \left|h_{current} - h_{neighbor(k,l)}\right| \tag{10}$$

where, $(k,l) \in N$ is the set of 3×3 neighbor grids excluding the grid of the robot's current location, $h_{current}$ is the height of the grid of the robot's current location, and $h_{neighbor(k,l)}$ is the height of the neighbor grid.

The unvisited grid search reward $r_{unvisited}$ is a reward for moving to an unvisited grid to improve the robot's search ability. Each time the robot moved to an unvisited grid, it was given a reward of 0.1.

The entropy reward $r_{entropy}$ was established to facilitate robotic exploration as follows

$$r_{entropy} = -0.00001 * \sum_{(i,j)\in M} H_{(i,j)} \tag{11}$$

where, $(i,j) \in M$ is the set of 47×47 grids centered on the grid of the robot's current location. From Eq. 3, entropy increases with the variance of the elevation probability distribution. As the robot searches, the accuracy of the surrounding elevations increases, so the variance and entropy decrease. Since the entropy reward is penalized when the entropy of the surrounding area is large, the robot learns to reduce entropy by exploring the surrounding area.

4.3. Learning Conditions

The learning environment was Windows 11, Intel Core i7-12700, 2.1GHz. Learning was performed by simulation because it is time-consuming to use a real robot. The simulator computes at 200Hz and the input is 10 Hz.

As shown in Fig. 6, for each episode, the robot's starting point was set at a random position and yaw angle on a circle of radius 4 m from $(-3,0)$. The goal was set at $(12,0)$. The end condition was set when the robot reached the goal area, when the robot's pitch angle exceeded 60° and it was considered to have tipped over, or when it exceeded 200 seconds.

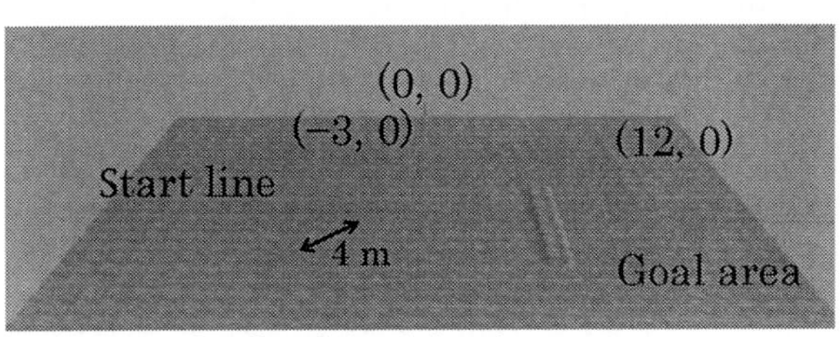

Fig. 6. Start and goal conditions in learning

5. Results

5.1. Straight line obstacle

Figure 7 shows the reward curves of the proposed method when learning to overcome a linear obstacle. In the visit count method, the robot can overcome an obstacle with a height of 10 cm up to episode 30, so it learned behaviors to reach the target point early and obtained high reward. From episode 90, the height of the obstacle was 25 cm, and the robot received a large negative reward when it avoided the obstacle, because it was stuck on the obstacle and could not move. The number of trials with high rewards increased after episode 200, indicating that the agent learned to receive a large positive reward for avoiding the obstacle and reaching the target. In the entropy method, learning was slower, but the number of trials with high rewards increased after episode 250. To compare the proposed methods, learning was also conducted under the condition that no search reward was given. However, the method failed to learn smooth avoidance behavior when the obstacles became higher after episode 30, and many trials resulted in low rewards.

In order to verify the learned behavior, we conducted running tests with the robot's starting point at $(-1,0)$, the target point at $(12,0)$, and the obstacle heights at 10 cm and 25 cm. The test results are shown in Fig.s 8 and 9.

When the obstacle height was 10 cm, the robot could reach the target in 60 seconds when no search reward was given, in 45 seconds with the visit count method, and in 50 seconds with the entropy method. In this case, the robot reached the target by overcoming the obstacle without avoiding it. However, to avoid getting stuck, the robot chose a path that allowed it to get in and out of the obstacle at an angle rather than perpendicular to it.

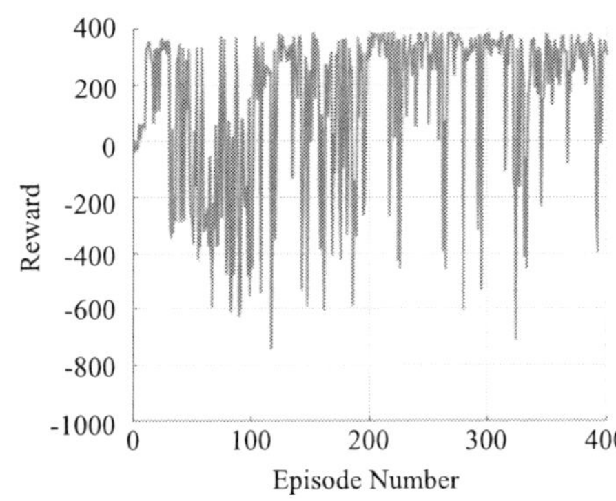

(a) Visit count method

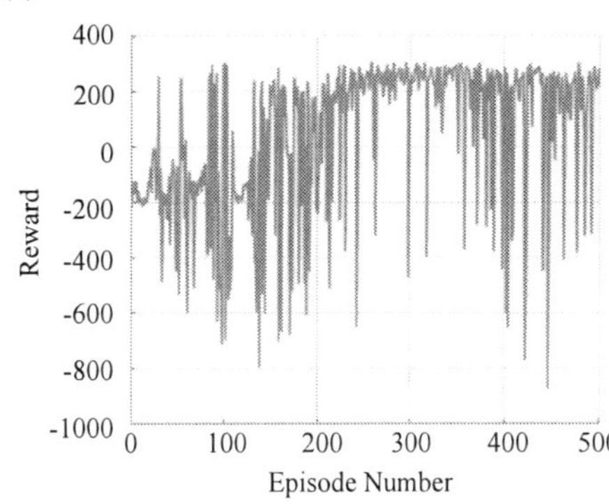

(b) Entropy method

Fig. 7. Results of reward on terrain with straight line obstacle

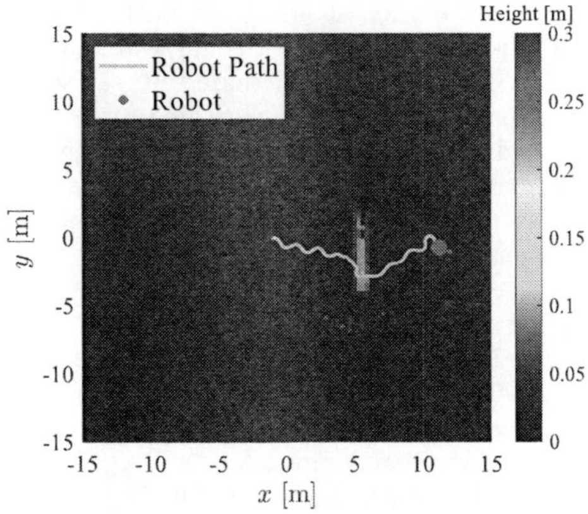

(a) Obstacle height 10 cm

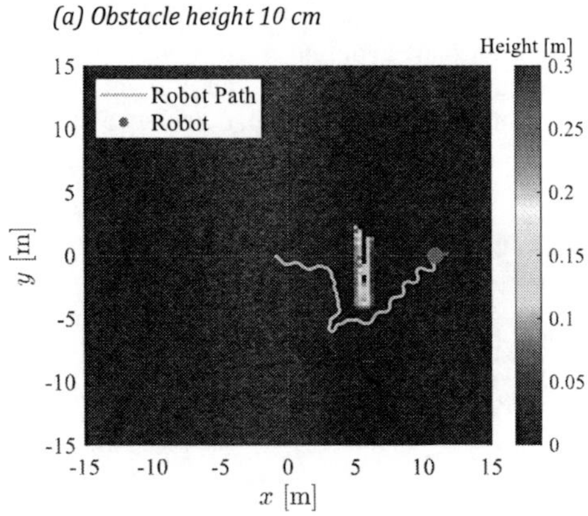

(b) Obstacle height 25 cm

Fig. 8. Results of height map and robot trajectory obtained by visit count method on terrain with straight line obstacle

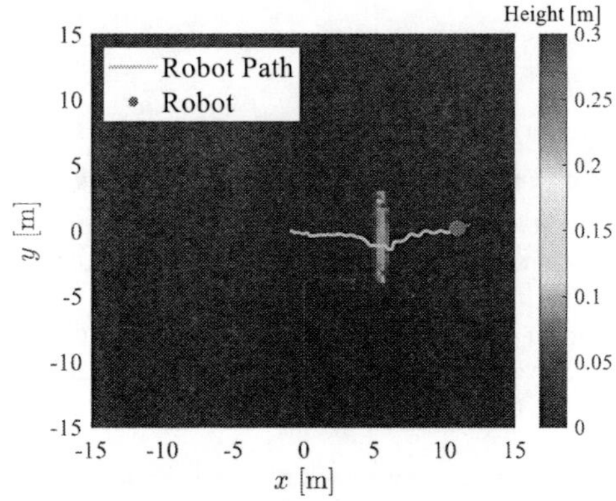

(a) Obstacle height 10 cm

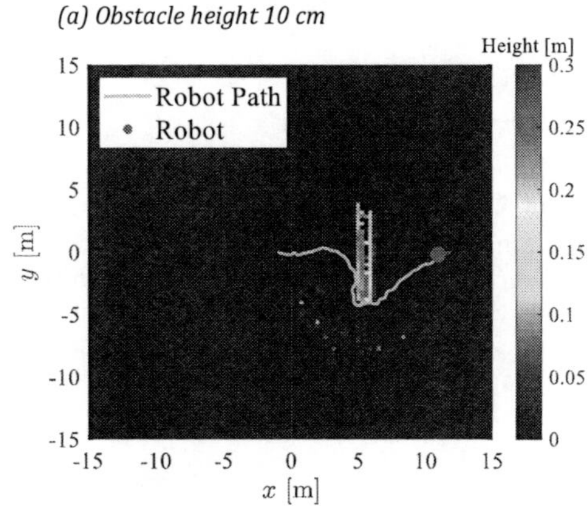

(b) Obstacle height 25 cm

Fig. 9. Results of height map and robot trajectory obtained by entropy method on terrain with straight line obstacle

When the obstacle height was 25 cm and no search reward was given, the robot was able to reach the goal area, but it gradually headed toward the edge and accidentally avoided the obstacle while colliding with it. On the other hand, the two proposed methods avoided obstacles more efficiently and reached the goal area. The approximate time required to reach was 160 seconds with no search reward, 60 seconds with the visit count method, and 70 seconds with the entropy method.

These results indicate that the robot acquired the behavior to explore the unknown environment by adding a search reward, and was able to efficiently find a safe path. On the other hand, the meandering of the robot increased with visiting count method. This may be due to the fact that the search reward for the unexplored grid was set larger than the angular velocity reward, which prioritized exploration over angular velocity suppression. The reason why the target arrival time was short despite the meandering was because the travel speed was higher than that of the entropy method. In addition, the entropy method sometimes stuck at the edge of obstacles when avoiding obstacles under the condition that the obstacle height is the limit height that can be overcome. This was due to an attempt to avoid the obstacle at the very edge of the obstacle. Therefore, it is necessary to further consider the weight of each reward for more efficient and safer traveling.

5.2. Dead-end

The visit count method, which performed well under the linear obstacle condition, was used to validate the method under the dead-end condition. Figure 10 shows the reward curve for learning to a dead-end using the visit count method. Without imitation learning, the robot reached the target only a few times in the flat terrain condition up to episode 100, and did not progress in this number of episodes. However, when imitation learning was used, the reward curve was higher because it was easy to follow the way point to the target point in the flat terrain condition. In contrast, in the environment with obstacles in the dead-end from episode 101, the agent learned to go straight to the target and initially received a large negative reward for colliding with the obstacle. However, the agent learned to avoid the obstacle through supervised behaviors and received high rewards even during unsupervised reinforcement learning around episodes 180 to 200. After episode 301, the environment was a large dead-end with obstacles, but the trials with high rewards were frequent, indicating that the robot learned to receive high success rewards by recognizing and avoiding obstacles in the dead-end. To compare the proposed method, we also conducted learning under the condition that no search rewards were given, but no avoidance behavior was acquired.

In order to verify the learned behavior, a running test was conducted with the starting point of the robot at $(-1, 0)$, the target point at $(12, 0)$, the height of 25 cm, and the side projection length of 2.5 m. Figure 11(a) shows the running path of robot without a search reward and Fig. 11(b) shows the running path with the proposed method. When the robot was not given a search reward, it went round and round in the dead-end and could not avoid the obstacle. On the other hand, using proposed method, when the robot encountered the dead-end and recognized the height of the dead-end as about 15 cm, it started to avoid the obstacle and was able to reach the goal area. Therefore, it was confirmed that the robot using proposed method can reach the target by exploring the surrounding environment even when it encounters an obstacle in a dead-end by considering a search reward.

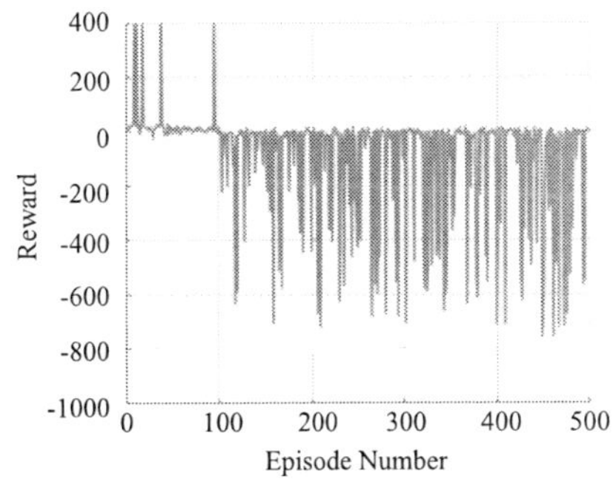

(a) *Not using imitative learning*

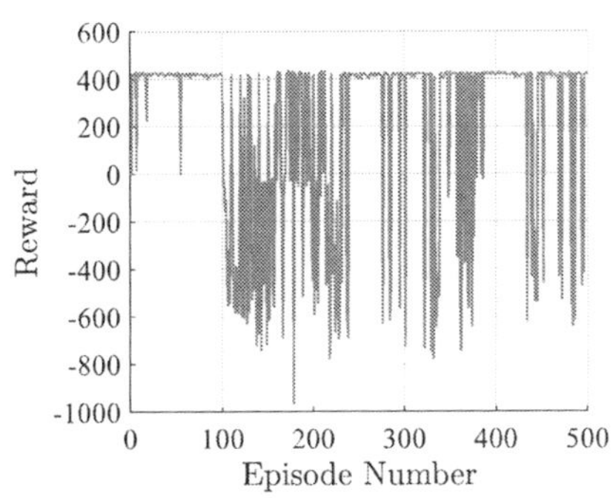

(b) Using imitative learning
Fig. 10. Results of reward using visit count method on terrain with dead-end

6. Conclusions

In this study, we verified whether the robot can avoid large obstacles by avoiding local solutions during local path generation by using a search reward based on the number of times the robot passes through the grid and search rewards based on the elevation probability entropy of an elevation map for reinforcement learning. The results showed that when the robot encountered an obstacle that was difficult to traverse, the robot was able to find a safer path by considering a search reward to go to an unknown environment.

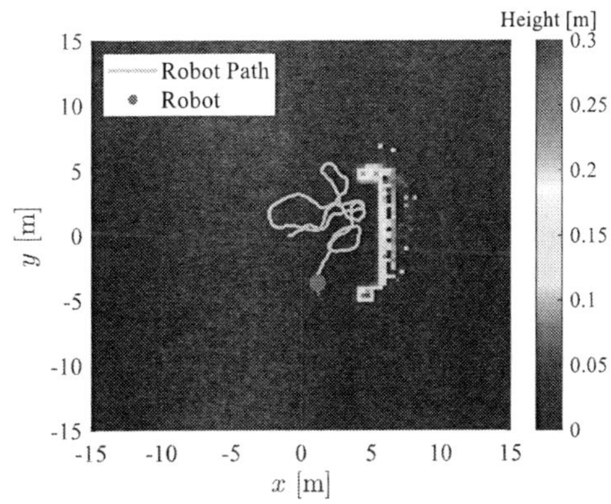

(a) Not using visit count method

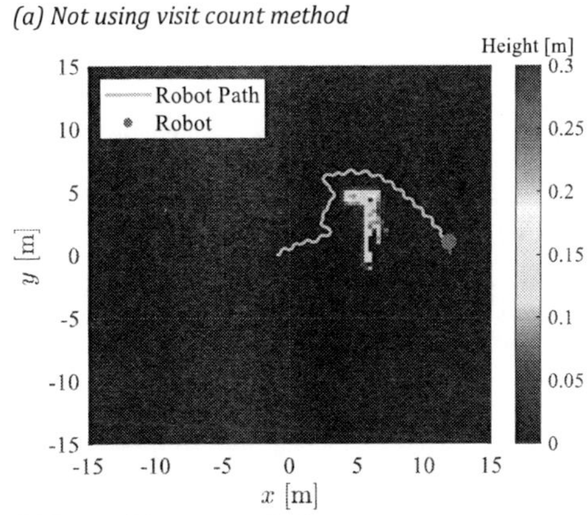

(b) Using visit count method
Fig. 11. Results of height map and robot trajectory on terrain with dead-end

7. References

Bengio, Y., Louradour, J., Collobert, R., Weston, J., 2009. Curriculum Learning. Proceedings of the 26th Annual International Conference on Machine Learning, pp. 41-48.
Bo, Z., Guobin, L., Qixin, Z., Yu, D., Awais, K., 2022. Path Planning for Wheeled Mobile Robot in Partially Known Rough Terrain. Sensors 22(14), 5217.
Dieter, F., Wolfram, B., Sebastian, T., 1997. The Dynamic Window Approach to Collision Avoidance. IEEE Robotics & Automation Magazine 4(1), pp. 23–33.
Konda, R. V., Tsitsiklis, N. J., 2000. Actor-Critic Algorithms. Advances in Neural Information Processing Systems 12, pp. 1008-1014.
Schulman, J., Wolski, F., Dhariwal, P., Radford, A., Klimov, O., 2017. Proximal Policy Optimization Algorithms, arXiv:1707.06347v2 [cs.LG].
Shannon, C.E., 1948. A Mathematical Theory of Communication. The Bell System Technical Journal 27(3), pp. 379–423.
Weerakoon, K., Sathyamoorthy, A. J., Patel, U., Manocha, D., 2021. TERP: Reliable Planning in Rough Outdoor Environments using Deep Reinforcement Learning. Proceedings of the IEEE/RSJ International Conference on Intelligent Robots and Systems (IROS), pp. 1–7.

2908| Drawbar pull experiment and analysis of groused-track unit on sandy terrain

Ryota Azuma [a,*], Tatsuya Maeda [b], Kosuke Horie [b], Shinichiro Miyai [b], Shunsuke Sugimura [b], Genya Ishigami [a]

[a] *Department of Mechanical Engineering, Keio University, Yokohama, Kanagawa, Japan*

[b] *Development Division, Komatsu Ltd, Hiratsuka, Kanagawa, Japan*

* *Corresponding author: r.azuma0293@keio.jp*

ABSTRACT

Tracked vehicles are commonly used on construction sites because of their high vehicle stability and traversability with large ground contact area. Grousers attached to the track shoe enhance vehicle stability and traversability, mitigating the risk of slipping on rough terrain. While many past studies investigated the mechanical effect of a single grouser shoe, experimental test of an entire groused-track unit remains limited due to non-uniform stress distribution of the track unit on the ground.

In this study, we developed an experimental apparatus with a groused-track unit and conducted drawbar pull experiment on sandy terrain. We measured the slip ratio and sinkage of the track unit with various traction load corresponding to the drawbar pull of the track unit. The relationship between the drawbar pull coefficient, which is the ratio of the drawbar pull to the vertical load of the track, and the slip ratio was examined. The experiments confirmed that the slip ratio remains constant when the drawbar pull coefficient is approximately 0.4 or lower. Within this range, the track unit maintains small sinkage and a stable driving state. However, when the drawbar pull coefficient exceeds approximately 0.45, the slip ratio sharply increases as larger traction load cases the track unit to an unstable driving state by repeating the move-and-stop motion. This trend was observed regardless of the varied vertical load. These findings suggest that the relationship of the slip-drawbar pull coefficient and the threshold for stable/unstable driving state are useful for evaluating the traversability of the groused-track unit.

Keywords
Groused-track unit
Drawbar pull coefficient
Slip ratio
Sandy terrain

1. Introduction

Enabling automatic or remote maneuvering of construction vehicles will improve safety in the construction industry, and it also helps rescue and recovery efforts in disaster areas. In recent years, construction vehicles with automatic/remote maneuvers are also considered to be used for space development. Expanding the area of lunar exploration and sustainable lunar activities require the construction of a lunar base. The construction machines need to be operated automatically or remotely since the harsh environments such as those in disaster and on the lunar surface make onboard operation infeasible. To achieve this, it is necessary to understand how the mobility and traversability of the construction vehicles behave on sandy terrain, and to predict whether they can successfully perform the required actions and how likely they are to become dangerous due to slippage.

Tracked vehicles are commonly used on construction site because of their high vehicle stability and traversability with large ground contact area. In addition, grousers are attached to the track shoe which is used on rough terrain because they enhance vehicle stability and traversability, mitigating the risk of slipping on rough terrain. Therefore, the effects of grouser should be taken into account for clarifying the driving characteristics of track.

Classical terramechanics have modeled the driving characteristics of tracked vehicles for many years. Bekker et al. (1956) conducted a vertical penetration test of a flat plate using a bevameter and proposed a semi-empirical vertical stress model using parameters of track width and sinkage into the soil. Reece (1965) pointed out that Bekker's model had a dependency between soil coefficients and proposed a more realistic estimation model by conducting experiments. Janosi et al. (1961) proposed a model for estimating shear stress from horizontal movement of a flat plate by considering Mohr-Coulomb yield criterion. However, these models are primitive since they assume the bottom of the track as a simple flat plate.

In recent years, there have been many studies on the drawbar pull exerted by grouser shoes. For example, Shaikh et al. (2021) conducted experiments to measure the pull exerted by a single grouser shoe at several different grouser heights and soil moisture contents. However, most of the previous studies on grouser shoes have been conducted on moist clay soil. This is because the current tracked vehicles are usually used in agricultural fields and construction sites on the earth, and these places have soil with a certain degree of moisture content. There are very few studies on grouser shoes on dry sand that can be potentially used as a reference for using tracked vehicles on the Moon.

On the other hand, Ishibashi et al. (2019) experimentally clarified that the stress distribution on the bottom surface of a tracked vehicle is not uniform. Therefore, the force exerted by a single grouser shoe should not be applied to the entire bottom surface of the track, and it is necessary to clarify the driving characteristics of the entire groused-track unit. However, few studies have conducted experiments using the entire groused-track unit. This is because experimental tests with real heavy construction vehicles or development of a large-scale measurement device cost too much time and money compared to experiments using a single grouser shoe.

The purpose of this study is to develop an experimental device that can simulate driving on sandy terrain using a reduced-scale model of the groused-track unit, and to experimentally analyze the driving characteristics of a groused-track unit.

https://doi.org/10.56884/Z4B59GCP

2. Drawbar Pull Estimation and Experimental Policy

Many studies have been conducted on the mechanical interaction between rough terrain and track in the field of terramechanics. In this section, the driving dynamics of tracked vehicles are estimated based on the conventional terramechanics theory in order to consider the experimental policy for analyzing the driving characteristics of a grousered-track unit.

Bekker (1956) conducted experiments using a bevameter to estimate the vertical stress acting on the bottom of the track. A flat plate as wide as the tracked vehicle was inserted into the soil, and the resistance force and sinkage were measured. Based on the results, a semi-empirical model that relates the sinkage z to the vertical stress σ was proposed using the following equation.

$$\sigma(z) = \left(\frac{k_c}{b} + k_\phi\right) z^n \tag{1}$$

Reece (1965) pointed out that Bekker's model had a dependency between soil coefficients and proposed a more realistic estimation model below.

$$\sigma(z) = \left(ck'_c + \rho bk'_\phi\right)\left(\frac{z}{b}\right)^n \tag{2}$$

Shear stress, which contributes greatly to the traction force of tracked vehicles, is caused by the friction between the soil and the track as the track deforms the soil. The slip ratio s is defined as Eq. 3 using the horizontal velocity of the vehicle v_x, the radius of driving sprocket of the track r, and the angular velocity ω.

$$s = 1 - \frac{v_x}{r\omega} \tag{3}$$

The amount of soil deformation $j(X)$ at a distance X from the top of the track ground surface can be expressed as follows using the slip ratio s defined in Eq. 3.

$$j(X) = sX \tag{4}$$

Janosi et al. (1961) focused on the exponential behavior of shear stress with respect to soil displacement and proposed the following model for shear stress occurring at the track ground surface.

$$\tau(j) = \tau_{max}\left(1 - e^{-\frac{j}{k}}\right) \tag{5}$$

$$\tau_{max} = (c + \sigma\, tan\phi) \tag{6}$$

Here, the maximum shear stress τ_{max} represents the yield strength of the soil by Mohr-Coulomb's yield criterion. These conventional terramechanics models are simplified ones that treat the ground surface of tracked vehicles as a single flat plate and do not consider the effects of rolling wheels and grousers. However, these models have been used in many previous studies.

Using these models, the drawbar pull exerted by tracked vehicles when driving can be roughly estimated. As shown in Fig.1, drawbar pull DP can be expressed by the following equation, which is obtained by subtracting the driving resistance R_x from the thrust force F_{h} generated at the track ground surface.

$$DP = F_{\mathrm{h}} - R_x \tag{7}$$

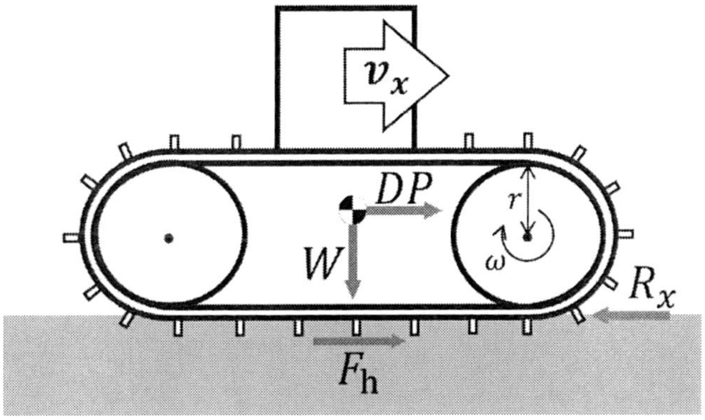

Fig. 1. Forces applied to the track unit

Assuming that the vertical stress distribution in the track ground surface is equal in the width direction, F_{h} can be obtained by integrating the shear stress (Eq. 5) with regard to the contact area of the track width b and the track ground length d_t.With discretized contact patch dX:

$$\begin{aligned} F_{\mathrm{h}} &= b\int_0^{d_t} \tau dX \\ &= (bc + \frac{W}{d_t}\tan\phi)\left\{d_t + \frac{k}{s}\left(e^{-\frac{s}{k}d_t} - 1\right)\right\} \end{aligned} \tag{8}$$

The traction resistance R_x is the force due to soil compaction by the track and ruts on soft ground with the track sinkage z_h. The resistance can be obtained from Eq. 5 as follows.

$$\begin{aligned} R_x &= b\int_0^{z_h} \sigma dz \\ &= \frac{b^{\frac{1-n}{n}}}{(n+1)\left(ck'_c + \rho bk'_\phi\right)^{\frac{1}{n}}}\left(\frac{W}{d_t}\right)^{\frac{n+1}{n}} \end{aligned} \tag{9}$$

Therefore, the drawbar pull DP exerted by the tracked vehicles when driving can be estimated from Eq.s 7, 8, and 9.

$$\begin{aligned} DP = &\left(bc + \frac{W}{d_t}\tan\phi\right)\left\{d_t + \frac{k}{s}\left(e^{-\frac{s}{k}d_t} - 1\right)\right\} \\ &- \frac{b^{\frac{1-n}{n}}}{(n+1)\left(ck'_c + \rho bk'_\phi\right)^{\frac{1}{n}}}\left(\frac{W}{d_t}\right)^{\frac{n+1}{n}} \end{aligned} \tag{10}$$

From this equation, the right-hand side is all soil parameters or track dimension parameters except for s and W, and DP varies with s. Therefore, the experimental policy of this study was to clarify the relationship between drawbar pull coefficient μ, which is DP divided by W, and slip ratio s, both of which are dimensionless parameters.

3. Experimental Apparatus

The experimental apparatus with a grousered-track unit developed in this study consists of a sand box and a traction unit, as shown in Fig. 2. The sand box is filled with Toyoura Sand, a dry sand. A grousered-track unit is attached to this experimental apparatus, which is grounded to simulate driving on a sandy terrain. The traction unit is equipped with horizontal and vertical linear guides which allow the track unit to sink as it moves forward.

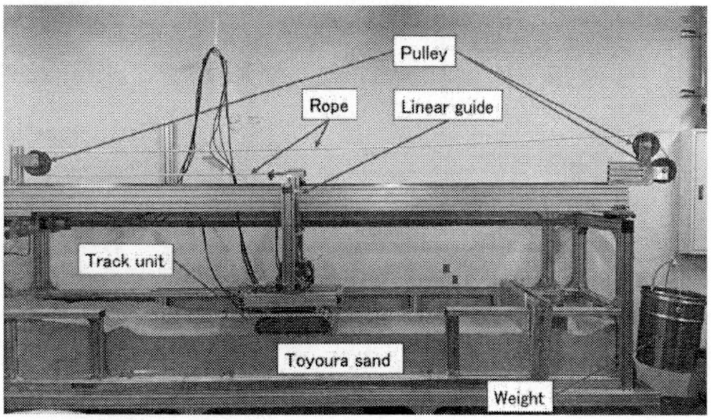

Fig. 2. Overview of track unit test bed

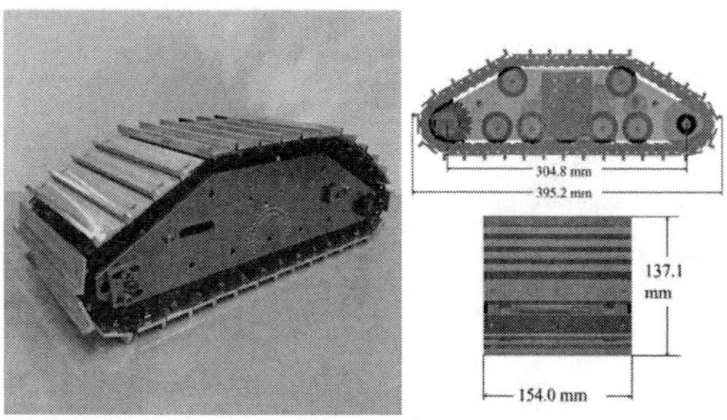

Fig. 3. Overview and dimension of track unit

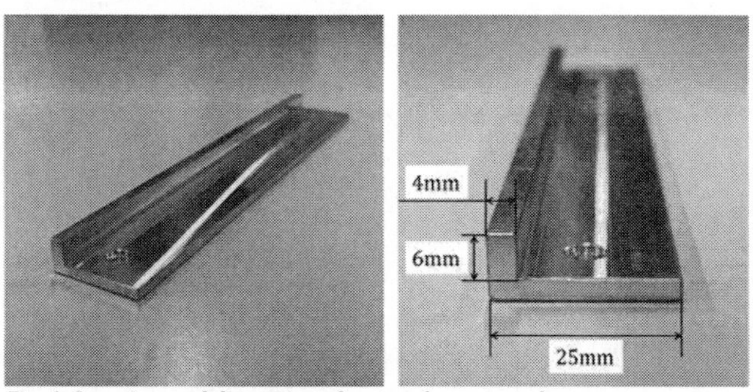

Fig. 4. Overview and dimension of rectangle grouser shoe

The track unit is mounted under the vertical linear guides and runs from left to right on Fig. 2. Each linear guide is equipped with a magnetic scale that can measure the horizontal and vertical travel distance. The drawbar pull exerted by the tracked unit during travel is simulated as a force applied in the opposite direction of the track through the pulley by the weight at the right of Fig. 2.

Fig. 3 shows an overview of the groused-track unit installed in this experimental apparatus. The track unit consists of a track part with rectangular grouser shoes attached to a chain and a drive transmission part with sprockets and rolling wheels that mesh with the chain and rotate. This track unit is based on Komatsu Ltd.'s PC18MR-3 mini-backhoe, with the track ground surface dimension reduced to 1/4 of the actual size. This was modified for this research from the one developed in the research of Ishibashi et al. (2019). Fig. 4 shows an overview of the rectangular grouser shoe attached to the track unit. The grouser shoe is made of metal, aluminum alloy (A6061), for mechanical simplicity since the elastic deformation of rubber makes experiments and analysis more complicated.

4. Drawbar Pull Experiment and Analysis

4.1. Experimental procedure

Because the soil in the traction test rig is soft, it is desirable to keep the soil condition constant for accurate comparison and reproducibility of the data obtained in each experiment. Therefore, the following procedures were used to prepare the soil before each run.

1. Sand was excavated by hand with a shovel to ensure uniform bulk density of the soil before each run.
2. The unevenness of the ground caused by the manual-shovel excavation was leveled using a blade. An acrylic plate was attached to the rail of the traction test apparatus via a slide guide, and the blade was moved horizontally in the direction of the wheel to level the soil surface.
3. After the soil was prepared according to steps 1 and 2, a vane shear test was conducted to confirm that the soil had been adequately prepared. The vane shear test is performed by inserting a torque driver with a blade at its tip vertically into the sand to a certain depth and rotating it in the axial direction to measure the shear strength of the soil and confirm soil compaction.

The track unit was driven for 15 seconds with a drawbar pull load applied, and the vertical velocity and sinkage were measured every 0.05 seconds using a magnetic scale. The same experiment was conducted with three different vehicle weights (21.285 kg for the track unit itself, 26.285 kg, and 31.285 kg) at various drawbar pull loads. The drawbar pull loads for each vertical load were adjusted so that the values translated into drawbar pull coefficient μ were approximately the same, referring to the drawbar pull loads for the vehicle weight of 21.285 kg shown in Table 1. In the experiment with a vehicle weight of 31.285 kg, the belt of the drive transmission part of the track unit could not bear a towing load larger than 14.5 kg, and the track was unable to run due to spinning caused by teeth skipping of a timing belt.

By conducting the above experiments, the track mobility of the groused-track unit at each drawbar pull load is discussed from the point of track velocity and sinkage, and the tractive performance of the groused-track unit are discussed based on the graphs of slip ratio and drawbar pull coefficient.

4.2. Processing of experimental data

This subsection describes how the horizontal velocity and sinkage data are processed. Fig. 5 shows a graph of time history of the horizontal velocity at a drawbar pull load of 0 kg, with a thin blue line. It can be seen that there is a high-frequency oscillation in

Table 1
Variation of load

Vertical load [kg]	Drawbar pull load [kg]
21.285	0, 2, 3, 4, 5, 6, 7, 8, 8.5, 9, 9.5, 10, 10.5, 11, 11.5
26.285	0, 3, 4, 5, 6.5, 7.5, 9, 10, 10.5, 11.5, 12, 12.5, 13, 14, 14.5
31.285	0, 3.5, 5, 6.5, 8, 9.5, 11, 12, 13, 13.5, 14.5

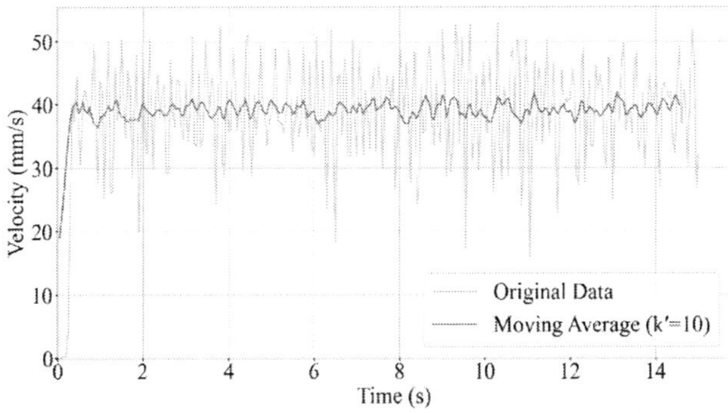

Fig. 5. Time history of horizontal velocity (Track weight is 21.285 kg, *Drawbar pull load is* 0.0 kg*)*

the speed every 0.05 seconds. This is thought to be due to the fact that the track consists of separated grouser shoes, so that a series of position changes occur periodically from the time one of the grouser shoes enters the soil until it makes full contact with the soil. In this study, since the behavior of the entire grousered-track unit is important, the graphs of the horizontal velocity and the sinkage are plotted using a moving average method, as shown by the red line in Fig. 5. The moving average method is a type of low-pass filter that removes high-frequency data, and outputs the value at a certain point as the average value of k' points in the neighborhood. In this experiment, the grouser shoe cycle of 0.5 seconds is used as a reference, where the distance between the grousers p = 25 mm is divided by the track rotation speed $r\omega$ = 52.9 mm/s. The average speed is output for 10 plots in a single plot every 0.05 seconds, which means that one plot is the average speed during the 0.5 seconds.

4.3. Mobility of the grousered-track unit

Fig. 6 shows the time history of horizontal velocity and sinkage when the drawbar pull load is between 0.0 kg and 8.0 kg. The time history of the drawbar pull load shows that, although there are some fluctuations, there are no large fluctuations after an acceleration time of approximately one second after the start of driving. Since the thickness of the grouser shoe is 3 mm, the sinkage is almost constant after 1 second. The sand between the grousers does not move and grousers stick the ground without causing soil failure, as seen when driving with drawbar pull loads in this range. As an example, Fig. 7 shows a capture of the track driving with a drawbar pull load of 4.0 kg. From these results, it can be said that the entire grousered-track unit is in a stable driving state when the drawbar pull load is in the range of 0.0 kg to 8.0 kg.

Fig. 8 shows the time history of horizontal velocity and sinkage at drawbar pull loads of 8.5 kg to 11.5 kg. To verify the reproducibility of the experimental results, similar measurements were taken twice each at 8.5 kg and 9.0 kg, and three times each from 9.5 kg to 11.0 kg. At 11.5 kg, the track was in full slip and it was buried in the sand as it sank, so measurements were taken only once to prevent machine failure. These results show that both horizontal velocity and sinkage fluctuate greatly, indicating that rapid acceleration and sinking are repeated irregularly. This behavior can be considered the Stick-Slip phenomenon. Stick-slip is a phenomenon in which friction causes alternating static and slippage at the contact surface of two objects, resulting in sudden acceleration and slowdown (Berman et al., 1996). In this experiment, the main reason for the stick-slip phenomenon is that the shear stress at the track ground surface reached the maximum shear stress shown in Eq. 6. In other words, as shown in Fig. 9,

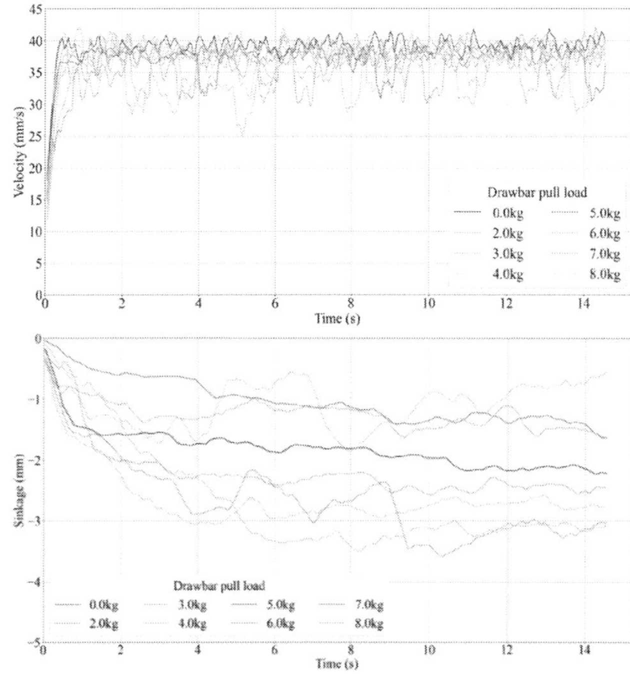

Fig. 6. Time history of horizontal velocity and sinkage (Track weight is 21.285 kg, *Drawbar pull load is from* 0.0 kg *to* 8.0 kg*)*

Fig. 7. Capture of track driving (Track weight is 21.285 kg, *Drawbar pull load is* 4.0 kg*)*

when the maximum shear stress is reached, shear failure occurs between the soil between the grousers and the soil underneath, and the slip phenomenon, in which the grousers and sand move together, and the stick phenomenon, in which the grousers stick the sand and move forward just before the maximum shear stress, are repeated irregularly.

Also, the following points are considered to be factors in the Stick-Slip phenomenon.

- In this experiment, soil bulk density was manually excavated using a shovel. Even though vane shear tests were conducted to confirm the bulk density at several points, the bulk density distribution was not completely uniform across the entire runway. Soil shear failure occurs in soft areas with low bulk density, causing the track to continue to sink, and when it hits an area with slightly higher bulk density, the soil shear failure is reduced and the track moves forward rapidly.
- As the drawbar pull load increased, tooth skipping of the timing belt was observed due to the increased load on the drive transmission part of the track unit. This may have caused a change in the condition of the sand on the ground surface due to a short period of drive stop.

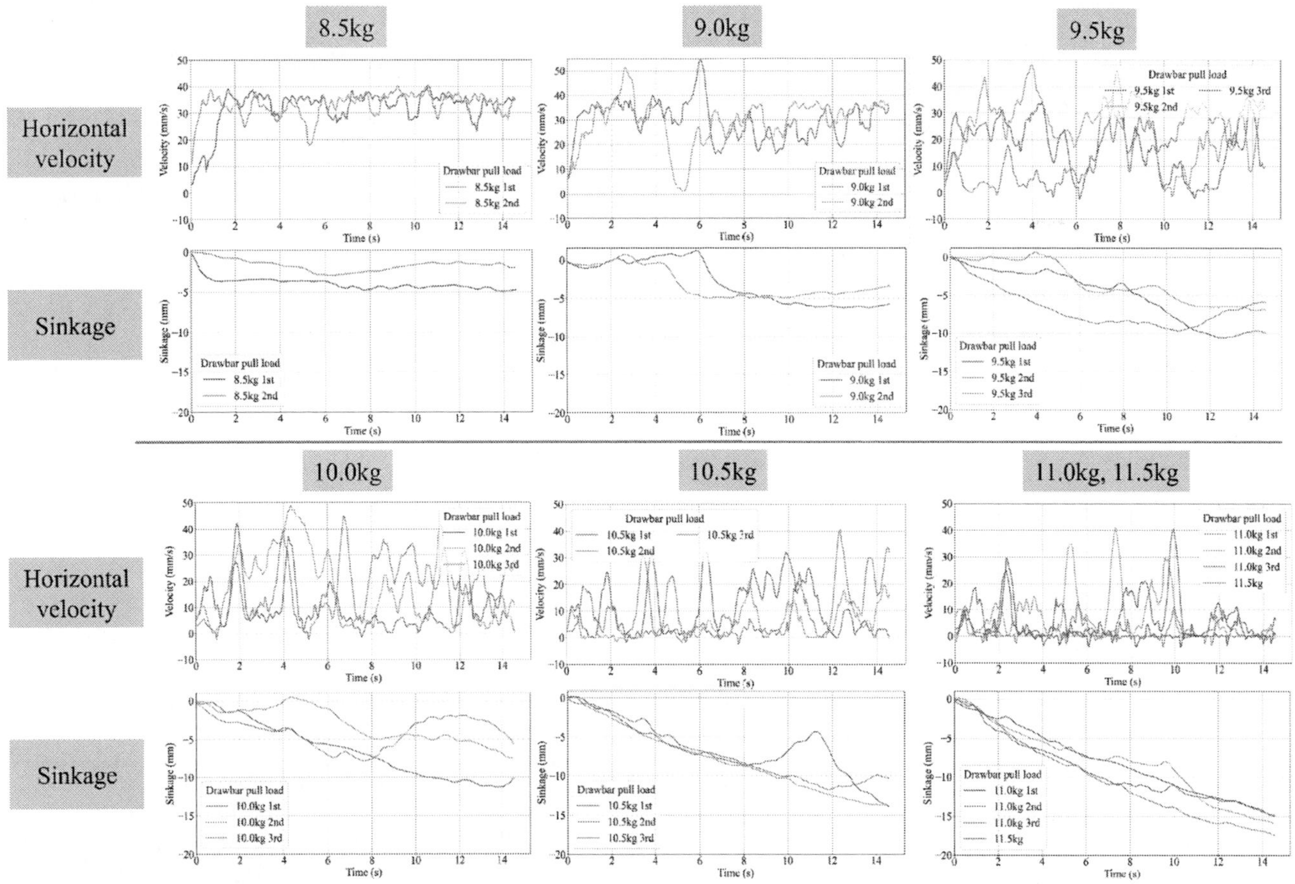

Fig. 8. Time history of horizontal velocity and sinkage (Track weight is 21.285 kg, *Drawbar pull load is from* 8.5 kg *to* 11.5 kg*)*

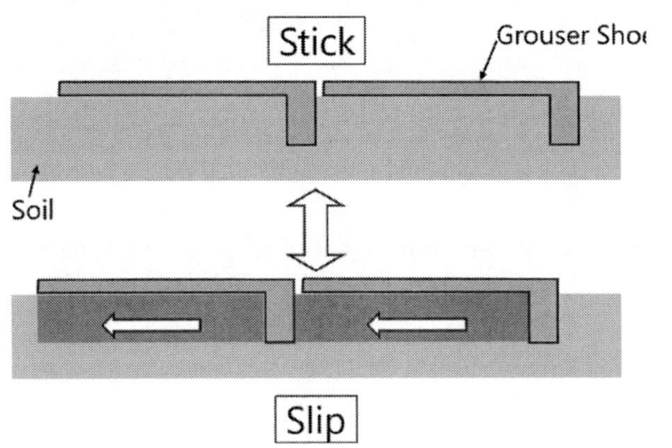

Fig. 9. Stick-Slip under grouser shoes

- The rope transmitting the drawbar pull load runs parallel to the ground and may be oscillating slightly due to the weight of the rope itself, which may also be causing minute oscillations in the drawbar pull load.

The above factors occur in combination when the drawbar pull load is between 8.5 kg and 11.5 kg, and the behavior of the track is considered to be unpredictable and unstable. In this range, the machine may be damaged or become unable to run.

Therefore, it is necessary to design the track to reduce the possibility of such an unstable driving state.

4.4. Relationship between μ and s

Fig. 10 shows a graph of the actual measured values of the drawbar pull coefficient μ and slip ratio s for the three vehicle weights shown in Table 1. The plots and error bars of slip ratio were calculated from the moving average value of the slip ratio from 2 to 14.5 seconds after the start of track travel. This range of time was based on the judgment that from Fig.5, the transition time from the start of track travel to the steady state is approximately one second. First, the experimental results for a vehicle weight of 21.285 kg show that the slip ratio increases rapidly from about $\mu = 0.45$, and it remains almost constant until then. The range of μ where s is almost constant corresponds to a drawbar pull load of up to 8.0 kg or 8.5 kg. In other words, the grousered-track unit maintains a nearly constant slip ratio under the stable driving conditions described in Section 4.3. Therefore, it is suggested that creating a $\mu - s$ graph by conducting drawbar pull experiments and confirming the range of μ in which track is in a stable driving state is an effective method for evaluating the tractive performance of the grousered-track units. In addition, comparing the experimental results with other vehicle weights in Fig. 10, the shapes of the graphs are almost the same. This indicates that the tractive performance of the grousered-track unit is less dependent on the vehicle weight. This further enhances the effectiveness of the evaluation method for grousered-track units using the $\mu - s$ graph described above.

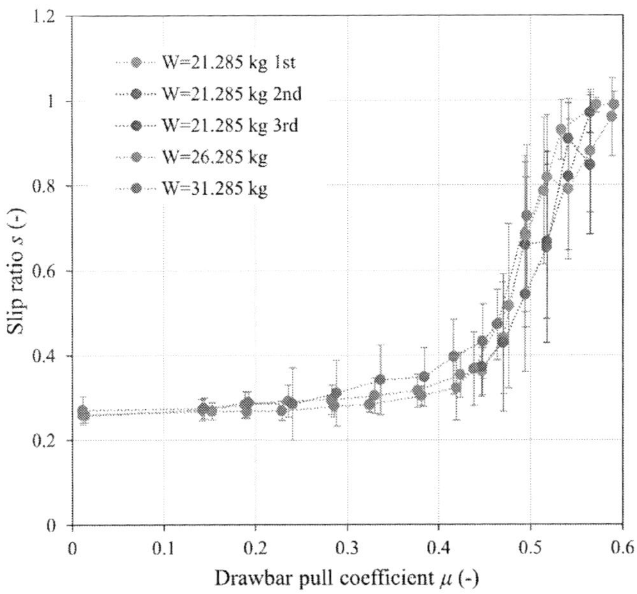

Fig. 10. Drawbar pull coefficient and slip ratio

5. Conclusions and future work

In this study, the drawbar pull experiment with the grousered-track unit was performed. The time history of horizontal velocity and sinkage of the track with varied drawbar pull loads and vehicle weights were measured. Furthermore, the drawbar pull characteristics have been analyzed based on the drawbar pull coefficient and slip ratio.

When the drawbar pull load was small, both horizontal velocity and sinkage fluctuated little, and the grouser stick the ground firmly, resulting in a stable driving state for the entire track unit. When the drawbar pull load was increased, both horizontal velocity and sinkage fluctuated greatly due to the stick-slip phenomenon, resulting in an unstable driving state.

In terms of the relationship between drawbar pull coefficient and slip ratio, it was confirmed that the slip ratio is almost constant in the stable driving state due to the effect of grousers. The graph shape of the drawbar pull coefficient and slip ratio was found to be less dependent on the vehicle weight. These results suggest that as a method for evaluating the tractive performance of grousered-track units, it is effective to generate a $\mu - s$ graph and check the range of μ within which the track unit can maintain a stable driving state.

In the future, we will conduct similar drawbar pull experiments using various shapes of grouser shoes to optimize the shape of the grouser. In addition, in this study, the dependence of this evaluation method on the rotational angular velocity of the track was not verified because the experiments were conducted with a fixed rotational angular velocity of the track. Future experiments should be conducted at other rotational angular speeds to further verify the validity of the evaluation method presented in this study.

Nomenclature

b	Track width	[m]
c	Adhesion of soil	[kN/m^2]
d_t	Track ground contact length	[m]
DP	Drawbar pull of track	[N]
F_h	Thrust force	[N]
j	Soil displacement	[m]
k	Soil shear deformation constant	[m]
k_c	Soil constant dependent on adhesion based on Bekker's model	[kN/m^{n+1}]
k_c'	Soil constant dependent on adhesion based on Reece's model	[-]
k_ϕ	Soil constant dependent on internal friction angle based on Bekker's model	[kN/m^{n+2}]
k_ϕ'	Soil constant dependent on internal friction angle based on Reece's model	[-]
n	Index of compaction and sinkage	[-]
r	Rotation radius of track	[m]
R_x	Driving resistance	[N]
s	Slip ratio	[-]
v_x	Horizontal velocity of track	[m/s]
W	Track weight	[N]
X	Distance from the front of the track contact surface	[m]
z, z_h	Sinkage of track	[m]
μ	Drawbar pull coefficient	[-]
ρ	Soil density	[kN/m^3]
σ	Vertical stress	[kPa]
τ	Shear stress	[kPa]
ϕ	Internal friction angle of soil	[rad]
ω	Angular velocity of track rotation	[rad/s]

Declaration of competing interests

The authors declare that they have no known competing financial interests or personal relationships that could have appeared to influence the work reported in this paper.

References

Bekker, M.G., 1956. Theory of Land Locomotion; the Mechanics of Vehicle Mobility, University of Michigan Press

Reece, A.R., 1965. Principles of Soil-Vehicle Mechanics, Proceedings of the Institution of Mechanical Engineers 180, 45-66

Janosi, Z., Hanamoto, B.,1961. The analytical determination of drawbar pull as a function of slip for tracked vehicles in deformable soils. Proceedings of the 1st International Conference on Terrain-Vehicle Systems, 707-736

Shaikh, S.A., Li, Y., Zheng, M., Chandio, F.A., Ahmad, F., Tunio, M.H., Abbas, I., 2021. Effect of Grouser Height on the Tractive Performance of Single Grouser Shoe under Different Soil Moisture Contents in Clay Loam Terrain. Sustainability 13, 1156 https://doi.org/10.3390/SU13031156

Ishibashi, S., Ishigami, G., 2019. Experimental Measurement and Analysis on Track-soil Interaction Stresses using In-track Sensor System. The Proceedings of the Symposium on the Motion and Vibration Control 16, A105 (in Japanese)

Berman, A.D., Ducker, W.A., Israelachvili, J.N., 1996. Origin and characterization of different stick-slip friction mechanisms. Langmuir 12(19) https://doi.org/10.1021/la950896z

3559 | Modeling light detection and ranging (LiDAR) response in snowy environments

Brian Brady *, Aaron Meyer, Sergey Vecherin, Mike Parker, Han Nguyen

Cold Regions Research and Engineering Laboratory, U.S. Army Corps of Engineers Research and Development Center, Hanover, NH

** Corresponding author: brian.r.brady@erdc.dren.mil*

ABSTRACT

The sensors used by a robust, off-road autonomous vehicle must be capable of detecting obstacles in any weather condition, but this becomes especially important during inclement weather with reduced visibility. Snowfall is no exception to this and presents extremely challenging conditions to vehicles exclusively employing Light Detection and Ranging (LiDAR) systems for object detection. Snowflakes may occlude "hard" targets in the vehicle's path of travel or introduce noise from "soft" detections of the snowflakes themselves. Since it is challenging to acquire experimental data for all possible snowfall conditions, numerical simulations are often employed to predict sensor performance. This study highlights preliminary results from a simulation that reproduces realistic snowfall effects within a Virtual Autonomous Navigation Environment (VANE). The simulation models snowflake size and concentration while accounting for sensor-specific modes of operation of the Velodyne HDL-32 LiDAR. VANE simulations provide vehicle designers with a rapid and affordable method to evaluate LiDAR perception and autonomy response in snowy conditions.

Keywords
LiDAR
snow
simulation
autonomous
vehicle

1. Introduction

During the winter of 2021-2022, the Michigan Office of Highway and Safety planning recorded 35,782 winter weather automobile crashes responsible for 78 fatalities (Michigan Traffic Crash Facts). During 2022, the Minnesota Department of public Safety recorded 20 fatalities and 1268 injuries due to automobile accidents occurring in snowy conditions. (Minnesota Motor Vehicle Crash Facts). Nationally, the U.S. Department of Transportation reports that falling snow is a contributing factor to automobile wrecks resulting in 76,000 injuries and 900 fatalities each year ("Snow & Ice - FHWA Road Weather Management 2025). Statistics like these indicate that snowy weather creates challenging and dangerous conditions for humans to operate ground vehicles in.

The automotive industry is actively working toward producing a "Level 5" vehicle capable of "full driving automation" (On-Road Automated Driving Committee) that could eliminate the human element from these winter accidents. The Waymo Corporation has performed extensive testing with self-driving taxicabs, logging over 20 billion simulated miles and 20 million road miles. But notably, their fleet is based out of Phoenix, San Francisco, Los Angeles, and Austin (Waymo Self-Driving Cars), all cities where snowfall events are exceedingly rare. Additional testing is required to demonstrate that safe and reliable results can be produced in snowy climates like Michigan and Minnesota.

Consideration must also be given to off-road environments during winter conditions. For example, in 2014 NATO soldiers travelled to Baffin Island in the Arctic Circle to participate in the Guerrier Nordique training exercise. A major finding from this exercise was that severe winter weather impeded the ability of ground-based supply convoys to travel from logistical hubs to isolated infantry platoons (Fry, 2014). Automated supply convoys are in development that could potentially assist with this problem (Bergin et al., 2023), but little research has been done to assess their performance in conditions like those encountered on Baffin Island.

The performance of an autonomous vehicle in winter weather depends, in part, on its ability to see through heavy snowfall to detect and avoid obstacles along its path of travel. Autonomous vehicles utilize a variety of sensors to perceive their surroundings including cameras, radar, and of interest to this study, Light Detection and Ranging (LiDAR) (Ignatious et al., 2022). In simple terms, a time-of-flight LiDAR unit is a laser tape-measure attached to a stopwatch. The laser emits a pulse of light into the environment and measures the amount of time required for the pulse to reflect off a target and return to the receiver. Distance between the receiver and the target is calculated using the measured flight time and known speed of light. Automotive LiDAR units can make thousands of measurements each second, which enables construction of a high-resolution, three-dimensional surface map of the vehicle's surroundings, commonly referred to as a "point cloud" (Royo and Ballesta-Garcia, 2019).

Snowfall causes LiDAR performance to degrade. Snowflakes falling through the laser frustum introduce noise into the point cloud and can potentially obscure obstacles in the vehicle's path of travel. The extent to which snow degrades LiDAR perception depends heavily on the make and model of the sensor in use. Laser parameters such as wavelength and pulse width, in conjunction with the optical geometry of receiver and transmitter all influence system performance. Additionally, signal processing algorithms embedded in the sensor's firmware and user-configurable modes of operation can also have significant impact. In the context of a winter autonomous driving simulation, a sensor-specific virtual

https://doi.org/10.56884/GPMZPXQX

LiDAR unit must faithfully reproduce effects introduced by both the physics of laser-snowflake interaction and the sensor's proprietary firmware.

This study presents an overview of how snowfall degrades performance of the Velodyne HDL-32 LiDAR and examines the impact of user-configurable settings on the sensor's behavior. Empirical data is compared to results from a physics-based ray-tracing simulation. This paper documents the discrepancies between the two results and provides evidence that sensor-specific firmware must be accounted for in the modelling effort. The paper concludes by showing how the model can be incorporated into a software-in-the-loop simulation for testing autonomous vehicle performance in off-road, snowy conditions.

2. Related Work

The impact of adverse weather on LiDAR performance has been well documented. Linnhoff et al., 2022 show how sunlight, rain, fog, and snow influence the performance of Blickfield, Ouster, and Velodyne sensors. Filgueira et al., 2017 study how rain attenuates the intensity and number of returns generated by a Velodyne VLP-16. Ronnback and Wernersson, 2008 examine performance of the SICK LMS200 and URG04-LX in falling snow. Bos and Weng, 2023 assess how blowing snow degrades the performance of automotive LiDAR.

Several adverse weather LiDAR datasets are available to scientific community. The Winter Adverse Driving Dataset (Kurup et. al 2022 and 2023) contains semantically labeled point clouds collected over multiple seasons. The aDverse wEather eNvironment Sensing systEm (DENSE) dataset (DENSE-Universität Ulm) provides access to sensor readings collected in a variety of weather conditions including fog (Bijelic et al., 2020) and snow (Kutila et al., 2020). The Canadian Adverse Driving Conditions dataset contains annotated LiDAR and camera output collected by an autonomous vehicle travelling through snowfall (Canadian Adverse Driving Conditions Dataset).

Datasets like these have been used to assist researchers in developing algorithms for de-noising snowy LiDAR point clouds. Refer to Kurup and Bos, 2021, Charron et al., 2018, Heinzler et al., 2020, and Park et al., 2020.

Significant efforts have also been devoted to the inverse of the point cloud de-noising problem, namely the generation of synthetic adverse weather data using predictive models. Goodin et al., 2019 present a method for predicting LiDAR range attenuation during rainfall. Rasshofer et al., 2011 provide LiDAR extinction coefficient models for rain, snow, and fog. Phillips et al. 2017 develop a model for predicting LiDAR response in dusty conditions. Rahm et al. 2021 present a model for laser transmission through snowfall. Hahner et al., 2022 and Roy et. al 2020 develop methods for predicting LiDAR response in snowfall.

This paper seeks to add to this collection of knowledge by examining the sensor-specific characteristics of the Velodyne HDL-32, modelling its performance in snowy conditions, and discussing reasons for discrepancies between the empirical and simulated results.

3. HDL-32 LiDAR degradation effects from snowfall

The Velodyne HDL-32 is a rotating, time-of-flight, automotive LiDAR. Selected operating characteristics are listed in Table 1.

Table 1
HDL-32 Characteristics (HDL-32E User's Manual and Programming Guide)

Parameter	Value
Wavelength	905 nm
Maximum Range	100 m
Number of Lasers	32
Vertical Field of View	-10 to + 30 deg
Horizontal Field of View	360 deg
Measurement Rate	21696 shots per beam per second

A common degradation effect that arises when operating a LiDAR like the HDL-32 in a snowstorm is the introduction of "spurious returns" (Hahner et al., 2022), "unwanted detections" (Rasshofer et al., 2011), or "false positives" (Bos and Weng, 2023) into the point cloud. These soft returns are due to laser reflections from snowflakes that cause the sensor to record distance measurements in the atmosphere between the transmitter and hard mobility impediments in the scene like tree trunks or boulders. Soft returns manifest as a collection of spurious points centered radially around the LiDAR hub. Figure 1 shows this effect in a point cloud collected with a Velodyne HDL-32 during a moderate snowfall event.

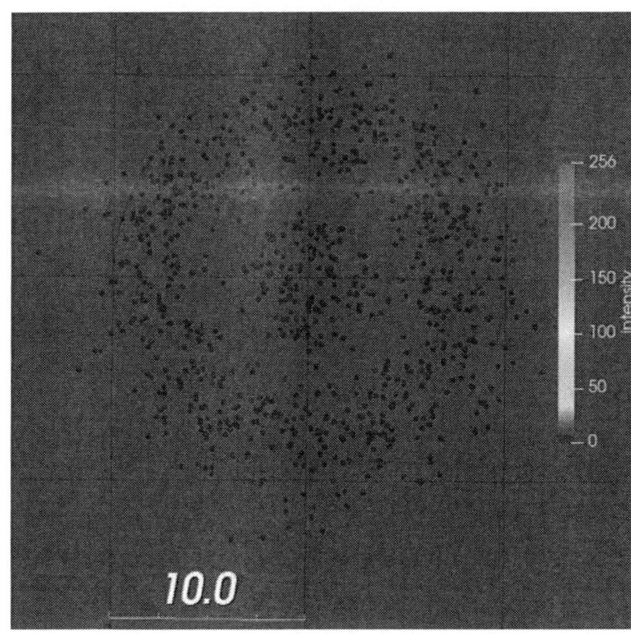

Fig. 1. Birds-eye view of a Velodyne HDL-32 point cloud showing blue soft returns from snowflakes. Grid spacing is 10 meters. Color scale corresponds to the calibrated reflectivity of the return.

This image shows that soft returns from the HDL-32 are distributed over distance in a double torus pattern. Several factors are responsible for this phenomenon. First, the HDL-32 uses a rectangular beam frustum with a horizontal divergence of 2.79 milliradians. At a range of 100 feet, divergence causes the size of the beam spot to grow to approximately four inches wide and two inches tall (HDL-32E User's Manual and Programming Guide). This means that the frustum generated by one laser pulse can capture reflections from multiple snowflakes. The HDL-32's internal signal processor decides which reflection to populate into the point cloud depending on a user-selectable mode of operation. When running in "strongest" mode, the HDL-32 only populates the most intense reflection from each shot to the final point cloud. In "last" mode, only the most distant reflection is populated. In "dual" mode, both the strongest and last returns are populated. (VLP-16 User Manual).

The point cloud depicted in Fig. 1 was captured in dual mode. The obvious hypothesis in this case is that outer torus of this figure corresponds to the last snowflakes detected, while the inner torus corresponds to the strongest snowflakes. But the double torus pattern also manifests in data collected with the sensor operating in both the strongest and last modes. To understand the root cause of this discrepancy, a point cloud collected in dual mode was loaded into Velodyne's analysis software, VeloView (VeloView: The Velodyne Lidar Viewer Based on Paraview Lidar). This program was used to process the point cloud and identify the dual returns in the data. Figure 2 shows the results of this analysis, with the dual returns highlighted in pink.

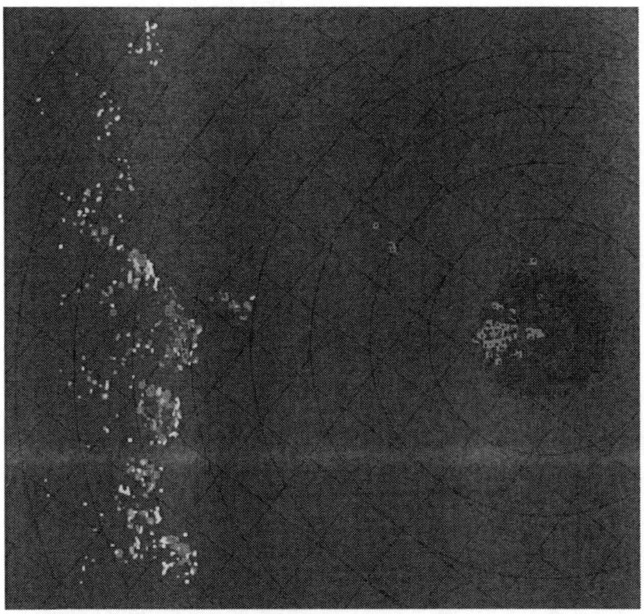

Fig. 2. Velodyne HDL-32 point cloud collected in dual return mode. The snowflake torus is visible on the right, and reflections from nearby vegetation are shown on the left. Dual returns are highlighted in pink.

The snowflake double torus is clearly visible in Fig 2. The points on the left-hand side of the image are returns from nearby vegetation. The pink dual returns almost exclusively correspond to pulses containing a hard target (i.e. vegetation) in the background and a snowflake in the foreground. Analysis of the strongest and last modes revealed no dual returns even though the same double torus pattern appeared. This implies that the presence of the double torus pattern is not due to the sensor's mode of operation.

Patent documents for the Velodyne HDL-64 provide some insight into the observed data pattern:

> Through the use of digital signal processor (DSP) control, a dynamic power feature allows the system to increase the intensity of the laser emitters if a clear terrain reflection is not obtained by photo detectors (whether due to reflective surface, weather, or other reasons), and to reduce power to the laser emitters for safety reasons if a strong reflection signal is detected by photo detectors. A direct benefit of this feature is that the lidar system is capable of seeing through fog and heavy rain by increasing laser power dynamically and ignoring early reflections (Hall, 2011).

Additional Velodyne Inc. documents suggest the sensor's firmware assigns a confidence level to each measurement based on the distance, reflectivity, and several continuity parameters of the return signal. The sensor discards points when the confidence level indicates they are likely to be spurious returns from snowflakes (Zhou et al., 2020).

It appears that regardless of the mode of operation selected, the HDL-32 firmware is selectively ignoring soft detections from snowflakes within a zone of exclusion around the sensor. This effect is clearly visible in the following histogram generated with data collected in strongest mode. Peaks corresponding to each ring of the double torus are clearly visible even though dual returns are not present.

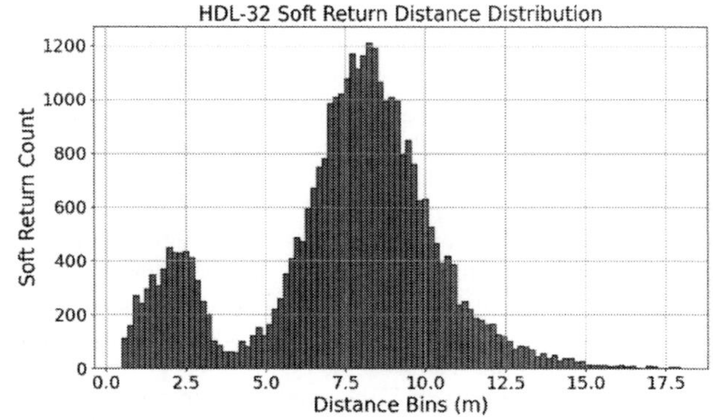

Fig. 3. Velodyne HDL-32 soft return distance distribution captured with the sensor operating in strongest mode.

Figure 3 also shows a minimum distance at which snowflake detections start registering. This is potentially due to the HDL-32 utilizing a bistatic receiver/transmitter configuration. In a bistatic LiDAR unit, the laser receiver and transmitter are not aligned coaxially. This means that there is a small region near the sensor housing where the receiver's field of view does not overlap with that of the transmitter. Inside this region detections are not possible (Phillips et al., 2017, Hahner et al., 2022). The HDL-32 manual does not provide details on whether a bistatic or coaxial configuration is employed, but it does note "the minimum return distance for the Velodyne HDL-32 is approximately 1 meter. Ignore returns closer than this." (HDL-32E User's Manual and Programming Guide). Ouster, the parent company of Velodyne, makes a similar statement in the firmware manual for their OS1 sensor:

> Minimum Range Threshold (cm) is a new feature added in FW v3.1 that reduces the minimum range of the sensor, enabling it to detect targets all the way up to the sensor window. In the past, Ouster sensors were subject to a minimum range limitation dictated by the hardware revision. However, Ouster has made significant strides in overcoming this limitation. Now, users have the capability to eliminate the blind zone and extend the field of view (FOV) to the sensor window (Ouster Firmware User Manual).

The data in Fig. 3 indicates that the minimum soft return threshold for the HDL-32 is approximately 0.5 meters.

In addition to knowing how soft returns are distributed as a function of distance, it is also important to understand how often they are generated as a function of time. The HDL-32 fires each of its 32 beams approximately 21696 times per second (HDL-32E User's Manual and Programming Guide). Even if the probability of detecting a snowflake on a single pulse is very low, the high rate at which the sensor fires could result in the introduction of thousands of spurious returns into the point cloud. To examine this effect, an angular sector was cropped from the snowflake torus shown in Fig. 1. The number of soft returns appearing within this sector was counted over the duration of the recording. The number of laser pulses fired through this sector was then calculated using the rotation rate of the sensor hub and the known laser firing rate. The probability of generating a soft return was calculated by taking the ratio of number of soft returns to number of shots fired. Table 2 shows the results of this calculation for three different snowfall conditions. The snowfall conditions in this table were based on qualitative assessments of the observer.

Table 2
HDL-32 Per-Shot Soft Return Probability

Snowfall Condition	Soft Return Probability
Light	0.0004
Moderate	0.014
Heavy	0.033

The sensor's mode of operation has a distinct influence on the probability of generating a soft return. Figure 4 shows a bird's-eye view of a point cloud collected in a typical suburban backyard during a snowstorm. The image was generated with the sensor set to run in last mode.

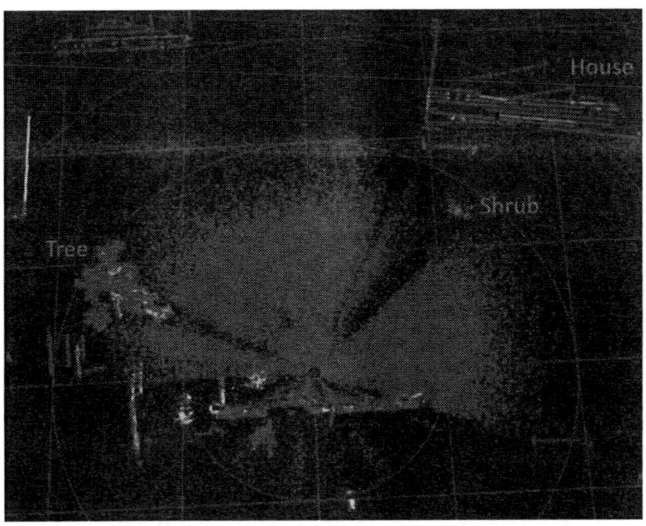

Fig. 4. Overlay of many HDL-32 frames collected in last mode.

When set to operate in last mode, the presence of a hard target in the background causes soft return detection probability in the foreground to drop toward zero. This results in distinct channels cut through the snow torus that align with the hard target. These channels are evident in Fig. 4 along azimuths coincident with the hard reflections from the tree, house, and shrub in the scene. Similar trends are visible in the following references: Real-Time Filtering of Snowfall from Lidar on Autonomoose Data 2018, Bos 2024a, and Bos 2024b.

This same trend is visible, although to a much lesser extent, when the sensor is set to operate in dual mode, as shown in Fig. 5.

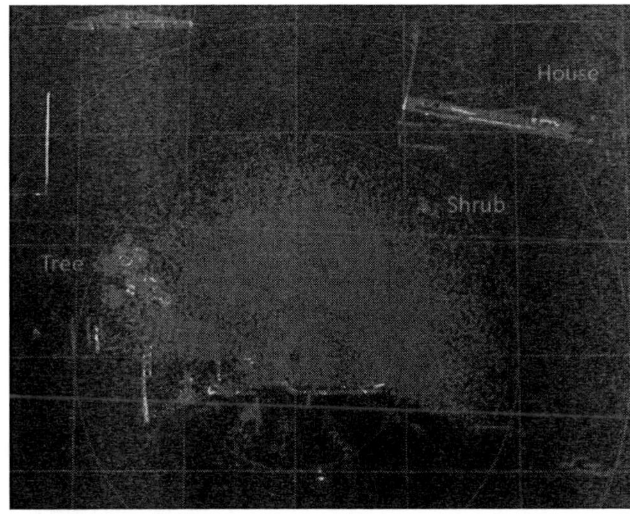

Fig. 5. Overlay of many HDL-32 frames collected in dual mode. Note the slight reduction in soft return probability on azimuths between the lidar hub and the tree.

The location of a soft return generated by the HDL-32 does not always appear to be completely randomized on each shot. Due to the laser's horizontal divergence and the finite width of beam spot, the same snowflake could be illuminated on sequential pulses. This could lead to streaks appearing in the point cloud as the same snowflake is detected across sequential azimuths. Streaks are visible in the data collected for this study, but surprisingly they appear to be oriented radially outward from the sensor hub, rather than perpendicular to the beam's flight path, as shown in Fig. 6.

Fig. 6. Soft return streaks in a snowy HDL-32 point cloud.

The presence of these radial streaks could be the result of a LiDAR phenomenon known as "mixed pixels" or "ghost points." This effect arises when the beam spot illuminates multiple targets in the same pulse. The LiDAR's internal signal processor then lumps the multiple reflections of the pulse into a single, intensity weighted, average distance. This distance measurement will fall between the minimum and maximum of the initially measured reflections and may not actually correspond to an object physically present in the scene (Carrillo et al., 2020). In the case of the soft returns depicted in Fig. 6, it is possible that a large snowflake exists near the start (foreground) of each radial streak. This large snowflake may be visible for several sequential pulses. But each pulse will contain a different amount of background snowflakes that cause the average distance reported by the sensor to change with each measurement. The streak continues until the large foreground flake is no longer illuminated by the beam spot.

The presence of snowflakes in the laser field of view also impacts the hard returns measured by the sensor. Work by Goodin et al., 2019 and Rasshofer et al., 2011 indicates that precipitation particles attenuate the intensity of hard returns and reduce the maximum range at which a LiDAR unit can detect an obstacle. This effect is demonstrated in Fig.7.

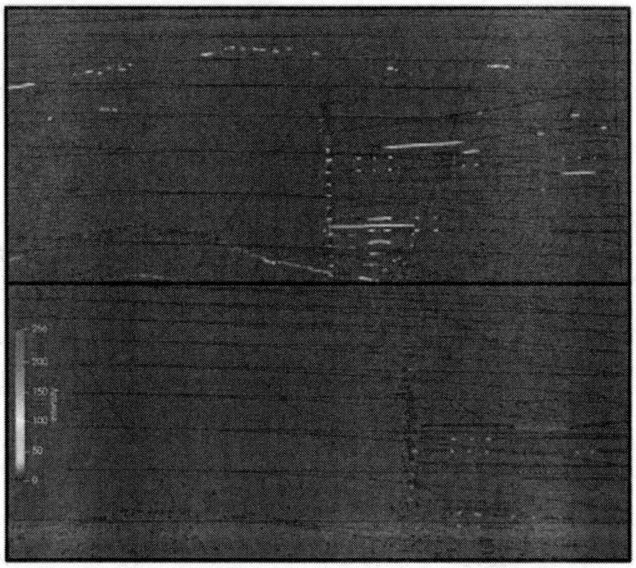

Fig. 7. Residential house scanned with an HDL-32 in two different weather conditions. Top: sunny weather, blue-sky conditions. Bottom: Heavy snowfall. Red dashed lines indicate an area cropped for intensity comparison. Color scale corresponds to reflectivity.

The top image was collected on a sunny, blue-sky day. The bottom image is the same scene but scanned during a heavy snowfall event. In the clear-skies image, bright green reflections are visible out to a range of nearly 100 meters. These reflections disappear during the snowstorm. In the top image, gutters and trim on the house appear as high reflectivity green stripes. During the snowstorm, these colors fade to a cooler shade of blue. To quantitively measure this effect, a small area to the right of the second story window was cropped from each image, as depicted by the dashed red outline. The average reflectivity of the hard returns in this region was computed for both scans. During sunny weather, the HDL-32 measured an average reflectivity of 7.1%. During the snowstorm, the reflectivity in this region deceased to 3.3%.

4. Test harness for generating synthetic snowfall data

The trends presented in the preceding section are not anchored to the particle size distributions or concentrations characterizing the snowfall events from which they were captured. These parameters likely have a strong impact on the behavior of the sensor. In the absence of disdrometer data, the authors developed a test harness to model the performance of a virtual HDL-32 in a variety of synthetic snowfall conditions.

The test harness utilizes physics-based ray tracing to model the laser pulse fired by the HDL-32. As noted in Section 2, this sensor has a rectangular beam "spot." Divergence causes the spot size to reach several centimeters in width and height at long ranges. Velodyne sensors utilize a unique beam pattern where each spot is composed of three horizontal bands of light (VLP-16 User Manual). The test harness models this effect by subsampling each pulse with many individual ray traces. Figure 8 demonstrates this concept.

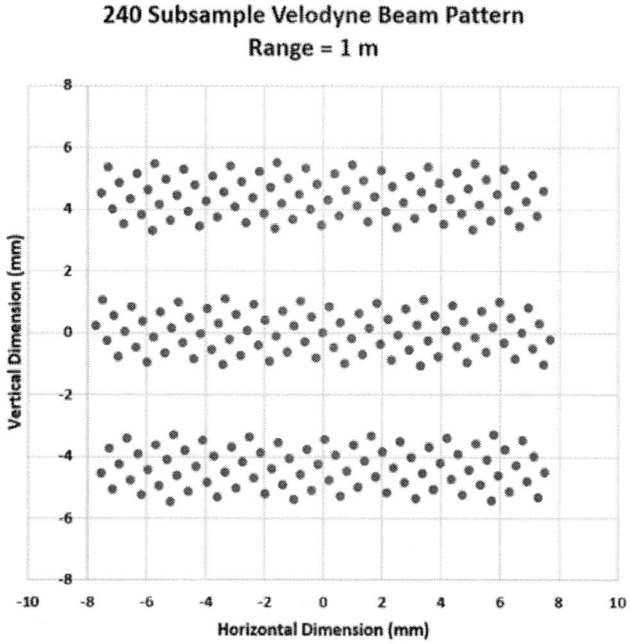

Fig. 8. Rectangular Velodyne beam pattern composed of three rows of 80 subsamples.

This beam pattern was fired through a virtual test chamber filled with synthetic snowflakes. The geometry of real-world snowflakes is highly variable. Dendrites, plates, columns, and graupel of lump, hexagonal, and conical varieties are all possible (Locatelli and Hobbs, 1974). In our ray tracing simulation, the reflectivity of an individual snowflake as measured by the LiDAR unit depends upon the angle of incidence at which the beam strikes the surface of the flake. To simplify geometry and avoid the necessity of tracking the orientation of thousands of particles, each synthetic flake was assumed to be spherical. Each flake was assigned a diffuse reflectivity of 80%. This value is close to the 905 nm reflectivity measured from newly fallen "partly decomposed particles" and "stellar dendrites" with diameters between 1.5 and 3 millimeters (Salvatori et al., 2005). Although this value corresponds to flakes on the ground, it was chosen as a reasonable value from which to begin developing a preliminary model.

Snowflakes were spatially frozen within the virtual test harness and populated with a variety of diameters and concentrations. Figure 9 depicts the geometry of the virtual setup.

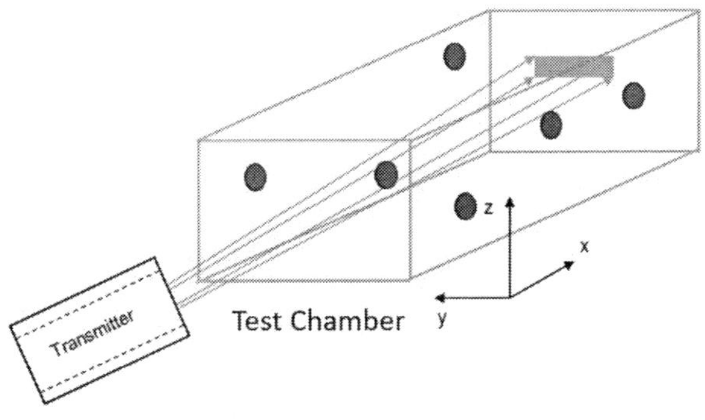

Fig. 9. Geometry of the HDL-32 virtual test harness.

As subsamples travel down the length of the test harness, the ray tracer may determine that they intersect with a spherical snowflake. When this occurs, the LiDAR equation (Goodin 2019) is used to calculate the amount of power reflected to the receiver. Due to beam divergence, one pulse could generate reflections with subsamples spread between several different snowflakes. After firing each shot, the distance and intensity of each intersection is loaded into a Digital Signal Processer (DSP). The DSP performs a convolution over the discrete subsamples and averages them into a smooth, analogue curve. The DSP then selects peaks from this curve to output as either the strongest or last return, as described in Section 3. If the peak of the analogue signal does not exceed the sensor's noise floor, the detection is ignored. Figure 10 illustrates this concept.

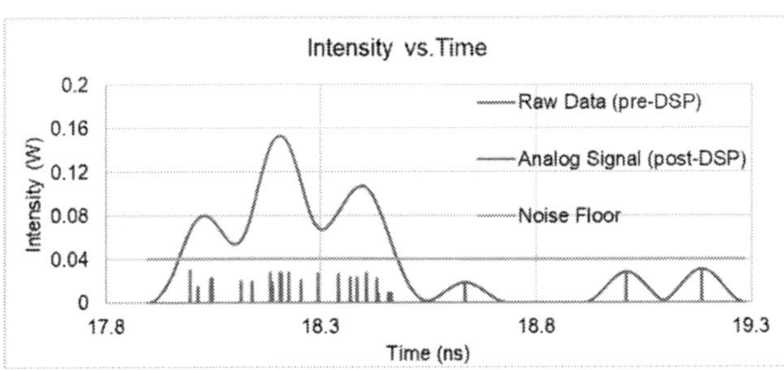

Fig. 10. DSP converting discrete subsamples into a continuous analog curve.

This test harness was used to run two different experiments. The first experiment aimed to characterize the soft returns generated by the DSP. The second experiment was focused on understanding how the presence of snowflakes impacted the ability of the sensor to detect a hard target.

4.1. Soft Return Characterization

This numerical experiment was designed to examine how the soft return distance distribution and detection probability varied as a function of snowflake diameter and concentration. The test harness was filled with particles of equal diameters at concentrations ranging from several hundred to several thousand particles per cubic meter. The origin of the transmitter was randomized on the y-z plane (see Fig. 9) between each shot, and rays were traced out to a range of 100 meters. The distance and intensity of the post-DSP returns were recorded. As expected, some of the shots resulted in soft returns, while others did not reflect enough power to exceed the DSP noise floor, resulting in a miss. After firing many shots, the probability of generating a soft return was calculated by taking the ratio of the number of hits to the number of shots fired.

An extinction coefficient was used to compare results between runs of different diameters and concentrations. The extinction coefficient, α, for each run was calculated using Eq 1. (Phillips et al., 2017).

$$\alpha = \sigma n \tag{1}$$

In Eq. 1 σ represents the cross-sectional area of the spherical snowflakes, and n represents the volumetric concentration at which they are populated in the test harness. The test harness was loaded with several different particle sizes, and runs were made across a range of extinction coefficients. Figure 11 shows how soft return probability varied as a function of these parameters.

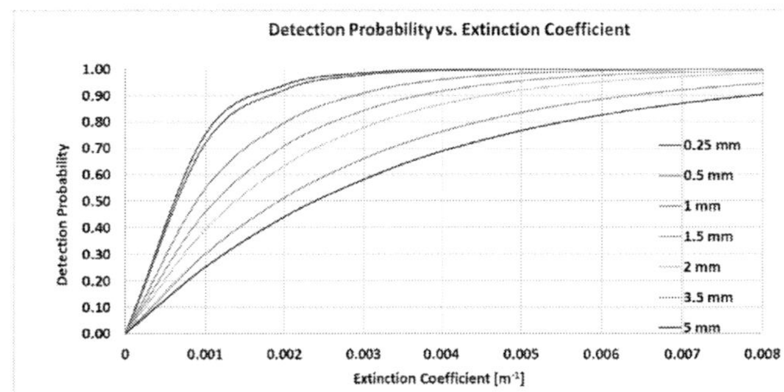

Fig. 11. Soft return probability varying as a function of extinction coefficient and particle diameter.

Figure 11 indicates that for a given extinction coefficient smaller particles are more likely to generate a soft return than large ones. This makes physical sense, as small particles need a much larger concentration to reach the given extinction coefficient than large particles.

Particle diameter and extinction coefficient also had a significant impact on how soft returns were distributed as a function of distance. Figure 12 plots these distributions for three different extinction coefficient/diameter pairings.

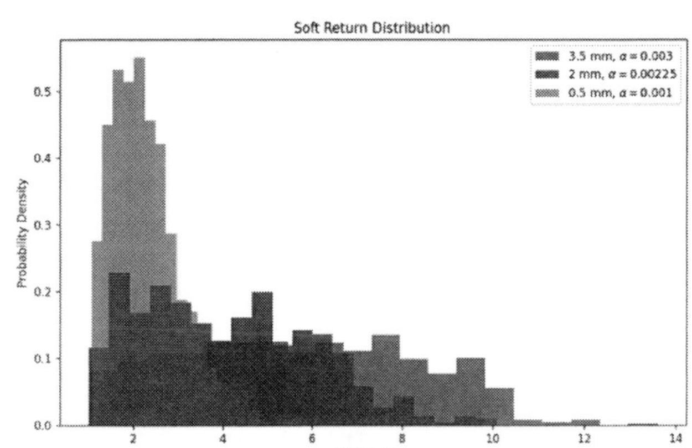

Fig. 12. Soft return probability varying as a function of extinction coefficient and particle diameter.

In both the synthetic and empirical data, the majority of the detections occur at distances less than 15 meters. However, the shape of the distributions in each case differ greatly from each other. The synthetic results do not contain the double torus effect seen in the real-world data. This may be due to internal signal processing in real-world HDL-32 resulting in the sensor selectively discarding points as discussed in Section 3.

It is also important to note that the real-world sensor made soft detections on less than 3% of the shots fired (refer to Table 2). In the synthetic test harness, probabilities this low are only obtained at extremely small extinction coefficients. This may indicate that the 80% reflective spherical snowflakes used in the test harness are not a good proxy for the complicated geometry found in natural flakes. Another possibility is that the HDL-32's internal signal processor is ignoring early returns from snowflakes (Hall 2011, 2014, and Zhou et al., 2020). The ray-tracing model described here does not attempt to account for processing done by the sensor's firmware. Additional computational tuning is required to calibrate the model. Carrillo et al. encountered similar issues modelling rainfall using a computational set up based on a similar ray tracing model to the test harness used in this study. They could not accurately replicate the sensor's maximum detection distance using solely extinction coefficient data collected from real-world rainfall events. The model had to be calibrated using an empirically determined correction factor (Carrillo et al., 2020).

4.2. Influence of Soft Returns on Hard Target Detection

In a second numerical experiment, a hard target (i.e. a solid wall with 80% diffuse reflectivity) was inserted into the test harness at varying ranges from the transmitter. Particle concentration and diameter were varied as previously described. Runs were made with the DSP set to output both the strongest and last detections. The probability of generating a hard or soft return was recorded for each mode. Figure 13 plots these probabilities as a function of distance between the transmitter and the hard target, x_{hard}, with the DSP operating in strongest mode.

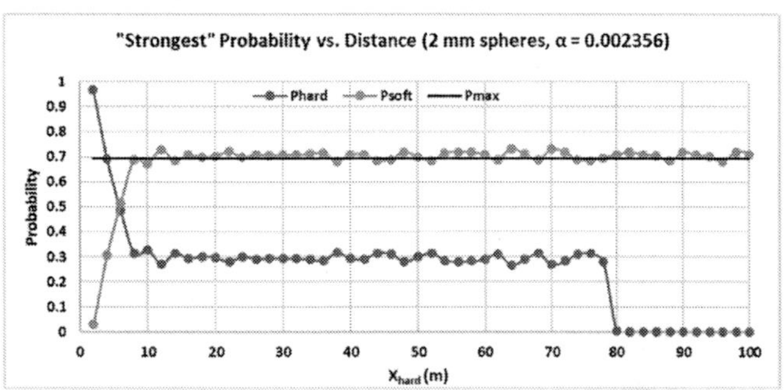

Fig. 13. Hard and soft return probability in the strongest mode varying as a function of distance to the hard target.

In this figure, P_{hard} and P_{soft} refer to the measured probability of generating a hard or soft return, respectively. P_{max} is the probability of generating a soft return in the absence of a hard target, as predicted by the curves shown in Fig. 11. When x_{hard} is small, almost all the detections correspond to the hard target. As x_{hard} increases, the DSP begins to detect snowflakes which causes P_{soft} to increase and P_{hard} to decrease, with the sum of both probabilities remaining constant at 1.0. When the hard target is approximately 10 meters from the transmitter, both P_{hard} and P_{soft} reach their fully developed values. When the hard target is approximately 80 meters from the sensor, P_{hard} drops to zero. This takes place because snowflakes have occluded a significant fraction of the subsamples in the beam pattern at this range. Therefore, the amount of power reflected from the hard target is not enough to exceed the sensors noise floor, causing the hard return to drop from view.

Related trends are visible when the DSP is set to operate in last mode, as shown in Fig. 14.

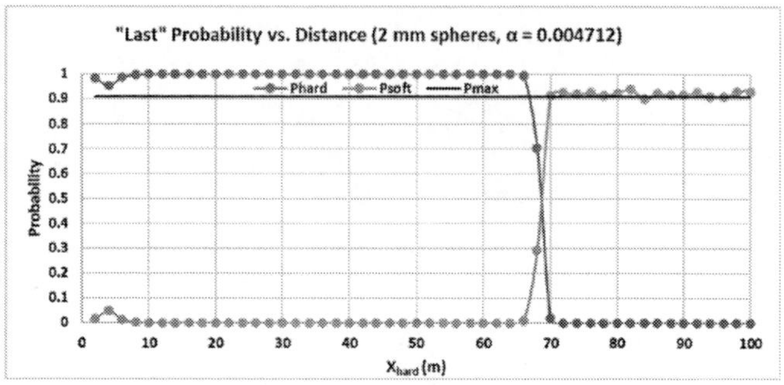

Fig. 14. Hard and soft return probability in last mode varying as a function of distance to the hard target. The spike in soft return probability at $x_{hard} = 4$ meters is due to mixed pixels or ghost points.

In most cases, the last (i.e. most distant) detection made by the DSP corresponds to the hard target. The exceptions to this rule occurs when the hard target is placed very far from the transmitter, or very close to the transmitter. As with strongest mode, at long distances snowflakes attenuate the amount of received power to levels that fall below the DSP noise floor. In this case, the last detection made by the DSP will always be a snowflake.

Note that a small peak in soft return probability develops when the hard target is positioned near the transmitter. Close examination of these returns revealed that the distances measured by the DSP did not correspond to either the range of the hard target or the distance to of any of the snowflakes intersected by beam pattern's subsamples. These soft returns are the ghost points or mixed pixels described in Section 3 that result from the DSP averaging the distance between snowflakes and the hard target.

5. Snowfall in the Virtual Autonomous Navigation Environment (VANE)

The Virtual Autonomous Navigation Environment (VANE) contains several software modules that simulate sensor perception and vehicle/terrain interaction. This provides robotics developers with the ability to connect an autonomy stack directly into a high-fidelity simulation environment. Using VANE, autonomy algorithms can be rapidly evaluated in conditions that are difficult, expensive, or dangerous to establish with physical testing.

First, a high-fidelity virtual scene is constructed within Unreal Engine (Unreal Engine Overview). Data from LiDAR point clouds, UAS photogrammetry, on-site imagery, and a variety of commercial and government repositories is used to construct a digital twin of the area of interest. This data is used to accurately characterize the terrain, structures, vegetation, and road networks present in the scene. Objects in the scene are assigned correct material attributes such as soil strengths, thermal properties, and optical textures (Holden et al., 2024). Figure 15 displays a VANE scene containing an unimproved track through a snow-covered forest.

Fig. 15. Unimproved track through a snow-covered forest in VANE.

As a vehicle navigates through a VANE scene, onboard sensors such as cameras, radar, GPS, and LiDAR interrogate the environment using physics-based raytracing. Each virtual sensor builds a data stream formatted identically to that of its real-world counterpart. These data streams are exported outside of VANE to an external customer's autonomy kit. The autonomy kit processes the sensor data and makes decisions on how the vehicle should navigate around obstacles detected by the sensors. These decisions are translated into vehicle control inputs, which are then sent back inside VANE's vehicle dynamics and terrain interaction module. This module uses the control inputs to update vehicle's position and orientation in the scene. The sensors then perform another scan, and the process is repeated until the vehicle reaches its desired destination (Holden et al., 2024).

VANE must be able to complete this entire process at real-time speed. For example, suppose the autonomy kit expects to receive 10 sensor data packets per second. VANE must be capable of outputting data at this rate, or the autonomy kit will assume the sensors are inoperative.

Generating synthetic LiDAR data at rates fast enough to meet the real-time threshold becomes challenging in snowy conditions. Automotive LiDAR units typically have a maximum range of approximately 100 meters. This means that in a VANE scene, snowflakes must be continually populated in a cylinder of 100-meter radius centered around the vehicle. During heavy snowfall events with high flake concentrations, this could easily lead to the generation of tens of millions of snowflakes. It is computationally inefficient to add and delete these particles as they move in and out of the sensor's field of view. Furthermore, runs with the virtual test harness revealed that small snowflakes (diameters of less than 0.25 millimeters) could not be resolved with ray traces of less than 1000 subsamples per pulse. Increasing the subsample count slows down both the ray tracer and increases the DSP workload. The result of these effects is that VANE cannot deliver a real-time LiDAR data stream in a scene filled with millions of snowflakes.

To circumvent this limitation, VANE LiDAR snowfall simulations are conducted in scenes completely devoid of snow particles. This allows the ray tracer and DSP to rapidly generate results corresponding to clear weather, blue-sky conditions. After ray tracing and signal processing are complete, statistical curve fits generated from data collected in the synthetic test harness (Section 4) are used to degrade the clear weather results. This is accomplished by introducing noise from soft returns and attenuating the intensity of the original hard reflections. Figure 16 depicts snowy point clouds generated using this algorithm.

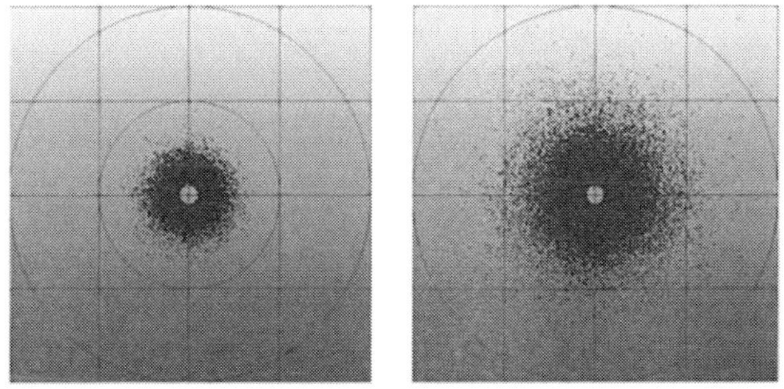

Fig. 16. Soft returns from snowflakes generated with a real-time LiDAR snowfall model in VANE. Left: 1 mm flakes. Right: 3.5 mm flakes. Both images were generated using identical extinction coefficients.

Figure 16 shows how the soft return distribution responds to both particle size and concentration. Notably, the double torus effect seen in Figs 1,2, 4, and 5 is not present in results from the VANE simulation. Additional work is needed to better understand how the internal signal processor used by the real-world HDL-32 is adjusting its power level and selectivity dropping returns to generate this effect.

The VANE LiDAR also replicates effects from both the strongest and last modes of operation. Figure 17 shows results with the virtual sensor running in last mode. A hard target (i.e. reflective wall) is situated in front of the sensor. When operating in last mode, the presence of the hard target cuts a channel through the soft return torus, just as seen in the real data shown in Fig. 4.

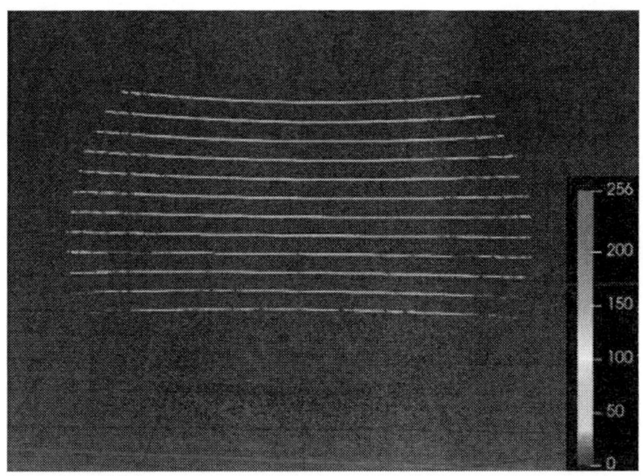

Fig. 17. VANE LiDAR operating in last mode. The presence of a hard target in the background causes soft returns to drop from the foreground of the point cloud.

The channel cut through the soft return torus does not manifest when the VANE LiDAR is set to run in strongest mode. In this mode, a complete snowflake torus is visible, but patchy gaps appear in the scan of the wall in the background, as seen below in Fig. 18.

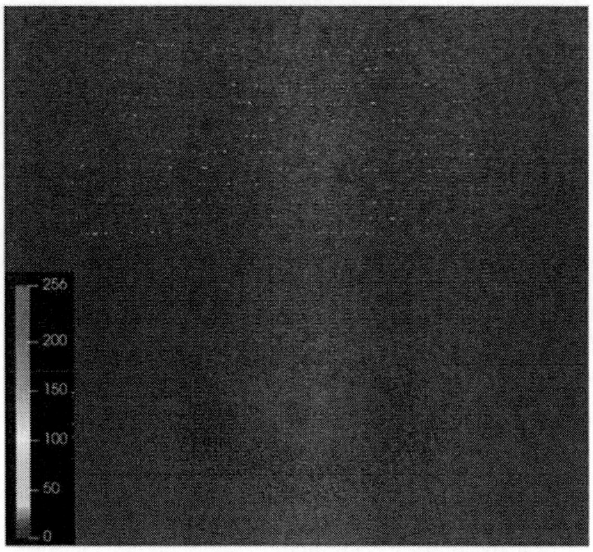

Fig. 18. VANE LiDAR operating in strongest mode. Reflections from snowflakes in the foreground frequently occlude the hard target in the background, resulting in the patchy coverage of the wall.

The patchy gaps across the wall arise when the strongest reflection measured by a single laser pulse corresponds to a snowflake rather than the wall itself. The probability of detecting a snowflake or the wall depends on the distance between the sensor and the hard target, as shown on Fig. 13. It is important to note that the strongest return does not always correspond to the brightest reflection, as indicated by the color scale. The color scale represents a calibrated reflectivity that normalizes the amount of power received according to the distance of the target. (VLP-16 User Manual). This means that snowflakes in the foreground could reflect more power than a hard target in the background but still have a lower overall calibrated reflectivity.

When operating in dual mode, the VANE lidar displays both the complete snow torus (as seen in strongest mode) and the smooth, uninterrupted scan of the hard target as seen in last mode. This is demonstrated below in Fig. 19.

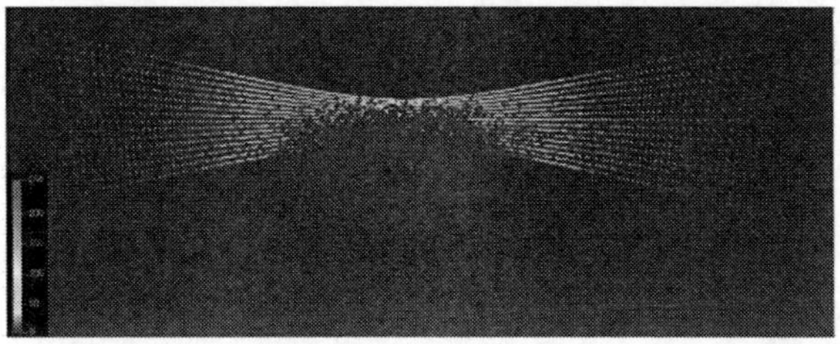

Fig. 19. VANE LiDAR operating in dual mode. Curvature visible on the wall is due to differences in the elevation angles of the beams as they rotate around the hub and project onto the flat surface.

The presence of snowfall in a virtual scene also attenuates both measured reflectivity of the hard targets and the maximum range to which they are detectable. Figure 20 demonstrates this effect with hard targets in both clear skies and heavy snowfall.

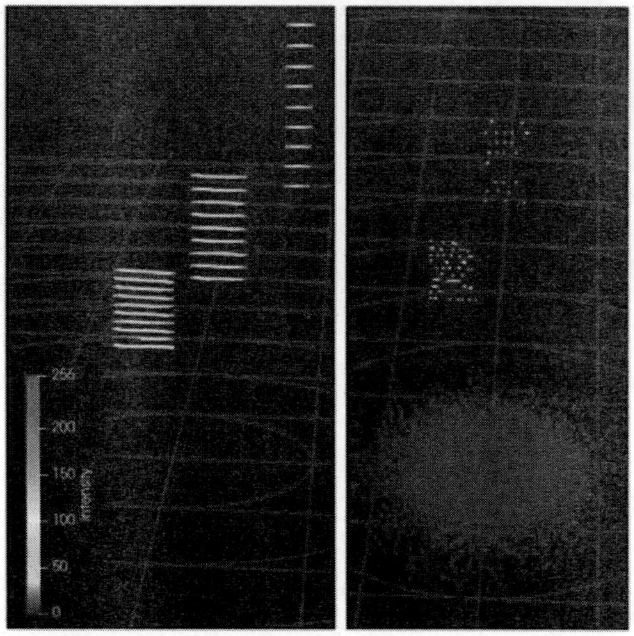

Fig. 20. Left: 100% reflective targets placed at ranges of 30, 50, and 90 meters scanned in a clear skies VANE simulation. Right: the same targets scanned during a heavy snowfall event in strongest mode.

Figure 20 displays three hard targets arranged at distances of 30, 50, and 90 meters from the LiDAR. The image on the left was scanned in sunny conditions, and the right image was taken during heavy snowfall. Snowfall causes the measured reflectivity of the targets to decrease from a yellow 100% in sunny weather to a green 75% in the snowy weather. In the snowy scene, the target positioned at 90 meters does not reflect enough power to exceed the sensor's noise floor, causing it to disappear.

Additionally, the VANE LiDAR model can reproduce the ghost points or mixed pixels described in Section 3. In VANE simulations, these points only appear near hard targets positioned close to the transmitter, as shown by the small spike in soft return probably on Fig. 14. In this case, the DSP reports an average distance between snowflakes and the hard targets, resulting in high intensity returns that appear several centimeters radially inward from the physical location of the hard target. Figure 21 shows a birds-eye view of ghost points forming near a hard target in a VANE scene.

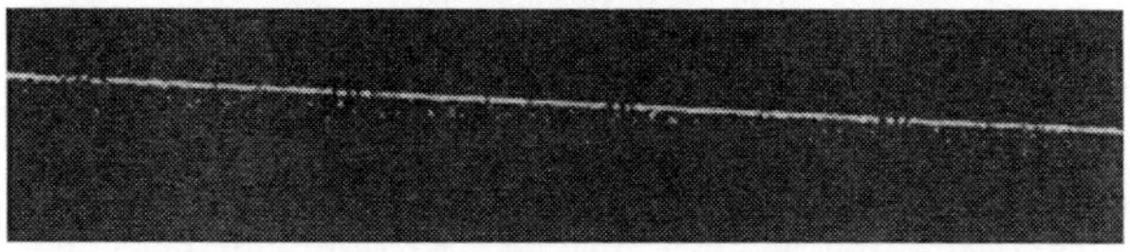

Fig. 21. Ghost points or mixed pixels generated near a hard target in a snowy VANE scene.

In contrast with Fig. 6, the mixed pixels generated in a VANE simulation do not appear as streaks within the torus of soft returns. Instead, they only seem to manifest when the snowfall has a large extinction coefficient and the hard target is in close proximity to the transmitter. Additional work is needed to reconcile this difference between the synthetic and real-world data.

6. Conclusions and future work

Falling snow degrades the performance of automotive LiDAR units by introducing noise and attenuating the maximum detection range of the sensor. This work provides qualitative examples of these effects in data collected with a Velodyne HDL-32 and introduces a framework to reproduce them in simulation using physics-based raytracing. This method allows researchers to predict the performance of a LiDAR unit at various snowfall intensities and examine how sensor performance changes as a function of the shape and divergence of beam pattern. Numerical experiments indicated that some features of LiDAR performance are captured by this approach, while others are not.

A significant amount of work remains to be completed. The preliminary results from the synthetic test harness described in Section 4 have not yet been validated against real-world snowfall particle size distributions and concentrations collected with a disdrometer. Additional effort is also needed to characterize the internal signal processing of the HDL-32. The data presented in Section 3 indicates that this sensor is dynamically adjusting the power of its laser and selectively ignoring soft detections. A purely physics-based approach to modeling will not capture effects generated by the sensor's proprietary signal processing algorithms. A hybrid approach using both physics-based modelling and empirical tuning will likely be required for each unique sensor make, model, and firmware version to achieve a high-fidelity match between numerical and experimental data.

7. Acknowledgements

The authors would like to thank the Mobility Systems Branch of the Geotechnical and Structures Laboratory in Vicksburg, MS for loaning us their HDL-32 for data collection during the winter of 2024-2025.

8. Declaration of competing interest

The authors declare that they have no known competing financial interests or personal relationships that could have appeared to influence the work reported in this paper.

9. References

Bergin, D, Carden, W., Huynh, K., Parikh, P., Bounker, P., Gates, B., Whitt, J., 2023. Tailoring the Digital Twin for Autonomous Systems Development and Testing. The ITEA Journal of Test and Evaluation 44(4). https://itea.org/journals/volume-44-4/tailoring-the-digital-twin-for-autonomous-systems-development-and-testing/.

Bijelic, M., Gruber, T., Mannan, F., Kraus, F., Ritter, W., Dietmayer, K., Heide, F., 2020. Seeing Through Fog Without Seeing Fog: Deep Multimodal Sensor Fusion in Unseen Adverse Weather. IEEE/CVF Conference on Computer Vision and Pattern Recognition (CVPR), 11679–89. https://doi.org/10.1109/CVPR42600.2020.01170.

Bos, J., 2024a. WADS Year 3, 22 Feb, APS to Michigan Tech (1/2). https://www.youtube.com/watch?v=rzzz8vbRJf4.

Bos, J., 2024b. WADS Year 3, 22 Feb, Mohawk to Calumet (2/3) Storm Peak. https://www.youtube.com/watch?v=J00sMCwrl4g.

Bos, J.P., Weng, W., 2023. Assessing the Effects of Severe Winter Weather on Automotive Lidar. Laser Communication and Propagation through the Atmosphere and Oceans XII, edited by D.T. Wayne, J.A. Anguita, and J.P. Bos. SPIE. https://doi.org/10.1117/12.2678571.

Canadian Adverse Driving Conditions Dataset, n.d. Accessed July 30, 2025. http://cadcd.uwaterloo.ca/index.html#overview.

Carrillo, J., Goodin, C., Fernandez, J., 2020. Sensor and Environment Physics in the Virtual Autonomous Navigation Environment (VANE)." Engineer Research and Development Center (U.S.), August 27. https://doi.org/10.21079/11681/37968.

Charron, N., Phillips, S., Waslander, S.L., 2018. De-Noising of Lidar Point Clouds Corrupted by Snowfall. 15th Conference on Computer and Robot Vision (CRV), 254–61. Toronto, ON, Canada: IEEE. https://doi.org/10.1109/CRV.2018.00043.

DENSE-Universität Ulm, n.d. Accessed July 31, 2025. https://www.uni-ulm.de/en/in/institute-of-measurement-control-and-microtechnology/research/data-sets/dense-datasets/#c811669.

Filgueira, A.,H. González-Jorge, S. Lagüela, L. Díaz-Vilariño, Arias, P., 2017. Quantifying the Influence of Rain in LiDAR Performance." Measurement 95: 143–48. https://doi.org/10.1016/j.measurement.2016.10.009.

Fry, N., 2014. Survivability, Sustainability, and Maneuverability: The Need for Joint Unity of Effort in Implementing the DOD Arctic Strategy at the Tactical and Operational Levels. Military Review 94(6): 54–62.

Goodin, C., Carruth, D., Doude, M., Hudson, C., 2019. Predicting the Influence of Rain on LIDAR in ADAS. Electronics 8(1): 89. https://doi.org/10.3390/electronics8010089.

Hahner, M., Sakaridis, C., Bijelic, M., Heide, F., Yu, F., Dai, D., Van Gool, L., 2022. LiDAR Snowfall Simulation for Robust 3D Object Detection. arXiv, https://doi.org/10.48550/ARXIV.2203.15118.

Hall, D., 2011. United States Patent No. US 7,969,558 B2, June 28. https://patentimages.storage.googleapis.com/3c/dd/91/8fd123ec4300f7/US7969558.pdf.

Hall, D., United States Patent No. US 8,767,190 B2, July 1. https://patentimages.storage.googleapis.com/b7/86/cb/e36b7056a2d7b2/US8767190.pdf.

HDL-32E User's Manual and Programming Guide, 2011. Velodyne LiDAR Inc. https://archive.org/details/manualzilla-id-6960639.

Heinzler, R., Piewak, F., Schindler, P., Stork, W., 2020. CNN-Based Lidar Point Cloud De-Noising in Adverse Weather. IEEE Robotics and Automation Letters 5(2): 2514–21. https://doi.org/10.1109/LRA.2020.2972865.

Holden, G., Aspin, Z., Monroe, J.G., McInnis, D., Price, P., Brad Hansen, B., 2024. Accelerating Autonomous Vehicle Development AND Evaluation with the VANE Simulation Tool Suite. Proceedings of the Ground Vehicle Systems Engineering and Technology Symposium (GVSETS), NDIA, Novi, MI, Aug. 13-15.

Ignatious, H.A., El-Sayed, H., Khan, M., 2022. An Overview of Sensors in Autonomous Vehicles." Procedia Computer Science, 12th International Conference on Emerging Ubiquitous Systems and Pervasive Networks / 11th International Conference on Current and Future Trends of Information and Communication Technologies in Healthcare, 198: 736–41. https://doi.org/10.1016/j.procs.2021.12.315.

Kurup, A., Bos, J.P., 2022. Winter Adverse Driving Dataset (WADS): Year Three." In Autonomous Systems: Sensors, Processing and Security for Ground, Air, Sea and Space Vehicles and Infrastructure 2022, edited by Michael C. Dudzik, Theresa J. Axenson, and Stephen M. Jameson. SPIE. https://doi.org/10.1117/12.2619424.

Kurup, A.M., Bos, J.P., 2023. Winter Adverse Driving Dataset for Autonomy in Inclement Winter Weather. Optical Engineering 62(03). https://doi.org/10.1117/1.OE.62.3.031207.

Kurup, Akhil, and Jeremy Bos. "DSOR: A Scalable Statistical Filter for Removing Falling Snow from LiDAR Point Clouds in Severe Winter Weather." arXiv:2109.07078. Preprint, arXiv, October 30, 2021. https://doi.org/10.48550/arXiv.2109.07078.

Kutila, M., Pyykonen, P., Jokela, M., Gruber, T., Bijelic, M., Ritter, W., 2020. Benchmarking Automotive LiDAR Performance in Arctic Conditions. 2020 IEEE 23rd International Conference on Intelligent Transportation Systems (ITSC), IEEE, September 20, 1–8. https://doi.org/10.1109/ITSC45102.2020.9294367.

Linnhoff, C., Hofrichter, K., Elster, L., Rosenberger, P., Winner, H., 2022. Measuring the Influence of Environmental Conditions on Automotive Lidar Sensors. Sensors 22(14): 5266. https://doi.org/10.3390/s22145266.

Locatelli, J.D., Hobbs, P.V., 1974. Fall Speeds and Masses of Solid Precipitation Particles." Journal of Geophysical Research 79(15): 2185–97. https://doi.org/10.1029/JC079i015p02185.

Michigan Traffic Crash Facts. Winter Weather October 2021-April 2022, 2022. Michigan Department of State Police. https://www.michigan.gov/msp/-/media/Project/Websites/msp/ohsp/1_September-2023/2022-Winter-Weather-Fact-Sheet-UMTRI.pdf?rev=e4423af5266c4caaafd99ffeeb4374a0&hash=AB81F304902439A5338B154CBEF6A105.

Minnesota Motor Vehicle Crash Facts, 2022. Minnesota Department of Public Safety, Office of Traffic Safety. https://s3.us-east-2.amazonaws.com/assets.dps.mn.gov/s3fs-public/migrated-files/divisions/ots/reports-statistics/Documents/2022a-crash-facts.pdf.

On-Road Automated Driving Committee, 2021. Taxonomy and Definitions for Terms Related to Driving Automation Systems for On-Road Motor Vehicles. SAE International. April. https://doi.org/10.4271/J3016_202104.

Ouster Firmware User Manual, 2024. Ouster Inc., July 29. https://data.ouster.io/downloads/software-user-manual/firmware-user-manual-v3.1.0.pdf.

Park, J.-I., Park, J., Kim, K.-S., 2020. Fast and Accurate Desnowing Algorithm for LiDAR Point Clouds. IEEE Access 8: 160202–12. https://doi.org/10.1109/ACCESS.2020.3020266.

Phillips, T.G., Guenther, N., McAree, P.R., 2017. When the Dust Settles: The Four Behaviors of LiDAR in the Presence of Fine Airborne Particulates." Journal of Field Robotics 34(5): 985–1009. https://doi.org/10.1002/rob.21701.

Rahm, M., Jonsson, P., Henriksson, M., 2021. Laser Attenuation in Falling Snow Correlated with Measurements of Snow Particle Size Distribution. Optical Engineering 60(09). https://doi.org/10.1117/1.OE.60.9.094102.

Rasshofer, R. H., Spies, M., Spies, H., 2011. Influences of Weather Phenomena on Automotive Laser Radar Systems." Advances in Radio Science 9: 49–60. https://doi.org/10.5194/ars-9-49-2011.

Real-Time Filtering of Snowfall from Lidar on Autonomoose Data, 2018. University of Waterloo. https://www.youtube.com/watch?v=2givFiLi2tQ.

Ronnback, S., Wernersson, A., 2008. On Filtering of Laser Range Data in Snowfall." 2008 4th International IEEE Conference Intelligent Systems, IEEE, September, 17-33-17–39. https://doi.org/10.1109/IS.2008.4670551.

Roy, G., Cao, X., Bernier, R., Tremblay, G., 2020. Physical Model of Snow Precipitation Interaction with a 3D Lidar Scanner." Applied Optics 59(25): 7660. https://doi.org/10.1364/AO.393059.

Royo, Ss, Ballesta-Garcia, M., 2019. An Overview of Lidar Imaging Systems for Autonomous Vehicles." Applied Sciences 9(19): 4093. https://doi.org/10.3390/app9194093.

Salvatori, R., Casacchia, R., Grignetti, A., Valt, M., Cagnati, A., 2025. Snow Surface Classification in the Western Svalbard Island. https://citeseerx.ist.psu.edu/document?repid=rep1&type=pdf&doi=786a2b4eb04618f5619373ba14a36f5b6f13e87c.

Snow & Ice - FHWA Road Weather Management, n.d. Accessed April 15, 2025. https://ops.fhwa.dot.gov/weather/weather_events/snow_ice.htm.

Unreal Engine Overview, n.d. Accessed April 23, 2025. https://www.unrealengine.com/en-US/home.

VeloView: The Velodyne Lidar Viewer Based on Paraview Lidar., n.d. Accessed October 22, 2024. https://www.paraview.org/veloview/.

VLP-16 User Manual, 2022. Velodyne Lidar, March 7. https://data.ouster.io/downloads/velodyne/user-manual/vlp-16-user-manual-revf.pdf.

Waymo Self-Driving Cars, n.d. Accessed April 22, 2025. https://waymo.com/waymo-driver/.

Zhou, X., Xiang, S., Zeng, Z., Shao, Z., 2021. Method for Identification of a Noise Point Used for LiDAR, and LiDAR System. United States Patent US10908268B2, filed March 23, 2020, and issued February 2. https://patents.google.com/patent/US10908268B2/en?oq=Patent+No.:+US+10%2c908%2c268+%D0%922.

4122 | DEM analysis of effect of interparticle forces on sand-outflow behavior from hopper under low-gravity environment

Shunsuke Muryokoji *, Shingo Ozaki

Yokahama National University, 79-5 Tokiwadai, Hodogaya-ku, Yokohama, Kanagawa 240-8501, Japan

* *Corresponding author: muryokoji-shunsuke-sd@ynu.jp*

ABSTRACT

Understanding the behavior of granular materials under reduced-gravity conditions is critical for planetary exploration, as it directly affects lander-ground interactions, rover mobility, and in-situ resource utilization. However, ground-based experiments using parabolic flights or drop towers are constrained by the short duration of microgravity, making it difficult to fully characterize granular behavior. To overcome these limitations, this study employs the discrete element method (DEM) to simulate sand outflow from hoppers under various gravity levels, using data from the Hourglass experiment conducted at the International Space Station (ISS). The experiment utilized an artificial gravity generator (0.06–2 G) to observe the flow and deposition of different granular materials. In our simulations, silica sands #5 and #8 were modeled with consideration of contact forces, rolling resistance, and electrostatic forces. The results revealed that the electrostatic force has almost no impact on silica sand #5, which has relatively larger particles, under all gravity conditions. In contrast, silica sand #8 showed substantial wall deposition, particularly under lower gravity. Furthermore, electrostatic forces alone were insufficient to fully reproduce the experimental results, suggesting that van der Waals forces must also be considered to achieve accurate modeling. These findings highlight the significance of interparticle forces in regolith behavior under low-gravity conditions and offer valuable insights for the development of robotic systems for planetary excavation, mobility, and construction.

Keywords
Discrete Element Method (DEM)
Electrostatic Force
Sand-flow
Hourglass

1. Introduction

A variety of surface operations involving soft extraterrestrial soils—such as lander ascent and descent, rover traversal, and subsurface drilling—are anticipated in current and planned satellite and planetary exploration missions (Michel et al., 2022; Zhang et al., 2022; Kawakatsu et al., 2023). To better understand the behavior of granular media such as regolith under reduced gravity conditions, experimental studies including parabolic flight tests (Kobayashi et al., 2010; Reiss et al., 2014) and drop tower experiments (Kumagai et al., 1994) have been conducted. However, these ground-based experiments are inherently limited by the short duration for which low-gravity conditions can be maintained, and it has been observed that the transition from normal gravity (1G) to reduced gravity can influence the initial state of the granular material. As a result, it remains challenging to fully characterize and model the interaction between granular media and mechanical systems solely based on terrestrial experiments.

In response to these limitations, the "Hourglass" experiment (Ozaki et al., 2023) was conducted in 2020 with the aim of investigating the gravity dependence of fundamental parameters of various granular materials—representative of regolith—to provide useful insights for the design of future exploration missions. The experiment utilized the artificial gravity generator installed in the Kibo module of the International Space Station (ISS), generating artificial gravity conditions ranging from 0.06 G to 2 G. A total of eight granular materials with differing physical properties were tested, and their dynamic behavior and deposition characteristics were recorded through onboard camera observations.

In this study, we apply the discrete element method (DEM), a widely used numerical technique (Briend et al., 2010; Obermayr et al., 2014; Zhao et al., 2014; Suzuki et al., 2019) in the field of terramechanics, to simulate the deposition and flow behavior of granular materials observed in the Hourglass experiment. In particular, we examine the competing influences of gravitational and inter-particle forces—especially electrostatic forces (Feng et al., 2003) — for sand particles of varying diameters. This focus is motivated by the fact that, in the case of fine granular materials such as regolith, inter-particle forces can dominate over gravity in low-gravity environments. Consequently, we demonstrate that parameter sets calibrated exclusively through terrestrial experiments may not be directly applicable to the pre-mission validation of exploration scenarios at low gravitational regimes.

2. Discrete element method analysis

The DEM is a numerical approach in which the system is modeled as an assembly of discrete particles. The method sequentially computes the motion of each particle by accounting for inter-particle interactions such as contact and sliding. In this study, the commercial software package Rocky DEM (ESSS Rocky, 2021) is employed to perform the simulations. This section outlines the contact force model and the electrostatic force model implemented in the analysis.

2.1. Contact force models

In this study, the Hertz–Mindlin contact model (ESSS Rocky, 2021) shown in Fig. 1 is employed. The normal contact force is modeled using a Hertzian spring-dashpot formulation, expressed as:

https://doi.org/10.56884/DZSIN0YU

$$F_n = \hat{K}_H s_n^{\frac{3}{2}} + \hat{C}_H s_n^{\frac{1}{4}} \dot{s}_n \tag{1}$$

where $\hat{K}_H$ is the normal stiffness coefficient, $\hat{C}_H$ is the normal damping coefficient, s_n is the contact normal overlap, and $\dot{s}_n$ is the time derivative of the normal overlap. The stiffness and damping coefficients for compressive interactions between spherical particles or between particles and container walls are computed as follows:

$$\hat{K}_H = \frac{4}{3} E^* \sqrt{R^*} \tag{2}$$

The equivalent Young's modulus E^* is given by:

$$\frac{1}{E^*} = \frac{1 - \nu_1^2}{E_1} + \frac{1 - \nu_2^2}{E_2} \tag{3}$$

where E_1 and E_2 are the Young's moduli of the two contacting particles or the particle and the boundary, and ν_1 and ν_2 are their respective Poisson's ratios. In addition, R^* in Eq. 2 represents the effective or equivalent radius, which is defined as follows:

$$\frac{1}{R^*} = \begin{cases} \dfrac{2}{L_1} + \dfrac{2}{L_2} & \text{for particle} - \text{particle collision} \\ \dfrac{2}{L} & \text{for particle} - \text{boundary collision} \end{cases} \tag{4}$$

Here, L_1 and L_2 denote the sizes of the two contacting particles, while L represents the size of the particle in particle–boundary interactions. The damping coefficient $\hat{C}_H$ is given by:

$$\hat{C}_H = \sqrt{5} \eta \sqrt{m^* \hat{K}_H} \tag{5}$$

where m^* is the effective mass during contact, defined as:

$$\frac{1}{m^*} = \begin{cases} \dfrac{2}{m_1} + \dfrac{2}{m_2} & \text{for particle} - \text{particle collision} \\ \dfrac{2}{m} & \text{for particle} - \text{boundary collision} \end{cases} \tag{6}$$

In this expression, m_1 and m_2 are the masses of the two particles in particle–particle contact, and m is the mass of the particle in contact with the boundary. The damping ratio η is related to the restitution coefficient e, and is determined by the following piecewise equation:

$$e = \begin{cases} \exp\left[-\dfrac{\eta}{\sqrt{1-\eta^2}}\left(\pi - \arctan\dfrac{2\eta\sqrt{1-\eta^2}}{1-2\eta^2}\right)\right] & \text{if } 0 \le \eta < \dfrac{1}{\sqrt{2}} \\ \exp\left(-\dfrac{\eta}{\sqrt{1-\eta^2}} \arctan\dfrac{2\eta\sqrt{1-\eta^2}}{2\eta^2 - 1}\right) & \text{if } \dfrac{1}{\sqrt{2}} \le \eta \le 1 \\ \exp\left(-\dfrac{\eta}{\sqrt{\eta^2-1}} \ln\dfrac{\eta + \sqrt{\eta^2-1}}{\eta - \sqrt{\eta^2-1}}\right) & \text{if } 1 < \eta \end{cases} \tag{7}$$

The tangential force F_τ is calculated using the Mindlin–Deresiewicz contact model (ESSS Rocky, 2021), and is expressed as:

$$F_\tau = -\mu F_n \left(1 - \zeta^{\frac{3}{2}}\right) \frac{s_\tau}{|s_\tau|} + \eta_\tau \sqrt{\frac{6\mu m^* F_n}{s_{\tau,max}}} \zeta^{\frac{1}{4}} \dot{s}_\tau \tag{8}$$

$$\zeta = 1 - \frac{\min(|s_\tau|, s_{\tau,max})}{s_{\tau,max}} \tag{9}$$

Here, μ is the friction coefficient, which is set to the static friction coefficient μ_s when no slipping occurs, and to the dynamic friction coefficient μ_d when slipping occurs. s_τ is the tangential relative displacement at the contact, and $\dot{s}_\tau$ is the tangential component of the relative velocity. The maximum relative tangential displacement at which particles begin to slide, $s_{\tau,max}$ is defined using the Poisson's ratios ν_1 and ν_2 of the two contacting bodies:

$$s_{\tau,max} = \mu \left(\frac{1-\nu_1}{2-\nu_1} + \frac{1-\nu_2}{2-\nu_2}\right)^{-1} s_n \tag{10}$$

The tangential damping ratio η_τ is given by:

$$\eta_\tau = -\frac{\ln \varepsilon}{\ln^2 \varepsilon + \pi^2} \tag{11}$$

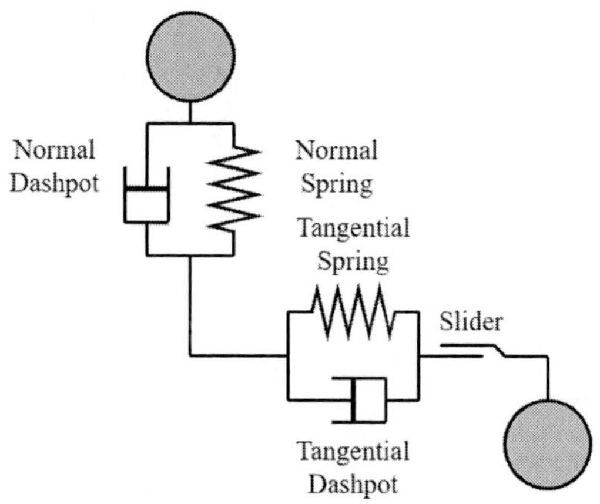

Fig. 1. Contact models.

2.2. Rolling resistance model

Rolling resistance accounts for surface roughness and non-spherical effects of particles. Among the two available rolling resistance models in Rocky DEM, the Type C: linear spring rolling limit model (ESSS Rocky, 2021) is adopted in this study. The rolling resistance moment is computed as:

$$M_{\mathrm{r}}^t = min\left(\left|M_{\mathrm{r,e}}^t\right|, M_{\mathrm{r,lim}}\right) \frac{M_{\mathrm{r,e}}^t}{\left|M_{\mathrm{r,e}}^t\right|} \tag{12}$$

Here, $M_{\mathrm{r,e}}^t$ denotes the rolling resistance moment under the assumption of purely elastic behavior. In this case, the rolling resistance moment would be updated incrementally in the following way:

$$M_{\mathrm{r,e}}^t = M_{\mathrm{r}}^{t-\Delta t} - K_{\mathrm{r}} \omega_{\mathrm{rel}} \Delta t \tag{13}$$

$M_{\mathrm{r}}^{t-\Delta t}$ represents the rolling resistance moment vector from the previous time step. The rolling stiffness K_{r} is defined as follows:

$$K_{\mathrm{r}} = R_{\mathrm{r}}^2 K_\tau \tag{14}$$

Here, K_{r} denotes the tangential stiffness, ω_{rel} is the relative angular velocity vector between the two contacting bodies, and Δt is the simulation timestep. The rolling radius R_{r} is defined as follows:

$$\frac{1}{R_{\mathrm{r}}} = \begin{cases} \frac{1}{|r_1|} + \frac{1}{|r_2|} & \text{for particle} - \text{particle collision} \\ \frac{1}{|r|} & \text{for particle} - \text{boundary collision} \end{cases} \quad (15)$$

r_1 and r_2 represent the rolling radius vectors of the two particles in contact during particle–particle interactions, while r denotes the rolling radius vector of the particle in particle–wall interactions. The rolling radius vector is defined as the vector connecting the particle's center of mass to the contact point. The limiting rolling resistance moment $M_{\mathrm{r,lim}}$ is defined as follows:

$$M_{\mathrm{r,lim}} = \mu_{\mathrm{r}} R_{\mathrm{r}} F_{\mathrm{n}} \quad (16)$$

Here, μ_{r} denotes the rolling resistance coefficient

2.3. Timestep

The simulation timestep Δt in the Hertzian spring-dashpot model is determined as follows (ESSS Rocky, 2021):

$$\Delta t = \frac{\pi}{2N_{\Delta t}^{l}} \min\left(\sqrt{\frac{m^*}{K_{\mathrm{H}}}}\right) \quad (17)$$

$$K_{\mathrm{H}} = \frac{4}{3} E^* \sqrt{R^*} s_{\mathrm{n}}^{\frac{1}{2}} \quad (18)$$

Here, $N_{\Delta t}^{l}$ is the minimum number of timesteps per loading cycle and is a user-defined input parameter. Among all particle–particle and particle–wall contact combinations, the minimum value of $\sqrt{\frac{m^*}{K_{\mathrm{H}}}}$ is selected to determine the time step.

2.4. Electrostatic Force

In the electrostatic force model, the magnitude of the interparticle force varies depending on the distance between two bodies. The model is described by the following equations:

$$F_{n,ele}(z) = \begin{cases} \frac{{q_p}^2}{4\pi\varepsilon z^2} & \text{particle} - \text{particle} \\ \frac{q_p q_w}{4\pi\varepsilon z^2} & \text{particle} - \text{boundary} \end{cases} \quad (19)$$

Here, $F_{n,ele}(r)$ is the electrostatic force, ε is the absolute permittivity, q_p is the point charge of a particle, q_w is the point charge of the wall, and z is the point-charge distance.

In Rocky DEM, the electrostatic force model involves four tunable parameters: Relative Permittivity, Initial Boundary Charge Density, Particle Charge, and Screening Distance. In this study, the absolute permittivity was set to that of vacuum, $\varepsilon = 8.85 \times 10^{-12}\ F/m$. Parametric tuning of the electrostatic interactions was performed by adjusting the values of Initial Boundary Charge Density, Particle Charge, and Screening Distance. Here, the Initial Boundary Charge Density represents the amount of electric charge per unit area on the wall surface. Particle Charge refers to the total charge carried by an individual particle. Screening Distance defines the maximum effective range over which electrostatic forces act and corresponds to the upper limit of z.

3. Analysis of Hourglass experiment

This study aims to reproduce the competitive effects between gravitational and interparticle forces observed in the Hourglass experiment through DEM simulations.

3.1. Hourglass experiment

As shown in Fig. 2, the Hourglass experiment was conducted using the Cell Biology Experiment Facility (CBEF) in the Kibo module at the ISS (Ozaki et al., 2023). Hourglass-shaped containers were mounted on a rotating turntable, where artificial gravity was simulated by varying angular velocity. During rotation, the hourglasses periodically flip to reset particle positions, allowing observation of the dynamic behavior and deposition characteristics of granular media.

A schematic of the Hourglass box is shown in Fig. 3. Each box includes an hourglass-shaped container, a graduated cylinder container, control units, a servo motor, a camera, and an accelerometer. The container material is aluminum alloy A2017, with a surface treated by electroless nickel plating.

The hourglass geometry features opening angles of 60° and 120°, an orifice width of 5 mm, a depth of 15 mm, and overall heights and widths as follows: 44 mm high, 29.2 mm wide on the 60° side, and 40 mm wide on the 120° side. The flip cycle and data recording were automated and repeated over a 7-hour period. Eight granular media were tested, including regolith simulants and natural silica sands.

CBEF can host four Hourglass Boxes simultaneously, allowing two batches of four experiments. Fig. 4 shows the operational timeline of artificial gravity during the experiment. Gravity levels from 0.06 G to 2G were simulated. At specific times (9, 20, and 32 minutes), 2G was applied to assist with unblocking the orifice in the event of clogging.

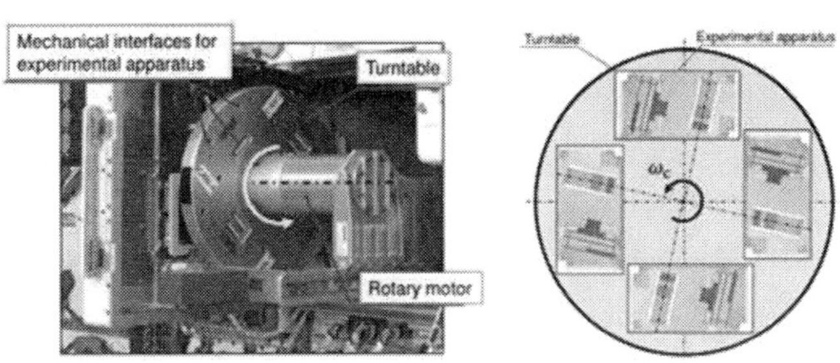

Fig. 2. Appearance of CBEF and turntable (Ozaki et al., 2023)

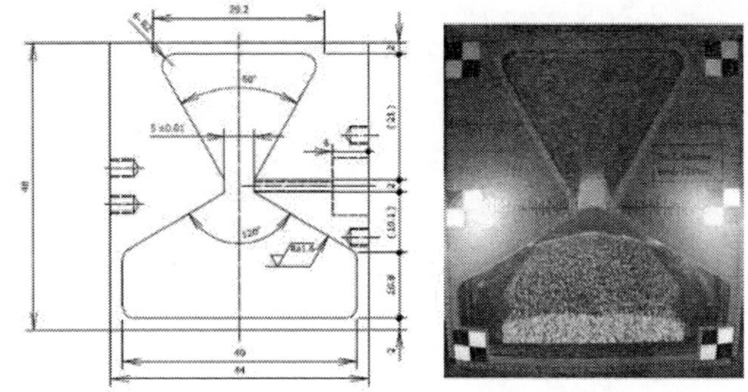

Fig. 3. Hourglass box and dimension of Hourglass (Ozaki et al., 2023)

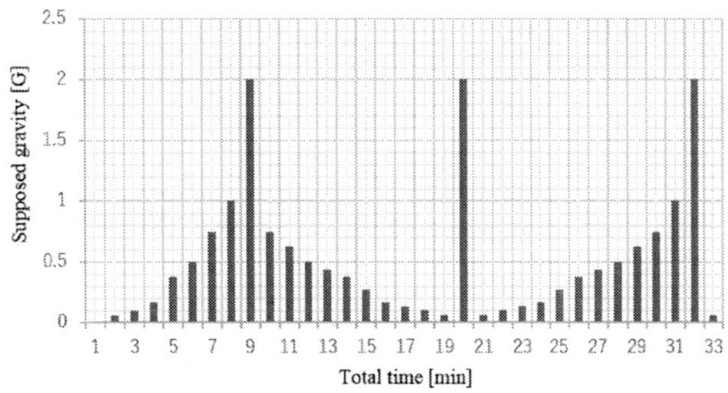

Fig. 4. Time history of controlled artificial gravity (Ozaki et al., 2023)

3.2. Analysis conditions

Table 1 lists the parameters (Ozaki et al., 2023) used in the DEM analysis, while Tables 2 and 3 show the particle size distributions of the tested silica sands #5 and #8, respectively. The "Numerical softening factor" in Table 1 is a coefficient used in Rocky DEM to reduce all stiffness values to shorten the computation time. Regarding the friction coefficient, the static and dynamic friction values are treated as equal in this study. The parameters presented in Tables 2 and 3 correspond to the actual samples used in the Hourglass experiment aboard the ISS.

Table 1
Parameters of DEM model.

Particle	Silica sand #5	Silica sand #8
Gravity[G]	0.063, 0.17, 0.38, 0.5, 1.0	
Young's modulus (particle) [MPa]	500	
Young's modulus (hourglass) [MPa]	500	
Numerical softening factor [-]	0.01	
Particle density [kg/m³]	2701	2634
Initial mass[g]	4.66	3.71
Static/Dynamic friction coefficient (particle-particle) [-]	0.3	
Static/Dynamic friction coefficient (particle-hourglass) [-]	0.2713	
Rolling resistance coefficient [-]	0.3	
Coefficient of restitution (particle-particle) [-]	0.8	
Coefficient of restitution (particle-hourglass) [-]	0.8	
Initial Boundary Charge Density [C/mm^2]	1.0×10^{-14}	
Particle Charge [C]	1.0×10^{-14}	
Screening Distance [m]	5.0×10^{-9}	

Table 2
Weight percentage for particle size of Silica sand #5.

Sieve size [um]	Percentage of passing weight [%]
1379	100
1224.31	99.8098
1086.97	98.3403
965.03	94.05
856.773	84.2228
760.661	71.2931
675.33	56.0284
599.572	40.7136
532.312	28.0358
472.597	17.5554
419.582	9.86431
372.513	4.55446
330.725	1.71532
293.624	0.4458
260.685	0

Table 3
Weight percentage for particle size of Silica sand #8.

Sieve size [um]	Percentage of passing weight [%]
599.572	100
532.312	99.6672
472.597	99.1125
419.582	98.3914
372.513	97.3375
330.725	95.9219
293.624	93.7876
260.685	91.0779
231.442	86.263
205.479	77.226
182.428	65.8718
161.963	52.8723
143.794	39.8091
127.664	28.4969
113.342	18.908
100.628	11.7886
89.3393	6.63126
79.3172	3.34155
70.4195	1.4269
62.5198	0.4725
55.5064	0.15066
49.2797	0

4. Analysis results and discussion

The section presents the simulation results and discusses their implications. A qualitative comparison with experimental results was performed based on the sedimentation patterns and flow behavior observed after inverting the hourglass. Figure 5 illustrates the relationship between gravitational and electrostatic forces under the conditions used in the numerical analysis.

Figures 6 and 7 compare the residual sand distributions after discharge between the experimental observations and simulation results for Toyoura sand No. 5 and No. 8, respectively. Figure 8 shows representative snapshots taken immediately after inversion of the hourglass to compare the initial flow behavior under the 0.063 G condition.

For Toyoura sand No. 5, both experimental and simulation results showed no significant sedimentation under any gravity condition. This agreement can be attributed to the relatively large particle size of No. 5 sand, where the effect of interparticle adhesion is comparatively small relative to gravitational force. Furthermore, good agreement was observed between the experiment and simulation regarding the onset time of discharge from the orifice and the termination time of flow in the upper hopper.

However, differences in flow patterns were observed. In the experiments, intermittent arch formation and collapse were noted during the flow, while the simulations exhibited a mass flow-like behavior with uniform discharge from the center and side walls, without forming any arch structures. This discrepancy is likely due to the simulation model considering only repulsive forces between particles, without accounting for attractive forces that are present in reality.

In contrast, Toyoura sand No. 8 exhibited stronger adhesion effects due to its finer particle size, resulting in significant wall adhesion and sidewall accumulation. Especially under lower gravity conditions, an increased amount of sedimentation was observed. The simulations demonstrated a similar gravity-dependent trend. When electrostatic forces were included in the simulation, sedimentation comparable to the experiments was reproduced under gravity levels of 0.38 G and above. However, under 0.17 G and 0.063 G, the simulated sedimentation amounts were lower than those observed experimentally.

Additionally, the duration of the flow tended to be longer in the simulations than in the experiments. This is likely because only repulsive forces were considered in the simulations, which allowed particles to flow more freely than real sand grains, thereby prolonging the discharge period.

These results suggest that it is difficult to fully reproduce the experimental behavior by modifying only the magnitude of electrostatic forces. It is therefore necessary to consider both van der Waals and electrostatic forces for more accurate modeling. Furthermore, although the Rolling Resistance Model was applied to spherical particles to approximate the influence of particle shape, more sophisticated modeling involving non-spherical particle geometries will be required for future analyses.

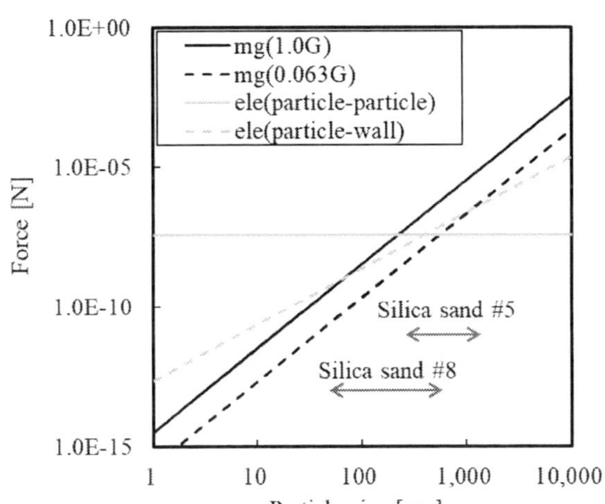

Fig. 5. Relationship between gravity and electrostatic forces to silica sand #5 and #8. Here, "ele" stands electrostatic forces and are plotted by the maximum value.

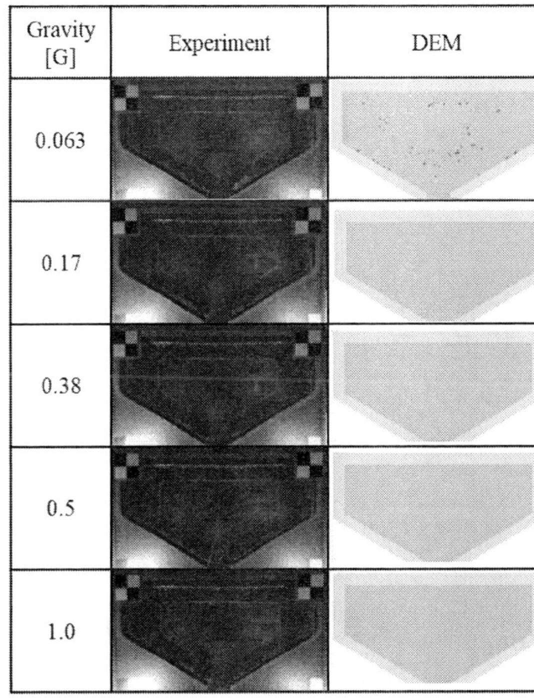

Fig. 6. Comparison of residual sand deposition of silica sand #5.

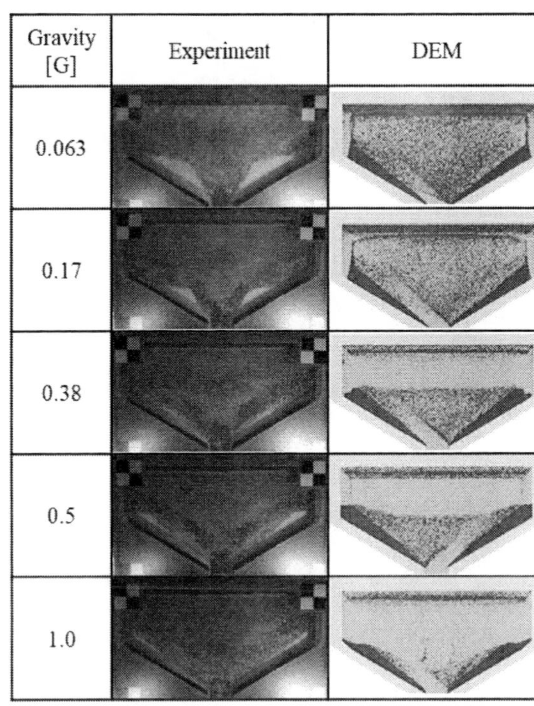

Fig. 7. Comparison of residual sand deposition of silica sand #8.

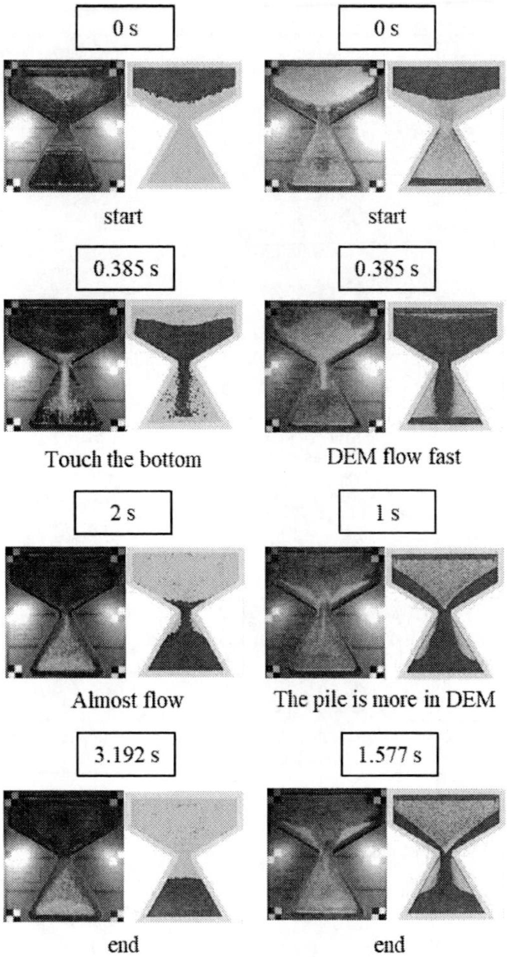

Fig. 8. Comparison of sand flow snapshots of silica sand #5(left) and silica sand #8(right).

5. Conclusions

In this study, we conducted DEM analysis to reproduce the deposition and flow behavior of granular media observed in the Hourglass experiment at the ISS. The DEM analysis aimed to reproduce the competitive relationship between gravity and electrostatic forces. The results demonstrated that appropriate tuning of electrostatic forces could replicate the gravity-dependent trends in flow characteristics and residual deposition of the sand. In particular, it was confirmed that for finer particles, electrostatic forces have a more dominant effect than gravity. However, to accurately reproduce the competitive relation between gravity and interparticle forces affecting sand flow characteristics, it is necessary to consider both electrostatic and van der Waals forces.

Acknowledgement

This study was supported by the Institute of Space and Astronautical Science and the Human Space Flight Technology Directorate, JAXA. The authors thank M. Otuski, G. Ishigami, M. Sutoh, and C. Kurosawa for their support in processing the experimental data. This study was also supported by the "Space construction innovation project" of the Ministry of Land, Infrastructure, Transport, and Tourism.

References

Briend, R., 2010. Modelling wheel-soil interactions using the discrete element method for tread shape optimization. Master's thesis, Department of Mechanical Engineering, McGill University, Montreal, Quebec.

ESSS Rocky, 2021. DEM Technical Manual Version 4.5.2.

Feng, J., Hays, D., 2003. Relative importance of electrostatic forces on powder particles. Powder Technol. 135–136, 65–75.

Kawakatsu, Y., Kuramoto, K., Usui, T., Sugahara, H., Ootake, H., Yasumitsu, R., Yoshikawa, K., Mary, S., Grebenstein, M., Sawada, H., Imada, T., Shimada, T., Ogawa, K., Otsuki, M., Baba, M., Fujita, K., Zacny, K., van Dyne, D., Satoh, Y., Tokaji, A., 2023. Preliminary design of Martian Moons eXploration (MMX). Acta Astronaut. 202, 715–728.

Kobayashi, T., Fujiwara, Y., Yamakawa, J., Yasufuku, N., Omine, K., 2010. Mobility performance of a rigid wheel in low gravity environments. J. Terramech. 47, 261–274.

Kumagai, M., Arai, S., Yamashita, Y., Suzuki, R., 1994. Speed-controllable drop tower for low gravity experiments. J. Kagaku Kogaku Ronbunshu. 20 (2), 295–298.

Michel, P., Ulamec, S., Böttger, U., Grott, M., Murdoch, N., Vernazza, P., Sunday, C., Zhang, Y., Valette, R., Castellani, R., Biele, J., Tardivel, S., Groussin, O., Jorda, L., Knollenberg, J., Grundmann, J., Arrat, D., Pont, G., Mary, S., Grebenstein, M., Miyamoto, H., Nakamura, T., Wada, K., Yoshikawa, K., Kuramoto, K., 2022. The MMX rover: performing in situ surface investigations on Phobos. Earth Planets Space. 74 (2), Dec.

Obermayr, M., Vrettos, C., Eberhard, P., Däuwel, T., 2014. A discrete element model and its experimental validation for the prediction of draft forces in cohesive soil. J. Terramech. 53 (1), 93–104.

Ozaki, S., Ishigami, G., Otsuki, M., Miyamoto, H., Wada, K., Watanabe, Y., Nishino, T., Kojima, H., Soda, K., Nakao, Y., Sutoh, M., Maeda, T., Kobayashi, T., 2023. Granular flow experiment using artificial gravity generator at International Space Station. npj Microgravity. 9, 61.

Reiss, P., Hager, P., Hoehn, A., Rott, M., Walter, U., 2014. Flowability of lunar regolith simulants under reduced gravity and vacuum in hopper-based conveying devices. J. Terramech. 55, 2014.

Suzuki, H., Katsushima, K., Ozaki, S., 2019. Study on applicability of RFT to traveling analysis of wheel with grousers: Comparison with DEM analysis as a virtual test. J. Terramech. 83, 15–24.

Zhang, J., Ding, Y., Chen, L., Song, A., Huang, F., Lu, X., 2022. A sweeping and grinding combined hybrid sampler for asteroid sample return mission. Acta Astronaut. 198, 329–346.

Zhao, C., Zang, M., 2014. Analysis of rigid tire traction performance on a sandy soil by 3D finite element–discrete element method. J. Terramech. 55, 29–37.

4217 | A computationally efficient plug-in for high-fidelity off-road vehicle dynamics studies on deformable terrains

Amir Khosravian [a,*], Sadegh Yarmohammadisatri [b], Corina Sandu [b], Yinglong He [a], Davide Tavernini [a]

[a] Centre for Aerodynamics, Aerospace and Automotive Engineering, School of Engineering, University of Surrey, Guildford, UK

[b] Department of Mechanical Engineering, Virginia Tech, Blacksburg, VA, USA

* Corresponding author: a.khosravian@surrey.ac.uk

ABSTRACT

In spite of advances in vehicle dynamics modeling, accurately simulating off-road vehicle performance on deformable terrains remains computationally challenging. In this paper, we develop an efficient yet high-fidelity off-road simulation toolbox, which incorporates the Hybrid Soft Soil Tire Model (HSSTM), a semi-empirical, experimentally validated off-road tire model, and the widespread simulation platform IPG CarMaker. This integration enables realistic simulations on deformable terrains and facilitates the development of off-road vehicle control algorithms for highly dynamic scenarios. The integrated vehicle-terrain model provides a tool for simulating the ride, handling, and mobility of off-road vehicles. HSSTM models the tire as discretized lumped masses with Kelvin-Voigt elements on a 3D deformable terrain. Unlike rigid-ring models, HSSTM captures tire flexibility, which is crucial for simulating tire dynamics on medium-hard to rigid terrains. To enhance computational efficiency, we adopt a Dynamic Terrain Adaptation (DTA) algorithm that models only a localized portion of the terrain around the vehicle, dynamically updating as the vehicle navigates. Simulations on deformable terrain in longitudinal and lateral scenarios (including force coupling) confirm a realistic representation of off-road dynamics based on the developed framework. The plug-in assesses contact patch pressure, normal and shear forces, elastic and plastic sinkage, and multi-pass effect.

Keywords
Off-road mobility
Tire-terrain interaction
Efficient off-road vehicle simulator
Deformable soil analyses
Off-road simulation plug-in

1. Introduction

Due to terrain variability, complex contact mechanics, and slip and traction limits, off-road vehicle dynamics are inherently more complex than their on-road counterparts. Unlike vehicles navigating on non-deformable surfaces, off-road vehicles ought to account for complex tire-terrain interactions that affect the mobility, handling, and stability of the vehicle. Tire material, the tread design, and the mechanical properties of the terrain all significantly impact the behavior of the vehicle, when simulating off-road vehicle dynamics. While the latter is most frequently considered for low-speed applications, such as agricultural, construction, or military vehicles, analyzing agile off-road vehicles has remained a less-explored territory.

In higher-speed off-road scenarios, tire-terrain interactions are more complex, requiring precise and responsive control strategies for handling the dynamics of the vehicle. However, despite such importance, researchers have yet to develop a computationally efficient framework to analyze off-road vehicle navigation and the interactions of tires with the terrain in a custom driving maneuver. Most existing vehicle dynamics simulators, such as CarSim ("Mechanical Simulation Corporation," n.d.) and IPG CarMaker ("CarMaker | IPG Automotive," n.d.), are inherently designed to simulate driving scenarios in structured environments.

There are simulators like Project Chrono ("Project Chrono - An Open-Source Physics Engine," n.d.), which are designed to handle vehicle simulations on deformable terrain; however, accurately modeling tire-terrain, especially in a long-distance driving scenario, can be computationally extensive and allocate an excessive amount of memory. Moreover, for a more computationally efficient simulation, Project Chrono adopts the Soil Contact Model (Krenn and Gibbesch, 2011), which models the tire as a solid ring, resulting in a noticeable inaccuracy on medium to hard terrain types due to imprecise estimation of the pressure distribution in the contact patch. Therefore, this research aims to strike a trade-off between high-fidelity modeling of off-road vehicle simulations over deformable terrains and the computational efficiency in a long-distance driving scenario.

Widely used tire models such as Magic Formula (MF) (Pacejka and Bakker, 1992) or MFSwift (Schmeitz and Versteden, 2009) cannot be directly adopted to capture accurate tire forces and moments on deformable terrains as they do not consider soil sinkage, bulldozing effect, and the pressure-sinkage relationship. On the contrary, terramechanics models incorporate the tire-terrain interaction either by means of curve fitting on experimental data or physics-based relationships.

More specifically, terramechanical models can be categorized into three main aspects: empirical, physics-based (analytical), and semi-empirical (Taheri et al., 2015). Pure empirical tire models (Allen et al., 1997; Hegazy and Sandu, 2013) frequently rely on extensive experimental tests and various curve-fitting techniques to capture the inherent patterns based on the measured inputs. While such models may be relatively efficient regarding computational demand, they may fail to capture important phenomena such as the bulldozing force, combined slip, and multi-pass effect. Moreover, the generalization capability of such models may be limited to the area of the data domain acquired from the conducted experiments. Consequently, the accuracy of the model may drop dramatically when employed on diverse terrains with different terramechanics properties that were not considered during the experimental data acquisition.

Conversely, physics-based models encompass physical principles, numerical analysis, or Finite Element Method (FEM) to

https://doi.org/10.56884/DZSIN0YU

quantify how the tire interacts with the terrain model (Fervers, 2004; Grujicic et al., 2010; Li and Schindler, 2013; Smith and Peng, 2013). The tire might have been modeled based on a solid ring, a 3D lumped mass, or pure FEM. In addition to that, the terrain might also range from discretized models up to FEM-based approaches, providing a wide trade-off between accuracy and computational complexity. However, such models may not be efficient enough to be incorporated into a vehicle dynamics simulator for assessing vehicle mobility due to the excessive computational demands and high memory allocation required in long-distance scenarios.

Considering the aforementioned trade-offs, in our research, we adopt the Hybrid Soft Soil Tire Model (HSSTM) (Corina Sandu et al., 2019; C. Sandu et al., 2019; Sandu et al., 2020), an experimentally validated semi-empirical tire model that provides an excellent trade-off between accuracy, computational efficiency, and generalization capability to various terrain types. Overall, semi-empirical tire models tend to take advantage of both experimental tests in conjunction with analytical approaches to balance the accuracy and computational effort of the model (Chan and Sandu, 2014; Harnisch et al., 2005; Krenn and Gibbesch, 2011; Vieira et al., 2022). HSSTM models the tire as 3D lumped masses of Kelvin-Voigt elements in the circumferential and lateral directions of the tire. This allows accurate and efficient modeling of the structural forces of the tire on the belt and sidewalls and allows HSSTM to capture the flexibility effect of the tire in the computation of the wheel center forces and moments. HSSTM employs a 3D Regular Grid Road (RGR) data model for terrain modeling, uniformly discretized with any desired resolution in X and Y directions. For deformable terrains, the terramechanics properties of the terrain are entirely user-defined based on the Bekker-Wong theory of terramechanics (Wong and Reece, 1967a, 1967b).

We integrate the high-fidelity HSSTM off-road tire model into IPG CarMaker to provide a customizable and efficient simulation plug-in for analyzing off-road vehicle mobility over deformable terrains on both low- and high-speed driving maneuvers. To enhance the computational efficiency of the simulation, we adopt the Dynamic Terrain Adaptation (DTA) algorithm. Instead of modeling the whole instance of the global terrain in the memory, the DTA only considers a limited portion of the terrain model around the vehicle. This localized terrain continuously slides and tracks the vehicle during the maneuver while maintaining the previous pressure distribution and elastic and plastic sinkage profiles. Such implementation allows our plug-in to efficiently and accurately simulate a long-distance driving scenario on deformable soil.

The main contributions of this paper can be summarized as follows:

- Development of a high-fidelity plug-in to simulate off-road mobility on deformable terrains for both low- and high-speed driving vehicles with a focus on the latter
- Implementing DTA to enhance the computational efficiency of the simulation for long-distance driving scenarios
- Development of a unified platform to support the design and assessment of chassis control strategies for off-road mobility
- Considering important characteristics of tire-terrain interaction in the integrated off-road plug-in, such as elastic and plastic sinkage, combined slip, bulldozing force, pressure distribution, and the multi-pass effect

The remainder of this paper is organized as follows: Section 2 focuses on the methodology development for the integrated plug-in between HSSTM and IPG CarMaker. It discusses the overview of the tire and terrain models utilized in this study to add off-road simulation capabilities to IPG CarMaker. Moreover, it presents how the integrated plug-in has been developed in the simulator and provides details of the communication protocol and data flow between tire, terrain, and vehicle plant models. Section 3 presents the results of the experiments and simulations. First, the experimental validation results are presented to demonstrate that the predictions of the developed tire model are in good agreement with other tire models relevant for non-deformable cases. Then, simulations on deformable terrain in both longitudinal and cornering scenarios are carried out to showcase the effectiveness of the proposed method in analyzing vehicle mobility on soft soil. Finally, Section 4 concludes the paper.

2. Methodology

The integration of the HSSTM tire and terrain model into the simulator is the key component of adding off-road simulation capability as a plug-in to IPG CarMaker. Therefore, this section first presents an overview of the HSSTM in both aspects of the tire and terrain models. Subsequently, the strategy to consolidate the tire and terrain model into the back-end of the simulator is discussed. Afterward, the data flow between the simulator, tire, and terrain models is presented. Finally, the simulation setup, discretization levels, and vehicle parameters are reported.

2.1. HSSTM tire-terrain interaction overview

According to the comparison between off-road tire models (Taheri et al., 2015), we have adopted HSSTM (Corina Sandu et al., 2019; C. Sandu et al., 2019; Sandu et al., 2020) to establish our off-road simulation plug-in. HSSTM accounts for key tire–terrain interaction phenomena, such as pressure-sinkage, combined slip, bulldozing force, and multi-pass effect, allowing accurate off-road simulation capabilities. HSSTM models the tire using 3D discretized lumped masses of Kelvin-Voigt elements on both the sidewalls and belt, providing the capability to model the tire's flexibility due to tire deflection and the effect of inflation pressure.

Unlike many existing models, HSSTM does not require a pre-defined normal pressure distribution approach for the contact patch. Instead, the tire envelopes the terrain and determines the stress distribution accordingly. The terrain model is based on a 3D RGR model, uniformly discretized in both X and Y directions with a user-defined segment resolution. Considering the fact that both tire and terrain models are 3D discretized, the approach for computing which nodes on the tire are in contact with the corresponding segments is a pivotal task. This ensures that the normal and tangential stress distributions are correctly determined for different terrain segments, which will eventually cause elastic and plastic sinkage on the terrain.

HSSTM adopts a contact formulation based on a master-slave approach, with tire-lumped mass nodes defined as the master and the ground segments as the slave (C. Sandu et al., 2019). In this method, the surfaces of two objects are in contact with each other, and each segment in the slave is continuously monitored to carefully detect penetration into the nodes of the master (Zavarise and De Lorenzis, 2009). For the contact algorithm in the deformable terrain, HSSTM leverages the Penalty method, which dictates that the contact forces are determined relative to the contact penetration of a specific node into the terrain without any extra degree of freedom (Taheri et al., 2014). The node penetration into the terrain can directly be mapped to the bulk modulus, volume, and surface area of the segment. It is important to highlight that since this modeling approach functions similarly to a force proportional to the deflection behavior of a spring, there is no zero-gap assumption between contact surfaces.

Figure 1 illustrates the overall tire-terrain interaction modeling of HSSTM. The tire starts building up the pressure distribution over the terrain segments based on the relative penetration of the lumped mass nodes, which continuously analyzes the contact between the tire and the terrain. Such normal loading generates elastic and plastic sinkages over the terrain segments, deforming them in the global Z direction. For tire-terrain interaction on deformable terrains, it is important to consider contact conditions in each iteration timestep and update the terrain profile according to the current condition, such as loading, unloading, and reloading.

For normal loading, the pressure-sinkage relationship is the primary cause of tire sinkage into the terrain and is heavily dependent on terrain characteristics and terramechanics properties. According to Bekker's theory of terramechanics (Bekker and Bekker, 1962), the pressure-sinkage relationship for mineral terrains can be expressed as:

$$p = \left(\frac{k_c}{b} + k_\phi\right) z^n = k_{eq} z^n \tag{1}$$

where p is the distributed pressure, b is the width of the contact area, k_c is the cohesive modulus of soil deformation, k_ϕ is the frictional modulus of soil compaction, z is the tire sinkage into the terrain (penetration depth), n is the sinkage exponent, and k_{eq} is the equivalent modulus value that considers both cohesive and frictional components.

Repetitive loading is a crucial component of off-road vehicle mobility. When the front tire passes a portion of the terrain, it unloads the terrain, and the stress and pressure distribution at that location will be reduced. It should be noted that, due to the hysteresis phenomenon, the force-deflection behavior of the terrain is different during the loading and unloading conditions. When the rear tire passes the same portion of the terrain and once more reloads the terrain segments at that location, the slope of the force-deflection path is approximately equal to the unloading phase. For mineral terrain types, the pressure sinkage relationship for repetitive loading conditions can be described as:

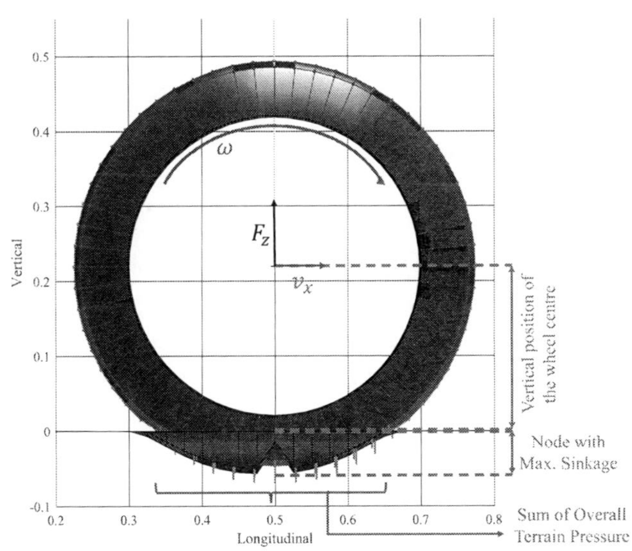

Fig. 1. HSSTM tire-terrain interaction modeling

$$p = \left(\frac{k_c}{b} + k_\phi\right) z_u^n - k_u(z_u - z) = p_u - k_u(z_u - z) \tag{2}$$

where k_u describes the terrain stiffness during loading and unloading as a function of unloading sinkage z_u.

$$k_u = k_0 + A_u z_u \tag{3}$$

k_0 and A_u can be determined using fitting techniques. The shear behavior of the tire is directly linked to the slip velocity of the tire tread elements. Just as the normal loading condition had different pressure-sinkage relationships for various terrain types, a similar pattern can be observed when investigating the terrain's response to shearing loads. This also results in unique and distinct shear stress-displacement behavior for each terrain type. The following expression can be adopted to describe shear stress distribution at the contact patch for mineral terrains (Yamakawa and Watanabe, 1961):

$$\frac{\tau}{\tau_{max}} = 1 - e^{-\frac{\lambda_s}{K}} \tag{4}$$

where τ_{max} is the maximum shear stress, K is the shear deformation modulus, which corresponds to the shear displacement value required to generate the maximum shear stress, and λ_s is the shear displacement value. Maximum shear stress can be expressed according to the Mohr-Coulomb equation.

$$\tau_{max} = c + p\tan(\phi) \tag{5}$$

where p is the normal stress, ϕ is the angle of shearing resistance, and c is the soil cohesion coefficient.

2.2. Integration of tire-terrain model in vehicle dynamics simulator

The incorporated vehicle dynamics simulator in this work is IPG CarMaker. Carmaker provides a comprehensive integration and test platform for the entire vehicle and its interaction with the environment. It includes a detailed multi-body simulation with a focus on real-time inference speed and Hardware-in-the-loop deployment. In addition to a robust and reliable vehicle dynamics model, CarMaker is also widely known for flexible environment modeling, which provides the capability to model complex road profiles, detailed 3D scenes, complex traffic congestion, and environmental conditions.

Figure 2 illustrates the communication data flow between IPG CarMaker as the vehicle dynamics simulator, the HSSTM tire model, and the 3D discretized terrain model. As illustrated in Fig. 2, first, user-defined parameters are utilized to initialize tire and terrain models. The initialization phase includes generating the initial instance of the global terrain model as well as calculating initial states ($q_{0_{Tire}}$) of all individual nodes on the belt and sidewalls based on the geometry of the tire as well as the adopted discretization level. IPG CarMaker provides the wheel kinetics and kinematics to the tire model at every simulation timestep. It is important to note that HSSTM employs Standard Tire Interface (STI) to form a well-known communication protocol with the vehicle plant model.

Since HSSTM directly envelopes the tire over the terrain to determine the pressure distribution of the tire over the contact patch, it requires receiving all its inputs in the global reference frame. Hence, as expected by the tire model, a pre-processing data handling block is considered to transfer simulator inputs to the global reference frame. The tire model employs this array of data to determine the tire kinematics, which includes the position and

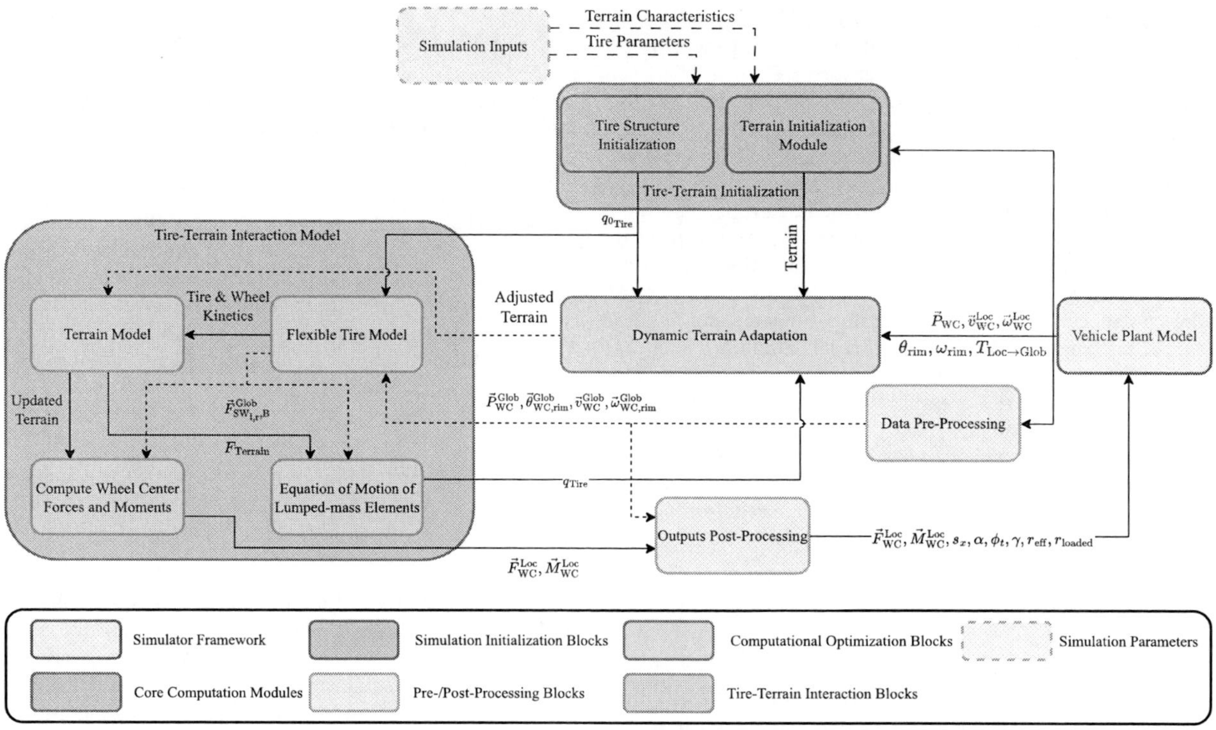

Fig. 2. Integrated plug-in flowchart between HSSTM (tire-terrain interaction model) and IPG CarMaker (vehicle dynamics simulator)

linear velocity components of every circumferential and lateral node of the tire, considering the tire deflection and the 3D structure of the flexible tire.

Instead of utilizing a fixed-size terrain model, the DTA algorithm has been exploited in the simulation loop. By doing so, DTA dynamically generates a localized terrain segment around the vehicle, based on the opted terrain discretization level. At each timestep, the DTA continuously monitors the positions of the Kelvin-Voigt elements on the tire model and compares them with the terrain boundary. Once it detects that the tire is nearing the edge of the active terrain, it generates a new small-scale terrain model centered on the updated wheel position. Finally, the pre-computed pressure, sinkage, and forces from the original terrain are then mapped onto the new terrain by tracking the corresponding segment positions. The DTA algorithm not only significantly increases the computational efficiency of the simulation by avoiding the allocation of excessive memory but also increases the inference speed of the simulation without harming the terramechanical relationships between the tire and the terrain.

After adjusting terrain boundaries, the tire model detects the nodes in contact with the terrain, and the terrain structure is updated using the computations from the tire model. Subsequently, the terrain response is calculated based on the normal and shear stresses from the tire-terrain interaction relationships. The tire model first calculates all individual Kelvin-Voigt elements' normal and tangential forces based on the modeled springs and dampers. After solving the structural forces, the terramechanical equations inside the contact patch of the tire are solved by considering the penetration of the nodes inside the contact patch into the terrain. This results in computing the pressure distribution, terrain forces, and elastic and plastic sinkage values of the terrain segments, which are updated and stored in the corresponding global array of the terrain model.

The computed tire and terrain forces are considered when solving the equation of motion for the tire nodes on the belt and sidewalls. It is important to note that, for enhancing computational efficiency, the nodes on the tire only have translational degrees of freedom, which is still sufficient to describe the flexible behavior of the tire accurately. Solving the equation of motion for tire nodes will provide tire states for the next iteration, which includes updating the rim, segments, and bristles states.

Finally, the overall forces and moments around the wheel center are computed using the structural forces of the left and right sidewalls. Subsequently, a post-processing block is used to calculate longitudinal slip ratio, slip angle, turn slip, and the loaded radius of the tire at each timestep, which, in conjunction with the wheel center forces and moments, are returned to the vehicle dynamics simulator. It is worth highlighting that the vehicle plant model requires only wheel center forces and moments to solve the kinetics of the vehicle and update its states accordingly. The other specified signals are only used for monitoring purposes.

It is essential to highlight that the whole pipeline illustrated in Fig. 2 is developed in a generalized manner and is not limited only to IPG CarMaker as the multi-body dynamics simulator. To be more specific, since the overall pipeline is designed based on the Tire STI communication protocol, it can be easily implemented on any multi-body dynamics simulator that supports the Tire STI data flow, with some potential minor adjustments to data pre- and post-processing to match the expected input/outputs of the plug-in correctly. Moreover, the DTA is designed to be compatible with standard RGR-based terrain and road models. Therefore, any desired tire-terrain interaction model that utilizes RGR-based terrain models can benefit from the DTA algorithm to enhance computational efficiency during the simulation.

2.3. Simulation setup

The entire plug-in is developed in the C programming language, which can execute significantly faster than high-level programming languages by being adequately compiled in the back-end of the IPG CarMaker and relying on pre-defined arrays and structures. A timestep of 4×10^{-5} seconds has been opted to carry out the experiments, proving accurate enough to perform well in the nonlinearities and complexities involved in deformable terrains. The number of lateral nodes on each segment is selected as three for properly modeling the belt as well as the left and right sidewalls. Furthermore, 80 segments have been opted for as the discretization level of the tire in the circumferential direction, which falls within the range recommended by the tire model developers (C. Sandu et al., 2019). Moreover, a terrain resolution of 5 cm was opted for in both terrain length and width directions to uniformize the terrain model.

Table 1
Specifications of the test vehicle for simulations

Parameter	Symbol	Value
Wheelbase	l (m)	2.97
Front/Rear load distribution	-	51/49
Front/Rear track	T_f/T_r (m)	1.66/1.7
Overall mass	m (kg)	2025
Sprung mass	m_s (kg)	1841.8
Wheel mass	m_w (kg)	20
Vehicle inertia around the x-axis	$I_{xx_{\text{body}}}$ (kg.m^2)	741.7
Vehicle inertia around the y-axis	$I_{yy_{\text{body}}}$ (kg.m^2)	2903
Vehicle inertia around the z-axis	$I_{zz_{\text{body}}}$ (kg.m^2)	3170.6
Wheel inertia around the x-axis	$I_{xx_{\text{wheel}}}$ (kg.m^2)	0.7
Wheel inertia around the y-axis	$I_{yy_{\text{wheel}}}$ (kg.m^2)	1.2
Wheel inertia around the z-axis	$I_{zz_{\text{wheel}}}$ (kg.m^2)	0.7
Motor power	P_{motor} (kW)	20
Maximum motor torque	T_{max} (Nm)	250
Maximum motor speed	ω_{max} (rpm)	8000
Gear ratio	n (-)	3.5
Motor efficiency	η (-)	0.95

Table 2
Tire characteristics used to carry out simulation experiments

Parameter	Symbol	Value
Tire radius	r (m)	0.33
Tire width	b (m)	0.215
Sidewall height	h_{sw} (m)	0.119
Tread thickness	t_{tread} (m)	0.0067
Wheel radius	r_w (m)	0.21
Inflation pressure	P (kPa)	240
Sidewall tangential stiffness	k_{st} (N/m)	2.9 e5
Sidewall radial stiffness	k_{sr} (N/m)	1.8 e3
Sidewall lateral stiffness	k_{sl} (N/m)	3.2 e5
Belt tangential stiffness	k_{bt} (N/m)	1.5 e6
Belt radial stiffness	k_{br} (N/m)	4.9 e5
Belt lateral stiffness	k_{bl} (N/m)	7.1 e6

Table 3
Terramechanical characteristics of the simulated terrain type

Parameter	Symbol	Value
Sinkage exponent	n (-)	0.8
Cohesive modulus	k_c (kN/m^{n+1})	16.54
Frictional modulus	k_ϕ (kN/m^{n+2})	911.4
Moisture content	- (%)	0
Cohesion	c (kPa)	3.71
Shearing resistance angle	ϕ (deg)	25.6
Shear deformation modulus	K (cm)	2.1
Initial stiffness modulus of repetitive loading	K_0 (kN/m^{n+1})	0
Empirical accumulative-based term of repetitive loading	A_u (kN/m^{n+2})	8.6 e4

A visualization plug-in has been developed in this research as a post-process sub-module to better study the vehicle-terrain interaction. It provides an animation tool to illustrate the terrain sinkage, tire deflection, longitudinal and lateral shear forces inside the contact patch, vehicle trajectory over the terrain, wheel center forces, and pressure distribution of the contact patch.

For simulation experiments, the employed vehicle is an All Wheel Drive (AWD) electric passenger vehicle with four on-board direct drive electric motors. Table 1 summarizes the main characteristics of the vehicle. The selected tire size in this study is 215/55R17, with parameters reported in Table 2. Finally, a soft-to-medium terrain model with mechanical properties presented in Table 3 has been employed for the terrain side.

3. Results and discussion

The previous section discussed the proposed framework to integrate tire and terrain models in the vehicle dynamics simulator. This section presents experimental validation results using a Flat-Trac machine first. Subsequently, simulation results with HSSTM on deformable terrain in longitudinal and combined slip lateral cornering scenarios are shown. Finally, the impact of the DTA algorithm on computational efficiency is discussed.

3.1. Tire model experimental validation results

In order to validate the accuracy of the HSSTM, a series of experimental tests were conducted using a Flat-Trac tire testing machine. These tests evaluate the capability of the tire model to faithfully estimate tire performance under combined, linear lateral, and nonlinear lateral slip scenarios. In these experiments, a wide range of critical input variables at the tire level, including inclination angle, tire slip angle, longitudinal slip ratio, and vertical load, were controlled as input parameters by the test machine. As indicated in Fig. 3 and Table 4, the performance of HSSTM is benchmarked against MF and CarSim Internal tire models. The input conditions and output responses across all models are visualized and compared in Fig. 3. For the sake of brevity, only the detailed results of the combined slip test have been presented in Fig. 3, as it imposes the most complexity to capture coupling force effects. Key results of the combined and pure lateral tests are represented in Table 4 by adopting the Root Mean Squared Error (RMSE) evaluation metric.

Table 4
Benchmarking the performance of HSSTM compared to MF and CarSim Internal models in Flat-Trac tire testing machine. The best and second-best performances are indicated by bold and underlined text, respectively.

Model	Signal	RMSE combined slip	RMSE linear lateral slip	RMSE nonlinear lateral slip
HSSTM	F_x (N)	474.2	**8.7**	**12.3**
	F_y (N)	**117.1**	268.5	327.4
	M_x (Nm)	**15.9**	8.1	10.7
	M_z (Nm)	25.9	9.2	14.4
MF	F_x (N)	**461.5**	**8.7**	**12.3**
	F_y (N)	265.9	**150.9**	**310.2**
	M_x (Nm)	30.4	**4.1**	**7.0**
	M_z (Nm)	69.9	**7.0**	**11.8**
CarSim Internal	F_x (N)	474.2	**8.7**	**12.3**
	F_y (N)	283.2	234.5	400.1
	M_x (Nm)	30.4	**4.1**	7.1
	M_z (Nm)	**20.8**	12.9	23.1

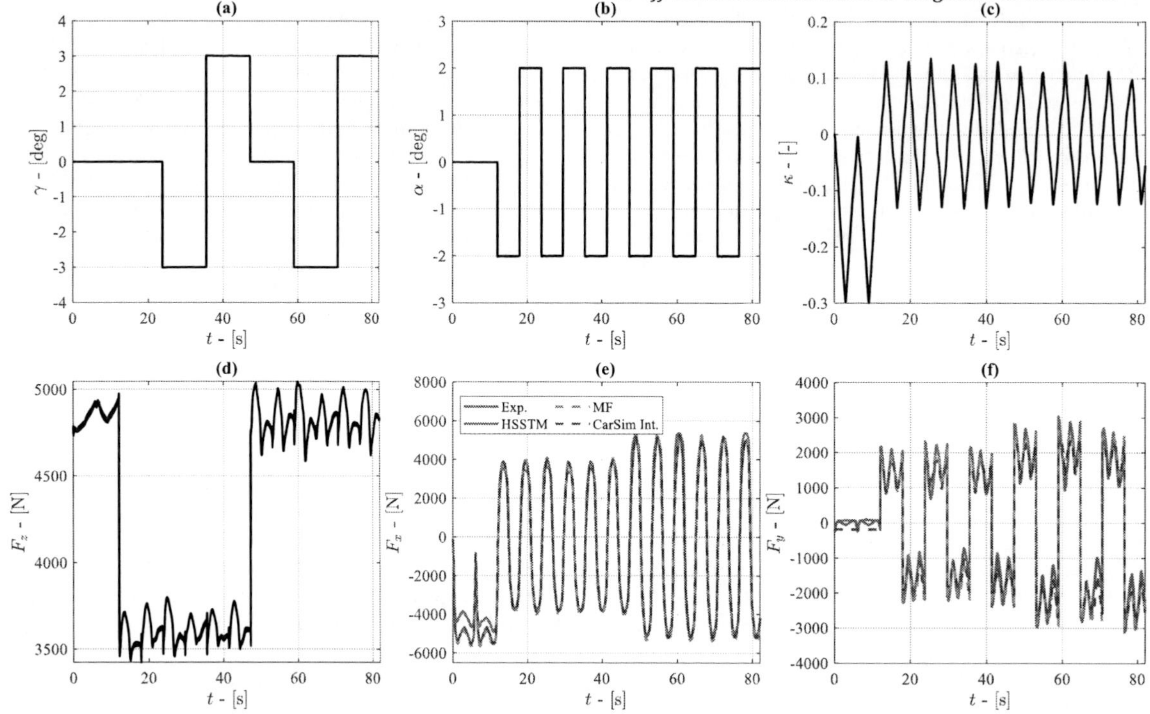

Fig. 3. Benchmarking comparison of HSSTM force prediction against other tire models in combined slip experiment using Flat-Trac data: (a) inclination angle, (b) tire slip angle, (c) longitudinal slip ratio, (d) vertical load, (e) longitudinal force, and (f) lateral force

In the combined slip test depicted in Fig. 3, the longitudinal slip ratio ranges between approximately −0.3 and +0.1, providing a full spectrum of braking and traction conditions. The HSSTM model effectively captures the transient dynamics of the longitudinal and lateral forces, particularly in reproducing the lateral force and overturning moment. The MF model overestimates transient peak values of longitudinal force, while the CarSim Internal model noticeably underestimates the lateral force. HSSTM, profiting from its Kelvin-Voigt-based flexible tire representation, demonstrates excellent compatibility with experimental data.

As indicated in Table 4, the MF model excels in pure lateral slip tests. Its optimized empirical structure accurately captures the relation between tire slip angle and lateral forces. However, a noticeable degradation in its performance is observed in the combined slip test. While MF does leverage a combined slip formulation, it often relies on weighting function to combine longitudinal and lateral forces, inherently limiting its capability to fully represent interacting forces and moments that are related to the flexible structure of the tire during combined slip.

To be more specific, in the combined slip scenario, HSSTM yields the lowest RMSE values for both overturning moment and lateral force compared to MF and CarSim Internal model. Notably, predictions of HSSTM for longitudinal force and aligning moment are quite similar to the best model in the benchmarking comparison.

In addition, in linear and nonlinear lateral slip scenarios — where tire lateral compliance plays a critical role— HSSTM maintains its accuracy and is in good agreement with the experimental data. Results indicate that HSSTM exceeds traditional models where lateral flexibility and dynamic force coupling are crucial. The experimental accuracy of HSSTM justifies its appropriateness as a high-accuracy off-road tire model in vehicle simulation.

3.2. Off-road simulation results: longitudinal maneuver

This subsection presents vehicle simulation results on deformable terrain in a longitudinal driving scenario. The longitudinal maneuver effectively showcases the traction and braking behavior of the vehicle on soft soil, considering the terrain sinkage and bulldozing force. Moreover, since the vehicle moves in a straight line, it enables us to visualize the multi-pass effect as the rear axle tires pass the same portion of terrain already deformed by the front ones.

The maneuver is defined as:

- Standstill of 2 seconds for the system to reach the initial steady-state conditions
- Traction scenario for 7.2 seconds, which includes building the traction torque from zero to 380 Nm with a linear ramp for five seconds, followed by driving with this constant torque for 2.2 seconds
- Regenerative braking for 2.2 seconds with linear ramp torque input (building up -170 Nm for rear tires and -350 Nm for the front tires)

Simulation results for the longitudinal driving scenario are presented in Fig. 4. Notably, only FL and RL tires are analyzed, considering that FR and RR tires behave similarly due to identical torque distribution in the powertrain system and homogenous terrain in all directions. The longitudinal forces predicted by the tire model are very responsive to the change of state in the applied torque during the traction build-up, constant torque phase, releasing the traction, and entering the regenerative braking phase.

It can be observed that during the traction maneuver, the longitudinal force on the RL tire is slightly lower compared to the FL tire, despite receiving identical traction torques at the wheel center level. This phenomenon can be attributed to the fact that

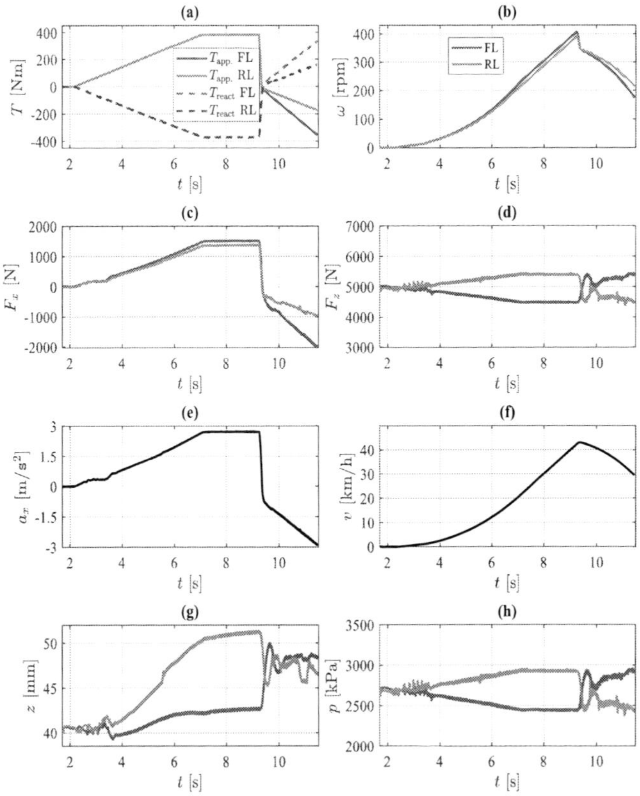

Fig. 4. Simulation results for longitudinal traction/braking maneuver on deformable terrain: (a) applied and reaction torques at wheel center, (b) angular velocity, (c) longitudinal force, (d) vertical force, (e) longitudinal acceleration, (f) vehicle velocity, (g) terrain sinkage, and (h) terrain pressure inside contact patch.

during the traction, there will be more load on the rear axle due to the load transfer effect, increasing the normal pressure on the terrain level. This leads to noticeably more sinkage on the RL tire compared to FL, as verified in Fig. 4. More sinkage is also followed by more bulldozing force, and thus more resistance torque at the wheel center level, decreasing the slip ratio on the RL tire. The effect of the bulldozing force is also illustrated in lower angular velocity for the RL tire despite the vehicle being AWD.

In addition to the accurate traction behavior, the model also performs well during the braking phase of the maneuver. Figure 4 illustrates that HSSTM reacts promptly when the applied torque is released, by dropping the longitudinal force. Once the vehicle enters the braking phase, more brake torque is applied to the FL tire due to the brake distribution ratio of the vehicle. HSSTM correctly captures this phenomenon by computing a greater braking force for the front axle tires. Additionally, the speed of the vehicle is slightly over 40 km/h when the vehicle enters the braking phase, which confirms that the developed simulation plug-in is not restricted to low-speed driving and is capable of accurately simulating agile off-road scenarios at higher speeds.

Vertical load and load transfer are also modeled and monitored during the simulation. Once the traction phase starts, the vertical load on the RL tire rises to 5400 N while the load on the FL tire drops to approximately 4400 N. During braking, the load transfer correctly shifts towards the front.

The overall normal terrain pressure inside the contact patch shows similar patterns to the vertical load, as both concepts are highly correlated. However, the sinkage of the tire into the terrain will behave in a nonlinear manner due to the sinkage exponent parameter. Figure 4 depicts that the terrain responds with roughly 40 mm of sinkage due to static load in the standstill phase. During the traction phase, both FL and RL tires sink even more into the terrain by a noticeable margin, a crucial phenomenon in off-road vehicle mobility studies. To be more specific, the overall sinkage for the FL and RL tires increases to 42.3 and 50.7 mm, respectively. This behavior can be explained by the fact that during the standstill phase, the tires only sink into the terrain due to the static load and normal pressure on the terrain. However, during the traction phase, the motion of the tire also builds shear forces and shear displacements, which disrupt the soil structure and reduce its load bearing capacity. In addition to that, it also results in local yielding and terrain failure, especially towards the leading edge. Therefore, since the terrain is now simultaneously under normal and shear stress, it yields more, resulting in higher sinkage than the standstill phase.

It is also worth highlighting the behavior of the vehicle at the beginning of the traction phase. In Fig. 4, during $t = 2 - 4$ s it can be observed that the tire longitudinal forces are generated at a much lower rate compared to when the vehicle builds up some speed. This phenomenon is also reflected in a chattery response to the normal pressure inside the contact patch and a sudden, temporary increase in sinkage of the tire.

More specifically, despite the linear increase of the traction torque, the tire first builds up the traction at a lower rate before stabilizing the rate of increase in the longitudinal force. This phenomenon is attributed to the fact that the soil beneath the tire is undisturbed in terms of shear forces in a standstill condition. After the start of traction, the soil has yet to develop shear bands. Tire sinkage gets altered since the terrain requires absorbing the vertical load, resisting longitudinal shearing, and initiating the soil failure mechanism.

It should be noted that shear stress in the soil does not immediately reach its peak value. Early on, due to low shear stress, the terrain cannot yet resist the traction. Therefore, it sinks more before the shear zone is fully developed. As sinkage increases, the contact patch grows, leading to increased shear force build-up area until it stabilizes the torque applied at the wheel center level.

Elastic and plastic contributions of terrain segments ought to be analyzed individually, and terrain deformation must be correctly investigated according to the loading condition. Figure 5 presents a snapshot of our visualization plug-in, which illustrates how such phenomena are considered in this work. Tire circumferential lumped-mass nodes apply normal pressure into terrain segments relative to their corresponding penetration into the terrain. This deformation is a mixture of elastic and plastic contributions based on the mechanical properties of the terrain.

When the unloading occurs, some portion of sinkage related to elastic deformation diminishes; however, the plastic deformation will not recover. The terrain contour visualized in red illustrates the remaining plastic sinkage after the front axle tires pass these portions of terrain. When the rear tires pass the exact location, existing soil compaction alters the tire-terrain interaction, including drawbar pull, bulldozing force, and shear stress distribution.

As visualized in Fig. 5, when the second passing occurs, the plastic sinkage profile completely differs from the first passing related to front axle tires. It is worth mentioning that this phenomenon is attributed to both the multi-pass effect and the higher vertical load on the rear axle tires.

3.3. Off-road simulation results: combined slip cornering maneuver

The primary purpose of the cornering maneuver in this subsection is to introduce a combined slip scenario to analyze

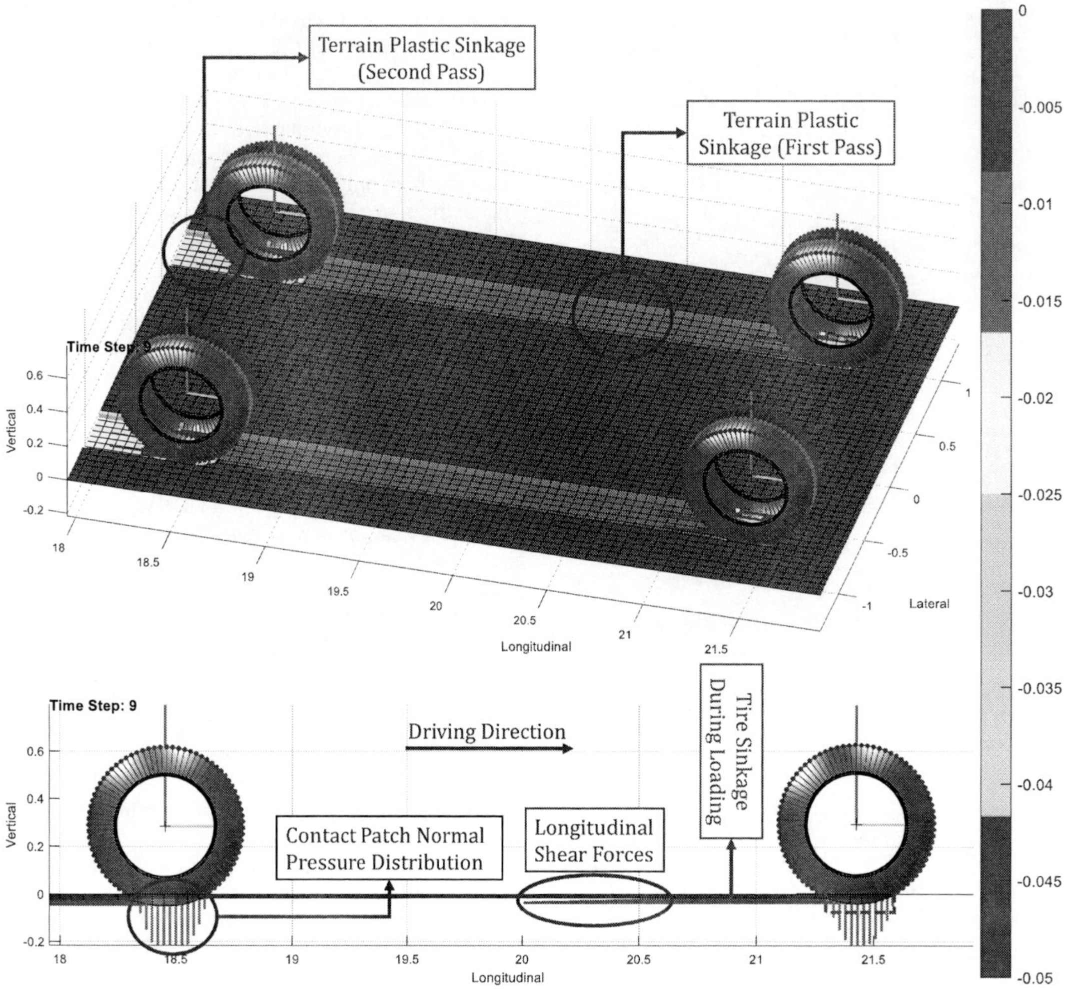

Fig. 5. Snapshot of tire-terrain interaction in longitudinal driving scenario to showcase the multi-pass effect.

vehicle-terrain interaction on deformable soil for a vehicle that is both accelerating and cornering simultaneously. Therefore, the tire should generate longitudinal and lateral forces, and we can illustrate how our plug-in captures tire force coupling and force redistribution.

Analyzing the vehicle behavior in a combined slip scenario will induce asymmetric terrain loading due to longitudinal and lateral load transfers. This will lead to asymmetrical sinkage on tires; therefore, shear band development differs for each tire as the soil beneath it fails differently. To analyze both vehicle and soil behavior under such a driving scenario, the following double-step steer maneuver is performed:

- Standstill of 2 seconds for dynamic settling
- Traction scenario for 7 seconds by increasing the applied torque from zero to 380 Nm in a duration of 4 seconds
- Decrease the traction torque to 190 Nm and maintain it constant for all four wheels during the rest of the maneuver
- A steering input of +90° is applied at a rate of +400°/s for 2 seconds
- A steering input of -90° is applied at a rate of -400°/s for 2 seconds
- Revert the steering input to 0° at a rate of +400°/s in 5 seconds.

Figure 6 illustrates the simulation results of the combined slip maneuver. As can be observed from the obtained results, the model accurately captures the load transfer effect both during the traction and during combined slip cornering maneuvers. Results demonstrate that the model not only computes higher vertical loads for outer tires but also exhibits a higher load on rear axle tires compared to front ones due to the positive acceleration. Therefore, due to different loads and normal pressure distribution for each individual tire, an asymmetric contact patch pressure exists for this combined slip cornering maneuver.

The tire sinkage is also unique for each tire. When the vehicle enters the first curve, the outer tires sink noticeably more into the terrain, and the sinkage decreases for the inner tires due to the unloading. In addition to that, the rear tires still have higher sinkage due to the presence of longitudinal acceleration. Higher sinkage leads to an increase in the contact patch area.

The plug-in correctly captures higher lateral forces for the outer tires during both turns. Tire lateral slip angles rise approximately to 12.5° and 9.5° for front and rear tires, respectively, confirming that the simulation is conducted in the nonlinear region of the tire force curve. However, a rise in both longitudinal and lateral tires can be observed from simulation results for the inner tires, specifying that although the scenario is combined slip, the tire is not operating near the friction limits and is capable of producing more force in both directions. A higher slip

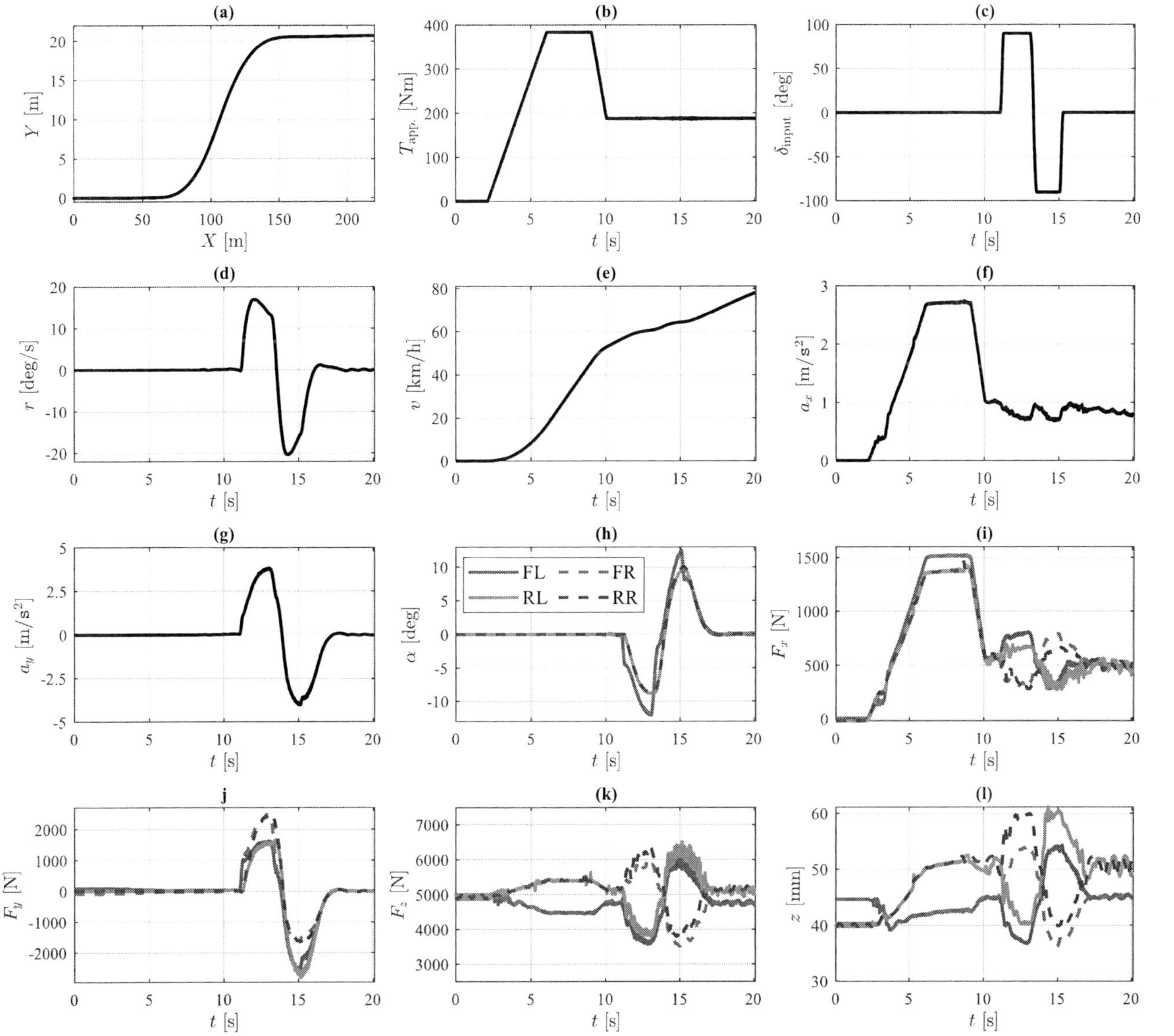

Fig. 6. Vehicle simulation results on combined slip maneuver on deformable terrain: (a) vehicle trajectory, (b) applied torque to all wheels, (c) steering input, (d) vehicle yaw rate, (e) vehicle velocity, (f) longitudinal acceleration, (g) lateral acceleration, (h) tire slip angle, (i) longitudinal force, (j) lateral force, (k) vertical force, and (l) terrain sinkage

angle for front tires in both turns is correlated to the steering angle and the fact that the simulated vehicle is front-wheel steering.

Interestingly, the behavior of longitudinal force indicates patterns that are significantly more noticeable in deformable terrains. As can be observed from Fig. 6, during cornering, the longitudinal forces for the inner tires increase. Meanwhile, they decrease for the outer ones despite the outer tires having a higher vertical load. Even though this phenomenon occurs in simulations on non-deformable terrain (for example, with the MFSwift model), the differences in longitudinal forces on inner and outer tires will not be as dominant as on such deformable terrain.

For the case of off-road simulations on a terrain with the provided mechanical properties, as sinkage increases for outer tires, more bulldozing force and resistive torques at the wheel center level are induced from the terrain, reducing the angular velocity on the outer tires. On the contrary, the unloading and decrease of sinkage for the inner tires contribute to a reduced resistance, further increasing the longitudinal slip ratio.

During combined slip cornering on soft soil, such a higher disparity between longitudinal forces generates an additional yaw moment directed outward of the turn, resisting the increase of the yaw rate. The relatively low yaw rate shown in Fig. 6 supports this. The phenomenon contributes to promoting understeering. Such an effect is even further amplified on softer terrains.

3.4. DTA Impact on Computational Efficiency

As previously mentioned, the DTA algorithm was proposed and implemented in our plug-in to significantly reduce computational complexity and provide long-distance simulation capability for the toolbox. To quantify its impact on the computational efficiency of the overall pipeline, we replicated the longitudinal traction maneuver (discussed in Section 3.2) and compared the total simulation time using the standard RGR-based terrain model against the RGR model combined with the DTA.

Table 5
Comparison of the overall computational time with and without adopting the DTA. Simulation maneuver is the longitudinal traction scenario with path distances of 20 and 30 meters long, with terrain resolutions of 2.5 cm and 5 cm.

Distance (m)	**Terrain resolution (cm)**	**Computational time (s)** RGR	RGR+DTA	**Gain (%)**
20	5	219.34	131.66	39.97
	2.5	563.64	171.48	69.57
30	5	370.75	164.38	55.66
	2.5	2554.12	222.84	91.27

It is important to note that replicating the whole maneuver without the DTA would not even be possible, as modeling such a vast terrain mesh with the required boundary would demand dramatically higher memory on RAM. Therefore, we simulated only the first 20 m and 30 m of the path. Additionally, we have duplicated the simulations with two terrain resolutions of 5 cm and 2.5 cm to better highlight the impact of DTA. The hardware used to carry out the simulation utilizes a 13th-generation Intel Core i5-1335U CPU with 16 GB of RAM. Table 5 demonstrates the computational time comparison of baseline RGR with RGR in addition to DTA.

As shown in Table 5, the DTA makes a significant impact on computational complexity with a noticeable margin in all cases. Additionally, its impact grows exponentially as the vehicle needs to cover a longer distance or in case of increasing the terrain resolution. That is due to the fact that modeling an entire mesh grid tensor for storing positions, sinkages, normal and shear pressures in both X and Y directions is such a memory-consuming process. Accessing data, performing matrix multiplications, and writing data on such large tensors cause significant simulation delays and substantially increase the overall computational overhead.

In contrast, by utilizing the DTA, the tensor sizes for the terrain model remain small. They depend only on the terrain resolution and the defined boundary of the small terrain patch modeled around the vehicle. This approach provides the capability to simulate the vehicle behavior on deformable terrain in a long-distance scenario.

4. Conclusion

This paper presents a new plug-in for analyzing vehicle-terrain mobility on deformable terrains by leveraging the Hybrid Soft Soil Tire Model and IPG CarMaker. While other simulation-based toolboxes might be applicable to slow-speed maneuvers that are more common for agricultural or military vehicles, our plug-in covers a wider range of user-defined maneuvers. Most significantly, it primarily targets agile high-speed driving on off-road terrain.

To achieve this, a flexible tire model in conjunction with a Regular Grid Road-based uniform terrain model has been adopted to accurately model the tire-terrain interaction according to the theory of Terramechanics. An accurate communication protocol between vehicle plant, tire, and terrain models has been developed, and it operates in IPG CarMaker, utilizing the Standard Tire Interface data communication protocol. Dynamic Terrain Adaptation algorithm has been developed to enhance computational efficiency and provide the capability of performing long-distance maneuvers, which works seamlessly in conjunction with the tire-terrain interaction model. A visualization toolbox to illustrate tire-terrain interactions, terrain sinkage profile, and pressure distribution inside the contact patch has also been developed to provide a deeper understanding for studying off-road vehicle-terrain interaction.

Experimental validation has been provided, and predicted forces of the tire model have been benchmarked against relevant tire models that operate on rigid surfaces. Simulation results in both longitudinal traction/braking and combined slip cornering maneuvers demonstrate the effectiveness of our simulation plug-in in accurately capturing tire-terrain interaction in highly dynamic nonlinear driving scenarios on deformable terrain.

Our future work includes conducting simulation experiments on various terrain types with a wide range of vehicle types featuring distinct powertrain systems and mass distributions. One limitation of the current Dynamic Terrain Adaptation algorithm is that when the vehicle has fully passed the terrain boundary, the prior terrain data is destroyed after the transition to the new boundary is complete. Future work will focus on further optimizing the algorithm to ensure it stores information in disk memory so that it can be retrieved and restored when needed. Such modifications make sure the multi-pass effect is always considered (for example, for simulating multi-agent systems on deformable terrains). Finally, we will focus on the integration of advanced control strategies utilizing the developed plug-in for applications such as torque vectoring and path-tracking control.

5. Declaration of competing interest

The authors declare that they have no known competing financial interests or personal relationships that could have appeared to influence the work reported in this paper.

6. References

Allen, R.W., Chrstos, J.P., Rosenthal, T.J., 1997. A tire model for use with vehicle dynamics simulations on pavement and off-road surfaces. Vehicle System Dynamics 27, 3–18. https://doi.org/10.1080/00423119708969663

Bekker, D.-I. M.G., Bekker, D.-I. M.G., 1962. Mechanics of Off-the-Road Locomotion. http://dx.doi.org/10.1243/PIME_AUTO_1962_000_009_02 16, 25–44. https://doi.org/10.1243/PIME_AUTO_1962_000_009_02

CarMaker | IPG Automotive [WWW Document], n.d. URL https://www.ipg-automotive.com/en/products-solutions/software/carmaker/ (accessed 2.12.25).

Chan, B.J., Sandu, C., 2014. Development of a 3-D quasi-static tyre model for on-road and off-road vehicle dynamics simulations: Part i-on-road flexible tyre model. International Journal of Vehicle Systems Modelling and Testing 9, 77–105. https://doi.org/10.1504/IJVSMT.2014.059157

Fervers, C.W., 2004. Improved FEM simulation model for tire–soil interaction. J Terramech. 41, 87–100. https://doi.org/10.1016/J.JTERRA.2004.02.012

Grujicic, M., Marvi, H., Arakere, G., Haque, I., 2010. A finite element analysis of pneumatic-tire/sand interactions during off-road vehicle travel. Multidiscipline Modeling in Materials and Structures 6, 284–308. https://doi.org/10.1108/15736101011068037/FULL/PDF

Harnisch, C., Lach, B., Jakobs, R., Troulis, M., Nehls, O., 2005. A new tyre–soil interaction model for vehicle simulation on deformable ground. Vehicle System Dynamics 43, 384–394. https://doi.org/10.1080/00423110500139981

Hegazy, S., Sandu, C., 2013. Experimental investigation of vehicle mobility using a novel wheel mobility number. J Terramech. 50, 303–310. https://doi.org/10.1016/J.JTERRA.2013.09.005

Krenn, R., Gibbesch, A., 2011. Soft Soil Contact Modeling Technique for Multi-Body System Simulation. Lecture Notes in Applied and Computational Mechanics 58 LNACM, 135–155. https://doi.org/10.1007/978-3-642-22167-5_8

Li, H., Schindler, C., 2013. Analysis of soil compaction and tire mobility with finite element method. Proceedings of the Institution of Mechanical Engineers, Part K: Journal of Multi-body Dynamics 227, 275–291. https://doi.org/10.1177/1464419313486627/ASSET/IMAGES/LARGE/10.1177_1464419313486627-FIG20.JPEG

Mechanical Simulation Corporation [WWW Document], n.d. URL https://www.carsim.com/ (accessed 2.12.25).

Pacejka, H.B., Bakker, E., 1992. THE MAGIC FORMULA TYRE MODEL. Vehicle System Dynamics 21, 1–18. https://doi.org/10.1080/00423119208969994

Project Chrono - An Open-Source Physics Engine [WWW Document], n.d. URL https://projectchrono.org/ (accessed 2.12.25).

Sandu, C., Taheri, Sh., Taheri, S., Els, S., Jimenez, E., 2020. Hybrid soft soil tire model (HSSTM). Part III: Model parameterization and validation. J Terramech. 88, 1–15. https://doi.org/10.1016/J.JTERRA.2019.11.002

Sandu, C., Taheri, Sh, Taheri, S., Gorsich, D., 2019. Hybrid Soft Soil Tire Model (HSSTM). Part II: Tire-terrain interaction. J Terramech. 86, 15–29. https://doi.org/10.1016/J.JTERRA.2019.08.004

Sandu, C., Taheri, Sh., Taheri, S., Gorsich, D., 2019. Hybrid Soft Soil Tire Model (HSSTM). Part I: Tire material and structure modeling. J Terramech. 86, 1–13. https://doi.org/10.1016/J.JTERRA.2019.08.002

Schmeitz, A.J.C., Versteden, W.D., 2009. Structure and Parameterization of MF-Swift, a Magic Formula-based Rigid Ring Tire Model. Tire Sci Technol 37, 142–164. https://doi.org/10.2346/1.3138768

Smith, W., Peng, H., 2013. Modeling of wheel–soil interaction over rough terrain using the discrete element method. J Terramech. 50, 277–287. https://doi.org/10.1016/J.JTERRA.2013.09.002

Taheri, Sh, Sandu, C., Taheri, S., Pinto, E., Gorsich, D., 2015. A technical survey on Terramechanics models for tire–terrain interaction used in modeling and simulation of wheeled vehicles. J Terramech. 57, 1–22. https://doi.org/10.1016/J.JTERRA.2014.08.003

Taheri, Sh., Sandu, C., Taheri, S., 2014. Finite Element Modeling of Tire Transient Characteristics in Dynamic Maneuvers. SAE International Journal of Passenger Cars - Mechanical Systems 7, 221–230. https://doi.org/10.4271/2014-01-0858

Vieira, D., Orjuela, R., Spisser, M., Basset, M., 2022. An adapted Burckhardt tire model for off-road vehicle applications. J Terramech. 104, 15–24. https://doi.org/10.1016/J.JTERRA.2022.08.001

Wong, J.Y., Reece, A.R., 1967a. Prediction of rigid wheel performance based on the analysis of soil-wheel stresses: Part II. Performance of towed rigid wheels. J Terramech. 4, 7–25. https://doi.org/10.1016/0022-4898(67)90047-X

Wong, J.Y., Reece, A.R., 1967b. Prediction of rigid wheel performance based on the analysis of soil-wheel stresses part I. Performance of driven rigid wheels. J Terramech. 4, 81–98. https://doi.org/10.1016/0022-4898(67)90105-X

Yamakawa, J., Watanabe, K., 1961. The Analytical Determination of Drawbar Pull as A Function of Slip for Tracked Vehicles in Deformable Soils. International Society for Terrain-Vehicle Systems, 1st Int. Conf. 707, 1720–1727. https://doi.org/10.1299/KIKAIC.68.1720

Zavarise, G., De Lorenzis, L., 2009. The node-to-segment algorithm for 2D frictionless contact: Classical formulation and special cases. Comput Methods Appl Mech Eng 198, 3428–3451. https://doi.org/10.1016/J.CMA.2009.06.022

4335 | Extended brush model for use with gravelly terrain

Devon Cox *, P. Schalk Els *

Department of Mechanical and Aeronautical Engineering,
University of Pretoria, Pretoria, South Africa

* *Corresponding authors: u20458208@tuks.co.za, schalk.els@up.ac.za*

ABSTRACT

This paper presents the development of a physics-based brush tyre model extended for applications beyond conventional smooth surfaces. The research investigates limitations in existing tyre models when simulating vehicle dynamics on non-smooth, hard terrain. A comprehensive literature survey informed the methodological approach, which combines a physics-based brush model with empirical tread and soil models implemented in MATLAB. Initial validation results demonstrate promising performance, indicating that the combination of empirical soil characterization with the physics-based brush model effectively predicts tyre forces on non-smooth terrain. The model provides a framework for predicting traction limits in off-road conditions, expanding the simulation capabilities for vehicle dynamics on complex surfaces.

Keywords
Extended brush tyre model
Off-road Mobility
Agricultural tyres
Tyre-soil interaction
Model validation

1. Introduction

Accurate tyre models have always been a part of simulation in the context of vehicle dynamics. Compared to the models developed for hard smooth roads, there exists a fundamental lack of data and or models that can simulate tyres over rough and deformable terrain. It is important to emphasize the difficulty and financial cost of developing and verifying these models over the unimaginably wide range of operating conditions. Over the last few decades, considerable research has gone into developing and understanding tyre models, such as the influential Magic formula presented by Bakker et al. (1989), to simulate realistic operating conditions. A more generalized 3D enveloping was introduced (Stallman, 2013; Klein, 2016; Farroni and Timpone, 2016; Behera, 2020) and is widely used in simulation software such as MSC ADAMS (Automated Dynamic Analysis of Mechanical Systems) to simulate tyres accurately. On the other end of the spectrum, one has Finite Element (FE) models that can be extremely accurate in their predictions (Ben Turkia et al., 2019; Gast et al., 2021), but computationally expensive to run and costly to parameterise.
The brush model differs from these models in that it is partly physics-based, unlike the magic formula, but does not model the interaction between contact forces, frictional force and the physics of the viscoelastic polymers as the FE models do. The brush model takes a lumped approach in the modelling of rubber and a distributed approach in the modelling of contact forces, which provides a good medium for interpretability, evaluation and parameterisation cost. The general brush theory has been widely discussed in literature but the most notable is the LuGre (Lundt-Grenoble) model developed by De Wit et al. (1995), which has been extended, improved and investigated numerous times (Svendenius and Wittenmark, 2003; Deur et al., 2004; Tsiotras et al., 2004; Deur et al., 2005; Wu et al., 2011; Spike, 2013; Davari et al., 2015; Lu et al., 2019a; Romano et al., 2023; Stocco et al., 2024). However, a significant shortcoming identified in the literature, is that the models were primarily described theoretically or verified on smooth surfaces such as glass plates or asphalt surfaces. This paper attempts to shed some light on these shortcomings by investigating the brush model's validity on rougher surfaces. For this reason, the authors reference the recent locally lumped model developed by Zhang et al. (2024) as it lends itself to easier implementation and faster computations than the distributed LuGre model.
The LuGre brush theory is briefly summarised, after which the model developed by Zhang et al. (2024) is discussed. Parallels are drawn between the two models and other works in the field of friction modelling. The implementation methodology is summarised in Section 88 and brief results are shown in Section 90. Commentary is made on the parameter estimation in brush models and the difficulties faced which can be seen in Section 90.

2. Literature Survey

2.1. Brief Review of LuGre Brush Theory

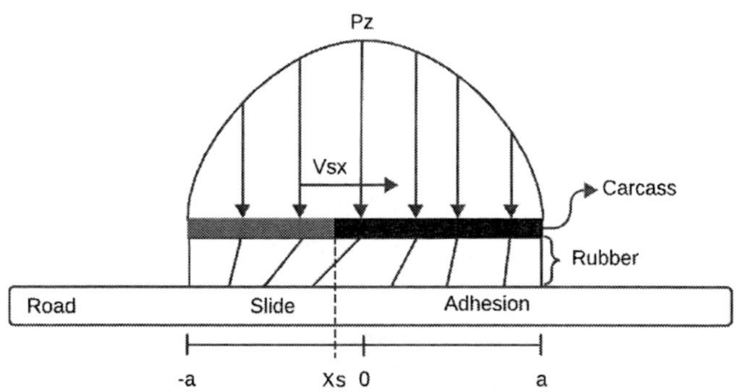

Fig. 1. Simple Brush model representation in 2D as described by Svendenius and Wittenmark, (2003). This representation can easily be extended to a two-dimensional contact patch.

The simplest of brush models tries to model the friction state between the road surface and the viscoelastic rubber found on the tyre tread blocks. The model describes the movement and interaction of brushes of infinitesimal length. As the tread block moves over the road surface, the brushes start displacing. This displacement in turn causes a force at the carcass of the tyre that is then transferred to the rest of the tyre.
As seen in Fig. 1, the brush model aims to model the deflection of brushes across the contact patch by taking the vertical pressure distribution and the relative velocity of the road w.r.t the tyre into account. The LuGre model aims to model these interactions through compact partial differential equations (PDEs), and the equations as summarised in Tsiotras et al. (2004), are given in Eq.s 1 through 5.

https://doi.org/10.56884/W40AXG3Q

$$\frac{dz_i(t,\zeta)}{dt} = \frac{\partial z_i}{\partial t} + |\omega r|\frac{\partial z_i}{\partial \zeta} \quad (1)$$

$$= v_{ri} + C_{0i}(v_r)z_i(t,\zeta), i = x, y \quad (2)$$

$$\mu_i(t,\zeta) = -\sigma_{0i}z_i(t,\zeta) - \sigma_{1i}\frac{\partial z_i(t,\zeta)}{\partial t} - \sigma_{2i}v_{ri}(t), i = x, y \quad (3)$$

$$F_i(t) = \int_0^L \mu_i(t,\zeta)f_n(\zeta)d\zeta\, , i = x, y \quad (4)$$

$$M_z(t) = -\int_0^L \mu_y(t,\zeta)f_n(\zeta)\left(\frac{L}{2} - \zeta\right)d\zeta \quad (5)$$

Here applicable boundary conditions are applied to ensure that the tyre has zero deflection at the leading and trailing edges of the contact patch.

2.2. Locally Lumped Model:

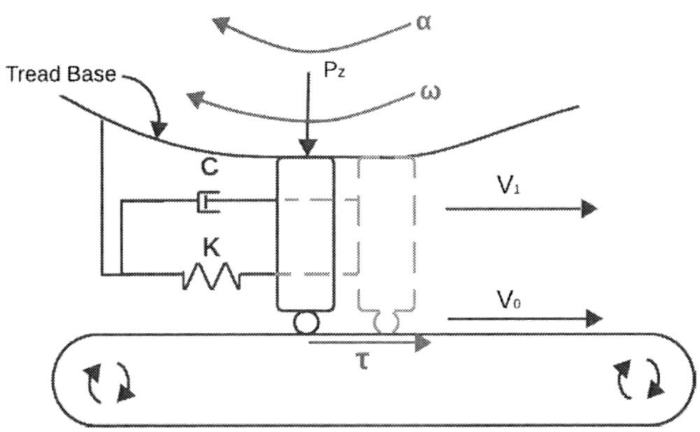

Fig. 2. Locally lumped brush model developed by Zhang et al. (2024). The wheel has an applied angular acceleration and velocity whilst the road has an associated velocity. The difference between the linearised wheel velocity and the velocity of the road element causes stresses to develop at the brush tip.

The author opted for a more recent implementation of the brush theory, as described by Zhang et al. (2024), for which the schematic can be seen in Fig. 2. The equations are summarised in Eq.s 6 through 11. The reader is encouraged to read Zhang et al. (2024) for a more in-depth explanation of the equations.

$$\boldsymbol{\tau}_{slide}(t) = \begin{cases} P_z\mu_d\cos(\theta_2) & for\ x\ direction \\ P_z\mu_d\sin(\theta_2) & for\ y\ direction \end{cases} \quad (6)$$

$$\boldsymbol{\tau}_{adh}(t) = \begin{cases} m\dfrac{dv_{rx}}{dt} + c_x v_{rx} + k_x\delta_x \\ m\dfrac{dv_{ry}}{dt} + c_y v_{ry} + k_y\delta_y \end{cases} \quad (7)$$

$$\boldsymbol{\tau}(t) = \begin{cases} \boldsymbol{\tau}_{adh} & if\ \sqrt{\tau_x^2 + \tau_y^2} \le \mu_0 P_z \\ \boldsymbol{\tau}_{slide} & otherwise \end{cases} \quad (8)$$

$$\mu_d = \left(\frac{P_z}{P_{ref}}\right)^{-q}\left[\mu_0 + (\mu_m - \mu_0)\exp\left(-\left\{h\log_{10}\left(\frac{\|\boldsymbol{v}_s\|}{v_m}\right)\right\}^2\right)\right] \quad (9)$$

$$\boldsymbol{F}(t) = \int_{\mathcal{P}} \boldsymbol{\tau}(t)dA \quad (10)$$

$$M_z(t) = -\int_{\mathcal{P}} \tau_y(t,x)\mathrm{COP_x}dA \quad (11)$$

Comparing the above equations to the ones presented in Eq.s 1 through 5, one can see that the LuGre model chooses a friction model for each direction, while Zhang et al. (2024) propose to model the brush dynamics in each direction. These two conventions are seen throughout the brush model literature where authors either must choose a friction model for the different directions or choose two sets of brush parameters for the different directions. This can be attributed to the velocity-dependent friction models and the inherent anisotropy of tyres. This is analogous to the magic formula tyre model, where there exists an entirely different set of parameters for the longitudinal, lateral and aligning moment models. The advantage of a brush model is that one must only parameterise once, and the forces and moments are extracted through the integration of the stress states. Here one can also see that by introducing a set of simpler ordinary differential equations (ODEs), logic for the two different states that the tyre can experience must be introduced. When developing a brush model, the author then has the choice to introduce the logic in the friction model or the brush dynamics. If the logic is introduced in the friction model, it leads to a discontinuity that lends itself to a less compact description and difficulties in solving the equations analytically.

2.3. Friction Model

Considering Eq. 3 and Eq. 7, one can see that they have similar forms. The model proposed by Zhang et al. (2024) aims to model the dynamics of the brushes with an analytical friction model, while the LuGre model aims to model the dynamics of the friction state rather than the brushes. The friction model proposed in Eq. 7 is a pressure-dependent and temperature-independent form of the friction master curve given in Kelly (2008). Expanding Eq. 2, one gets the following:

$$C_{0i}(v_r) = \frac{|v_r|\sigma_{0i}}{g(v_r)}, i = x, y \quad (12)$$

$$g(v_r) = \mu_0 + (\mu_m - \mu_0)\exp\left(-\frac{|v_r|}{v_s}\right)^{\gamma} \quad (13)$$

Here we can see a similar form of the friction model as described in Eq. 7. As discussed in section 2.1, simplifying the description leads to a discontinuity in the equations brought on by the added logic for the two different sliding conditions. This is where the LuGre model has an advantage, as the PDEs lead to a more elegant and continuous description of the friction coefficient, compared to Eq. 7. Zhang et al. (2024) describe the pressure dependence of the force generation that tyres exhibit and then include a saturation term to avoid excessively large values for the friction coefficient at low-pressure values.

2.4. Pressure Distribution

Generally, one finds a form of analytical quadratic pressure distribution, for which an example is given in Eq.s 14 - (18). The problem is that this is unrealistic for a general tyre, as they usually have some type of tread pattern, and especially for off-road tyres, as these usually have large lugs. Then there is the issue of the pattern that the road element transfers as this is usually random in nature and not easily modelled by the analytical quadratic expression. Although the pressure distribution usually takes a quadratic or trapezoidal shape to satisfy the boundary conditions, the interior is mostly random.

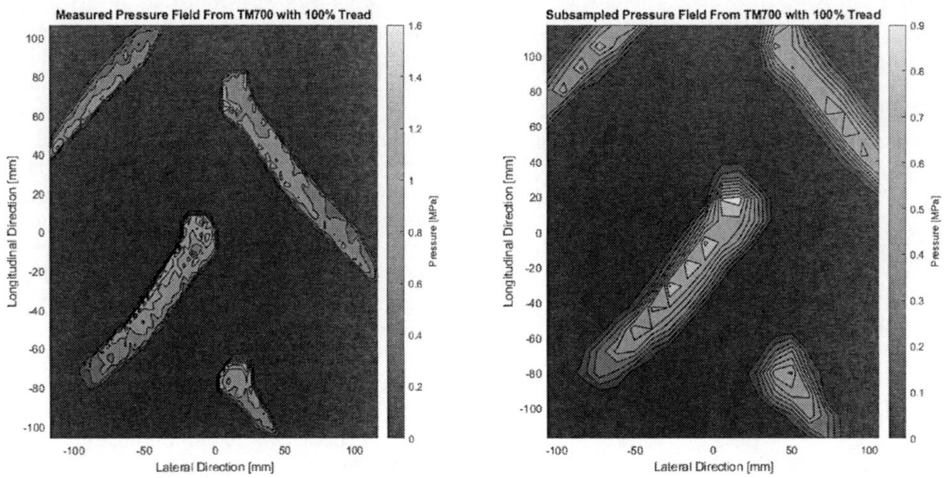

Fig. 3. Measured Pressure Distribution obtained for the Trelleborg TM700 agricultural tyre. On the left-hand side, one can see the full resolution obtained by the Tekscan pressure mat. On the right-hand side, one can see the subsampled pressure used for the model input.

$$f_z(x) = \frac{F_z}{2a} \cdot g_x\left(\frac{x}{a}\right) \tag{14}$$

$$g_i(\zeta_i) = \mathrm{A}_i\left(1 - \zeta_i^{2n_i}\right)\left(1 + \lambda_i \zeta_i^{2n_i}\right)\left(1 - \mathrm{B}_i \zeta_i\right), i = x, y \tag{15}$$

$$\mathrm{A}_i = \frac{(2n_i + 1)(4_{n_i} + 1)}{2n_i(4n_i + 1 + \lambda_i)}, i = x, y \tag{16}$$

$$\mathrm{B}_i = \frac{3(2n_i + 3)(4n_i + 3)}{(2n_i + 1)(4n_i + 1)(4n_i + 3 + 3\lambda_i)} \cdot \frac{i_e}{a}, i = x, y \tag{17}$$

$$P_z(x, y) = \frac{f_z(x)}{2w(x)} \cdot g_y\left(\frac{y}{w(x)}\right) \tag{18}$$

As mentioned, this distribution is good for testing the model numerically but not suitable for realistic simulations. Generally, The pressure distribution is measured in situ using a pressure-sensitive mat, such as the Tekscan products (Tekscan, 2024). This gives good results on the smooth terrain, that the brush model is usually used for, but for rougher surfaces, this is unfortunately not possible, as the sensors in the mat are sensitive and easily damaged. Another problem is the resolution of these mats, as they can be very coarse, and depending on the size of the wheel used, will need some imputation or interpolation scheme. There are relatively easy methods to measure the contact area on rougher surfaces, but explicitly measuring the pressure distribution remains a problem.

3. Methodology: Simulation

The author implemented the model proposed by Zhang et al. (2024) in MATLAB as there is a suite of built-in tools that can be accessed with minimal coding. Since the model describes the dynamics of each brush, the author opted for an efficient Object-Oriented Programming (OOP) approach. This allows one to organize methods and efficiently keep track of the brush properties, such as displacement, velocity and induced stresses. MATLAB is optimised for matrix and vector operations; therefore, one class instance can be used to store information about the whole contact patch. Since computation time is always important in numerical simulations, the author decided to make use of MATLAB's built-in code generation capabilities which builds the MATLAB code to C++, which runs more efficiently. This alone decreases the computation time by a factor of 2.

Since the model is effectively two models joined by one state that keeps track of whether the induced stresses are greater than dry friction allows for or not, the code was split into 2 sections. One section deals with the simpler adhesion region, and the second section deals with the more complex sliding region.

3.1. Adhesion Region:

For the adhesion region, the stresses are computed as in Eq. 7. The author opted for a 2nd order finite difference method to estimate the derivative of the relative velocity since one does not always have analytical inputs, so it does not make sense to take analytical derivatives of this function.

3.2. Sliding Region:

For the sliding region, one needs to calculate the sliding velocity at the current time step. This was done by solving the constraint equations given by Zhang et al. (2024). Since these are implicit equations, the author opted to discretise the equations, which led to the following:

$$v_{si,n} = v_{i,(n-1)} - v_{ri,n} \;, i = x, y \tag{19}$$

$$v_{s,n} = \sqrt{v_{sx,n}^2 + v_{sy,n}^2} \tag{20}$$

$$\theta_{1,n} = \operatorname{atan}\left(\frac{v_{sy,n}}{v_{sx,n}}\right) \tag{21}$$

$$\theta_{2,n} = \theta_{1,n} - \pi \tag{22}$$

This formulation allows one to compute the sliding velocities without the need for a root-finding algorithm, which adds computation time. Notice though, that one needs the deformation velocity $\boldsymbol{v}_{(\boldsymbol{n-1})}$. This property is calculated in the previous time step using a Velocity Verlet approach. Zhang et al. (2024) opted for an implicit Euler integration scheme, as this tends to underestimate the values and in turn does not cause the integration of long-time duration simulations to diverge numerically. The disadvantage of the implicit Euler scheme is that it requires a root-finding algorithm, usually a Newton-Raphson implementation, which again increases the computation time. By using a Velocity Verlet integration scheme, one decreases the global error to be of order 2, compared to order 1 for an Euler scheme, and foregoes the need for the use of an iterative solver. The Verlet integration method is usually considered in the graphics programming community as it has less error accumulation over time, compared to the Euler method. This is due to it being part of a class of geometric integrators called symplectic integrators that exhibit numerical conservation of the Hamiltonian, should the analytical one be conserved. The Velocity Verlet implementation is summarised below:

$$v_{i,\left(n-\frac{1}{2}\right)} = v_{i,(n-1)} + 0.5\, a_{i,(n-1)}\Delta t \ , \ i = x,y \tag{23}$$

$$\delta_{i,n} = \delta_{i,(n-1)} + v_{i,\left(n-\frac{1}{2}\right)}\Delta t + 0.5a_{i,(n-1)}\Delta t^2,\ i = x,y \tag{24}$$

$$a_{i,n} = \frac{1}{m}\left(\tau_{i,n} - k_i\delta_{i,n} - c_i v_{i,\left(n-\frac{1}{2}\right)}\right),\ i = x,y \tag{25}$$

$$v_{i,\left(n+\frac{1}{2}\right)} = v_{i,\left(n-\frac{1}{2}\right)} + 0.5\left(a_{i,n} + a_{i,(n-1)}\right)\Delta t,\ i = x,y \tag{26}$$

3.3. Pressure Distribution:

The model was investigated using two different pressure distributions, namely a uniform, quadratically shaped pressure distribution and measured distribution of a Trelleborg TM700 agricultural tyre, as seen in Fig. 3 (Becker and Els, 2022).

3.4. Model Inputs:

For much of the work so far, the author used a simple input of different angular velocities and a constant slip value. The velocity of the road element was then calculated using the percentage slip that was chosen. This was done to test and debug the model whilst it was being written. In an experiment, the slip value may not link angular velocity and road velocity due to road surface, temperature, tyre condition, and applied torque. Although the initial simulations may not be extremely realistic, they do provide confidence that the model should be able to handle more complex inputs, as the simulations ran smoothly without any unexpected behaviour. A simple parameterised smooth step function (Ebert et al., 2002) was used for the angular velocity input and can be seen in Eq. 27. This function was chosen for its smooth transition between the on and off states, which results in better numerical stability for the brush model simulation compared to the traditional step input. Three different values were chosen to demonstrate that the model can be used for a range of applications namely: 1 m/s, 5 m/s and 10 m/s.

$$smooth\ step(t) = \begin{cases} 0 & t < 0 \\ 6t^5 - 15t^4 + 10t^3 & 0 \le t \le 1 \\ 1 & 1 < t \end{cases} \tag{27}$$

To evaluate model performance on rougher terrains, the Trelleborg TM700 pressure distribution was used, due to its discrete, high-pressure locations. These discrete locations represent a similar condition to a smooth tyre contacting a rough substrate. To use the TM700 pressure distribution, it had to be spatially resampled to the same resolution as the brush model. After running a sensitivity study on the brush resolutions, it was found that a brush model resolution of 20x20 (400 brushes) achieves a good compromise between solution times and model convergence.

An initial guess for the model parameters was used, as the parameters proposed by Zhang et al. (2024) in their Grosch wheel model, resulted in numerically unstable results if the time step was not set to extremely small values such as $\Delta t \le 10^{-6}s$. Considering that one would like to parameterise the model using available data, this is unreasonable as most of the data the author has access to has been measured with a sampling frequency of 1000Hz, which results in a timestep of $\Delta t = 10^{-3}s$. These findings suggest that, while the absolute magnitudes may not be quantitatively representative of a realistic tyre, the model remains valid as a proof of concept by capturing the trends one expects. A later discussion on parameter identification follows, but due to time constraints, the parameterised model could not be validated with

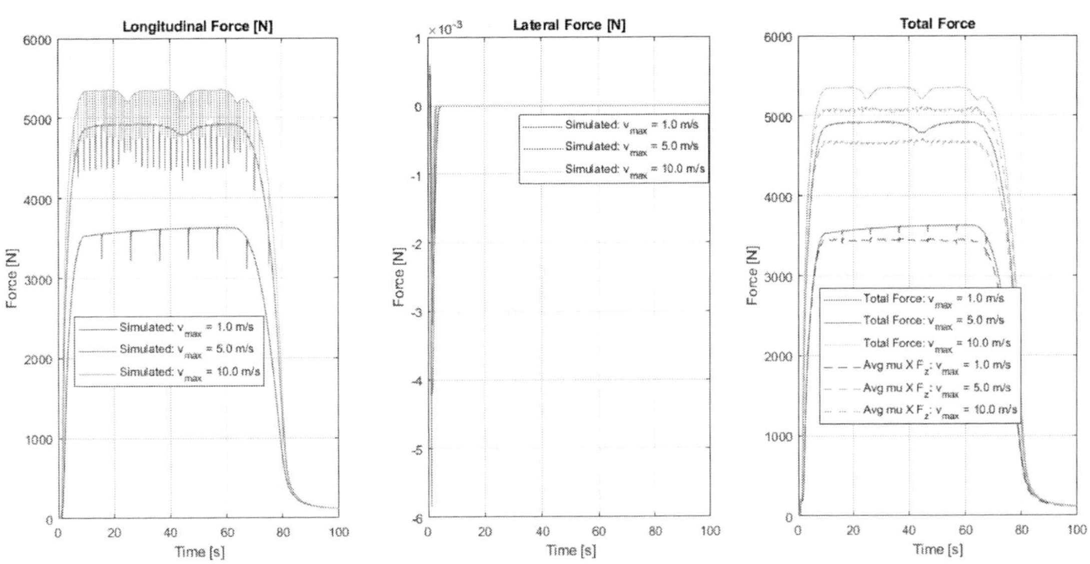

Fig. 4. Rolling tyre simulations using a uniform pressure distribution on a smooth surface. The higher speeds result in larger sliding velocities which induce larger forces. It should be noted that a constant slip ratio was used, which is quite unrealistic, but shows the trend for the outputs one can expect. In a realistic sense, one would ideally measure or estimate the velocity of the road element.

dynamically measured tyre data. This is discussed further in Section 6.

4. Results and Discussion: Simulation

4.1. Rolling Tyre test:

As mentioned in Section 3 above, the rolling tests consisted of three speeds. The range was used with a uniform and a more realistic measured pressure distribution. The results can be seen in Fig. 4. From these results, one can see that the higher speeds result in higher forces, this is due to the higher sliding velocities which results in a higher friction coefficient. Take note that these simulations were run with a constant slip ratio of 10%, meaning that the road element has a linear velocity 10% lower than that of the wheel. This is quite unrealistic and ideally, one would measure or estimate either the slip ratio or the velocity of the road element. The measurement can be done with a simple Laser Surface Velocimeter (LSV), which is frequently used in automotive testing. Considering the Total Force graph, one can notice that the conventional notion of the friction force as a function of friction coefficient and vertical force results in a slightly lower value for the force. The author attributes this to the unrealistic nature of assigning a single slip ratio value for the entirety of the simulation, which is a dynamic property depending on external factors, which in turn means that the model output for force will be higher as the brushes displace more. Considering the Longitudinal Force graph, one can see high and low-frequency intermittent peaks. The high-frequency peaks are an artefact of the numerical simulation and the author attributes this to the assigning of a fixed slip ratio, as in reality, the tyre would have a smoother transition into and out of the contact patch. The low-frequency undulations are due to the pressure distribution being interpolated across the domain to simulate a tyre rolling. This was done for pure numerical convenience, as it proved difficult to simulate the movement of the individual brushes.

4.2. Trelleborg TM700 Simulations:

From the TM700 rolling simulation results, it was concluded that a similar trend to the tyre with a uniform pressure distribution was found, indicating that the model can successfully take a more complex measured pressure distribution as an input. This shows potential in that the model can be used to simulate more complex, gravelly terrain which is the goal of this model.

4.3. Sliding Tyre Test:

In addition to rolling simulations, simulations with the tyre sliding were investigated. The tyre angular velocity was set to zero, whilst the road element still had some velocity. Again, three different values for the road velocity were chosen.

The author decided to model the tyre sliding in the interest of parameter estimation which is inherently part of the modelling workflow. Since it is physically and financially less expensive to conduct static testing of tyres, sliding tyre data is more abundant than dynamically measured data. The fact that the tyre is sliding, will also induce more deflection from the brushes, which in theory, should make estimating the parameters for the model easier as the dynamics for more easily excited.

5. Parameter Identification

5.1. Statistical Outlook:

For a model to be useful, the model parameters are required. In the interest of offline parameter estimation, an optimisation-based approach was used. Making use of a simple Bayesian inference approach, one defines a likelihood function over the measured data and a prior distribution over the parameters. This allows one to calculate the Maximum A Posteriori (MAP) estimate by minimising the negative log of the resulting probability distribution. Assuming the measurement noise takes a Gaussian distribution, and that the parameters are also normally distributed one can calculate the covariance of the posterior and therefore the covariance of the predictive distribution (Bishop and Nasrabadi, 2006), which gives the likelihood of the specific model based on the data used. This is useful in quantifying the uncertainty in the parameters.

$$\mathcal{L} = -\frac{1}{2}\sum_{i=1}^{N}(x_i - y_i)^T \Lambda (x_i - y_i) - \frac{1}{2}(\beta - \mu_\beta)^T \Lambda_\beta (\beta - \mu_\beta) - \frac{N}{2}\log(2\pi) - \frac{1}{2}\log(2\pi) - \frac{ND}{2}\log(|\Lambda^{-1}|) - \frac{K}{2}\log(|\Lambda_\beta^{-1}|) \quad (28)$$

The loss function is defined Eq. 28, which corresponds to $\log(P(\beta|Y))$ when a Gaussian Likelihood and Gaussian Prior are used. For this to be effective, one generally needs an informative prior, which is usually difficult to define since the parameters, or their scale is unknown beforehand. When the mean vector, for the parameters, is chosen as the zero vector, this promotes smaller parameter values, which is generally preferred in parameter and system identification (Brunton and Kutz, 2019).

5.2. Optimisation Method:

Generally, one would prefer a gradient-based optimisation procedure, as this results in fast convergence. The built-in MATLAB optimisation toolbox uses a 2nd order method known as the BFGS or L-BFGS method, which approximates the Hessian matrix w.r.t the parameter vector. This method guarantees faster convergence than first-order methods but is computationally more expensive.

After extensive experimentation and data preparation, a meta-heuristics-based Particle Swarm Optimisation (PSO) algorithm was decided to be the best, similar to Lu et al. (2019b), which opted for a variant of the widely used genetic algorithm known as GA-PS (genetic algorithm and pattern-search). The PSO algorithm is less likely to get stuck in local minima as there are several particles scattered throughout the parameter space. This makes finding the global minimum more likely. It was found that the PSO algorithm converged faster than the BFGS-based one, as it was not necessary to implement a finite difference scheme at every iteration. The PSO algorithm can be a constrained optimisation algorithm, which simplifies the proposed loss function in Eq. 28, as usually one would need to define a strictly positive distribution over the parameters, which in turn complicates the calculation of the posterior predictive covariance. The most common implementation of PSO is to sample the particle locations from a uniform distribution, but a more efficient way to sample a high dimensional space is known as Latin Hypercube Sampling (LHS). By implementing this method, one can reduce the computation time significantly as the particle locations are more space-filling and do not oversample a specific area which can be the case with

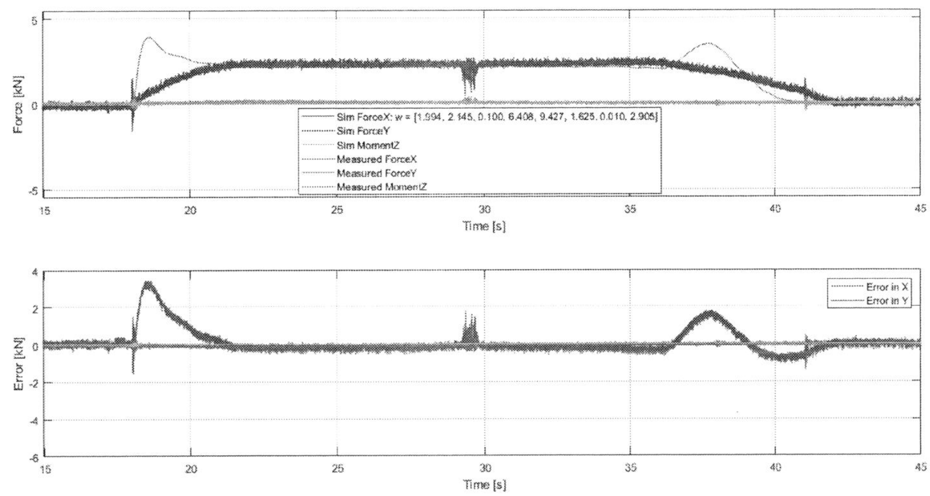

Fig. 5. Result from the parameter estimation using experimental data gathered on a Static Tyre Test Rig (STTR). The top graph shows the model output in blue, red and yellow, while the experimental data is in purple green and light blue. The model parameters are summarised in Table 1

sampling from a uniform distribution (Chen et al., 2013; Afzal et al., 2017).

5.3. Discussion and Conclusion

Data measured on a static tyre test rig (STTR) (Becker and Els, 2022) was used in the parameter identification study. The results after 20 iterations can be seen in Fig. 5. One can see that with the estimated parameters; the model output follows the same trend as the experimental data. The initial and final parameters used are summarised in Table 1. From Fig. 5, one can see that the model accurately captures the steady state behaviour of the tyre. In the transient state, one can see a significant overshoot of the model. This can be attributed to the optimisation procedure not having enough iterations. As the iterations are increased, it is believed that the model will converge to a more realistic solution.

Table 1
Results from parameter estimation.

Parameters	Initial Guess	Final Estimated
k_x [N/mm³]	6.865	1.994
k_y [N/mm³]	5.177	2.145
c_x [Ns/mm³]	0.654	0.100
c_y [Ns/mm³]	0.215	6.408
m [kg/mm²]	4.942	9.427
h [-]	0.999	1.625
q [-]	0.114	0.010
v_m [mm/s]	3.477	2.905

Since only longitudinal data was used in this parameter identification study, there is a lot of uncertainty pertaining to the correctness of the parameters in the y-direction. Although only longitudinal data was considered in this study, the parameter identification study proved that it is possible to estimate the model parameters from sliding data gathered using an STTR. This has the potential to significantly reduce the time required to parameterise a tyre model.
In conclusion, the gradient-free optimisation procedure has been proven to be sufficiently accurate in estimating the brush model parameters from experimental data. The hyperparameters for the optimisation procedure and the brush model must be tuned to improve the accuracy, but the model is more than capable of predicting trends in either sliding or rolling simulations.

6. Conclusions and future work

This paper presented an implementation of an extended brush model developed by Zhang et al. (2024). The use of brush models has seen an increase as the computational power of computers increases. They have been used to understand the complex phenomenon that occurs while a tyre is rolling and deflecting, to aid in the investigation of tyre efficiencies (Davari et al., 2015) and how to improve them. This shows that the brush model is still an applicable physics-based modelling technique that researchers can use in understanding the complex and partially unknown physics of tyre dynamics.
Although the brush model is widely used, there exists a fundamental lack of implementation on rougher or off-road conditions in literature. This paper presented the initial results of the brush model being implemented with more complex pressure distributions, which showcases the model's ability to capture complex tyre dynamics. A discussion on parameter estimation in brush models was included and the difficulty in using traditional gradient-based learning methods was addressed by making use of a more general meta-heuristic approach. All of this was implemented in MATLAB which made the programming easier and execution times were optimised by making use of the built-in code generation capabilities.

7. Nomenclature

A	Contact area	[mm²]
A_i	Pressure Shape Constant 1	[-]
$\boldsymbol{a}$	Deformation acceleration	[m/s²]
a	Halfwidth of the contact patch	[mm]
B_i	Pressure Shape Constant 2	[-]
b	Half-length of the contact patch	[mm]
c_i	Brush Damping Coefficient	[Ns/mm³]

C_{0i}	Steady-state friction function	[-]
D	The dimensionality of the measured signal	[-]
F	Force	[N]
g	Friction coefficient function	[-]
h	Friction master curve shape parameter	[-]
k_i	Brush Spring Constant	[N/mm³]
$\mathcal{L}$	Loss function	[-]
M_i	Moment	[Nm]
m	Brush Inertia	[kg/mm²]
N	Number of data points in the signal	[-]
P_z	Vertical Pressure	[MPa]
$\boldsymbol{v_r}$	Relative velocity	[mm/s]
$\boldsymbol{v_s}$	Sliding velocity	[mm/s]
$\boldsymbol{v}$	Deformation velocity	[mm/s]
w	Halfwidth function	[mm]
z	Internal friction state	[-]
$\boldsymbol{\alpha}$	Angular acceleration	[rad/s²]
$\boldsymbol{\beta}$	Parameter vector	[-]
$\boldsymbol{\delta}$	Brush deflection	[mm]
$\boldsymbol{\Delta}$	Difference operation	[-]
ζ	Local coordinates	[m]
θ_1	The angle between the sliding velocity vector and the positive x-axis	[rad]
θ_2	The angle between the shear stress vector and the positive x-axis	[rad]
λ	Pressure shape constant 2	[-]
Λ	Probability distribution precision	[-]
μ	Friction coefficient	[-]
π	Transcendental constant	[-]
$\sigma_0, \sigma_1, \sigma_2$	The inverse of transient stiffnesses for the LuGre model	[m/N]
$\boldsymbol{\tau}$	Shear stresses developed in contact patch	[MPa]
$\boldsymbol{\omega}$	Angular velocity	[rad/s]

8. Acknowledgements

This research is part of a larger project funded by the South African Department of Science and Innovation (DSI). Funds were received through the Council for Scientific and Industrial Research (CSIR) as part of their Capability-building initiative in Artificial Intelligence (AI) and Robotics ("CSIR robotic platform supports further research at local universities | CSIR," n.d.).

9. Declaration of competing interest

The authors declare that they have no known competing financial interests or personal relationships that could have appeared to influence the work reported in this paper.

10. References

Afzal, A., Kim, K.-Y., Seo, J., 2017. Effects of Latin hypercube sampling on surrogate modelling and optimization. International Journal of Fluid Machinery and Systems 10, 240–253.

Bakker, E., Pacejka, H., Lidner, L., 1989. A New Tire Model with an Application in Vehicle Dynamics Studies. Autotechnologies 1989 Proceedings 98, 13. https://doi.org/10.4271/890087

Becker, C., Els, S., 2022. Effect of surface roughness on tyre characteristics. Journal of Terramechanics 102, 27–48.

Behera, A., 2020. Improved accuracy of tyre obstacle enveloping model. Thesis for: Master's in automotive technology, 2020. Doi: 10.13140/RG. 2.2

Ben Turkia, S., Wilke, D.N., Pizette, P., Govender, N., Abriak, N.-E., 2019. Benefits of virtual calibration for discrete element parameter estimation from bulk experiments. Granular Matter 21, 1–16.

Bishop, C.M., Nasrabadi, N.M., 2006. Pattern recognition and machine learning. Springer.

Brunton, S.L., Kutz, J.N., 2019. Data-Driven Science & Engineering.

Chen, R.-B., Hsieh, D.-N., Hung, Y., Wang, W., 2013. Optimizing Latin hypercube designs by particle swarm. Statistics and computing 23, 663–676.

CSIR robotic platform supports further research at local universities | CSIR [WWW Document], n.d. URL https://www.csir.co.za/csir-robotic-platform-supports-further-research-local-universities (accessed 5.13.25).

Davari, M.M., Jerrelind, J., Trigell, A.S., Drugge, L., 2015. A multi-line brush-based tyre model to study the rolling resistance and energy loss. Presented at the Proceedings of 4th International Tyre Colloquium: Tyre Models for Vehicle Dynamics Analysis, Guildford, UK.

De Wit, C.C., Olsson, H., Astrom, K.J., Lischinsky, P., 1995. A new model for control of systems with friction. IEEE Transactions on Automatic Control 40, 419–425.

Deur, J., Asgari, J., Hrovat, D., 2004. A 3D brush-type dynamic tyre friction model. Vehicle System Dynamics 42, 133–173.

Deur, J., Ivanović, V., Troulis, M., Miano, C., Hrovat, D., Asgari, J., 2005. Extensions of the LuGre tyre friction model related to variable slip speed along the contact patch length. Vehicle System Dynamics 43, 508–524.

Ebert, D.S., Musgrave, F.K., Peachey, D., Perlin, K., Worley, S., 2002. Texturing and modelling: a procedural approach. Elsevier.

Farroni, F., Timpone, F., 2016. A test rig for tyre envelope model characterization. Engineering Letters 24, 45–49.

Gast, R.G., Els, P.S., Kok, S., Wilke, D.N., Botha, T.R., 2021. Identifiability of tyre force contact prediction from deformation measurements. Presented at the MATEC Web of Conferences, EDP Sciences, p. 00027.

Kelly, D.P., 2008. Lap time simulation with transient vehicle and tyre dynamics.

Klein, P., 2016. A 3D envelopment procedure for tyre belt radiated noise level prediction, in: INTER-NOISE and NOISE-CON Congress and Conference Proceedings. Institute of Noise Control Engineering, pp. 3823–3834.

Kruger, R., Els, P.S., Hamersma, H.A., 2023. Experimental investigation of factors affecting the characterisation of soil strength properties using a Bevameter in-situ plate sinkage and shear test apparatus. Journal of Terramechanics 109, 45–62. https://doi.org/10.1016/j.jterra.2023.06.002

Lu, Y., Zhang, J., Yang, S., Li, Z., 2019a. Study on improvement of LuGre dynamical model and its application in vehicle handling dynamics. Journal of Mechanical Science and Technology 33, 545–558.

Romano, L., Bruzelius, F., Jacobson, B., 2023. An extended LuGre-brush tyre model for large camber angles and turning speeds. Vehicle System Dynamics 61, 1674–1706.

Spike, J., 2013. Adaptive Tire Model for Dynamic Tire-Road Friction Force Estimation. University of Waterloo.

Stallman, M.J., 2013. Tyre model verification over off-road terrain. University of Pretoria.

Stocco, D., Biral, F., Bertolazzi, E., 2024. A physical tyre model for real-time simulations. Mathematics and Computers in Simulation 223, 654–676.

Svendenius, J., Wittenmark, B., 2003. Brush tyre model with increased flexibility. Presented at the 2003 European Control Conference (ECC), IEEE, pp. 1863–1868.

Tekscan, 2024. TireScan [WWW Document]. URL https://www.tekscan.com/products-solutions/systems/high-speed-tirescan (accessed 5.15.25).

Tsiotras, P., Velenis, E., Sorine, M., 2004. A LuGre tyre friction model with exact aggregate dynamics. Vehicle System Dynamics 42, 195–210.

Wu, X.D., Zuo, S.G., Lei, L., Yang, X.W., Li, Y., 2011. Parameter identification for a LuGre model based on steady-state tyre conditions. International Journal of Automotive Technology 12, 671–677.

Zhang, M., Unrau, H.-J., Gießler, M., Gauterin, F., 2024. A detailed tyre tread friction model considering dynamic friction states. Tribology International 193, 109342.

4342 | Cornering performance of rigid wheel in granular media using coarse-scale DEM models

Aidan Dickerson [a], Bohumir Jelinek [a,*], George L. Mason [a], Thomas Skorupa [b], Michael Cole [b], Jody D. Priddy [c]

[a] *Center for Advanced Vehicular Systems, Mississippi State University, Mississippi State, Mississippi, United States*

[b] *U.S. Army Combat Capabilities Development Command Ground Vehicle Systems Center, Warren, Michigan, United States*

[c] *U.S. Army Engineer Research and Development Center, Geotechnical and Structures Laboratory, Vicksburg, Mississippi, United States*

* *Corresponding author: bj48@cavs.msstate.edu*

ABSTRACT

Understanding interactions of wheels and tracks with granular media under variable loading conditions, including longitudinal and side slip, is critical for prediction of mobility for wheeled and tracked vehicles in off-road environments. The discrete element method (DEM) is routinely used for modeling interaction of soil with track and tires, but the method's accuracy needs to be better established.

In this work, two DEM models from the Generic EDEM Material Model (GEMM) database from Altair®'s EDEM™ software package, which were identified (Jelinek et al., 2025) as the best match to physical experiments, are used to calculate tractive performance of a rigid wheel in sand under braked, towed, and powered conditions with side slip. The simulations follow the experiments by Shinone et al. (2010), examining a 165/60R13 wheel with constant circumferential velocity of 97.6 mm/s and vertical contact load of 980 N operating in powered conditions under forward slips in the range of -5.9% to 54.8%. The steady-state tractive forces on the wheel in sand were evaluated under the same forward slip conditions as in the experiment, except with added side slip of 3, 6, and 12 degrees. The side-slip effects on net traction, gross traction, and sinkage of the wheel in sand are evaluated by comparison with forward-slip-only results, as well as analytical solutions for lateral forces. Lateral forces and overturning moments were then calculated and compared with relevant relationships derived from physical measurement and other simulation methods. The results indicate a good match to experimental trends, encouraging further use and calibration of the DEM soil-wheel interaction models.

Keywords
Off-road mobility
Tractive performance
Discrete element method
Rigid wheel
Side slip

1. Introduction

Prediction of vehicle mobility for vehicles in off-road environment presents a challenge for agriculture, construction, military, and space exploration. The subgrade materials that the off-road vehicle operates on may include rock, gravel, sand, silt, clay soil, or their mixtures with various degrees of saturation. DEM-based numerical models of soils provide truthful physical characterization of particulate nature of soil subgrade; however, due to large number of particles in fine grain soils, a compromise need to be found between sufficiently accurate representation of the particulate nature of the soil and capabilities of the high-performance computing.

Multiple studies have used DEM to characterize the interaction of granular subgrades with rigid wheels, nevertheless only a few of them include side slip. The tractive performance of off-road wheels on dry sand including steering was analyzed by Du et al. 2017a, 2017b, 2018. A DEM simulation along with comparison with analytical results were presented (Hu et al., 2021). These models, however, require careful calibration (Coetzee, 2017; Zeng et al., 2020a, 2020b), and, in many cases, non-spherical particles are used to better match the experiments (Coetzee, 2020).

This work presents recent progress on extension of our previous effort (Jelinek et al., 2025) to side slip. The goal is to assess capability of coarse grain DEM models to predict lateral forces and torques acting on a smooth wheel operating in dry sand under braked and powered conditions with side slip.

2. Discrete element method background

A particle bed to simulate wheel-soil interactions was generated using particle-factory functionality of the Altair®'s EDEM™ software. The particle bed was assembled from five identical material blocks repeated along the longitudinal direction. Only the particles in a single block were settled under periodic boundary conditions to lower the simulation time needed to populate the bed with particles. A smooth-surface cylindrical wheel without tread was constructed from triangular facets. Figure Fig. 1 illustrates the wheel and enclosing box while Table 1 specifies their dimensions.

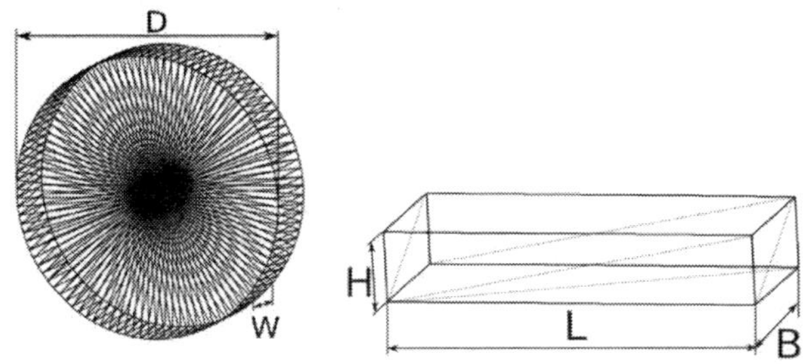

Fig. 1. Triangular facet representation of the wheel and enclosing box

Table 1
Wheel and Box Dimensions

Property	Symbol	Value	Units
Wheel diameter	D	53.49	cm
Wheel width	W	16.99	cm
Wheel weight	F_G	980.33	N
Box length	L	301.5	cm
Box width	B	48.0	cm
Box depth	H	60.5	cm
Soil depth	h	60	cm
Number of particles	N	97152	-

https://doi.org/10.56884/FV0FI0QP

All particles have a multi-spherical shape. Individual spheres had a radius of 9.5mm with a fixed overlap. Fig. 2 visualizes the multi-sphere particles.

Fig. 2. Multi-spherical particle shape

The Hertz-Mindlin (no slip) and Hertz-Mindlin JKR (Johnson et al., 1971) models were used to represent DEM contact interactions. The GEMM database does not include material models with elasto-plastic properties such as Edinburgh Elastic-Plastic-Adhesive models (Jones, 2003; Jones et al., 2004; Thakur et al., 2014), nor Type C rolling friction models (Ai et al., 2011). Hertz-Mindlin (Mindlin and Deresiewicz, 1953) is the default model used in EDEM due to its accurate and efficient force calculation. The Hertz-Mindlin JKR model was used for particles with a non-zero JKR coefficient. For Hertz-Mindlin, the normal force is a function of normal overlap δ_n:

$$F_n = \frac{4}{3} E^* \sqrt{R^*} \delta_n^{\frac{3}{2}} \tag{1}$$

where the E^* is the equivalent Young's Modulus and R^* is the equivalent radius. They are defined as:

$$\frac{1}{R^*} = \frac{1}{R_i} + \frac{1}{R_j} \tag{2}$$

$$\frac{1}{E^*} = \frac{(1 - v_i^2)}{E_i} + \frac{(1 - v_j^2)}{E_j} \tag{3}$$

with E^i, v^i, R^i, and E^j, v^j, R^j being the Young's Modulus, Poisson ratio, and radius of each sphere in contact. Additionally, there is a damping force, $F_n{}^d$, given by:

$$F_n^d = -2\sqrt{\frac{5}{6}} \beta \sqrt{S_n m^*}\, \boldsymbol{v}_n^{rel} \tag{4}$$

$$m^* = \left(\frac{1}{m_j} + \frac{1}{m_i}\right)^{-1} \tag{5}$$

$$\beta = \frac{-\ln e}{\sqrt{ln^2 e + \pi^2}} \tag{6}$$

$$S_n = 2E^* \sqrt{R^* \delta_n} \tag{7}$$

where m* is the equivalent mass, $v_n{}^{rel}$ is the normal component of the relative velocity, β, and S_n (the normal stiffness). The constant e is the coefficient of restitution, the tangential force, F_t, depends on the tangential overlap δ_t and the tangential stiffness S_t.

$$F_t = -S_t \delta_t \tag{8}$$

$$S_t = 8G^* \sqrt{R^* \delta_n} \tag{9}$$

Here, G^* is the equivalent shear modulus. Additionally, tangential damping is defined as:

$$F_t^d = -2\sqrt{\frac{5}{6}} \beta \sqrt{S_t m^*} \boldsymbol{v}_t^{rel} \tag{10}$$

where v_t^{rel} is the relative tangential velocity. The tangential force is limited by Coulomb friction, $\mu_s F_n$, where μ_s is the coefficient of static friction. The rolling friction (Sakaguchi et al., 1993) is implemented by applying torque

$$\tau_i = -\mu_r F_n d_i \Omega_i \tag{11}$$

to the contacting objects, where μ_r is the coefficient of rolling friction, d_i is the distance of the i-th contact point from the center of mass and Ω_i is the unit angular velocity vector of the object at the contact point.

The Hertz-Mindlin contact model with JKR cohesion (Johnson et al., 1971) allows modelling of cohesive materials. It calculates the normal force as a function of overlap δ_n and the surface energy parameter γ as follows:

$$F_{JKR} = -4\sqrt{\pi \gamma E^*} a^{3/2} + \frac{4E^*}{3R^*} a^3 \tag{12}$$

$$\delta_n = \frac{a^2}{R^*} - \sqrt{\frac{4\pi\gamma a}{E^*}} \tag{13}$$

where E^* is the equivalent Young's modulus, a is contact overlap radius, and R^* is the equivalent radius from Eq. 2. Wheel slip and circumferential velocity were used as an initial condition to each system. Wheel slip is defined as:

$$i = 1 - \frac{v_f}{\omega r} \tag{14}$$

Equation 14 defines i as wheel slip in terms of the wheel forward velocity v_f, wheel angular speed ω, and the wheel radius $r=D/2$. Post-processing values such as net traction and gross traction were calculated using Eq. 15 and 16 respectively. Sinkage amount was based on the initial position of the suspended wheel.

$$F^x = \sum_j F_j^x \tag{15}$$

$$T = \sum_j F_j^t d_j \tag{16}$$

F^x is the forward component of the contact force, F^t is the wheel-surface-tangential component of the contact force, D is the wheel diameter and d_j is the distance from the contact point to the wheel axis. F^t has negative/positive sign if the force acts to increase/decrease wheel rotation.

DEM nomenclature is summarized in Section 7.

Table 2
Particle properties for two DEM models selected from GEMM database due to their best match to measurements from Shinone et al. (2010).

DEM Mat. #	ν	ρ	G	e^p	μ_s^p	μ_r^p	γ^p	e^g	μ_s^g	μ_r^g	γ^g
1311	0.25	3300	10	0.55	0.68	0	0	0.5	0.45	0.15	0
1111	0.25	3300	10	0.55	0.44	0	0	0.5	0.45	0.15	0

*a superscript of p implies particle-particle interaction while g is a particle-geometry interaction

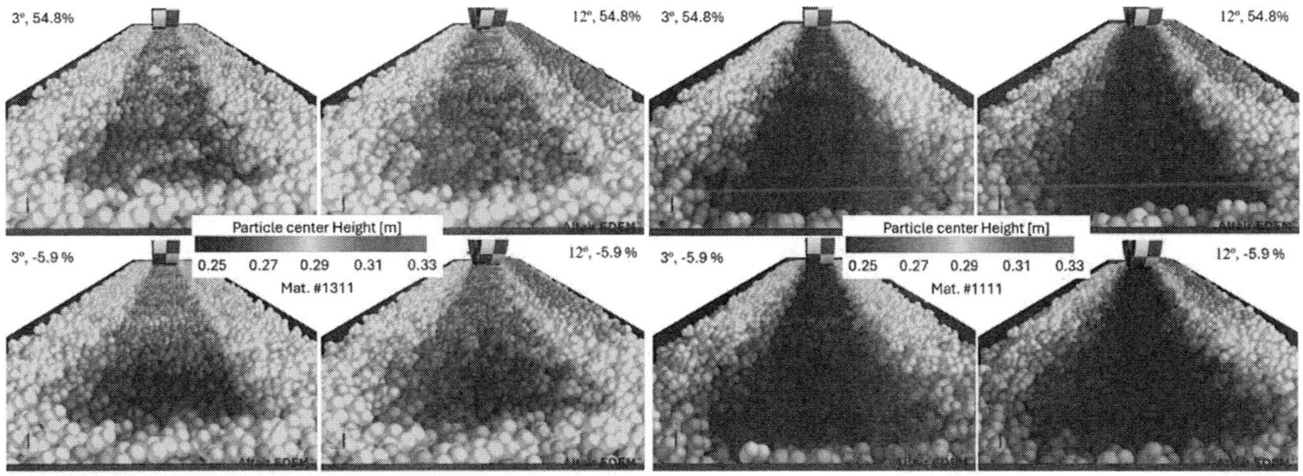

Fig. 3. Final configurations of the wheel at 54.8% slip (upper row) and -5.9% slip (lower row) with 3° (odd columns) and 12° (even columns) side slip and particle bed material # 1311 (left half) and 1111 (right half).

3. DEM material model selection

In Jelinek et al. (2025), ten DEM models from the GEMM database were selected and evaluated for their ability to match the experimental measurement of tractive performance of a wheel in dry sand performed by Shinone et al. (2010). GEMM database provides three selection criteria to select models with desired material properties. These are 1) particle size, 2) bulk density, and 3) an angle of repose. To match material characterization of the dry sand, the following material-selection criteria were applied in Jelinek et al. (2025): 1) the smallest available particle size, 2) a bulk density range of 1500-2000 kg/m^3, and 3) an angle of repose of 34°. Given these three selection criteria, the GEMM database returns ten materials. The evaluation in Jelinek et al. (2025) found that two out these ten materials, the DEM models with GEMM reference numbers 1311 and 1111, resulted in the closest match to tractive performance measurements by Shinone et al. (2010). These two materials differ only in a coefficient of static friction, and, similarly to dry sand, they don't exhibit a significant cohesion. They were selected in the present study for further evaluation of their ability to predict tractive performance of a wheel with side slip. The present study did not perform any direct calibration of the DEM materials, but instead selected the two materials with the best match (Jelinek et al. 2025) to forward-slip-only tractive performance measurements performed by Shinone et al. (2010).

4. Simulation results

The DEM simulation procedure follows the experiment by Shinone et al. (2010), with only difference being that side slip angles of 3°, 6°, and 12° were prescribed. Final configurations of the wheel from simulations at -5.9% and 54.8% forward slip with 3° and 12° side slips for the two materials are shown in Fig. 3. It is evident that material number 1111 with lower static friction coefficient resulted in higher wheel sinkage than material 1311 with higher static friction coefficient.

4.1. Side slip effects

Figure Fig. 4 compares drawbar pull (F_x), gross tractive effort (M_y/R) and sinkage results for materials 1311 and 1111 at varying side slip. *X*-axis of the coordinate system points forward, *y* points left, and *z* points up. The forward component of the drawbar pull F_x decreases with increased side slip. Material 1111, with static friction lower than material 1311, produces less forward pull and more sinkage at each side slip. Interestingly, gross tractive effort is higher for lower-friction material 1111, pointing out even higher losses when wheel operates with higher sinkage. The most significant difference in all three variables is seen when the side slip changes from 6° and 12°.

Figure Fig. 5 compares lateral force (F_y), overturning moment (M_x/R) and self-aligning moment (M_z/R) for the two materials. Side slip force F_y increases with side slip angle, as expected. The same is true for the magnitude of overturning torque M_x and for self-aligning torque M_y, with self-aligning torque having much lower value than overturning torque. A significant change is seen in all three variables even when the side slip changes from 0° to 3°.

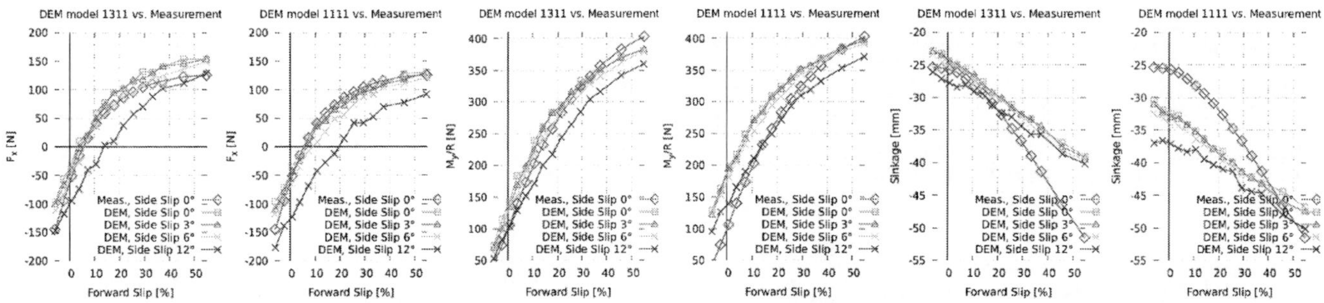

Fig. 4. Drawbar pull, gross tractive effort, and sinkage from DEM models 1311 and 1111 at varying side slip, along with measurement values at zero side slip.

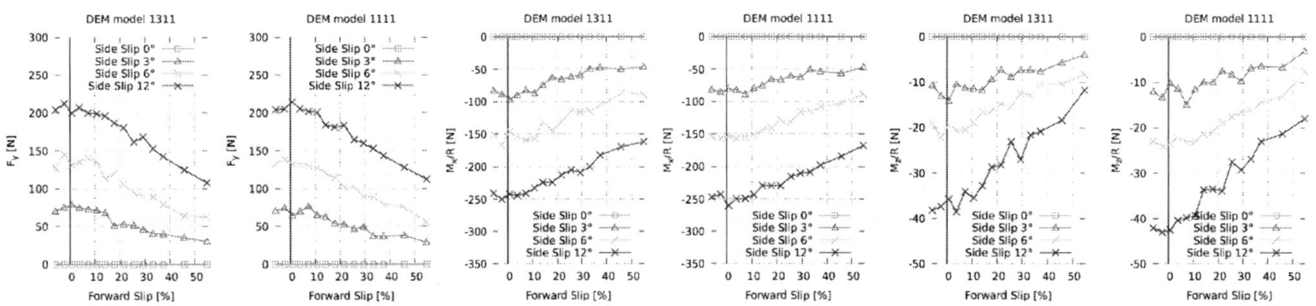

Fig. 5. DEM simulation results for lateral force, self-aligning torque, and overturning torque for varying side slip, showing increase in magnitude with increasing side slip as expected.

4.2. Stress profiles and relative velocities along the wheel-soil contact patch

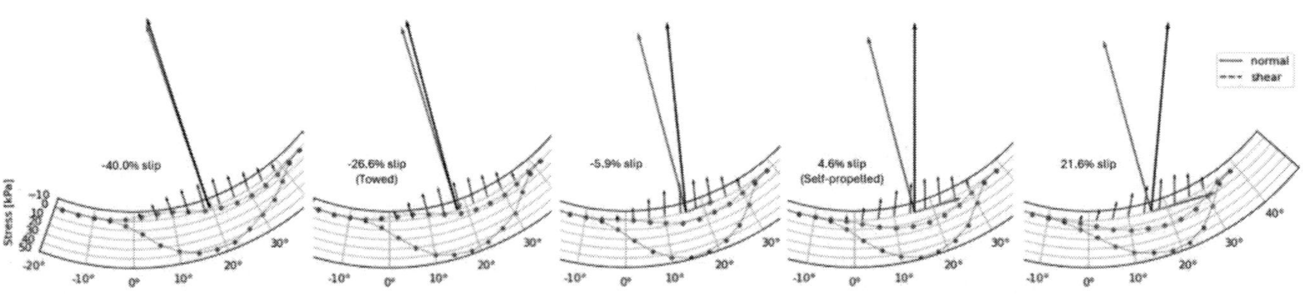

Fig. 6. Normal and shear stress profiles along the wheel-soil contact patch for straight-driving wheel in DEM material 1111

Figure Fig. 6 shows normal and sheer stress profiles along the wheel-circumference contact with soil for three prescribed slips (-40%, -5.9%, and 21.6%) by forcing decreasing forward velocities and keeping the circumferential wheel velocity of v_c=0.0976 m/s (Shinone et al., 2010), towed (forcing v_x with the value of v_c), and self-propelled (prescribed v_c=0.0976 m/s). With prescribed slip, the wheel is only allowed to move freely in vertical (z) direction. Towed and self-propelled conditions allow one extra degree of freedom: free angular velocity ω_y in towed, and free horizontal velocity v_x in self-propelled conditions. The largest black arrow in Fig. 6 shows the total force exerted by soil subgrade on the wheel. Its vertical component balances the wheel weight, which is 980 N. The green arrow shows the sum of normal forces, and the red arrow the sum of shear forces exerted by soil on the wheel.

Relative velocity of soil particles at the wheel-soil contact surface with respect to wheel circumference is shown in Fig. 7. The transition from negative to positive shear stress roughly corresponds to the transition from negative to positive relative shear velocity of particles with respect to the wheel surface. Entry and exit angles were estimated by approximating the normal stress profile with a squared "tilted" sine function shown in Eq.s 17 an(18) 18.

$$\tau(\theta') = \left(\frac{1}{t}\arctan\left(\frac{t\sin(\theta'-\pi)}{1-t\,\cos(\theta'-\pi)}\right)\right)^2 \tag{17}$$

where θ' is

$$\theta' = \pi(\theta-\theta_2)/(\theta_1-\theta_2) \tag{18}$$

Here, θ_1 and θ_2 are the angles at which the wheel enters and exits the soil. The approximation in Eq.s (17) and 18 offers an alternative to connected halves of Gaussian distributions used by

Horiko and Ishigami (2020). The shear stress profile in Fig. 6 resembles those measured by Horiko and Ishigami (2020), however the shear stress transition point in the present study does not correspond the location of maximum normal stress. Figure Fig. 8 shows maximum normal stress value along with entry angle, exit angle, and angle of maximum normal stress angle at varying slips. Interestingly, the present study show a minimum in the dependence of the entry angle on slip, while Wong and Reece (1967) show increasing trend only. The maximum normal stress being largest close to zero slips can be justified by contact patch being shortened at close-to-zero slips, as indicated by lower difference between the entry and exit angle for close-to-zero slips.

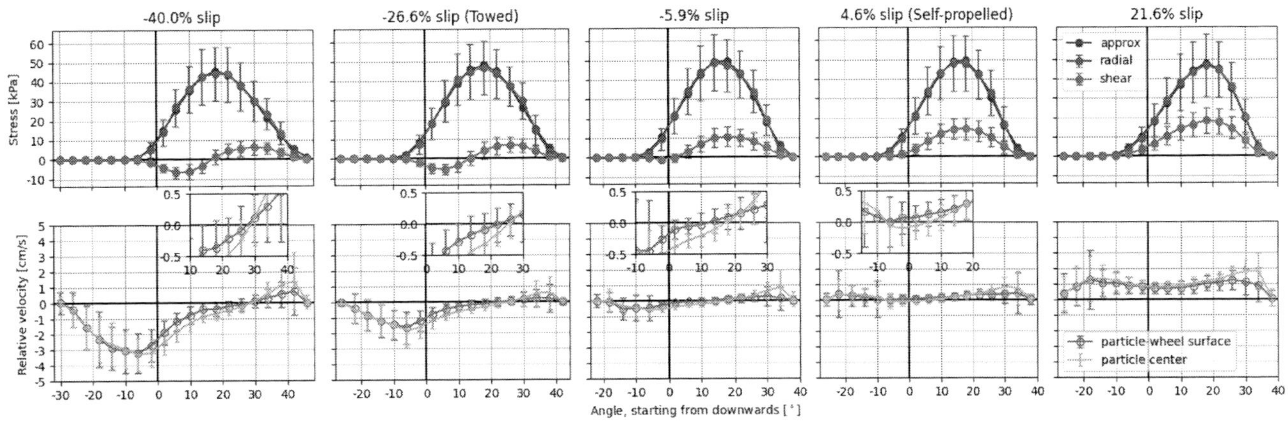

Fig. 7. Velocity of particles relative to the wheel circumference at the particle centers and at the contact point (lower row) aligned with normal and shear stress profiles along the contact patch (upper row).

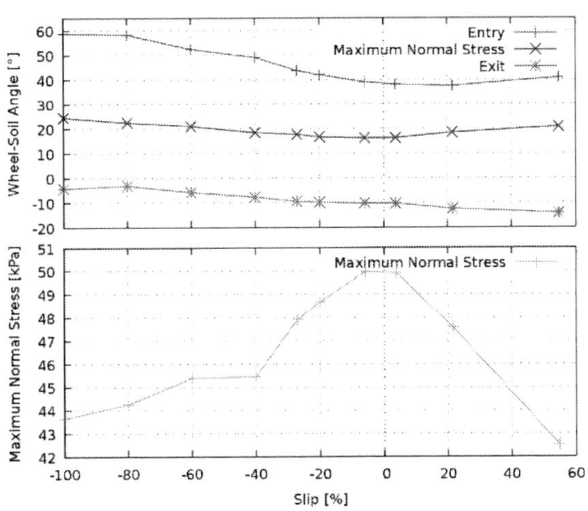

Fig. 8. Maximum normal stress values along with entry angle, exit angle, and maximum normal stress angle at varying slips.

Figure Fig. 9 shows average particle-center velocities under the wheel. The forward velocity of the wheel at each slip is indicated by the topmost black arrow. The rim velocity is indicated by three arrows: first on the left side of the rim, second on the right slide of the rim, and the velocity of the bottom-most point of the rim is indicated by an arrow placed between just above the bottom-most point of the rim. It is important to note that velocities of soil particles were averaged over multiple steps of the simulation. The averaging started 7 seconds after the wheel started moving forward to allow the wheel-soil system to reach a steady state and finished when the wheel center reached the distance of 1 meter from the center of the box, what is approximately 0.5 meter away from the end of the box to avoid effects of box ending wall.

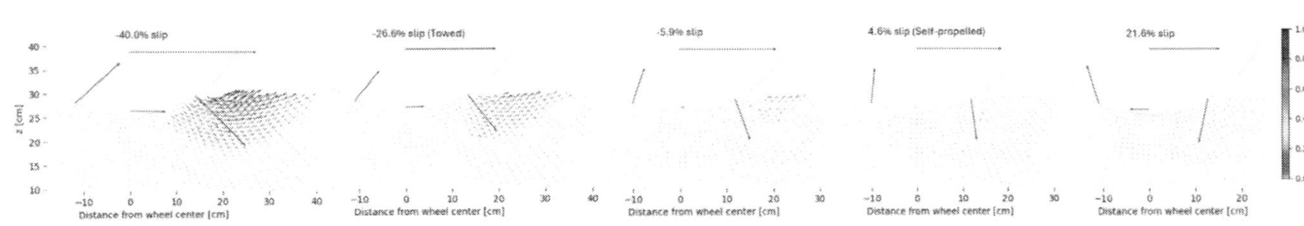

Fig. 9. Averaged particle velocities under the wheel. Forward and wheel surface velocities are shown as large black arrows

5. Comparison with analytical expressions

To help analyze the results of the simulation and evaluate their accuracy, an analytical model of tractive performance was created using a combination of the Bekker and Janosi terramechanics models (Bekker, 1956; Janosi and Hanamoto, 1961). The Bekker model is used to evaluate sinkage, while the Janosi model is used for evaluating drawbar pull and gross tractive effort. The equations used for this analytical solution are derived from previous work (Hu et al., 2021; Li and Bingham, 2022).

$$\sigma_{Bekker} = \left(\frac{k_c}{b} + k_\varphi\right) * |z|^n \tag{19}$$

$$z_{equillibrium} = P + Q - \sigma * A_{modified} \tag{20}$$

$$A_{modified} = \sqrt{2 * r * |z| - |z|^2} * b * \cos(\beta) * (1-i)^\alpha \tag{21}$$

$$Q = D * i * e^{-E*i} * P \tag{22}$$

Equations 19, 20, 21 and 22 above represent the normal stress σ_{Bekker}, the sinkage equilibrium, the modified contact area, and a load mimicking the bulldozing effect, respectively. Table 3 below defines the parameters used in these equations.

Table 3
Nomenclature for Bekker / Janosi terramechanics models

Symbol	Definition	Units
z	Sinkage	[m]
b	Wheel width	[m]
k_c	Cohesive modulus	[N/m^{n+1}]
k_φ	Friction modulus	[N/m^{n+2}]
σ	Normal stress, Bekker	[Pa]
P	Load	[Pa]
Q	Additional load from bulldozing effect	[Pa]
r	Wheel radius	[m]
β	Side-slip angle	[rad]
i	Forward slip	[]
D	Additional load scaling parameter	[]
E	Shear scaling for additional load	[]
n	Bekker exponent	[]
c	Soil cohesion	[Pa]
φ	Internal friction angle	[rad]
K_x	Longitudinal shear modulus	[m]
K_y	Lateral shear modulus	[m]
μ_{rr}	Running resistance friction	[]
α	Slip exponent	[]
A	Contact area	[m^2]

The above parameters constitute the Bekker and Janosi soil characterization along with units for calculation in their corresponding equations. Some parameters, such as sinkage, are converted to other units after calculation to improve clarity in displaying their results.

$$\tau_x = (c + \sigma_{Bekker} * \tan(\varphi)) * (1 - e^{-\frac{r*i}{K_x}}) \tag{23}$$

$$\tau_y = (c + \sigma_{Bekker} * \tan(\varphi)) * (1 - e^{-\frac{|r*\tan(\beta)|}{K_y}}) \tag{24}$$

$$DBP = A_{modified} * (\tau_x * \cos(\beta) - \tau_y * \sin(\beta)) \tag{25}$$

Equations 23, 24, and 25 show how the shear stresses in each direction are calculated, then the calculation of the drawbar pull.

$$RR = P * \mu_{rr} \tag{26}$$

$$GTE = DBP + RR \tag{27}$$

Equations 26 and 27 above show how to calculate running resistance and gross tractive effort, respectively. Together, each of these equations can be used to generate an analytical solution for the effects of side slip.

The fitting of DEM results to Bekker / Janosi parameters was done by using a Python script utilizing several tools from the SciPy library, which is a library focused on scientific work with extensive functionality for solving complex systems of equations. The issue when using these calculations is that some of the necessary parameters are not included in the reference material from Shinone et al. (2010) on which the original work involving only forward slip was based. To compensate for this, SciPy tools such as *fsolve*, *least squares*, and *differential evolution* were used to fit the solution to Shinone et al. (2010) data using a 0° side-slip angle. That provided approximations for the missing parameters through several stages of fitting. Documentation of these tools can be found on SciPy's website ("SciPy User Guide," 2025).

Example results of fitting results from simulations of the DEM material number 1311 with 12° side slip to Bekker / Janosi soil model are shown in Fig. 10, Fig. 11, and Fig. 12, demonstrating a very good fit for the drawbar pull and gross traction and a reasonable fit for sinkage. Bekker and Janosi parameters obtained by fitting 12° side slip response from both 1311 and 1111 DEM materials are listed in Table 4.

Table 4
Bekker and Janosi parameters fitted to DEM results mat. # 1311 and 1111

Parameter	Mat. # 1311	Mat. # 1111	Units
k_c	1.40	1.20	[N/m^{n+1}]
k_φ	9,920	31,500	[kN/m^{n+2}]
n	1.48	2.00	[]
K_x	0.164	0.206	[m]
K_y	0.00124	0.00147	[m]
D	0.703	0.240	[m]
E	1.00	1.00	[]
μ_{rr}	0.207	0.255	[]
α	0.684	0.869	[]

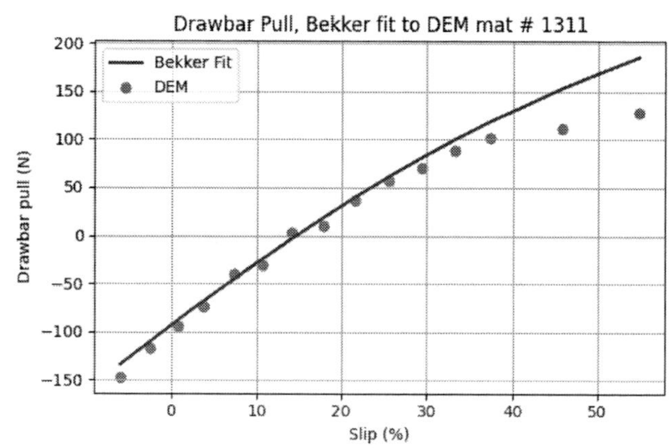

Fig. 10. Drawbar pull- Bekker / Janosi fit to DEM material # 1311

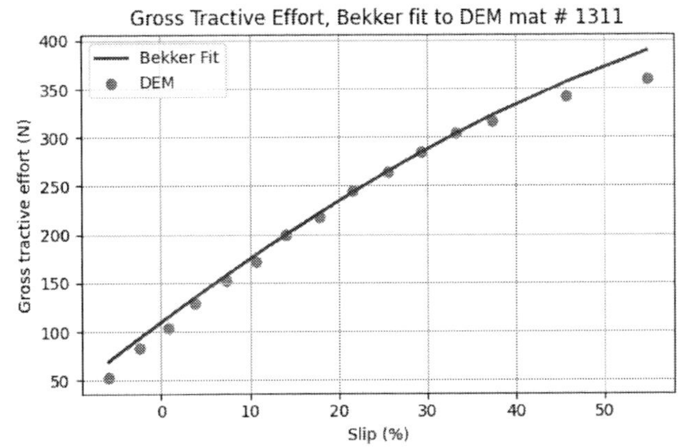

Fig. 11. Gross tractive effort - Bekker / Janosi fit to DEM material # 1311

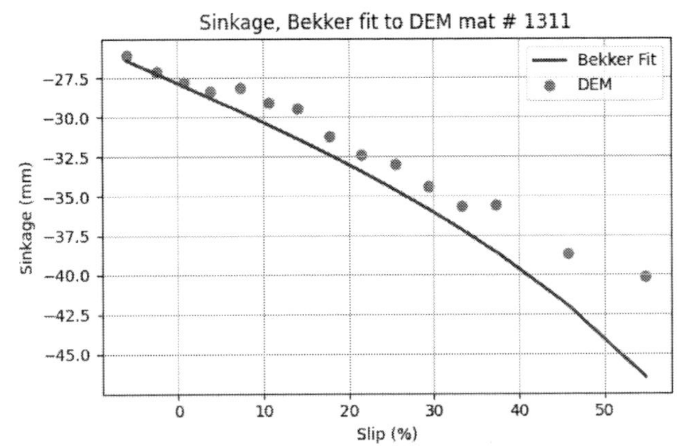

Fig. 12. Sinkage – Bekker / Janosi fit to DEM material # 1311

6. Conclusions and future work

This work analyzes the results of DEM simulations including side slip. A new approximation of the normal stress profile along the wheel-soil contact patch for a straight-driving wheel is proposed. The results were fitted to Bekker / Janosi terramechanics relations. A reasonable agreement with Bekker theory is found.

In the future, further calibration of DEM material parameters can be performed to better match the experimental data. It would add value significantly if the side slip results were compared and additionally calibrated with an experiment. Approximation for the shear stress profile, allowing to predict drawbar pull and gross traction, will be investigated.

7. Nomenclature

a	Contact radius	[m]
B	Box width	[m]
D	Wheel diameter	[m]
d_i	Distance from the i-th contact point from the center of mass	[m]
d_j	Distance from the j-th particle wheel contact point to wheel axis	[m]
e	Coefficient of restitution	[]
E	Young's modulus	[m]
F^d	Damping force	[m]
F_G	Wheel weight	[N]
F_t	Tangent component of contact force	[N]
F_n	Normal component of contact force	[N]
F^x	Forward component of force on wheel	[N]
G	Shear modulus	[MPa]
H	Box depth	[m]
h	Soil depth	[m]
i	Wheel slip	[]
L	Box length	[m]
m	Mass	[kg]
N	Number of particles	[]
R	Hertz-Mindlin radius	[m]
R_i	Radius of sphere in contact	[m]
S	Stiffness	[N/m]
T	Gross tractive effort	[Nm]
t	Parameter of the normal stress approximation function	[]
v_f	Forward component of wheel velocity	[m/s]
v_n^{rel}	Relative normal velocity	[m/s]
v_t^{rel}	Relative tangential velocity	[m/s]
W	Wheel width	[m]
α	Contact overlap	[m]
β	Damping ratio	[]
γ	JKR surface energy parameter	[J/m^2]
δ	Overlap	[m]
μ	Coefficient of friction	[]
μ_r	Coefficient of rolling friction	[]
μ_s	Coefficient of static friction	[]
ν	Poisson's ratio	[]
ρ	Solid Density (material without voids)	[kg/m^3]
ω	Angular speed	[rad/s]
τ_i	Rolling friction torque for i-th contact	[Nm]

8. Acknowledgements

This material is based upon research conducted under contract W912HZ-22-C0004 with the U.S. Army Engineer Research and Development Center (ERDC). The views and conclusions contained herein are those of the authors and should not be interpreted as necessarily representing the official policies or endorsements, either expressed or implied, of ERDC or the U.S. Government.

9. Declaration of competing interest

The authors declare that they have no known competing financial interests or personal relationships that could have appeared to influence the work reported in this paper.

10. References

Ai, J., Chen, J.-F., Rotter, J.M., Ooi, J.Y., 2011. Assessment of rolling resistance models in discrete element simulations. Powder Technology 206, 269–282. https://doi.org/10.1016/j.powtec.2010.09.030

Bekker, M.G., 1956. Theory of Land Locomotion: The Mechanics of Vehicle Mobility, 9690401st ed. University of Michigan Press, Ann Arbor, MI. https://doi.org/10.3998/mpub.9690401

Coetzee, C., 2020. Calibration of the discrete element method: Strategies for spherical and non-spherical particles. Powder Technology 364, 851–878. https://doi.org/10.1016/j.powtec.2020.01.076

Coetzee, C.J., 2017. Review: Calibration of the discrete element method. Powder Technology 310, 104–142. https://doi.org/10.1016/j.powtec.2017.01.015

Du, Y., Gao, J., Jiang, L., Zhang, Y., 2018. Development and numerical validation of an improved prediction model for wheel-soil interaction under multiple operating conditions. Journal of Terramechanics 79, 1–21.

Du, Y., Gao, J., Jiang, L., Zhang, Y., 2017a. Numerical analysis on tractive performance of off-road wheel steering on sand using discrete element method. Journal of Terramechanics 71, 25–43.

Du, Y., Gao, J., Jiang, L., Zhang, Y., 2017b. Numerical analysis of lug effects on tractive performance of off-road wheel by DEM. Journal of the Brazilian Society of Mechanical Sciences and Engineering 39, 1977–1987.

Horiko, S., Ishigami, G., 2020. Experimental study on wheel-soil interaction mechanics using in-wheel sensor and particle image velocimetry part II. Analysis and modeling of shear stress of lightweight wheeled vehicle. Journal of Terramechanics 91, 243–256. https://doi.org/10.1016/j.jterra.2020.07.002

Hu, C., Gao, J., Song, X., Zhang, M., Tan, X., 2021. Analytical modeling and DEM analysis of soil–wheel interaction under cornering and skidding conditions in off-road vehicles. AIP Advances 11, 085122. https://doi.org/10.1063/5.0057046

Janosi, Z., Hanamoto, B., 1961. The analytical determination of drawbar pull as a function of slip for tracked vehicles in deformable soils, in: International Society for Terrain-Vehicle Systems, 1st Int. Conf.

Jelinek, B., Card, A., Mason, G.L., Grebner, K., Dickerson, A., Skorupa, T., Cole, M., Priddy, J.D., 2025. Tractive performance of rigid wheel in granular media using coarse-scale DEM models. Journal of Terramechanics 117, 101016. https://doi.org/10.1016/j.jterra.2024.101016

Johnson, K.L., Kendall, K., Roberts, A.D., 1971. Surface energy and the contact of elastic solids. Proc. R. Soc. Lond. A 324, 301–313. https://doi.org/10.1098/rspa.1971.0141

Jones, R., 2003. From Single Particle AFM Studies of Adhesion and Friction to Bulk Flow: Forging the Links. Granular Matter 4, 191–204. https://doi.org/10.1007/s10035-002-0122-6

Jones, R., Pollock, H.M., Geldart, D., Verlinden-Luts, A., 2004. Frictional forces between cohesive powder particles studied by AFM. Ultramicroscopy 100, 59–78. https://doi.org/10.1016/j.ultramic.2004.01.009

Li, Z.Q., Bingham, L.K., 2022. NASA White Paper - Terramechanics for LTV Modeling and Simulation.

Mindlin, R.D., Deresiewicz, H., 1953. Elastic Spheres in Contact Under Varying Oblique Forces. Journal of Applied Mechanics 20, 327–344. https://doi.org/10.1115/1.4010702

Sakaguchi, H., Ozaki, E., Igarashi, T., 1993. Plugging of the Flow of Granular Materials during the Discharge from a Silo. International Journal of Modern Physics B 07, 1949–1963. https://doi.org/10.1142/S0217979293002705

SciPy User Guide [WWW Document], 2025. URL https://docs.scipy.org/doc/scipy/tutorial/index.html

Shinone, H., Nakashima, H., Takatsu, Y., Kasetani, T., Matsukawa, H., Shimizu, H., Miyasaka, J., Ohdoi, K., 2010. Experimental Analysis of Tread Pattern Effects on Tire Tractive Performance on Sand using an Indoor Traction Measurement System with Forced-slip Mechanism. Engineering in Agriculture, Environment and Food 3, 61–66. https://doi.org/10.1016/S1881-8366(10)80020-5

Thakur, S.C., Morrissey, J.P., Sun, J., Chen, J.F., Ooi, J.Y., 2014. Micromechanical analysis of cohesive granular materials using the discrete element method with an adhesive elasto-plastic contact model. Granular Matter 16, 383–400. https://doi.org/10.1007/s10035-014-0506-4

Wong, J.-Y., Reece, A.R., 1967. Prediction of rigid wheel performance based on the analysis of soil-wheel stresses part I. Performance of driven rigid wheels. Journal of Terramechanics 4, 81–98.

Zeng, H., Xu, W., Zang, M., Yang, P., 2020a. Calibration of DEM-FEM model parameters for traction performance analysis of an off-road tire on gravel terrain. Powder Technology 362, 350–361. https://doi.org/10.1016/j.powtec.2019.12.006

Zeng, H., Xu, W., Zang, M., Yang, P., Guo, X., 2020b. Calibration and validation of DEM-FEM model parameters using upscaled particles based on physical experiments and simulations. Advanced Powder Technology 31, 3947–3959. https://doi.org/10.1016/j.apt.2020.06.044

4368 | Improving the expanding/contracting speed of a push-rolling rover on loose surfaces with steep slopes

Daisuke Fujiwara [a,*], Hiroto Fujisawa [a], Tasuku Hoshino [a], Kojiro Iizuka [b]

[a] *Department of Mechanical and Electrical Engineering, Faculty of Engineering, Suwa University of Science, Japan*

[b] *Department of Machinery and Control Systems, System Engineering and Science, Shibaura Institute of Technology, Japan*

* *Corresponding author: fujiwara@rs.sus.ac.jp, iizuka@shibaura-it.ac.jp*

ABSTRACT

The wheeled rover has played an important role in planetary or lunar exploration. The planet and moon surfaces are covered with granular materials, which causes the wheeled rover to get stuck. To suppress wheel slippage, the rover, which utilizes shear force beneath the anchored wheel, can effectively climb a loose surface with steep slopes. Previous studies have developed various types of push-rolling and wheel-walking rovers. Furthermore, our previous study developed a minimal configuration rover and confirmed that it can climb loose surfaces with steep slopes while suppressing slippage.

During its movement, the rover changes its wheelbase length and moves its wheels using the resistance forces of the anchored wheels. However, the variable speed of the wheelbase length remains constant at 20 mm/s, and the effect of this speed on the rover's traveling performance is still unclear. The traveling speed of the rover with small wheels tends to slow down; therefore, understanding the potential to improve its speed is important. The purpose of this paper is to validate the impact of the changing speed of the wheelbase length on the climbing ability. This paper presents a testbed rover that can change the speed of the expanding/contracting wheelbase length and performs the climbing experiment on loose soil with steep slopes. The experimental results suggested that driving performance was maintained even when the expanding/contracting speed increased three times faster than the conventional speed.

Keywords
Mobile robot
Push-rolling motion
Loose soil
Planetary explorations

1. Introduction

For direct planetary explorations, various wheeled rovers have been developed. The previous study confirmed that the wheeled rover that uses shear forces beneath a locked wheel can suppress the rover's slip on the loose soil with slopes. These kinds of rovers are called push-rolling or inching locomotion, push-pull locomotion rovers (Cao et al., 2023, Cordes and Babu, 2016, Creager et al., 2015, Kemurdjian et al., 1992, Wilcox et al., 2007). These rovers lock one pair of wheels to the ground, then reposition the other pair of wheels. The repositioned methods of the wheels are two. The one is a walking motion (Wilcox et al., 1992, Cordes et al., 2016), and the second is driving while expanding and contracting (Cao et al., 2023, Creager et al., 2015). The second one is a relatively simple configuration and is designed with relatively few actuators and is small.

Our previous study confirmed that a small and minimal push-rolling rover can climb loose soil on a steep slope of approximately around 30 degrees with low slip (Fujiwara et al., 2025). In terms of transportation costs, the small rovers are beneficial. In fact, the ball-shaped small rover with a diameter of 80 mm has been used in planetary exploration (JAXA, 2024). However, the traveling speed of the rover with small wheels is low, and the additional motion of the push-pull moving decreases the rover's traveling speed.

The important factor of the push-rolling rover developed in the previous study is the slip state of the rotational wheel during expansion/contraction motion. The slip ratio of the rotational wheel relates to the drawbar pull force of the wheels. The previous study indicated that the drawbar pull force becomes large when the slip ratio is large (Wong and Reece, 1967); therefore, when the slip ratio of the driving wheel is high, the push-rolling rover suppresses the rover's slip. The slip ratio is calculated by the expansion/contraction speed and the speed of a rotational wheel. The arbitral slip ratio is realized by the arbitrary push-pull speed and the speed of a rotational wheel. Hence, if the effect of the increased push-pull speed on the traveling performance is small, the rover can still climb the loose soil with steep slopes even at high traveling speeds. To confirm the effect of the push-pull speeds on the rover's performance, this study investigates the traveling performance of the small push-pull rovers at different push-pull speeds.

2. Locomotion mechanism of the push-pull rover

Many configurations of the push-rolling method have been developed in previous studies. The Scarab and the robot developed by (Creager et al., 2015) rotates the wheel legs centering on the arbitrary point of the body; therefore, the robot's center of mass changes during the push-pull motion. In addition, the robot developed by (Shrivastava et al., 2020) added the steering motion to the motion of the leg wheels. This motion is similar to the squirming motion in the soil. Cordes and Babu (2016) and Wilcox et al. (1992) developed a multiple-wheel-legged robot, Sherpa and ATHLETE, that can raise their legs from the ground and reposition them. The robot developed in our group (Fujiwara et al., 2025) and one method described in (Bouton et al., 2023) expands/contracts its wheelbase along the body direction, and the height of the robot's body is unchanged. To design a small robot, this study focuses on the motion of the driving wheel that moves along the robot's body without changing the height of the center of mass as indicated in Fig. 1

https://doi.org/10.56884/KCKUMG76

In Fig. 1, first, the rover expands the front wheels. The speed of the rotational wheel is $r\omega$, and the expansion/contraction speed is v_l. F_x means the drawbar pull force of the rotational wheels, and F_r is the resistance force of the locked wheels from the ground. Second, the rover contracts the rear wheels. The rover repeats these motions and moves in a longitudinal direction. The forces from the ground acting on the rigid wheels have been investigated in the terramechanics field. Figure 2 shows the wheel-soil contact model. Furthermore, F_x is as follows (Wong and Reece, 1967):

$$F_x = rb \int_{\theta_r}^{\theta_f} \{\tau(\theta)\cos\theta - \sigma(\theta)\sin\theta\}\, d\theta \tag{1}$$

where r and b are the wheel radius and width, respectively. $\tau(\theta)$ and $\sigma(\theta)$ represent the shear forces and normal forces beneath the wheel. Furthermore, θ_r, θ_f are the contact angles between the wheel and the ground. In these parameters, $\tau(\theta), \sigma(\theta)$ relate to the slip ratio of the wheel. In the case of the push-rolling rover, the wheel slip ratio is represented as follows:

$$s = \begin{cases} 1 - v_l/r\omega & r\omega > v_l \\ r\omega/v_l - 1 & r\omega < v_l \end{cases} \tag{2}$$

where v_l is the contracting/expanding speed of the rover. r is the wheel radius, and ω is the angular velocity of the rotational wheel. The value range of the slip ratio is -1 to 1. When $r\omega > v_l$, the rotational wheel moves with high slip. The drawbar pull forces increase with an increasing slip ratio. Meanwhile, when $v_l > r\omega$, the wheel moves by pushing the ground. When s is close to -1, the wheel becomes the locked state. This means that the forces calculated by Eq. 1 are close to F_r.

Herein, s is a ratio; therefore, the slip ratio can be set by arbitrary v_l and v_w.

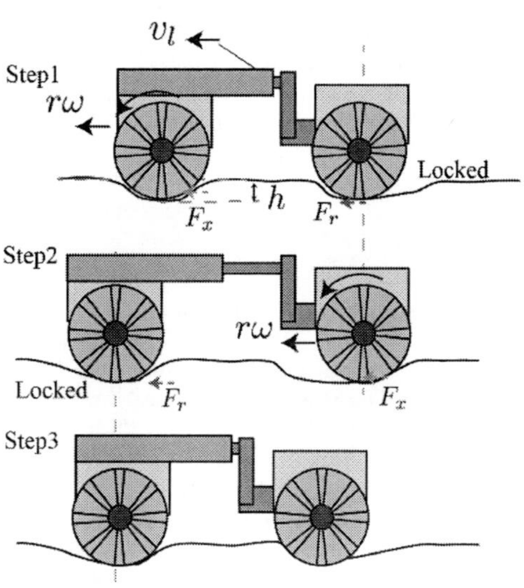

Fig. 1. Push-rolling locomotion

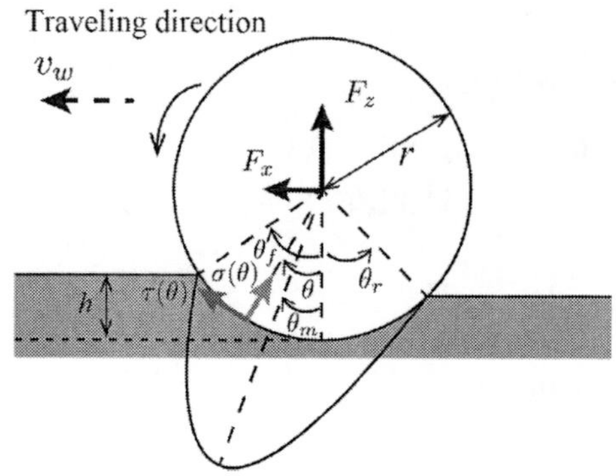

Fig. 2. Wheel-soil contact model

3. Testbed rover

Figures 3 and 4 show the testbed rover developed in this study. The wheel is 100 mm in diameter, including lug height. The number of lugs is 21. The number is decided by the dimension of the failure area of the soil mass acting on the lug (Sutoh et al., 2011). The size of the testbed is 202 mm in width and 111 mm in height. The wheelbase length is from 185 mm to 285 mm. This length is changed by the linear actuator. The rover mass is 1.03 kg. The angular speed of the wheel is controlled based on the pulses of the encoder. To realize different push-pull speed, this paper adopts the linear actuator with stepper motor (Actuonix S20 100 mm). The maximum speed of the linear actuator v_l is 120 mm/s. To adjust the slip ratio of the driving wheel, its angular velocity is regulated based on the linear actuator speed of the rover.

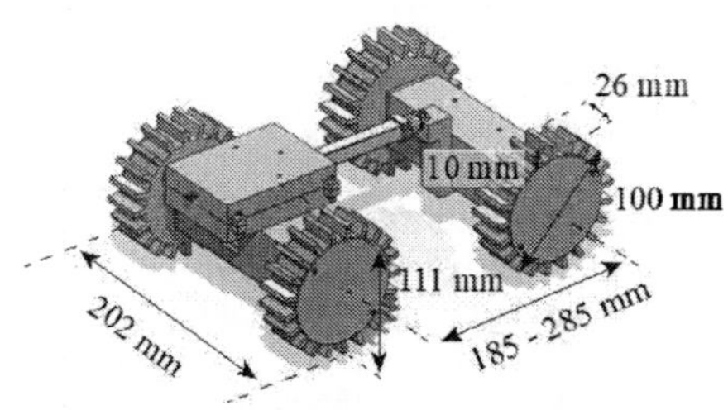

Fig. 3. CAD image of the testbed rover

Fig. 4. Actual image of the testbed rover

4. Experiment

4.1. Environment

Figure 5 shows the soil box for the climbing test. The soil is the Silica sand No. 5 (Table 1). The size of the soil bin is 1016 mm in length, 480 mm in width, and 110 mm in height. The motion capture system measures the displacement of the rover. The maker position is placed on the side of the wheel as shown in Fig. 4.

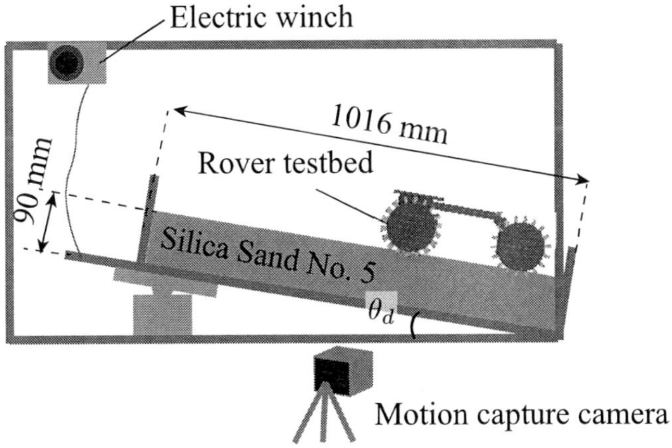

Fig. 5. Experimental environment

Table 1
Parameters of Silica sand No. 5

Description (Unit)	Value
Bulk density (g/cm^3)	1.29-1.31[a]
Angle of repose (°)	33[b]
Internal friction angle (°)	22.3[c]
Cohesion (N/m^2)	761.8[c]
Mean particle size (mm)	≈ 0.59 [a]

a: data of supplier, b: measured, c: (Noriaki, 2013)

4.2. Experimental conditions

Table 2 shows the experimental conditions for the traveling test. The slope angles are 10°, 20°, and 30°. The slip ratios of the driving wheel are 0.2, 0.6, and 0.8. v_l is set at 20, 40, 60, and 100 mm/s. The angular velocity of the wheel is adjusted to realize the slip ratios at each v_l. Table 3 indicates the angular velocity at each v_l.

The evaluation index is the traveling ratio as indicated in Eq. 3.

$$T_r = d_a/d_i \tag{3}$$

where d_a and d_i are the actual distance and traveling distance when the rover does not slip. For example, the telescopic distance of the testbed rover is 100 mm. In this case, d_i is 100 mm.

Table 2
Experimental conditions for the traveling tests

Description (Unit)	Value
Slope angle (°)	10, 20, 30
v_l (mm/s)	20, 40, 60, 100
Slip ratio s (-)	0.2, 0.6, 0.8
Trial at each condition (times)	5

Table 3
Angular velocity at each v_l

Telescopic speed v_l(mm/s)	Slip ratio	Angular velocity ω (rad/s)
	0.2	0.5
20	0.6	1.0
	0.8	2.0
	0.2	1.0
40	0.6	2.0
	0.8	4.0
	0.2	1.5
60	0.6	3.0
	0.8	6.0
	0.2	2.0
80	0.6	4.0
	0.8	8.0
	0.2	2.5
100	0.6	5.0
	0.8	10.0

4.3. Results and discussions

Figure 6 shows the actual images of the traveling experiments on the loose soil with a slope angle 30°. The rover repeats expansion and contraction motion and climbs the slope.

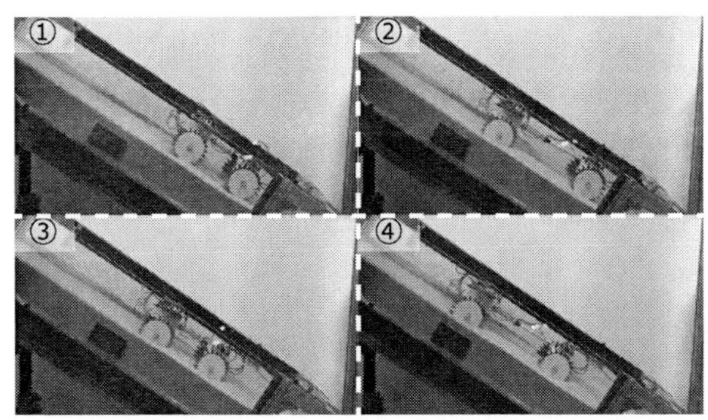

Fig. 6. Actual images of the traveling test. Slope angle is 30°. Telescopic speed is 60 mm/s. Slip ratio is 0.8.

Figures 7, 8, 9, and 10 show the raw data of the displacement at different telescopic speeds and slip ratios. The displacement is calculated from the maker position of the front wheel. The graph shows the rover repeats intermittent motion. The steady part of the graph indicated the interval of the motion. Note that the traveling test at high v_l, for example, 80 or 100 mm/s, in the slip ratio 0.8, does not exist due to the mechanical limitation of the linear actuator. Each graph shows that the gradient of the increasing part rises with an increasing v_l. This indicates that the rover's climbing velocity increased.

Figures 11, 12, and 13 show the traveling ratio of the rover at different slip ratios of the driving wheels and push-pull speeds. Note that the result in the case of $v_l = 100$ mm/s and slope angles of 10° and 20° is omitted. In this condition, the linear actuator is stopped. This is because the driving force of the linear actuator decreases as the rotational speed increases.

When the slip ratio is 0.6 and 0.8, the effect of v_l on the traveling ratio is small. Meanwhile, when the slope angle is 20° and the slip ratio is 0.2 and 0.6, the traveling ratio slightly decreases with an increasing v_l. The drawbar pull forces decrease as the slip ratio decreases. This leads to unstable running on the loose soil with slopes. Furthermore, high push-pull speeds induce complex destruction of the soil due to the high rotation of the wheels. Therefore, the driving performance at low slip ratios is reduced.

In summary, the results suggested that the rover can increase climbing speeds while suppressing slip even at high slope angles.

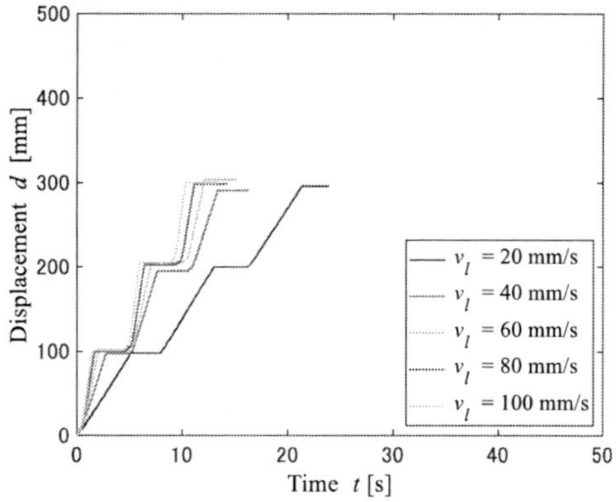

Fig. 7. Displacement of the rover on a slope angle of 10° at different telescopic velocity. The slip ratio of the driving wheel is 0.2.

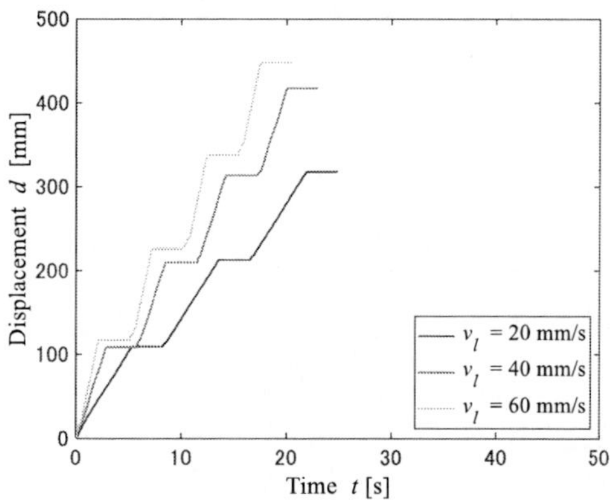

Fig. 8. Displacement of the rover on a slope angle of 10° at different telescopic velocity. The slip ratio of the driving wheel is 0.8.

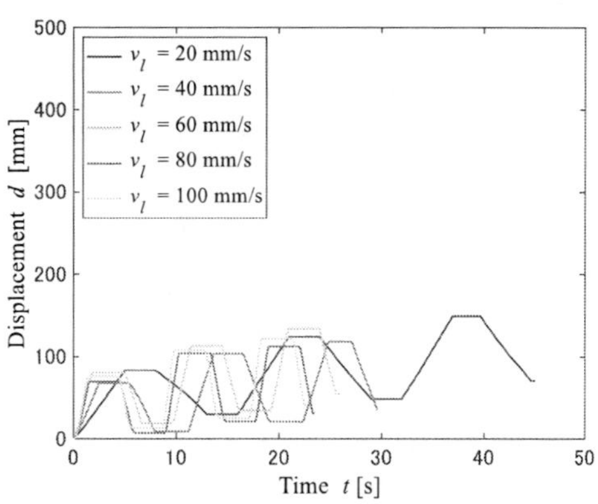

Fig. 9. Displacement of the rover on a slope angle of 30° at different telescopic velocity. The slip ratio of the driving wheel is 0.2.

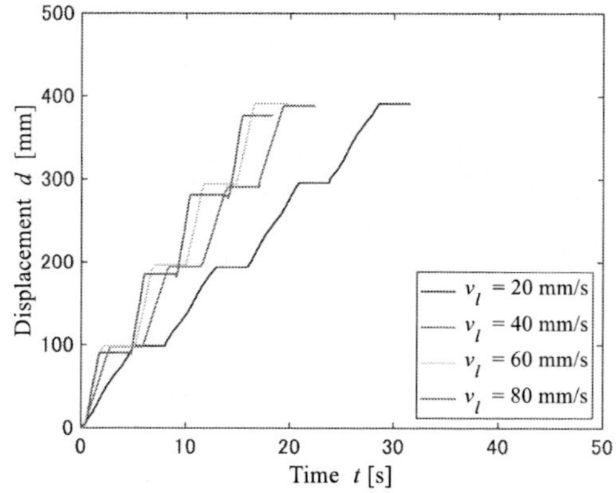

Fig. 10. Displacement of the rover on a slope angle of 30° at different telescopic velocity. The slip ratio of the driving wheel is 0.8.

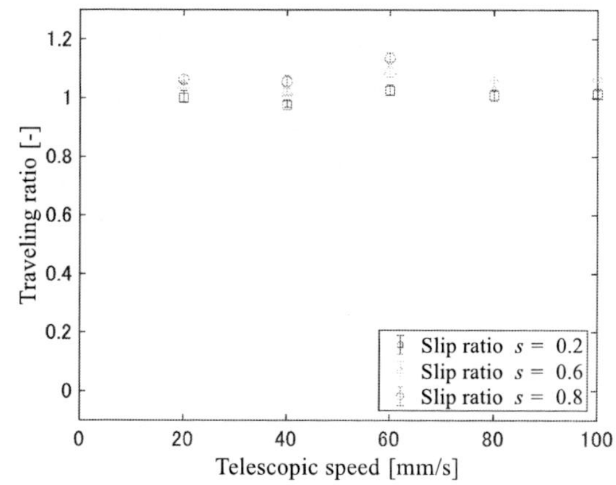

Fig. 11. Traveling ratio at a slope angle of 10°.

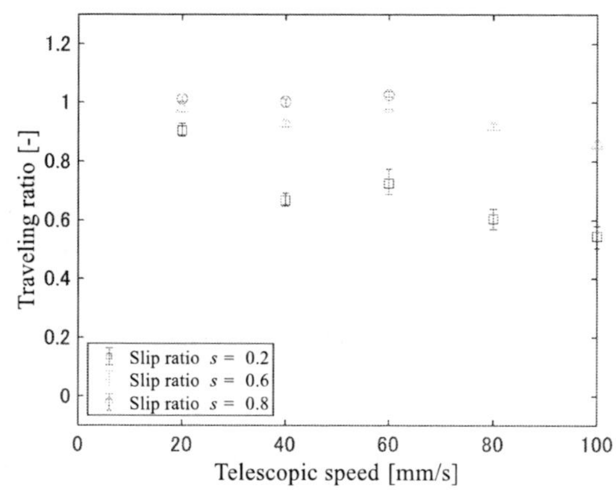

Fig. 12. Traveling ratio at a slope angle of 20°.

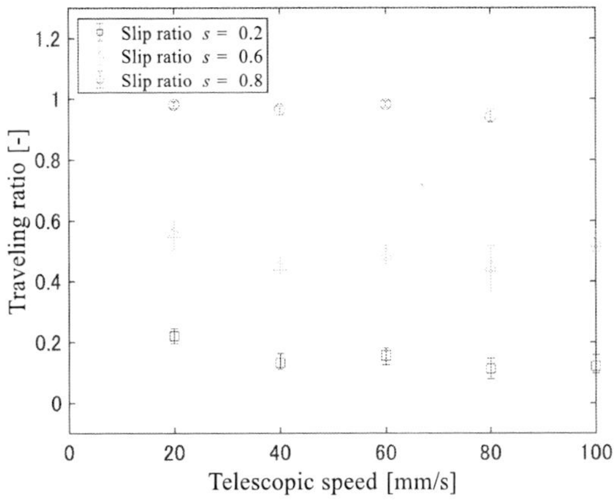

Fig. 13. Traveling ratio at a slope angle of 30°.

5. Conclusions and future work

The purpose of this study is to investigate the effect of the push-pull speed on the climbing performance of the small rover at different slope angles and slip ratios of the driving wheels. To change the push-pull speeds, this paper adopted a linear actuator with a stepper motor that can adjust telescopic speeds.

Then, this paper performed a climbing experiment on the loose soil with slopes. The experimental results indicated that the effect of the telescopic speeds of the rover on the climbing ability is small when the driving forces of the rover are enough. This suggests that a small push-rolling rover can improve the traveling speed by suppressing slip on the steep slopes.

This paper mainly focused on the climbing test of the rover. Future works should investigate the effect of the driving speed on the forces acting on a single wheel under the condition of the same driving length of 100 mm. Furthermore, future studies consider the mechanical limitation of the linear actuator when the rotational speed is high.

6. Nomenclature

b	Wheel width	[mm]
d_a	Actual traveling distance	[mm]
d_i	Ideal traveling distance (No slip)	[mm]
F_r	Resistance force	[N]
F_x	Drawbar pull force	[N]
h	Sinkage	[mm]
r	Wheel radius	[mm]
s	Slip ratio	[-]
v_l	Telescopic speed	[mm/s]
v_w	Traveling speed of the wheel	[mm/s]
θ_d	Slope angle	[°]
θ_f, θ_r	Wheel contact angle	[°]
$\sigma(\theta)$	Normal stress	[N/m^2]
$\tau(\theta)$	Shear stress	[N/m^2]
ω	Angular velocity	[rad/s]

7. Declaration of competing interest

The authors declare that they have no known competing financial interests or personal relationships that could have appeared to influence the work reported in this paper.

8. References

Bouton, A., Reid, W., Brown, T., Daca, A., Sabzehi, M., Nayar, H., 2023. Experimental study of alternative rover configurations and mobility modes for planetary exploration, In: 2023 IEEE AeroConf, pp. 1–13. doi:10.1109/AERO55745.2023.10115624.

Cao, C., Moon, D., Creager, C., Lieu, D.K., Stuart, H.S., 2023. Push-pull locomotion: Increasing travel velocity in loose regolith via induced wheel slip. J. Terramech. 110, 87–99. doi:https://doi.org/10.1016/j.jterra. 2023.08.005.

Cordes, F., Babu, A., 2016. Sherpatt: A versatile hybrid wheeled leg rover. In: Proceedings of the 13th i-SAIRAS 2016.

Creager, C., Johnson, K., Plant, M., Moreland, S., Skonieczny, K., 2015. Push pull locomotion for vehicle extrication. J. Terramech. 57, 71–80. doi:https://doi.org/10.1016/j.jterra.2014.12.001.

Fujiwara, D., He, Q., Iizuka, K., 2025. Climbing loose surfaces with steep slopes using a small, lightweight push-rolling rover with minimal configuration. J. Terramech. 118, 101043. URL: https://www.sciencedirect.com/science/article/pii/S0022489824000855, doi:https://doi.org/10.1016/j.jterra.2024.101043.

JAXA, 2024. Jaxa, slim mission. URL: https://global.jaxa.jp/.

Noriaki, M., 2013. Study on Development Locomotion Performance based on Terramechanics for Wheeled Rover on Soft Ground, Ph.D. thesis (SOKENDAI).

Shrivastava, S., Karsai, A., Aydin, Y.O., Pettinger, R., Bluethmann, W., Ambrose, R.O., Goldman, D.I., 2020. Material remodeling and unconventional gaits facilitate locomotion of a robophysical rover over granular terrain. Sci, Robot. 5, eaba3499. doi:10.1126/scirobotics.aba3499.

Sutoh., M., Nagatani, K., Yoshida, K., 2011, Evaluation of influence of surface shape of wheel on traversability of wheeled mobile robots on loose soil, In: Proceedings of the Movic, B108.

Wilcox, B.H., Litwin, T., Biesiadecki, J., Matthews, J., Heverly, M., Morrison, J., Townsend, J., Ahmad, N., Sirota, A., Cooper, B., 2007. Athlete: A cargo handling and manipulation robot for the moon. J. Field Robot. 24, 421–434. doi:https://doi.org/10.1002/rob.20193.

Wong, J.Y., Reece, A., 1967. Prediction of rigid wheel performance based on the analysis of soil-wheel stresses part i. performance of driven rigid wheels. J. Terramech. 4, 81–98. URL: http://dx.doi.org/10.1016/0022-4898(67)90105-x, doi:10.1016/0022-4898(67)90105-x.

Kemurdjian, A., Gromov, V., Mishkinyuk, V., Kucherenko, V., Sologub, P., 1992. Small marsokhod configuration, In: Proceedings 1992 IEEE ICRA, pp. 165–168 vol.1. doi:10.1109/ROBOT. 1992.220318.

4417 | The status of soil physical, mechanical, and chemical properties two years after deforestation in a military training area

Kersti Vennik [a,*], Tõnu Tõnutare [b]

[a] *Department of Academic Studies, Estonian Military Academy, Estonia*
[b] *Chair of Soil Science, Estonian University of Life Sciences, Estonia*

* *Corresponding author: kersti.vennik@mil.ee*

ABSTRACT

Military training areas are unique in their function and usage. These areas must endure intensive and repeated movement of wheeled and tracked military vehicles, as well as dismounted troops. Depending on the training requirements of defense forces, military training areas can undergo rapid land-use changes. For instance, when new firing ranges or maneuvering grounds are needed, forests may be cleared to create open areas. Deforestation has a significant impact on soil properties. Logging and uprooting activities displace soil layers and alter soil structure. The primary goal of training land managers is to restore the usability of deforested areas as quickly as possible, making the restoration of disturbed ground a key focus.

The objective of this study was to assess the status of soil physical properties, chemical composition, and trafficability conditions two years after deforestation. As deforestation contributes to the release of carbon dioxide (CO_2) into the atmosphere, it plays a significant role in climate change. A key objective of this study was to evaluate changes in soil CO_2 levels following deforestation. Measurements were conducted in 2020 at a military training ground in northern Estonia. The study included bulk density and moisture content measurements, as well as the determination of California Bearing Ratio (CBR) and Cone Index (CI) values for both disturbed and undisturbed areas. The plant-available nutrient content (P, K, Mg, Ca) was determined using the ammonium acetate-lactate (AL) method. Soil carbon and nitrogen content were measured using the Dumas method. Measurement results are presented, highlighting the effects of disturbance and recovery on soil properties.

Keywords
Soil shear strength
Cone index
Trafficability
Soil chemical elements

1. Introduction

Field exercises are an essential component of military training. To support such activities, designated training areas, occupying approximately 6% of the Earth's terrestrial surface, have been established (Zentelis and Lindenmayer, 2015). Typical military training areas include sections for tank and infantry fighting vehicle maneuvers, impact sites, infantry tactical training grounds, and firing ranges (Havlíček et al., 2018). These various sections are commonly used both by military vehicles and dismounted infantry. Consequently, all designated areas are subject to intensive use and heavy load; during a single training cycle, a given area may be traversed by up to 1000 infantry personnel and 100 vehicles. Military training activities typically include vehicle maneuvering, dismounted movement, live-fire exercises, and activities such as trench digging (Zentelis et al., 2017). Soils of varying structure and texture must be capable of supporting all equipment used while also recovering as rapidly as possible, regardless of specific soil or climatic conditions.

Driven by changes in military-strategic objectives and the acquisition of new equipment, military training areas often undergo rapid land-use transformations. According to Havlíček et al. (2018), land-use change processes within military training areas may differ from global land-use trends and are primarily influenced by the strategic planning and management decisions of military authorities. Typical modifications include the rapid establishment of aircraft landing zones or the construction of new roads with high bearing capacity requirements. Additionally, changes in vegetation cover, such as increasing forested or grassland areas, are frequently required. In the latter case, the creation of open areas necessitates deforestation, which involves cutting down trees, removing stumps and roots, surface milling, and ground leveling. These activities can disturb soils and alter natural horizons to a depth of 20 cm or more (Smith and Wass, 1989).

For land managers of military training grounds, a key challenge is the long recovery time of disturbed land, which typically has initially a low bearing capacity. Tobias et al. (2008) noted that soil dislocation and restoration disrupt soil structure, making it highly susceptible to compaction. For example, most Swiss soil restoration guidelines recommend limiting traffic to light vehicles for up to five years post-restoration (Tobias et al., 2008). However, such a prolonged restriction is not optimal for military training areas, where intensive use is often required shortly after land disturbance.

One strategy employed to accelerate the restoration of soil bearing capacity is the sowing of herbaceous plants in disturbed areas. Several studies (Trükmann et al., 2009; Goldsmith, 2006; Machado et al., 2015; Ali and Osman, 2008) have demonstrated that plant roots can significantly enhance soil shear strength. Goldsmith (2006), for instance, found that the shear strength of soil containing roots of the tall grass species *Panicum virgatum* (switchgrass) was five times greater than that of soil without roots. Ali and Osman (2008) investigated the influence of herbaceous plant roots grown under laboratory conditions on shear strength, with root-soil samples tested to depths of up to 0.9 meters. Their results indicated that within just six months, the plants developed root systems strong enough to significantly improve soil shear strength, with further increases observed one year after planting. Although these experiments tested soil-root complexes to a depth of approximately one meter, the majority of the roots are concentrated within the top 0–10 cm of soil (Brown et al., 2010), where their contribution to soil strength is greatest. This is particularly relevant because the shear strength of the 0–15 cm

https://doi.org/10.56884/7ZENE0MP

soil layer is critical for determining trafficability (Shoop et al., 2015).

Soil strength is most commonly assessed using a cone penetrometer and/or a Dynamic Cone Penetrometer (DCP). In contrast, bulk density and porosity provide direct information about the soil's compaction status. Soil shear strength, in relation to vehicle trafficability, is often determined using the cone penetrometer method. In this method, soil strength is measured using a penetrometer (comprising a 50 cm rod with a 30° circular cone and a base area of 3.23 cm^2), and the resulting parameter is known as the Cone Index (CI). CI is typically measured in 15 cm increments down to a depth of 45 cm (FM 5-430-00-1, 1994). For estimating trafficability, i.e. the general risk of vehicles getting stuck, the CI value for the 15–30 cm soil layer is commonly used and compared against the strength requirements of specific vehicles (vehicle cone index, VCI) (FM 5-430-00-1, 1994). Different VCI values are defined for various soil types (coarse-grained, fine-grained, and peat soils) and vehicle types (wheeled or tracked). In this study, fine-grained soils were analyzed using the corresponding VCI values. VCI can be calculated for a single vehicle pass (VCI_1) or for repeated passes, typically 50, along the same track (VCI_{50}). The formulas for calculating VCI_1 and VCI_{50} are provided by Priddy (1995) and are not reproduced here.

The California Bearing Ratio (CBR) test method was developed in the 1920s within the field of road construction to assess the strength of subgrade layers beneath pavements. In the 1940s, the United States Army Corps of Engineers (USACE) adopted the CBR method for evaluating the suitability of natural terrains for airfield construction. Beyond its application to military airfields, the CBR index is also used to assess soil strength at potential sites for military roads. Based on the measured bearing capacity, the required thickness of the surface layer is determined (TM 5-822-12, 1990). The standard CBR test can be conducted either in laboratory conditions or directly in the field. Due to its resource- and time-intensive nature, the USACE has also adopted the use of the Dynamic Cone Penetrometer (DCP) as a more efficient alternative. The DCP measurements are subsequently converted into CBR values (Shoop et al., 2008).

In the context of forest harvesting, the release of carbon dioxide (CO_2) into the atmosphere and its contribution to climate change have become subjects of increasing attention (Peng et al., 2023). Forests act as carbon sinks, storing carbon not only in biomass but also, often in higher concentrations, in forest soils. The uppermost few centimeters of the soil, known as the litter or duff layer, are particularly carbon-rich. If followed by a raw humus layer, this can further contribute significantly to carbon storage. However, when soils are disturbed, the carbon stored in these layers is released into the atmosphere. From a climate change perspective, it is therefore crucial to understand how soil mixing affects carbon content and how this varies throughout the soil profile.

In the framework of vegetation recovery in disturbed soils, it is essential to assess the concentrations of chemical elements necessary for plant development. The most critical macronutrients for plant growth are nitrogen (N), phosphorus (P), and potassium (K). Additionally, calcium (Ca) and magnesium (Mg) play important roles. If any of these nutrients are present in insufficient concentrations, plant growth may be significantly inhibited.

This research focused on the changes occurring in a military training area following deforestation and soil disturbance activities. The main objective was to assess the status of soil physical properties, chemical composition, and trafficability conditions two years after deforestation and the sowing of herbaceous seeds. Although the study was conducted with the aim of better understanding bearing capacity recovery and the effects of deforestation in a military training context, its findings contribute to the broader understanding of soil structural recovery processes, with principles that are applicable beyond military training areas as well.

2. Methods

A field study, including soil measurements, sampling, and military vehicle trafficability experiments, was conducted in 2020 at a military training area located in northern Estonia, near the city of Tapa. Large-scale deforestation was carried out in 2018 to establish a new firing range. Grass seed had been sown on half of the deforested area in 2018. Additionally, a small section of the deforested area was milled again in 2020 due to overgrowth. That same year, another wooded area was cleared as part of the firing range expansion. Unfortunately, the type of stump grinder used and its working depth are unknown. Soil condition measurements and sampling were conducted on July 3 and November 21, 2020, in non-freezing conditions. The exact locations of the measurement areas are shown in Fig. 1, while the status, activities carried out at each area, and the soil type are summarized in Table 1. According to the Unified Soil Classification System (USCS), all areas are characterized by SM type soils (Estonian Soil Map Application, 2025; García-Gaines and Frankenstein, 2015). Trafficability experiments were conducted on August 18, 2020, in Area 1.

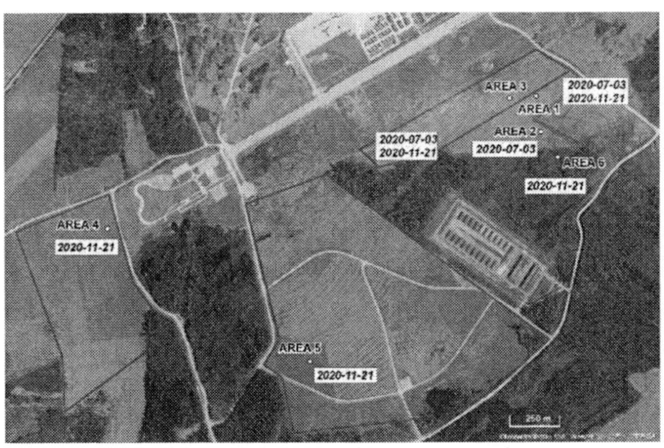

Fig. 1. Location of measured areas and measurement dates

Table 1
Soil type (according to the World Reference Base, WRB, (World reference base ..., 2022)) and management activities at the measurement areas

Area and soil type (by WRB)	Date and type of activities
Area 1: Mollic Cambisols Area 2: Gleyic Cambisols	2018-07 – deforestation (including cutting of trees, milling) 2018-09 – ground leveling and sowing grass seeds
Area 3: Mollic Cambisols	2020-06 – deforestation
Area 4: Calcaric Cambisols	Undisturbed natural grassland (co-called natural reference area)
Area 5: Mollic Cambisols	2018-06 - deforestation 2020-10 – additional soil surface milling up to depth of 5 cm
Area 6: Gleyic Cambisols	Natural forest

The main focus area of the study was a section of the new firing range, specifically, a site cleared in 2018 where herbaceous plant seeds had been sown. Since two different soil subtypes, Mollic

Cambisols and Gleyic Cambisols (Table 1), occur within this area, samples were collected from two separate locations to identify potential differences. These locations, referred to as Area 1 and Area 2 on the map shown in Fig. 1, differ topographically: Area 1 is situated on visually higher ground, while Area 2 lies in a relatively flat, gently sloping depression. In this study, Area 3 represents soil conditions immediately following the completion of earthworks, whereas the soils in Areas 1 and 2 allow for conclusions to be drawn regarding the changes that may have occurred during a two-year recovery period. The summer soil conditions in Area 1 and Area 3 are shown in Fig. 2. As of 2020, military training activities had not yet commenced in the newly established firing range.

Fig. 2. Ground conditions in Areas 1 and 3 on July 3, 2020

Area 4 (Tabel 1, Fig. 1) represents a natural grassland where no direct soil-disturbing activities have taken place, apart from occasional mowing and crossings by armored vehicles. This area serves as a reference to assess the load-bearing capacity and condition of the soil in its natural state. Again, the measurements in Area 5 allow for conclusions to be drawn regarding changes in soil strength additional round of surface milling. Soil samples were collected from Area 6 to determine the carbon and nutrient content of forest soils, with the aim of comparing these values to those of soils formed as a result of deforestation. The soil sampling and measurement plan are presented in Table 2.

For bulk density sampling, cylinders with a height of 4 cm and a diameter of 5.3 cm were used. To determine bulk density, the samples were oven-dried at 105 °C for 24 hours and weighed. The density was then calculated based on the soil volume (88.2 cm^3). In addition, the gravimetric water content of the soil samples was calculated, i.e., the ratio of the mass of water to the mass of dry soil. Penetrometer readings were recorded with a Humboldt penetrometer at depths of 0 cm, 2.5 cm, 5 cm, 7.5 cm, 10 cm, 12.5 cm, 15 cm, 23 cm, 30.5 cm, 38 cm, and 45.5 cm, or until the penetrometer reached its maximum value of 300 psi.

Based on the values obtained from the DCP measurements, the CBR was calculated using Eq. 1, where DCP represents the penetration (in mm) per blow. A graph of CBR values versus depth was then compiled based on the results.

$$CBR = \frac{292}{DCP^{1.12}} \quad (1)$$

The content of mobile plant nutrients (P, K, Ca, Mg) in the soils was determined using the ammonium acetate-lactate (AL) method (Egner et al., 1960). In the AL method, nutrient extraction is performed with a solution of 0.1 M (i.e. mol/L) ammonium lactate and 0.4 M acetic acid at a pH of 3.7. The extraction lasted for four hours, after which the resulting suspension was filtered, and the nutrient content was determined using microwave plasma atomic emission spectrometry (MP-AES). An Agilent MP-4200 atomic emission spectrometer was used for the analysis. Results are presented as the average of three measurements. For the evaluation of soil nutrient content, a classification developed for Estonian conditions was applied (Table 3).

Table 2
Soil property measurements and sampling timeline

Area	Sampling and measurements date 2020-07-03	2020-11-26
Area 1	(1)Cone indexes in 20 repetitions (2)DCP in 3 repetitions (3)Samples for bulk density and chemical composition determination from depth of 0, 5, 10, 15, 20 and 30 cm, in 4 repetitions (4)Samples for porosity determination from 0 cm, in 5 repetitions	(1)Cone indexes in 20 repetitions (2)DCP in 3 repetitions
Area 2	(1)Cone indexes in 20 repetitions (2)DCP in 3 repetitions (3)Samples for bulk density and chemical composition determination from depth of 0, 5, 10, 15, 20 and 30 cm, in 4 repetitions (4)Samples for porosity determination from 0 cm, in 5 repetitions	-
Area 3	(1)Cone indexes in 10 repetitions (2)Samples for bulk density determination from depth of 0, 5, 10, 15, 20 and 30 cm, in 4 repetitions	(1)Cone indexes in 10 repetitions (2)DCP in 3 repetitions
Area 4	-	(1)Cone indexes in 20 repetitions (2)Samples for bulk density determination from depth of 0, 5, 10, 15, 20 and 30 cm, in 4 repetitions (3)Samples for porosity determination from 0 cm, in 5 repetitions
Area 5	-	(1)Cone indexes in 20 repetitions (2)DCP in 3 repetitions (3)Samples for bulk density determination from depth of 0, 5, 10, 15, 20 and 30 cm, in 4 repetitions (4)Samples for porosity determination from 0 cm, in 5 repetitions
Area 6	-	(1)Samples for chemical composition determination from depth of 0, 5, 10, 15, 20 and 30 cm, in 4 repetitions

Table 3
Classification of mineral soils based on plant-available phosphorus (P), potassium (K), and magnesium (Mg) content according to the AL method (Nurmekivi, 2002)

Class	Phosphorus [mg/kg]	Potassium [mg/kg]	Magnesium* [mg/kg]
Very low	<20	<50	-
Low	20–40	50–100	<140
Moderate	41–80	101–200	150–250
High	81–120	>200	>250
Very high	>120	-	-

*Applies to soils with a humus content of 8–12% and pH_{KCl}> 5.6

The soil carbon and nitrogen contents were determined using the Dumas method with an Elementar Macro CNS analyzer. In this method, the sample is combusted in an oxygen atmosphere at 1000° C. The amounts of CO_2 and N_2 produced during combustion are measured, and the carbon and nitrogen contents of the analyzed material are calculated accordingly.

Soil porosity was determined using computed tomography. For tomographic analysis, undisturbed monolith samples with a diameter of 10 cm and a height of up to 20 cm were taken from the study area. Three monolith samples were collected from each site. The samples were extracted using a 25 cm long PVC plastic tube, which was used both for transportation to the laboratory and for conducting the tomographic analysis. The scans were performed using a Yxlon FF35CT tomograph and analyzed with Volume Graphics VGStudio Max 3.2 software. Cylinders with a diameter of 8 cm and a length of 15 cm (10 cm for one site) were selected from the tomographic images for porosity analysis. To determine changes in porosity along the soil profile, the cylinders were divided into 5 cm segments, and porosity was calculated separately for each segment. The results from three samples were used to characterize each area. Statistical analysis were carried out in MS Excel.

3. Results

3.1. Bulk density, moisture, and porosity of the measured sites

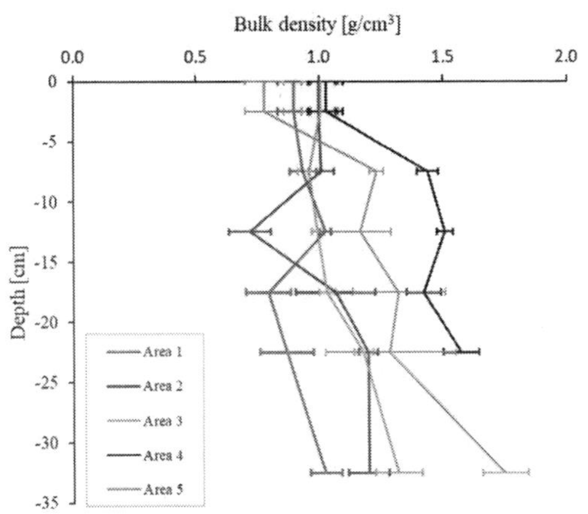

Fig. 3. Bulk density for measured areas. Horizontal bars indicate standard deviations.

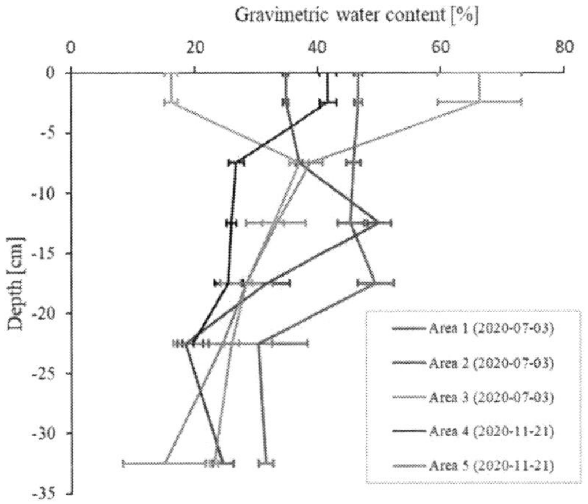

Fig. 4. Gravimetric water contents of studied areas. Horizontal bars indicate standard deviations.

The bulk densities (Fig. 3) determined for Area 1 and Area 2 in the 0–10 cm soil layer were similar, with an average value of 0.96 g/cm³. However, the gravimetric water content (Fig. 4) in the surface layer of Area 1 was approximately 10% higher. In Area 2, the relatively dry topsoil layer is followed by a significantly less dense (0.72 g/cm³) and moister intermediate layer at a depth of 10–15 cm. In the deeper layers, bulk density increased while moisture content decreased, which is characteristic of typical soil bulk density profiles. In Area 1, soil moisture content was higher and bulk density values were significantly lower in the layers deeper than 15 cm compared to Area 2. At a depth of 30–40 cm, the bulk density was 1.03 g/cm³, which is considered relatively low. According to Astover (2005), the typical bulk density of arable soils in Estonia ranges between 1.4 and 1.5 g/cm³, while similar forest soils may exhibit slightly lower values. In her PhD thesis, Vennik (2019) reported a bulk density of 1.8 g/cm³ at a depth of 30 cm in a soil with similar texture and soil type under natural grassland conditions. Hence, the bulk density of 1.2 g/cm³ at 30–40 cm depth in Area 2 can also be regarded as unusually low. This suggests that the soil disturbance caused by deforestation likely extended deeper than 20 cm and that the soil has not yet regained its natural density within the two-year recovery period.

The recovery of a soil's natural compaction state is a long-term process, driven by natural phenomena such as freeze–thaw and wetting–drying cycles (Hillel, 1998). Closer to the surface, soils are subject to more intense changes due to exposure to wind, temperature fluctuations, and moisture variation compared to deeper layers. Based on the bulk density profiles of Areas 1 and 2 (Fig. 3), it can be observed that in some locations, deeper layers are denser than the upper layers, while in others, the deeper layers are in fact looser than the surface soil. According to the moisture content values presented in Fig. 4, the initial visual estimation, i.e., that Area 1 was drier and higher and Area 2 wetter and lower, did not hold true. In all layers except at 12.5 cm depth, Area 2 exhibited lower moisture content compared to Area 1. In summary, Areas 1 and 2 are similar, both representing soils with a typical mixed structure, where bulk density varies considerably randomly with depth.

By comparing the bulk density values at different depths from Areas 1 and 2 with those from Area 3 (Fig. 3), it can be concluded that no significant changes in density have occurred in Areas 1 and 2 during the two-year period. The values measured in Area 3 are more or less identical to those in Areas 1 and 2. In fact, in the deeper layers (15–33 cm), the bulk density in the newly established Area 3 is even higher than in the adjacent area that has remained undisturbed for two years. One possible reason for the higher values is the greater amount of stones in the soil in Area 3, which may also have influenced the depth of milling. Whitecotton et al. (1999), referencing Stewart and Cameron (1992), noted that even a small amount of stones in the soil can reduce compaction, as stones help distribute the pressure.

Despite the autumn conditions, typically associated with higher precipitation (at the time of measurement, the soil surface was covered by a 5 cm snow layer) and limited evaporation, the moisture content across the soil profile in the reference natural grassland area (Area 4) on November 21 was lower than during the summer measurements in the newly established firing range, i.e. Area 1 (Fig. 4). The bulk density of the soil in Area 4 at the surface was equal to that recorded in the new firing range, i.e. 1.03 g/cm^3, and already reached 1.57 g/cm^3 just a few centimeters deeper (Fig. 3). Due to the high stoniness of the soil, it was not possible to collect samples from depths beyond 20 cm. In comparison to the natural reference area (Area 4), Areas 1, 2, and 3 were in a noticeably looser and wetter condition, extending down to a depth of 32.5 cm. The higher soil bulk densities observed in the uppermost 2.5 cm layer of Area 1, 2 and 3 can likely be associated with surface levelling operations conducted after forest clearing. In the case of Area 5, the deeper soil layers, starting from 17.5 cm, were already comparable to those in the natural area (Area 4). However, the recent clearing had clearly affected the bulk density of the surface layer (Fig. 3).

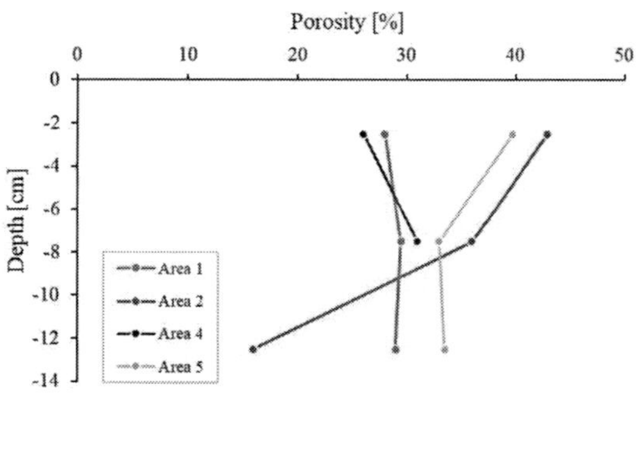

Fig. 5. Porosity of the measured areas

Figure 5 illustrates the porosity values of the measured soils. The porosity of Area 1 is quite similar to that of reference area - Area 4. The low total pore volume and high water content in Area 1 (Fig. 4) indicate that the soil is predominantly composed of fine pores, which have a high water retention capacity. In contrast, Area 4 is probably characterized mainly by larger pores, which retain less water. Based on porosity and moisture content, it can be inferred that in Area 2, the uppermost 7.5 cm layer is dominated by large pores, while at a depth of 12.5 cm, there is a layer with lower porosity, composed mainly of fine pores but also exhibiting lower bulk density. Thus, although the surface bulk density of Areas 1 and 2 is similar, their pore systems, and consequently their water-holding capacities differ significantly. Overall, there is considerable variability in density, porosity, and water retention over a short distance. Such areas affected by land clearing represent a much more complex physical environment and pose different challenges to load-bearing capacity compared to soils with an undisturbed natural structure and compaction. The measurement results from Area 5 suggest that the most recent clearing cycle increased the total pore volume. (Fig. 5).

3.2. Cone index values and their relation to trafficability

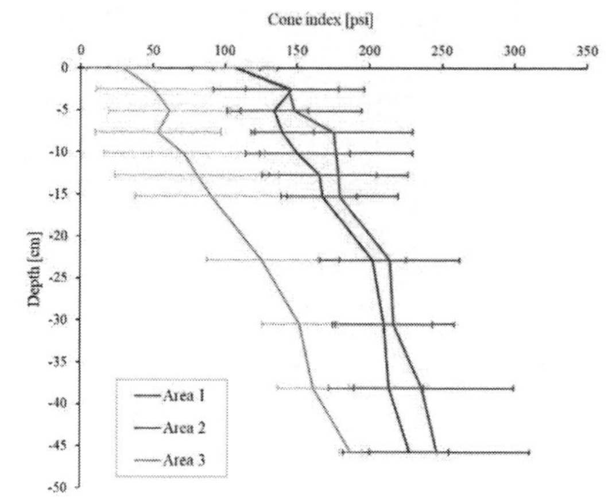

Fig. 6. Measured cone index values as a function of depth on 2020-07-03 in Areas 1, 2, and 3. Horizontal bars indicate standard deviations.

Figure 6 shows that the strength profiles determined in Areas 1 and 2 are similar, with the average cone index increasing almost linearly from 107 psi at the surface to 247 psi at a depth of 45 cm. However, analysis of individual penetrometer thrusts reveals the presence of many abnormal strength profiles, i.e. cases where cone index (CI) values in deeper layers are lower than those at the surface. The high standard deviation of mean CI values indicates considerable variability in strength, which is typical of soil layers that have been recently disturbed and have not yet reconsolidated. On the freshly milled surface in Area 3, CI values are statistically significantly lower than those in Areas 1 and 2 ($p<0.05$). Furthermore, as shown in Fig. 4, this lower strength cannot be attributed to differences in moisture content.

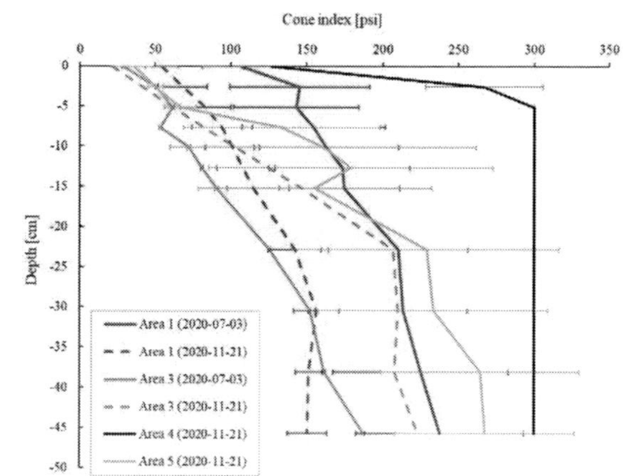

Fig. 7. Measured cone index values as a function of depth on 2020-07-03 and 2020-11-21 in Areas 1, 3, 4, and 5. Horizontal bars indicate standard deviations.

All CI values (except for those from Area 2) are presented in Fig. 7. The autumn measurements in the main study area, Area 1, indicate that strength values in all layers are, on average, approximately 75 psi lower than in summer. In contrast, autumn strength values in Area 3 are higher than the summer values. This difference may be explained by local variability within Area 3, i.e. the autumn penetrometer measurements were likely conducted on a stonier segment than those in summer. The most stable and strongest subsoil was found in Area 4, a natural grassland area, where the penetrometer reached its maximum value at a depth of only 5 cm. Across the entire measured profile, Area 5 generally showed higher strength values than Area 1 in November; however, the high variability (i.e., large standard deviation) diminishes the significance of these two means. Once again, additional milling just before the measurements significantly decreased the CI values in the top 5 cm layer of Area 5. To assess the soil strength conditions of the measured areas from the perspective of vehicle trafficability using the penetrometer method, it is essential to present the average CI values for the 15–30 cm soil layer. The corresponding values are presented in Table 4.

Table 4
Cone index values in depth of 15-30 cm

Area	2020-07-03	2020-11-21
Area 1	202 psi	138 psi
Area 2	194 psi	-
Area 3	135 psi	189 psi
Area 4	-	300 psi
Area 5	-	200 psi

For comparison, the ground strength requirements for the tacked military combat vehicle CV90 (with a total weight of 32.8 t, track length of 4.1 m, and track width of 0.53 m) to perform one and fifty passes VCI_1 and VCI_{50} are 22 psi and 50 psi, respectively. According to the 2020 measurements (Table 4), all studied areas had sufficient strength in the critical soil layer to support at least 50 passes by the above-mentioned armored vehicle.

As discussed above, disturbed soils present a unique environment, characterized by high variability, including layers with differing porosity and water-holding capacities. This high heterogeneity may seriously affect actual trafficability. To test the applicability of the penetrometer method and its assessment capability for disturbed soils, trafficking experiments were conducted in August, 18 in Area 1. For comparison also up to five passes on Area 4 in November were carried out with CV90.

Fig. 8. Tracks left behind by the CV90 on 2020-08-18: after 50 passes (left) and after a 90-degree turn (right) in Area 2.

Fig. 9. CV90 trafficability tests on 2020-11-21: CV90 on the field (left) and tracks left behind after five passes (right) in Area 4.

Figure 8 presents the tracks left behind by the CV90 after 50 passes and after performing a 90-degree turn. All the tracks on the ground were homogeneous and even. Measured rut depths were shallow, ranged from 2 to 7 cm after 50 passes and from 2 to 9 cm after turning. The rut depths were similar to the Area 4 results (Fig. 9). The tests conducted indicated the applicability of the penetrometer method regardless of soil disturbance.

To analyze the potential influence of plant root strengthening on soil shear strength in the surface layer, Table 5 presents the average strength values of the topsoil layer (up to 7.5 cm thick) separately.

Table 5
Cone index values in depth of 0-7.5 cm

Area	2020-07-03	2020-11-21
Area 1	144 psi	75 psi
Area 2	134 psi	-
Area 3	49 psi	51 psi
Area 4	-	248 psi
Area 5	-	71 psi

While the summer measurements showed more than a twofold difference in surface layer strength between Areas 1 and 2 compared to Area 3, the difference observed in the autumn measurements was only several tens of psi (Table 5). This difference was statistically significant ($p < 0.05$), indicating that the freshly cleared open area established in 2020 to expand the firing area has a weaker surface layer than the disturbed area that has recovered over two years and developed a grass cover.

3.3. CBR values of measured sites

The CBR values for the measured areas are presented in Fig.s 10 and 11.

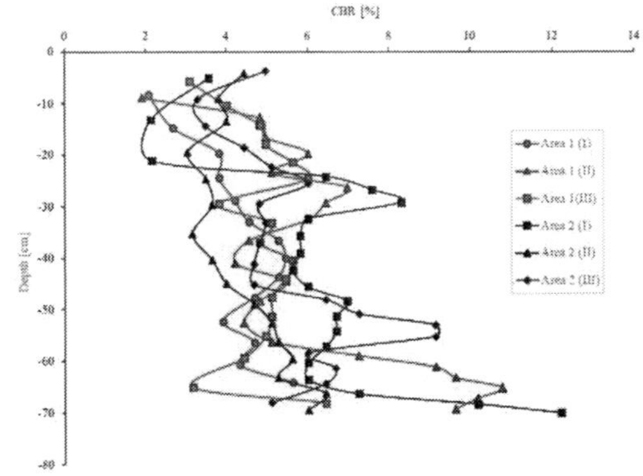

Fig. 10. CBR values on 2020-07-03 in Areas 1 and 2.

Although the summer CBR profile lines presented in Fig. 10 show some horizontal curvature, indicating the presence of stronger and weaker layers within the soil in Area 1 and 2, the overall strength values remain low, with only minor variations. A slight increase in bearing capacity is observed in the deeper layers. The CBR values in Areas 1 and 2 across the entire study range from 1.9% to 12.2% for multiple DCP measurements, with values exceeding 10% found only below 60 cm. The average CBR value throughout the studied depth range in these areas is 5.4%. According to Shoop et al. (2008), CBR values below 10% are typical of fine-grained soils with low strength under moist conditions. Even lower CBR values are found in peat soils, where values typically fall below 1%. Based on U.S. Army engineer guidelines for expedient road construction and considering the movement of a 40-ton tank up to 40 times annually over a 25-year period, a CBR value of 5% implies that a 44 cm layer of crushed rock and gravel must be added to the subgrade to achieve the necessary soil strength and stability (TM 5-822-12, 1990).

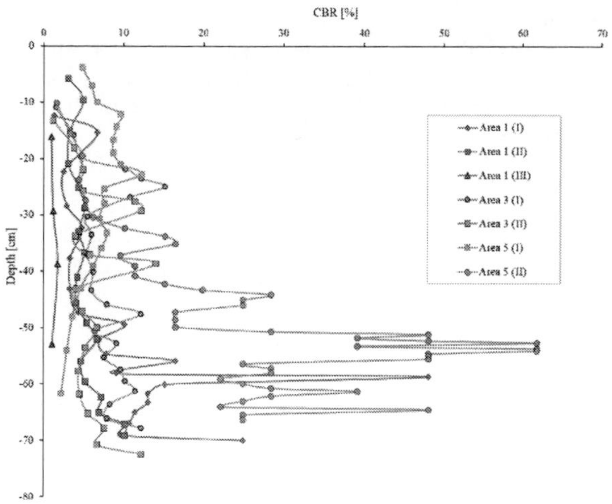

Fig. 11. CBR values on 2020-11-21 in Areas 1, 3 and 5.

From the measurements taken in November, one strength profile at Area 1 (Fig. 11, Area 1 (III)) showed that the CBR values did not exceed 2%. The results from the other two measurements were similar to the summer values (Fig. 10), remaining within the range of 1.3% to 15.5% at various depths. A high CBR value of 48% at a depth of approximately 60 cm suggests the displacement of a stone that remained in the cone path. Based on the autumn measurements, it can be concluded that there are no significant differences in load-bearing capacity between Area 3 and the study areas. The presence of sharp peaks in the characteristic curves (CBR up to 15.5%) in Area 3 is most likely due to the presence of a large number of stones. The load-bearing capacity heterogeneity in the soil profile, resulting from milling and mixing of the soil cover, is well represented by the results from Area 5. According to one CBR profile (Area 5 (I)), the deeper layers of the ground (starting from 40 cm) are weaker than the surface layers: the average CBR for the 0–40 cm depth is 8%, while for the 40–62 cm depth, it is only 3.4%. However, according to a profile measured a few meters away (Area 5 (II), the load-bearing capacity values at depths greater than 40 cm ranged from 20% to 60%. In the strongest soil, Area 4, CBR measurements could not be performed due to the high content of stones and slate slabs in the soil.

3.4. Content of chemical elements and conclusions on environmental impacts

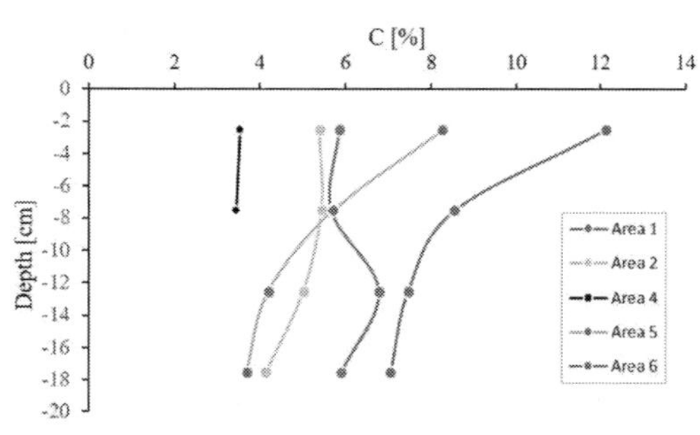

Fig. 12. The content of total carbon (C) of studied areas as a function of depth

The total carbon content in the analyzed soils ranged from 3.4% to 12.3% (Fig. 12). The highest carbon content was found in the upper soil layer of Area 6 (forest site). In the milled areas (Areas 1, 2, and 5), carbon content was lower across all measured layers compared to the undisturbed forest soil (Area 6). In some cases within the milled areas, surface layers exhibited lower carbon concentrations than deeper layers, while in others, the opposite pattern was observed. Measurement results suggest that milling has homogenized the organic matter content, at least down to a depth of 25 cm.

In the context of climate change, the observed change in carbon content can be regarded as a natural consequence of human activity. The findings show that carbon content in the soil decreased in all milled areas but remained higher than in the natural grassland (Area 4). Similar changes in carbon content are also observed in agricultural soils as a result of cultivation. Monitoring results from Estonia show that the average carbon loss from agricultural soils is 2.14 t/ha from the plough layer, resulting in 7.8 t CO_2 ha^{-1} a^{-1} emissions (Eesti maaelu arengukava 2014–2020). Based on the same study, it can be concluded that although intensive oxidation of loose milled material will continue in the near future, its rate will decrease over time, and with the development of a dense grass cover, the area will begin to act as a carbon sink.

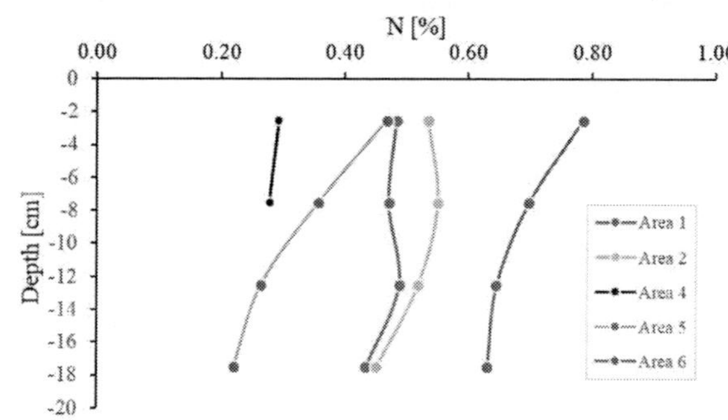

Fig. 13. The content of total nitrogen (N) of studied areas as a function of depth

The total nitrogen content of the soil samples collected from the study areas ranged from 0.25% to 0.8% (Fig.13). The highest nitrogen content was observed in Area 6 (forest area) throughout the entire 25 cm profile, although it decreased to 0.57% with depth. In the disturbed areas, specifically Areas 1 and 2, nitrogen levels were lower than in the forest, with particularly low concentrations recorded in Area 5. This supports the findings from the carbon data, suggesting that milling has homogenized the organic matter within the upper 25 cm of soil. However, due to the heterogeneity of the original material, localized zones of higher or lower nutrient content may still be present within the disturbed layer.

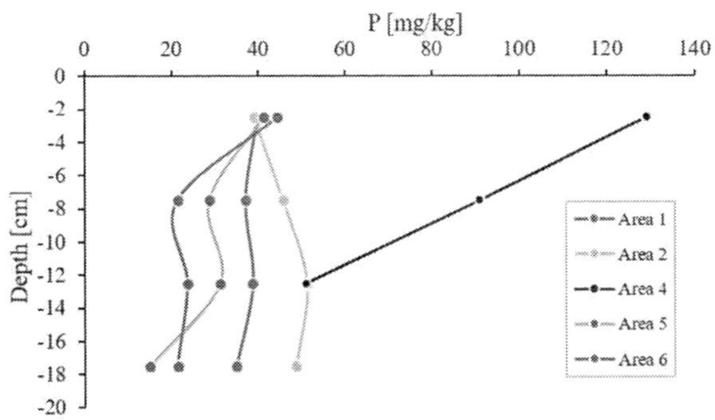

Fig. 14. The content of phosphorus (P) of studied areas as a function of depth

Soils are classified into five categories based on their phosphorus (P) content (Table 3). All P content classes, from very low to very high, are represented within the study areas (Fig. 14), with concentrations ranging from 11 to 129 mg/kg. High and very high P levels were found in the natural grassland area (Area 4). Naturally, the soils in these areas are relatively low in P. To clarify the reasons for the observed high P content, follow-up sampling and investigation over a larger area would be needed. P content was very low in the forest area (Area 6). In the disturbed areas, P content generally fell into the low to moderate classes, and such low concentrations are likely to act as limiting factors for plant growth.

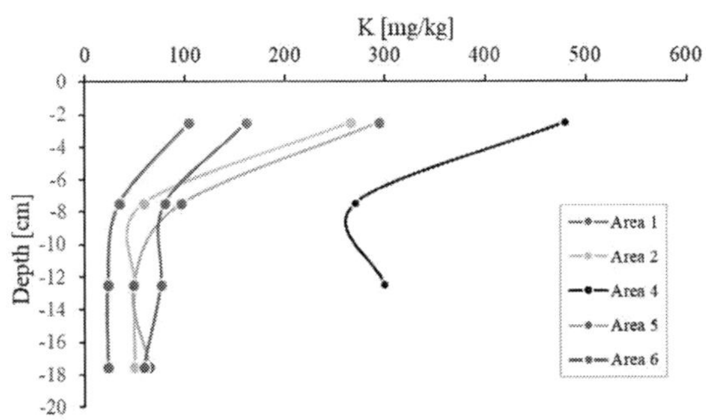

Fig. 15. The content of potassium (K) of studied areas as a function of depth

Based on the proposed potassium (K) classes (Table 3), it can be estimated (Fig. 15) that in the upper 5 cm soil layer, K content in all areas is either high or moderate, followed by a sharp decline in plant-available potassium across all test sites. In the 5–10 cm layer, most soils in the test areas already fall into the low potassium category, with concentrations that may be limiting to plant growth. Area 4 stands out with significantly higher potassium levels, where even deeper layers exhibit high K concentrations. No differences attributable to land use or treatment were observed across the measured areas, and no discernible variation was found in the potassium content profiles between milled and un-milled sites.

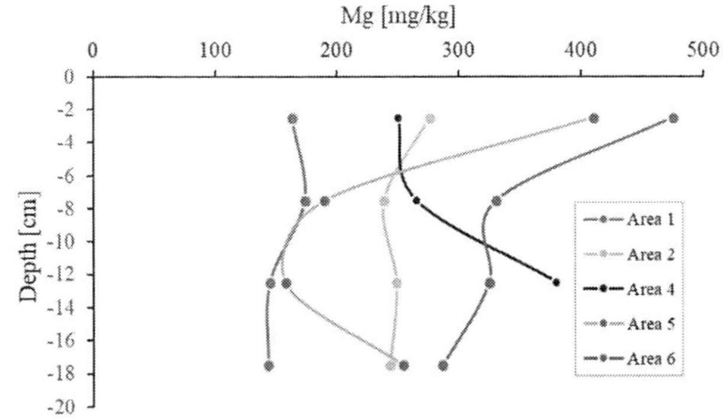

Fig. 16. The content of magnesium (Mg) of studied areas as a function of depth

The magnesium (Mg) concentration (Fig. 16) is high in forest soils (Table 3) and was likely similarly high in the study areas prior to deforestation. As a result of soil disturbance, the Mg content decreased significantly; however, the current values still fall within the moderate range.

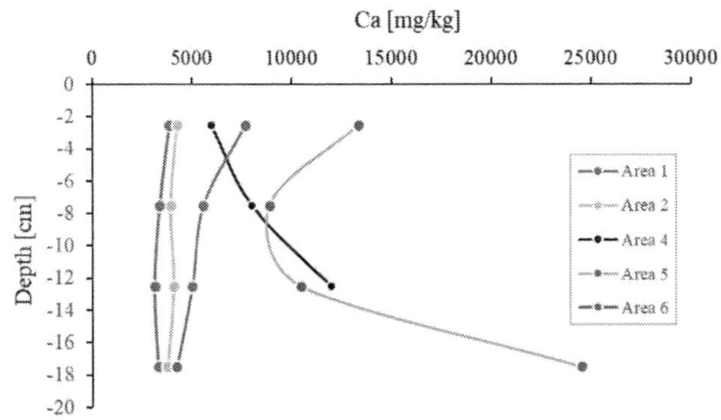

Fig. 17. The content of calcium (Ca) of studied areas as a function of depth

The extractable calcium content in the soils of the test areas, as determined by the AL method, varies up to tenfold, ranging from 2,500 to 25,000 mg/kg (Fig. 17). As the soils are calcareous, a high calcium content is naturally expected. Calcium is a relatively immobile element, and variations in its content within the soil profile result from long-term leaching processes, which are also influenced by the thickness of the soil layer overlying the carbonate-rich subsoil. Due to the mixing of different soil layers during milling, the calcium content has become homogenized throughout the profile, and given the slow rate of leaching, noticeable stratification has not yet developed. Since both calcium and magnesium form bicarbonates in the soil solution, their leaching under local climatic conditions is inevitable. However, leaching is reduced in grass-covered soils, as the presence of herbaceous vegetation with a longer growing season decreases the amount of water percolating through the soil profile.

4. Conclusions

Based on bulk density measurements conducted in the newly established firing range, it can be concluded that, at the soil surface, there is no significant difference in density between disturbed and natural soils, except in the case of very recently milled soils (treated less than one month prior). Significant differences in bulk density become evident in deeper soil layers, i.e. 5–25 cm. No substantial changes in the bulk density of treated soils were observed over a two-year period, and the density remained nearly as low as that of the surface layer. Wet and loosely packed intermediate layers, including most probably larger but also thinner pores, were found between stronger and drier horizons. Penetrometer measurements and trafficability tests suggest that the soil in the new firing range provides sufficient strength to support repeated vehicle traffic. Nonetheless, it is necessary to conduct trafficability tests under wet conditions as well. The CBR test results further confirm the earlier observations: alternating strong and weak layers are present within the soil profile, and from a road construction perspective, the subgrade can be considered relatively weak.

Soil milling has induced changes in the chemical element composition of the soil. The primary effect has been the homogenization of nutrient and carbon concentrations throughout the soil profile. In milled areas, lower concentrations of key nutrients such as phosphorus and potassium have been recorded. The carbon content in the surface layer of undisturbed forest soil is approximately twice as high as that in milled areas. In the context of climate change, the observed reduction in soil carbon can be considered a typical anthropogenic impact. A comparable loss of carbon occurs annually in cultivated agricultural fields. Overall, these changes are rather unfavorable in the context of cross-country trafficability. Previous research indicates that a higher organic matter content (up to a certain limit) improves trafficability conditions by making the soil behave more elastically and enhancing its recovery after traffic.

In summary, while the milled areas still exhibit characteristic features of mixed soils after 2-years recovery period, they are generally in good condition from the perspective of trafficability. The sowing of grass seed has clearly contributed to the rapid re-establishment of vegetation cover and increase in surface layer strength.

5. References

Ali, F.H., Osman, N., 2008. Shear strength of a soil containing vegetation roots. Soils Found. 48 (4), 587–596.

Astover, A., 2005. Eesti mullastik ja muldade kasutussobivus. Tartu. https://maad.maad.ee/avalik_leht/img/Eesti-mullastik.pdf

Brown, R.N., Percivalle, C., Narkiewicz, S., DeCuollo, S., 2010. Relative rooting depths of native grasses and amenity grasses with potential for use on roadsides in New England. HortScience 45 (3), 393–400.

Eesti maaelu arengukava 2014-2020 4. JA 5. Prioriteedi püsihindamine. Mulla orgaanilise süsiniku varu muutus ja CO2 emissioon erinevate toetusmeetmete ning maakasutuse puhul. Töö teostaja: Põllumajandusuuringute Keskus, mullaseire büroo. https://pmk.agri.ee/sites/default/files/2019-11/Kliimamuutused_2016.pdf

Egner, H., Riehm, H., Domingo, W.R., 1960. Untersuchungen über die chemische Bodenanalyse als Grundlage für die Beurteilung des Nahrstoffzustandes der Boden, II : Chemische Extractionsmetoden zu Phosphor- und Kaliumbestimmung. Kungliga Lantbrukshügskolans Annaler, 26, 199–215.

Estonian Soil Map Application, 2025. https://xgis.maaamet.ee/xgis2/page/app/mullakaart

FM 5-430-00-1, 1994.Planning and Design of Roads, Airfield, and Heliports in the Theater of Operations – Road design, Headquarters, Department of the Army, Department of the Air Force, Washington DC, Aug. 1994.

García-Gaines, R., A., Frankenstein, S., 2015. USCS and the USDA Soil Classification System. Development of a Mapping Scheme. ERDC/CRREL TR-15-4, US Army Corps of Engineers, Cold Regions Research and Engineering Laboratory.

Goldsmith, W., 2006. Soil strength reinforcement by plants. In: Proc. of the 37th Conference of the International Erosion Control Association, February 20–24, 2006, Long Beach, California, 3–18.

Havlíček, M., Skokanová, H., Dostál, I., Vymazalová, M., Pavelková, R., Petrovič, F., 2018. The consequences of establishing military training areas for land use development—A case study of Libavá, Czech Republic. Land Use Policy 73, 84–94.

Hillel, D., 1998. Enironmental soil Physics. Elsevier (USA).

Machado, L., Holanda, F.S.R., da Silva, V.S., Maranduba, A.I.A., Lino, J.B., 2015. Contribution of the root system of vetiver grass towards slope stabilization of the São Francisco River. SEMINA: CIENCIAS AGRARIAS. August 2015, 2453-2463.

Nurmekivi, H., 2002. Põllukultuuride väetamine. Põllumajandusministeerium.

Peng, L., Searchinger, T.D., Zionts, J., Waite, R., 2023. The carbon costs of global wood harvests. Nature 620 (7972), 1–6, DOI: 10.1038/s41586-023-06187-1

Priddy, J., 1995. Stochastic vehicle mobility forecasts using the NATO reference mobility model. Report 3. Database Development for Statistical Analysis of the NRMM II Cross-Country Traction Empirical Relationships. Technical Report GL-95-8. https://scholar.google.com/citations?view_op=view_citation&hl=en&user=9Jx-n3MAAAAJ&citation_for_view=9Jx-n3MAAAAJ:u5HHmVD_uO8C

Shoop, S.A., Diemand, D., Wieder, W.L., Mason, G., Seman, P.M., 2008. Opportune Landing Site Program. Predicting California Bearing Ratio from Trafficability Cone Index Values. ERDC/CRREL TR-08-17.

Shoop, S.A., Coutermarsh, B., Cary, T., Howard, H., 2015. Quantifying vegetation biomass impacts on vehicle mobility. J. Terramech. 61, 63–76.

Smith, R.B., Wass, E.F., 1989. Soil displacement in stump-uprooting equipment trials on a root rot-infested cutover. J. Soil Water Conserv. 44, 351–352.

Stewart, D.P.C., Cameron, K.C., 1992. Effect of trampling on the soils of the St. James Walkway, New Zealand. Soil Use Manage 8, 30–36.

TM 5-822-12, 1990. Design of aggregate surfaced roads and airfields. Headquarters Department Of The Army. Washington, DC, 28 September.

Tobias, S., Haberecht, M., Stettler, M., Meyer, M., Ingensand, H., 2008. Assessing the reversibility of soil displacement after wheeling in situ on restored soils. Soil Tillage Res. 98, 81–93.

Trükmann, K., Horn, R., Reintam, E., 2009. Impact Of Roots On Soil Stabilisation In Grassland. ISTRO 18th Triennial Conference Proceedings, June 15-19, 2009 Izmir-TURKEY.

Vennik, K., 2019. The effect of military vehicles on rut formation on Estonian soils and natural recovery of the ruts. PhD Thesis. Estonian University of Life Sciences. https://dspace.emu.ee/xmlui/handle/10492/4859

Whitecotton, R.C.A., David, M.B., Darmody, R.G., Price, D.L., 1999. Foot Traffic Effects on Grassland Soil Properties at the U.S. Air Force Academy, Colorado. US Army Coprs of Engineers Engineer Research and Development Center. CERL Technical Manuscript 99/98 Decemmer 1999.

World reference base for soil resources 4th edition, 2022. https://www.isric.org/sites/default/files/WRB_fourth_edition_2022-12-18.pdf

Zentelis, R., Lindenmayer, D., 2015. Bombing for Biodiversity—Enhancing Conservation Values of Military Training Areas. Conservation Letters, July/August 2015, 8(4), 299–305. doi: 10.1111/conl.12155

Zentelis, R., Banks, S., Roberts, J.D., Dovers S., Lindenmayer, D., 2017. Managing military training-related environmental disturbance. J Environ Manage Dec 15: 486-493. doi: 10.1016/j.jenvman.2017.09.029.

4558 | In-situ soil-property estimation and Bayesian mapping with a simulated compact track loader

W. Jacob Wagner [a,b,*], Ahmet Soylemezoglu [b], Katherine Driggs-Campbell [a]

[a] *Department of Electrical and Computer Engineering, University of Illinois at Urbana-Champaign, Urbana, IL, United States*

[b] *Construction Engineering Research Laboratory, Engineer Research and Development Center, United States Army Corps of Engineers, Champaign, IL, United States*

* *Corresponding author: William.J.Wagner@erdc.dren.mil*

ABSTRACT

Existing earthmoving autonomy is largely confined to highly controlled and well-characterized environments due to the complexity of vehicle-terrain interaction dynamics and the partial observability of the terrain resulting from unknown and spatially varying soil conditions. To overcome these restrictions and facilitate development of more robust autonomous earthmoving, a soil-property mapping system is proposed to extend the environmental state.

A GPU-accelerated elevation mapping system is extended to incorporate a blind mapping component that traces the movement of the blade through the terrain to displace and erode intersected soil, enabling separately tracking undisturbed and disturbed soil. Each interaction is approximated as a flat blade moving through a locally homogeneous soil, enabling modeling of cutting forces using the fundamental equation of earthmoving (FEE). Building upon our prior work on in situ soil-property estimation, a method is devised to extract approximate geometric parameters of the model given the uneven terrain, and an improved physics-infused neural network (PINN) model is developed to predict soil properties and uncertainties of these estimates.

A simulation of a compact track loader (CTL) with a blade attachment is used to collect data to train the PINN model. Post-training, the model is leveraged online by the mapping system to track soil property estimates spatially as separate layers in the map, with updates being performed in a Bayesian manner. Initial experiments show that the system accurately highlights regions requiring higher relative interaction forces, indicating the promise of this approach in enabling soil-aware planning for autonomous terrain shaping.

Keywords
Autonomous Earthmoving
Soil-Property Estimation
Physics-Infused Neural Networks
Robotics
Terrain Mapping

1. Introduction

The construction equipment manufacturing industry has advanced equipment automation over the last few decades (Saidi et al., 2016). Most earthmoving equipment can now either be retrofitted with or come directly from the factory with advanced operator assist and machine control capabilities (Jackson, 2017; Ries, 2019). The most advanced versions of this technology not only provide feedback to the operator about the state of the current terrain compared to the design surface but also automate portions of the control to increase inexperienced operator efficiency and to ensure that soil is not unnecessarily disturbed (Azar and Kamat, 2017; Hayashi et al., 2013).

Given a desire to increase machine productivity, the growing labor shortage for skilled equipment operators, and advances in autonomy in self-driving, there is recent increased interest in the development of autonomous construction equipment (Nguyen and Ha, 2023; Mathur et al., 2024). In the domain of autonomous earthmoving, heavy equipment original equipment manufacturers have focused on developing autonomy for large-scale mining operations, including autonomous dump truck navigation (Bellamy and Pravica, 2011) and execution of slot-dozing operations autonomously (D'Adamo, 2023). Several startups have also entered the market and are working to bring autonomy to other heavy equipment, including excavators, and to other applications such as solar-farm construction and earthmoving for construction. For a more comprehensive treatment of autonomous construction, see these reviews: Nguyen and Ha, 2023; Dadhich et al., 2016; Ha et al., 2019.

While in some applications, such as surface mining, it is possible to have a significant degree of control over the environmental conditions (D'Adamo, 2023), in many applications of interest, the environment is much more unstructured (Soylemezoglu et al. 2021, Zhu et al., 2023). The environment is partially observable because existing sensors are only capable of producing information regarding the surface characteristics of the terrain. The shape of the terrain can be provided a priori, or mapped by inferring ground height from machine position and/or using onboard depth sensing (Jud et al., 2019; D'Adamo, 2023). Semantic segmentation methods can provide some information regarding the broad class of material on the surface but provide limited information regarding the physical properties of the soil relevant to earthmoving. Given that soil conditions can substantially vary with both lateral position and depth, surface measurements are not adequate to inform soil-aware planning and tool control.

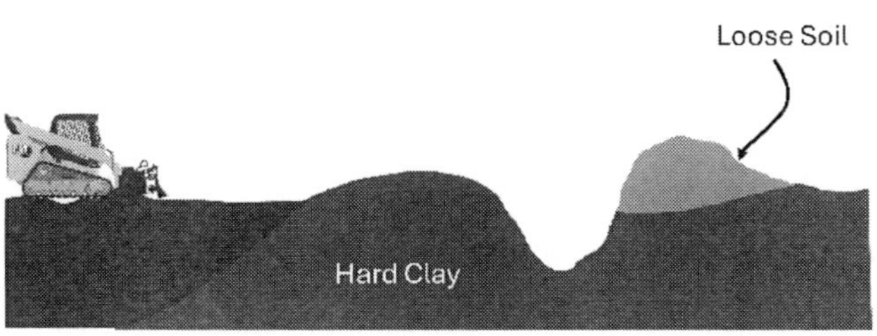

Fig. 1. Example earthmoving scenario used to illustrate the need for soil-aware planning.

It is generally recognized that some level of adaptation is necessary at the control level for autonomous earthmoving to enable robust and efficient execution of earthmoving cycles in environments with unknown and spatially varying soil conditions (Dadhich et al., 2016). However, adaptation of earthmoving plans to soil conditions has not been thoroughly investigated.

https://doi.org/10.56884/M9KLDUH0

To illustrate the need for adaptation to soil conditions at the planning level, consider the following scenario, depicted in Fig. 1. The task is for a bulldozer to fill in a ditch and ensure the resulting terrain surface is smooth enough for a wheeled vehicle to traverse the region. The planner's job is to select a cut start location and a desired relative cut-depth for the blade controller. It is important to note that the soil strength, arising from the soil properties, determines the maximum allowable cut depth to avoid machine stalling, given a finite bulldozer tractive force. If the soil properties are unknown, it is impossible to select the dig parameters optimally.

A soil-ignorant planner can act conservatively by selecting a cut start location far from the deposit point and a small depth of cut, but this is only optimal for very high-strength soils. Alternatively, the planner can act more aggressively by selecting a dig point near the deposit location and a large depth of cut. This approach is optimal for uniform low-strength soils but leads to insufficient accumulated material volume when an adaptive low-level blade controller is utilized to avoid stalling. In fact, this scenario is encountered by Egli et al. (2022) in the development of a learned adaptive scooping policy, where it is remarked that for hard soils the controller is only able to partially fill the bucket, even when the user defined dig-point is selected to be at the maximum extent of the excavator arm.

Soil conditions are often not spatially homogeneous, which can further complicate planning. Consider a slightly modified scenario where far from the ditch, a low-strength loam is present, but near the ditch the soil is made of high-strength clay. If a soil-ignorant planner is tuned aggressively to make deep cuts near the ditch, this will result in a much higher number of cuts to fill the ditch than a soil-aware planner which can more quickly accumulate material by choosing to cut the lower-strength material further away from the ditch and then carry it over the higher-strength material to the deposit location. The possibly obvious but important lesson here is that spatially varying soil conditions result in varying forces required to shear soil, which subsequently affect earthmoving plan feasibility and the resulting cut trajectory, further affecting accumulation and deposition of material. The implication is that to support development of soil-aware earthmoving planners, it is first necessary to develop an approach to obtaining knowledge of soil conditions in situ and to devise a method for keeping track of how these conditions vary spatially, which is the focus of this work.

1.1. Soil-Aware Autonomous Earthmoving

Early approaches to autonomous earthmoving focused on the development of low-level control systems to coordinate machine movement throughout the phases of a dig cycle. One approach is to formulate the control problem in terms of kinematic trajectory tracking. This is intuitive, especially for excavation applications, given that an excavator arm is a serial manipulator and can be treated similarly to robotic manipulators, enabling re-use of motion planning tools for trajectory generation and task-space control. However, these kinematic trajectory control approaches fail if the provided trajectory is unable to be executed accurately when machine force limits are reached. The force required to move the tool through the ground depends not only on the dynamics of the machine, but on the resistance of the soil to shearing, which is dependent on the soil properties.

Rule-based methods and fuzzy logic controllers were initially explored to enable adaptation to soil conditions but require elicitation of rules from expert operators which can be challenging and have fallen out of favor recently (Bradley et al, 1998; Lever, 2001). Another approach is to use compliant control to balance the trajectory tracking objective with machine force constraints (Dobson et al., 2016; Ha et. al, 2000). A challenge with using these compliant methods, such as impedance control, is that the impedance of the environment is unknown and variable. If a fixed environmental impedance is assumed, tracking performance will vary significantly across soil types. The open-loop bandwidth of hydraulic systems on heavy equipment is limited, and high gains can excite low resonant modes of the tool, leading to dangerous oscillations (Maeda et al., 2015). One way of addressing this tuning problem is to adapt controller parameters, e.g. stiffness and damping, online (Feng, et al., 2022; Ha et al., 2000). Sotiropoulos and Asada (2019) propose an alternative compliant control formulation based on maximization of power transmitted to the soil, which aims to match the impedance of the excavator with the impedance of the load, but it is not clear how this approach can be extended for tasks that require achieving a desired terrain shape. Maeda et al. (2015) suggest taking advantage of the cyclical nature of the excavation dig-cycles by using iterative learning control to compensate for the interaction forces encountered in the subsequent digging pass. Jud et al. (2017) discard the idea of tracking a desired kinematic trajectory and instead detail a force trajectory control approach to digging that supports constraining positions, velocities, and forces while avoiding penetration of a design surface. This scheme results in the production of various kinematic trajectories as the system reacts to differing soil conditions, but significant retrofitting of the machine hydraulic system is necessary to attain sufficient force tracking performance (Jud et al., 2019).

More conventional compliant manipulation approaches can also be used successfully and can be combined with methods that predict interaction forces and/or disturbances to improve tracking and ensure selection of feasible trajectories (Maeda et al., 2015; Singh, 1995a; Mononen et al., 2022). The main idea is that during earthmoving, when a system deviates from the model-derived expected behavior, this is caused by unmodeled soil-tool interaction forces and therefore, tracking errors contain information regarding the soil conditions. These errors are evident at the force level but are also reflected at the trajectory and velocity level (Bradley and Seward, 1995; Maeda, 2013), which helps to explain how human operators can infer information about soil conditions without knowledge of interaction forces.

Rather than treating interaction forces as disturbances, they can be predicted and incorporated in a feed-forward manner to enhance control (Cannon and Singh, 2000). Singh's (1995a) pioneering work introduced the concept of soil property identification to enable adaptive tactical planning, allowing for efficient excavation while staying within machine force limits. By estimating soil properties iteratively, it becomes possible to optimize digging trajectories under force constraints, thereby preventing actuator saturation. These properties are inferred from the soil-tool interaction using a combination of terrain mapping, known tool trajectories, and force measurements collected during execution (Luengo et al., 1998; Singh, 1995b; Cannon, 1999). For a comprehensive review of inverse terramechanics methods encompassing both traditional techniques and machine learning approaches, see Arreguin et al. (2021).

Many semi-empirical models of soil-tool interaction exist with varying degrees of complexity and different parametrizations, but the fundamental equation of earthmoving (FEE) is a common choice (Blouin et al., 2001). This model is derived by assuming a Mohr-Coulomb model of soil shear strength and performing a force balance on a system with a flat angled blade failing a wedge of soil (Mckyes, 1989; Reece, 1964; Holz et al., 2013; Wagner, 2025), as depicted in Fig. 2.

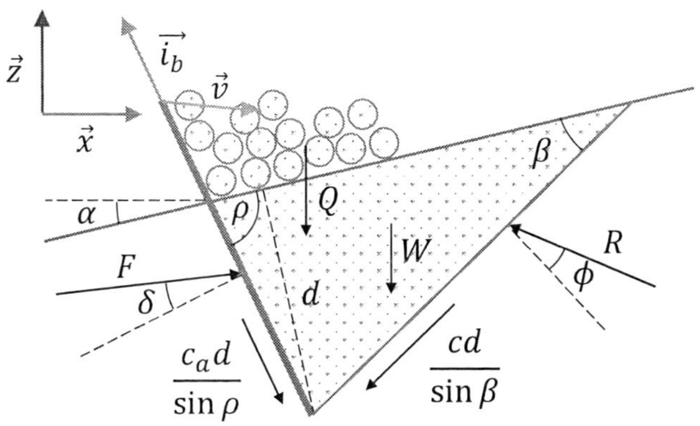

Fig. 2. Force diagram for the fundamental equation of earthmoving (FEE) with a surface inclined at angle α.

The resulting basic equation is given by

$$f = f_{FEE}(\Theta) = \gamma d^2 w N_\gamma + cdwN_c + QN_Q + c_a dwN_a \qquad (1)$$

$$F(\Theta) = f \cdot [\,\sin(\rho + \delta - \alpha),\ \cos(\rho + \delta - \alpha)] \qquad (2)$$

$$N_\gamma = \frac{[\cot(\rho)+\cot(\beta)]\sin(\alpha+\phi+\beta)}{2\sin(\eta)}, \quad N_c = \frac{\cos(\phi)}{\sin(\beta)\sin(\eta)} \qquad (3\text{-}4)$$

$$N_Q = \frac{\sin(\alpha+\phi+\beta)}{\sin(\eta)}, \quad N_a = \frac{-\cos(\rho+\phi+\beta)}{\sin(\rho)\sin(\eta)} \qquad (5\text{-}6)$$

where the soil properties are the cohesion c, internal angle of friction ϕ, moist unit weight γ, adhesion c_a, and soil-tool friction angle δ. The surcharge force is defined as Q, and the remaining parameters describe the geometry with $\eta = \delta + \rho + \phi + \beta$ defined for notational convenience.

Existing soil property estimation methods for earthmoving applications are aimed at excavation or scooping operations and assume the measurement of multiple (3-8) full dig cycles in the same spatial region prior to producing an estimate (Luengo et al., 1998; Tan et al., 2005; Yu et al., 2023). This is not adequate for bulldozing operations where soil-tool interaction takes place over a large area, sometimes with only partial overlap between cuts. By combining a fast physics-infused neural network (PINN) approach to soil property estimation with an elevation mapping system, estimates can be tracked spatially and fused over time via Bayesian updates, the details of which are laid out in Section 2.

1.2. Mapping for Autonomous Earthmoving

Building and maintaining a detailed persistent representation of the environmental state can be beneficial at both control and planning levels, enabling consideration of approaching terrain shape in machine blade control and allowing planning over large regions outside of the machine's immediate field of view. To represent the topography of terrain for earthmoving, a raster format is often used, where spatial data in the horizontal plane is discretized and elevation at each location is represented continuously. This is a common approach to mapping in robotics (Cremean and Murray, 2005; Fankhauser et al., 2014; Miki et al., 2022) that extends simple occupancy grid representations, and these maps are often referred to as elevation maps, heightmaps, heightfields, digital elevation models (DEMs), digital surface models (DSMs), or gridmaps. For clarity, we will adopt the heightmap nomenclature and use gridmaps when describing the extension to additional layers. Mononen et al. (2022) use a Bayesian mapping approach to produce online updated heightmaps from range measurements obtained from a forward-facing LiDAR sensor mounted to a bulldozer chassis. A smooth spline is fit to the approaching surface and adjusted based on blade load and track slip to produce the target blade trajectory.

In place of or in addition to mapping the terrain with vehicle-mounted depth sensors, the elevation of the tractor can be used to update the terrain height under the machine tracks, as is done on some commercial systems (Komatsu, 2022). Further, given knowledge of the trajectory that the blade takes, intersections between the blade and map can be determined and used to move material within the map (D'Adamo, 2023, Brewster, 2018), i.e., proprioceptive (Jud et al., 2019) or blind mapping. By modeling material movement, real-time estimation of the terrain changes in areas not visible from machine-mounted sensors due to obscuration from the blade and terrain is enabled. This approach can facilitate tracking earthmoving efficiency via estimation of material cut and fill volumes. These maps provide a natural representation on which to plan earthmoving sequences.

D'Adamo (2023) describes a method for estimating the maximum pushable volume for a bulldozer bulk-earthmoving task where geometric, tractive, and engine limits are determined from blade and track geometry, known fixed soil properties, and engine torque curves. These limits are used to help compensate for vertical blade localization errors when mapping terrain, but it is implied that a similar model is used by the Caterpillar SATS planner when sequencing push actions.

A heightmap can be extended to include additional information by simply adding additional raster layers to represent other qualities relevant to navigation and earthmoving. We will refer to these extended representations as gridmaps. This may include tracking of loose soil separately from undisturbed terrain (D'Adamo, 2023), categorical information produced by a semantic segmentation system (Guan et al., 2022; Erni et al., 2023), traversability scores (Miki et al., 2022), or even material properties (Chen, 2024). Guan et. al. (2022) leverage gridmaps to build a traversability estimation pipeline that combines geometric features with semantic information to enable improved planning for excavator navigation at a worksite.

2. Methods

In our prior work (Wagner et al., 2023), the core physics-infused neural network (PINN)-based soil property estimation system was developed, and simulation-based experiments demonstrated the merits of this approach for the task of rapidly estimating soil properties and strength using the bulldozer itself as a sensor. While promising, this approach has limited direct applicability as it relies on the assumption of flat terrain to obtain the geometric parameters of the FEE (α, ρ, d, w) and it presumes knowledge of privileged information such as the surcharge force Q obtained using a virtual soil mass sensor, of which there is no analog for on a real bulldozer. Additionally, all prior experiments were performed using the simulation of a simple bladed vehicle model lacking a drivetrain and constrained to move along a vertical plane along a flat, level surface.

To enable soil property estimation in more realistic earthmoving scenarios and to enable future soil-aware earthmoving planning, several significant advancements are required. The major contributions of this work therefore include:

- Development and demonstration of a system performing soil property estimation in-situ on an advanced simulation of a compact track loader (CTL) with articulated bulldozer blade performing earthmoving operations.

- Extension of the physics infused neural network approach to estimate parameter uncertainties without requiring knowledge of true soil properties during training
- Development of a swept-volume based approach to heightmap deformation and concurrent extraction of FEE parameters, including surcharge, when model assumptions are met
- Development of a Bayesian estimation method for merging multiple soil property estimates and gridmap-based approach for tracking these estimates spatially

2.1. Physics Infused Neural Network for In-Situ Soil Property Estimation

A key component of the soil property estimation approach taken here is to bake-in knowledge of soil-tool interaction physics to the network architecture in the form of Eq. 1, referred to as a physics infused neural network (PINN) (Wagner et al., 2023). In comparison to the previously developed model, a number of improvements have been made to both the network architecture and training of the system to generalize the approach to uneven terrains, increase accuracy, and facilitate Bayesian mapping including:

- Training on a more realistic and diverse dataset
- Building in additional physics using error propagation, and leveraging a negative-log likelihood (NLL) loss term to enable estimation of soil property estimate uncertainty Σ_{Θ_u}
- Expanding estimated soil properties to include γ and c_a and improving estimates by making multi-step force predictions throughout a sweep to reduce ambiguity in parameter estimation

As outlined in Fig. 4, the network ingests a history of vehicle states and actions $\mathbf{o}_{[t:t+T]}$ along with a subset of known FEE parameters $\Theta_k = [\rho, \alpha, w, \boldsymbol{d}_{T_P}, \mathbf{Q}_{T_P}, \mathbf{v}_{T_P}]$ over the prediction horizon $\boldsymbol{T}_P = [t+T-P: t+t]$. It produces, as intermediate outputs, an estimate of the unknown soil properties $\widehat{\Theta}_u = [\hat{c}, \widehat{\phi}, \hat{c}_a, \hat{\delta}, \hat{\gamma}]$ and a predicted variance for each of these estimates $\Sigma_{\Theta_u} = \mathrm{diag}[\widehat{\sigma}_c, \widehat{\sigma}_\phi, \widehat{\sigma}_{c_a}, \widehat{\sigma}_\delta, \widehat{\sigma}_\gamma]$, as well as predicted variances for a subset of the known parameters. These estimates are then provided to the embedded FEE block to produce the predicted force $\boldsymbol{F}^{\mathrm{FEE}}_{T_P}$ and propagated uncertainty Σ_F. A residual force F^r is also directly produced by the network and summed with $F^{FEE}_{T_P}$ to produce the final force prediction $\boldsymbol{F}_{T_P}$. This residual force serves the purpose of compensating for forces not well captured by the FEE model such as those caused by penetration and blade acceleration. An example rollout of both the estimated parameters and the predicted forces is shown in Fig. 3.

For brevity, some of the details of the loss function definition are omitted here; similar formulations can be found in our prior work (Wagner et al., 2023). The two major changes made here are the averaging of the loss over the multi-step prediction horizon T_P and the inclusion of an additional negative log-likelihood (NLL) loss term

$$\mathcal{L}_{\mathrm{NLL}}(\boldsymbol{F}, \widehat{\boldsymbol{F}}, \boldsymbol{\Sigma}_F) = \frac{1}{2}\left[(\boldsymbol{F}-\widehat{\boldsymbol{F}})\boldsymbol{\Sigma}_F^{-1}(\boldsymbol{F}-\widehat{\boldsymbol{F}})^{\top} + \ln(\det(\boldsymbol{\Sigma}_F))\right] \qquad (7)$$

which is derived by formulating the optimization of the distribution $F \sim \mathcal{N}(F(\Theta), \Sigma_F(\Theta, \Sigma_\Theta))$ as a minimum NLL problem. To reduce notational clutter, the subscript T_P indicating a quantity is a vector over the time window has been removed,

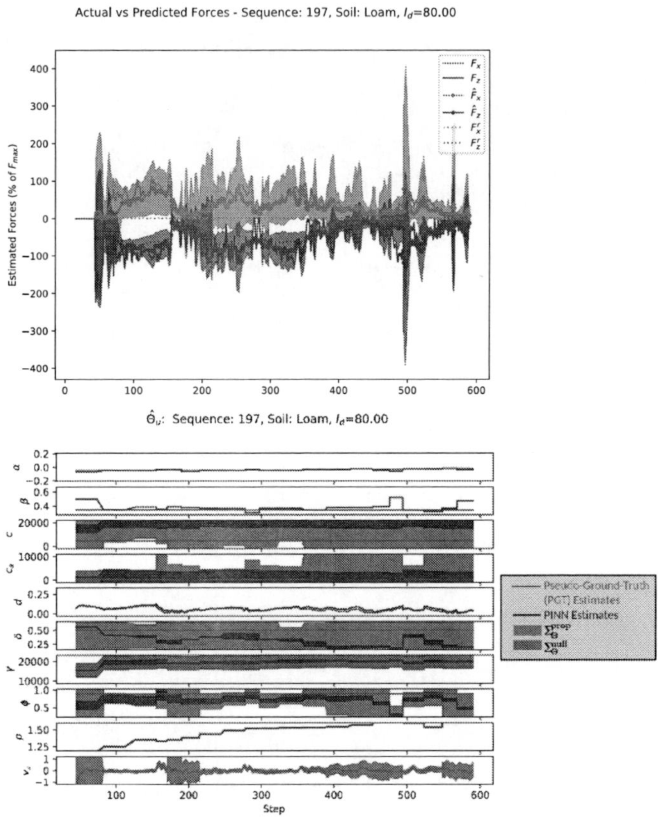

Fig. 3. PINN predictions for a single rollout in $I\_d = 80\%$ loam soil. (top) Predictions of soil properties $\widehat{\Theta}_u$ (blue line), estimated parameter uncertainties Σ_Θ (blue shading), and nullspace projected uncertainties Σ_Θ^n (gray shading) overlaid on PGT soil properties Θ_u (green line). Note, accelerations can lead to observed forces transiently exceeding F_{max}. (bottom) Corresponding force predictions for the same rollout.

but the bold facing remains where required. The challenge here is that a standard NLL loss will only enable determination of uncertainties for the quantity being regressed upon, i.e. the cutting force F, but we are interested in obtaining parameter estimate uncertainties Σ_Θ not Σ_F. Since the true parameter values are not assumed to be known, the trick employed here is to require the network to produce Σ_Θ and propagate that uncertainty through the Eq. 2 to obtain Σ_F as a function of Θ and Σ_Θ. This is accomplished here using a local expansion method which is suitable when uncertainties are relatively small and where the function is not highly non-linear. The first order Taylor series expansion of f_{FEE} around the estimate $\widehat{\Theta}$ is

$$F \approx f_{\mathrm{FEE}}(\widehat{\Theta}) + J(\Theta - \widehat{\Theta}) \qquad (8)$$

where $J_{ij} = \frac{\partial F_i^{\mathrm{FEE}}}{\partial \Theta_j}$ is the Jacobian. The propagated force covariance matrix can then be defined as $\Sigma_F = J\Sigma_\Theta J^T$. In practice, the soil parameter uncertainty is not estimated directly by the network. Instead, the network produces$S_{\Theta_u} = \ln(\Sigma_{\Theta_u})$ resulting in the modified error propagation $\Sigma_F = J\exp(S_\Theta)J^T$. This change is suggested to avoid the potential division by zero in Eq. 7, because it is more numerically stable and doesn't require resolving network outputs to the positive domain (Kendall and Gal, 2017).

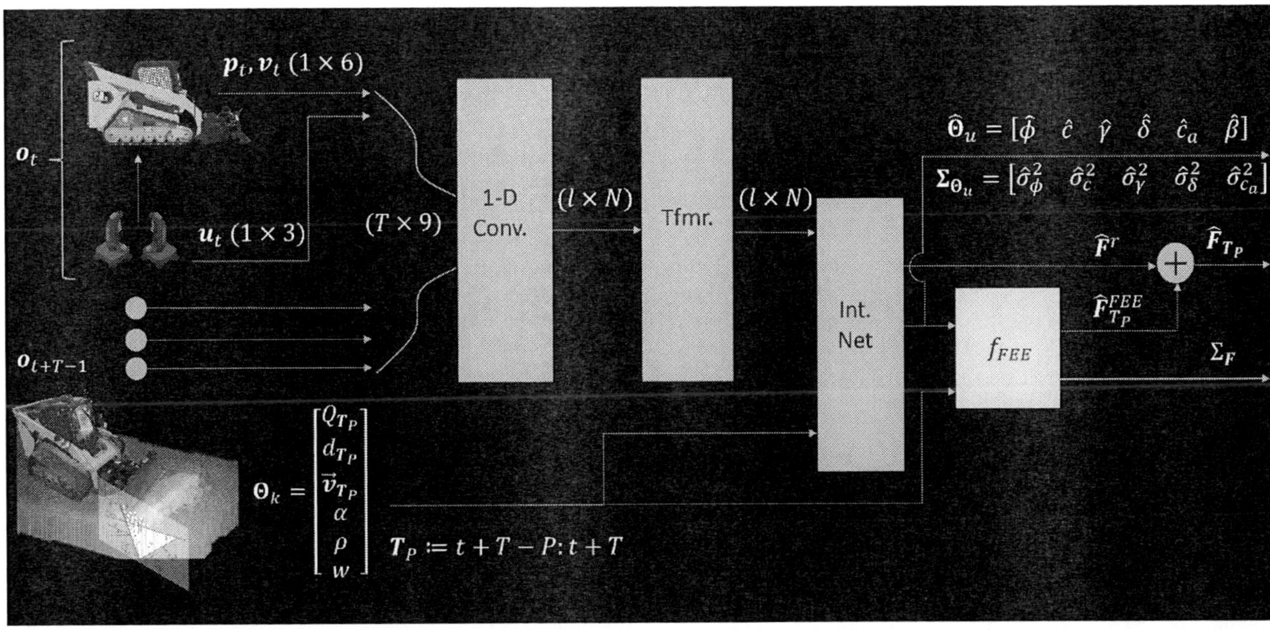

Fig. 4. PINN-based soil property estimation system diagram. Inputs flow from the left to produce the estimates on the right. The PINN block is highlighted in in green.

Overall, this error propagation approach enables us to train the network to estimate parameter uncertainties via back-propagation through Eq. 7 without requiring knowledge of the true Θ_u. Uncertainties obtained in this manner are aleatoric and reflect the uncertainty of the underlying noisy force measurements. This will not capture epistemic uncertainty of the parameter estimates, which may arise during inference when, for example, testing on a soil that was not present in the training dataset.

During training, it was observed that some parameter estimates were overly confident. Upon closer inspection, it was observed that under certain conditions, the Jacobian $\boldsymbol{J}$ is not well conditioned, i.e., has very small singular values. When this occurs then the operator has a non-zero null space, meaning that changes in parameters and parameter variances on this null space do not affect changes in F and Σ_F. The PINN relies upon kinematic observations and the action history $\mathbf{o}_{[t:t+T]}$, not the force, to make the parameter estimates, but the force is the only supervisory signal used to train the network. Therefore, when the Jacobian is deficient, during training the supervision of the parameters that lie within the null space collapses, and during inference the network is unlikely to produce reasonable estimates of those parameters under these conditions.

Similar problems have been observed in other soil property estimation efforts in which non-linear optimization methods are used to solve the inverse terramechanics problem, i.e. estimate Θ from F. Luengo et al. (1998) perform a sensitivity analysis using true parameters as a baseline and individually perturbing each value to observe the corresponding change in predicted force. For the soil considered and assuming a fixed dig angle, it was found that the cohesion c was the least sensitive to perturbation out of the considered parameters $[\phi, \delta, \beta, c]$ where a 50% change in c resulted only in a 25% change in the force. Alternatively, Tan et al. (2005) investigate the sensitivity of the parameter estimation when the measured force is perturbed, and find that ϕ is impacted minimally, while γ is much more easily influenced, e.g. changing by a factor of 23% for a 5% perturbation of the force.

While these analyses are somewhat specific to the optimization methods used, they indicate that certain parameters may be more difficult to estimate accurately due to the structure of the FEE. This occurs acutely when parameters lie within the null space of the Jacobian. In the mapping task outlined here, downstream of estimation, soil-property estimates are fused together in a Bayesian manner. In order to avoid the fused estimates from collapsing erroneously due to an artificially confident individual estimate, it is preferable to inflate Σ_Θ so that if J is well conditioned on a subsequent pass, the likely more accurate parameter estimates will be given more weight. A simple way of inflating the covariances when J is singular is to project a fixed covariance into the null space

$$\Sigma^n_\Theta = \left(I - J^{\dagger J}\right)\Sigma^a_\Theta \tag{9}$$

where Σ^n_Θ is the projected nullspace covariance, Σ^a_Θ is a fixed value referred to here as the ambiguity covariance, and † refers to the pseudo-inverse operation. The projected null space covariance is then combined with the estimated covariance as $\Sigma_{\Theta_u} = \min\left(\Sigma_{\Theta_u}, \Sigma^n_{\Theta_u}\right)$ to ensure that the uncertainties do not become erroneously too small.

Depending on the data and the number of timesteps for which a set of predicted parameters is assumed constant, i.e. P, J_{T_P} may or may not have a nullspace. However, it almost always has some very dominant singular values and some much smaller, near zero ones. When performing the pseudo-inverse, both absolute a_{tol} and relative tolerance r_{tol} can be defined to set the threshold for selecting which singular values and corresponding singular vectors should be removed to produce a low-rank approximation of the inverse. In the current implementation, these values are set to $a_{\text{tol}} = 1{,}500$ and $r_{\text{tol}} = 0.05$, however, the resulting Σ^n_Θ is not particularly sensitive to these values. This is related to the magnitude of the parameter values and methods for scaling this projection or modifying the underlying optimizer may improve overall performance (Thuerey et al., 2021).

The overall loss function is expressed as

$$\begin{aligned}\mathcal{L}\left(\boldsymbol{F},\widehat{\boldsymbol{F}},\widehat{\boldsymbol{\Theta}},\boldsymbol{F}^r,\frac{\partial N_\gamma}{\partial\beta},\boldsymbol{\Sigma_F}\right) &= \\ &\lambda_F\mathcal{L}_{\text{wMAE}}\left(\boldsymbol{F},\widehat{\boldsymbol{F}},w_{xz}\right) + \lambda_{NLL}\mathcal{L}_{\text{NLL}}^{\Sigma_\Theta\leftharpoondown}\left(\boldsymbol{F},\widehat{\boldsymbol{F}},\boldsymbol{\Sigma_F}\right) + \\ &\lambda_{\partial N_\gamma/\partial\beta}\mathcal{L}_{\text{MAE}}^{\beta\leftharpoondown}\left(0,\frac{\partial N_\gamma}{\partial\beta}\right) + \sum_{\hat{\theta}\in\Theta^{\text{rng}}}\left[\lambda_\theta\mathcal{L}_{\text{ReLU}}\left(\hat{\theta},l^\theta\right)\right] + \\ &\sum_{\hat{\theta}\in\Theta^{\text{res}}}\left[\lambda_{\theta^{res}}\mathcal{L}_{\text{ReLU}}\left(0,\hat{\theta}^{res}\right)\right] + \lambda_{F^r}\mathcal{L}_{\text{wMAE}}\left(\boldsymbol{0},F^r,w_{xz}\right)\end{aligned} \tag{10}$$

The first term is a weighted mean average error (MAE) loss based on the predicted force error and is the primary supervised loss component. The second term is the supervised NLL loss from Eq. 8, but gradients with respect to this loss are only backpropagated through Σ_Θ as indicated by the superscript $\Sigma_\Theta \leftharpoondown$. The remaining four terms are regularization losses whose purpose is to enable consistent estimation of β, enforce estimated parameters to lie with physically meaningful ranges l^θ, allow for small residual corrections of the FEE parameters extracted via the mapping process $\hat{\theta}_u = \theta_k + \hat{\theta}^{\text{res}}$ for some subset $\Theta^{\text{res}} \in \Theta_k$, and ensure that the residual force F^r remains small to compensate for mismatch between the FEE and the true physics. The λ terms refer to weights of each loss component and the loss functions containing the argument $w_{xz} = [1.0, 0.5]$ are used to weight the x and z components separately. It is also important to note that the parameter uncertainty values produced by the network are also enforced to lie within a reasonable range using the ReLU regularization $\Sigma_\Theta \in \Theta^{\text{rng}}$. When any of the following parameters exceed their clipping limits $[\eta, \beta, \rho]$ or if $\hat{d} = d + \hat{d}^{res} < 0$, the FEE is not valid and therefore all losses except for F^r regularization are invalidated. This loss is averaged over the prediction horizon P helping to reduce the size of the Jacobian null space, and mini-batches of size 100 are used to help stabilize training.

To further improve estimation of the soil failure angle, a nominal value of β as a function of ϕ is obtained to enable the network to estimate the residual angle β^{res} as opposed to determining it directly. This expression is derived by considering a frictionless $\delta = 0°$ vertical blade $\rho = 90°$ cutting through a flat terrain $\alpha = 0°$ comprised of a cohesionless soil $c = 0, c_a = 0$ (Miedema, 2014), yielding $\beta = \pi/4 - 0.5\,\phi$.

2.2. Terrain Mapping

To enable use of this soil property estimation approach on a realistic compact track loader (CTL) simulation performing earthmoving operations on an uneven terrain while articulating the blade to vary its roll, pitch, yaw, and height, a method is devised to obtain the required network input parameters Θ_k via tracking of blade movement through a terrain map. The basic idea is to represent the terrain as a heightmap and then track the movement of the blade as it intersects the heightmap in order to both update the map by displacing soil and to and extract approximate FEE parameters Θ_k from the geometric relationship between the blade and map. Algorithm 1 outlines this procedure and Alg. 2-5 provide more of the details and are available in the appendix. The open-source Elevation Mapping cupy software provides an implementation of a GPU accelerated height mapping approach used in legged robotics applications and offers support for custom semantic mapping and plugins (Miki et al., 2022; Erni et al., 2023). This package acts as a foundation that is modified and extended to enable the desired earthmoving applications.

In addition to tracking the terrain height, a gridmap layer containing disturbed/loose soil is also maintained, similar to the method proposed by D'Adamo (2023), enabling computation of surcharge volume V^Q and application of heightmap erosion to loose soil alone. Given the focus of this work is on using mapping as a means to obtain Θ_k, some additional factors must be considered to enable accurate soil property estimation. In derivation of the FEE, a number of assumptions are made that must be satisfied including: the blade is in contact with the soil, the soil is failing as a result of the applied force, the forces are balanced i.e. in static equilibrium, the blade is flat, and the cut geometry is uniform along the width of the blade. By carefully designing the methods used to determine if the blade is in contact with the heightmap and how soil is displaced when contact occurs, these requirements can be met.

A volume, represented as a polygonal mesh, is swept out by the movement of the blade once a minimum translational or rotational displacement has occurred, and subsequently this swept volume is checked for intersection with rays emanating from the center of each heightmap cell, see Fig. 5. This differs from the blind mapping terrain update described by D'Adamo in which a fixed solid volume representing the Ground Engaging Tool (GET) geometry is intersected with the heightmap at a fixed rate (D'Adamo, 2023). Due to the discretization of the terrain in the horizontal plane implicit with a heightmap representation, it is possible for intersections between the map and a fixed GET volume to be missed, depending on the GET velocity, GET thickness, and update rate, although this was not reported. Additionally, it isn't obvious how such an approach could be utilized to ensure the FEE assumptions are being met. For example, using this solid volume intersection approach, it would be difficult to automatically differentiate between the following states: active cutting, a stalled blade still exerting force on the soil, and a blade merely making contact with the terrain but not exerting a force on the soil. By instead considering the intersection between a GET swept volume and the heightmap, any detected intersections necessarily imply contact and soil failure i.e. the active cutting scenario, thereby satisfying two of the FEE assumptions.

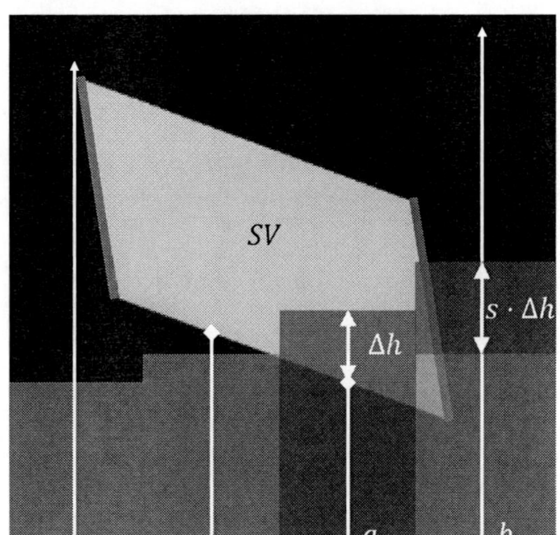

Fig. 5. A profile view of the swept volume (SV) intersecting the elevation map M. The ray casting procedure is illustrated with the white lines emanating from the bottom of the cells with intersections marked with white diamonds. The displaced terrain height Δh is determined from difference between the cell height and intersection height and the volume is then swelled and deposited in cell b.

In simulation of soil-tool interaction, Holz et al. (2013) similarly rely upon swept volumes for soil displacement, but instead leverage a discretely sampled signed distance function to represent the terrain surface more continuously. This work also hints at a method for satisfying the FEE assumption of a flat blade with uniform cut geometry, by breaking the cutting surface up into multiple thin slices where these assumptions are close to valid. The cutting force for each slice can then be computed independently using the known soil properties and summed across the blade to produce an aggregate force and torque applied to the blade by the soil.

The objective here is the inverse, to obtain soil property estimates from observation of cutting force, but this slicing concept can be leveraged to back out Θ_k as laid out in Alg. 2. For each cell intersected by the swept volume, points on the terrain surface along the direction of GET translation are obtained using a Digital Differential Analyzer (DDA) line rasterization algorithm. During this process, the volume of surcharge lying in front of the blade V^Q is obtained. A line is then fit to these points using weighted least squares to better fit points close to the blade, and the geometric parameters of the FEE for the slice (α, ρ, d) are obtained.

While it may be possible to develop a soil property estimation algorithm that directly consumes FEE parameters for each slice, for simplicity we have chosen to obtain a single parameterization of the FEE for each sweep. The cutting force, Eq. 1, is strongly sensitive to d and therefore slice FEE parameters are averaged together with a weight determined by d. Averaging these values is somewhat crude, as Equation 1 is non-linear, but in practice it has been sufficient. Although not shown in Algorithm 2, to enable the model to compensate for poor extracted Θ_k, average goodness of fit metrics the line fit and parameter variances are computed and provided as inputs to the integration network. To obtain the blade width w each intersection point is projected onto the blade along the translation direction $\hat{t}$ and the range of the points along the direction perpendicular to the translation $\perp \hat{t}$ is determined to be the equivalent blade width. During partial contact between the blade and the ground w will be less than the full blade width and during cuts where the blade is yawed to one side the width is reduced in proportion to the cosine of the yaw angle. This entire FEE parameter approximation and extraction procedure is necessary, as it is important for the system being designed to support uneven terrains along with blade roll and yaw because the goal is to enable soil property estimation during performance of typical earthmoving tasks when these configurations will be encountered.

Once Θ_k has been obtained, updating the map as a result of the sweep can proceed and is described in Algorithm 3. The physics of terrain deformation from soil-tool interaction is very complex, and while the Vortex Studio simulation used for all of the experiments presented here captures some of this behavior, it is still an approximation of reality. The methods described here are further simplified and rely upon several heuristics but have proven adequate for the application. For simplicity, it is assumed that when soil is displaced from one cell a it is deposited entirely in another cell b. To determine which cell should receive the displaced material, a DDA rasterization starting at the cell a is performed along the direction $\hat{m}$ to find the first cell lying outside of the swept volume. The search direction $\hat{m}$ is defined to be between the starting blade normal $\hat{n}$ and blade center translation direction $\hat{t}$. This process is depicted in Fig. 6. Both intersected and deposit cell heights and loose soil values are updated, accounting for swell of the undisturbed material assuming a fixed swell ratio s. The height uncertainties of both cells are also updated conservatively by considering a 1 σ range of possible cell heights

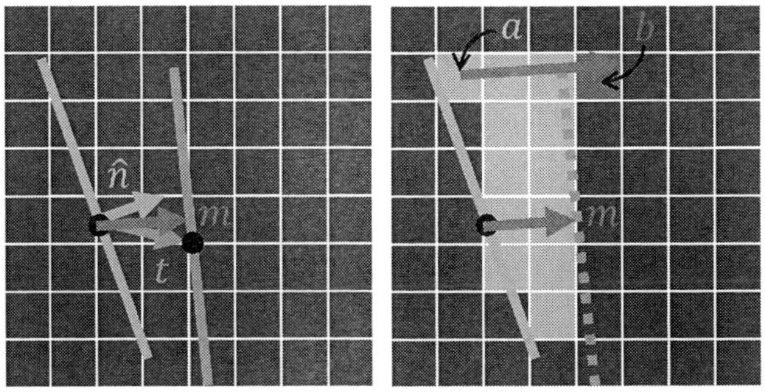

Fig. 6. (Left) depiction of the method used for determining the displaced cell deposit direction $\hat{m}$ from the blade normal $\hat{n}$ and translation vector t. Note that $\hat{m} = m/||m||$. (Right) Once a deposit direction is selected, a DDA algorithm is used to search along the line beginning at the center of an intersected cell a with the orientation of $\hat{m}$ to find the first cell b that is not intersected in the sweep where the soil is deposited.

given the prior uncertainties and uncertainty of the swept volume height σ_{SV}. The change in loose soil volume lying in front of the blade from the start of the sweep to the end of the sweep ΔV^Q is determined by accounting for loose soil not being displaced, loose soil flowing below the blade, and newly disturbed and swelled soil.

The parameter extraction described in Alg. 2 yields a single set of values Θ_k for the entirety of the sweep although it takes place over a period of time in which surcharge is accumulating on the blade and the depth of cut is varying. To produce a denser dataset for training the network with multiple predictions of cutting force per sweep, the depth of cut d and surcharge volume V^Q are interpolated. If it is assumed that the disturbed soil has a fixed density γ_l the surcharge force Q can then be obtained from the interpolated surcharge volume. During training, the extracted parameters Θ_k are logged along with the observation history o from the vehicle's sensors. The ground truth soil properties Θ_u can also be logged if available from the simulation or from manual testing but are not required for training the network.

Once the model is trained, these extracted parameters can be fed to the network to produce the soil property estimates $\hat{\Theta}_u$ and estimate uncertainties Σ_{Θ_u}. These estimates can then be used to perform a Bayesian update of the soil properties as described in Alg. 4. Given that soil properties are spatially varying and that these transitions can occur abruptly, e.g. from buried objects or soil strata, we conservatively choose to apply the updates to the region of the map corresponding to the horizontal extent of the FEE soil wedge while inflating covariances for cells further from the blade.

As discussed in Sec. 2.1, the accuracy of the estimated soil properties will vary based on both the underlying true soil properties and the geometric configuration of the FEE in relation to the Jacobian J. For example, a shallow cut through a region may result in poor estimation of ϕ and c as compared to a deeper cut. By leveraging Bayesian updates to combine soil property estimates from multiple passes across the same spatial region, the ambiguity inherent in the estimates can be reduced and the accuracy of the estimates can be improved while still accommodating for the spatial variance of soil properties.

The final step of updating the map following a sweep is to perform soil erosion. This step is necessary to produce realistic looking heightmaps and to account for spill around the edges of the blade during a push. As discussed earlier, when soil is displaced, it is moved from one cell to another, and over the course of multiple sweeps, loose soil accumulates in the cells lying before the blade edge, producing a high wall of surcharge. In reality, as soil is displaced it spreads out and flows until reaching an equilibrium slope, the angle of repose, that is dependent on the soil properties of the disturbed soil.

The erosion method used in Algorithm 5 lines 17-18 is largely based on established methods for soil grid erosion (Li, 1993; Holz, 2009). Soil slips when the shear stress force τ' caused by the weight of the soil W becomes greater than the shear strength force s' that is based on the cohesion c_l and internal friction angle ϕ_l of the loose soil. A factor of safety F_s can be defined

$$F_s(\alpha) = \frac{s'}{\tau'} = \frac{cL + W\cos(\alpha)\,tan(\phi)}{W\sin(\alpha)} \tag{11}$$

that indicates slip will occur when $F_s < 1$. The weight W is a function of the grid resolution and loose soil density γ_l, and α is the soil failure angle whose range depends on the relative height between adjacent cells. By minimizing F_s over α, it can be determined if slip will occur. A slip velocity can then be obtained using Newton's second law, and Euler integration can be used to compute the soil slip height h_{slip}. In practice, the erosion operation is performed multiple times after each sweep with a small timestep δt to ensure smooth erosion and avoid over-sized h_{slip}.

To avoid soil from slipping through the blade, a mask is defined indicating where the swept volume overhangs a given cell. The height of the bottom of the swept volume at the masked cell centers is also provided to allow for soil to flow from other cells to beneath the blade. This enables surcharge to slip beneath the blade when it is raised above the compacted soil surface as during a spreading operation. To leverage the GPU to perform these slip calculations in parallel, the slip calculation is applied in a tiled fashion where for each considered cell out of a block of 4 cells, two neighbors at 90° apart are considered for soil exchange, see Fig. 7. This is applied for each cell in the block and is only performed for a region of interest (ROI) near the blade to reduce computation. The direction considered for erosion is selected to be in line with the deposit vector $\hat{m}$ so that soil is allowed to erode away from the blade.

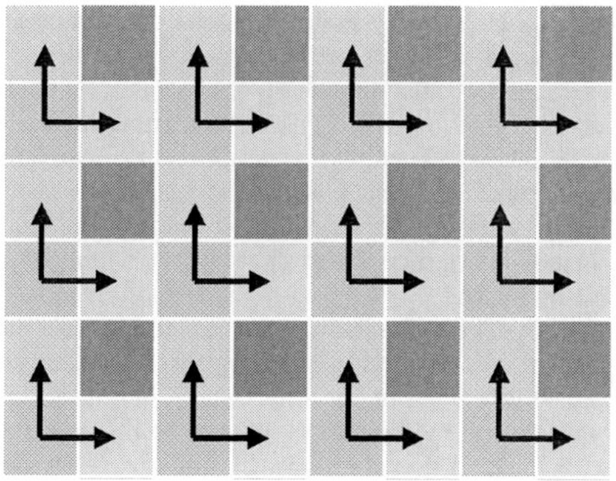

Fig. 7. The four different grid colors indicate the four groupings of cells for which slip calculations are performed in parallel and the black arrows depict one of the four possible sets of directions for the cells considered for soil exchange, selected using the deposit direction vector $\hat{m}$.

3. Results and Discussion

All experiments have been performed using a simulated compact track loader (CTL) modeled after a CAT 299 D3 XE outfitted with an articulated bulldozing blade, see Fig. 8. The simulation was constructed by CM Labs within Vortex Studio, which supports simulation of multi-body dynamics, real-time hybrid soil-tool interaction, low-fidelity hydraulics systems modeling, various virtual sensors (CM Labs, 2016). Very few specifications of the of the system are available publicly, making development of an accurate simulation challenging. A mix of physical measurements and photographs were used to produce an approximate CAD model of the system with a focus on proper machine kinematics. This CAD model, public high-level machine specifications, and hydraulic specifications obtained from the manufacturer informed the simulation development. Realistic dynamics were prioritized over cosmetic fidelity, but no analysis regarding the accuracy of the model to the real-world system has yet been performed.

The CTL has a higher center of gravity and a shorter wheelbase than a standard bulldozer, which makes it more sensitive to the terrain underfoot and less stable around the pitch axis in particular. Oscillations in pitch must be properly compensated for by varying blade height to ensure the creation of smooth terrain profiles. The CTL also has a lower drawbar-pull than even small dozers, meaning that the cutting force it can apply to the soil is lower and therefore so is the maximum depth of cut. Due to these differences, bulldozing with a CTL is a more challenging task than with a conventional bulldozer, but is considered a sufficient analog platform for this work.

Using the Vortex Studio Python API, the simulation can be configured with randomized terrain, parameters for the soil simulation can be specified, commands can be provided to the vehicle actuators and sensor observations can be obtained at a rate of 60 Hz. The measurements consist of chassis and blade pose and twist, hydraulic cylinder lengths and velocities, track velocities, engine speed, soil cutting force, and pointclouds from front and rear mounted single beam LiDAR scanners.

To enable automatic data collection, cascaded control architecture is developed where at the lower-level feed-forward PID velocity controllers are tuned for each arm joint and the tracks. A combination of individual joint position control and rudimentary task-space control is implemented using multiple PID controllers to enable tracking of blade height and roll relative to the surface while maintaining a desired pitch joint position and taking into account the loss of manipulability when the lift arm reaches its lowest position. To keep the vehicle from stalling and still collect a dataset with varying depths of cut, a controller that raises the blade in response to poor tracking of the desired forward velocity is developed. This approach works reasonably well, but is not optimally tuned and some oscillation occurs, particularly upon a substantial step change in the height command.

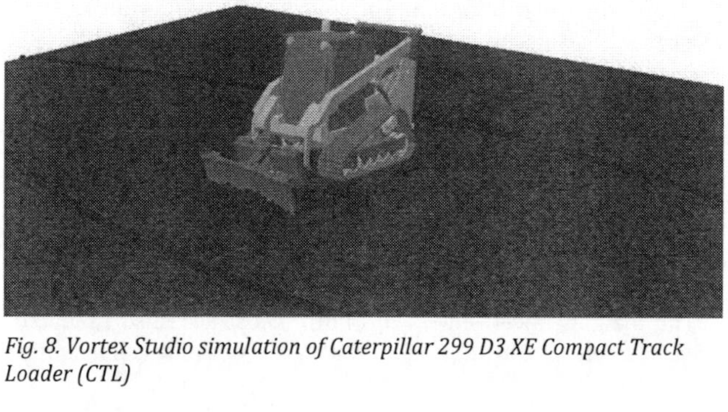

Fig. 8. Vortex Studio simulation of Caterpillar 299 D3 XE Compact Track Loader (CTL)

Fig. 9. Illustration of how the blind mapping system updates the terrain height in front of the blade during a cut through a berm.

The terrain mapping system described in Sec. 2.2 is initialized using the starting heightmap from the simulation. The mapping system is provided with pointcloud measurements from both LiDARs at each timestep as well as the blade pose to be used for blind mapping using blade swept volumes as described in Algorithm 1. See Fig. 9 for a visualization of how the blind mapping enables updating the map in the region directly in front of the blade that is not visible from the front LiDAR.

To validate this approach, tests were performed in which the pointcloud observations were disabled, and only the blade movement was used to update the map from its starting state. Fig. 10 depicts one such experiment. There are some discrepancies between the estimated height and the true height in two main areas: along the edges of the cut path that was taken by the CTL and where the soil was deposited. Errors of this nature are expected and are a result of a mismatch between the way that soil particles flow in simulation and the approximation of this process using soil erosion, see Alg. 5. The primary issue is that the soil erosion parameters are assumed to be fixed $c_l = 25\,Pa, \phi_l = 0.26\,rad, s = 1.2$ when in reality they vary with the undisturbed soil properties. The reason for this is that outside of simulation these values are not known a priori. With some tuning, the blind mapping error can be reduced, but this is not necessary to achieve good soil property estimation performance. Additionally, when pointcloud measurements are enabled, these mapping errors are corrected when in the field of view of the sensor.

3.1. Soil Property Estimation Accuracy

A dataset of vehicle observations o, extracted FEE parameters Θ_k, and pseudo ground truth (PGT) soil properties Θ_u is collected using the data collection controller and performing a sweep over controller height and forward velocity setpoints. A total of 268, 10-second-long episodes are collected across four pre-tuned soil types (Clay, Loam, Gravel, Sand) where the relative density I_d of the soil is varied from 1%-100%. Some configuration changes were made to the Vortex Studio soil simulation to ensure more realistic behavior for bulldozing operations (Wagner, 2025). To improve the ability of the system to perform in various environments, the terrain heightmap is randomized to small hills and dips using a mixture of Gaussians approach. Additionally, the blade controller roll and yaw setpoints along with the relative height, forward velocity, and chassis yaw angle setpoints are made to drift throughout an episode using a random walk.

The soil property estimation network is then trained on this dataset. The model is used to make predictions of across the entire dataset and then averaged across values for fixed Θ_u corresponding to changes in relative density I_d. The averaged soil property predictions are plotted with the PGT values against increasing I_d for each of the four soil types, see Fig. 11. Before analyzing these results it is important to explain why the soil properties obtained from the simulation Θ_u are referred to as *pseudo* ground truth (PGT). Vortex Studio simulates soil cutting forces use a hybrid heightmap-DEM method that utilizes the FEE to determine forces when the shear strength has been overcome by the tool and then spawns particles to enable more realistic dynamic behavior. A variety of heuristics have also been implemented to improve realism from the perspective of human operators. So, although the values Θ_u are obtained directly from the simulation, for the purposes of evaluating the performance of the soil property estimation system, they can only be used as a reference and not absolute ground truth values.

With this in mind, it is useful to compare trend of the estimated parameters $\widehat{\Theta}_u$ with relative density I_d to the trend of the PGT properties Θ_u. Increasing values of $\widehat{\phi}, \hat{c}$, and $\hat{\gamma}$ are observed for both of the cohesive soils, and ϕ also displays a positive trend for

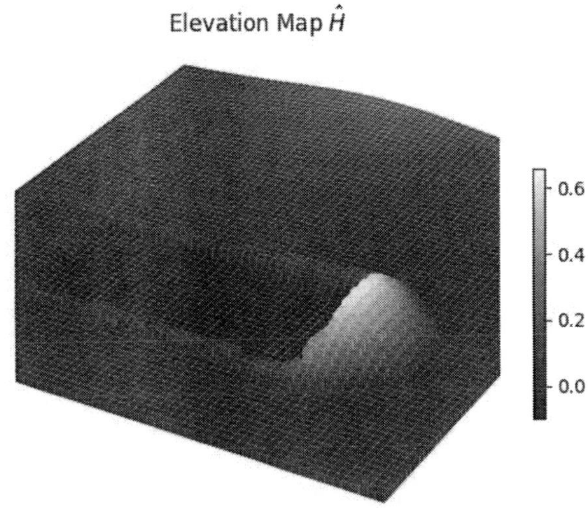

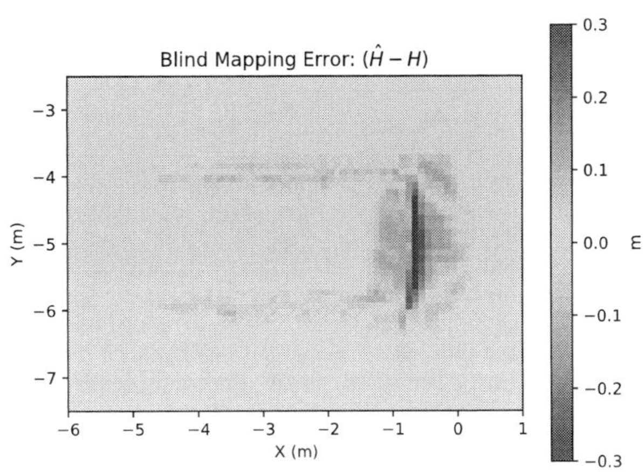

Fig. 10. (top) A 3D rendering of a heightmap $\hat{H}$ following a cut when using only blind mapping. (bottom) The resulting heightmap error.

the cohesionless soils. The friction ϕ is estimated particularly well, whereas the estimated density is not terribly accurate. This aligns with the Jacobian J having very small singular values for γ and may be improved with some modifications to the underlying gradient descent method. In agreement with our prior results, $\hat{c}$ is consistently significantly larger than the PGT value and thought to be related to additional forces from DEM. Similarly, for $I_d \in [0-60]$, $\hat{\delta}$ is consistently lower, but jumps up to near the PGT value for higher relative densities. Correspondingly, the average depth of cut across the episode decreases drastically and residual forces F^r increase drastically. This is reflective of an odd behavior,

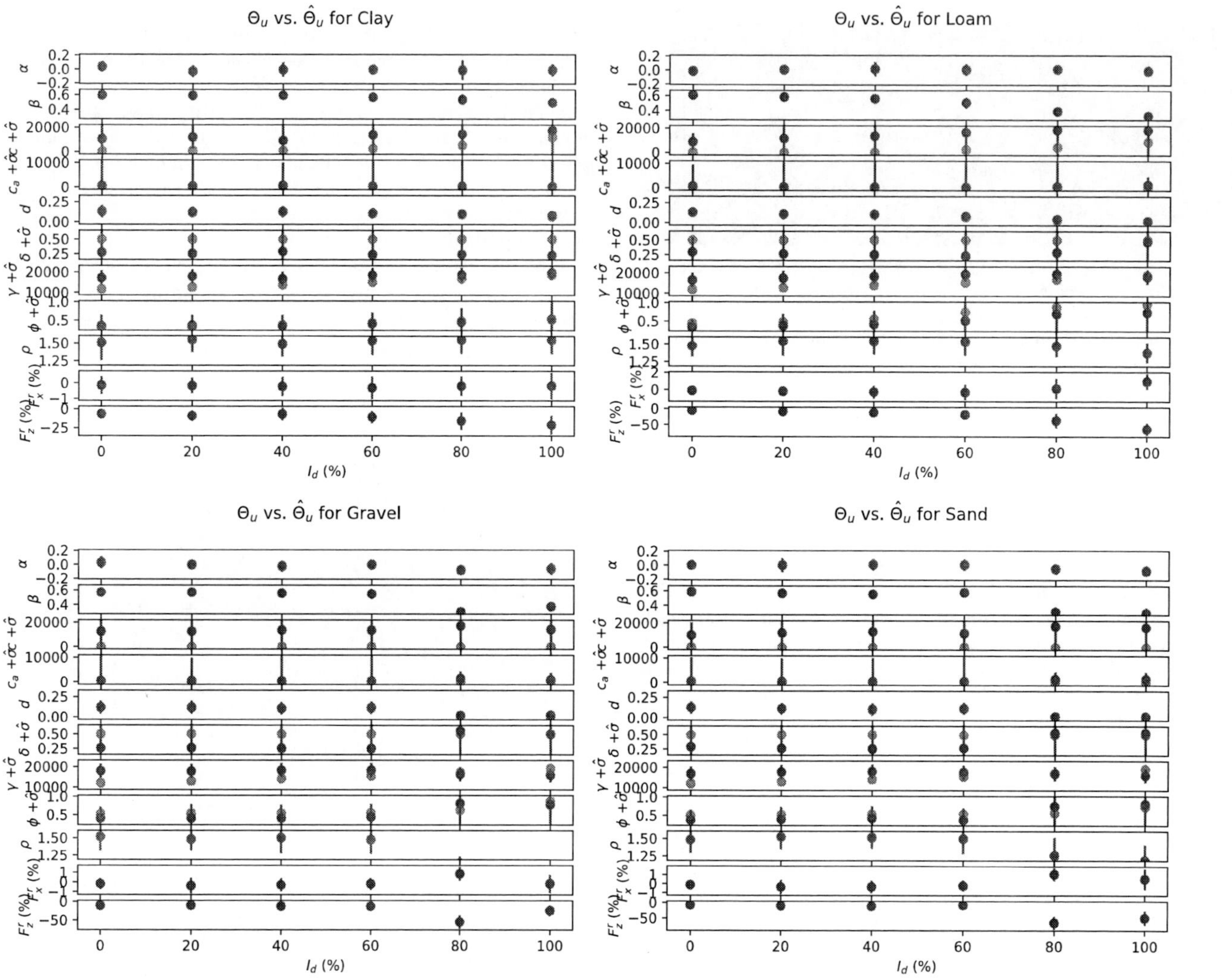

Fig. 11. PINN soil property estimates $\hat{\Theta}_u$ aggregated over the dataset and binned according to the relative density I_d are shown in blue and PGN values Θ_u are shown in green. The vertical lines indicate the average predicted standard deviation $\Sigma_u^{-1/2}$

observed during simulation with height relative density, where the blade was incapable of penetrating the soil surface even when the entire weight of the vehicle is applied. While penetration forces are expected to increase with relative density, at slightly lower relative densities the soil can be penetrated and the maximum depth of cut that can be achieved without stalling increases drastically as well.

Without more detailed documentation regarding the Vortex Studio soil simulation, it is difficult to say what the cause of this behavior is, but the important take away is that at higher relative densities $I_d \in \sim [70 - 100]$, poor soil property estimates are more reflective of issues within the dataset than of the estimation approach. It is interesting to point out that when this penetration issue occurs, the network produces a large F_z^r to compensate for the non-FEE compatible behavior as intended. The other important thing to note is that the uncertainty estimates for the parameters also increase in this case which is a desirable feature for the downstream Bayesian mapping. Taking a closer look at the rollout in Fig. 4, the projected force uncertainty is shown to increase when a step change in height is observed at around the 500-step mark.

3.2. Soil Property Mapping Accuracy

To qualitatively evaluate the ability of the combined soil property estimation and mapping system to handle spatially varying soil properties, a simple experiment was performed in which a strip of soil with a high $I_d = 80\%$ is present in a flat terrain with relatively low $I_d = 10\%$. The vehicle is then operated by the author to perform a sequence of cuts across the region, where the cut begins on the left in the lower density soil and moves to the right across the strip of higher density soil. The individual estimated soil properties $\widehat{\Theta}_u$ are tracked as separate layers in the map and fused together using Bayesian updates as outlined in Alg. 4. The fused soil properties from each layer are then used to compute what we define as the FEE index $F^{\text{idx}} = f_{\text{FEE}}(\Theta^{\text{idx}})$, which corresponds the the force that would be required to shear the soil under a specific geometric scenario where $\Theta^{\text{idx}} = [\alpha = 0°, \rho = 80°, w = 1.85\ m, d = 0.2\ m,\ \phi = \bar{\phi},\ c = \bar{c},\ \delta = \bar{\delta},\ c_a = \bar{c}_a,\ \gamma = \bar{\gamma}]$. The resulting map for this test is shown in Fig. 12, where the denser soil region is outlined.

The FEE index within the green outline is shown to be substantially higher than the index outside for most of the sweeps. There are some erroneously higher index regions outside of the

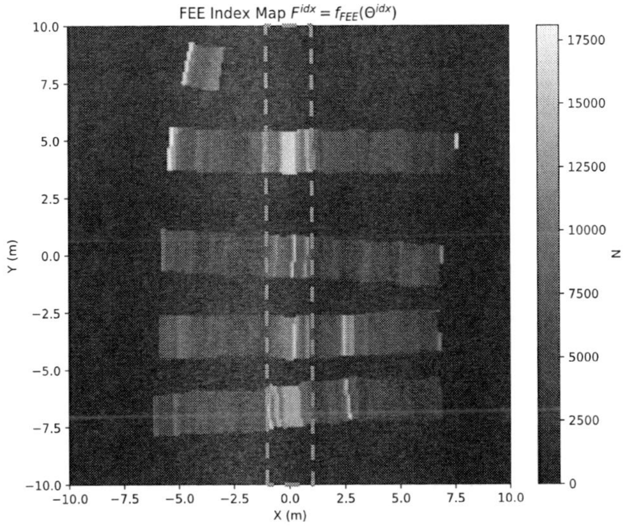

Fig. 12. FEE Index map for multi-cut experiment. Clay soil was specified with $I_d = 10\%$ everywhere except inside the dashed green line where $I_d = 80\%$.

higher-density strip. These errors sometimes occur at the beginning of the push which is a time when the blade is moving somewhat dynamically as the controller adjusts the arm position to arrive at the setpoint. In this case the assumption of static equilibrium is violated which may explain the inaccuracy to some extent. The soil property estimation network was designed to partially compensate for this through prediction of the residual force F^r (fixed across the prediction horizon T_P) which is added to the multi-step predicted force $\hat{F}_{T_P}^{FEE}$. If there exist any oscillations (due to poor controller tuning), then this fixed value residual compensation may provide adequate compensation for the blade acceleration. Oscillations of the blade both at the beginning of many cuts and when transitioning out of the higher strength soil.

Overall, the results of this test show that our approach to soil property mapping is largely successful. To achieve this result, multiple trials were performed and in that process some issues were uncovered. The primary problem was that to perform a multi-pass test, some minor modifications were made to the controller to enable a person to control the track velocity and chassis angle setpoints. Additionally, a toggle was implemented to enable the operator to raise the blade when reversing to the following cut start location. This resulted in some slight differences in the blade pitch angle after lifting, that were not immediately obvious. While the blind mapping system performed as expected, the soil property estimation system produced poor estimates. The working theory is that this is because the training dataset did not include any observations with this blade angle and therefore the data was out of distribution. This can be remedied by improving the blade controller to achieve more consistent, less oscillatory motions and by expanding the training dataset to cover the domain more fully. However, given the complexity of such a system and the large number of degrees of freedom, it may be challenging to ensure full coverage over the domain. Therefore, this approach might additionally benefit from development of a method for epistemic uncertainty estimation to ensure that when conditions not encountered during training are observed, a high uncertainty is produced so as to not negatively influence the fused map states.

4. Conclusions

This work has advanced the state of autonomous earthmoving by extending previous methods for soil property estimation to handle non-flat terrain geometries. A key contribution is the integration of a terrain mapping system, in which a heightmap of the environment is constructed from onboard LiDAR data, and a blind mapping component uses blade kinematics to infer terrain displacement. An improved PINN-based model is introduced to estimate a broader set of FEE soil parameters and perform uncertainty quantification. These estimates are fused into soil property layers of a gridmap using a Bayesian update mechanism, enabling continuous soil-aware terrain modeling during operation.

Initial experiments indicate that this system accurately identifies regions requiring higher interaction forces, showing promise for its application in soil-aware planning and control. Remarkably, despite the increased complexity of operating over uneven terrain, the system achieves better estimation performance than previous approaches limited to flat surfaces. To our knowledge, this is the first system to perform real-time, in-situ soil property mapping. This capability establishes a foundation for intelligent, soil-aware earthmoving systems capable of optimizing cut strategies, avoiding failure conditions, and autonomously responding to subsurface anomalies.

Several avenues remain for future work. The mapping framework could benefit from techniques to mitigate aliasing artifacts, for example by projecting GET motion to apply gradual cell updates. Incorporating a priori soil class knowledge or image-based classification could further constrain the estimation space and improve accuracy (Kasaragod et al., 2022). More sophisticated update strategies, such as inflating covariance with depth or applying observations via Gaussian processes, could enhance robustness in stratified soils.

While this study was conducted in simulation, transitioning to real-world platforms will require new methods for force estimation—either via hydraulic pressure sensing or by leveraging dynamics models to infer force from kinematics. Exploring these directions will be critical for validating the approach outside of controlled environments. Ultimately, while perfect soil property estimation may be infeasible due to the inverse terramechanics problem, producing estimates that yield accurate force predictions under varying blade configurations may be sufficient for autonomous planning. Where higher fidelity estimates are needed for engineering tasks, improved uncertainty modeling will be necessary.

5. Declaration of competing interest

The authors declare that they have no known competing financial interests or personal relationships that could have appeared to influence the work reported in this paper.

6. References

Arreguin, A.L., Montenegro, S., Dilger, E., 2021. Towards in-situ characterization of regolith strength by inverse terramechanics and machine learning: A survey and applications to planetary rovers. Planet. Space Sci. 204, 105271.

Azar, E.R., Kamat, V.R., 2017. Earthmoving equipment automation: A review of technical advances and future outlook. J. Inf. Technol. Constr. 22(13), 247–265.

Bellamy, D., Pravica, L., 2011. Assessing the impact of driverless haul trucks in Australian surface mining. Resour. Policy 36(2), 149–158.

Blouin, S., Hemami, A., Lipsett, M., 2001. Review of resistive force models for earthmoving processes. J. Aerosp. Eng. 14(3), 102–111.

Bradley, D.A., Seward, D.W., 1995. Developing Real-Time Autonomous Excavation—The LUCIE. In: Proc. Conf. Decision and Control, 3028–3033.

Bradley, D.A., Seward, D.W., 1998. The Development, Control and Operation of an Autonomous Robotic Excavator. J. Intell. Robot. Syst. 21(1), 73–97.

Brewster, H.J., 2018. An analysis of the accuracy of varying terrain modelling methods for semi-autonomous bulldozing. Ph.D. dissertation, The University of Queensland.

Cannon, H.N., 1999. Extended Earthmoving with an Autonomous Excavator. Ph.D. dissertation, Carnegie Mellon University.

CM Labs Simulations, 2016. Theory guide: Vortex software's multibody dynamics engine, Online

Cannon, H., Singh, S., 2000. Models for automated earthmoving. In: Amato, N. (Ed.), Experimental Robotics VI. Springer, London, pp. 163–172.

Chen, J., Frey, J., Zhou, R., Miki, T., Martius, G., Hutter, M., 2024. Identifying terrain physical parameters from vision – Towards physical-parameter-aware locomotion and navigation. IEEE Robot. Autom. Lett.

Cremean, L.B., Murray, R.M., 2005. Uncertainty-based sensor fusion of range data for real-time digital elevation mapping (RTDEM). In: IEEE Int. Conf. Robot. Autom.

D'Adamo, T., 2023. Building and maintaining real-time terrain maps for autonomous bulldozer mining. Ph.D. dissertation, The University of Queensland.

Dadhich, S., Bodin, U., Andersson, U., 2016. Key challenges in automation of earth-moving machines. Autom. Constr. 68, 212–222.

Dobson, A.A., Marshall, J.A., Larsson, J., 2016. Admittance Control for Robotic Loading: Underground Field Trials with an LHD. J. Field Robot. 34(1).

Egli, P., Gaschen, D., Kerscher, S., Jud, D., Hutter, M., 2022. Soil-Adaptive Excavation Using Reinforcement Learning. IEEE Robot. Autom. Lett. 7(4), 9778–9785.

Erni, G., Frey, J., Miki, T., Mattamala, M., Hutter, M., 2023. MEM: Multi-modal elevation mapping for robotics and learning. In: IEEE/RSJ Int. Conf. Intell. Robot. Syst., pp. 11011–11018.

Fankhauser, P., Bloesch, M., Gehring, C., Hutter, M., Siegwart, R., 2014. Robot-centric elevation mapping with uncertainty estimates. In: Int. Conf. Climbing Walking Robots, pp. 433–440.

Feng, H., Song, Q., Yin, C., Cao, D., 2022. Adaptive Impedance Control Method for Dynamic Contact Force Tracking of Robotic Excavators. J. Constr. Eng. Manag. 148(11).

Guan, T., He, Z., Song, R., Manocha, D., Zhang, L., 2022. TNS: Terrain traversability mapping and navigation system for autonomous excavators. In: Robotics: Sci. Syst.

Ha, Q.P., Nguyen, Q.H., Rye, D.C., Durrant-Whyte, H.F., 2000. Impedance control of a hydraulically actuated robotic excavator. Autom. Constr. 9(5–6), 421–435.

Ha, Q.P., Yen, L., Balaguer, C., 2019. Robotic autonomous systems for earthmoving in military applications. Autom. Constr. 107.

Hayashi, K., Shimada, K., Ishibashi, E., Okamoto, K., Yonezawa, Y., 2013. Development of D61EXi/PXi-23 Bulldozer with automatic control system of work equipment. Komatsu Tech. Rep. 59(166), 1–8.

Holz, D., Azimi, A., Teichmann, M., 2013. Real-time simulation of mining and earthmoving operations: A level set-based model for tool-induced terrain deformations. In: ISARC.

Holz, D., Beer, T., Kuhlen, T., 2009. Soil deformation models for real-time simulation: A hybrid approach. In: VRIPHYS Workshop, pp. 21–30.

Jackson, T., 2017. Side-By-Side: Comparing Cat, Deere, Komatsu and Topcon mast-less dozer GPS systems. Equipment World, Apr.

Jud, D., Hottiger, G., Leemann, P., Hutter, M., 2017. Planning and Control for Autonomous Excavation. IEEE Robot. Autom. Lett. 2(4), 2151–2158.

Jud, D., Leemann, P., Kerscher, S., Hutter, M., 2019. Autonomous free-form trenching using a walking excavator. IEEE Robot. Autom. Lett. 4(4), 3208–3215.

Kasaragod, A., Addepalli, A., Vuyyuru, A., et al., 2022. Image-based soil granularity classification for robotic excavation. In: IEEE Int. Conf. Robot. Autom.

Kendall, A., Gal, Y., 2017. What uncertainties do we need in Bayesian deep learning for computer vision? Adv. Neural Inf. Process. Syst. 30, 5575–5585.

Komatsu, 2022. Intelligent machine control features iMC 2.0 – Dozers. https://www.komatsu.eu/en/intelligent-machine-control-features-imc-20-dozers.

Lever, P.J.A., 2001. An automated digging control for a wheel loader. Robotica 19, 497–511.

Li, X., 1993. Physically-based modeling and distributed computation for simulation of dynamic terrain in virtual environments. Ph.D. dissertation.

Luengo, O., Singh, S., Cannon, H., 1998. Modeling and identification of soil-tool interaction in automated excavation. In: IEEE Int. Conf. Intell. Robot. Syst. 3, 1900–1906.

Maeda, G.J., Manchester, I.R., Rye, D.C., 2015. Combined ILC and disturbance observer for rejection of near-repetitive disturbances, with application to excavation. IEEE Trans. Control Syst. Technol. 23(5), 1754–1769.

Maeda, G.J., 2013. Learning and reacting with inaccurate prediction: Applications to autonomous excavation. Ph.D. dissertation, The University of Sydney.

Mathur, A., Singh, V.K., Shivam, et al., 2024. A comprehensive review of bulldozers in modern construction. J. Sci. Res. Rep. 30(5), 337–342.

Mckyes, E., 1989. Chapter 8. Soil cutting and tillage. In: Soil Cutting Mechanics. Elsevier, pp. 192–221.

Miedema, S.A., 2014. The Delft sand, clay and rock cutting model. IOS Press.

Miki, T., Wellhausen, L., Grandia, R., Jenelten, F., Homberger, T., Hutter, M., 2022. Elevation mapping for locomotion and navigation using GPU. In: IEEE/RSJ Int. Conf. Intell. Robot. Syst., pp. 2273–2280.

Mononen, T., Kolu, A., Mattila, J., 2022. Semi-autonomous bulldozer blade control using real-time terrain mapping. In: IEEE/ASME Int. Conf. Adv. Intell. Mechatron., pp. 775–782.

Nguyen, H.A., Ha, Q.P., 2023. Robotic autonomous systems for earthmoving equipment operating in volatile conditions and teaming capacity: A survey. Robotica 41(2), 486–510.

Reece, A.R., 1964. The fundamental equation of earth-moving mechanics. Proc. Inst. Mech. Eng. 17, 16–22.

Ries, R., 2019. With intelligent excavators ruling the marketplace, here's how to find machine controls that match your needs. Equipment World, Mar.

Saidi, K.S., Bock, T., Georgoulas, C., 2016. Robotics in construction, in: Siciliano, B., Khatib, O. (Eds.), Springer Handbook of Robotics. Springer, pp. 1493–1520.

Simulations, C.L., 2016. Theory guide: Vortex software's multibody dynamics engine, pp. 1–65.

Singh, S., 1995a. Synthesis of tactical plans for robotic excavation. Ph.D. dissertation, Carnegie Mellon University.

Singh, S., 1995b. Learning to predict resistive forces during robotic excavation. In: Proc. IEEE Int. Conf. Robot. Autom., pp. 2102–2107.

Sotiropoulos, F.E., Asada, H.H., 2019. A model-free extremum-seeking approach to autonomous excavator control based on output power maximization. IEEE Robot. Autom. Lett. 4(2), 1005–1012.

Soylemezoglu, A., Williams, D., Ellison, C.C., Klein, J.D., Lopez, I.J., Netchev, A., Nottage, D.S., William, J., 2021. Robotics for engineer operations. Eng.: Prof. Bull. Army Eng. 2021 Annu. Issue, 58–60.

Tan, C.P., Zweiri, Y.H., Althoefer, K., Seneviratne, L.D., 2005. Online soil parameter estimation scheme based on Newton-Raphson method for autonomous excavation. IEEE/ASME Trans. Mechatron. 10(2), 221–229.

Thuerey, N., Holzschuh, B., Holl, P., Kohl, G., LIno, M., Liu, Q., Schnell, P., Trost, F., 2021. Physics-based deep learning. https://physicsbaseddeeplearning.org.

Wagner, W.J., 2025. Deriving the fundamental equation of earthmoving and configuring Vortex Studio earthmoving simulation for soil property estimation experimentation. arXiv:2505.19330.

Wagner, W.J., Nottage, D., Soylemezoglu, A., Driggs-Campbell, K., 2023. In situ soil property estimation for autonomous earthmoving using physics-infused neural networks. In: Proc. 16th Eur.-Afr. ISTVS Conf.

Yu, S., Song, X., Sun, Z., 2023. On-line prediction of resistant force during soil–tool interaction. ASME J. Dyn. Syst. Meas. Control 145(8).

Zhu, Y., Thangeda, P., Ornik, M., Hauser, K., 2023. Few-shot adaptation for manipulating granular materials under domain shift. In: Robotics: Sci. Sys

Appendix

Algorithm 1 Update Map with GET Movement

$M \leftarrow$ elevation map
$\boldsymbol{T}_{0:1} \leftarrow$ Blade pose history for current sweep
$\sigma_{SV} \leftarrow$ Blade height uncertainty during sweep
$\gamma_l \leftarrow$ Fixed loose soil moist unit weight
$\boldsymbol{o} \leftarrow$ Observation history
update, $\hat{t}, \hat{n} \leftarrow \textsc{CheckMovement}(T_0, T_1)$
if update **then**
  $SV \leftarrow \textsc{GenerateSweptVolume}(T_0, T_1)$
  $\boldsymbol{A}, \boldsymbol{\Delta H}, \boldsymbol{G}, \boldsymbol{h}^G \leftarrow \textsc{IntersectHeightmap}(SV, M)$
  $\alpha, \rho, d, w, V_0^Q, \boldsymbol{P}, \boldsymbol{X}^{\hat{t}}, \boldsymbol{Z} \leftarrow \textsc{ExtractFEEParams}(M, T_0, \boldsymbol{A}, \boldsymbol{\Delta H}, \hat{t}, \hat{n})$
  $\boldsymbol{B}, \Delta V^Q \leftarrow \textsc{CutAndDeposit}(M, \boldsymbol{A}, \boldsymbol{\Delta H}, \hat{t}, \hat{n}, \sigma_{SV})$
  $\mathbf{d}, \boldsymbol{V}^Q \leftarrow \textsc{InterpolateFEEParams}(\alpha, \rho, d, \boldsymbol{T}_{0:1}, V_0^Q)$
  $Q \leftarrow \gamma_l \boldsymbol{V}^Q$
  ▷ Collect Known FEE Parameters
  $\boldsymbol{\Theta}_k = [\alpha, \rho, w, \mathbf{d}, \boldsymbol{Q}]$
  if training **then**
    $\textsc{LogData}(\Theta_k, \boldsymbol{o})$
  else
    $\hat{\Theta}_u, \Sigma_{\Theta_u} \leftarrow \textsc{EstimateSoilProperties}(\boldsymbol{\Theta}_k, \boldsymbol{o})$
    $\textsc{UpdateSoilLayers}(M, \hat{\Theta}_u, \Sigma_{\Theta_u}, \boldsymbol{A}, \boldsymbol{P}, x_{\min}^{\hat{t}})$
  end if
  $\textsc{PerformSoilErosion}(M, T_1, \hat{m}, \boldsymbol{G}, \boldsymbol{h}^G)$
end if

Algorithm 2 ExtractFEEParams

$C_x \leftarrow$ Line fit distance exponential weight
$C_d \leftarrow$ Parameter average depth exponential weight
function $\textsc{ExtractFEEParams}(M, T_0, \boldsymbol{A}, \boldsymbol{\Delta H}, \hat{t}, \hat{n})$
    $\boldsymbol{P}, \boldsymbol{X}^{\hat{t}}, \boldsymbol{Z}, V^Q \leftarrow \textsc{GetSurfacePoints}(\boldsymbol{A}, M, \hat{t}, \hat{n}, T_0)$
    for all $a, \Delta h, \boldsymbol{x}^{\hat{t}}, \boldsymbol{z} \in \boldsymbol{A}, \boldsymbol{\Delta H}, \boldsymbol{X}^{\hat{t}}, \boldsymbol{Z}$ **do**
        ▷ Fit Line to surface points & Intersect with blade surface & Extract slice FEE parameters
        $\boldsymbol{W}_a = \exp(-C_x \boldsymbol{x}^{\hat{t}})$
        $\boldsymbol{\alpha}[a], \boldsymbol{\rho}[a], \boldsymbol{d}[a] \leftarrow \textsc{WLSLineFitIntersect}(\boldsymbol{x}^{\hat{t}}, \boldsymbol{z}, \boldsymbol{W}_a, M, \Delta h, \hat{n}, \hat{t})$
    end for
    ▷ Average FEE parameters over slices
    $\boldsymbol{W}_d \leftarrow \exp(C_d \boldsymbol{d})$
    $\alpha, \rho, d \leftarrow \textsc{WAvg}(\boldsymbol{\alpha}, \boldsymbol{\rho}, \boldsymbol{d}, \boldsymbol{W}_d)$
    $\boldsymbol{P}^{\perp \hat{t}} = \textsc{ProjectPointsOnToBladeApprox}(\boldsymbol{A}, \hat{t})$
    $w = \max(\boldsymbol{P}^{\perp \hat{t}}) - \min(\boldsymbol{P}^{\perp \hat{t}})$
    return $\alpha, \rho, d, w, V^Q, \boldsymbol{P}, \boldsymbol{X}^{\hat{t}}, \boldsymbol{Z}$
end function

Algorithm 3 CutAndDeposit

$\lambda_n \leftarrow$ Deposit direction normal weight
$s \leftarrow$ Soil swell ratio
function $\textsc{CutAndDeposit}(M, \boldsymbol{A}, \boldsymbol{\Delta H}, \hat{t}, \hat{n}, \sigma_{SV})$
    ▷ Get deposit direction
    $m \leftarrow \lambda_n \hat{n} + (1 - \lambda_n)\hat{t}$
    $\boldsymbol{B} \leftarrow \textsc{GetDepositCell}(\hat{m}, \boldsymbol{A}, M)$
    $\hat{m} \leftarrow \frac{m}{||m||}$
    $H := M[\text{elevation}]$
    $L := M[\text{loose}]$
    $\sigma := M[\sigma]$
    for all $a, \Delta h, b \in \boldsymbol{A}, \Delta H, \boldsymbol{B}$ **do**
        ▷ Update intersected cells
        $H[a]^+ \leftarrow H[a] - \Delta h$
        $L[a]^+ \leftarrow \max(L[a] - \Delta h, 0)$
        $\sigma[a]^+ \leftarrow \max\left(\max(\sigma[a] - \Delta h, 0), \sigma_{SV}\right)$
        ▷ Update deposit cells
        $H[b]^+ \leftarrow H[b] + s\Delta h$
        $L[b]^+ \leftarrow L[b] + s\Delta h$
        $\Delta h_{\max} \leftarrow \textsc{Get1SigmaMaxDisplacement}(\sigma_{SV}, \Delta h, \sigma[a], s)$
        $\sigma[b]^+ \leftarrow \sqrt{\sigma[b]^2 + \left(\frac{\Delta h_{\max}}{2}\right)^2}$
    end for
    $\Delta V_Q \leftarrow \textsc{GetDeltaVolume}(H, L, H^+, L^+)$
    return $\boldsymbol{B}, \Delta V_Q$
end function

Algorithm 4 Update Soil Layers

$C_s \leftarrow$ Soil wedge distance exponential weight
$x^{\hat{t}}_{\min} \leftarrow$ Minimum soil wedge marking distance
function UPDATESOILLAYERS($M, \hat{\Theta}_u, \Sigma_{\Theta_u}, \boldsymbol{P}, \boldsymbol{X}^{\hat{t}}$)
    ▷ Obtain distance to end of wedge
    $x^{\hat{t}}_{\max} \leftarrow \frac{d}{\cos(\alpha)(\tan(\alpha+\beta)-\tan(\alpha))}$
    $\boldsymbol{\Theta_u} := M[\Theta_u]$
    $\boldsymbol{\Sigma}_{\Theta_u} := M[\Sigma_{\Theta_u}]$
    for all $\boldsymbol{p}, \boldsymbol{x}^{\hat{t}} \in \boldsymbol{P}, \boldsymbol{X}^{\hat{t}}$ **do**
        ▷ Inflate estimated variances
        $\boldsymbol{v} \leftarrow \boldsymbol{x}^{\hat{t}} < x^{\hat{t}}_{\max} \vee \boldsymbol{x}^{\hat{t}} < x^{\hat{t}}_{\min}$
        $\boldsymbol{C}_w \leftarrow \exp(C_s \frac{\boldsymbol{x}^{\hat{t}}}{x^{\hat{t}}_{\max}})$
        $\Sigma^{+}_{\Theta_u} \leftarrow \boldsymbol{C}_w \boldsymbol{\Sigma}_{\Theta_u}$
        ▷ Perform Bayesian update for valid wedge cells
        $\boldsymbol{\Sigma}^{-}_{\Theta_u} \leftarrow \boldsymbol{\Sigma}_{\Theta_u}[\boldsymbol{v}]$
        $\boldsymbol{\Theta}^{-}_{u} \leftarrow \boldsymbol{\Theta}_u[\boldsymbol{v}]$
        $\boldsymbol{\Theta}_u[\boldsymbol{v}] \leftarrow \frac{\Sigma^{+}_{\Theta_u} \boldsymbol{\Theta}^{-}_{u} + \boldsymbol{\Sigma}^{-}_{\Theta_u} \Theta_u}{\boldsymbol{\Sigma}^{-1}_{\Theta_u} + \Sigma^{+}_{\Theta_u}}$
        $\boldsymbol{\Sigma}_{\Theta_u}[\boldsymbol{v}] \leftarrow \frac{\Sigma^{+}_{\Theta_u} \boldsymbol{\Sigma}^{-}_{\Theta_u}}{\boldsymbol{\Sigma}^{-1}_{\Theta_u} + \Sigma^{+}_{\Theta_u}}$
    end for
    return
end function

Algorithm 5 Perform Soil Erosion

$c_l, \phi_l, \gamma_l \leftarrow$ Fixed loose soil properties for erosion
$D_{\text{ROI}} \leftarrow$ Soil erosion ROI dimensions
$g \leftarrow$ Heightmap resolution
$E \leftarrow \mathbf{0}$ Blade and swept volume mask
$Z \leftarrow \mathbf{0}$ Mask heights
$\delta t \leftarrow$ Erosion timestep
function PERFORMSOILEROSION($M, T_1, \hat{m}, \boldsymbol{G}, \boldsymbol{h}^G$)
  $\boldsymbol{R}_a, \boldsymbol{R}_b \leftarrow$ GETROITILES($T_1, \hat{m}, D_{\text{ROI}}$)
  $H := M[\text{elevation}]$
  $L := M[\text{loose}]$
  $\sigma := M[\sigma]$
  ▷ Fill Erosion Mask
  $E[\boldsymbol{G}] \leftarrow 1$
  $Z[\boldsymbol{G}] \leftarrow \boldsymbol{h}^G$
  for all $\boldsymbol{r}_a, \boldsymbol{r}_b \in \boldsymbol{R}_a \boldsymbol{R}_b$ **do**
    $\boldsymbol{\Delta h} \leftarrow \min(H[\boldsymbol{r}_a] - H[\boldsymbol{r}_b], L[\boldsymbol{r}_a])$
    $F_{\min}, \alpha^* \leftarrow$ MINIMIZESLIPSAFTEYFACTOR($\boldsymbol{\Delta h}, c_l, \phi_l, \gamma_l, g$)
    $\boldsymbol{h}_{\text{slip}} \leftarrow$ COMPUTESLIP($\alpha^*, \boldsymbol{\Delta h}, c_l, \phi_l, \gamma_l, g, \delta t$)
    ▷ Limit slip if masked
    $\boldsymbol{e}_a \leftarrow E[\boldsymbol{r}_a]$
    $\boldsymbol{z}_a \leftarrow Z[\boldsymbol{r}_a]$
    $\boldsymbol{e}_b \leftarrow E[\boldsymbol{r}_b]$
    $\boldsymbol{z}_b \leftarrow Z[\boldsymbol{r}_b]$
    $\boldsymbol{h}_{\text{slip}}[\boldsymbol{e}_b] \leftarrow \min(\boldsymbol{h}_{\text{slip}}[\boldsymbol{e}_b], \max(\boldsymbol{z}_b - H[\boldsymbol{r}_b], 0))$
    $\boldsymbol{h}_{\text{slip}}[\boldsymbol{e}_a] \leftarrow 0$
    ▷ Update map
    $H[\boldsymbol{r}_a] \leftarrow H[\boldsymbol{r}_a] - \boldsymbol{h}_{\text{slip}}$
    $L[\boldsymbol{r}_a] \leftarrow H[\boldsymbol{r}_a] - \boldsymbol{h}_{\text{slip}}$
    $\sigma[\boldsymbol{r}_a] \leftarrow \sigma[\boldsymbol{r}_a] - \boldsymbol{h}_{\text{slip}}$
    $H[\boldsymbol{r}_b] \leftarrow H[\boldsymbol{r}_b] + \boldsymbol{h}_{\text{slip}}$
    $L[\boldsymbol{r}_b] \leftarrow H[\boldsymbol{r}_b] + \boldsymbol{h}_{\text{slip}}$
    $\sigma[\boldsymbol{r}_b] \leftarrow \sigma[\boldsymbol{r}_a] + \boldsymbol{h}_{\text{slip}}$
  end for
  return
end function

4579 | Analysis of passenger tire-salted pavement interaction: a traction perspective

Erin Fenton *, Zeinab El-Sayegh

Department of Automotive and Mechatronics Engineering, Ontario Tech University, Oshawa, Canada

** Corresponding author: erin.fenton@ontariotechu.net*

ABSTRACT

This paper introduces a novel approach to investigate the interaction between a passenger car tire and salted pavement. A passenger car tire size 235/55R19 was modelled within a Finite Element Analysis (FEA) software called Virtual Performance Solution (VPS) using 28 layers including solid, layered membrane, beam, and shell elements. The modelled tire was validated in static and dynamic domains using manufacturer-provided data. The salt used to sit on top of the pavement was modelled using Smoothed-Particle Hydrodynamics (SPH) techniques and calibrated using the shear-strength test. The salted pavement–tire interaction required the use of a node-to-segment contact algorithm with edge treatment to ensure no penetration between the salt particles and the tire elements. The focus of this paper is to investigate the tractive efforts of the driven tire over the salted pavement under different operating conditions. The operating conditions include vertical loads of 3.5 kN, 5 kN, and 8 kN, and a nominal inflation pressure of 228 kPa at constant angular velocities equivalent to longitudinal speeds of 10 km/h, 50 km/h, and 100 km/h. To understand the impact salt has on the driving conditions of icy roads several simulations including a 57 mm, 114 mm, and 170 mm salt depth were selected to understand the impacts of the amount of salt on a surface.

This study provides insight into the use of road salt in the winter months in Canada. As road salt is crucial in winter climates it is important to understand the implications that salting pavement will have on the performance of passenger car tires through advanced computational techniques. By understanding the implication salt has on different pavements affected by winter a deeper understanding of safety, and tire traction performance can be achieved.

Keywords
Tire Mechanics
Vehicle Dynamics
Terramechanics
Smoothed Particle Hydrodynamics
Finite Element Analysis

1. Introduction

Winter in Canada includes changing weather conditions that greatly increase the risk of accidents as road conditions become unstable. Road maintenance is extremely important to keeping the conditions of the roads as stable as possible while weather conditions change. To do so snowplows, de-icing agents and road salt are used to maintain stable conditions. With variability in driving conditions, it is important to understand how road salt can influence the traction between passenger car tires and pavement.

By understanding the tractive forces between car tires and salted pavement an optimized solution for the use of road salt can be provided to control the influence that road salt has on habitation near the roads. The goal is to find the optimal solution for salting pavement that encourages the safest conditions for vehicles without increasing the risk of hyper-saline bodies of water (Radosavljevic et al., 2024).

Extensive research on tire-pavement interaction over the last two decades has evolved significantly between physical and modelled tire testing, of which, the results all capture the same characteristics of tires whether performed on the roads or in a lab. Methods for performing research include computational approaches to modelling across many software's for validating tire forces and moments in tire science. Many researchers evaluated the impacts of snow on tire performance and are used to predict the salt requirement of the roads.

Wang et al. (2023) studied traction of three types of tires at different inflation pressures and SPH modelled snow in an FEA software. The results provided a correlation between the traction property and wheel load that determined the lower the inflation pressure the better the contact with the ground. This research however lacked the use of varying speeds in tests in different snow depths. Zeng et al. (2022) measured the tractive interaction between an off-road tire and a granular gravel/sand terrain using a large flat track indoor bin. The results concluded the tractive effort increased linearly with granular size though the results lacked application to winter analysis the granular size of the gravel and sand does provide insight to impact of salting icy roads. Kurczynski et al. (2022) analyzed the impacts that invisible ice had on road surfaces tractive forces by evaluating acceleration capability. Using a Ford Transit Van, they determined the influence the type of ice has on the acceleration that can be acquired. The results showed black ice provided the least tractive force but did not account for the increase in acceleration possibility when the pavement was salted for added traction.

Vieira et al. (2021) evaluated the impact five different types of tires have on rolling resistance on two road surfaces. Using a rolling resistance measurement trailer the results showed the highest rolling resistance was from studded winter tires and winter tires. Evaluating the rolling resistance on ice provides insight to free rolling tires but does not evaluate the increase in tractive effort in driven tires. Ruzinskas et al. (2021) performed tire traction performance tests with mixtures of snow and water, and crushed ice with water. Using an internal drum test bench tire, traction tests were performed at 4 kN vertical load, 220 kPa inflation pressure with 0-degree camber angle at 50 km/h. These tests lacked application of salt that is used to limit the standing snow, ice, water on road surfaces. In the same year, Carlson et al. (Carlson, 2021) studied the impacts of snow and rainwater on rolling resistance of aircraft wheels at three different speeds with a constant tire pressure. The results showed that increased snow depth and lower tire pressure gave higher rolling resistance. As the study aimed to understand the fuel efficiency of vehicles, they did not bear in mind the importance of the traction for these vehicles.

https://doi.org/10.56884/JF3254YB

Xu et al. (2020) examined the tractive efforts between off-road tires and granular terrain in LS-DYNA to determine how tread pattern, vertical load, and inflation pressure can influence the tractive effort. The granular terrain used was gravel in a soil bin and increased the tractive efforts of tires. The granular terrain though different from salt provides the insight this paper is providing. This work, however valuable, should also be applied to passenger car tires rather than off-road tires on granular surfaces. Giudici et al. (2019) performed tests on free rolling tires over snow mixed with salt to understand the concentration of salt has on tire performance. This research allows for the practical application of salt to ensure the lowest density snow that will be evacuated by passing traffic. These tests were only conducted on compacted snowy roads and not icy roads which could provide insight to the best practices of salting icy roads and braking up the solid ice.

Ekinci et al. (2015) evaluated the tractive performance of radial tires on off-road vehicles with different lug heights, axle loads and inflation pressures. The surfaces used were clay, silt, and sand. The findings showed that efficiency in tractive effort increased when inflation pressure was reduced. The findings are beneficial when evaluating the use of salt on pavement as sand and salt act to increase tractive forces in the winter months for roads that cannot be salted. This research however provides the opposite advice to always have your vehicle tires inflated to the appropriate pressure and cannot be applied to winter driving on salted icy roads. Bhoopalam et al. (2014) conducted research with free rolling tires at varying normal loads, inflation pressures, and tread depth on ice. This research used free rolling tires for testing, therefore, the traction effort of tires was not measured. The research however provides insight to depth of ice influencing performance of tires in winter. Bhoopalam et al. (2014) provided a state-of-the-art mathematical modelling review of tire performance on ice. This research provided insight to ten consistencies of ice providing a coefficient of friction that varies. This review however does not provide insight to what salt or snow can do to this coefficient of friction.

Heggli et al. (2010) used numerical simulations in LS-DYNA using Finite Element Methods (FEM) to predict the tire traction between compacted snow and a passenger car tire size 195/65R15 at inflation pressure 200 kPa and vertical load 4 kN. This testing only predicted the tractive effort, but they did not evaluate any additional surfaces such as any mixed terrain with these predictive numerical simulations. Nakajima (2003) developed an analytical model to estimate longitudinal traction of a passenger tire on snow with good reliability. This work however did not consider any measurements outside of static traction on a different surface other than snow. Scheibe (2002) provided insight into the use of studded tires and winter tires for winter climates that encourage the use of studded tires versus regular tires for winter performance as traction was far higher with the use of snow tires on ice and snow. This research performed physical testing of all tires and recorded the tire traction by time it took to reach 32.2 km/h but did not use any forces to determine the tractive effort of the tires. Raad et al. (2000) performed research to understand the tractive performance of transit and paratransit vehicle tires on snowy and icy surfaces. As a result, they found when vehicle tire traction forces decrease the stopping distances increase and decrease controllability of the vehicle. The team did not, however, consider how salt that is consistently used in winter can increase the tractive forces seen under these transit vehicle tires.

The research presented in this paper provides an innovative solution to understanding the tractive forces experienced by passenger car tires through a hybrid computational approach. An FEA passenger car tire sized 235/55R19 was modelled and validated in VPS alongside an SPH salted pavement model that was calibrated using a shear-strength test. The research work below provides insight to salt particles on an icy road surface and the influence they have on increased traction of passenger car tires on low friction surfaces. The interaction was analyzed using three different depths of salt on the same 0.2 low coefficient of friction surface to compare how tractive forces change with depth of salt applied to the pavement. By understanding the influence road salt has on increased tractive forces refined techniques of salting and maintaining roads during the Canadian winter months can be applied.

2. Tire-Salted Pavement Model

A four-groove Passenger car tire size 235/55R19 Continental CrossContact LX Sport model seen in Fig. 1 was used to demonstrate the interaction between tires and road salt. All interactions were captured using a hybrid FEA-SPH approach in the software called VPS.

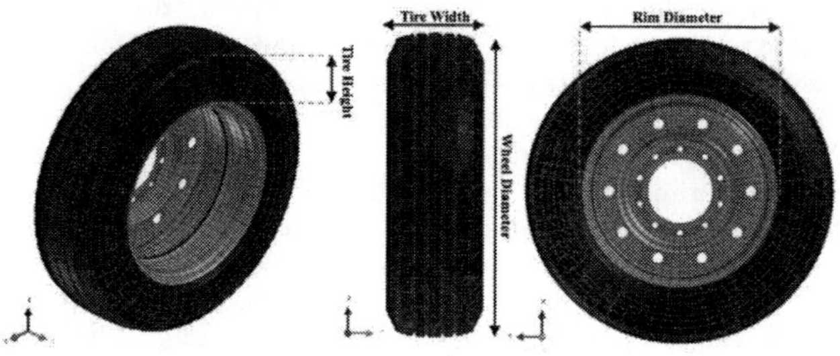

Fig. 1. VPS Modelled Tire

The tire modelled in this research work included a rim assembly with the layers to best capture the material behaviour exhibited by the rubber compounds found in the tire (Fathi, 2024). To capture the behaviours of rubber the Mooney-Rivlin hyperelastic model was used for the nonlinear deformation that occurs in incompressible rubber material used in the layers of the tire tread.

2.1. Wheel Material Behaviour and Modelling

The behaviour of the FEA tire model is derived from the Strain Energy Function (SEF) proposed by Rivlin. The model is expressed as a linear function in Eq. (1 with the two invariants for incompressible materials.

$$W = \sum_{i=0,j=0}^{\infty} C_{ij}(I_1 - 3)^i (I_2 - 3)^j \tag{1}$$

Where C_{ij} are representative of the material constants.

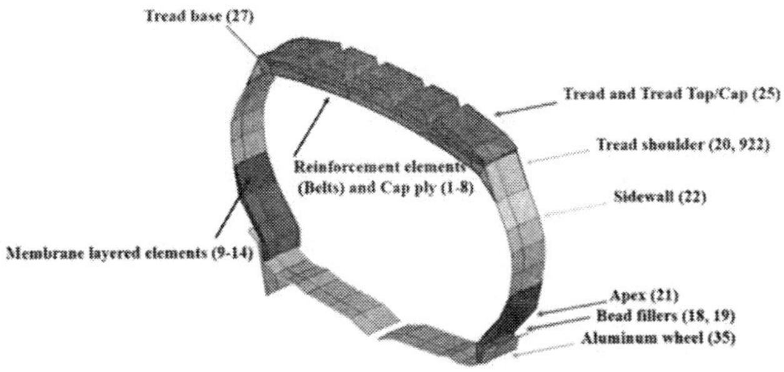

Fig. 2. Cross Section of Tire Components with Labels (Fathi, 2024)

The generalized Mooney-Rivlin model is seen in (2) and best utilized for the tread modelling of the tire seen in Fig. 2.

$$W = C_{10}(I_1 - 3) + C_{01}(I_2 - 3) \tag{2}$$

In Eq. 2 C_{01} represents the loading coefficient and C_{10} corresponds to the unloading coefficient. Figure 2 represents the cross section of the tire with labelled parts. Table 1 provides the Mooney-Rivlin properties for the tread and under tread tire components.
The rim was modelled as an aluminum alloy to reflect the increase in stiffness and change in mass distribution than the layers of the tire. This stiffness of the alloy reflects the assumption the rim is a solid material without deflection during the operation of the testing. An additional assumption during the operation of the tire-rim assembly is there is a negligible amount of tire-rim slip that occurs. These assumptions help to reduce the complexity of the problem and provide a solid foundation for which all simulations occur.

Table 1
Mooney-Rivlin Tire Tread Component Properties (Fathi, 2024)

Tire component	C_{01}(MPa)	C_{10} (MPa)	Poisson's Ratio
Tread	2.49	0.67	0.473
Under tread	0.51	1.86	0.409

Numerical calibration was used for the wheel assembly geometry of the rim weight and diameter along with the tire weight and diameter seen in Table 2. Simulation values were compared against Continentals manufacturer data. These values correspond to the tire geometry when the tire is inflated before loading conditions were applied.

Table 2
Wheel Component Comparison between Manufacturer Specifications and Simulation Results (Fathi, 2024).

Wheel Assembly	Tire Geometry	Manufacturer Data*	Simulation Results	Units
Rim	Weight	11.4	11.05	kg
	Diameter	482.6	445.27	mm
Tire	Weight	13.65	14.81	kg
	Diameter	741.68	739.69	mm

*Manufacturer data taken from the Continental website (Continental, 2024).

2.2. Wheel Validation

Validation of the tire is required to ensure the results of simulation match those of physical tire testing completed by the tire manufacturer. To ensure appropriate results static and dynamic tests were conducted and compared against manufacturer data. The static tests used were lateral, longitudinal, and vertical stiffness. The dynamic tests were rolling resistance, cornering stiffness and critical vertical frequency. In Table 3 below the evaluated tests are presented to compare the simulation results to the manufacturer data of the tire and confirm the validity. There is however a discrepancy between the longitudinal stiffness of the tire and the manufacturer data, and the tire would therefore reach maximum traction sooner than the physical tire would.

Table 3
Static and Dynamic Tire Verification (Fathi, 2024).

Dynamic Testing	Manufacturer Data*	Simulation Results	Units
Rolling Resistance Coefficient	0.015–0.03	0.006–0.01	N/A
Cornering Stiffness	1200–1800	1057.26	N/degree
Critical Vertical Frequency	70–90	74	Hz
Static Testing	Manufacturer Data*	Simulation Results	Units
Vertical Stiffness	181.1	235.86	N/mm
Lateral Stiffness	115–140	116.66	N/mm
Longitudinal Stiffness	200–250	373.11	N/mm

The simulation validation results in the above Table 3 were all performed at a nominal load of 5 kN and an inflation pressure of 228 kPa.

2.3. Salt Modelling and Calibration

The salt particles used in the testing were modelled as SPH particles. To model a material as an SPH particle the material properties of the salt were gathered and validated. The material properties used were density, shear modulus, yield stress, tangent modulus and an equation of state represented by Eq. 3;

$$p = c_0 + c_1\mu + c_2\mu_2 + c_3\mu_3 + (c_4 + c_5\mu + c_6\mu_2)E_i \tag{3}$$

In Eq. (3 c_0 and c_1 represent material constants, μ is the ratio of current over initial mass density and E_i is the internal energy of the material.

To calibrate the SPH salt particles with the found material properties of salt a shear box test was performed. The shear box test loaded six identical boxes filled with the SPH material with increasing pressures from 0 kPa to 50 kPa in increments of 10 kPa. Once the tests completed the salt particle from each box with the greatest shear strength value would be plotted to show a shear-strength relationship. The relationship between the shear-strength and pressure values would then be fitted with a line of best fit. Using the Mohr-Coulomb failure criterion in Eq. (4 the shear box test was calibrated using a linear line of best fit to determine the internal angle of friction and cohesion which was validated against literature values (Kolano et al., 2024) .

$$\tau_{max} = c + \sigma \tan\phi \tag{4}$$

Here, the c represents cohesion, and ϕ represents the angle of internal friction. Where the angle of internal friction is evaluated as $\tan^{-1}(\phi)$. As seen in Fig. 3 the accuracy of the VPS calibration provided an internal friction angle of 47.02 degrees where literature values gave an angle of 44 ± 0.85 degrees. The VPS values were within 6% of the physical testing properties so the model was determined to be validated. The cohesion was also calibrated against literature values as the y-intercept from the equation given provides the cohesion to be 257 kPa.

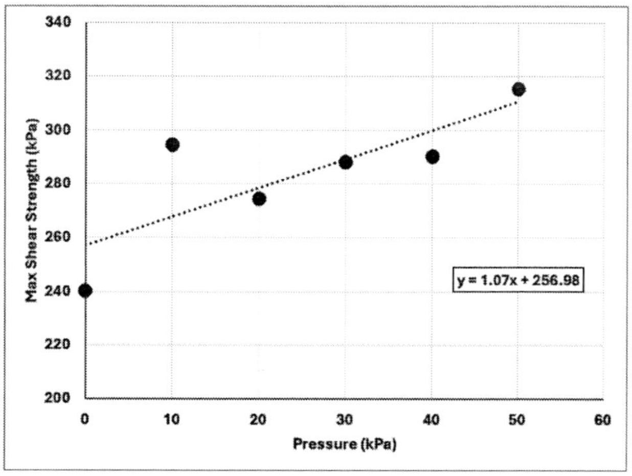

Fig. 3. Shear-Strength Results for the Salt Particle Calibration

3. Salted Pavement Tire Interaction

The salted pavement tire interaction simulations were evaluated at varying conditions of depth of salt, angular velocity equivalent to longitudinal speed, and vertical loading. To evaluate the effect salt has on increased tractive effort coefficients a 0.2 coefficient of friction was applied to the road surface. The simulation consists of the FEA tire model appended on to an icy rigid road topped with SPH salt particles. A node-to-segment contact with edge treatment algorithm was applied to the salt and road as the salt is a slave to the road. The same node-to-segment contact with edge treatment algorithm is also applied to the tire and the road with the tire being a master to the road. These contact algorithms are applied so the particles are forced to interact with each other in a controlled manner where the particles do not pass through each other. The tire, salt and road are shown in Fig. 4 where the x-direction is the direction the tire moves in, the negative z-direction is the direction in which the vertical load is applied, and the y-direction is the direction where the angular velocity is applied around.

Within an evaluated simulation the tire has an inflation pressure applied to the interior of the tire sidewall, the vertical load is applied to the center of the rim, and the tire is then settled on the surface of the road before the angular velocity is applied in the y-direction. This allows the tire to be settled on the road surface before the angular velocity is applied so no additional forces are added to the simulation.

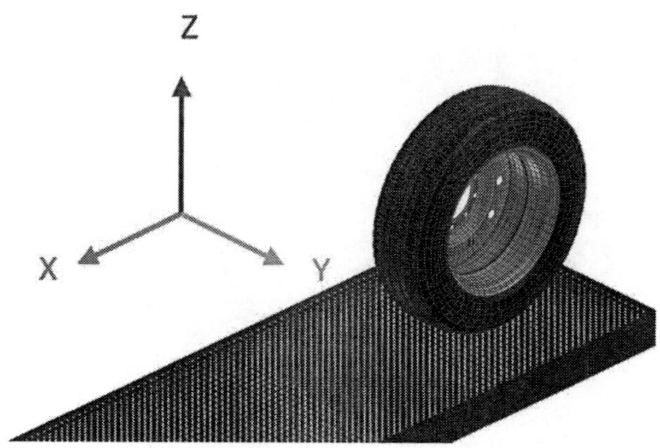

Fig. 4. Tire on Top of Salted Pavement

The numerical conditions applied to the simulations used angular velocities equivalent to longitudinal speeds at 10 km/h, 50 km/h and 100 km/h, with vertical loading from 3.5 kN, 5 kN and 8 kN over the three salt depths of 57 mm, 114 mm and 170 mm. Within the simulations all the tires were inflated with the same pressure of 228 kPa. To evaluate the interaction between the modelled tire and salt on the road surface tractive effort and longitudinal slip were evaluated at several operating conditions.

Tractive effort evaluation of a passenger car tire required the use of two forces in Kilonewtons; the longitudinal force acting in the x-direction (F_X) of the tire and the vertical load applied in the z-direction (F_z) of the tire from the center of the rim. As seen in Eq. 5 the tractive effort was determined as the ratio of the generated longitudinal force to the applied vertical force, the forces were taken as an average after steady state was reached at 0.3 seconds until the duration of the two second test.

$$TE = \frac{F_x}{F_z} \tag{5}$$

The test was taken in a two second interval as the tire reaches steady state around 0.3 seconds and then the characteristics of the interaction beyond this give results that are all average. A longer test would require greater computer power and would not yield more accurate results than needed at two decimal places of importance to evaluate the tractive forces of the tire.

To validate the use of tractive effort testing for all simulations longitudinal slip during driving was provided for each angular velocity and vertical load. Longitudinal slip over the salt surface provides an understanding of the amount of tractive effort that is gained by the tire as it moves over the surface of the road. For the tire to gain the most traction the longitudinal slip of the tire reaches the peak value between 20-40% seen in Fig. 5.

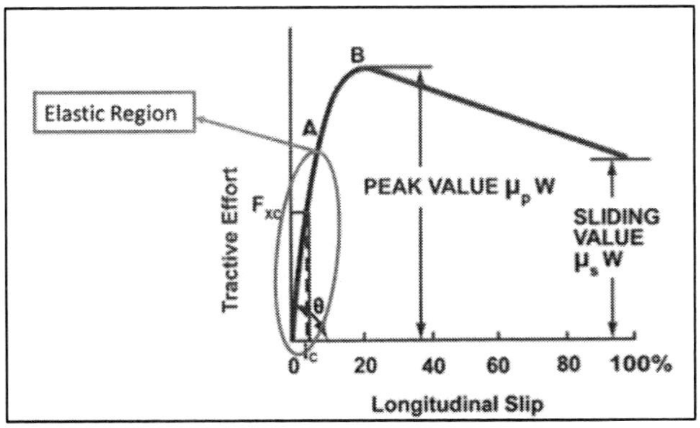

Fig. 5. Longitudinal Slip vs Tractive Effort (Wong, 2022)

Longitudinal slip in Eq. 6 is defined as the percentage difference between the translational speed of the vehicle and the speed of the tire.

$$s\,(\%) = \left(1 - \frac{v}{\omega * r}\right) * 100 \tag{6}$$

Here, v represents longitudinal speed (mm/s) in the x-direction, ω represents angular velocity (rad/s) in the y-direction, and r represents the effective loading radius of a tire (mm) between the center of the rim to the center of the contact patch. To calculate the longitudinal slip using (6) the tires average longitudinal speed and angular velocity were taken after the tire reached steady state at 0.3 seconds over the length of the two second test. To evaluate the effective rolling radius of the tire a single measurement at 0.69 seconds was taken where maximum deflection occurred seen in Fig. 6. Using the measurement feature in VPS the deflection of the tire between the center of gravity node of the tire and the middle of the contact patch was chosen. This point however not perfectly aligned with the center of the contact patch was measured using only the z-axis component for accuracy.

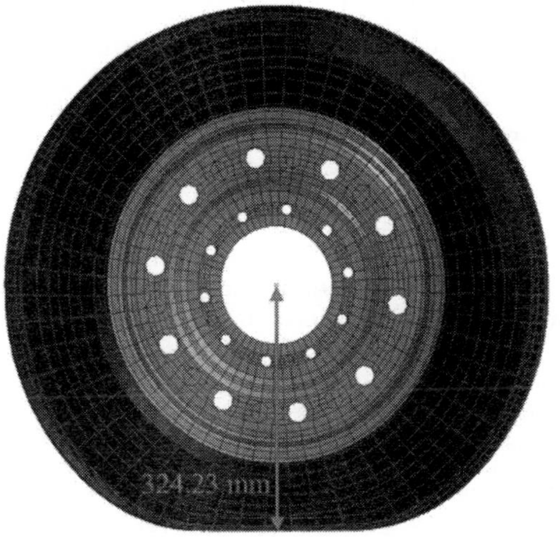

Fig. 6. Effective Rolling Radius of Tire at 50 km/h and 8 kN Load

Using the longitudinal slip equation, Eq. 6, in conjunction with Eq. 5 the tractive effort values were validated using the longitudinal slip percentage. When the longitudinal slip percentage is around its peak value between 20-40% the tire is in the best situation to provide adequate tractive effort. When longitudinal slip is much lower than the peak value around 20% the tire is essentially sliding over the road surface and not providing enough tractive effort to be measured and the tire acts as though it is free rolling.

4. Results and Discussion

The simulations were performed at several operating conditions including 3.5 kN, 5 kN and 8 kN applied vertical load and applied angular velocity equivalent to longitudinal speeds of 10 km/h, 50 km/h and 100 km/h were used in junction with a nominal inflation pressure of 228 kPa. In addition, the simulation was also repeated for several salt depths including 57 mm, 114 mm and 170 mm to analyze the effect of salt depth on tire-salted pavement interaction characteristics.

At the salt depth of 57 mm, longitudinal slip and tractive effort were evaluated at four speeds of 10 km/h, 50 km/h, 80 km/h and 100 km/h. As seen in Fig. 7, of the four tests the 10 km/h tires show the least percentage of longitudinal slip where the other tires provide an adequate amount of slip. The tractive effort of this tire could not be created as the 10 km/h test had far too low longitudinal slip, so the contact patch was sliding over the salt. The other three tires however show adequate amounts of tractive effort over the salt as they are producing much higher amounts of longitudinal slip ranging between 35 and 77%. For this reason, the tractive effort across the 50 km/h, 80 km/h and 100 km/h tests have similar ranges of tractive effort as reflected in Fig. 7.

Table 4
Tire Slip Percentage for Driving Over 57 mm Salt Depth

Slip (%)		Loads (kN)		
		3.5	5	8
Speed (km/h)	10	2.415	3.812	6.464
	50	71.723	34.720	51.839
	80	71.723	59.617	36.639
	100	77.518	71.113	51.435

As seen in Table 4 and Fig. 7 the longitudinal slip most closely correlated to increased tractive effort are the simulations with peak longitudinal slip between 20-40%. Two values that fall between this range are at 80 km/h and 8 kN with 37% longitudinal slip and 50 km/h and 5 kN with 34% longitudinal slip. These two tires both provided the highest tractive effort coefficients of 0.192 and 0.206, respectively.

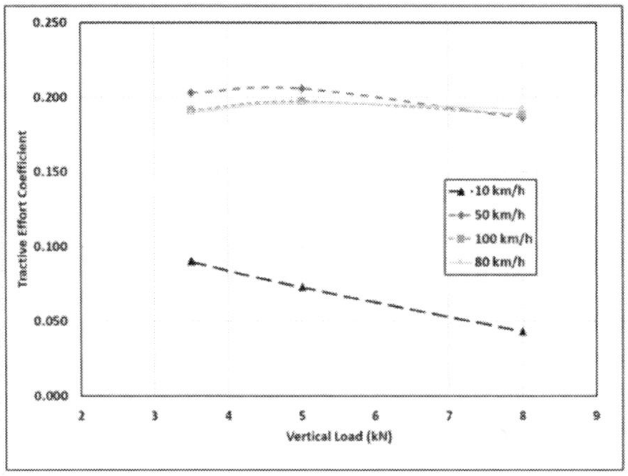

Fig. 7. Tractive Effort Coefficient -57 mm Salt Depth

For the 114 mm salt depth simulations were run at 10 km/h, 50 km/h and 100 km/h for three vertical loads of 3.5 kN, 5 kN and 8 kN. Of these similar to the 57 mm salt depth it was determined that in Table 5 the longitudinal slip of the 10 km/h tire is far too low to provide a significant amount of tractive effort to the tire and it remains as a driven tire acting like a free rolling tire. The other tires at higher speeds do however provide an adequate amount of slip to induce additional tractive effort over the salt.

Table 5
Tire Slip Percentage for Driving Over 114 mm Salt Depth

Slip (%)		Loads (kN)		
		3.5	5	8
Speed (km/h)	10	1.364	2.812	5.148
	50	52.630	35.769	52.630
	100	77.069	67.934	50.747

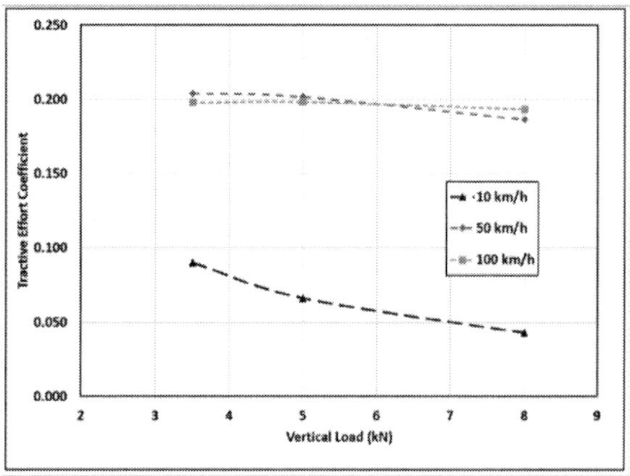

Fig. 8. Tractive Effort Coefficient-114 mm Salt Depth

In Fig. 8 the tractive effort coefficients of the 10 km/h tire are incredibly low as provided the tire does not create enough longitudinal slip to gain traction on the salty surface like the higher speed tires. As seen in the 50 km/h and 100 km/h tires there were higher amounts of tractive effort when driven as they have adequate amounts of longitudinal slip. As seen in Table 5 and Fig. 8 above the longitudinal slip most closely correlated to increased tractive effort are the tests with the peak longitudinal slip between 20-40%. One value that fell between this range was at 50 km/h and 5 kN with 36% longitudinal slip. The resulting tractive effort coefficient was 0.201.

At the salt depth of 170 mm longitudinal slip and tractive effort were evaluated at three equivalent longitudinal speeds of 10 km/h, 50 km/h and 100 km/h. As seen in Table 6 of the three tires loaded with 10 km/h they show the least percentage of longitudinal slip, and the contact patch slides over the salt. The other tires however show adequate amounts of tractive effort over the salt surface as they are producing much higher amounts of longitudinal slip between 35 and 77%. For this reason, the tractive effort across the 50 km/h and 100 km/h tests have similar amounts of tractive effort reflected in Fig. 9.

Table 6
Tire Slip Percentage for Driving Over 170 mm Salt Depth

Slip (%)		Loads (kN)		
		3.5	5	8
Speed (km/h)	10	0.061	1.416	4.909
	50	52.422	35.249	52.422
	100	77.191	67.647	50.008

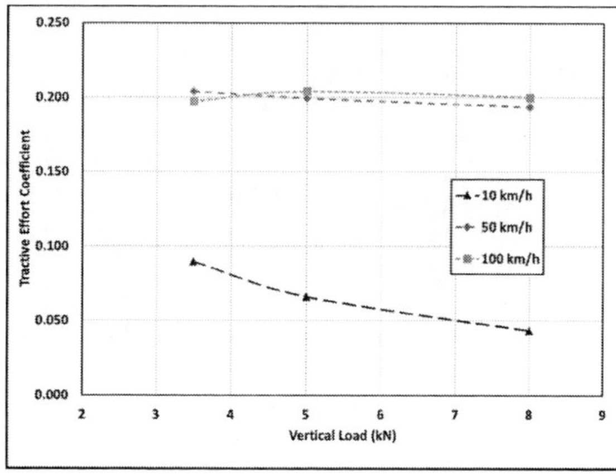

Fig. 9. Tractive Effort Coefficient-170 mm Salt Depth

In Fig. 9 the tractive effort coefficients of the 10 km/h are incredibly low where the tire does not create enough longitudinal slip to gain traction on the salty surface like the higher speed tires. At higher speeds of 50 km/h and 100 km/h there were higher amounts of tractive effort when driven as they have adequate amounts of longitudinal slip. As seen in Table 6 and Fig. 9 above the longitudinal slip that can provide near peak tractive effort are the tires with longitudinal slip between 20-40%. One tire produced longitudinal slip in this range, and it was at 50 km/h and 5 kN with 35% longitudinal slip. The resulting tractive effort coefficient was 0.199.

As a result of this study the highest tractive effort was seen on the icy surface with 57 mm of salt depth. Any salt above this amount does not increase tractive performance as the performance threshold was met with the lowest salt depth. As the testing was performed at three loads and three salt depths the next steps will be to understand the influence inflation pressure has on increasing the tractive performance at this salt depth.

4.1. Contact Stress Analysis

To understand how the contact patch is influenced by pressure building from the tire-road surface interaction the stress pressure was evaluated in the contact patch at the end of the two second simulation shown below in Fig. 10, Fig. 11 and Fig. 12. The simulation was set at 50 km/h, 57 mm of salt depth, 228 kPa and all three vertical loading conditions. As seen in Fig. 10, Fig. 11 and Fig. 12 the contact patches have an even and symmetrical pressure exerted by the road surface in all three loading conditions. As seen the trend in pressure of the contact patch increased as load increased. The highest contact pressure in the 3.5 kN loaded tire was 1.52 MPa. While the highest contact pressure in the 5 kN loaded tire was 2.64 MPa, and that in the 8 kN loaded tire was 3.27 MPa. The area where the contact pressure is most prominent does not change but an increase in pressure is seen.

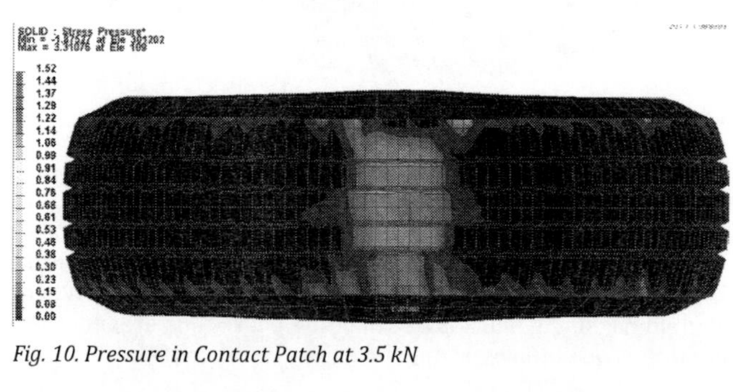

Fig. 10. Pressure in Contact Patch at 3.5 kN

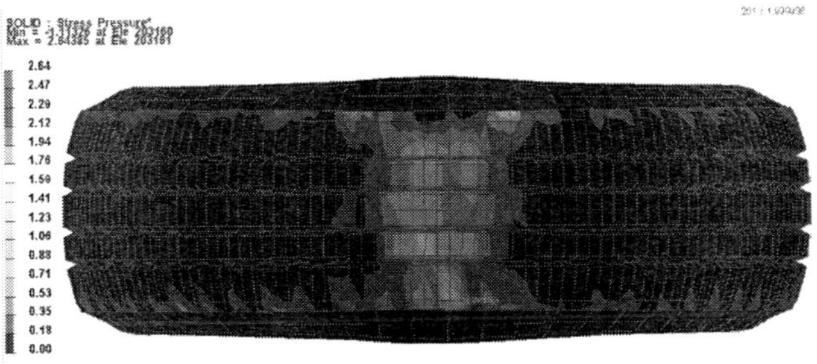

Fig. 11. Pressure in Contact Patch at 5 kN

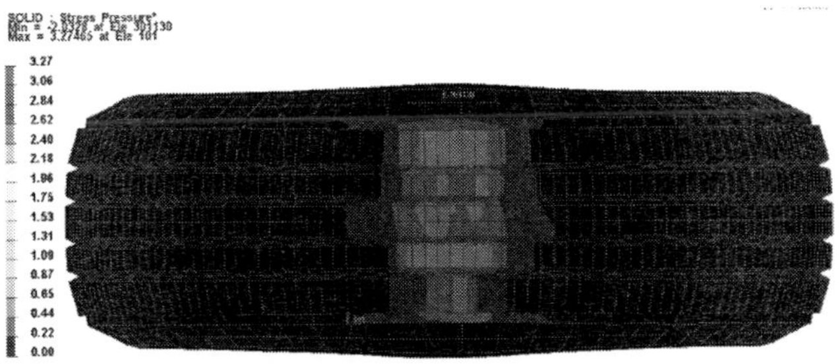

Fig. 12. Pressure in Contact Patch at 8 kN

As a result of this study, it can be understood that with increased loading the contact patch will have an increased influence of building pressure on the tread where the tire interacts with the road. This is consistent with what is seen in literature, as the vertical load increases the contact area increases as well which results in further stress generation. This will in turn result in increased wear because of increased pressure influence where the tire meets to road surface.

5. Conclusions and Future Work

Road salt is an integral component of road maintenance during the winter months in North America. Without it, road conditions could become unstable and the formation of ice on service roads would increase the risk of accidents. Using a Continental CrossContact Sport LX 235/55R19 tire, the interaction between the tire and salted surface was investigated. The application of the salt on icy roads was explored at varying salt depths of 57 mm, 114 mm and 170 mm with the tire having a nominal inflation pressure of 228 kPa and loaded at three vertical loads of 3.5 kN, 5 kN, and 8 kN.

This research provided insight to how longitudinal slip influences the tractive effort coefficient. As seen the tire with adequate amounts of longitudinal slip between 20-40% can gain traction on a surface and the tractive effort coefficient represents a driven tire. Below 20% longitudinal slip the tire does not create adequate tractive effort, and the tire will function as though it is a free rolling tire. This research overall provides the trend that once the road surface has been salted there is a limited amount of additional tractive effort that can be provided. With the data it was determined that longitudinal slip at given conditions can reach a peak value of 37% at 80 km/h and 8 kN for 57 mm salt depth, at 114 mm salt depth the peak longitudinal slip reaches 36% at 50 km/h and 5 kN, and at 170 mm peak longitudinal slip is 35% at 50 km/h and 5 kN. These three loading conditions gain the most tractive effort coefficients of 0.206, 0.201, and 0.199 respectively. This provides insight that once the peak amount of longitudinal slip is reached the tractive effort coefficient cannot be increased much past 0.2. Though additional salt on top of 170 mm will not create additional tractive effort the initial salting of the road surface does increase the tractive effort values seen by ~10% at operating conditions of 50 km/h at 5 kN load with 57 mm salt depth.

In the near future, works including icy surfaces in combination with sand, snow, and water in any combination will be investigated. As the materials may interact with each other during the winter months it is of importance to understand the variety of agents that influence the safety of the drivers on the road. To further increase this knowledge braking and accelerating will be applied to understand how normal road behaviours are influenced in these conditions.

6. Nomenclature

c	Cohesion constant of terrain	(kPa)
ϕ	Internal shear resistance	(deg)
τ_{max}	Maximum shear stress	(kPa)

7. Acknowledgements

The authors express their gratitude to the Natural Sciences and Engineering Research Council of Canada (NSERC Discovery Grant) for their continuous support during this research work.

8. Declaration of competing interest

The authors declare that they have no known competing financial interests or personal relationships that could have appeared to influence the work reported in this paper.

9. References

Bhoopalam, A.K., Sandu, C., 2014. Review of the state of the art in experimental studies and mathematical modeling of tire performance on ice. Journal of Terramech. 53(1), 19–35. https://doi.org/10.1016/j.jterra.2014.03.007

Carlson, A., 2021. The effect of water and snow on the road surface on rolling resistance. In Swedish National Road and Transport Research Institute. https://doi.org/10.13140/RG.2.2.12407.75686

Continental, 2024. CrossContact LXSport. https://continentaltire.com/tires/crosscontact-lx-sport

Ekinci, Ş., Çarman, K., Kahramanli, H., 2015. Investigation and modeling of the tractive performance of radial tires using off-road vehicles. Energy 93, 1–11. https://doi.org/10.1016/j.energy.2015.10.070

Fathi, H., 2024. Modeling and Analysis of All-Season Passenger Car Tire using Finite Element Analysis. Vehicles 6, 0–133.

Giudici, H., Wåhlin, J., Klein-Paste, A., 2019. Tire–Pavement Interaction in Contact with Salted Snow: How an Aqueous Solution Affects Compression, Squeeze-Out, and Compaction of Snow. Journal of Cold Regions Engineering 33(4). https://doi.org/10.1061/(asce)cr.1943-5495.0000192

Heggli, M., Jaggi, M., Löwe, H., Schneebeli, M., Seta, E., Nakai, T., Szabo, D., 2010. Prediction of tire traction on a compacted snow road. https://www.researchgate.net/publication/292792235

Kolano, M., Cała, M., Stopkowicz, A., 2024. Mechanical Properties of Rock Salt from the Kłodawa Salt Dome—A Statistical Analysis of Geomechanical Data. Materials 17(14). https://doi.org/10.3390/ma17143564

Kurczyński, D., Zuska, A., 2022. Analysis of the Impact of Invisible Road Icing on Selected Parameters of a Minibus Vehicle. Sensors 22(24). https://doi.org/10.3390/s22249726

Nakajima, Y., 2003. Analytical model of longitudinal tire traction in snow. Journal of Terramech. 40(1), 63–82. https://doi.org/10.1016/j.jterra.2003.09.003

Raad, L., Lu, J.J., 2000. Traction Performance of Transit and Paratransit Vehicles in Winter. Journal of the Transportation Research Board 1731(1).

Radosavljevic, J., Slowinski, S., Rezanezhad, F., Shafii, M., Gharabaghi, B., Van Cappellen, P., 2024. Road salt-induced salinization impacts water geochemistry and mixing regime of a Canadian urban lake. Applied Geochemistry 162. https://doi.org/10.1016/j.apgeochem.2024.105928

Ružinskas, A., Giessler, M., Gauterin, F., Wiese, K., Bogdevičius, M., 2021. Experimental investigation of tire performance on slush. Eksploatacja i Niezawodnosc 23(1), 103–109. https://doi.org/10.17531/EIN.2021.1.11

Scheibe, R. R., 2002. An Overview of Studded and Studless Tire Traction and Safety.

Vieira Tiago, Sandberg, U., & Erlingsson, S., 2021. Rolling resistance evaluation of winter tyres on in-service road surfaces. Tire Science and Technology 49(2).

Wang, D., Wang, H., Xu, Y., Zhou, J., Sui, X., 2023. Numerical Simulations of the Driving Process of a Wheeled Machine Tire on a Snow-Covered Road. Machines 11(6). https://doi.org/10.3390/machines11060657

Wong, J. Y., 2022. Theory of Ground Vehicles (5th ed.). John Wiley & Sons, Inc.

Xu, W., Zeng, H., Yang, P., Zang, M., 2020. Numerical analysis on tractive performance of off-road tire on gravel road using a calibrated finite element method–discrete element method model and experimental validation. Proceedings of the Institution of Mechanical Engineers, Part D: Journal of Automobile Engineering 234(14), 3440–3457. https://doi.org/10.1177/0954407020930175

Zeng, H., Xu, W., Zang, M., 2022. Experimental investigation of tire traction performance on granular terrain. Journal of Terramech. 104, 49–58. https://doi.org/10.1016/j.jterra.2022.09.002

4755 | Relationship between seasonal terrain properties and vehicle rut depths

Susan Frankenstein *, Allan Wheeler, Bruce Elder

Cold Regions Research and Engineering Laboratory, Hanover, New Hampshire, United States

** Corresponding author: susan.frankenstein@erdc.dren.mil*

ABSTRACT

Laboratory and field studies were conducted over three years on fine grained and peat soils measuring the effects of soil properties on vehicle performance, namely drawbar pull, and motion resistance. Tests were conducted for unfrozen, frozen, and thawing states using eight different vehicles both wheeled and tracked. In the laboratory, soil moisture and temperature could be controlled. In both the field and laboratory, the soils were tilled then smoothed before testing. The soil was characterized pre and post testing using a variety of strength instruments. Rut depths associated with motion resistance tests were recorded. This paper discusses the effects of measured soil properties on rut depth as a function of soil state as well as compares results to existing algorithms. Soil properties measured using different techniques are also compared.

Keywords
Soil freeze/thaw
Peat
Rut depth

1. Introduction

The NATO (North Atlantic Treaty Organization) reference mobility model (NRMM) calculates sinkage for fine- and coarse-grained soils as a function of cone index, or rating cone index, of the critical layer (Priddy, 1995). These equations are empirical, based on hundreds, if not thousands, of measurements. In NRMM, the soil is broken into 2 layers, an upper one from 0-6 in (0-15 cm) and a lower one from 6-12 in (15-30 cm) (Ahlvin and Haley, 1992). The critical layer is the one that can support the vehicle, starting from the surface. There is no sinkage equation for peat.

Affleck (2005) measured rut depths in Alaska in peat and fine-grained soils in unfrozen, frozen, and thawing conditions for heavy wheeled and tracked vehicles. She found that the NRMM predictions under estimated rut depths in thawing soil for higher cone index (greater than 100 psi). Affleck (2005) found the opposite for weak soils. In winter, little or no rut depths were found on the frozen soil. Notably, no correlation was found between water content and rut depth for thawing soils.

Vennik et al. (2017) measured single and multi-pass rut depths in fine-grained soil for both a wheeled and tracked vehicle then compared them to the model developed by Willoughby and Turnage (1990), and implemented in NRMM. They used the same approach for peat even though the model was not intended for these soils. Vennik et al. used the cone index of two critical layers, 7.5-23 cm and 15-30 cm. If the strength of the upper critical layer was used, the modelled single-pass ruts over predicted the measured ones for ruts less than 7.5 cm and under predicted rut depth for deeper ruts. The model under predicted the measured rut depths for all ruts using the lower critical layer strength.

Vennik et al. (2019) measured single and multi-pass rut depths from a wheeled vehicle on natural grasslands. Soil type ranged from various loams to peat. Similar to this study, they also measured soil properties such as density and moisture content. They used *SoilFlex* (Keller et al., 2007) with the uniaxial strain rate option to predict rut depth. The *SoilFlex* model overestimated rut depth in peat soils.

As stated previously, none of these models are intended for use with peat soils. They are also not meant to predict rut depth for frozen or thawing soils. As with unfrozen soils, the bearing capacity of frozen and thawing terrain depends on several properties including, but not limited to, wetness, density, organic content, and vegetation. It is hard to capture all soil states needed to formulate new off-road mobility predictors, particularly for winter and shoulder conditions. Multiple soil types, moisture levels, freeze-thaw states, and vehicles are needed. These are hard to control in natural environments. Therefore, field-scale experiments under controlled conditions were conducted in the Frost Effects Research Facility, FERF, in Hanover, NH as well as at the Keweenaw Research Center, KRC, Calumet, MI test facility. The results of the rut depth measurements will be discussed in this paper.

2. Test Set-Up

2.1. Vehicles

Eight different vehicles were tested over the course of the project, five in the FERF and three at KRC. They ranged from wheeled and tracked all-terrain to medium tracked vehicles. General descriptions, test location, and vehicle parameters pertinent to this study are listed in *Table 1*.

https://doi.org/10.56884/H9K33U27

Table 1
Test vehicle information. NGP is nominal ground pressure, VCI1 is the one pass vehicle cone index calculated using the NRMM VCI equations (Priddy, 1995), FG is fine-grain, and MK is muck. In cells where there are two entries, the top is for the front axle while the bottom is for the rear.

vehicle	test location	total vehicle weight (kg)	*NGP* (kPa)	*VCI1FG* (psi)	*VCI1MK* (psi)
light wheeled	FERF	3075	114 133	18.3 21.2	28.8 31.5
wheeled light all-terrain	FERF	1579	95 111	16.1 19.6	24.4 28.2
tracked light all-terrain	FERF	1851	9.5 12.6	8.2 9.6	13.2 13.2
tracked light all-terrain	FERF	2077	10.8 13.9	8.8 10.1	13.2 13.2
tracked articulated amphibious	FERF	7452 5014	19 13	8.2 6.8	13.3 13.2
light wheeled	KRC	6080	268 228	30.2 27.8	40.9 49.3
medium tracked	KRC	10324	48	15.5	13.9
tracked light all-terrain	KRC	2569	5.8 7.1	8.9 10.5	13.1 13.1

2.2. Soils

Laboratory experiments were conducted in the FERF between November 2023 and March 2025 over two test periods, FY2024 (three test soil types) and FY2025 (two test soil types) where FY is fiscal year. The test layouts for each period are shown in *Fig. 1*.

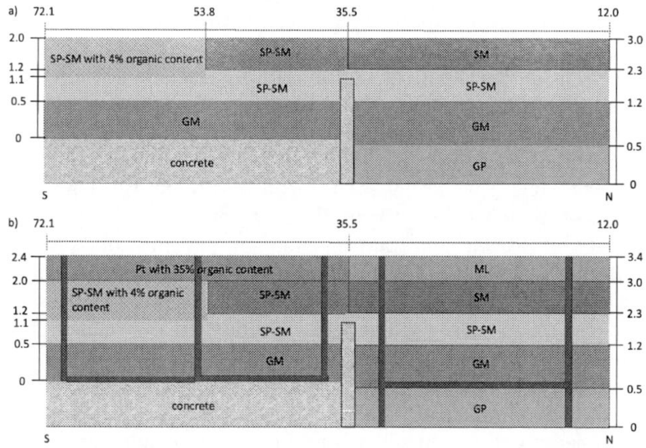

Fig. 1. Side view of test section layouts. 0 ft (0 m) is at the North end of the test bed (right side of drawings). All measurements shown are in feet. a) FY2024; b) FY2025. Vertical blue lines in the bottom diagram are the syphons used to introduce water to the bottom of the sections.

The FERF testing lane is over 76 m (250 ft) long. In FY2024 the test lane was divided into three, approximately equal, partitions each with a different frost susceptible soil. The section lengths were, in-part, determined by the presence of a concrete vertical wall (top diagram *Fig. 1*). Initial USCS (Unified Soil Classification System) identification labeled them as ML, low plasticity silt; SM, silty sand; and Pt, peat. The soils were reclassified after testing and instead found to be SM, SP-SM, a mix of poorly graded (i.e. uniform) sand and silty sand, and SP-SM with 4% organics (*Fig. 1* top). The organic content was measured using a burn technique as outlined in ASTM D2974-20 (2020). The plastic limit (PL) and liquid limit (LL) were 17 and 19 for the SM soil. They were 0 for the other soils. Larger gravel and cobble size stones were present in each section but were especially prevalent in the SM portion farthest north. Below the top, were layers of increasingly better drained soils to allow for water infiltration from both the top and bottom to incorporate a water table and more natural soil conditions.

In FY2025 another layer was added to the FY2024 test configuration. Instead of three different soil types, the test lane was divided into two test soils of approximately equal partitions as shown in the bottom of *Fig. 1*. In the USCS schema they are ML and Pt with 35% organics (*Fig. 1* bottom). The peat is a mix of manufactured sand and peat generated from natural sphagnum moss as a starter. The silt is also manufactured.

Sixteen tests were conducted in FY2024 and thirteen in FY2025 still, the data is too limited to determine relationships between rut depth and soil properties, especially for frozen and thawing soils. Therefore, data collected at KRC in their outdoor peat and fine-grained, ML, soil test sections between September 2020 and March 2021 were incorporated with the two years of data collected in the FERF. The organic content of the KRC peat is 28%. No rut measurements are available for the fine-grained soils in the unfrozen soil state, but the other tests provide valuable data for evaluating rut depths under different seasonal conditions.

Prior to testing, the soils at both the FERF and at KRC were ripped to a depth of approximately 0.5 m and then the top 15 cm was tilled to a uniform consistency. Soil wetness and freeze/thaw depths were varied in the FERF beginning with dry conditions in FY2024. In FY2025, water was added from above and below. The volumetric water content over the two testing years ranged from 2-30% while at KRC it was 25-45%. The frost depth for all three test seasons was 5-40 cm. The thaw depth was about 5 cm in the FERF and 20 cm at KRC. Seven different instruments were used to characterize soil conditions. These will be discussed in next.

3. Soil Measurements

At both KRC and in the FERF soil measurements were made before vehicle passage then in the vehicle rut after a single pass motion resistance test. No measurements were made in the drawbar pull ruts due to override of the holdback vehicle. Only the pretest measurements are used for rut analysis as the models are intended to be predictive. In the FERF, the full suite of measurements was taken at two locations for each soil type while at KRC there were five locations for each vehicle test lane.

3.1. Surface soil moisture

Average surface volumetric soil moisture (VWC = volume water/volume soil) was measured using the FieldScout TDR (time domain reflectometry) 350 soil moisture meter. The Standard soil setting was used for all FieldScout moisture readings. Instrument resolution is 0.1 % and accuracy is ± 3 %. Soil saturation is typically around 50 % VWC (Spectrum Technologies, 2022a). Four tine lengths are available: 3.8, 7.6, 12.2, and 20.3 cm (1.5, 3.0, 4.8, and 8.0 in). The NRMM bins the soil into 6 in (15.2 cm) layers to predict vehicle performance. Thus, for the unfrozen soils, the 7.6 cm (4.8 in) tines were used as these are closest to the surface layer thickness. For the frozen soil, tine length was determined based on penetration ability while for the thawing soil, tine length was chosen to best characterize the thawing layer. Five readings were taken at each measurement location, from which average, standard deviation, minimum, and maximum were calculated for both the location and soil type, FERF, or test lane, KRC.

3.2. Soil Density Profile

In situ (wet) density (total mass/total volume) averaged over two-inch (5.1cm) increments from the surface to a depth of 20.3 cm (8 in) was measured using the Troxler EGauge™ Model 4590 surface density gauge. Instrument resolution is 0.3 lb/ft³ (4.8 kg/m³) and accuracy is ± 0.5 lb/ft³ (8.0 kg/m³) (Troxler, 2016). If the soil moisture is known, then at each depth the dry (bulk) density (soil mass/total volume), gravimetric moisture content (wet density - dry density), and percent moisture (gravimetric moisture content/dry density) are also determined. Troxler provides the model 6760 Moisture Probe which is calibrated to a single EGauge™. This was done for the FERF tests but unavailable available at KRC. The moisture measurement depth is approximately 3.5 in (9 cm; Troxler, 2022) and is assumed to be constant over the depth of the density range measured. A set of measurements takes approximately 10 minutes. Average, standard deviation, minimum, and maximum were calculated for each soil type in the FERF and test lane at KRC.

3.3. Surface shear

The surface shear was measured using the Geonor H-60 Hand-Held Vane Tester instruments which was designed to measure the undrained shear strength of soft to stiff clays (Geonor, 1995). The Geonor has three vanes, 16, 20, and 25.4 mm in diameter. There is only one scale and so a multiplicative factor (20, 10, 5 respectively) is applied to obtain the maximum shear value. Resolution and accuracy before applying the factor are 0.2 kPa and ± 0.1 kPa respectively. The vane size was soil and freeze/thaw state dependent. For the unfrozen soil, the 25.4 mm vane was used while for the frozen soil only the 16 mm vane could be used. As with the soil moisture, five measurements were made from which the average, standard deviation, minimum, and maximum were calculated for both location and soil type, FERF, or test lane, KRC.

3.4. Surface California Bearing Ratio - Clegg Impact Hammer

The Clegg impact hammer is often used to determine the strength of manmade turf surfaces. A cylinder (the hammer) is dropped from a given height down a guide tube and the peak deceleration of the hammer on impact is measured. Four drops at a given point consist of one test (Lafayette Instrument Co, 2005). Further test details can be found in ASTM D5874-16 (2016). Three replicates were taken at a location to allow for basic statistical analysis both at the location and for the soil type or test lane. The hammer is available in three sizes, 4.5, 2.25, and 0.5 kg. For all tests, the lightest-weight, 0.5 kg hammer was dropped from a 30 cm height. Output units are in Clegg Impact Value (*CIV*), *CIV* = 10 gravity units = 98.1 m/s². *CIV* is then converted to *CBR* (California Bearing Ratio, 0 - 100 %) using a hammer dependent algorithm. For the 0.5 kg hammer the algorithm in Shoop et al. (2019) was used: $CBR = 0.55CIV - 1$.

3.5. California Bearing Ratio profile - Dynamic Cone Penetrometer

The California Bearing Ratio profile was measured using a dynamic cone penetrometer (DCP). Test procedure is outlined in ASTM D6951/D6951M-09 (2015). For all of the sites the 4.6 kg (10.1 lb) sliding weight was used along with the standard 60-degree, 20 mm diameter cone tip. Depth of the tip as a function of the number of drops was recorded as millimeter per blow, known as the DCP index. This is then converted to a *CBR* value ($CBR = 292(2 \cdot \text{DCP index})^{-1.12}$). Total penetration depths ranged from 0.5 to 1.0 m. Diverging from the standard procedure, at the start of the test, the penetrometer was allowed to sink under its' own weight and the depth recorded before the first drop. This is referred to as the zero drop depth later in the paper.

3.6. Static Cone Index Profile

The cone penetrometer is used extensively by the U.S. Army to quantify soil penetration strength, combined shear and compressive. In part, this is due to its portability and ease of use. The unit of strength is the cone index (*CI*; psi, kPa). The penetrometer should be pushed into the soil at a constant rate of 3 cm/s (0.1 ft/s). Two tip sizes are available, 0.5 in (1.27 cm) and 0.75 in (1.91 cm) diameter. The cone angle is 30°. The 0.5 in (1.27 cm) cone was used for this study at both locations. The FieldScout SC900 Soil Compaction Meter was used in these tests. It continuously measures the cone index in pounds per square inch (psi) from the surface to a depth of 18 in (45.72 cm) recording values every 2.54 cm (1 in) (Spectrum Technologies, 2022b). Instrument resolution is 1 in (2.54 cm) and 35 kPa (5 psi) with an accuracy of ± 0.5 in (1.3 cm) and ± 103kPa (± 15psi). A minimum of three measurements to a minimum depth of 6 in (15.24 cm) were done at each grid point. This depth corresponds to the top layer thickness of NRMM (Ahlvin and Haley, 1992). From these, the average, standard deviation, minimum, and maximum from 0-6 in (0-15.2 cm), 6-12 in (15.2-30.5 cm), 12-18 in (30.5-45.7 cm), as well as the maximum at the surface and over every 3 in (7.6 cm) increment from 0-18 in (0-45.7 cm) were calculated. The later calculations facilitate comparison with measurements taken using the dial cone penetrometer (Vennik et al., 2017). For thawing soils, statistics were calculated just for that layer as well as over the NRMM layers.

4. Inter-Instrument Comparisons

The volumetric water content is measured directly by the FieldScout TDR and indirectly with the EGuage™. Similarly, penetration is obtained using both the static cone penetrometer and dynamic DCP as well as the Clegg Impact Hammer. Inter-instrument comparisons for these two properties will be discussed next. For the purposes of this study, we set a goodness of fit criteria of $r^2 \geq 0.75$ after removal of outliers which were defined as points laying beyond one standard deviation from the initial curve fit. Relationships that meet this cut-off are considered meaningful.

4.1. Water content: EGuage™ and FieldScout

In FY2024 and FY2025 water content was measured using the FieldScout TDR and EGauge™. The EGauge™ dry, or bulk, density was used to convert the TDR volumetric water content to gravimetric and to convert the EGauge™ gravimetric to volumetric.

The results for all soil types for unfrozen soil are shown in *Fig. 2*. The tine length for the comparisons was 12.2 cm (4.8 in). Goodness of fits, r^2, assuming a linear relationship for all soils as well as for each individual soil type were investigated. Goodness of fit using all available points as well as after eliminating outliers defined as any measurements greater than one standard deviation from the fit curve. Fits for which $r^2 \geq 0.75$ were found for gravimetric and volumetric water content using all soil types (55 points) as well as for the silt, ML, and peat, Pt, soils in FY2025 (14 points). At larger water contents, the TDR water contents are lower than the EGuage™ values for peat. Overall, the fits are good.

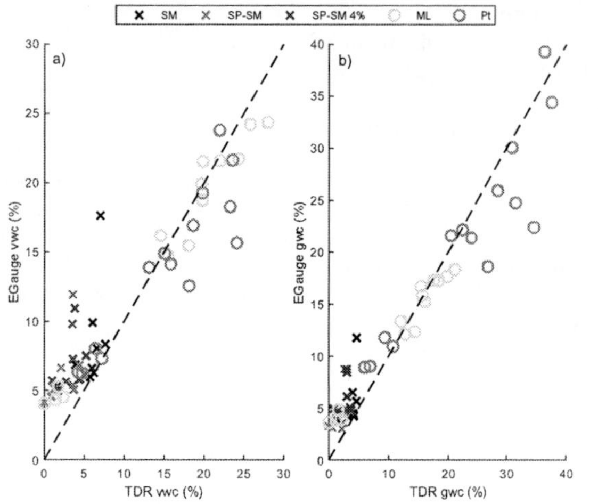

Fig. 2. Comparison between FieldScout TDR and EGauge™ unfrozen water content. a) volumetric; b) gravimetric. The dashed line is one-to-one.

4.2. CBR: Clegg impact hammer and cone penetrometer

Shoop et al. (2008) as part of the Opportune Landing Site program investigated several relationships between *CI* and *CBR*. They found that the best relationship followed a power law, $CBR = aCI^b$ where a and b are soil type dependent. Mason and Baylot (2016) also found that a power law was the best approach. However, instead of *CI*, they used *RCI*. All relations are for unfrozen soil. The fits for Shoop et al. (2008) coarse-grained, fine-grained, and ML are plotted in *Fig. 3a* while the fits for Mason and Baylot (2016) are in *Fig. 3b* for all soil types and ML. For the purposes of this comparison, Pt was considered fine-grained.

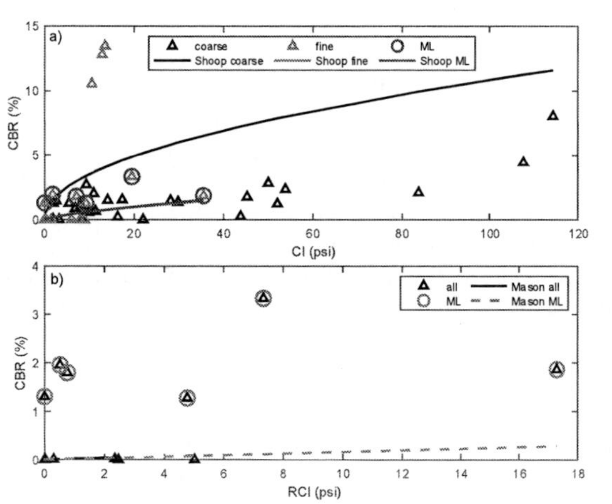

Fig. 3. Comparison between the Clegg drop 4 CBR and cone penetrometer cone indexes for unfrozen soil. a) curve fits from Shoop et al. (2008), b) curve fits from Mason and Baylot (2016).

From *Fig. 3b* it is evident that the measured ML *CBR* values are all greater than both the ML and all-soil power curves proposed by Mason and Baylot (2016). The peat soils are all below the curves. These curves are essentially the same, and thus indistinguishable in the figure for the low *RCI* values. This is also true for the fine-grained and ML curves of Shoop et al. (2008). As with the Mason and Baylot fits, the measured ML values are larger than the predicted ones while the Pt ones are smaller. The coarse-grained curve of Shoop et al. (2008) is greater than all of the corresponding measurements. It should be noted that both approaches were developed using natural soils and not manipulated ones like in this study.

4.3. Cone Index: DCP and cone penetrometer

Although designed for different purposes and soil states, some researchers, for example Hibbard et al. (2025), have used DCP measurements in lieu of the cone penetrometer. Since both sets of measurements were made for this study, it was decided to investigate the viability of this approach. To do so, the relation between the cone index calculated from the DCP zero and first drop depths and the corresponding cone penetrometer values were compared for unfrozen and frozen soils. The zero drop depths ranged from 12 - 34 cm while for the first drop it was 15 - 38 cm. The 'Dutch Method', calculates resistance into the soil from the DCP using an energy approach (Stolf et al., 1998; Herrick and Jones, 2002; Minasny, 2005). Hibbard et al. (2025) use this as a proxy for the cone index here after known as method 1 (m1) for both drop depths. The Army Field Manual (FM) 5-430-00-1 in Figure H-1, Appendix H compares the cone index, *CBR*, and soil shear strength (Army, 1994). From it the relationship $CI = 40 \cdot CBR$ is found. This is referred to as method 2 (m2) for drop one. Method 2 for the zero drop DCP is calculated from a force calculation assuming a friction coefficient of 0.2.

The DCP calculated *CI* were compared to both maximum and average cone penetrometer *CI* values. These were calculated over the corresponding DCP penetrations. The results for the linear fits that meet our criteria, $r^2 \geq 0.75$, after outliers were removed, are presented in *Fig. 4* for unfrozen and *Fig. 5* for frozen soils. The total number of points for each fit ranged from 5 - 12. The dashed line is the one-to-one fit.

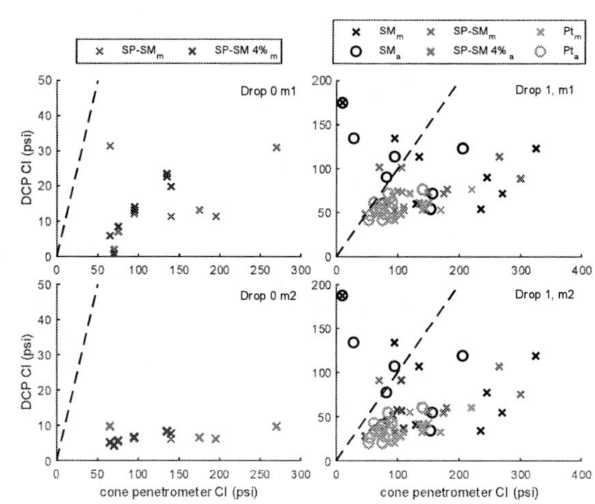

Fig. 4. Comparison between the DCP zero and first drop and cone penetrometer cone indexes for unfrozen soil, using methods 1 (m1) and 2 (m2). Subscripts 'm' and 'a' are for maximum (x) and average (o) respectively. The dashed line is the one-to-one curve.

Before discussing the results, it should be noted, that the DCP is mainly intended for measuring subsurface surface soil strength and that surface values are less reliable. A first look at *Fig. 4* and *Fig. 5* it is evident that, except for the m1 drop 0 comparison, the DCP calculated *CI* are smaller than the cone penetrometer measurements regardless if the maximum or average values are used. Although not shown, better fits were found for drop 1 than drop 0 for both unfrozen and frozen soils. Linear fits with $r^2 \geq$

0.75 were only found for the two SP-SM soils for drop 0. These soil types also meet the criteria for the drop 1 fits. For drop 1 and unfrozen soil, both peat and SM soils also meet this cut-off for unfrozen soil. The only scenario where the ML soil meets this criterion is for frozen soil using m1. Based on these results, caution should be used if the DCP is used instead of the cone penetrometer. Although a linear relation is applicable for some soils, it is not one-to-one.

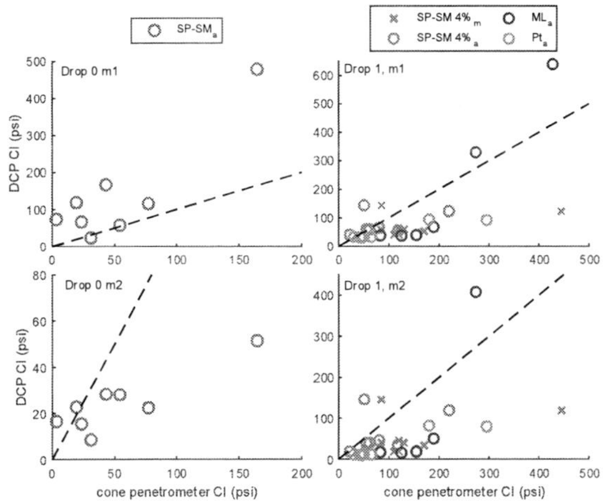

Fig. 5. Comparison between the DCP zero and first drop and cone penetrometer cone indexes for frozen soil, using methods 1 (m1) and 2 (m2). Subscripts 'm' and 'a' are for maximum (x) and average(o) respectively. The dashed line is the 1-to-1 curve.

5. Rut Analysis

5.1. Rut depth dependence on measured soil properties

To begin the analysis, the relationship between rut depth (cm) and a subset of the various measured physical soil parameters, x_n, was investigated. The only vehicle property used was the nominal ground pressure, v. The soil properties used, by subscript, were

1. dry density (g/cm^3) 0-2" (0-5.1 cm),
2. wet density (g/cm^3) 0-2" (0-5.1 cm),
3. cone index (psi) from 0-6" (0-15.24 cm),
4. Clegg 4th drop *CBR* (%),
5. surface shear (kPa),
6. volumetric water content (%),
7. DCP 0 drop depth (cm),
8. carbon content (% total by weight), and
9. frost/thaw depth (cm).

The dependent variable subscript numbers are used in *Table 2*. For frozen soil tests where the instruments could not be inserted into the soil, maximum or minimum values were used for all values. Affected instruments include the shear vane, 260 kPa, the cone penetrometer, 750 psi, and the VWC, 0 %. This affects one test in FY2024, three in FY2025, and the KRC February 2021 measurements.

Goodness of fit values for rut depths as a linear function ($a + b \cdot z$) of twelve combinations were investigated. The dependent variable combinations, z, used were:

- x_n,
- $x_n \cdot v$,
- $x_n \cdot \ln(v)$,
- x_n / v,
- $\ln(x_n)$,
- $v \cdot \ln(x_n)$,
- $\ln(v) \cdot \ln(x_n)$,
- $\ln(x_n)/v$,
- $1/x_n$,
- v/x_n,
- $\ln(v)/x_n$, and
- $1/(x_n \cdot v)$

The vehicle *NGP*, v, are listed in *Table 1*. In addition, combinations of $x_n + v$ in their three different forms (x_n, v; $\ln(x_n)$, $\ln(v)$; $1/x_n$, $1/v$) were also tried but abandoned due to poor fits. The fit results were sorted based on: all soil states, unfrozen, frozen, and thawing using the nine soil parameters for each of the twelve fits.

The best fits, based on highest r^2 value after removing outliers, for nine combinations of vehicle and soil types (all vehicles, all soil types; all vehicles, non-peat soils; all vehicles, peat; wheeled, all soil types; wheeled, non-peat soil, wheeled, peat; tracked, all soil types; tracked, non-peat soil; tracked, peat) as well as four soil states (all, unfrozen, frozen, thawing) were then extracted and the best of these 36 fits compared. From these 36 fits, a final curve was chosen as the optimum for each of the nine vehicle and soil type groups for each soil state defaulting to the simplest x_n and v combinations whenever possible. The fit equations are listed in *Table 2*. The soil parameter that results in the best fits most often is the DCP 0 drop depth (x_7) which accounts for 15 of the 36 possible followed by Clegg surface *CBR* (x_4, 6/36), cone index from 0-6" (x_3, 5/36), freeze or thaw depth (x_9, 4/36), dry and wet density (x_1 and x_2, 2/36 each), and finally shear strength and volumetric water content (x_5 and x_6, 1/36 each). Carbon content did not have any impact on rut depth at all, most likely due to the fact that so few levels were tested. Although somewhat biased due to the fact that the simplest forms were chosen when there were comparable r^2 values and similar number of fitting points, fits that were a function of $x_n \cdot v$ were most common (13/36), then v/x_n (8/36), $x_n \cdot \ln(v)$ and $\ln(v)/x_n$ (4/36), $v \cdot \ln(x_n)$ (3/36), x_n, (2/36), and $\ln(x_n)/v$ and $1/(x_n \cdot v)$ (1/36).

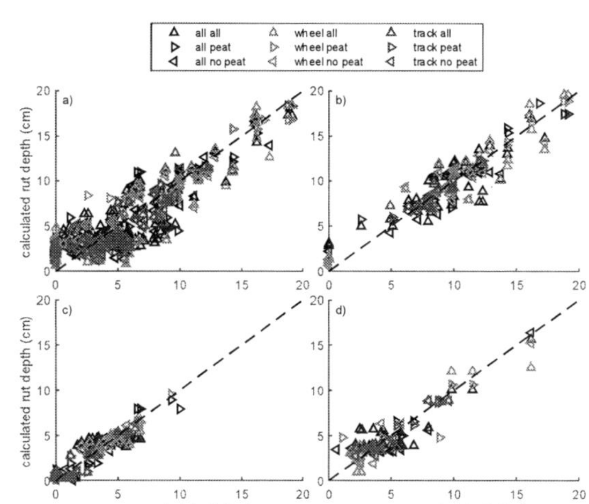

Fig. 6. Calculated rut depths using the equations in Table 2 versus measured ones. a) all soil states, b) unfrozen, c) frozen, and d) thawing.

Table 2
Rut depth fit equations resulting in the highest r^2. v is vehicle nominal ground pressure (kPa). x_n is measured soil property.

soil state	r^2	number of fit points	rut depth fit equation (cm)	vehicle type	soil type
all	0.705	123	$2.584x_7v + 0.004$	all	all
all	0.703	40	$3.620x_7v + 0.004$	all	peat
all	0.791	87	$1.139x_7\ln(v) + 0.111$	all	no peat
all	0.800	71	$0.650x_7v + 0.004$	wheel	all
all	0.850	16	$10.766\ln(x_4)v - 0.010$	wheel	peat
all	0.786	53	$0.942x_7v + 0.004$	wheel	no peat
all	0.673	57	$1.872x_7v + 0.016$	track	all
all	0.774	23	$2.159x_7v + 0.015$	track	peat
all	0.725	35	$9.666\ln(x_3)/v - 16.700$	track	no peat
unfrozen	0.744	40	$2.338x_7\ln(v) + 0.108$	all	all
unfrozen	0.887	11	$5.316\,v/x_5 + 0.095$	all	peat
unfrozen	0.784	27	-1118.721	all	no peat
unfrozen	0.865	26	$0.777x_7v + 0.005$	wheel	all
unfrozen	0.990	5	$7.284x_7v + 0.003$	wheel	peat
unfrozen	0.879	20	$-0.015x_7\ln(v) + 0.124$	wheel	no peat
unfrozen	0.784	14	$7.291\ln(x_4)v - 0.030$	track	all
unfrozen	0.941	4	$19.493\ln(v)/x_4 - 0.069$	track	peat
unfrozen	0.923	8	$-3.772\,v/x_2 + 1.184$	track	no peat
frozen	0.728	47	$4.838x_3 - 0.006$	all	all
frozen	0.870	14	$-1.314x_9\ln(v) + 0.020$	all	peat
frozen	0.858	28	$-0.210\,v/x_1 + 0.039$	all	no peat
frozen	0.742	21	$3.336\ln(x_7)v + 0.006$	wheel	all
frozen	0.997	5	$10.481x_4v - 0.001$	wheel	peat
frozen	0.787	15	$2.797x_7v + 0.002$	wheel	no peat
frozen	0.858	26	$0.072\ln(v)/x_3 + 41.468$	track	all
frozen	0.838	12	$0.201\ln(v)/x_3 + 38.892$	track	peat
frozen	0.883	14	$-0.093\,v/x_3 + 9.328$	track	no peat
thawing	0.750	26	$3.721x_7v + 0.003$	all	all
thawing	0.897	10	$4.366\ln(v)/x_4 + 1.413$	all	peat
thawing	0.805	22	$3.252\,v/x_4 + 0.014$	all	no peat
thawing	0.873	9	$-1.472\,v/x_9 + 8.382$	wheel	all
thawing	0.564	5	$-69.905x_1 + 80.093$	wheel	peat
thawing	0.937	9	$-3.639x_6v + 0.003$	wheel	no peat
thawing	0.887	13	$2.742\,v/x_9 + 20.537$	track	all
thawing	0.886	4	$5.677x_9v + 0.0004$	track	peat
thawing	0.919	13	$1.704\,v/x_2 + 0.273$	track	no peat

The calculated rut depths, using the equations listed in *Table 2*, versus the measured ones are plotted in *Fig.6*. Looking at the tightness of points around the 1-to-1 line in each subplot, the frozen has the best fits. Combining all soil states results in some of the worst fits, *Fig* 6a. Overall, the calculated rut depths tend to be smaller than the measured ones although the fits are, for the most part, within the range of measurement error.

5.2. Comparison with other rut models

According to Affleck (2005) and Richmond et al. (1995), Willoughby and Turnage (1990) developed semi-empirical equations to predict rut depth, sinkage, in fine-grained soil for wheeled and tracked vehicles. These algorithms are implemented in NRMM. All of the tests in this study classify as either fine-grained or muck, i.e. peat. The NRMM algorithms assume that there is no sinkage for muck soils. The sinkage per axle (z_i, in) in fine-grained soil for wheeled vehicles is quantified as

$$z_i = \frac{5d_i}{(xn_i s^{0.2})^{5/3}}$$
$$xn_i = \max\left(CI_l\, n_i b_i d_i / W_i / \left(1 - {\delta_i}/{h_i}\right)^{1.5}, 0.7\right) \tag{1}$$

while for tracked vehicles it is

$$z_i = 0.00443 l_i e^{5.887/xn_i}$$
$$xn_i = \max\left(CI_l \frac{2b_i l_i}{W_i}, 0.7\right) \tag{2}$$

where the subscript *i* is per axle, *d* is the tire diameter (in), *b* is tire/track width (in), *h* is tire section height (in), δ is tire deflection (in), *n* is the number or tires, *W* is the weight (lb), *l* is the track length (in), CI_l (psi) is the layer cone index, and *s* is the tire slip. Sullivan and Anderson (2000) suggest using 20 % slip when not measured. The xn_i minimum is to limit the sinkage to $5.3d_i$ and $20l_i$ for wheeled and tracked vehicles respectively for low *CI*. The layer cone index used depends on the vehicle cone index, *VCI1FG* which are listed in *Table 1*. If $CI_{0\text{-}6"} < VCI1FG$ and $CI_{6\text{-}12"} > VCI1FG$ then $CI_l = CI_{6\text{-}12"}$ else $CI_l = CI_{0\text{-}6"}$. Liu et al. (2008) modified the Willoughby and Turnage method to incorporate turning effects. Their approach requires vehicle and location specific fitting parameters. One factor that they found important regardless of turning radius is vehicle speed.

Comparison between the measured rut depth for the fine-grain, fg, and peat, pt, soils and those calculated using Eq. 1 and (2) are shown in *Fig. 7* for unfrozen soil. Calculated rut depths for which $CI_{6\text{-}12"} < 2$ psi are not shown as the sinkage was greater than 350 cm (138 in) for both soil types. Even though Eq. 1 and 2 are intended for fine-grained soils, better agreement between measured and calculated results are found for peat. The number of points however is very low for these fits.

Vennik et al. (2017) measured rut depths for wheeled military vehicles natural grasslands and arable fine-grained soil. One of the grasslands had peat. For the arable locations they found that the Willougby and Turnage model overpredicts rut depth if $CI_{0\text{-}6"}$ is used in the calculations and under predicts when $CI_{6\text{-}12"}$ is used. They also calculated rut depth using $CI_{3\text{-}9"}$ and found mixed results. At the one peat location the predicted ruts were greater than the measured ones for $CI_{0\text{-}6"}$. Affleck (2005) also found that the Willougby and Turnage model overpredicts rut depth. Similar results were found in this study for the wheeled vehicle in both the fine-grained and peat soils as can be seen in *Fig. 7*. This is encouraging as our tests were on prepared sites while Vennik et al.'s (2017) were in the field on both disturbed and natural locations.

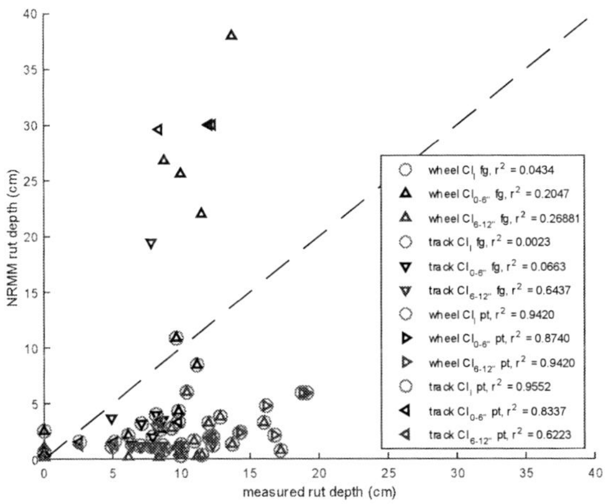

Fig. 7. Measured versus calculated rut depths for tracked and wheeled vehicles for fine-grained (fg) and peat (pt) soil as functions of $CI_{0\text{-}6''}$, $CI_{6\text{-}12''}$, or CI_l. r^2 values are for $rut_{calc} = m \cdot rut_{meas} + b$ fits. The dashed line is 1-to-1.

For frozen peat Shoop (1994) states that the minimum frost depth (m) needed to support a vehicle is $\text{minimum}\left(0.5, \sqrt{P/0.35}\right)$ for dry soil and $\text{minimum}\left(0.5, \sqrt{P/0.86}\right)$ for wet soil where P (MN) is the vehicle load. These calculated frost depths are interpreted as frozen soil thickness needed for which no break-through into the underlying unfrozen soil occurred. This was interpreted as no measurable sinkage for this study. In *Fig. 8* the measured frost depth, for which the rut depths are negligible, is compared to both the wet and dry calculated minimum frost depths for the peat soil tests. The Shoop (1994) depths average about 5 times larger for dry conditions and 3 times bigger for dry peat regardless of vehicle type. Of the 16 tests on frozen peat, vehicle break-through occurred for four of them, three of which were for the light wheeled vehicle where the frost depths ranged from 5-7 cm and the soil was considered wet. For the non-peat soils, the only vehicle break-through instances occurred with the wheeled light all-terrain vehicle when the frost depth was approximately 4.5 cm and the volumetric soil moisture before freezing was less than 10 %.

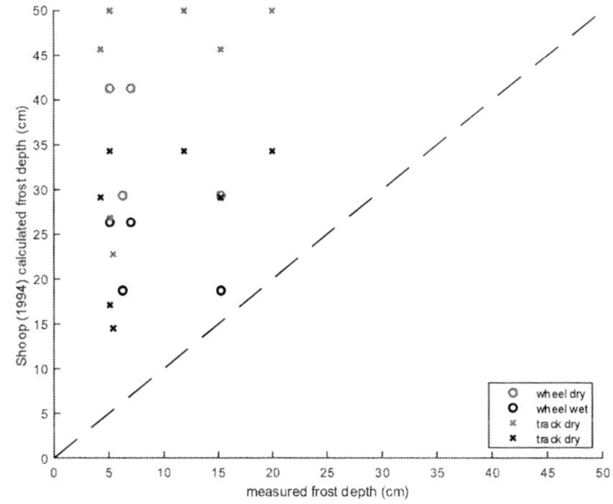

Fig. 8. Measured peat frost depth versus the wet and dry depth calculations based on Shoop (1994) for wheeled and tracked vehicles.

6. Conclusions and recommendations

Based on the results found in this study, Vennik et al. (2017) and Affleck (2005) the current fine-grained sinkage equations in NRMM do not adequately predict rut depth for either wheeled or tracked vehicles in northern soils. Overall, they underpredict measured values. The opposite is true for the frost depth needed to support vehicles on peat soils. The equations suggested by Shoop (1994) are overly conservative by a factor of 3-5 for wet and dry peat respectively.

Findings from this study also suggest that the DCP 0 drop depth followed by the Clegg *CIV* are better predictors of rut depth when coupled with the vehicle nominal ground pressure rather than the cone, or rating cone, index. Relationships found indicate that one equation may be used for all soil states, unfrozen, frozen, and thawing, although the number of data points is limited for certain conditions.

Finally, caution should be used when using DCP calculated cone index in lieu of cone penetrometer measurements.

7. Declaration of competing interest

The authors declare that they have no known competing financial interests or personal relationships that could have appeared to influence the work reported in this paper.

8. References

Affleck, R.T., 2005. Disturbance Measurements from Off-Road Vehicles on Seasonal Terrain. ERDC/CRREL TR-05-12.

Ahlvin, R.B., Haley, P.W., 1992. NATO Reference Mobility Model Edition II, NRMM II User's Guide. U.S. Army Waterways Experimental Station, Geotechnical Laboratory, Vicksburg, Mississippi, TR GL-92-19

Army, 1994. FM 5-430-00-1, Planning and Design of Roads, Airfields, and Heliports in the Theater of Operations. Washington, D.C.

ASTM D2974-20, 2020. Standard Test Methods for Determining the Water (Moisture Content, Ash Content, and Organic Material of Peat and Other Organic Soils. ASTM International. West Conshohocken, PA.

ASTM D5874-16, 2016. Standard Test Methods for Determination of the Impact Value (IV) of a Soil. ASTM International. West Conshohocken, PA.

ASTM D6951/D6951M-09, 2015. Standard Test Method for use of the Dynamic Cone Penetrometer in Shallow Pavement Applications. ASTM International. West Conshohocken, PA.

Geonor, Inc., 1995. Instructions for use Inspection Vane Tester, H-60. Milford, PA.

Herrick, J.E., Jones, T.J., 2002. A Dynamic Cone Penetrometer for Measuring Soil Penetration Resistance. Soil Sci. Soc. Am. J. 66, 1320-1324.

Hibbard, S., Sion, B., Peters, J., Sabol, D., Houseman-Lehman, S., Bieszczad, J., Eylander, J., 2025. Field Validation of GeoWATCH soil moisture downscaling model: Implications for soil strength predictions and cross-country mobility. GSA. In Press.

Keller, T., Defossez, P., Weisskopf, P., Arvidsson, J., Richard, G., 2007. SoilFlex: A model for prediction of soil stresses and soil compaction due to agricultural field traffic including a synthesis of analytical approaches. Soil and Tillage Res. 93(2), 391-411. https://doi.org/10.1016/j.still.2006.05.012.

Lafayette Instrument Company, 2005. Clegg Manual. Lafayette, IN.

Liu, K., Ayers, P., Howard, H., Jones, R., Anderson, A., 2008. Prediction of Rut Depth During Military Vehicle Turning Maneuvers Using a Modified Sinkage Numeric. Transactions American Soc. of Agricultural and Biological Engineers, 53(4), pp. 1019-1024.

Mason, G.L., Baylot, E.A., 2016. Predicting Soil Strength in Terms of Cone Index and California Bearing Ratio for Trafficability. ERDC/GSL TN-16-1.

Minasny, B., McBratney, A.B., 2005. Comments on "Simultaneous Measurement of Soil Penetration Resistance and Water Content with a Combined Penetrometer-TDR Moisture Probe" and "A Dynamic Cone Penetrometer for Measuring Soil Penetration Resistance". Soil Sci. Soc. Am. J. https://doi.org/10.2136/ssaj2005.0925.

Priddy, J., 1995. Stochastic Vehicle Mobility Forecasts Using the NATO Reference Mobility Model, Report 3 Database Development for Statistical Analysis of the NRMM II Cross-Country Traction Empirical Relationships. US Army Corps of Eng. Waterways Experiment Station, Vicksburg, Mississippi, Tech Rep. GL-95-9.

Richmond, P., Shoop, S., Blaisdell, G., 1995. Cold Regions Mobility Models. U.S. Army Cold Regions Research and Engineering Laboratory, Hanover, NH, report 95-1.

Shoop, S.A., Wieder, W.L., Elder, B.C., Beal, S.A., Deeb, E.J., 2019. Assessment of Field Methods for Measuring Mechanical Properties of Snow. ERDC/CRREL TR-19-17.

Shoop, S., 1994. Vehicle bearing capacity of frozen ground over a soft substrate. Canadian Geotech. J. 32, 552-556.

Shoop, S., Diamond, D., Seman, P.M., Wieder, W.I., Mason, G., 2008. Predicting California Bearing Ratio from Trafficability Cone Index Values. ERDC/CRREL TR-08-17.

Spectrum Technologies, Inc., 2022a. FieldScout TDR 350 Soil Moisture Meter Product Manual.

Spectrum Technologies, Inc., 2022b. FieldScout SC 900 Soil Compaction Meter Product Manual.

Stolf, R., Cassel, D.K., King, L.D., Reichardt, K., 1998. Measuring mechanical impedance in clayey gravelly soils. Rev. Bras. Cienc. Solo. 22, 189-196.

Sullivan, P., Anderson, A., 2000. A methodology for estimating Army Training and Testing Area Carrying Capacity (ATTACC) vehicle severity factors and local condition factors. U.S. Army Engineer Research and Development Center, Construction Engineering Research Laboratory, Champaign, Illinois, ERDC/CERL TR-00-02.

Troxler Electronic Laboratories, Inc., 2016. Manual of Operation and Instruction, EGauge™ Model 4590 Surface Density Gauge. Research Triangle Park, NC.

Troxler Electronic Laboratories, Inc., 2022. Model 6760 Moisture Probe Quick Reference Guide, Edition 1.6. Research Triangle Park, NC.

Vennik, K., Kukk, P., Krebstein, K., Reintam, E., Keller, T., 2019. Measurements and simulations of rut depth due to single and multiple passes of a military vehicle on different soil types. Soil and Tillage. 186, 120-127.

Vennik, K., Keller, T., Kukk, P., Krebstein, K., Reintam, E., 2017. Soil rut depth prediction based on soil strength measurements on typical Estonian soils. Biosystems Eng. 163, 78-86.

Willoughby, W.E., Turnage, G.W., 1990. Review of a procedure for predicting rut depth. Internal Working Document.

5493 | Effects of electronic control systems on autonomous vehicle performance in granular terrain

Bohumir Jelinek [a,*], Aidan Dickerson [a], Angela Card [a], Greg Henley [a], George L. Mason [a], Ethan Salmon [a], C. Michael Gibson [a], Tyler Hannis [a], Thomas Skorupa [b], Michael Cole [b], Jody D. Priddy [c], Miriam Figueroa-Santos [b], Sara Boyle [b], Jeremy Mange [b]

[a] *Center for Advanced Vehicular Systems, Mississippi State University, Mississippi State, Mississippi, United States*

[b] *U.S. Army Combat Capabilities Development Command Ground Vehicle Systems Center, Warren, Michigan, United States*

[c] *U.S. Army Engineer Research and Development Center, Geotechnical and Structures Laboratory, Vicksburg, Mississippi, United States*

* *Corresponding author: bj48@cavs.msstate.edu*

ABSTRACT

Understanding the effects of electronic control systems, such as Antilock Braking System (ABS) and Electronic Stability Control (ESC), and their interaction with the steering control algorithm is critical for improving mobility and maneuverability of autonomous wheeled vehicles in off-road environments.

In this work, we evaluate the effects of wheel-speed based ABS and yaw-rate-following ESC on the performance of an autonomously-driven Polaris vehicle performing a double-lane-change (DLC) maneuver in granular terrain. The performance is evaluated in terms of the maximum speed at which the vehicle can pass the DLC test. Two lateral speed controllers are compared: a simple PID and a Stanley autonomous steering controller, along with the longitudinal speed controller that accelerates the vehicle to identify the maximum DLC passing speed. The simulations were conducted using the Chrono simulation package with the Soil Contact Model (SCM). The SCM parameters based on dry sand soil bin measurements were chosen from published literature. The ABS effects were found to be negligible or detrimental, while the ESC augmented the lateral steering control algorithm allowing the vehicle to achieve higher DLC passing speeds.

Keywords
Off-road mobility
Soft terrain
Electronic stability control
Antilock braking system
Double lane change

1. Introduction

Electronic control systems allow vehicles to better handle slippery road-surface conditions and improve the safety of occupants during emergency driving maneuvers. These systems therefore became standard for on-road vehicles. Improving vehicle mobility and driver safety for vehicles in off-road environments is less explored, presenting challenges for agriculture, construction, military, and space exploration.

This work examines the effect of Antilock Braking System (ABS) and Electronic Stability Control (ESC) on the ISO-3888-1 test performance in sand-like terrain.

2. Double lane change test

The double lane change test (DLC) for road vehicles is described by the ISO 3888-1 standard. The test aims to evaluate the safety of the vehicle during an aggressive lane change maneuver. The vehicle driving path during the test is constrained as shown in Fig. 1. Details for the vehicle studied in this work are shown in Fig. 4. Widths of the individual sections of the path are proportional to the width of the vehicle being tested, with the consecutive sections increasing in width. The ISO 3888-1 test targets passenger cars and light commercial vehicles with limited length, therefore the horizontal lengths of the path sections are constants not depending on the vehicle length. Lateral boundaries of sections 1, 3, and 5 are marked by cones that create boundaries for acceptable vehicle trajectories. The vehicle passes the test only if it stays within the boundaries.

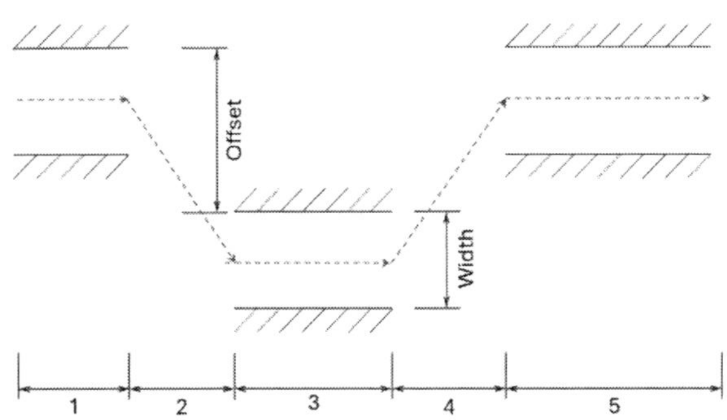

Fig. 1. ISO-38881 double lane change maneuver

3. Terrain: Soil Contact Model

The soil contact model (Krenn and Hirzinger, 2009), specifically its efficient implementation in Chrono (Serban et al., 2023), was used to model the interaction between the soil and the vehicle's wheels. The model is based on Bekker's terramechanics law

$$p = \left(\frac{K_c}{b} + K_\phi\right) z^n \tag{1}$$

(Bekker, 1956). This expression relates the normal pressure p along the contact patch for a wheel with the width b to the contact sinkage z, in terms of the coefficient K_c related to soil cohesion, the coefficient K_ϕ related to soil stiffness, and the soil sinkage exponent n. The shear stress in SCM is expressed following the formulation by Janosi and Hanamoto (1961)

https://doi.org/10.56884/1EBEV4Q4

$$\tau = \tau_{\max}\left(1 - e^{-j/k}\right) \quad (2)$$

where j is accumulated shear and k is the Janosi parameter. The $\tau_{\max}$ is given by Mohr–Coulomb failure criterion

$$\tau_{max} = c + p\ tan\ (\phi) \quad (3)$$

where c is the soil cohesion, where p is the normal pressure from Eq. 1, and ϕ is the soil internal friction angle. This study applied SCM parameters for the dry sand from Guo et al. (2019) listed in Table 1.

Table 1
Parameters of the SCM soil model (Guo et al., 2019)

Parameter	Value	Unit
K_c	1000	N/m^{n+1}
K_ϕ	1528600	N/m^{n+2}
n	1.08	-
k	0.024	m
c	200	Pa
ϕ	27	°

4. Chrono Polaris 3D vehicle dynamics model

The effect of control systems was studied on a 3D Polaris AWD vehicle model as implemented in the Chrono software package (Serban et al., 2019; Tasora et al., 2015). The model consists of a rigid chassis, a double wish-bone suspension in front, a three-link independent rear arm suspension in the back, and a powertrain with automatic transmission characterized by lookup tables. Most significant parameters of the vehicle model are listed in Table 2.

Table 2
Parameters of the Chrono Polaris vehicle model

Parameter	Value	Unit
Mass	1378	kg
Wheelbase	2.72	m
Max steering angle	27	°
Vehicle width	1.51	m

5. Control systems

Four control systems were subject of this study: antilock braking system (ABS), electronic stability control (ESC), along with two steering controllers: a proportional steering controller with look-ahead and a Stanley steering controller.

5.1. Antilock braking system

The ABS algorithm following the one presented in Pacejka and Besselink (2012) was applied in the present work. As described in Jelinek et al. (2024) and illustrated in Fig. 2, the ABS algorithm applied in this study examines the wheel speed, which is designated by $r_{eo}\Omega$ in the upper plot in Fig. 2, to modulate the braking effort by either 1) applying reduction gradually lowering the braking effort requested by the driver, 2) keeping the net braking effort at the constant level, or 3) increasing the braking effort gradually back to the level requested by the driver. When braking is required, the driver requests an increase in braking torque M_B at the driver's desired rate up to a desired maximum value. The ABS will allow the braking torque to increase until point B is reached. After reaching point B, the torque is reduced by the ABS at a given rate R_2 until point C is reached, as shown in the middle plot in Fig. 2. Then, the braking torque remains constant in the interval between points C and E. At point E, when the wheel velocity has caught up sufficiently with the vehicle velocity, the braking pressure is pumped up at a rate R_1 until it reaches the original driver's brake request (state Z).

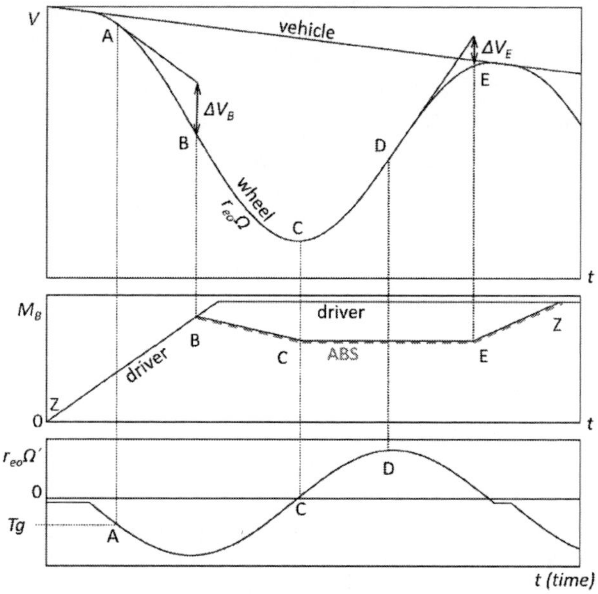

Fig. 2. ABS algorithm

5.2. Steering controllers

Two steering controllers implemented in Chrono software suite were examined in this study.

The first steering controller was a path-following controller with look-ahead and feedback proportional to the error in projected trajectory. This path-following steering controller first constructs a cubic Bezier curve located between the DLC path boundaries as depicted by the dashed line in Fig. 4. The controller sets the steering command so that the projected path of the vehicle at the look-ahead distance (sentinel point) follows a point on the Bezier curve at the look-ahead distance (target point). For this controller, only the proportional feedback coefficient K_P of the full proportional-integral-derivative (PID) version available in Chrono was tuned, while the integral and derivative feedback coefficients K_I and K_D were set to zero in this work to avoid the need for extensive tuning. The constant parameters of the path-following proportional steering controller and the intervals of values for the parameters varied in this study are shown in Table 3.

Table 3
Parameters of path-following proportional steering controller

Parameter	Value / <range>	Unit
Look-ahead distance	5.0	m
K_P	<0.1, 256>	-
K_I	0.0	-
K_D	0.0	-

The second steering controller examined in this study was the Stanley path-following steering controller (Hoffmann et al., 2007). This controller was developed by a team from Stanford University for a vehicle that won the 2005 DARPA Grand Challenge (Thrun et al., 2006). It corrects both directional and lateral (cross-track) path errors, allowing the vehicle to return to the desired path even in the presence of large disturbances. The desired steering angle δ is given by

$$\delta(t) = \psi(t) + \arctan \frac{K_S e(t)}{v(t)} \quad (4)$$

Here, the ψ is the directional path error, the e is the lateral (cross-track) path error, and the $v = |\boldsymbol{v}|$ is the vehicle speed as illustrated in Fig. 3. The parameter K_S quantifies the corrective strength of the feedback for the lateral error.

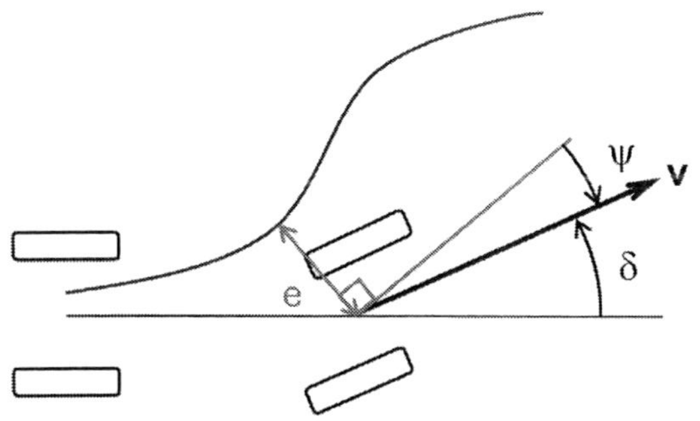

Fig. 3. Stanley steering control variables

The intervals of values for the parameters that were varied in this study are listed in Table 4. Additional features of the Stanley controller, as time delay, reset, dead zone, and tire dynamics, were not applied here.

Table 4
Parameters of Stanley steering controller

Parameter	Value / <range>	Unit
K_S	<0.1, 256>	-

5.3. Electronic stability control

The yaw-rate following ESC algorithm applies braking command correction proportional to the difference between the desired and actual yaw rate. The desired yaw rate ω_D is calculated from the steering angle δ, a forward component of the vehicle velocity v_x and from the vehicle wheelbase L as follows.

$$\omega_D = \frac{v_x \tan(\delta)}{L} \quad (5)$$

The actual yaw rate ω_A is the angular speed of a vehicle about its vertical axis. The corrective braking command c_F for the front wheels and c_R for the rear wheels, both limited within the interval <0, 1> corresponding to zero braking torque and maximum braking torque, are calculated from the steering delta $\Delta = \omega_A - \omega_D$ as follows.

$$c_F = K_F \Delta, \quad c_R = K_R \Delta \quad (6)$$

The corrective braking is applied only on the left (right) side when the desired yaw rate is larger (lower) than the actual yaw rate. Yaw rate is considered positive for the left turn. The different correction coefficients K_F and K_R can be applied for the wheels on the front (steered) axle and the rear (not steered) axle. The ESC will activate only after Δ exceeds the threshold of 0.01 radians per second, and it will deactivate when Δ is lower than this threshold for more than one second. The ESC parameters and intervals within which the parameters were varied are listed in Table 5.

Table 5
Parameters of Electronic Stability Control

Parameter	Value / <range>	Unit
Activation threshold	0.01	rad/s
Deactivation delay	1.0	s
Front correction coefficient K_F	<0.01, 32>	1/rad
Rear correction coefficient K_R	<0.01, 32>	1/rad

6. Results

This section summarizes simulation results, showing how the controllers were tuned and assessing the effects of electronic control systems on the maximum speed at which the vehicle can pass the DLC test.

6.1. ABS effects on DLC test

ABS significantly improves control of the vehicle when braking on a hard surface. The goal of the present ABS algorithm is to reduce braking effort to avoid the wheel locking (skid) and to regain the grip with the road surface. On granular terrain, represented by the SCM model in this study, the reduction of the braking command resulting from ABS action was found either negligible – when the braking was not sufficiently strong to cause excessive wheel deceleration, or detrimental – when the ABS reduction in braking interfered with the braking requested by ESC. To avoid this interference, the ABS was turned off during all ESC tests on granular terrain presented next.

6.2. ESC effects on DLC test without tuned steering

Tuning of ESC parameters for the vehicle to achieve the maximum DLC-passing speed was first performed with the default values kept for the steering parameters.

The path-following proportional steering controller had the default value of $K_P = 0.5$. The DLC tests with combinations of K_F and K_R values of 0, 2, 4, 8, 16, 32 were executed. The vehicle passed the DLC test at the maximum speed of 10 m/s, with K_F:K_R combinations of 4:2, 8:4, 8:2, 16:8, 16:4, 16:2 and 32:3, 32:4, 32:2. Locking of the front wheels (but not rear) when the steering delta was large was observed in each of these cases. Wheel trajectories along with target and sentinel point locations for the DLC test of the vehicle using the proportional steering controller with $K_P = 0.5$ are shown in Fig. 4. Without ESC (upper plot), the vehicle fails the DLC test due to the front right wheel leaving the course. The lower plot shows the vehicle passing the DLC test with $K_F = 4$ and $K_R = 2$. The corresponding actual and desired yaw rate, steering delta, and braking commands for the front left, front right, rear left, and rear right wheels are shown in Fig. 5.

For the Stanley steering controller with the default $K_S = 0.5$, the maximum speed for the vehicle to pass the DLC test was found to be 10 m/s. The only K_F:K_R combination that passed the DLC test at the maximum speed 10 m/s was 32:2, which resulted in the front wheel being locked when the most intense steering was applied.

6.3. ESC effects on DLC test with tuned steering

Tuning the proportional steering controller revealed that the K_P values of 16 or higher resulted in the maximum DLC passing speed of 15 m/s even without ESC. When ESC was applied at 15 m/s, the K_P value of 8 was sufficient with the K_F:K_R combination of 0.01:0.01. An increased maximum DLC passing speed of 16 m/s was then achieved with the help of ESC, but only with $K_P \geq 32$ and small K_F:K_R combinations, as 0.02:0.02 or 0.04:0.02. The vehicle failed the test with the K_F:K_R combinations of 0.01:0.01 and 0.04:0.04. Notably, the maximum steering angle is reached even when performing the DLC test at lower speeds. For example at 13 m/s, the minimum required $K_P = 4$ resulted in applying maximum steering angle - specifically when the vehicle turns to return to the original lane (x = 80 m in Fig. 4) and when it turns to straighten the direction of travel after returning to the original lane (x = 100 m in Fig. 4).

Tuning the Stanley steering controller led to the maximum DLC passing speed of 11 m/s - with the K_S values in the interval <110, 256>. The maximum passing speed of 11 m/s was achieved without ESC, and engaging ESC did not help enough to pass the DLC test at 12 m/s.

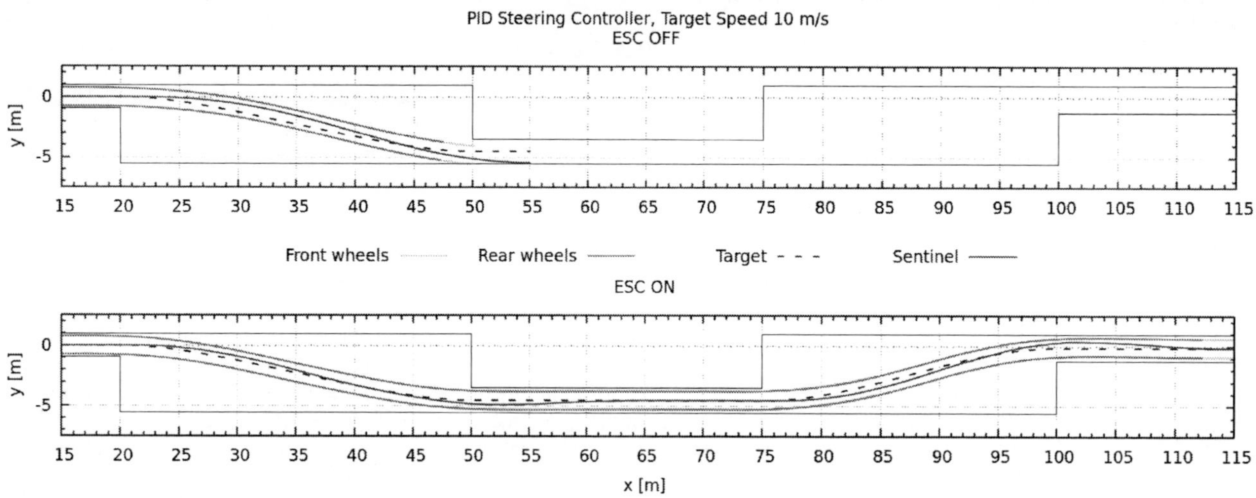

Fig. 4. DLC with ESC versus without ESC

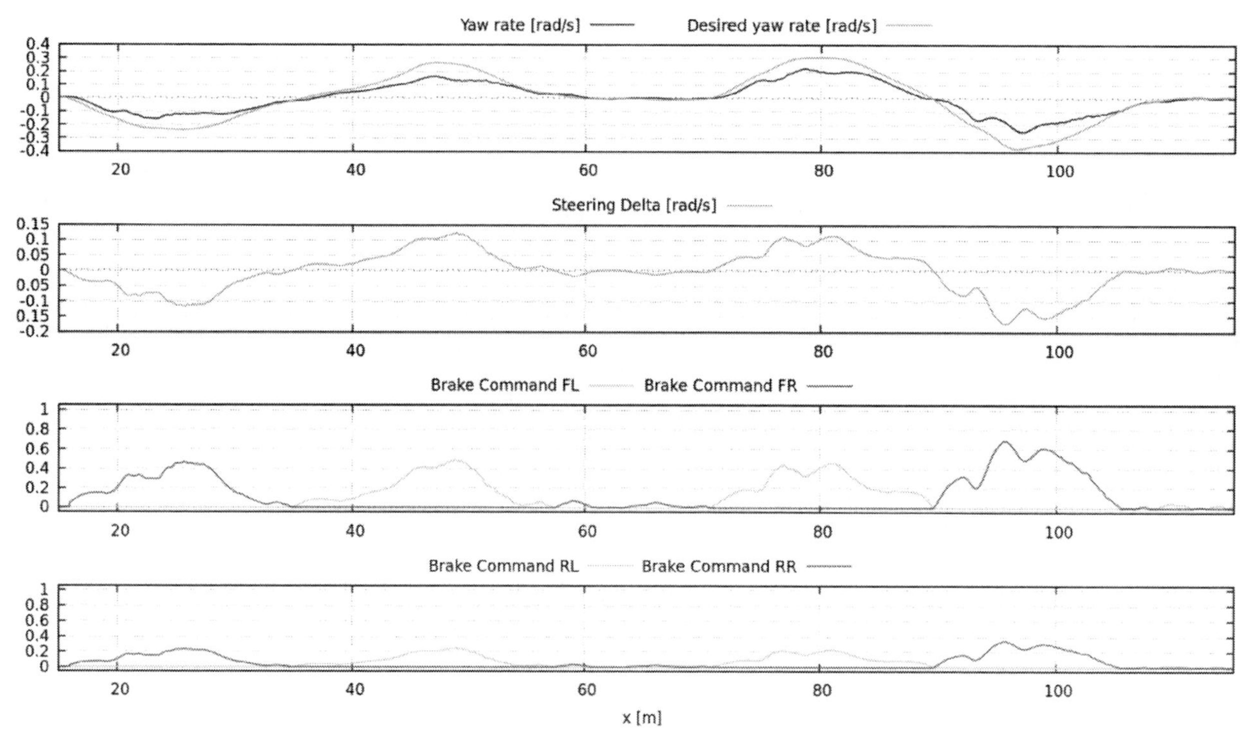

Fig. 5. Actual and desired yaw rates, steering delta, and corrective brake commands (front left, front right, rear left, rear right) initiated by ESC

6.4. Comparison of DLC tests without ESC on hard surface vs. SCM

The maximum passing speeds and tuned parameters for the proportional and Stanley steering controller when the vehicle drives on an SCM surface and on a hard surface are listed in Table 6.

The maximum passing speed is always lower on the SCM terrain than on the hard surface. To achieve maximum DLC passing speeds on either surface requires optimization of the steering controller's K_P or K_S values.

For the soft surface, it proved more difficult to find a single optimal value than for the hard surface. Numerous runs for the Stanley controller were examined to identify the range of parameters where the vehicle passed the test. Just outside the boundaries of this range, some of the tests have failed and some passed. Further from the boundaries, all tests failed.

With the Stanley controller, the optimal K_S for soft surface is much larger than for the hard surface.

Table 6
Max passing speed without and with ESC on SCM surface vs. hard surface

Steering controller	Hard surface		SCM surface		
	Tuned parameter	Max DLC passing speed	Tuned parameter	Max DLC passing speed	With ESC
Proportional	K_P =20	24 m/s	$11 \le K_P \le 64$	15 m/s	16 m/s
Stanley	K_S =0.2	19 m/s	$110 \le K_S \le 430$	11 m/s	11 m/s

7. Conclusions and future work

The present study revealed that only minimal ESC intervention was required when the proportional steering feedback coefficient K_P was tuned to achieve the highest DLC speed on the sand-like SCM surface. However, the ESC helped to increase the maximum DLC passing speed when the steering was not tuned, compensating for the low value of the steering feedback coefficient K_P. With the Stanley steering controller, ESC did not help to improve maximum DLC passing speed.

The proportional path-following steering controller allowed the vehicle to pass the DLC test at significantly higher speeds than the Stanley steering controller, demonstrating that robustness of Stanley controller comes with a cost of accuracy in the case of a tight-boundary DLC test on a homogeneous SCM surface. With increased speed, the two steering controllers required higher values of K_P and K_S, respectively.

For future work, the effects of the look-ahead distance parameter should be studied. The derivative and integral feedback should also be evaluated for improvement of the steering stability. Other steering controllers, for example simple-realistic and pure-pursuit controllers, need to be examined and a wider range of testing scenarios should be deployed. Given that the maximum steering angle was reached in many of the SCM tests, simulations with increased maximum steering angle could be performed to check for increased maximum DLC passing speed.

8. Acknowledgements

This material is based upon research conducted under contract W912HZ-22-C0004 with the U.S. Army Engineer Research and Development Center (ERDC). The views and conclusions contained herein are those of the authors and should not be interpreted as necessarily representing the official policies or endorsements, either expressed or implied, of ERDC or the U.S. Government.

9. Declaration of competing interest

The authors declare that they have no known competing financial interests or personal relationships that could have appeared to influence the work reported in this paper.

10. References

Bekker, M.G., 1956. Theory of Land Locomotion: The Mechanics of Vehicle Mobility, 9690401st ed. University of Michigan Press, Ann Arbor, MI. https://doi.org/10.3998/mpub.9690401

Guo, T., Guo, J., Huang, B., Peng, H., 2019. Power consumption of tracked and wheeled small mobile robots on deformable terrains–model and experimental validation. Mech. Mach. Theory 133, 347–364. https://doi.org/10.1016/j.mechmachtheory.2018.12.001

Hoffmann, G.M., Tomlin, C.J., Montemerlo, M., Thrun, S., 2007. Autonomous Automobile Trajectory Tracking for Off-Road Driving: Controller Design, Experimental Validation and Racing. IEEE, pp. 2296–2301. https://doi.org/10.1109/ACC.2007.4282788

Janosi, Z., Hanamoto, B., 1961. The analytical determination of drawbar pull as a function of slip for tracked vehicles in deformable soils, in: International Society for Terrain-Vehicle Systems, 1st Int. Conf.

Jelinek, B., Henley, G., Card, A., Hannis, T., Gibson, M., Priddy, J., Boyle, S., Figueroa-Santos, M., Mange, J., 2024. Vehicle-Level Control Systems Framework for Use in Create-GV Vehicle Dynamics Simulations. Presented at the 2024 NDIA Michigan Chapter Ground Vehicle Systems Engineering and Technology Symposium, Novi, Michigan, United States, pp. 2024-01-4107. https://doi.org/10.4271/2024-01-4107

Krenn, R., Hirzinger, G., 2009. SCM – A SOIL CONTACT MODEL FOR MULTI-BODY SYSTEM SIMULATIONS, in: Proceedings of the 11th European Regional Conference of the ISTVS.

Pacejka, H.B., Besselink, I., 2012. Tire and vehicle dynamics, 3rd edition. ed. Butterworth-Heinemann Elsevier, Oxford Waltham.

Serban, R., Taves, J., Zhou, Z., 2023. Real-Time Simulation of Ground Vehicles on Deformable Terrain. J. Comput. Nonlinear Dyn. 18. https://doi.org/10.1115/1.4056851

Serban, R., Taylor, M., Negrut, D., Tasora, A., 2019. Chrono::Vehicle: template-based ground vehicle modelling and simulation. Int. J. Veh. Perform. 5, 18. https://doi.org/10.1504/IJVP.2019.097096

Tasora, A., Serban, R., Mazhar, H., Pazouki, A., Melanz, D., Fleischmann, J., Taylor, M., Sugiyama, H., Negrut, D., 2015. Chrono: An open source multi-physics dynamics engine, in: Kozubek, T. (Ed.), International Conference on High Performance Computing in Science and Engineering, Lecture Notes in Computer Science. Springer, pp. 19–49.

Thrun, S., Montemerlo, M., Dahlkamp, H., Stavens, D., Aron, A., Diebel, J., Fong, P., Gale, J., Halpenny, M., Hoffmann, G., Lau, K., Oakley, C., Palatucci, M., Pratt, V., Stang, P., Strohband, S., Dupont, C., Jendrossek, L., Koelen, C., Markey, C., Rummel, C., van Niekerk, J., Jensen, E., Alessandrini, P., Bradski, G., Davies, B., Ettinger, S., Kaehler, A., Nefian, A., Mahoney, P., 2006. Stanley: The robot that won the DARPA Grand Challenge. J. Field Robot. 23, 661–692. https://doi.org/10.1002/rob.20147

5658 | Winter route planning algorithm

Mark Bodie *, Michael Parker

US Army Engineer Research and Development Center
Cold Regions Research and Development Center, Hanover, NH

* *Corresponding author: mark.o.bodie@erdc.dren.mil*

ABSTRACT

Cold regions represent an especially challenging environment for maneuver. In addition to standard obstacles which occur in less extreme environments, the surface itself can become an obstacle. Winter surfaces are characterized by low strength, resulting in deep vehicle sinkage with high motion resistances. In this work, we developed a winter route planning (WRP) algorithm that identifies low risk routes that minimizes the energy required by the vehicle to complete a mission. The algorithm fuses weather, terrain, and vehicle information to generate a unique on and off-road graph of the area of interest. A winter surface algorithm is used to predict the current snowpack's depth and density using weather and terrain information. This information along with vehicle data, is used to generate the vehicle's motion resistance and work associated with surface compaction, tree push over force, and gravity forces. Search requests are initiated, and optimal routes are displayed through Keyhole Markup Language (kml) files, enabling the use of existing GIS software to interface with the WRP algorithm. Experiments were conducted at the Ethan Allen Firing Range (EAFR) to validate the algorithm. Driven and predicted routes and mission completion times were compared with favorable results. Additional algorithm enhancements are identified to improve route accuracy and in-field operation.

Keywords
Route planning
Graph
Winter mobility

1. Introduction

Maneuvering in cold regions poses significant challenges due to the inherent limitations of winter terrain. Low surface strength leads to deep vehicle sinkage and elevated motion resistance, while reduced friction and shear strength further diminish tractive force, severely limiting mobility compared to temperate environments. Moreover, volatile weather conditions, such as heavy snowfall, can swiftly render previously passable routes impassable, underscoring the importance of timely surface and vehicle data for identifying low risk routes.

This work presents a Winter Route Planning (WRP) algorithm designed to identify low-risk routes which avoids areas prone to vehicle immobility. It integrates geo-spatial and weather data, which are processed by meteorological models to estimate snow depth and density across cold-region surfaces. Vehicle-specific attributes—such as ground pressure, gross traction , and ground clearance—are then used to evaluate mobility over predicted terrain conditions. By combining detailed environmental data with vehicle capabilities, the algorithm effectively determines optimal low risk routes.

Drivers interface with the WRP algorithm through the Tactical Assault Kit (TAK) software. TAK was developed by the U. S. Air Force Research Laboratory and is currently maintained by the Joint Product Center; it is a geospatial infrastructure and situational awareness software. TAK is used by 15 DoD Programs with over 250k users from the US military, federal and state government agencies, and US partner nations (Usbecka et al., 2015). By coupling the WRP algorithm with TAK we enable evaluation and coordination of optimal cold region routes in the field. Rapid route modification in the field allows real-time information to be integrated into route planning, enabling soldiers to assess alternative routes and targets. The ability to quickly modify routes in the field is also applicable to civilian applications. Time critical containment and search and rescue missions can benefit from the fast evaluation of changing conditions in the field. Updated cold regions routes can be quickly generated and shared with other responders, enabling efficient mission completion.

The remainder of the document will provide greater detail of the WRP algorithm and TAK interface. Experimental results are also discussed with comparisons between predicted and driven routes. Additional algorithm enhancements are discussed in the conclusion and recommendation section.

2. Winter Route Planning Algorithm

The Python-based WRP algorithm provides drivers with optimal routes which minimize vehicle energy or mission time given a start and target location. Route generation relies on a transportation graph for the area of interest (AOI), where vertices define geographic points and edges define the connections between the vertices. The edges have attributes such as length, mission time, and energy cost. The algorithm evaluates the graph to identify the route with the lowest cumulative attribute value connecting the start and target points.

The WRP algorithm integrates three core components—winter surface generation, graph generation, and graph search—to fuse weather, terrain, and vehicle data for optimal route planning (Fig. 1). The winter surface generation algorithm utilizes weather and terrain data to estimate snowpack depth and density across the area of interest (AOI). Using the resulting snowpack characteristics, along with terrain and vehicle-specific attributes, the graph generation algorithm constructs unique transportation

https://doi.org/10.56884/HKXFJ43Z

graphs tailored to each vehicle and daily snowpack conditions. Graphs are developed for both on-road and off-road environments. The graph search algorithm then determines the route that minimizes either vehicle energy consumption or total mission time.

Drivers interact with the WRP algorithm via the TAK software, where they define the start location, target destination, and any no-go zones. This input is encoded in a Keyhole Markup Language (KML) file, which the WRP algorithm parses to initiate graph-based routing. Upon computing the optimal route, the algorithm outputs a KML file displaying the selected route. When reloaded into TAK, the software presents both the optimal route and the vehicle's current location, offering real-time feedback to guide the driver in maintaining the planned route.

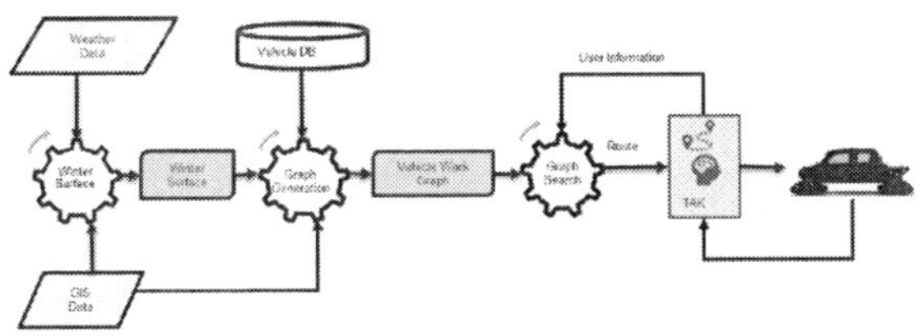

Fig. 1. Winter Route Planning Algorithm

2.1. WRP TAK Interface

Drivers interact with the WRP algorithm via the TAK software interface. Start and target points are designated using the point TAK dropper tool, where the start is marked as SP and the target as VDO (vehicle drop-off). No-go zones are defined through the TAK drawing tool and excluded from the graph search to ensure the computed route avoids restricted areas. Once these inputs are set, the TAK Overlays tool is used to export the route parameters—start, target, and no-go boundaries—as a Keyhole Markup Language (KML) file, which is then processed by the WRP algorithm. Fig. 2 illustrates a typical TAK input scenario, displaying the start and target locations along with a defined no-go region.

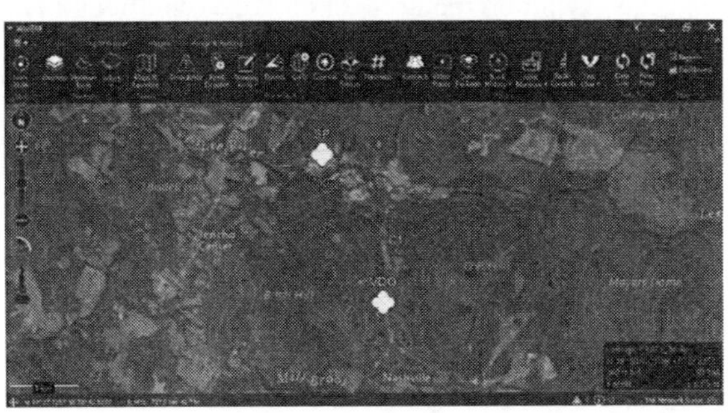

Fig. 2. Driver TAK Inputs

The WRP algorithm enables drivers to evaluate multiple optimal routes. The initial route corresponds to the path with the minimum cumulative attribute—such as time or energy—between the defined start and target locations. For subsequent evaluations, drivers can exclude the on-road segments, off-road segments, or both from the initial optimal route. The graph is then solved, minimizing the cumulative attribute, based on the modified graph. Each computed route is exported as a Keyhole Markup Language (KML) file, defining the route's geographic position. These files are then imported into the TAK software's Routes Tool, which visually displays the optimal path alongside the vehicle's real-time position, enabling the driver to course-correct to the optimal route.

Figure 3 illustrates two computed optimal routes generated via the TAK Routes Tool. The initial route, depicted in red, minimizes vehicle work required to traverse from the designated start point (SP) to the target point (VDO), and includes both on-road and off-road segments. Once this route is removed from the routing graph, a second-best route—shown in yellow—is computed. This alternate path is longer and primarily follows on-road segments. Importantly, both routes are calculated to avoid the constrained no-go area labeled C1.

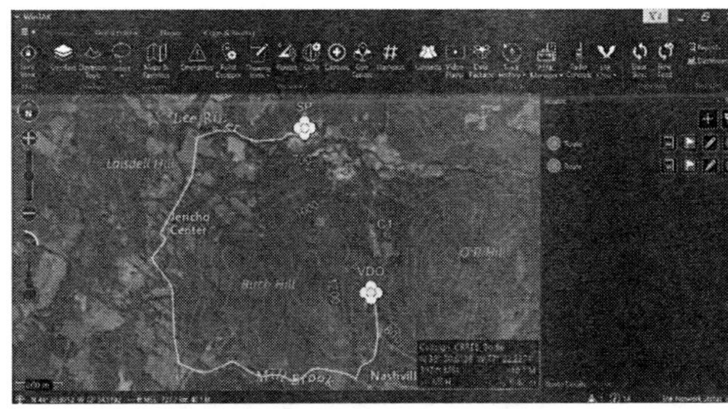

Fig. 3. Optimal Routes Displayed in TAK

2.2. Winter Surface Generation Algorithm

The Winter Surface Generation algorithm utilizes modified versions of established meteorological models—specifically Noah-MP (Niu et al., 2011) and SnowModel (Liston et al., 2006)—to generate predictive data on winter surface conditions. These models ingest digital terrain models (DTM), Land Cover, and weather data to estimate snowpack depth and density across the area of interest (AOI). The models can use commercial or government weather data, although highest predictive accuracy is achieved using data from a dedicated weather station located within the AOI. Enhancements to the base meteorological models allow for snowpack prediction at mobility-relevant spatial resolutions, with resolutions of 10 meters. The localized meteorological framework accounts for terrain-induced snow dynamics, including wind-driven snow redistribution, solar radiation effects on slope orientation and shading, as well as subsurface thermal gradients between soil and snowpack layers. These variables are processed to deliver high-fidelity estimations of snowpack properties necessary for mobility analysis.

The Noah-MP, Snowmodel, and weather data process workflow has been consolidated into a Docker container, enabling automation and cross platform compatibility. The integrated workflow monitors atmospheric conditions at 3-hour intervals, collecting key meteorological variables such as air temperature, humidity, wind speed, wind direction, precipitation rates, and longwave radiation. These inputs facilitate daily updates to the snowpack prediction, ensuring that model outputs remain aligned with weather fluctuations. This dynamic monitoring enables the Winter Surface Generation algorithm, to track evolving snow conditions.

In addition to depth and density predictions, the models provide snow stratigraphy predictions such as gain size, bond size, and coordination number. While these stratigraphy parameters are not currently integrated into the WRP algorithm, future implementation is planned to improve snow strength estimation based on formulations proposed by Golubev et al. (2000).

Incorporating stratigraphic strength predictions will enable more accurate assessments of vehicle motion resistance, thereby refining route optimization within cold regions. A full description of the modified metrological models can be found in found in Letcher et al. (2023).

For the Ethan Allen Firing Range (EAFR) case study a dedicated weather station was erected within EAFR (44.472180° latitude, -72.938930° longitude). The collected weather data was used to predict the snowpack's depth and density. Fig. 4 through Fig. 7 show the predicted snowpack depth and density from March 10-12, 2024. The black dot illustrates the location of the weather station. The figures show the change in the snowpack in a short time duration and illustrate the need to generate timely winter surfaces based on weather conditions.

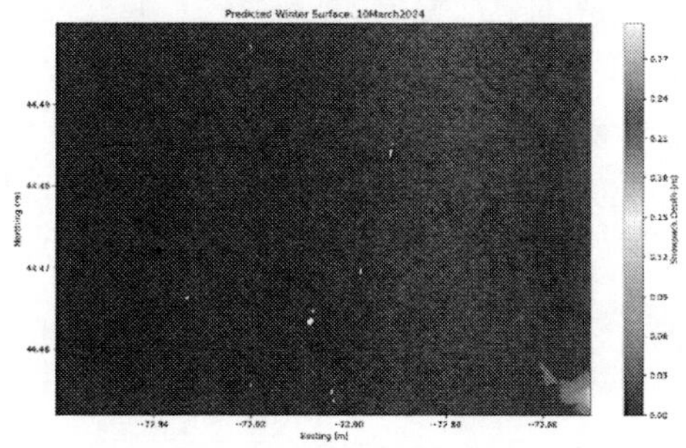

Fig. 4. Predicted Snow Depth: CEATS 10 March 2024

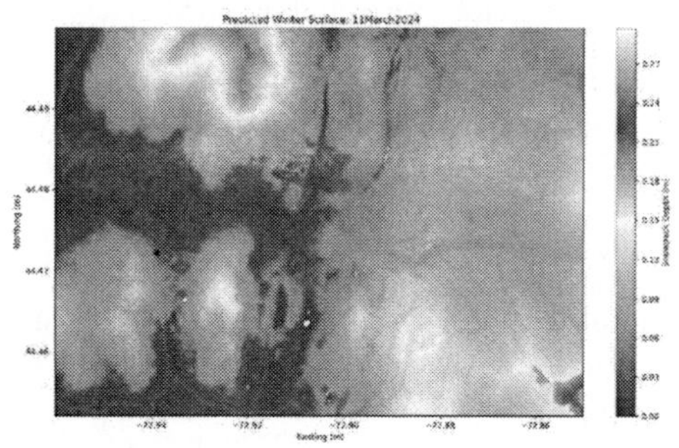

Fig. 5. Predicted Snow Depth: CEATS 11 March 2024

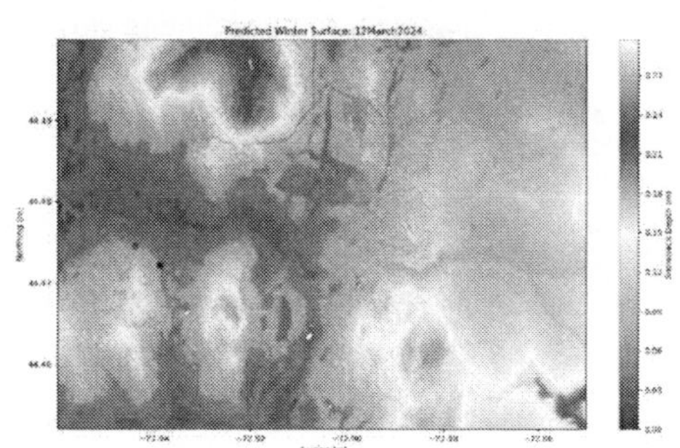

Fig. 6. Predicted Snow Depth: CEATS 12 March 2024

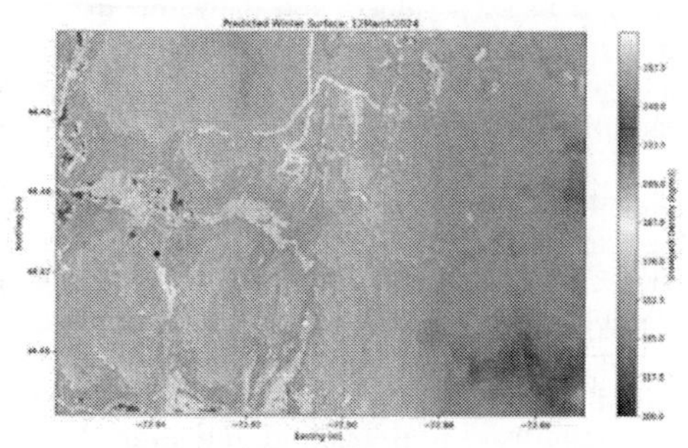

Fig. 7. Predicted Snow Density: CEATS 12 March 2024

To validate the predicted winter surface, we dug a snow pit on March 12th, 2024, at latitude 44.475° and 72.9441° which is illustrated by the red dot on Fig. 6 and Fig. 7. The air temperature was 7.5 °C and the snow temperature was 2 °C. The measured snow depth was 17.5 cm with a density variation of 0.171 g/cc in the top 10cm and 0.277 g/cc in the bottom 10 cm. The meteorological models predicted a snow depth of approximately 9.0 cm and a density of 0.17 g/cc at the snow pit location, resulting in a difference of 8.5 cm in depth and 0.054 g/cc in bulk density. Additional validation of the meteorological models can be found in Letcher et al. (2023).

2.3. Graph Generation Algorithm

The graph generation algorithm generates a unique graph for each vehicle and winter surface. The algorithm inputs the winter surface, DTM, DSM, Open Street Maps (OSM) road centerlines, and vehicle data to develop a graph that defines the vehicle energy and time required to transverse between connected vertices (Open Street Maps, 2015). The algorithm uses the Python import iGraph to build and solve the graph (Csardi, 2006).

The algorithm merges on and off-road graphs into a combined graph. The combined graph allows for large area searches while retaining high resolution required for off-road conditions. In Fig. 8 the on-road graph of northern Vermont is approximately 1500 km^2 and the off-road graph is approximately 50 km^2. The combined graph allows for graph searches that result in non-intuitive routes, which can result in a longer on-road route that requires lower energy or reduced mission time when compared to a more direct off-road route (Fig. 3).

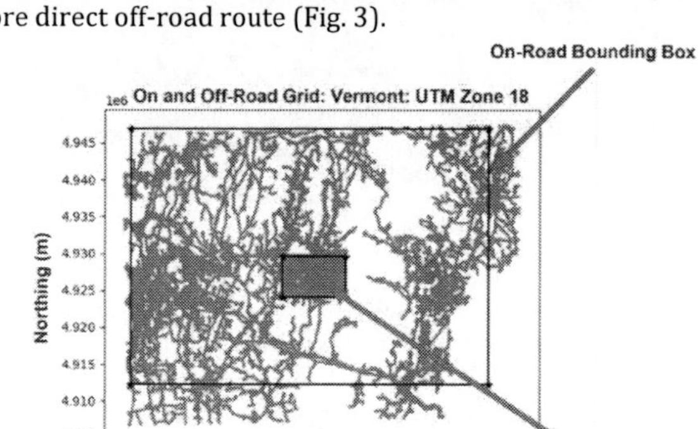

Fig. 8. Combined Graph, Northern Vermont

The on-road graph is developed from the OSM road-centerlines. The road centerlines identify the connection between road segment and define road attributes such as geographic location, road name, maximum speed, road type, etc. In addition to the road centerlines, OSM provides building and water features. If desired, the features are implemented as no-go areas and removed from the graph. The NASA STRM data product is used to generate the road centerline elevations (Farr et al., 2000). Both OSM and the NASA STRM provide a wide coverage of the earth. On-road graphs are generated for plowed and unplowed road conditions. The algorithm allows roads not listed in OSM Maps to be added using a kml file. The kml file provides names and geographic locations for the unmarked roads.

The algorithm uses a gridded graph to develop the off-road graph. The off-road graph is generated with resolutions between 3 to 10m. The gridded graph provides edge connections to the eight surrounding gridded vertices (Fig. 9).

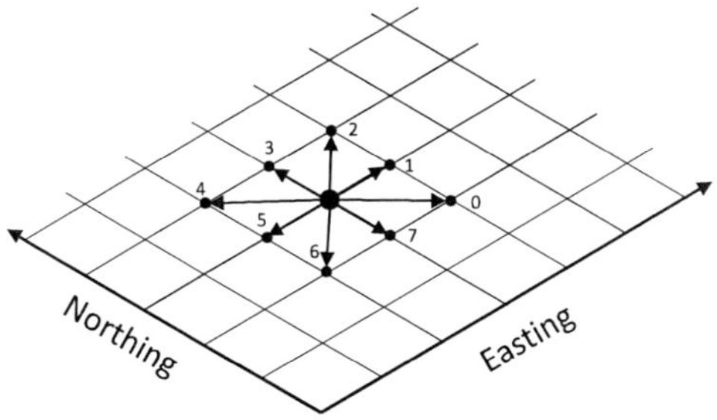

Fig. 9. Off-Road Gridded Graph

When the road centerlines are within the off-road bounding box, the road centerline elevations are generated using the finer resolution off-road DTM. The off-road tree canopy is determined by subtracting the DSM from the DTM and further refined with an edge detection algorithm. The edge algorithm identifies unmarked trails using a spatial high-pass filter on the DSM to identify sharp transitions in the surface. If the transition is beyond a calibratable value, the canopy tree height for the pixel is eliminated. The tree canopy height is used to calculate the tree push over force and the energy required for the vehicle to transverse across the graph edge. Figure 10 shows the ability of the edge algorithm to identify unmarked trails in an area of northern New York state.

Fig. 10. Tree Canopy Edge Detection

A combined graph is generated from the on and off-road graphs. Within the off-road bounding box, transition edges, at the off-road graph resolution, are generated between the on-road edges and the nearest off-road vertices. The combined graph with the transition edges allows the on-road network to access locations within the off-road grid.

The WRP algorithm populates the graph with required energy and time edge attributes. The required energy attribute is calculated by multiplying the vehicle's motion resistance by the edge length, and the required time attribute is calculated by multiplying the vehicle velocity by the edge length. The remainder of the section describes the method used to calculate vehicle motion resistance and velocity.

2.3.1. External Motion Resistance and Event Constraints Determination

The vehicle's external motion resistance associated with each of the graph edges is calculated based on snow compression, snowplowing force, tree push over force and elevation changes. The motion resistance associated with the elevation changes, is determined by projecting the heading vector, defined by the edge's start and end vertex positions, on the vehicle's gravity vector. The vertex positions are defined by their easting and northing position and deformed surface elevation. The deformed surface elevation is determined by subtracting the vehicle sinkage, described in the next section, from the vertex's winter surface elevation.

$$\vec{V_x} = (v_{sx} - v_{ex})\vec{\iota} + (v_{sy} - v_{ey})\vec{\jmath} + (v_{sz} - v_{ez})\vec{k} \tag{1}$$

where $\vec{V_x}$ is the heading vector, v_{sx} is the start vertex easting position, v_{ex} is the end vertex easting position, v_{sy} is the start vertex northing position, v_{ey} is the end vertex northing position, v_{sz} is the start vertex deformed surface elevation, and v_{ez} is the end vertex deformed surface elevation.

Next, we generate the lateral vector by rotating the normalized heading vector, $\vec{V_x}$, counterclockwise 90° and determine the deformed surface elevation at the lateral position,

$$v_{lx} = -t\frac{(v_{sy} - v_{ey})}{|\vec{V_x}|} + v_{sx} \tag{2}$$

$$v_{ly} = t\frac{(v_{sx} - v_{ex})}{|\vec{V_x}|} + v_{sy} \tag{3}$$

$$\vec{V_y} = (v_{lx} - v_{sx})\vec{\iota} + (v_{ly} - v_{sy})\vec{\jmath} + (v_{lz} - v_{sz})\vec{k} \tag{4}$$

where $\vec{V_y}$ is the lateral vector, v_{lx} is the lateral easting position, v_{ly} is the lateral northing position, and v_{lz} is the lateral deformed surface elevation. Next, the normal vector, $\vec{V_z}$, is generated by taking the cross product of the heading, $\vec{V_x}$, and lateral, $\vec{V_y}$, vectors.

$$\vec{V_z} = \vec{V_x} \times \vec{V_y} \tag{5}$$

The heading, lateral, and normal vectors makeup the vehicles local coordinate system, with the heading vector defining the x-axis, the lateral vector the y-axis, and normal vector the z-axis. The gravity vector is then projected on the vehicle's local coordinate system, this is done by taking the dot product between the heading, lateral, and normal vectors with the gravity vector.

$$\vec{F_g} = 0\vec{\iota} + 0\vec{\iota} + mg\vec{k} \tag{6}$$

$$\vec{F}_{xg} = \vec{V_x} \cdot \vec{F_g} \tag{7}$$

$$\vec{F}_{yg} = \vec{V_y} \cdot \vec{F_g} \tag{8}$$

$$\vec{F}_{zg} = \vec{V_z} \cdot \vec{F_g} \tag{9}$$

To determine the vehicle work associated with deforming the cold regions surface, we used procedures outlined in Shoop et al. (2020) and in Welling et al. (2022). To calculate the vehicle's motion resistance, the vehicle sinkage is first determined using the equation below,

$$z = z_w \left[1 - e^{\left(-P_{gp}/C_r\rho_o\right)}\right] \quad (10)$$

where z is the vehicle sinkage (m), P_{gp} is the vehicle's ground pressure (kPa), ρ_o is the initial snow density (g/cm^3), C_r is a regression constant (48.36 kPa/g/cm^3), and z_w is the maximum asymptotic sinkage given by the following equation,

$$z_w = h\left[1 - \frac{\rho_o}{\rho_f[1 + bs/b]}\right] \quad (11)$$

where h is the snow depth (m), b is the track or tire width (m), ρ_f is a regression constant (0.416604 g/cm^3), and bs is a regression constant (0.614817 m). The calculated sinkage, z, is subtracted from the winter surface elevation to determine the deformed surface elevation. The deformed surface elevation is used in the calculation of the slope motion resistance described above.

If the sinkage is not greater than the vehicle's ground clearance, the following equation from Shoop et al. (2020) is used to calculate the vehicle's deformable surface motion resistance,

$$MR = 2bz_wC_r\rho_oD_rN_a\left[\left(1-\frac{z}{z_w}\right)\ln\left(1-\frac{z}{z_w}\right) - \left(1-\frac{z}{z_w}\right)+1\right] \quad (12)$$

where MR is the vehicle's tire/track deformable surface motion resistance (N), N_a is the number of axles, and D_r is a regression constant (1.633). If the vehicle sinkage exceeds the vehicle's ground clearance an additional motion resistance term from Welling et al. (2022), associated with snow plowing, is added to the motion resistance term above,

$$MR_{ds} = MR + A_p\frac{1.11e4\ (N\frac{g}{cm^5})}{0.5\frac{g}{cm^3} - \rho_o} \quad (13)$$

where MR_{ds} is the total deformable surface motion resistance (N) and A_p is the plowing area (m^2). Ap is calculated as,

$$A_p = w(z - GC) \quad (14)$$

where w is the vehicle's width (m) and GC is the vehicle's ground clearance (m).

In addition to the surface motion resistance, an additional tree push over motion resistance term is also added. The tree push over force calculation uses a simple linear relationship between the tree's height and push over force (Fig. 11). The minimum tree height and the slope of the push over force, are user provided parameters.

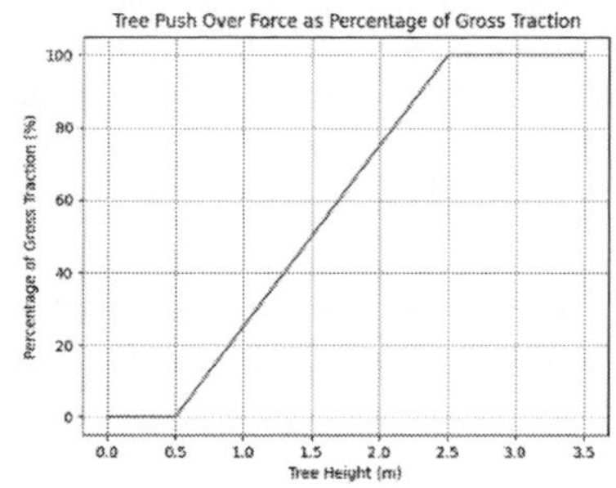

Fig. 11. Tree Push Over Force

The four motion resistance terms snow compression, snowplowing force, tree push over force and elevation changes are added to generate the external motion resistance term. The external motion resistance is compared to the vehicle's gross traction, if the external motion resistance is greater than the gross traction, this represents a no-go condition, a condition where the vehicle will become immobile. If this condition occurs, the edge connection will be removed.

Checks are conducted to assess if vehicle sliding, topple or rollover events can occur. To assess vehicle sliding, the longitudinal and lateral gravity forces are compared to a user defined percentage of the available traction. If the gravity force exceeds the vehicle's sliding constraint, the edge connection is removed from the combined graph.

Vehicle topple occurs when the vehicle rotates 90 degrees or more about the lateral vehicle axis, y-axis. We use a simple static analysis to determine if a topple event is possible. In the analysis if either the front or rear axle normal force is zero a topple event is possible. This is determined by taking the moments around the front or rear axle and setting the normal force to zero on the opposite axle (Fig. 12). If the longitudinal force is outside the bounds of Eq. 9 a topple condition could occur.

$$F_{zg}\,{}^{b}/_{h_{cg}} < F_{xg} < -F_{zg}\,{}^{a}/_{h_{cg}} \quad (15)$$

where F_{xg} is the gravity force in the longitudinal direction, F_{zg} is the gravity force in the vertical direction, h_{cg} is the center of gravity (cg) height, b is the distance from the cg to the rear axle, and a is the distance from the cg to the front axle.

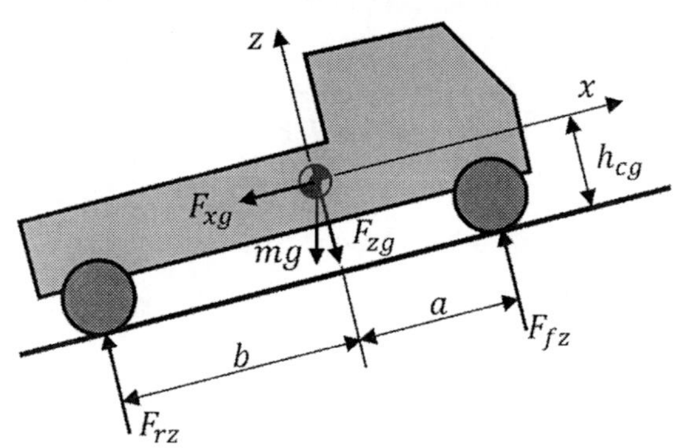

Fig. 12. Vehicle Topple Force Balance

As with the sliding constraint, a user defined percentage of the topple force is used to determine if the constraint is exceeded. If the longitudinal gravity force exceeds the topple constraint, the edge connection is removed from the combined graph.

A similar analysis is conducted to identify if a vehicle rollover event is possible. Vehicle rollover occurs when the vehicle rotates 90 degrees or more about the longitudinal vehicle axis, x-axis (Gillespie, 1994). If the lateral force is outside the bounds of Eq. 10 a rollover condition could occur, and the edge connection is removed from the combined graph.

$$F_{zg} {}^{t}/_{2h_{cg}} < F_{yg} < -F_{zg} {}^{t}/_{2h_{cg}} \qquad (16)$$

where t is the vehicle track width.

2.3.2. Vehicle Velocity and Internal Motion Resistance Determination

The vehicle's edge velocity is determined from the minimum of the edge speed limit and the vehicle capable speed. The vehicle's thrust versus speed curve is used to identify the vehicle capable speed. The vehicle's thrust versus speed curve is determined from the engine torque curve, gear ratios with losses and all internal motion resistance terms,

$$T_{fd} = T_e(GR_t\eta_t)(GR_{fd}\eta_{fd}) \qquad (17)$$

$$Thrust = {}^{T_{fd}}/_{R_{rl}} - mr_i - F_d \qquad (18)$$

where T_{fd} is the final drive torque, T_e is the engine torque at the given engine speed, GR_t is the transmission gear ratio, η_t is the transmission gear efficiency, GR_{fd} is the final drive gear ratio, η_{fd} is the final gear efficiency, mr_i is the internal motion resistance, and R_{rl} is the ratio of rotational to linear motion.

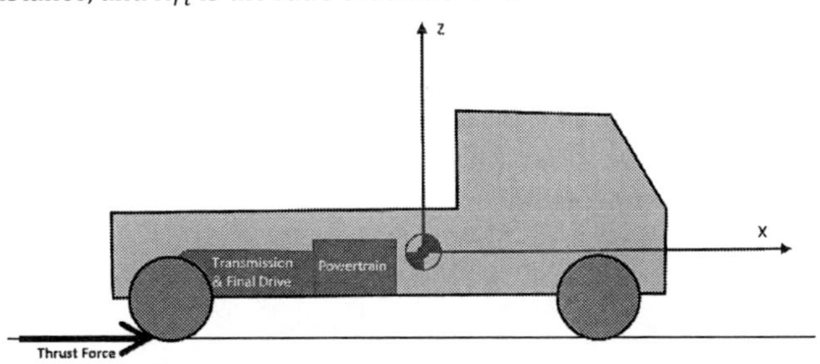

Fig. 13. Vehicle Thrust Force

The vehicle speed is given by,

$$V_{spd} = \frac{\omega_e R_{rl}}{GR_t GR_{fd}} \qquad (19)$$

where V_{spd} is the vehicle speed and ω_e is the engine rotational speed. Fig. 14 shows an example thrust velocity curve for a conceptual vehicle. The vehicle capable speed is the speed where the external motion resistance equals the vehicle thrust.

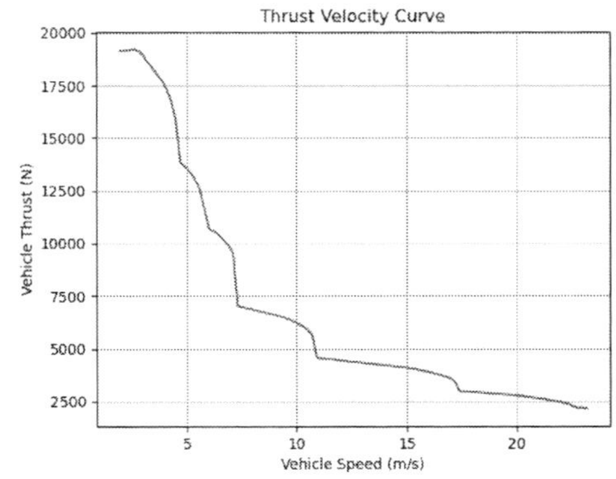

Fig. 14. Thrust Velocity Curve

Speed constraints are assigned to each edge segment based on its classification as either on-road or off-road. For on-road segments, maximum allowable vehicle speed is derived from OSM attributes associated with each road segment. Off-road segments utilize a user-defined maximum speed parameter, which can be calibrated to reflect terrain-specific mobility limitations. The operational vehicle speed for each segment is computed as the lesser of the vehicle's maximum capable speed and the applicable segment speed limit. This speed value is used to estimate total mission completion time.

The required time to complete an edge is determined by multiplying the edge speed by the edge length. The total energy required to complete an edge is the sum of the external and internal motion resistances multiplied by the edge length.

2.4. Graph Search Algorithm

Upon receiving a driver-requested KML file, the graph search algorithm is triggered to compute the optimal route between defined start and target points. The optimal route is defined by its geospatial latitude and longitudinal coordinates along with performance metrics including estimated mission completion time and fuel consumption. Mission completion time is calculated by summing the traversal times across each edge in the optimal route, while fuel burn is estimated by summing the edge energy values in the optimal route. The computed route is exported as a KML file and imported into the TAK software's Routes Tool, which visually displays the route alongside the vehicle's real-time position, providing continuous navigation feedback and facilitating course adjustments.

Users can select routes optimized for mission completion time, vehicle energy, and weighted vehicle energy. The weighted energy attribute incorporates a user-defined weighting factor that varies according to route segment type. This factor is multiplied by the vehicle's energy requirement for each edge, generating the weighted energy attribute. By tuning the weighting parameter, users can prioritize larger on-road segments and penalize off-road segments.

The Graph Search Algorithm computes optimal routes across on-road areas covering up to 1500 km^2 and off-road areas of 50 km^2 in under 10 seconds. This rapid performance is achieved through the separation of the computationally intensive Graph Generation Algorithm from the runtime Graph Search algorithm. In operational settings where server access is not possible, forecasted weather can be used to generate graphs before the mission. The pre-generated graphs can then be stored on the in-field device and used by the Graph Search algorithm to identify optimal routes. By pre-generating graphs, the WRP algorithm enables operators in the field to quickly generate routes, supporting agile decision-making in dynamic operational environments.

The Graph Search Algorithm employes the iGraph get_shortest_path() method, an implementation of the Dijkstra algorithm, to identify the minimum cumulative attribute cost from the start to target vertices (Csardi et al., 2006). The Dijkstra algorithm (Dijkstra, 1959) initially assigns all graph vertices high-cost values, except for the start vertex, which is set to zero to indicate the starting point. It then evaluates all adjacent vertices by calculating the sum of the current vertex's cost and the weight of the connecting edges. If this computed cost is lower than the existing cost of the adjacent vertex, the vertex is updated with the new cost, and the current vertex is stored as its predecessor. The process continues iteratively until the target vertex is reached. Upon completion, the shortest path is reconstructed by tracing backward from the target vertex to the start vertex, following the sequence of recorded predecessors.

3. Winter Route Planning Evaluation

On 10 March 2025, a series of field experiments were conducted at the EAFR facility using a tracked Polaris MRZR Alpha to evaluate and refine the performance of the WRP algorithm (Fig. 15). The evaluation focused on route prediction between a designated start and target location under two conditions: one scenario constrained access to W. Hill Road, while the other allowed its inclusion within the search area.

Fig. 15. Polaris MRZR Alpha: https://military.polaris.com/en-us/mrzr-alpha

Fig. 16 and 17 illustrate the initial conditions, where SP denotes the vehicle's start position and VDO indicates the target location. The transparent red overlay identifies driver-defined constraint zones, while grey lines represent road centerlines. Optimal routes are depicted in red, Fig. 16 for the constrained scenario and Fig. 17 for the unconstrained scenario.

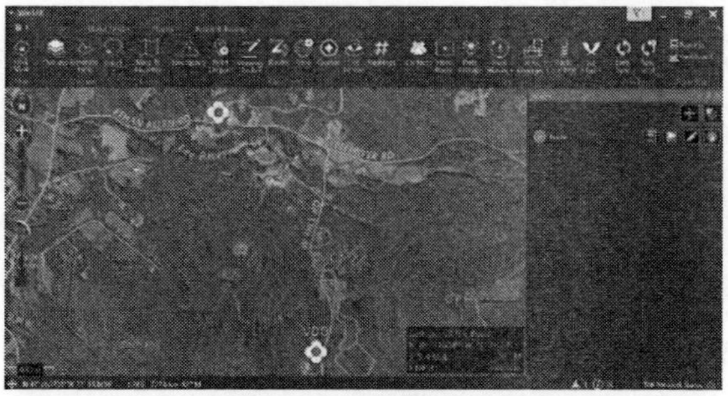

Fig. 16. Optimal Vehicle Work Path, Main Road Constrained

Fig. 17. Optimal Vehicle Work Path, Main Road Unconstrained

In the unconstrained condition, the search algorithm selected W. Hill Road as the lowest-risk route. Route computation and mission time estimation were completed in 3 seconds, with a predicted mission duration of 6 minutes. This route was subsequently traversed by the instrumented Polaris MRZR Alpha, which collected GPS position, velocity, IMU data, and driver inputs. As shown in Fig. 18, the driven path (cyan) closely overlaid the predicted route (red), with the actual mission time matching the predicted duration.

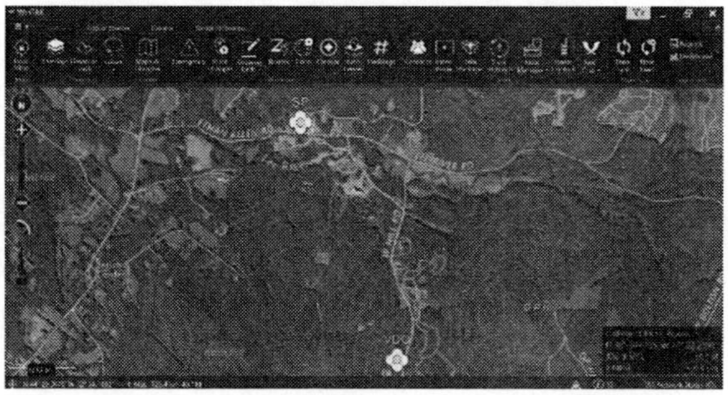

Fig. 18. Comparison of Unconstrained Road Route

In the constrained scenario, W. Hill Road was excluded from the search area. The algorithm selected a secondary road equivalent to a level 2 to level 3 logging road before transitioning to an off-road route to reach the target. At a latitude of 44.465° and longitude of -72.941°, an off-road route was selected. With the main road constrained off, transversing to the target location required an off-road route through a dense woodblock. Field reconnaissance identified only one viable route: a seasonal drainage area beneath a sparse tree canopy. The WRP algorithm successfully selected this route, demonstrating minimal deviations to avoid natural obstacles such as downed trees. Figure 19 highlights the accurate transition from road to forested terrain, with only minor route discrepancies observed at latitude 44.464°

and longitude -72.9398°. Route computation in this scenario was completed in 5 seconds, with a predicted mission time of 9 minutes. The actual traversal took 11 minutes.

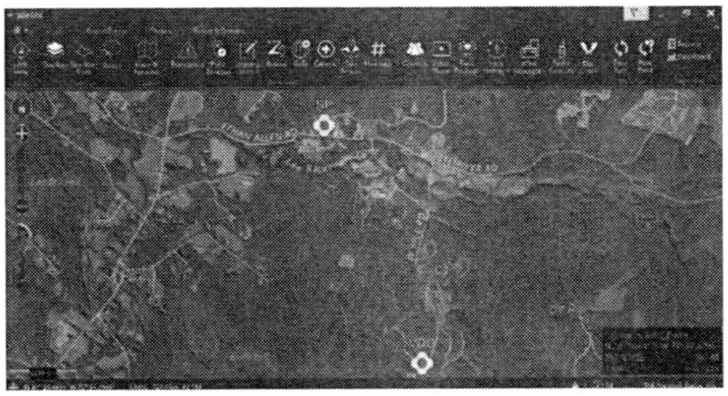

Fig. 19. Comparison of Constrained Road Route

Overall, the WRP algorithm showed good correlation between the predicted and driven routes. When the main road was included in the search area, the algorithm correctly selected W. Hill Road as the low-risk route. The predicted and actual mission completion times compared well with one another, with both driven and predicted time being 6min. When the road was constrained from the search area, the algorithm selected a secondary road. When required to proceed off-road, the algorithm selected the only viable off ramp into the forested area. Once on the off-road route, the predicted route selected areas with limited tree cover.

4. Conclusions and Recommendations

Due to the rapidly changing environment and low strength surfaces, maneuver in cold regions can be especially challenging. In this work we have developed a WRP algorithm that fuses current weather, terrain, and vehicle data to generate optimal routes. The WRP algorithm interfaces with drivers through the TAK software enabling quick generation of optimal routes in the field. Meteorological models fuse weather and terrain data to predict the current snowpack's depth and density. Road centerlines from OSM are used to generate an on-road graph of the AOI. An off-road graph is generated using detailed DTM and DSM products. The on and off-road graphs are combined into a single graph allowing large areas searches with the road network connected to the off-road areas. Edge weights representing the vehicle's energy consumption and traversal time for each segment are computed. Rollover, topple and sliding events are evaluated. If the events are possible, the edge connection is removed from the combined graph. The combined graph is generated daily to ensure that the current winter surface is used in the graph development. The combined graph is searched using the Dijkstra algorithm provided by the iGraph python import. Optimal routes, based on used selected parameters, are determined by the algorithm in less than 10 seconds for an on-road area of 1500km^2 and an off-road area of 50km^2. The quick search times enable the generation of multiple routes in dynamic field operations depending on what parameters are important to the user (time to target, minimizing energy consumption, on or off-road).

Validation testing of the WRP algorithm was conducted at the EAFR on 10 March 2025. The algorithm showed good correlation between the predicted and driven routes. When the main road was included in the search area, the algorithm correctly selected the main road, W. Hill Road, as the low-risk route. When the road was constrained from the search area, the algorithm selected a secondary road. When required to proceed off-road, the algorithm selected the only viable off ramp into the forested area. Once on the off-road route, the predicted route selected a seasonal drainage area with limited tree cover. The predicted mission completion times and the driven route times compared well. For the prominently on-road route, both the driven and predicted mission times were 6 minutes. For the on and off-road route, the WRP algorithm predicted a 2-minute faster completion time than the driven route.

Future algorithm enhancements will focus on three areas: 1) an improved driver interface, 2) improved route accuracy, and 3) continued algorithm validation. To improve the driver interface with the WRP algorithm, we will develop a TAK plugin for the algorithm. To improve route accuracy, we will investigate methods to improve off-road resolution without increasing search times. The finer resolution will allow for improved discovery of off-road obstacles. Also, we will use snow stratigraphy to improve predictions of vehicle motion resistance and immobility. To improve algorithm validation, we will conduct route testing at multiple locations and compare Lidar scanned and predicted winter surfaces.

5. Acknowledgements

The authors extend their gratitude to Theodore Letcher for generously providing the meteorological models utilized in the Winter Surface Generation Algorithm, as well as for his valuable guidance in applying these models. We also thank Eric Jerez for his contributions in streamlining the workflow by containerizing the meteorological models using Docker. Additionally, we acknowledge Jason Olivier for providing the GIS data employed in this study.

6. References

Csardi, G., Nepusz, T., 2006. The igraph software package for complex network research. InterJournal, Complex Systems, 1695.

Dijkstra, E.W., 1959. A note on two problems in connexion with graphs. Numer. Math. 1, 269–271. https://doi.org/10.1007/BF01386390

Farr, T.G., Kobrick, M., 2000. Shuttle Radar Topography Mission produces a wealth of data. Eos Trans. AGU, 81:583-583.

Gillespie, T., 1994. Fundamentals of Vehicle Dynamics. Society of Automotive Engineers, Warrendale, PA.

Golubev, V.N., Frolov, A.D., 2000. Model of structure and mechanical properties of dry granular snow. Annals of Glaciology 31:434-438. doi:10.3189/172756400781819815

Letcher T. Parno, J., 2023. Incorporating Advanced Snow Microphysics and Lateral Transport into the Noah-Multiparameterization (Noah-MP) Land Surface Model. US Army Engineer Research and Development Center, Cold Regions Research and Engineering Laboratory TR-23-10. Hanover, NH.

Liston, G., Elder K., 2006. A Distributed Snow-Evolution Modeling System (SnowModel). American Meteorological Society.

Niu, G.-Y., Yang, Z.-L., Mitchell, K.E., Chen, F., Ek, M.B., Barlage, M., Kumar, A., et al., 2011. The Community Noah Land Surface Model with Multiparameterization Options (Noa h-MP): 1. Model Description and Evaluation with Local-Scale Measurements. Journal of Geophysical Research: Atmospheres 116 (D12). https://doi.org/10.1029/2010JD015139.

OpenStreetMap contributors, 2015. Retrieved from https://planet.openstreetmap.org

Shoop, S.A., Bodie, M.O., Frankenstein, S., Bigl, M.F., 2020. Recent improvements to snow mobility algorithms, Cold Regions Research and Engineering Laboratory Technical Report, MP-20-9. doi:10.21079/11681/37821.

Usbecka K., Gillena M., Loyalla J., Gronoskya A., Sterlingb J., Kohlerc R., Hanlon K., 2015. Improving Situation Awareness with the Android Team Awareness Kit (ATAK). SPIE Conference on Defense and Security. Baltimore, MD.

Welling, O., Shoop, S., Letcher T., Elder, B., Bodie, M., 2022. Estimation of Plowing Forces on Vehicle Driving Through Deep Snow. Journal of Terramech. 104: 25-29.

6001 | Modeling and simulation of autonomous vehicular systems at ERDC – II Soft soil model support in vehicle dynamics

Peilin Song *, Burhman Gates, Garrett Holden, Brent Towne, Joshua Fairley

Mobility System Branch, Engineer Research and Development Center, Vicksburg, MS, USA

** Corresponding author: peilin.song@erdc.dren.mil*

ABSTRACT

As modeling and simulation (M&S) become increasingly vital in the development of autonomous vehicular systems, the fidelity of vehicle dynamics models has emerged as a critical factor—especially for military vehicles operating in off-road environments dominated by varying types of soft soils. Accurate simulation of mobility in these conditions requires reliable models that reflect real-world interactions between vehicles and terrain.

The U.S. Army Engineer Research and Development Center (ERDC), through its Geotechnical and Structures Laboratory (GSL), has been a leader in off-road mobility research for decades. ERDC/GSL has conducted extensive testing on a wide range of vehicles and terrains, leading to the development of empirical formulations for traction, resistance, and sinkage of tractive elements such as tires and tracks. These formulations have been encapsulated in the Vehicle-Terrain Interaction (VTI) API, which supports both wheeled and tracked vehicles and serves as a foundational tool for many off-road mobility simulation programs.

This paper presents the integration of the VTI API as a plugin within Unreal Engine. This integration aims to enhance the realism and accuracy of mobility simulations by enabling high-fidelity modeling of vehicle-terrain interactions within immersive simulation environments. By leveraging the VTI API, the resulting simulations more effectively represent vehicle performance on soft soils, thereby advancing the analysis and development of autonomous and military vehicles in complex off-road conditions.

Keywords
Off-road mobility
Soil characterization
Soft soil model
Vehicle Terrain Interaction
Unreal Engine

1. Introduction

Modeling and simulation (M&S) techniques are essential in the development of autonomous vehicle systems. ERDC's Geotechnical and Structures Laboratory (GSL), in collaboration with Army partners, has applied M&S across key programs such as Virtual Engineering Evaluation Test (V-EET), Autonomous System Test Capability (ASTC), and PDM-RASContinuous Autonomy Simulation Test Laboratory Environment (CASTLE). These efforts feature high-fidelity sensor simulation, RTK-based navigation stacks, and advanced vehicle dynamics models developed with GVSC.

The effectiveness of these simulations has been demonstrated in real-world scenarios. For example, lidar simulations have revealed unexpected behaviors, such as vehicle acceleration in fog—mirroring test observations. Object detection and avoidance (ODOA) testing has informed safety tuning, highlighting the impact of obstacle size on stopping distance.

Accurate vehicle dynamics modeling is also essential in the M&S of autonomous systems, as it directly impacts the realism and effectiveness of autonomy development. Unreal Engine is used in our simulation as physical engine for vehicle dynamics. Unreal Engine's native physics engine provides a capable foundation for general vehicle dynamics simulation, making it suitable for basic mobility analysis. However, its vehicle-terrain interaction model is limited and primarily tuned for rigid, hard-surface environments. This simplification reduces simulation fidelity when applied to off-road or soft-soil conditions, often leading to unrealistic behaviors such as oscillations and instability that are difficult to interpret or correlate with physical phenomena. These limitations highlight the need for integrating more advanced, physics-based mobility models to accurately capture terrain effects and improve simulation reliability for autonomous system development.

A cornerstone of ERDC's mobility modeling is the Vehicle-Terrain Interaction (VTI) software, developed through decades of off-road mobility research. VTI incorporates empirical models of tractive element behavior in soft soil and other terrains, enabling predictive performance analysis. Integrating VTI with vehicle dynamics models supports accurate simulation of wheeled and tracked vehicles.

This paper focuses on the integration of ERDC's soft soil modeling capability into Unreal Engine via the VTI API. Section 2 introduces the soft soil model—an algorithmic framework for predicting steady-state tractive performance across terrain types. Section 3 details the Unreal Engine integration, outlining plugin architecture and the vehicle template interface. The final section presents example applications and outlines future directions for extended implementation.

2. Vehicle Terrain Interface (VTI)

The NATO Reference Mobility Model (NRMM) is a well-established computational tool used to predict the mobility performance of a vehicle operating on a specified terrain under defined environmental conditions. First released in Hailey (1979), the model has undergone numerous updates and enhancements over the years, including the addition of new algorithms and expanded scenario support. A major revision in 1992 resulted in the release of NRMM II, which featured a modernized codebase (Ahlvin, 1992).

NRMM implements a set of equations and algorithms that estimate a vehicle's steady-state mobility across various terrains. These models are empirically derived from decades of laboratory

https://doi.org/10.56884/606GXXY9

and field-testing data (see Fig. 1), making NRMM a foundational tool in the analysis and planning of vehicle mobility in military and engineering contexts.

Fig. 1. Field Test and Lab Test

At the core of NRMM is a set of empirical relationships that describe tractive force, resistance force, and sinkage as functions of slip, soil type, vegetation, and moisture content—all evaluated at the tractive element level. Tractive elements refer to wheels for wheeled vehicles and tracks for tracked vehicles. To support dynamic simulation of off-road military vehicles, ERDC encapsulated these empirical relationships into a software library known as the Vehicle-Terrain Interaction (VTI) library. This library is designed for easy integration and can be seamlessly invoked by simulation programs or used within a hardware-in-the-loop (HIL) architecture. One of the early applications of the VTI is in the N-Post Simulator in which VTI is used to evaluate traction and resistance force on the test rig so that it could be driven properly (Fig. 2).

Fig. 2. N-Post-Simulator at GVSC

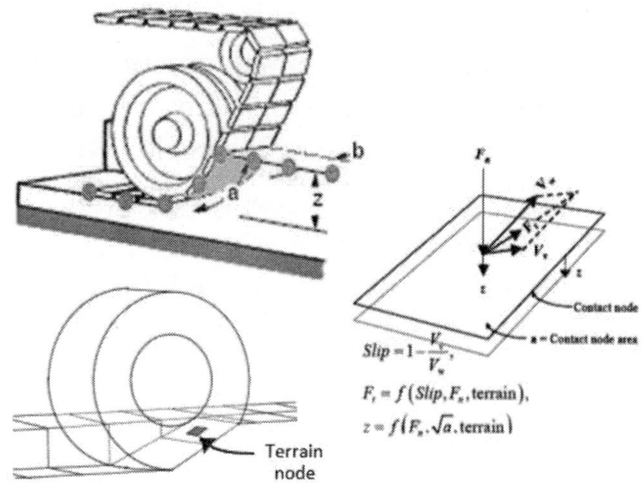

Fig. 3 Physics of the VTI

2.1. Tractive Elements: Physics and Equations

The Vehicle-Terrain Interaction (VTI) library captures the relationship between the forces and sinkage experienced by a tractive element on a given terrain under specific conditions. These forces include tractive or resistance forces in the longitudinal direction, depending on whether the tractive element is active (powered) or passive (unpowered), as well as resistance forces in the lateral direction, which influence vehicle stability and maneuverability. In addition, VTI also estimates the sinkage of tractive elements, providing critical input for assessing vehicle mobility and ground pressure distribution. Figure 3 illustrates the contact physics of the tractive elements with terrain.

From the Fig. 3, we could see that the compound force Ft and sinkage z are the functions of tractive slip, vehicle configuration and terrain information.

In general, VTI is composed of the following four functions:

1. Traction as a function of slip and slip angle
2. Resistance as a function of slip angle
3. Sinkage
4. Slip as a function of traction and slip angle

2.2. Key VTI Terrain Parameters

As is well understood, terrain characteristics significantly influence the tractive forces, resistance forces, and sinkage experienced by tractive elements. These effects are governed by several key factors, including surface condition, soil type, Cone Index (RCI), snow depth, frost depth, and moisture content.

In particular, soil type, as classified by the Unified Soil Classification System (USCS), plays a central role. The USCS categorizes soils based on grain size, plasticity, and gradation into groups such as well-graded gravels (GW), silty sands (SM), lean clays (CL), and fat clays (CH). Each soil class exhibits distinct mechanical behavior that affects vehicle-terrain interaction. For example, coarse-grained soils like sands and gravels typically provide better traction but may have variable sinkage depending on compaction and moisture, while fine-grained soils like clays and silts can lead to increased resistance and significant sinkage, especially when wet or poorly consolidated.

Understanding and incorporating these terrain attributes is critical for accurate mobility prediction and effective dynamic simulation in off-road environments.

Table 1
Key VTI Terrain Parameters

Parameter	Type	Allowable values	units
Terrain Surface Condition – determined/derived by terrain file reader or spreadsheet used to produce file and based on all other terrain state parameters. These codes are for use in a "CASE" statement, with the intention of reducing the number of if statements (but not completely).	char*3	NFG = Normal Fine Grain[1] soil NCG = Normal Course Grain[2] soil FFG = Frozen Fine Grain soil FCG = Frozen Course Grain SNO = Any condition with Snow ICE = Ice (without a snow cover) RD = Primary or Secondary Road OTH = use soil type/conditions, specifically for deep water, shallow water, PT, RK	none
Soil Type	char*4	0 = UN = Unknown 1 = GW 2 = GP 3 = GM 4 = GC 5 = SW 6 = SP 7 = SM 8 = SC 9 = ML 10 = CL 11 = OL 12 = CH 13 = MH 14 = OH 15 = PT 16 = RK = Rock 17 = SR = secondary road 18 = PR = primary road 19 = WS = shallow water 20 = WD = deep water	none
RCI 0-6 – note that RCI values are actual values, and are not adjusted by the vehicle terrain interface code (i.e. frozen soil RCI is not adjusted from normal nor is the condition of Slippery coarse grain soils)	float	0-700	psi
RCI 6-12 (see note for RCI 0-6), not currently used	float	0-700	psi
Surface cover depth (if density = 1.0, the soil is water covered with this depth. If soil type is shallow or deep water, this is the water depth, otherwise it is snow depth, also snow density < 1.0	float	>= 0	inches
Snow Density	float	>= 0 and < 1.0 snow density >= 1.0 water covered soil	g/cc
Frost Depth	float	>= 0	inches
Thaw Depth	float	>= 0	inches
Moisture Content Code (Note these codes are specific to the above conditions, some combinations should not occur)	char*3	'DRY', 'AVG', 'SAT' – used if soil is frozen and coarse grained. 'NOR', 'SLP' – normal or slippery conditions of fine grain soil only 'ICE' there is a layer of ice present over the given soil type, and there can be a layer of snow over the ice 'DRY', 'WET' – used on non-ice covered roads only	none

Notes:
[1]The fine-grained soils for the current mobility models are USCS soil types: GM, GC, SM, SC, ML, CL, OL, CH, MH, and OH.
[2]The coarse-grained soil for the current mobility models are USCS soil types SP, SW, GP and GW, which are soils in which less than 5% pass a No. 200 sieve (grain size is greater than 0.074 mm).

Table 1 show a comprehensive list of the factors VTI output. A detailed list of input parameters and variables to VTI model could be found in Priddy (1995).

2.3. VTI and High-Fidelity Mobility Prediction

The Vehicle-Terrain Interaction (VTI) model, developed by ERDC, is a data-driven, empirical framework for simulating how ground vehicles interact with a variety of terrains. Built on extensive field test data collected over decades, VTI captures the practical behavior of vehicles under diverse off-road conditions. Rather than relying on first-principles physics, it uses empirically derived relationships to deliver highly realistic mobility predictions.

At a high level, VTI:

- Predicts essential mobility parameters such as traction, rolling resistance, and sinkage
- Uses empirical relationships based on actual test data, incorporating key variables like soil type, moisture, vehicle mass, configuration, and slip
- Delivers the highest fidelity in off-road mobility simulation by directly reflecting ground-truth performance
- Interfaces with simulation platforms, such as Unreal Engine, to drive vehicle dynamics
- Serves as a reference standard, rather than being subject to Verification and Validation (V&V) itself. Since its algorithms are derived from ground-tested data, VTI acts as a baseline for validating other terramechanical models or formulations

By relying on proven, real-world performance data, VTI ensures trusted mobility predictions and provides a robust foundation for simulation-based autonomy development, vehicle design, and test planning.

3. VTI in Vehicle Dynamics in Unreal 4

In Unreal Engine 4 (UE4), vehicle simulation is structured around a template-based architecture, built on the WheeledVehicle class and powered by the PhysX physics engine. This architecture enables efficient development of ground vehicle simulations by providing a modular and extensible framework.

3.1. Core Architecture with VTI

At the heart of UE4's vehicle simulation is the VehicleMovementComponent, specifically UWheeledVehicleMovementComponent, which:

- Governs all low-level vehicle dynamics, including engine torque, braking force, steering angles, and gear shifting
- Allows configuration of drivetrain behavior such as front/rear/all-wheel drive
- Provides access to real-time vehicle telemetry such as wheel slip, throttle input, and gear status
- Supports customization through Blueprints or C++ for advanced control or integration

This component is tightly coupled with:

- Suspension and Wheel Models: Defined through the vehicle's skeletal mesh and wheel classes, enabling independent control over radius, friction, damping, and whether a wheel is steerable or powered
- Chassis Setup: Uses physical material and collision settings to interact with the terrain mesh
- Input System: Connects user or AI inputs to drive throttle, steering, and brake commands

However, the contact mechanics between terrain and tractive elements (wheels and tracks) are still based on the hard surface and friction circle.

For high-fidelity simulation, the VehicleMovementComponent can be extended or overridden for more sophisticated soft soil applications by integrating

- Vehicle-Terrain Interaction (VTI), which computes traction, sinkage, and resistance from real-world empirical data
- Terrain and environment definition, providing complex surface properties for soft soil.

The integration is conducted in ProjectGL. Figure 4 illustrates the file structure of the vehicle movement component plugin that supports VTI. This structure will be further detailed in the subsequent section.

3.2. ProjectGL and VTI Support

ProjectGL, a suite of Unreal Engine plugins developed by GVSC, offers customized components for autonomous simulations. This suite enables the modeling and simulation of both wheeled and tracked vehicles by inheriting and extending the VehicleMovementComponent class. The classes extended from VehicleMovementComponent in ProjectGL are UCruxWheeledVehicleMovement for wheeled vehicles and UCruxTrackedVehicleMovement for tracked vehicle.

Nevertheless, the new movement component classes still rely on a simple contact model based on hard surfaces and a friction circle. To enable VTI support in ProjectGL, the UCruxWheeledVehicleMovement and UCruxTrackedVehicleMovement classes could be inherited, with VTI integrated to estimate traction, resistance, and sinkage.

Figure 4 list all the header files of VTI, vehicle movement component classes supporting VTI and template files for wheeled and tracked vehicles with VTI movement components.

- VTI: VTI is provided as header files (TerrainStateData.h, VTI_internal.h, and vti_mob.h) and a library. For simulations, a terrain soft property file is also required, and the utility file asc_reader.h is included to read information from this file.
- Vehicle movement component class for wheeled vehicle: UCruxSoftSoilWheeledVehicleMovementComponent in header file CruxSoftSoilWheeledVehicleMovement ComponentUE4.h.

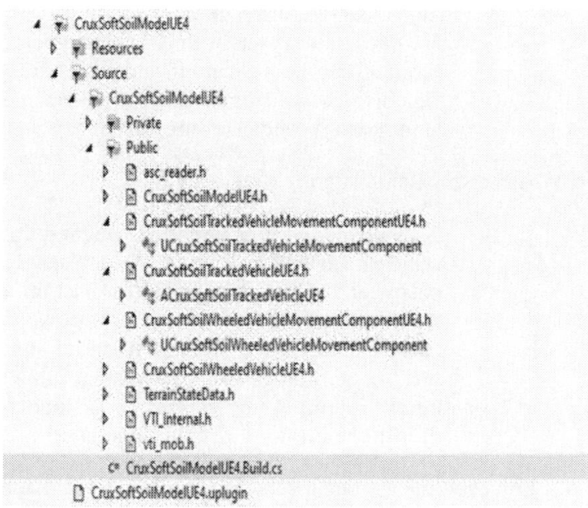

Fig.4. ProjectGL Plugin for Vehicles with VTI Support

- Vehicle movement component class for tracked vehicle: UCruxSoftSoilTrackedVehicleMovementComponent in header file CruxSoftSoilTrackedVehicleMovement ComponentUE4.h.
- Template class supporting VTI for wheeled vehicles UCruxSoftSoilWheeldVehicle in CruxSoftSoil WheeledVehicleUE4.h
- Template class supporting VTI for tracked vehicles UCruxSoftSoilTrackedVehicle in CruxSoftSoil TrackedVehicleUE4.h

VTI Movement Component of Wheeled Vehicle

3.3. VTI Movement Component of Wheeled Vehicle

```
class CRUXSOFTSOILMODELUE4_API UCruxSoftSoilWheeledVehicleMovementComponent : public UCruxWheeledVehicleMovementComponent
{
    GENERATED_BODY()

public:

    virtual void GenerateTireForces(class UVehicleWheel* Wheel, const FTireShaderInput& Input,
        FTireShaderOutput& Output) override;
    float GetFrictionCoff(float coffs, float ShapeWidth, float ShapeRadius, int WheelIndex,
        float offset, double slip, double _load, FVector loc);
```

Fig. 5. Implementation of VTI in wheeled movement component class

VTI model for wheeled vehicle is embedded in member function *void GetFrictionCoff(...)* of class UCruxSoftSoilWheeledVehicleMovementComponent. This function is called by *GenerateTireForces(...)* which overrides the function with the same name in the parent class of UCruxSoftSoilWheeledVehicleMovementComponent so that VTI model is effective in tire force computation (Fig. 5).

3.4. VTI Movement Component of Tracked Vehicle

```
class CRUXSOFTSOILMODELUE4_API UCruxSoftSoilTrackedVehicleMovementComponent : public UCruxTrackedVehicleMovementComponent
{
    GENERATED_BODY()

public:

    virtual void GenerateTireForces(class UVehicleWheel* Wheel, const FTireShaderInput& Input, FTireShaderOutput& Output) override;
    double GetFrictionCoff_(float coffs, float ShapeWidth, float ShapeRadius, int WheelIndex,
        float offset, double slip, double _load, FVector loc);
    double GetFrictionCoff(float coffs, float ShapeWidth, float ShapeRadius, int WheelIndex,
        float offset, double slip,  FVector loc);
```

Fig. 6. Implementation of VTI in tracked movement component class

VTI model for tracked vehicle is embedded in member function *void GetFrictionCoff(...)* of class UCruxSoftSoilTrackedVehicleMovementComponent. This function is called by *GenerateTireForces(...)* which overrides the function with the same name in the parent class of UCruxSoftSoilTrackedVehicleMovementComponent so that VTI model is effective in tire force computation (Fig. 6).

3.5. Data Source for VTI Model

The data required by VTI for traction, resistance and sinkage are from several sources:

- Vehicle states from dynamics engine of Unreal/PhysX: vehicle speed, vehicle slip, vertical force on tractive elements.
- Vehicle configuration from Unreal: vehicle type, vehicle width between tractive elements, dimension of tractive elements and so on.
- Terrain Information from terrain soil property file: soil type and soil strength. However, these two pieces of information could be overwritten from user input.
- Scenario information from user input: Weather (snow, rain and ice), moisture contents.

User could define scenario information and overwrite the soil type and strength with the user input panel in Unreal (Fig. 7).

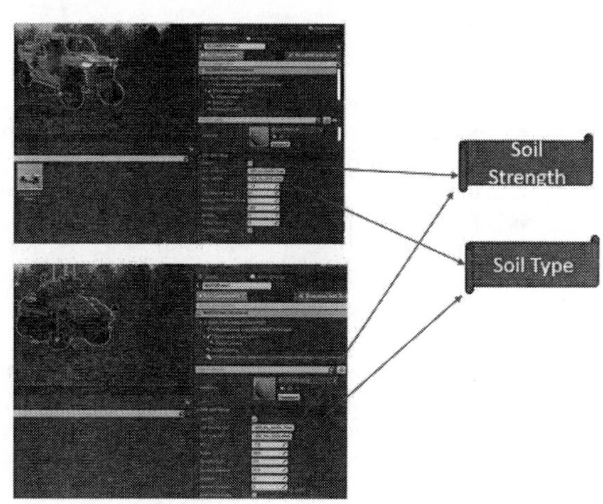

Fig. 7. Terrain and Scenario Information Input Panels

In general, the soil type and strength are mesh-wisely defined in terrain soil property files. However, soil type and property could be overwritten with a single value from the interface (Fig. 7). The default values for soil type and property are -1.0. Any positive setting will overwrite.

4. Simulation Improvement with VTI-enabled Vehicle Dynamics

Using a military combat test vehicle, simulation runs were conducted to evaluate the impact of VTI support on vehicle model performance. The results clearly demonstrate that VTI support produces significantly smoother motion, closely matching real-world test observations. VTI also enhances simulation stability, reducing the need for human intervention and eliminating the jump-stop phenomenon encountered without VTI.

A comparison of simulated vehicle speed profiles, with and without VTI support, is shown in Fig. 8. The vehicle in both simulations followed the same mission, guided by an identical navigation stack. The blue profile represents the speed of the VTI-supported vehicle, while the red profile shows the speed of the vehicle simulated with the original Unreal Engine tire model.

Crucially, the VTI model is an empirical model, derived from test data and considered ground truth. Therefore, it is not subject to validation but, rather, provides the benchmark against which other models are validated. This establishes the validity of the improvements achieved through its implementation.

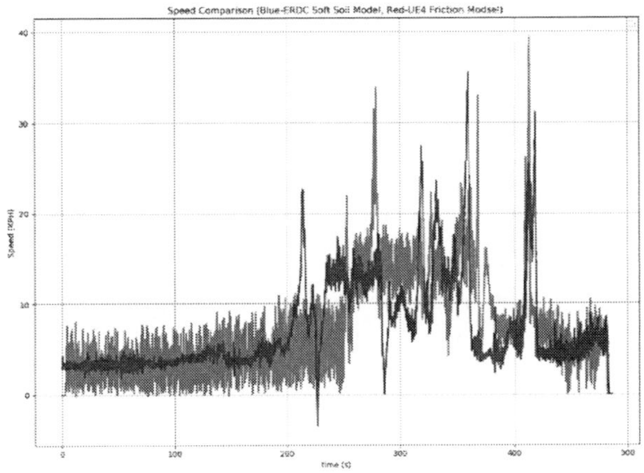

Fig. 8. Speed Comparison between Two Simulations with and without VTI Support

5. Conclusions and future work

This paper presented our efforts to enhance mobility simulation fidelity in ProjectGL through VTI model support. The effectiveness and utility of the VTI model in simulations are highly encouraging (Rhodes, 2025), and we are actively integrating this contact model into other projects, such as ASTC and PDM-RAS CASTLE. We anticipate further valuable findings from our continued work with the VTI model.

Our future plans include a more comprehensive integration of the VTI model within Unreal Engine. Currently, Unreal Engine's tire model relies on an oversimplified friction circle, which compromises the fidelity of lateral motion by relating longitudinal and lateral forces. In the next phase, we will replace this simplified tire model with a full VTI model, enabling accurate force calculations in all directions.

6. Reference

Ahlvin, R., Haley, P., 1992. NATO Reference Mobility Model, Edition II, NRMM II User Guide. TR GL-92-19, WES, 1992.

Haley, P., Jurkat, M., Brady, P., 1979. NATO Reference Mobility Model, Edition I, NRMM I User Guide. TR 12503, TACOM.

Priddy, J.D., 1995. Stochastic Vehicle Mobility Forecasts Using the NATO Reference Mobility Model; Report 3, Database Development for Statistical Analysis of the NRMM II Cross-Country Traction Empirical Relationships. TR GL-95-8, ERDC/GSL.

Rhodes, M., Gaucher, B., Zitiello, T., Thompson, B., 2025. Autonomous Ground Vehicle Simulation: An Agile Research Study, Department of Systems Engineering, United States Military Academy, West Point, New York 10996. Internal report.

6010 | Validation of mechanistic DEM calibration

Zamir Syed *

Singularity Solutions, Goleta, California, USA

** Corresponding author: zamir@cavator.ai*

ABSTRACT

Proper Discrete Element Method (DEM) calibration is essential for guiding reliable engineering decisions on the optimal design of tools and wheels interacting with soil. Optimally designed tools and wheels save time, reduce costs and lower emissions.

The paper highlights and validates two overlooked contributions to the DEM calibration literature: the relationship between physical and DEM relative density, and the scale-invariant effect of friction on critical state mechanics. These contributions significantly reduce calibration iterations and enhance the accuracy of soil simulations.

The research includes direct shear tests on Michigan 2NS sand and blade mixing experiments to validate the DEM calibration. The results show that mechanistic, non-iterative approach to DEM calibration can yield an accurate DEM parameter set that can be further refined to meet specific application needs.

Keywords
Discrete element method
DEM calibration
Soil modeling
Blade mixing

1. Introduction

The study of tool-soil interaction is crucial for various engineering applications, including agriculture, construction, and mining. Understanding the dynamics between tools and soil can lead to more efficient and effective tool designs, reducing wear and tear on equipment and improving overall productivity. Mechanistic and generalizable Discrete Element Method (DEM) calibration is essential for accurately simulating the behavior of soil particles when interacting with tools, especially over large strains. Proper DEM calibration ensures that simulations reflect real-world conditions leading to more reliable engineering decisions. This is particularly important for modeling excavation, where soil cutting imparts very large strains on the soil, often beyond the calibrated range. Therefore, it is very important to choose DEM calibration methods that generalize to a wide variety of mechanical soil responses (e.g., stability, shearing, compaction, and flow).

Optimizing tools through accurate DEM calibration can significantly enhance efficiency, resulting in cost savings and reduced emissions. By designing tools that perform better and last longer, industries can lower operational costs and minimize their environmental impact. In the case of mobility, optimizing wheels and tires using properly calibrated DEM models can improve vehicle performance on various terrains, enhancing mobility and reducing fuel consumption.

2. Background

2.1. DEM Calibration

Two thorough reviews, Coetzee et al. (2022) and Fang et al (2022), describe DEM calibration methods in great depth. However, two important contributions to the calibration literature were overlooked and necessitate elaboration here. The first overlooked contribution is the relationship between physical relative density and DEM relative density established by Salot et al. (2009). This relationship provides the mechanism for peak strength development of a granular assembly that is analogous to real soils. In addition, Salot et al.'s method places the calibrated DEM model at the correct relative density ensuring realistic soil volume change in simulation. The second contribution is the scale invariant effect of friction on critical state mechanics (Syed et al., 2017). Combining these deterministic steps leads to very significant reduction in the calibration iterations described in Fang et al.

2.2. Soil-tool interaction in DEM

Many soil-tool interaction problems have been simulated using DEM including cutting, blade sweeping, and cone penetration (Shmulevich, 2010). Shmulevich et al. (2007) performed blade cutting simulations on various blade geometries using a calibrated soil model from Asaf et al. 2007. Using similar methods, Sadek et al. (2011) performed soil calibration using laboratory direct shear tests over a range of moisture contents. A tool-soil problem simulated by Obermayr et al. (2011) used triaxial test data as a basis for calibration as well as the relative density model proposed by Salot et al. (2009). Correlation between experimental and DEM model showed that the relative density method of calibration finds applications outside of standard geotechnical testing.

https://doi.org/10.56884/LHDE3WFS

3. Methods

3.1. Direct shear test

Six specimens of Michigan 2NS sand underwent direct shear testing (ASTM3080) at three normal loads. The residual shear strength of each specimen was recorded and plotted (Fig. 1) for calculation of the residual friction angle, ϕ.

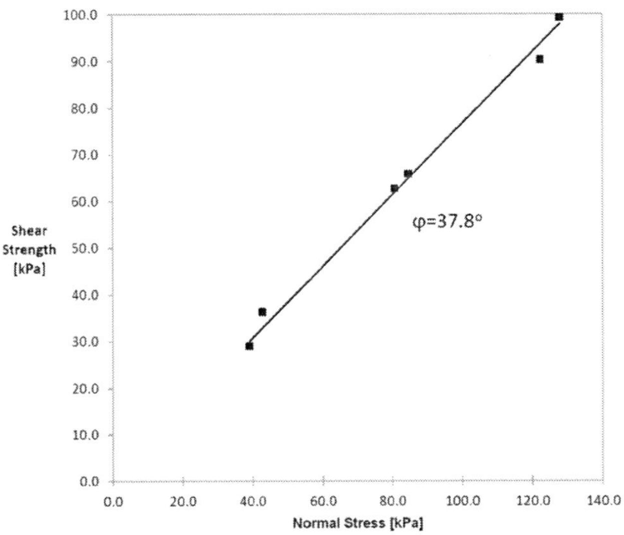

Fig. 1. Mohr-Coulomb failure envelope of residual shear strength for Michigan 2NS sand from direct shear tests

The minimum and maximum void ratios of the Michigan 2NS sand was also determined at emin = 0.54 and emax = 0.82 [ASTM 4253,4254, and 854-06].

3.2. Blade Mixing

A blade mixing apparatus was built from commercial-off-the-shelf (COTS) materials consisting of a bucket on a base plate, two bearings, a shaft and a blade (Fig. 2). The shaft's angular position was measure using a pulley-coupled string potentiometer attached to the aluminum shaft. Shaft torque was measured using a torsional load cell fabricated by outfitting the hollow, square aluminum shaft with 4 strain gages in a full Wheatstone bridge configuration (Fig. 3). It was calibrated by applying a known moment load to the shaft and measuring voltage drop across the Wheatstone bridge circuit.

A rectangular (100 mm x 23 mm) steel mixing blade was also bolted to the shaft. The shaft-blade assembly was lowered into a bucket while a bearing constrained all movement except cylindrical rotation. The bucket was filled with 2NS sand and lightly tamped and leveled (Fig. 4). The void ratio of the prepared sand specimen was 0.65 (relative density, Dr = 40%). The bulk density of the sand specimen was 1600 kg/m3. The bulk volume of the sand was approximately 3750 cm3.

After the sand was leveled and the data acquisition equipment was active, the blade shaft was manually rotated to sweep the blade through the sand. Shaft torque and blade angle was measured through three trials of the blade sweep.

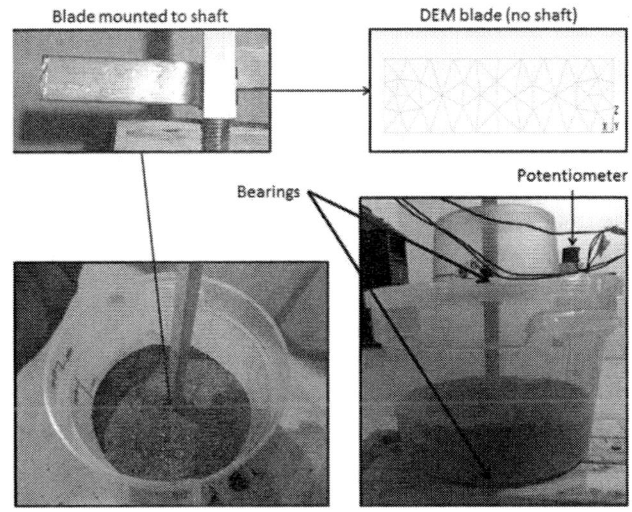

Fig. 2. Blade mixing apparatus

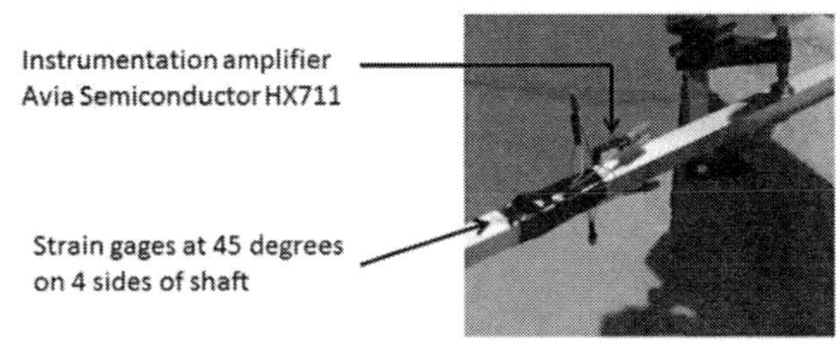

Fig. 3. Full Wheatstone bridge torque cell mounted on shaft

3.3. DEM Blade Mixing

A cylindrical bin of radius of 15 cm was initialized with an elastic wall of Young's modulus 200 GPa and Poisson's ratio of 0.3. Bottom surface of the bin was also modeled by the same elastic wall, but the top of the bin was left open. The interaction between the walls and the particles were modeled by the Hertzian contact law. The sliding and rolling friction coefficients acting between the particles and wall were 0.3 and 0.155, respectively. The steel blade was modeled as a tri-surface mesh with Young's modulus of 200 GPa and Poisson's ratio of 0.3. The shaft was omitted from the simulation. However, the blade's rotation was centered on the rotation axis of the shaft. In addition, the reaction forces acting on the blade were transformed to a reaction torque acting on the shaft axis. The blade was rotated at 2 deg/sec.

The physical sand specimen, after leveling, was approximately 70 mm deep. Therefore, the cylindrical DEM specimen was initialized at a depth of 70 mm. The bulk density of the physical sand was matched in DEM at 1.596 g/cm^3, but only after the specimen was brought to equilibrium at a desired initial state. Therefore, the particle density was highly dependent on the void ratio immediately before the simulation started.

3.4. DEM Calibration Methodology

DEM calibration was performed using a non-iterative procedure combining closed-form frameworks established by Salot et al. (2009) and Syed et al. (2017). The coefficients of sliding and rolling friction are identified by the effective friction angle method in Syed et al. (2017) based on the critical state strength of the 2NS sand (Fig. 1). Particle-scale Young's modulus was chosen at $1x10^8$ Pa to keep the timestep size above 10-6 seconds.

Poisson's ratio was arbitrarily chosen to be 0.3. The particle radius was set to 2 mm so that reasonable computation times could be achieved on available computing hardware (AMD Opteron 6274, 32 cores @ 2.2 GHz). The relative density of the physical specimen was estimated to be 40%. However, the shaft blade geometry and the irregular bin shape create some error in the determination of physical relative density. In addition, precise prescription of virtual initial state is often confounded by tool geometry; especially in cases where a blade is represented by shell surface. Regardless, the virtual relative density was approximately prescribed at 40%. The actual virtual relative density after specimen preparation was approximately 51%. Using the initial void ratio and physical bulk density, the particle scale density was prescribed at 2606 kg/m3. The coefficient of restitution was arbitrarily set to 0.1. Absolute damping was not active in this simulation. Table 1 summarizes the calibrated DEM input parameters.

Table 1
DEM parameters after calibration

DEM Input Parameters	Value
Sliding friction, μ_s	0.300
Rolling friction, μ_R	0.155
Particles radius, R	2.0 mm, mono-disperse
Young's modulus, Y	$1x10^8$ Pa
Target virtual relative density, D_r	Estimated 40%
Intermediate/latent DEM Params	**Value**
Maximum void ratio, e_{max}	0.873
Minimum void ratio, e_{min}	0.446
Initial void ratio, e_0	0.655
Actual virtual relative density	51%
Particle-scale density	2606 kg/m^3
Number of particles	87884

4. Results

4.1. Physical blade mixing

The reaction torque on the shaft during the blade mixing is reported in Fig. 4.

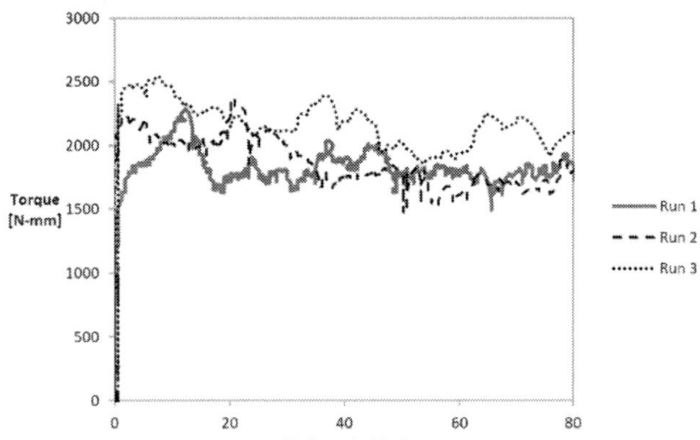

Fig. 4. Blade shaft reaction torque with respect to blade angle

Each of the three blade-sweeps were rotated at different speeds to assess the sensitivity of shaft reaction torque on shaft velocity. The shaft velocity curves for each test are shown in Fig. 5.

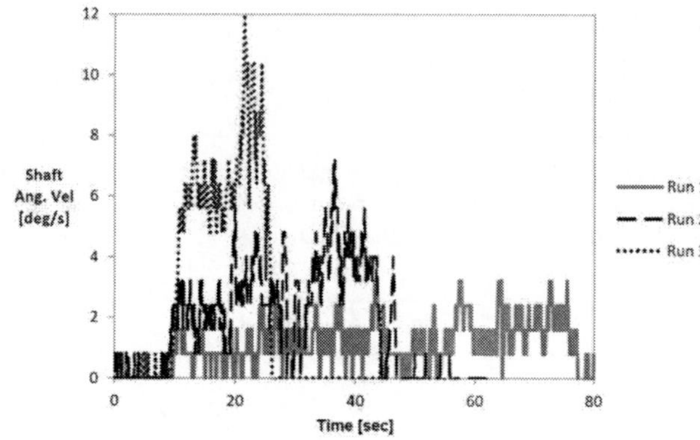

Fig. 5. Shaft velocity as a function of time during blade mixing

From **Errore. L'origine riferimento non è stata trovata.** and **Errore. L'origine riferimento non è stata trovata.**, shaft reaction torque was sensitive to shaft velocity greater than 4 deg/s. Therefore, the DEM simulation of the blade sweep system was performed at a shaft velocity of 2 deg/s.

4.2. DEM vs. Experiment

Runs 1 and 2 from the physical testing were compared to the DEM simulation of blade mixing (Fig. 6). The DEM results were smoothed with a moving average and the two physical runs were averaged. The resulting comparison is shown in Fig. 7.

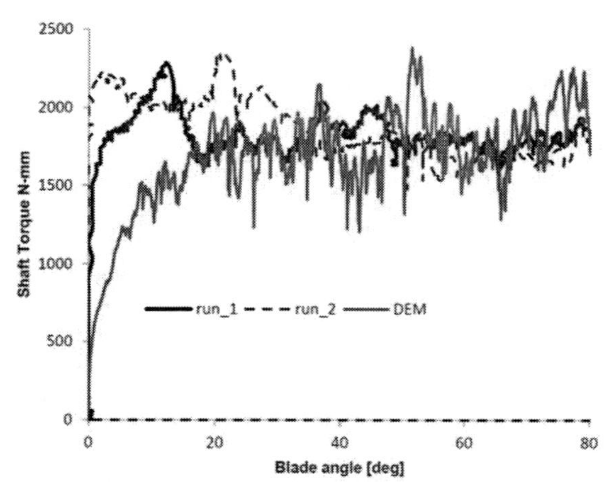

Fig. 6. Raw blade mixing reaction torque vs blade angle overlay

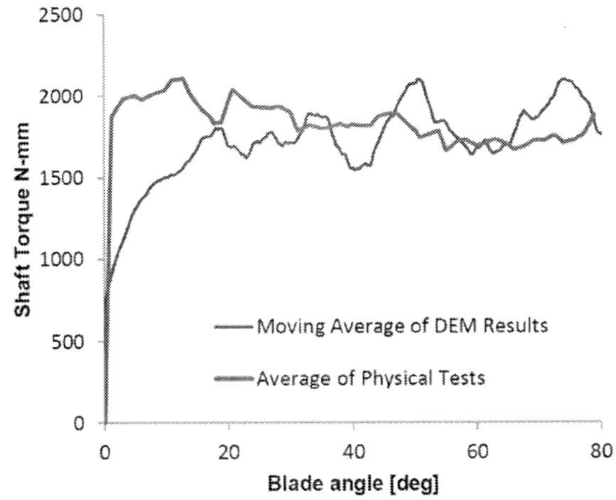

Fig. 7. Average physical vs moving average DEM blade mixing

Indeed, from **Errore. L'origine riferimento non è stata trovata.**, the lagging nature of the DEM simulation compared to the physical testing is still apparent, even after smoothing and averaging operations. Additional analysis revealed that the bulk tangent modulus of the virtual material was less stiff than the experimental tangent modulus. Therefore, additional work will be needed to calibrate the early onset of the stress-strain response of the soil.

When the blade speed was increased to 6 deg/s, the calibrated DEM model under-predicted the shaft reaction torque. It seemed that dynamics became a significant contributor to shaft torque at 6/deg/s. It was hypothesized that the coefficient of restitution would improve the agreement between virtual and physical experiments in the dynamic case. Coefficient of restitution was uniformly varied over 5 blade mixing simulations and shaft torque was recorded as a function of blade rotation. The results of the virtual experiment are shown in Fig. 8.

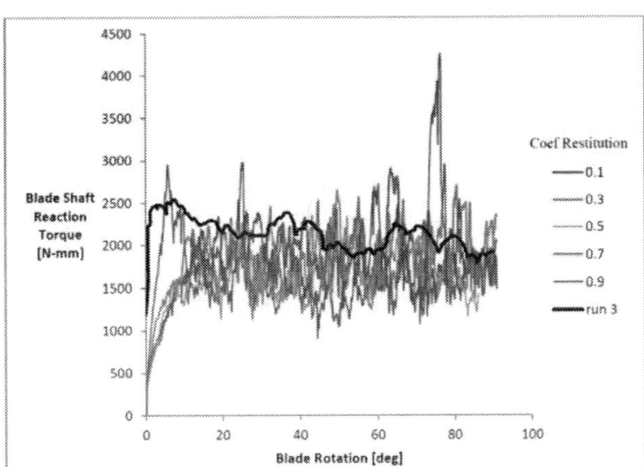

Fig. 8. Blade mixing at shaft velocity 6 deg/s. Run 3=physical experiment. Coefficient of restitution was varied from 0.1 to 0.9 in virtual experiments

Errore. L'origine riferimento non è stata trovata. shows that increasing coefficient restitution did not improve shaft reaction torque prediction. Indeed, the signal became noisier as the coefficient of restitution was increased.

5. Conclusion and Future Work

DEM calibration using a one-step, non-iterative procedure produces good accuracy for dry non-cohesive soil simulation in the context of blade mixing.

Future work is needed to add structure to the DEM calibration process. Just as Syed et al. (2017) and Salot et al. (2009) provided anchors for friction and volume change, additional research is needed to robustly describe the relationships between void ratio, particle stiffness and bulk tangent modulus of material in triaxial test or direct shear configuration. In addition, significant attention needs to be paid to the volumetric response of virtual soils, especially in the simulation of mobility systems so that sinkage, shear sinkage, and bulldozing effects can be accurately reflected in analysis.

6. Acknowledgements

The author would like to thank Prof. Mehari Tekeste (Iowa State University) for providing guidance on this project.

7. Declaration of competing interest

The author declares that they have no known competing financial interests or personal relationships that influence the work reported in this paper.

8. References

Asaf, Z., Rubinstein, D., Shmulevich, I., 2007. "Determination of discrete element model parameters required for soil tillage." Soil and Tillage Research 92 (1):227-242.

ASTM, 1994. "D3080-90: Standard test method for direct shear test of soils under consolidated drained conditions." Annual Book of ASTM Standards 4:290-5.

ASTM, D. 4253, 2006. Standard Test Methods for Maximum Index Density and Unit Weight of Soils Using a Vibratory Table.

ASTM, D. 4254, n.d. Standard Test Methods for Minimum Index Density and Unit Weight of Soils and Calculation of Relative Density (Unified Soil Classification System). ASTM International, West Conshohocken, PA.

Coetzee, C. J., Scheffler, O.C., 2022. The calibration of dem parameters for the bulk modelling of cohesive materials. Processes, 11(1), 5.

Fang, W., Wang, X., Han, D., Chen, X., 2022. Review of material parameter calibration method. Agriculture, 12(5), 706.

Salot, C., Gotteland, P., Villard, P., 2009. Influence of relative density on granular materials behavior: DEM simulations of triaxial tests. Granular matter 11 (4):221-236.

Syed, Z., Tekeste, M., White, D., 2017. A coupled sliding and rolling friction model for DEM calibration. Journal of Terramech., 72, 9-20.

Shmulevich, I., 2010. State of the art modeling of soil–tillage interaction using discrete element method. Soil and Tillage Research 111 (1):41-53.

Shmulevich, I., Asaf, Z., Rubinstein, D., 2007. Interaction between soil and a wide cutting blade using the discrete element method. Soil and Tillage Research 97 (1):37-50.

6540 | KRC bevameter development and mobility data collection

Russell Alger *, Peter Alger

Keweenaw Research Center, Michigan Technological University, Houghton

* *Corresponding author: rgalger@mtu.edu*

ABSTRACT

The Keweenaw Research Center (KRC) at Michigan Tech University has designed and built an automated Bevameter to measure in-situ soil strength and derive Bekker-Wong mobility parameters. Testing has been performed in a number of soils including high organic soils (peat), silts, sands, gravels, and lunar regolith, all at various moisture contents. A data set containing derived strength parameters is being built and results that are non-proprietary and non-controlled are shared with permission on the KRC website.

KRC has performed several vehicle mobility tests both in the field and in the lab on the KRC single wheel tire tester. A large amount of this collected data is also available on the website with permission. All testing includes cone penetrometer, nuclear density, moisture, and surface friction when possible.

This paper will be an overview of the Bevameter and the data collected to date, as well as an overview of other mobility testing that has been performed.

Keywords
Bevameter
Bekker-Wong
Vehicle Mobility
Traction
Sinkage

1. Introduction

The Bevameter is an instrument used to measure soil strength parameter in situ. The measurements from a Bevameter are used to calculate Bekker-Wong parameters (Bekker, 1956; Wong, 1978). These parameters are used as soil parameters in various simple terramechanics mobility models.

KRC designed and built the test unit described in this paper with the purpose of measuring in-situ soil strength, shear and compression, and to characterize soils during vehicle testing, but more importantly to develop a library of soil types and the corresponding characteristics including Bekker-Wong values. The intent was, and is going forward, to build a reference set of soil characteristics that can be shared with the vehicle mobility modelling community. KRC has developed a shared drive that contains a large amount of vehicle testing results and data that can be accessed with permission.

2. The KRC Bevameter

The KRC Bevameter is a self-contained unit that can be transported in the field by a number of different prime movers. It is designed to take in-situ soil strength measurements, usually in conjunction with mobility testing and modeling. The KRC Bevameter consists of three stations; a plate sinkage device, and two annular shear devices. For most testing, the two shear types used are a metal grouser and a rubber faced annulus. A steel face annulus can also be used to analyze belly drag. Fig. 1 is a photo of the Bevameter in the field, and Fig. 2 shows the Annuli as well as two plates used for plate sinkage testing.

Fig. 1. Bevameter operation in an open field

https://doi.org/10.56884/M0DWTB9K

Fig. 2. Shear annuli and sinkage plates

The unit is normally mounted on a vehicle or moved with a forklift from site to site during testing. It has removeable stability pads that allow it to be operated in a free-standing mode also. Figure 3 shows the Bevameter on an excavator.

Fig. 3. Bevameter in transport via excavator

2.1. Site Preparation

Test sites are generally prepared in three different ways. Some areas are tested without changing or removing any of the vegetation, as shown in Fig. 4. On occasion, if the vegetation is too high or thick, the grass is cut down to a mowed surface, as shown in Fig 5. Vegetation can also be carefully scraped away for some tests to get readings on the "pure" soil, as shown in Fig. 6. Care must be taken to make sure the soil layer is thick enough to make the full root bed removal possible.

Fig. 4. Bevameter operation on vegetated surface

Fig. 4. Bevameter operation on mowed surface

Fig. 5. Bevameter operation on soil surface

Fig. 6. Vegetation layer removal using an excavator bucket

2.2. Annular Shear Testing

The two annular shear devices are mechanically identical. The face areas and dimensions of the annuli are also the same for the grouser and rubber coated rings. The outside and inside diameter of the rings are 34 cm and 27 cm. The area of the annular ring is 335.4 cm^2. The rubber covered ring has smooth rubber glued to it. The grouser ring is steel, with 12 grousers spaced every 30°. The grousers are the width of the ring, 1 cm tall, and 5 mm thick.

The two shear baskets are mounted on the ends of ball splines that are designed to move freely in the vertical axis. The mechanisms are raised and lowered with cable winches that are disconnected during a test to allow free movement of the ball splines. The splines are gear driven to rotate them. The shear baskets are designed to carry plates in order to change the vertical load for the testing. For most test scenarios, three different loads are used resulting in pressures under the annulus of 8.62 kPa, 17.24 kPa, and 25.86 kPa. To achieve these three loads the basket is tested empty (293.6 N), with two weights (587.2 N), and with four weights (880.8 N) applied. Each test of a different load is performed on a virgin surface, i.e., the unit is moved each time. Figure 4 and Fig. 5 show some of the variations of weights during testing. Figure 8 shows the computer and control panel for the Bevameter.

Fig. 7. Bevameter control panel

To run a test, the Bevameter is placed in an area that looks fairly flat and is representative of the desired soil type and condition. Some minor smoothing of the surface under the cage is occasionally performed prior to the test if it is necessary. The computer is set up for the test and automatically lowers the first basket onto the surface and the winch cable is disconnected. The start button is pushed and the annulus is spun through 90° at an angular velocity of 2.95 degrees/sec. When the test is complete, the winch is reconnected and the cage is lifted. Once the first cage is in the up position, the second cage is lowered and the process is repeated. After the two shear tests are run, a plate sinkage test is usually performed in-place between the shear readings. This will be explained in the next section of this report.

When the 3 tests are complete, the weights are moved to test the next scenario, and the bevameter is moved to a new location. The tests are then repeated with the new load configuration. This process is repeated a third time to get the three different load readings and responses need for deriving the Bekker-Wong parameters. During a test, location, time, angular location, torque, and basket sinkage are recorded.

2.3. Plate Sinkage Testing

The actuator used for the demonstration in Trier is capable of 6672 N of compressive force. KRC also has a system that delivers a 16 kN of force to allow testing at higher loads. This is only possible if enough dead load is available to counteract the force supplied by the actuator.

KRC has a number of plate sizes that are used for plate sinkage tests. The plates used for the Estonian tests were 10.16, 15.24, and 30.40 cm. The smaller diameter plate was used for stronger soils and the largest for soft soils. In most cases, two plates were used for each soil. Multiple tests were performed for each plate at most sites.

To perform a test, the plate is lowered close to the terrain surface using a manual switch. Once this has been done, the test is started on the control box and proceeds automatically at a speed of 1.45 cm/sec. The plate is pushed into the soil until the load reaches 2224 N. At this point the load is quickly released and the compression immediately starts again up to 4448 N, released and reloaded until the load reaches 6672 N and the test is complete.

The test is stopped here and the device is retracted and made ready for another test. At this point the same plate can be run again, or another size plate can be installed and tested. Figure 9 and Fig. 10 are photos of the plate sinkage during a test. During a test, location, time, sinkage and force are recorded.

Fig. 8. Plate sinkage device at starting position

Fig 9. Plate sinkage device stopped at maximum force threshold

3. Data Reduction

There are numerous ways to reduce the data collected by the device. For this work, a MatLab script developed at KRC is used to allow reduction of the measured parameters. The MatLab script can be followed to see how the data was reduced to arrive at these particular Bekker-Wong values.

3.1. Shear Data Analysis

The general method to reduce the shear data is to determine the maximum available shear strength for development of Mohr-Coulomb theory. Figure 11 is a plot of three annular shear tests performed on a soil at KRC. There are a number of ways to reduce this data to derive the required maximum shear stresses. The shapes of the curves, zero starting points, initial slope, etc., can all be examined to adjust the outcome.

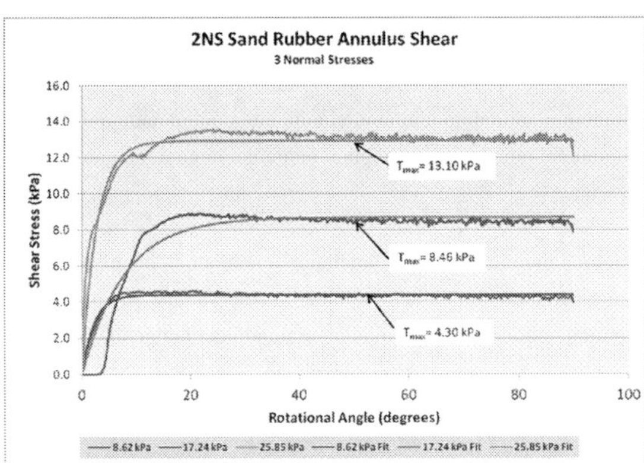

Fig. 10. Example shear test for visual analysis

From the shear data plot, the Mohr-Coulomb Shear vs Normal Stress relationship can be derived as shown in the plot in Fig. 12. For this analysis, three shear tests were used. Further reduction can be made by adding more tests in a given soil into the fit, but other in situ soil measurements, including but not limited to moisture content and density, must be similar between all tests.

The process is repeated for both annuli. Very simply, the maximum traction available from a certain soil can be surmised from these two plots.

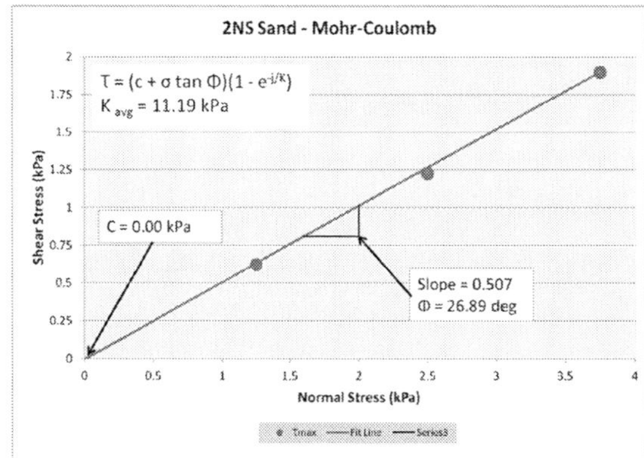

Fig. 11. Example Mohr-Coulomb fit of Fig. 11's data

3.2. Plate Sinkage Analysis

Figure 13 is a plot of a set of plate sinkage tests on the same KRC soil. For this particular test, the plate sizes were 10 cm and 15 cm in diameter. The Bekker-Wong derivations can be made using a single plate size. This graph shows the two plates plus the average. As with the shear reduction, tests in a like soil can be averaged, mixed and matched, etc. There are several other ways to get results.

The plots in Figure 13 also show the load - unload - reload curves. These can be used to look at soil rebound and multi-pass analyses.

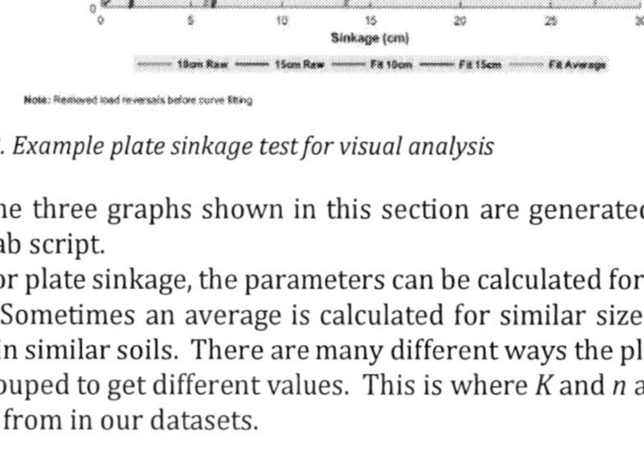

Fig. 12. Example plate sinkage test for visual analysis

The three graphs shown in this section are generated by the MatLab script.

For plate sinkage, the parameters can be calculated for a single test. Sometimes an average is calculated for similar sized plates used in similar soils. There are many different ways the plates can be grouped to get different values. This is where *K* and *n* averages come from in our datasets.

4. Simple Models

The following are some simple relationships for Mobility parameters.

The following are basic equations for shear strength (Perloff and Baron, 1976):

$$\tau = (c + p\tan\varphi)\left(1 - e^{\frac{-j}{K_s}}\right) \quad (1)$$

$$\tau_{max} = c + p\tan\varphi \quad (2)$$

Where τ=shear strength, p=normal load, τ_{max}=maximum shear, c=cohesion, φ=friction angle, K_s=average K_{avg}, and j=angular displacement.

The following are the basic equations for pressure/sinkage (Wong, 2010):

$$p = K_{eq} z^n \quad (3)$$

$$K_{eq} = \frac{K_c}{b} + K_\varphi \quad (4)$$

Where p=pressure, z=sinkage, and b=plate diameter, while n, K_c, and K_φ are constants within the Bekker pressure sinkage equation.

5. Data Sharing

As mentioned previously, KRC has developed a shared data site located at www.mtukrc.org. At this site, select the "Data Sets" tab and a drop-down menu is available with numerous different data sets and mobility tests. Bekker-Wong parameters have been measured in a number of soils, including lunar simulant. Whenever Bevameter tests are performed, other measurements including density, moisture content, cone penetrometer, Clegg hammer, vane shear, and others are made at the same time and are also included in the data sets. Most of the soils have been measured at varying moisture content and in some cases bulk density. There is also a tab that contains the design data and CAD for the bevameter.

Each tab requires asking for permission to receive the contents.

6. Acknowledgements

Some of the data collection was funded by research contracts. KRC thanks the various sponsors that helped to develop the data included in the shared drive.

7. Declaration of competing interest

The authors declare that they have no known competing financial interests or personal relationships that could have influenced the work reported in this paper.

8. References

Bekker, M.G., 1956. Theory of Land Locomotion, The University of Michigan Press, Ann Arbor.
Perloff, W.H. and Baron, W., 1976. Soil Mechanics, John Wiley & Sons, New York.
Wong, J.Y., 1978. Theory of Ground Vehicles, John Wiley & Sons, New York.
Wong, J.Y., 2010. Terramechanics and Off-Road Vehicle Engineering, Elsevier, New York.

6650 | Reliability of gravity offload for lunar rover wheel performance

James Hurrell *, Keisuke Takehana, Kentaro Uno, Kazuya Yoshida

Department of Aerospace Engineering, Tohoku University, Sendai, Miyagi, Japan

**Corresponding author hurrell.astro@dc.tohoku.ac.jp*

ABSTRACT

Purpose: The impact of gravity and the use of gravity offload on rover tractive performance is evaluated and compared between Earth and lunar environments. Single-wheel experiments and discrete element method (DEM) simulations of a micro-rover wheel modelled on the Rashid-1 rover designed for the Emirates lunar mission, are conducted. The wheel's interaction with Toyoura sand and FJS-1 lunar regolith simulant is studied. Consideration is made for why the use of gravity offloading has been prominent and other options that are available.

Methods: Slip conditions, traction coefficient and grouser pitch for single-wheel experiments are measured for a range of fixed slip values at the expected rover load of 24.5 N at a nominal 0.02 m/s, termed a low-speed regime. Single-wheel simulations have been previously verified by comparison with experimental results. Lunar simulations can predict lunar gravity performance.

Results: Wheel performance regarding traction coefficient and tractive efficiency under lunar gravity matches that under Earth gravity. Traction performance in tractive force and resistive torque is reduced by the ratio of Earth to lunar gravity, 1/6, for the same mass. Earth gravity results for 4.1 N are understood to over-predict the traction coefficient compared to lunar conditions. Using load matching of 24.5 N in lunar gravity reduces performance.

Conclusion: Gravity offloading is unsuitable. Depending on the aspect desired to study the influence of gravity on the rover, consideration should be made to use either Earth conditions, granular scaling laws, or a full-scale reduction of the rover. This may change with a transition to high-speed wheel motion.

Keywords
Terramechanics
Single-wheel test
Lunar rover
Grouser wheel
Lunar regolith simulant

1. Introduction

The focus of extraplanetary terramechanics has been on lunar and Martian rover systems, especially prominent now with several attempted and successful private lunar landings. Typically, this concerns small, slow-moving autonomous robotic vehicles. Drive systems and suspension are uniquely designed, with a limit to performance due to weight and power requirements. Some level of automation is required, with an expected lack of direct human interaction if issues arise. An often-used example of the requirement to study wheel-soil interaction is the Spirit and Opportunity rovers, which experienced severe mobility issues and highlighted the need to prepare well for extraplanetary missions. Furthermore, given the limits on power availability, maximising and understanding wheel performance aids in motor selection and can be used to improve rover efficiency (Wong, 2022).

The lunar surface is predominantly composed of lunar regolith, a heterogeneous layer of unconsolidated, fine-grained, and angular particulate material interspersed with larger parts such as rocks, breccias, and boulders. Lunar regolith exhibits a wide range of particle sizes, with a significant fraction measuring less than 100 microns in diameter and the majority being under 1 cm in size (Heiken et al., 1991).

There are two materials considered in this study, with Toyoura sand a fairly homogenous sand and Fuji Japan Simulant 1 made from crushed volcanic rock from Mt. Fuji, replicating the size distribution and aspects of chemical composition and material properties from Apollo 14 samples (Kanamori et al., 1998).

Under mechanical interaction, such as wheel-soil contact from a planetary rover, granular material may undergo local fluidisation, leading to wheel sinkage, increased slip ratios, and reduced net traction forces. These mobility constraints are particularly critical in planetary exploration scenarios, as excessive slip or sinkage can significantly impair rover traversability or result in complete immobilisation in severe cases. Understanding and modelling the interaction mechanics between rover locomotion systems and the lunar regolith is therefore vital for the development of robust, terrain-adaptive mobility platforms.

Rover wheel performance is often analysed using a full rover or a single-wheel testbed. Our research group has developed a single-wheel testbed to study the forces exerted on the wheel during motion, which can then be further explored through simulation or numerical methods.

In addition to the traditional terramechanics approaches, such as the Bekker-Wong model, other numerical techniques, like finite element analysis and soil contact modelling, simulate wheel-soil interactions by modifying a meshed surface. While these methods replicate bulk behaviour, particle-level interactions and motion can only be studied using particulate methods, for example Discrete Element Method (DEM) (Sullivan, 2011). DEM simulates granular materials in various industries, from pharmaceuticals to material processing, capturing the interactions between particles based on contact models and soil properties.

A key challenge in numerical modelling, especially for granular systems, is determining the baseline fidelity of the simulation, particularly in wheel-soil interactions. This work investigates this by simplifying the system, modelling spherical, homogeneous particles for both Toyoura sand and FJS-1, with parameters kept as similar as possible across the two materials.

Considering force representation, particle size, and gravity effects, it's expected that apparent cohesion becomes more

https://doi.org/10.56884/5A0PXCI7

Fig. 1. The test wheel, a like-ness of the Rashid-1 rover design but closed not skeletonized, for PIV study

Fig. 2. Comparison of the simulated and experimental wheel and testbed at 24.5 N, 30% slip

influential at reduced gravity levels, overshadowing frictional effects. For this study, Toyoura sand is modelled without apparent cohesion, while FJS-1 includes cohesion, with both materials calibrated to their respective AoR values from experiments.

The influence of gravity on granular behaviour can be studied through parabolic flight experiments or numerical models, particularly in simulating lunar gravity. The angle of repose (AoR) is a common method to assess shear strength and material behaviour under varying gravity conditions (Beakawi, 2018). While most studies indicate that static AoR remains invariant with changes in gravity (Marshall, 2018; Karapiperis, 2020; Nakashima, 2011), some report an increase in static AoR during failure in a rotating drum, or a decrease with lower gravity for dynamic AoR (Kleinhans, 2011). These discrepancies likely arise from differences in experimental methods, such as the material pile generation techniques, which were reported in studies (Beakawi, 2018).

Changes in dynamic AoR with gravity are more understandable due to the gravity-driven nature of material flow (Reiss, 2014; Ozaki, 2023). However, whether dynamic AoR increases or decreases with reduced gravity is still debated, as noted in studies discussed in Hoff *et al.* (2009) and Kleinhans *et al.* (2011), potentially due to variations in experimental setups and materials.

In this study, static AoR is taken to be invariant across gravity changes, acknowledging that reduced gravity decreases shear strength and bearing capacity, making the material prone to flow with less force required but fall more slowly.

Several studies have explored the effects of lunar gravity on wheel motion, both experimentally and numerically (Nakanishi, 2020; Niksirat, 2020; Daca, 2021; Daca, 2023; Kobayashi, 2010). However, differences persist, particularly between large rovers with small or no grousers and the micro-rover studied here, as well as variations in the dimensionality of DEM models (Wong, 2022; Jiang, 2018). Capturing lunar conditions experimentally remains challenging due to limited data and controlled conditions with lunar surface performance.

This study aims to establish a DEM simulation of a micro-rover's single-wheel performance, based on the Rashid-1 wheel, Fig. 1, with independently calibrated soil parameters using AoR. These simulations will be validated against experimental results, focusing on the traction coefficient and dynamic sinkage. Figure 2 shows a visual comparison of the simulation and experiment. After successfully comparing with Earth-based results, simulations will be extended to lunar gravity conditions. Full details of this process are given in Hurrell, et al. (2025), with the overall process outlined here.

2. Methodology

2.1. Overview

The generation of DEM simulations for single-wheel performance and the development of lunar predictions have four main steps, including simulation and experiment. First, the calibration of simulation parameters is performed independently of the wheel experiment data. Wheel simulations can then be generated based on single wheel experiments. Using the experimental data, simulation parameters are then validated against wheel experiment data. Following successful validation, lunar predictive simulations are performed. In addition, the wheel studied is based on the Rashid-1 rover and the associated Emirates Lunar Mission project.

2.2. Simulation

The simulations were developed using a previous methodology developed in Hurrell, et al. (2023) and finalised and validated as shown in Hurrell, et al. (2025). An outline is provided for the methodology as follows.

The single-wheel experiment was replicated using the Discrete Element Method (DEM). This approach simulates individual soil particles interacting via contact models to reproduce bulk soil behaviour. The Johnson-Kendall-Roberts (JKR) contact model captures cohesive forces between particles, reducing to the Hertz-Mindlin model when cohesion is set to zero (Coetzee, 2023). Additionally, the Type-C rolling resistance model accounts for particle shape effects when using spherical particles developed in Ai et al. (2011). For simplicity, particles are modelled as homogeneous spheres with a fixed diameter of 2 mm, scaled from the actual sub-millimetre size of Toyoura sand and FJS-1 particles. Scaling is ten times the median Toyoura particle size and thirteen times the FJS-1 particle median size.

Parameter calibration was conducted using the static angle of repose (AoR) test, a commonly used method for isolating and tuning key simulation parameters before validation with single-wheel test data (Smith, 2014). The calibration employed a design of experiments (DoE) approach using Altair's HyperStudy software (Douglas, 2013). Initial screening was performed using a fractional factorial design of resolution IV, allowing evaluation of seven parameters through sixteen simulations. Least squares regression and analysis of variance (ANOVA) identified key factors influencing the result. These were further refined using the Box–Behnken technique to optimise parameter values for matching the experimental AoR. Final calibration values were checked using five repeated AoR simulations, with means and standard deviations reported in Table 1. The methodology follows the approach outlined in Hurrell, et al. (2023) and Hurrell, et al. (2025).

Key parameters identified for Toyoura sand were static and rolling friction between particles, rolling friction between particles and surface, while cohesion was additionally calibrated for FJS-1 to capture its increased angularity and the influence of the fine particle content (Matsushima, 2014; Ozaki, 2023; Coetzee, 2023). Less influential parameters—including Poisson's ratio, coefficient of restitution, and Young's modulus—were held constant at nominal values, consistent with prior DEM studies (Gu, 2013; Lommen 2014; Ono 2013; Bui, 2009; Nakanishi, 2020).

The simulation replicates an experimental setup by filling a 0.65×0.14×0.14 m box with particles, which settle under gravity to a depth of approximately 0.12 m. A wheel is then introduced with a specified load, allowed to settle, and driven forward at controlled translational and rotational velocities. Wheel forces are measured at each timestep to compute the traction coefficient, tractive efficiency, sinkage, and resistive torque. Soil flow behaviour and wheel traces are also captured.

For suitable replication of bulk densities, Toyoura sand simulations used the actual particle density, while FJS-1 required a higher particle density to achieve an appropriate minimum bulk density due to the lack of size distribution in the model. Although the full experimental bulk density range of FJS-1 could not be captured, the simulations sufficiently represented the lower range observed in the experiment.

Simulations were performed using EDEM software (Altair Engineering), with calibration via HyperStudy. All simulations were GPU-accelerated using an NVIDIA GeForce RTX 4090. A dynamic domain approach was implemented, where only the region surrounding the wheel's path was active, significantly reducing computational cost. For 1.38 million particles simulated over a 30-second run time, it required approximately seven hours.

Table 1
Particle simulation parameter summary

Parameter	Toyoura	FJS-1
Particle diameter (mm)	2	2
Particle number	1,382k	1,382k
Time step (s)	8e-6	8e-6
Density (kg/m³)	2650	3230 (2860)
Average bulk density (kg/m³)	1440	1740
Bulk density range (kg/m³)	1380–1630	1620–1870
Poisson's ratio	0.25	0.25
Young's Modulus (N/m²)	5e7	5e7
Coefficient of Restitution	0.3	0.3
Static friction	0.65	0.65
Rolling friction	0.15	0.15
Simulated cohesion (kN/m²)	0	0.08
Angle of repose (°)	34 ± 0.5	38 ± 1

2.3. Experiment Validation

The simulations were based on a single wheel experiment using the wheel as shown in Fig. 1 and Fig. 2.

The wheel used in the experiments is a model of the Rashid-1 Rover wheel as part of the first Emirates lunar mission. The experimental wheel is shown in Fig. 1 and has a radius of 100 mm, a width of 80 mm, 14 grousers with a length of 20 mm and a mass of 1.18 kg. The wheel was entirely closed on one side to aid in visually studying soil flow under the wheel. This differs from the final rover's wheel design, which is skeletonised with a grouser length of 15 mm. Furthermore, the mass and material of the wheel are different from those of the flight wheel, but vertical load adjustment of the testbed maintains the determined load for the wheel.

The single-wheel testbed allows for one wheel to move on a fixed axis in the testbed. It consists of a wheel drive unit, a passive vertical moving unit, a translational drive unit and the sandbox. Further details on the components and experimental method are in Takehana (2025).

The aim is to evaluate the wheel's performance with fixed slip conditions for an expected vertical load. As such, load force, angular velocity and horizontal velocity are set at fixed values for each experiment run. This was a 24.5 N load including wheel mass and a nominal 0.2 rad/s wheel rotation speed. In addition, the packing condition of the testbed is maintained so that each experiment can be matched in the simulation.

The packing condition studied is that of loose material found approximately at a relative density of 25% by measurement with a hand vane and calibration against samples of known bulk density. Full details of this process are in Hurrell, et al. (2025).

2.4. Measured Quantities

Several aspects of the wheel motion are measured or set. Slip ratio, sinkage, traction coefficient, bulk density and the wheel trace can be considered for the experimental and simulated sand bed.

In both the experiment and simulation, the rotational and horizontal velocity of the wheel can be set. Setting fixed values of the rotational and horizontal velocity of the wheel gives a fixed slip ratio characterising the wheel's travelling state, or wheel rotation drives the wheel motion, giving a measurable slip. Originally, for simplification in the experiment, slippage of a wheel without grousers was given, but this was updated after driving motion was studied due to negative slippage being observed unless the grouser length was considered. This is defined in Eq. 1, where originally for fixed slip conditions r_s was treated as zero, but now updated with r_s with a grouser length as 0.02 m: For example, with no grouser, a slip of 20% then with the grouser length included in the calculation, the slip ratio becomes 33%.

$$i = \frac{r_w + r_s * \omega - v_x}{r_w * \omega} \tag{1}$$

r_w is the wheel radius, r_s is the grouser length, ω is the angular velocity, and v_x is the translational velocity. The slip ratio ranges between 0% and 100%. Zero indicates a wheel moving forward without any slippage, whereas a slip ratio of 100% indicates the wheel cannot move forward because of slippage (Ding, 2011}.

Dynamic sinkage is measured in the experiment and replicated in the simulation. This is sinkage measured from when the wheel is at rest given by Eq. 2., d, is the wheel z-coordinate at rest, s_w is the soilbed height and r_w is the wheel radius.

$$s_d = d - (s_w - r_w) \tag{2}$$

In the simulation, total wheel sinkage is determined by including the initial static sinkage with the dynamic sinkage.

The force of the particles on the wheel is measured and can be split into the force along each axis. The net tractive force in the x direction is F_x; dividing this value by the load force, F_z gives the traction coefficient, μ shown in Eq. 3. For wheel performance, the

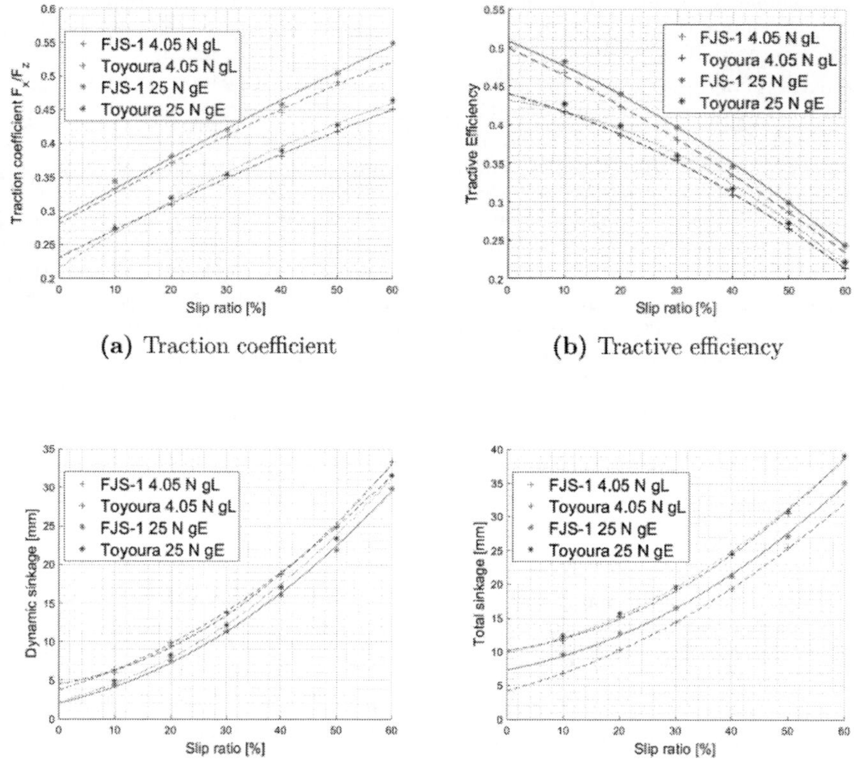

(a) Traction coefficient (b) Tractive efficiency

(c) Dynamic wheel sinkage (d) Total wheel sinkage

Fig. 3. Traction coefficient, tractive efficiency dynamic sinkage and total sinkage against slip ratio simulation results, at 4.05 N under Lunar gravity and 24.5 N under Earth gravity for FJS-1 and Toyoura.

traction coefficient is equivalent to the drawbar pull coefficient, which is the ratio of drawbar pull to rover weight (He, 2020).

$$\mu = \frac{F_x}{F_z} \tag{3}$$

In addition, the wheel motion can be evaluated as tractive efficiency, η, by the required torque to maintain the angular velocity needed or resistive torque at each slip condition. This is then determined as Eq. 4 where F_x is net tractive force, v_x is the horizontal (David, 2019; Ding, 2011).

$$\eta = \frac{F_x v_x}{T\,\omega} \tag{4}$$

The bulk density of each material and the relative bulk density are used to assess the testbed state and allow the simulation to match the experimental conditions. The minimum and maximum values of the bulk density were experimentally measured as summarised in Table 1.

3. Results

3.1. FJS-1 and Toyoura Lunar Comparison

Results of the validation against experimental data are covered in Hurrell, et al. (2025), the focus here is to discuss the lunar results. It was found that the traction coefficient and dynamic sinkage for experiment and simulation matched between Toyoura and FJS-1 results, although higher correlation for the Toyoura results.

An initial comparison can be made between the two materials studied, Toyoura sand and FJS-1, with the lunar gravity condition and the equivalent mass on the wheel as on Earth. This gives the 4.05 N load in lunar gravity equivalent to 24.5 N on Earth with the same mass of 10 kg across four wheels.

Figure 3 shows the traction coefficient, tractive efficiency, dynamic wheel sinkage and total wheel sinkage for FJS-1 and Toyoura sand at 4.05 N load with lunar gravity. In Fig. 3a) and b), the traction coefficient and tractive efficiency, respectively, for FJS-1 is consistently higher than for Toyoura matching behaviour for 24.5 N Earth gravity. Interestingly, the dynamic wheel sinkage is similar for both as seen in Fig. 3c), while due to the difference in static sinkage, FJS-1 has a higher bearing capacity and lower static sinkage than Toyoura, resulting in the wheel consistently sinking deeper overall as Fig. 3d).

If Toyoura sand is used instead of a lunar simulant to estimate lunar performance, as represented by the 4.1 N lunar FJS-1 result. In that case, performance will be underestimated. The total sinkage may be overestimated by 3-4 mm, the traction coefficient by 15-20% lower and the tractive efficiency by 5-10% lower.

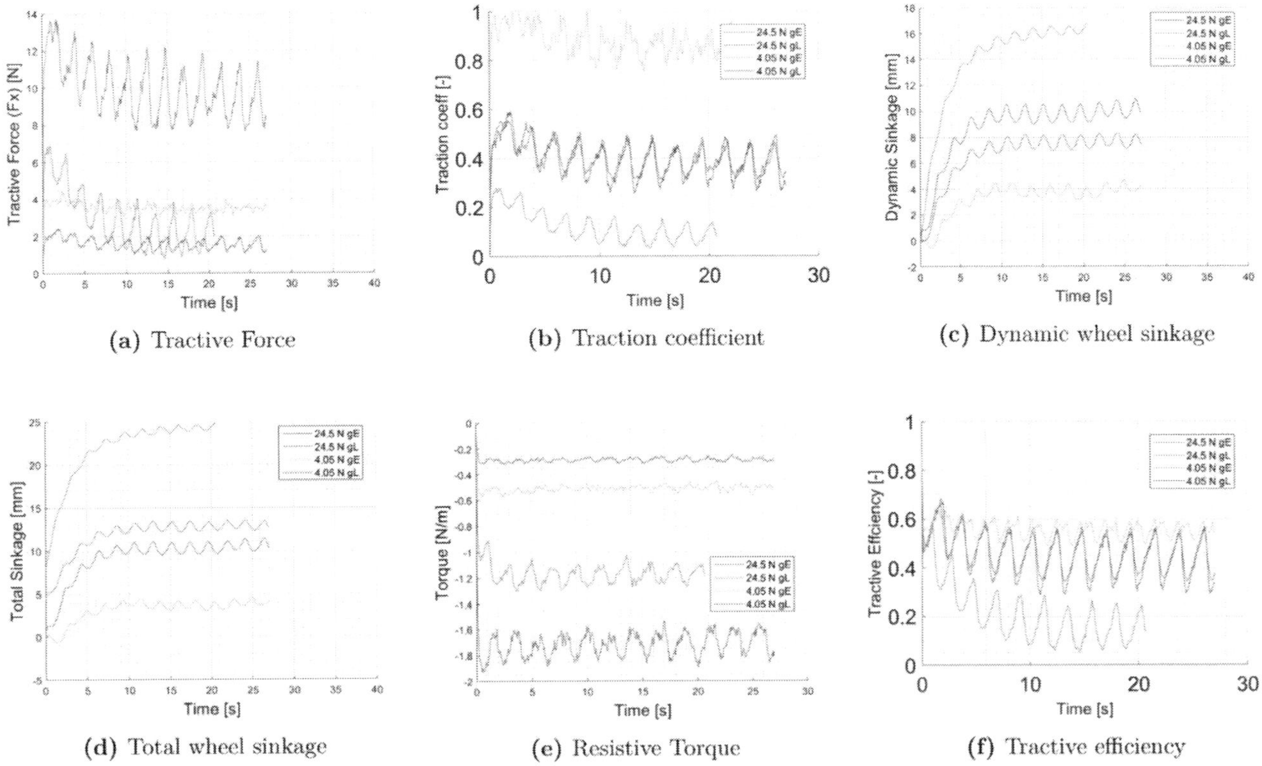

(a) Tractive Force (b) Traction coefficient (c) Dynamic wheel sinkage

(d) Total wheel sinkage (e) Resistive Torque (f) Tractive efficiency

Fig. 4. Tractive force, tractive coefficient, tractive efficiency dynamic sinkage and total sinkage against slip ratio simulation results, at 20% slip comparison of 4.05 N and 24.5 N load for Earth and Lunar gravity for FJS-1 results

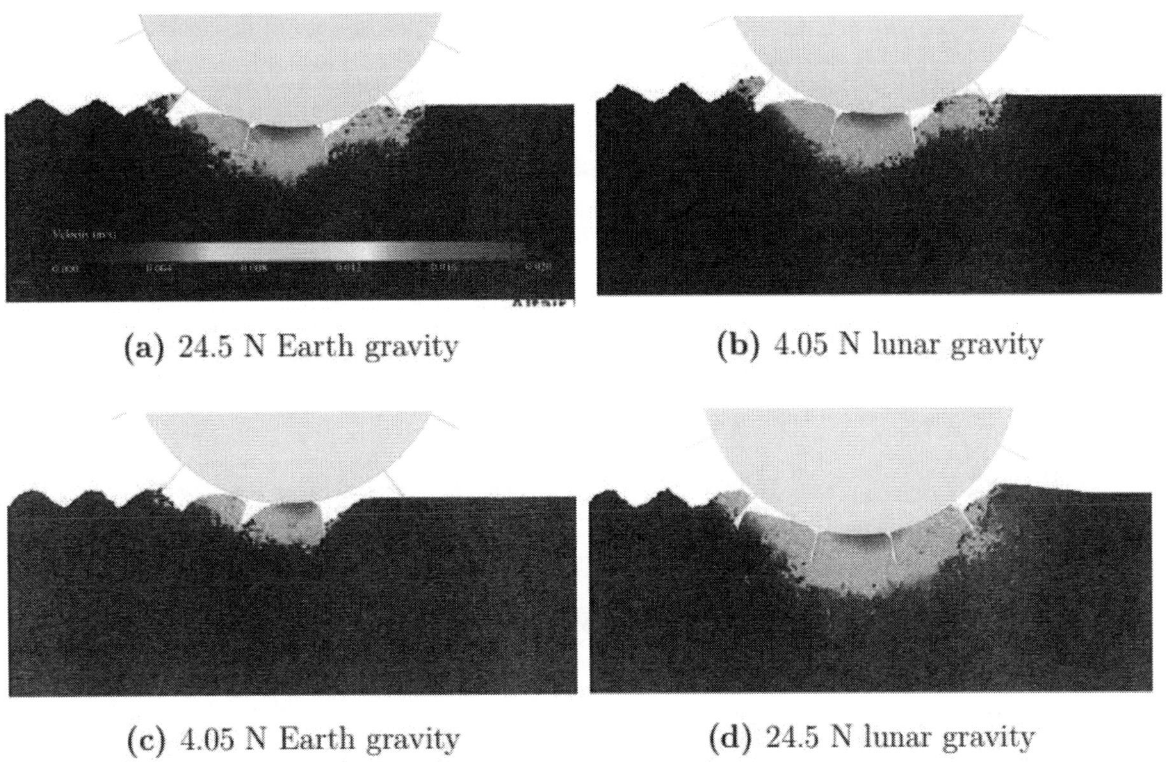

(a) 24.5 N Earth gravity (b) 4.05 N lunar gravity

(c) 4.05 N Earth gravity (d) 24.5 N lunar gravity

Fig. 5 Comparison of material motion, specifically particle velocity magnitude, under the wheel at 24.5 N and 4.05 N under Earth and lunar gravity for FJS-1 at 20 seconds, maximum Fx, at 33% slip

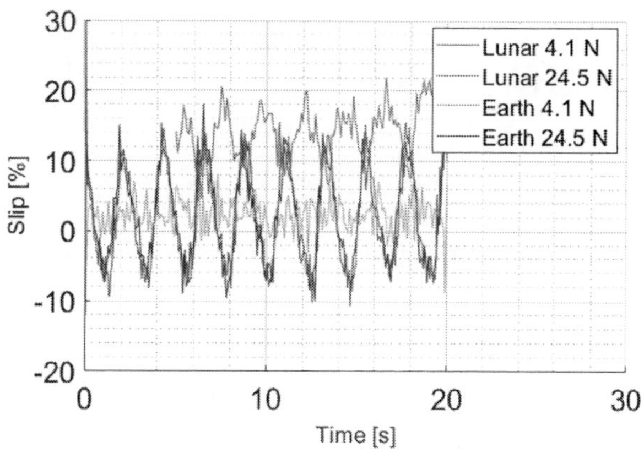

Fig. 6. Slip ratio over time for 24.5 N and 4.05 N under Earth and lunar gravity for FJS-1 over time with driving motion

3.2. FJS-1 Gravity and Load Study

The results for the 33% ratio in the tractive force, traction coefficient, dynamic and total wheel sinkage, as well as resistive torque and tractive efficiency are shown in the graphs in Fig.s 4 a), b), c), and d). Ultimately, visualising the soil behaviour and motion under the wheel in these four cases can aid in understanding the similarities and differences as shown in Fig. 5. The particle motion under the wheel for 4.1 N Earth gravity is small, while the area under the wheel for 24.5 N lunar gravity is significantly larger.

This can be explained because the motion of particles increases for the same wheel slip and mass due to the reduced shear strength of the soil as a result of the reduced gravity. Meanwhile, a reduced load reduces the amount of particle motion.

The overall result and combination of these two influences is a similar area of particle motion for the nominal Earth gravity and lunar gravity results, as seen comparing Fig.s 5a) and 5b).

Furthermore, this behaviour is not limited to 33% slip ratio but can be seen across the simulation results for 0% to 50%. The ratio for traction coefficient Earth to lunar is close to one, as can be seen in Fig. 4f) with them matching or, and the ratios for tractive force and resistive torque are close to 1/6. This ratio of 1/6 is similar to the ratio for Earth to lunar gravity, 9.81 1.62 = 1/6. The tractive force and resistive torque are reduced in the lunar gravity for the same resultant motion. As the traction coefficient accounts for the difference in load, Fz, and lunar gravity is 1/6 of Earth's, this cancels the 1/6 tractive force reduction and results in the traction coefficients and tractive efficiency matching.

When the rover's performance on Earth is compared to lunar gravity, keeping mass fixed, the traction coefficient and tractive efficiency are the same. Total wheel sinkage at low slip, where lunar sinkage is lower, has a 2 mm difference at 10% slip, reducing to less than a 1 mm difference at 50% slip. This difference was not observed in comparisons of Toyoura and with gravity variation, suggesting the apparent cohesion model could be responsible.

3.3. Influence of Gravity on Slip Ratio

In addition to the fixed slip ratio considered for the experiment and replicated in simulation, the wheel can be allowed to drive the motion and the subsequent slip ratio of the wheel can be measured. The influence of gravity and load can be considered with the same four cases across 4.1, or 24.5 N and Earth or lunar gravity.

Summarised in Fig. 6, it was found that 4.1 N lunar and 24.5 N Earth have matching trends in the wheel slippage data, although 4.1 N Earth is similar on average but doesn't match the large variation of slip with 4.1 N lunar. These average values are summarised in Table 2.

Table 2
Driving wheel or free slip condition results for Earth and lunar gravity with 4.1 and 24.5 N loads

Condition	Load [N]	Slip Ratio [%]
Earth	4.1	2.6
Earth	24.5	2.3
Lunar	4.1	2.1
Lunar	24.5	12.3

When looking at the material motion in Fig. 7 at the same point, due to the matching slip, it can be seen as with fixed slip, with driving wheel motion the effect on the material motion influences the same area, there is however slight difference in the grouser trace behind the wheel with a sharper peak, likely due to the influence of cohesion under lower gravity maintain a more rigid shape. Not shown here but again the area for 24.5 N lunar is much higher and for 4.1 N Earth much lower reinforcing that gravity offload may be unsuitable.

3.4. Literature Comparison

Comparison can be made to prior literature, namely the reduced gravity experiments and DEM studies of Kobayashi, et al. (2010) and Nakashima, et al. (2014) and followed by modified Bekker-Wong analysis and discussion of the DEM results and modified Bekker-Wong study in Wong, et al. (2022). In both the experiment and simulation studies, the wheel's performance in terms of reactive force is similarly reduced by a factor of the gravity ratio, 1/6, supporting the simulation results here (Wong, 2022).

Another key study is that of Niksirat, et al. (2020), although the load force on the wheel is kept constant across the gravity level, leading to a different response. Which was performed here in simulation to see if the same behaviour occurs and clearly 24.5 N on the lunar surface gives a very different result. Ultimately, the ideal study will come from a study of rover performance on the lunar surface, which is a topic of ongoing interest with the Rashid rovers and ELM project.

Furthermore, studies such as the Kuroda, et al. (2005) paper are interesting, where the results of a fully reduced-in-scale rover were studied under reduced gravity. Alongside examples such as the granular scaling law studies presented initially by Slonaker, et al. (2017) and developments with a study in reduced gravity presented in the Daca, et al. (2023) paper, the conclusions of the Bekker-Wong predictions covered in Wong (2022) book chapter 2.11 on the application of terramechanics to the study of mobility and extraterrestrial rovers and their running gears, with reduced gravity studies and DEM simulations in Kobayashi, et al. (2010), it supports that gravity offloading of a rover alone may not be suitable.

Approaches such as the full reduced scaling of the rover or the use of granular scaling laws are required to appropriately assess a rover's performance in reduced gravity when testing under Earth gravity. However, there could be differences at high-speed motion of the wheel, where the material is highly fluidised. Depending on what aspect of performance is required to study it may be best to use the nominal load on Earth conditions to suitably predict behaviour on the lunar conditions.

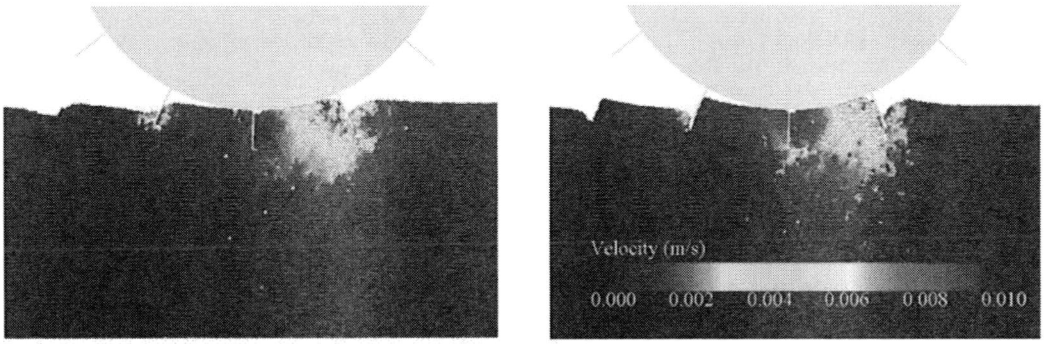

(a) 24.5 N Earth gravity (b) 4.05 N lunar gravity

Fig. 7 Comparison of material motion, specifically particle velocity magnitude, under the wheel at 24.5 N and 4.05 N under Earth and lunar gravity for FJS-1 at 16.6 seconds, with driving motion

The motion and shear strength of the granular material are affected by reduced gravity but balanced with the reduction of load force.

4. Conclusions and Future Work

Challenges in calibrating regolith material partly arise from its intermediate nature between soil and powder, affecting its bulk behaviour. Additionally, separating the roles and influence of cohesion, shape, and friction is complex and may not even be possible or necessary given the framework of DEM and current geomechanical understanding.

The study suggests avoiding gravity offloading in microrover testing, as it may overestimate performance. Simulations should also explore smaller particles' behaviour to replicate lunar conditions better. The reasoning for the notion that gravity offloading is suitable has been challenged, and this study adds to the growing evidence that the influence of reduced gravity on the granular material, as well as the rover itself, is of critical importance to a suitable study of the rover performance and likely any interaction behaviours. Reducing the weight to match that expected in the lunar environment is unsuitable, and other techniques should be used or approximated by testing on Earth. Some aspects may differ in control techniques and the direct forces involved, but the motion of the granular material will be similar at low speeds. High-speed conditions with greater fluidisation of the material may have differences.

Prior study established a validated material model and calibration process for studying the interaction of FJS-1 and Toyoura sand with rover wheels at a low-speed regime (0.02 m/s) and loose testbed packing (approximately 25% relative bulk density). Several avenues of further investigation and study can then be considered from expanding to study with variation in the testbed bulk density, in-depth study of slope or variation in the surface, developing towards full rover simulations and complex wheel designs.

Additionally, developing more complex simulations using coupling with multi-body dynamics to simulate more complex behaviour and motions. For example, suspension in the rover system and how that then affects the wheel behaviour or metal deformable wheels, which allow increased contact area and improved traction over rigid wheels. This study, alongside high speed, will be critical to crewed lunar missions and future lunar and other planetary rovers. Coupling between EDEM and Motionsolve was already performed with a study on gripper interaction with a granular surface. This could then be extended to a deformable wheel or suspension system.

Future work could include refining DEM calibration for small rovers (<10 kg), understanding high-speed interactions, and exploring agglutinate behaviour in lunar regolith simulations. Experimental setups should enable consistent bulk density control and account for external factors affecting regolith behaviour, like ionisation. More cohesive or tribocharged layers in the lunar regolith could further impact performance predictions and can be studied. In general, including agglutinate distributions in simulations may be critical for accurately modelling lunar surface interaction, as this differs from many lunar regolith simulants' compositions. The high-speed regime of wheel motion consists of an area likely above 0.2 m/s in nominal linear velocity, 10x higher than the 0.02 m/s nominal linear velocity studied in this thesis. Granular scaling laws (GSL) also indicate a differing regime and support the gravity invariance of material motion under the wheel at the low-speed regime. Further simulation studies alongside an understanding of the granular scaling laws would be beneficial to understanding differences in wheel performance due to speed and gravity variation.

Ultimately and ideally, it will be required to perform dedicated terramechanical or geotechnical experiments on the lunar surface to best understand the lunar regolith characteristics in situ and its influence on tractive performance.

5. Acknowledgements

Thank you to the members of the Space Robotics Lab for the support and facilities to continue this research.

6. Declaration of competing interest

The authors declare that they have no known competing financial interests or personal relationships that could have appeared to influence the work reported in this paper.

7. References

Beakawi Al-Hashemi, H.M., Baghabra Al-Amoudi, O.S., 2018. A review on the angle of repose of granular materials. Powder Technology, 330, 397–417. https://doi.org/10.1016/j.powtec.2018.02.003

Bui, H.H., Kobayashi, T., Fukagawa, R., Wells, J.C., 2009. Numerical and experimental studies of gravity effect on the mechanism of lunar

excavations. Journal of Terramechanics, 46(3), 115–124. https://doi.org/10.1016/J.JTERRA.2009.02.006
Coetzee, C.J., Scheffler, O.C., 2023. Review: The Calibration of DEM Parameters for the Bulk Modelling of Cohesive Materials. Processes, 11(1). https://doi.org/10.3390/pr11010005
Daca, A., Tremblay, D., Skonieczny, K., 2021. The Relationship between Cone Penetration Resistance and Wheel-Soil Interactions in Lunar Gravity. IEEE Aerospace Conference Proceedings, 2021-March. https://doi.org/10.1109/AERO50100.2021.9438503
Daca, A., Tremblay, D., Skonieczny, K., 2023. Expansion and Experimental Evaluation of Scaling Relations for the Prediction of Wheel Performance in Reduced Gravity. Microgravity Science and Technology, 35(6), 59. https://doi.org/10.1007/s12217-023-10087-4
Ding, L., Gao, H., Deng, Z., Nagatani, K., Yoshida, K., 2011. Experimental study and analysis on driving wheels' performance for planetary exploration rovers moving in deformable soil. Journal of Terramechanics, 48(1), 27–45. https://doi.org/10.1016/J.JTERRA.2010.08.001
Heiken, G.H., Vaniman, D.T., French, B.M., 1991. Lunar Sourcebook. Cambridge University Press. https://www.lpi.usra.edu/publications/books/lunar_sourcebook/
Gu, X., Yang, J., Huang, M., 2013. Laboratory measurements of small strain properties of dry sands by bender element. Soils and Foundations, 53(5), 735–745. https://doi.org/https://doi.org/10.1016/j.sandf.2013.08.011
Hofmeister, P.G., Blum, J., Heißelmann, D., 2009. The Flow Of Granular Matter Under Reduced-Gravity Conditions. AIP Conference Proceedings, 1145(1), 71–74. https://doi.org/10.1063/1.3180028
Hu, C., Gao, J., Song, X., Zhang, M., Tan, X., 2021. Analytical modeling and DEM analysis of soil–wheel interaction under cornering and skidding conditions in off-road vehicles. AIP Advances, 11(8), 085122. https://doi.org/10.1063/5.0057046
Hurrell, J., Takehana, K., Tanaka, T., Uno, K., Busoud, A.K., Yoshida, K., 2025. Traction Performance Evaluation for a Rashid-1 Rover Wheel. Space Science Reviews. https://doi.org/10.1007/s11214-025-01164-8
Hurrell, J., Takehana, K., Uno, K., Yoshida, K., 2023. Lunar rover discrete element method study and calibration. Proceedings of the 16th European-African Regional Conference of the ISTVS. https://doi.org/10.56884/XUAC7188
Wong, J.Y., 2022. Mechanics of Vehicle–Terrain Interaction. In Theory of Ground Vehicles (pp. 77–219). https://doi.org/https://doi.org/10.1002/9781119719984.ch2
Kanamori, H., Udagawa, S., Yoshida, T., Matsumoto, S., Takagi, K., 1998. Properties of Lunar Soil Simulant Manufactured in Japan. Space 1998, 462–468. https://doi.org/10.1061/40339(206)53
Karapiperis, K., Marshall, J. P., Andrade, J. E., 2020. Reduced Gravity Effects on the Strength of Granular Matter: DEM Simulations versus Experiments. Journal of Geotechnical and Geoenvironmental Engineering, 146(5), 06020005. https://doi.org/10.1061/(ASCE)GT.1943-5606.0002232/SUPPL_FILE/SUPPLEMENTAL_DATA_GT.1943-5606.0002232_KARAPIPERIS.ZIP
Kleinhans, M.G., Markies, H., de Vet, S.J., In't Veld, A.C., Postema, F.N., 2011. Static and dynamic angles of repose in loose granular materials under reduced gravity. Journal of Geophysical Research: Planets, 116(E11), 11004. https://doi.org/10.1029/2011JE003865
Kobayashi, T., Fujiwara, Y., Yamakawa, J., Yasufuku, N., Omine, K., 2010. Mobility performance of a rigid wheel in low gravity environments. Journal of Terramechanics, 47(4), 261–274. https://doi.org/https://doi.org/10.1016/j.jterra.2009.12.001
Lommen, S., Schott, D., Lodewijks, G., 2014. DEM speedup: Stiffness effects on behavior of bulk material. Particuology, 12(1), 107–112. https://doi.org/10.1016/J.PARTIC.2013.03.006
Matsushima, T., Ishikawa, T., 2014. Particle Grading Effect on Mechanical Properties of Lunar Soil Simulant FJS-1. Earth and Space 2014, 60–68. https://doi.org/10.1061/9780784479179.008
Nakanishi, R., Nakashima, H., Miyasaka, J., Ohdoi, K., 2020. Tractive performance analysis of a lugged wheel by open-source 3D DEM software. Journal of Terramechanics, 92, 51–65. https://doi.org/https://doi.org/10.1016/j.jterra.2020.10.001
Nakashima, H., Kobayashi, T., 2014. Effects of gravity on rigid rover wheel sinkage and motion resistance assessed using two-dimensional discrete element method. Journal of Terramechanics, 53(1), 37–45. https://doi.org/10.1016/J.JTERRA.2014.03.004
Nakashima, H., Shioji, Y., Kobayashi, T., Aoki, S., Shimizu, H., Miyasaka, J., Ohdoi, K., 2011. Determining the angle of repose of sand under low-gravity conditions using discrete element method. Journal of Terramechanics, 48(1), 17–26. https://doi.org/10.1016/J.JTERRA.2010.09.002
Niksirat, P., Daca, A., Skonieczny, K., 2020. The effects of reduced-gravity on planetary rover mobility. The International Journal of Robotics Research, 39(7), 797–811. https://doi.org/10.1177/0278364920913945
Ono, I., Nakashima, H., Shimizu, H., Miyasaka, J., Ohdoi, K., 2013. Investigation of elemental shape for 3D DEM modeling of interaction between soil and a narrow cutting tool. Journal of Terramechanics, 50(4), 265–276. https://doi.org/10.1016/J.JTERRA.2013.09.001
O'Sullivan, C., 2011. Particulate Discrete Element Modelling. CRC Press. https://doi.org/10.1201/9781482266498
Ozaki, S., Ishigami, G., Otsuki, M., Miyamoto, H., Wada, K., Watanabe, Y., Nishino, T., Kojima, H., Soda, K., Nakao, Y., Sutoh, M., Maeda, T., Kobayashi, T., 2023. Granular flow experiment using artificial gravity generator at International Space Station. Npj Microgravity 2023 9:1, 9(1), 1–13. https://doi.org/10.1038/s41526-023-00308-w
Reiss, P., Hager, P., Hoehn, A., Rott, M., Walter, U., 2014. Flowability of lunar regolith simulants under reduced gravity and vacuum in hopper-based conveying devices. Journal of Terramechanics, 55, 61–72. https://doi.org/10.1016/J.JTERRA.2014.04.005
Rodríguez-Martínez, D., van Winnendael, M., Yoshida, K., 2019. High-speed mobility on planetary surfaces: A technical review. Journal of Field Robotics, 36(8), 1436–1455. https://doi.org/10.1002/ROB.21912
Slonaker, J., Motley, D.C., Zhang, Q., Townsend, S., Senatore, C., Iagnemma, K., Kamrin, K., 2017. General scaling relations for locomotion in granular media. Physical Review E, 95(5), 52901.
Takehana, K., Kizaki, S., Tanaka, T., Uno, K., Yoshida, K., 2025. Comparison of lunar rover wheel performance in soils with different cohesive properties. Journal of Terramechanics, 117, 101011. https://doi.org/10.1016/J.JTERRA.2024.101011

6775 | Modeling and simulation of autonomous vehicular systems at ERDC - I Software integration laboratory: development and applications

Peilin Song *, Burhman Gates, Garrett Holden, Brent Towne, Joshua Fairley

Mobility System Branch, Engineer Research and Development Center, Vicksburg, MS, USA

* *Corresponding author: peilin.song@erdc.dren.mil*

ABSTRACT

Unmanned Ground Vehicles (UGVs) are becoming critical assets for the Army, offering enhanced capabilities for mission execution. However, the complexity of autonomous systems presents challenges that traditional analysis tools are ill-equipped to handle. To address this, ERDC has developed a Software Integration Laboratory (SIL) to support the modeling and simulation (M&S) of UGVs in virtual environments. The SIL provides a digital testbed for performance prediction and risk mitigation across both typical and edge-case scenarios.

As part of ongoing test and evaluation (T&E) endeavors, ERDC's SIL supports a wide range of activities that complement physical field trials. This includes the creation of high-fidelity virtual environments that integrate vehicle dynamics, terramechanics, terrain models, and sensor simulations. These environments enable realistic testing of autonomous navigation systems—such as the Robotic Technology Kit (RTK) and its variants—in off-road and contested conditions.

This paper highlights the design and capabilities of ERDC's SIL and its application to autonomous ground vehicle testing and development. It demonstrates how digital experimentation using the SIL supports the integration of autonomous systems into manned-unmanned teams and helps inform future mission planning. The paper also presents the application of the SIL in various projects, including both single autonomous vehicle operations and the coordinated motion of multiple vehicles in team-based scenarios.

Keywords
UGV
Autonomous Systems
Modeling and Simulation
Software Integration Laboratory
Test and Evaluation
Manned-Unmanned Teaming

1. Introduction

Modeling and Simulation (M&S) plays a critical role in advancing autonomous driving systems, especially in military contexts where vehicles often operate in unstructured, off-road, or uncharted environments. By enabling extensive virtual testing, M&S significantly reduces development costs and time by minimizing the need for expensive field trials. It allows engineers to simulate complex terrains, adverse weather, and mission-specific scenarios that are difficult or dangerous to replicate in real life—helping validate performance under extreme conditions. This is particularly valuable for military applications, where encountering rare edge cases like improvised obstacles or unexpected terrain features is common. Through high-fidelity simulations, developers can identify and correct system vulnerabilities early, mitigating operational and safety risks. As a result, M&S accelerates deployment, enhances system reliability, and supports mission readiness while maintaining a strong emphasis on safety and cost-efficiency.

ERDC has been engaged in the research and development of M&S of autonomous vehicles fort a few decades to support the test and evaluation (T&E). In this paper, we outline ERDC 's endeavors in development of Software Integration Laboratory (SIL) to support the evaluation of navigation capabilities of autonomous vehicles for military applications. We present the development of simulation components such as terrain model, sensor model, vehicle dynamics and their integration into a simulation platform enabling plug-N-play of navigation stack to be evaluated. The paper ends with some examples of practical applications in supporting T&E of autonomous vehicles.

This is the first of the three papers presented. In this conference. The second paper (Song 2025a) addresses the soft soil modeling and the third one, (Song 2025b) presents the verification and validation efforts at both system and component level.

2. Background and Related Works

M&S of autonomous driving is not new in research; it is also prevailing in industry. Of the notables, Waymo and Tesla are forerunners in M&S development and applications. Waymo (2021) developed a virtual environment *Simulation* City to its vehicles' interactions with various road users and scenarios. This platform allows for the simulation of complex and rare situations in a controlled setting. Tesla (Toolify.ai, 2023) builds the complex urban environments like San Francisco to refine their navigation stack.

While simulation technologies developed by civilian companies such as Tesla and Waymo represent significant advancements in autonomous vehicle (AV) development, they fall short of meeting the rigorous Test and Evaluation (T&E) requirements necessary for military applications. These civilian platforms are typically designed around well-mapped, structured environments—such as urban roads and highways—and are supported by vast amounts of real-world driving data. This combination ensures high-fidelity simulation tools that are highly effective for civilian use, enabling accurate modeling of traffic behaviors, road interactions, and regulatory compliance. However, this same reliance on structured environments and extensive datasets limits their applicability to military contexts, where AVs must operate in unstructured, GPS-denied, and often unmapped terrains. Moreover, the scarcity of operational data in such domains makes it challenging to generate representative scenarios. As a result, while civilian M&S excels in predictable,

https://doi.org/10.56884/N1CGA6U2

controlled settings, it lacks the adaptability and resilience required to evaluate systems in the complex, dynamic, and high-risk conditions typical of defense operations.

To address the limitations of civilian simulation tools in meeting military Test and Evaluation (T&E) requirements, the U.S. Army Engineer Research and Development Center (ERDC) has developed a specialized Modeling and Simulation (M&S) environment tailored for autonomous ground systems in defense contexts. Instead of relying on pre-mapped, structured environments and large volumes of real-world driving data—as is common in civilian platforms—ERDC's environment is designed to overcome data scarcity and operational unpredictability. It incorporates high-fidelity, physics-based models capable of generating accurate simulated data independent of existing infrastructure. Additionally, the platform enables extensive simulation of diverse edge cases, including degraded environmental conditions such as fog, heavy rain, and snow, as well as off-road terrain variability, GPS denial, sensor failures, and adversarial interference. This comprehensive scenario coverage ensures that autonomous systems are rigorously tested and validated under the full spectrum of challenges expected in real-world military operations.

This paper presents the design, implementation, and application of a high-fidelity Modeling and Simulation (M&S) environment developed by the U.S. Army Engineer Research and Development Center (ERDC) to support the testing and evaluation of autonomous ground systems in military contexts. Chapter 2 provides an overview of the overall M&S architecture, highlighting its goals, core design principles, and system-level capabilities. Chapter 3 delves into the key components that compose the environment, including terrain modeling, physics-based simulation engines, sensor emulation, and scenario orchestration tools. Finally, Chapter 4 illustrates how this M&S environment is applied to evaluate autonomous vehicle performance across a range of mission-critical scenarios, including off-road navigation, degraded environmental conditions, and GPS-denied operations. Together, these chapters demonstrate the environment's utility in bridging critical gaps left by civilian simulation technologies and ensuring readiness for real-world military deployment.

3. Simulation Integrated Lab Architecture for Autonomous Navigation Testing

The Modeling and Simulation (M&S) environment, structured as a Simulation Integrated Lab (SIL), is a modular and extensible framework designed to support the development, testing, and evaluation of autonomous ground systems under realistic operational conditions. The primary purpose of the SIL is to test and evaluate the navigation capability of autonomous vehicles; therefore, the navigation stack—including perception, localization, planning, and control algorithms—is not an inherent part of the SIL itself but is instead integrated externally for evaluation within the simulated environment.

At its core, the SIL integrates three principal components: a vehicle dynamics module, terrain modeling engine, and a sensor simulation suite. The vehicle dynamics module employs high-fidelity physics-based models to accurately represent the mechanical behavior of various vehicle platforms, including powertrains and tire-terrain interactions. This enables realistic simulation of mobility and then, control responses across diverse driving conditions.

The Terrain Modeling Module provides the environmental context in which autonomous vehicles are evaluated, enabling the simulation of complex, real-world operating conditions. This module is responsible for generating high-resolution terrain representations that capture both natural and man-made features relevant to autonomous mobility and navigation tasks.

To support realistic testing, ERDC has conducted extensive surveys of the U.S. Army's major test sites using advanced geospatial technologies. These surveys have produced high-density mesh terrain models that accurately capture the geometric and physical characteristics of diverse test environments. The resulting digital terrain data includes rich details such as elevation changes, surface roughness, buildings, vegetation, water bodies, and other structural features commonly encountered in military operations.

The sensor simulation module emulates the behavior of common autonomous vehicle sensors such as LiDAR, radar, EO/IR cameras, GPS, and IMUs. These simulated sensors include realistic physical effects like noise, occlusion, latency, and environmental degradation from conditions such as fog, rain, and snow. This ensures accurate perception emulation for validating the navigation stack under a wide range of operational scenarios.

These components are synchronized within a unified simulation loop, allowing for repeatable, scalable testing of autonomy algorithms across normal and edge-case scenarios.

The initial implementation of the SIL is deployed on a cluster of networked desktop machines connected via a Local Area Network (LAN), providing a flexible and scalable development environment. As the system matures and demands for higher computational throughput grow—particularly for running large-scale, high-fidelity simulations—the SIL architecture is designed to be ported to high-performance computing platforms, including supercomputers, to support more advanced experimentation and analysis.

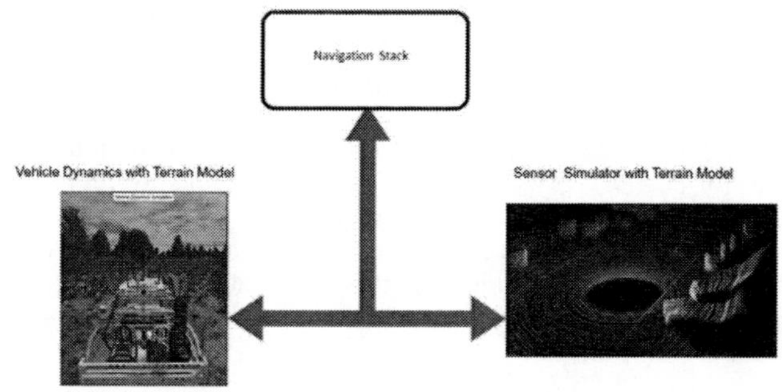

Fig. 1. Simulation Integrated Laboratory Diagram

Figure 1 shows the diagram of the SIL. The vehicle dynamics with terrain model embedded simulate and generate the vehicle position and orientation, which are fed to sensor simulator.

The sensor simulator updates the positions and orientations of various sensors simulated, calculates the physics of sensors and send the simulated signals via different protocols to navigation stack for signal fusion, cost map generation and path planning. Eventually, the planned motion is turned into the throttle and steering signals, which are sent to vehicle dynamics model for real-time vehicle update.

4. Implementation of Components of the SIL

This chapter presents the key software components used to build and operate the Simulation Integrated Lab (SIL) for testing autonomous ground vehicle navigation. Each component is selected for its high-fidelity modeling capability and ability to integrate into a synchronized simulation framework. The main elements of the SIL include:

- A Vehicle Dynamics Module implemented using Unreal Engine's physics engine.
- A Terrain Modeling Tool developed with ERDC's in-house VEGAS system.
- A Sensor Simulation Framework built around the Environment Sensor Engine (ESE).

These components work in concert to produce a realistic, physics-aware simulation environment suitable for evaluating perception, decision-making, and control algorithms in both structured and unstructured operational settings.

4.1. Vehicle Dynamics for Mobility Analysis in Unreal Engine

A key enabler of realistic autonomous ground vehicle simulation is the accurate modeling of vehicle-terrain interaction, commonly referred to as terramechanics. This field encompasses the study of how terrain properties—such as soil strength, surface roughness, and moisture—affect vehicle behavior, including traction, sinkage, rolling resistance, and mobility degradation. Terramechanics modeling is essential for analyzing vehicle performance in off-road and unstructured environments where terrain unpredictability poses operational challenges.

At the core of the SIL's terramechanics implementation is the Vehicle-Terrain Interaction (VTI) model, developed by the U.S. Army Engineer Research and Development Center (ERDC). ERDC has led terramechanics research for several decades, and its findings have been systematically captured in the VTI framework. This model serves as the authoritative foundation for predicting vehicle mobility in complex terrain (Fig. 2).

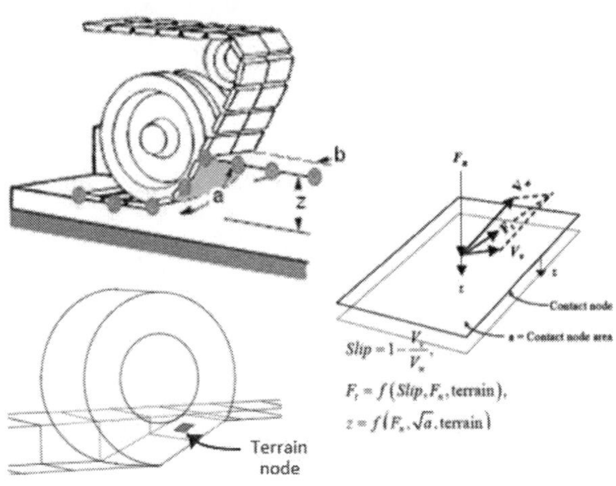

Fig. 2. Vehicle-Terrain Interaction Model

The VTI model quantitatively predicts key metrics such as:

- Traction forces
- Soil resistance
- Wheel or track sinkage

These predictions are based on a range of input parameters including:

- Soil type and geophysical properties
- Weather conditions (e.g., wet/dry, snow)
- Vehicle mass and geometry
- Configuration of tires or tracks
- Slip ratios and contact pressure

The VTI model ensures high-fidelity, physics-based mobility predictions, making it a critical element for simulation-based testing and evaluation (T&E) of autonomous systems in military contexts.

To bring these analytical capabilities into a real-time, interactive simulation setting, the VTI model is implemented within Unreal Engine, which serves as the platform for the vehicle dynamics module. The details of mobility analysis with VTI in Unreal will be discussed in another paper (Song, 2025a). The vehicle dynamics model in Unreal receives throttle and steering commands from the navigation stack to drive the vehicle(s) during simulation runs. It continuously updates the vehicle's position and orientation, which are sent to the sensor simulation module to ensure accurate and synchronized sensor outputs. This feedback loop is essential for evaluating closed-loop navigation performance in realistic environments.

4.2. Terrain Modeling and Environmental Features

Realistic simulation of autonomous vehicle behavior requires not only accurate vehicle dynamics but also high-fidelity terrain environments that reflect the complexity and variability of real-world operational areas. To support rigorous test and evaluation (T&E), the SIL includes a suite of detailed terrain models, developed specifically to replicate several of the U.S. Army's major test sites.

These terrain models are constructed using high-resolution spatial data, with grid resolutions of less than 25 cm, allowing for precise representation of micro-terrain features such as:

- Surface roughness and soil variation;
- Small obstacles and ditches;
- Ruts, erosion patterns, and irregularities that affect vehicle traction and stability.

In addition to raw elevation data, the terrain models incorporate rich environmental features, including:

- Buildings and man-made structures, accurately placed and dimensioned.
- Vegetation layers, such as grass, brush, and complex tree canopies.
- Trees modeled individually with biologically plausible geometry and density.

To further enhance realism, a tree growth model is employed during scene generation. This model simulates natural growth patterns over time, resulting in more authentic forest distributions and vegetation layering. This enables simulation of not only current conditions but also projected or degraded environments for mission planning and long-term system performance analysis.

These comprehensive terrain representations are essential for capturing the operational nuances of off-road navigation and for introducing variability into edge-case simulations. They also provide consistent geospatial context for the Vehicle-Terrain Interaction (VTI) model and sensor simulation components, ensuring coherent multi-domain simulation.

To streamline the process of building high-fidelity simulation environments, the U.S. Army Engineer Research and Development Center (ERDC) developed a toolset called Virtual Environment Generation Automation Scripts (VEGAS). VEGAS enables automated and scalable creation of complex virtual scenes, significantly reducing the manual effort required to build and maintain terrain and environmental models for simulation.

VEGAS is designed to:

- Ingest and process geospatial and survey data from diverse sources.
- Fuse multiple data layers, including elevation, imagery, vegetation, building footprints, and infrastructure.
- Generate simulation-ready terrain and asset files for both the Unreal Engine and the sensor simulation environment.
- Support quick modification and regeneration of scenes to reflect different mission requirements or environmental conditions.

Through VEGAS, ERDC can rapidly produce scene files that maintain the resolution and fidelity of manually created environments, while enabling automation at scale. The tool also supports integration with the tree growth model, allowing dynamic vegetation scenarios to be represented in different seasons or climate conditions. Figure 3 illustrates the process to apply the vegetation from LIDAR signal to a scene.

By coupling high-resolution input data with procedural generation capabilities, VEGAS ensures that terrain models are both physically representative and simulation-ready, enabling a continuous pipeline from field data to synthetic test environments.

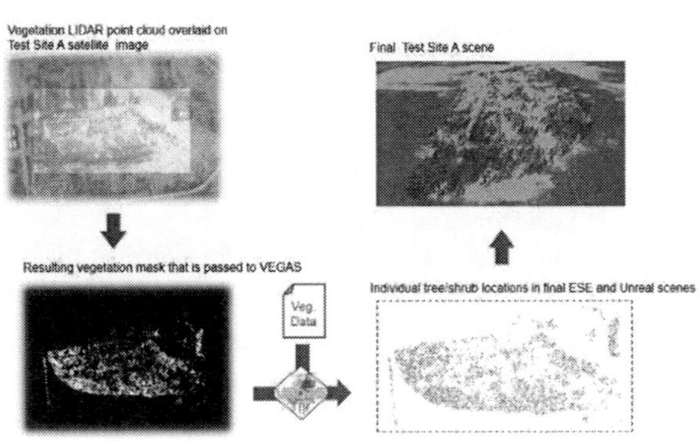

Fig. 3. VEGAS Vegetation Placement

5. Environment Sensor Engine (ESE)

The Environment Sensor Engine (ESE) is a core component of the Simulation Integrated Lab (SIL), responsible for producing high-fidelity, physics-informed synthetic sensor data. ESE plays a critical role in the testing, evaluation, and training of autonomous systems by replicating how real-world sensors perceive complex and dynamic environments.

ESE is designed to operate over the same high-resolution terrain and virtual environments generated by VEGAS and visualized in Unreal Engine, providing a tightly integrated pipeline from physical modeling to perception simulation.

Key Capabilities of ESE

- Realistic Sensor Simulation:

  ESE uses sophisticated ray tracing and casting technologies to simulate various sensor modalities, embedding physics-based models that replicate real-world interactions between electromagnetic waves and the environment. Supported sensors include:
  1. Cameras (RGB, Hyperspectral, Stereo)
  2. LiDAR
  3. Radar
  4. GPS and IMUs
- Environmental Fidelity:

  ESE integrates environmental effects that impact sensor performance in the real world, such as rain, snow, fog and dust

- High-Performance Computing & Parallelism:

  To support large-scale and high-fidelity simulation, ESE is built for parallel computing architectures. It leverages multi-threading and GPU acceleration on local clusters, and utilizes Message Passing Interface (MPI) on high-performance computing (HPC) platforms to scale simulations across multiple nodes for faster execution and batch processing.

Figure 4 show a comparison of simulated and actual pitures.

Fig. 4. Pictures (Simulated-Left, Actual-Right)

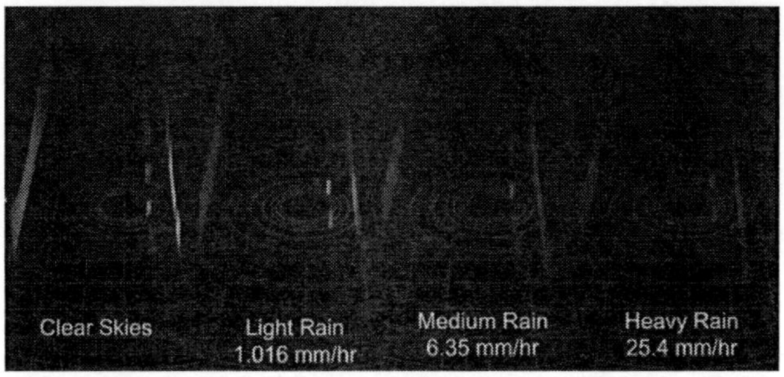

Fig. 5. Simulated Lidar Point cloud (Clear Sky vs Rains)

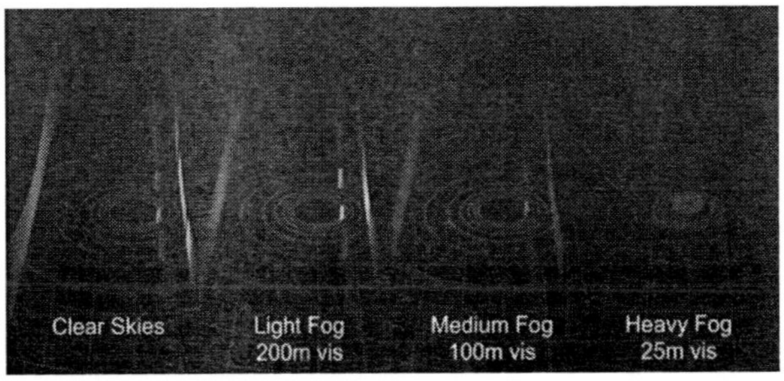

Fig. 6. Simulated Lidar Point cloud (Clear Sky vs Fog)

Figures 5 and 6 show the weather effects on the simulated lidar reflections. It is worth noting that the in Fig. 6 that the donut-shape artifact is shown, which is well known in in sensor physics and autonomous vehicle testing Liu (2003).

6. Applications and Conclusions

The Simulation Integrated Lab (SIL) has proven to be a versatile and powerful tool for evaluating autonomous systems under a wide range of operational scenarios. Built with high-fidelity terrain modeling, advanced vehicle dynamics, and realistic sensor simulation, the SIL environment enables rapid, cost-effective, and repeatable testing that is difficult or impossible to achieve in the physical world.

6.1. Applications Across Programs

The SIL has been successfully deployed across multiple government-sponsored programs, supporting both single-vehicle autonomy and multi-vehicle convoy operations. These applications have demonstrated the environment's flexibility and effectiveness in modeling realistic mission conditions, including:

- Navigation in complex off-road terrain
- Obstacle avoidance in unstructured environments
- Convoy behavior under GPS-denied conditions
- Sensor degradation in adverse weather (e.g., fog, rain, and snow)
- Edge-case testing, including degraded terrain, occlusions, and dynamic obstacles

By allowing rapid iteration and stress-testing of navigation stacks, SIL has accelerated development timelines while reducing reliance on costly and logistically intensive live testing.

6.2. Understanding Complex Vehicle Behaviors

In addition to validating autonomy, the SIL environment has played a key role in diagnosing and understanding unexpected vehicle behaviors. Using high-fidelity simulation, developers have been able to demystify strange phenomena, such as:

- Rattling gear shifts under varying traction conditions
- Overly aggressive throttle or braking behaviors in adverse weather
- Navigation anomalies due to sensor interference or poor terrain feedback

By recreating these behaviors virtually and modifying terrain, environmental, or vehicle parameters, engineers gain insight into root causes and can fine-tune control algorithms without risking physical assets. We will discuss more in Song (2025b)

6.3. Simulation Benefits Realized

Several concrete benefits have emerged from SIL-based testing:

- Risk mitigation through pre-mission validation of autonomy software
- Reduction in cost and time compared to live testing, especially for edge and failure cases
- Repeatability of complex scenarios, enabling fair benchmarking and debugging
- Scalability, with support for extensive batch simulations and multi-vehicle interactions

6.4. Conclusion

The SIL environment developed by ERDC represents a mature and scalable simulation platform for autonomous vehicle research. Its successful use in both single-vehicle and convoy autonomy programs has validated its design goals and affirmed its importance in modern autonomy development workflows.

As autonomy continues to evolve, SIL will remain a critical asset for modeling, simulation, and evaluation, enabling the Department of Defense and research communities to develop more robust, capable, and resilient autonomous systems for complex and unstructured environments.

7. References

Gulino, C., Fu, J., Luo, W., Tucker, G., Bronstein, E., Lu, Y., Harb, J., Pan, X., Wang, Y., Chen, X., Co-Reyes, J.D., Agarwal, R., Roelofs, R., Lu, Y., Montali, N., Mougin, P., Yang, Z., White, B., Faust, A., McAllister, R., Anguelov, D., Sapp, B., 2023. Waymax: An Accelerated, Data-Driven Simulator for Large-Scale Autonomous Driving Research. arXiv.2310.08710v1. https://doi.org/10.48550/arXiv.2310.08710

Li, Y., Duthon, P., Michele Colomb, M., Ibanez-Guzman, J., 2003. What happens to a ToF LiDAR in fog? arXiv.2003.06660v4. https://arxiv.org/pdf/2003.06660

Song, P., Gates, B., Holden, G., Fairley, J., 2025a. Modeling and Simulation of Autonomous Vehicular Systems at ERDC - II, Soft Soil Model Support in Vehicle Dynamics, 55th Conference of the International Society for Terrain-Vehicle Systems, October 6–9, Lebanon, New Hampshire, USA

Song, P., Gates, B., Holden, G., Fairley, J., 2025b. Modeling and Simulation of Autonomous Vehicular Systems at ERDC-III, Verification and Validation of M&S for Autonomous Vehicles, 55th Conference of the International Society for Terrain-Vehicle Systems, October 6–9, Lebanon, New Hampshire, USA.

Toolify.ai, 2023. Experience Tesla's Self-Driving Car Simulation Learning, https://www.toolify.ai/ai-news/experience-teslas-selfdriving-car-simulation-learning-28863, Dec.

Waymo Team, 2021. Simulation City: Introducing Waymo's most advanced simulation system yet for autonomous driving, https://waymo.com/blog/2021/07/simulation-city, July.

6905 | Investigating data collection and processing strategies for UAV-based terrain models in vehicle dynamics simulation

Herman A. Hamersma *, J. Christian van Aswegen, A. Glenn Guthrie, P. Schalk Els, Carl M. Becker

Department of Mechanical and Aeronautical Engineering, University of Pretoria, Pretoria, South Africa

* *Corresponding author: hermanh@up.ac.za*

ABSTRACT

Accurate modeling of tire–terrain interaction is critical for effective vehicle dynamics simulations, particularly in off-road environments. A previous study demonstrated the feasibility of generating three-dimensional (3D) terrain models from images captured by an unmanned aerial vehicle (UAV) and identified several areas for further investigation—chief among them the influence of UAV flight altitude and image overlap ratio. This study investigates these parameters, along with the impact of the interpolation resolution used to regularize point cloud data.

New UAV imagery of a Belgian paving test track was collected, and baseline measurements from a traditional mechanical profilometer were used for direct comparison. While constructing 3D terrain models from UAV data is relatively straightforward, our findings highlight several critical factors – such as flight altitude, image overlap, and interpolation method – that can significantly affect model accuracy. Based on these results, we propose a practical, step-by-step approach for developing accurate terrain models suitable for use in vehicle dynamics simulations.

The UAV-based technique is broadly applicable, from smooth man-made surfaces (e.g., ISO 8608:2016 Class C roads) to rough, natural off-road terrains, enabling new opportunities for mobility research across a wide range of challenging environments.

Keywords
Terrain modelling
Road profiling
Photogrammetry
UAV surveying
Off-road mobility
Vehicle dynamics simulation

1. Introduction

In our previous study (Hamersma et al., 2024), we explored the feasibility of using an unmanned aerial vehicle (UAV) equipped with a wide-lens camera to generate full three-dimensional (3D) terrain models for vehicle dynamics simulations. This UAV-based measurement technique was benchmarked against data obtained using a traditional mechanical terrain profilometer – the Can-can machine (Becker and Els, 2014) – on the same test track, enabling a direct comparison. The UAV approach offered several advantages: rapid data acquisition, non-contact operation (eliminating integration drift on uneven terrain), and minimal disturbance to soft surfaces.

That initial investigation focused on a Belgian paving test track at the Gerotek Test Facilities (GTF) in South Africa (Armscor Defence Institutes SOC, n.d.). The Belgian paving test track is approximately 100 m long and 4 m wide, and consists of ±200 mm by ±130 mm cobblestones with gaps of ±35 mm between the cobblestones. Orthographic and oblique aerial images were captured to construct a 3D terrain model, with variations in flight altitude caused by obstacles around the test track, such as vegetation. Somewhat unexpectedly, orthographic imaging produced more accurate terrain models than oblique imaging. However, that study used default image overlap settings across all surveys and did not vary flight parameters systematically.

The present study builds on this foundation by examining how key UAV survey parameters and post-processing – namely flight altitude, image overlap ratio, and interpolation resolution – influence the accuracy of the resulting 3D terrain models. As before, the Belgian paving track at GTF is used as the test site, and UAV results are compared against baseline data from the Can-can machine.

2. Methodology

The methodology consists of six steps:

1. Survey the terrain with an UAV.
2. Develop the 3D model as a point cloud.
3. Post-process the point cloud to crop unwanted areas and ensure correct orientation.
4. Regularize the point cloud into an evenly spaced grid through interpolation.
5. Calculate the displacement power spectral density (PSD) as described by ISO 8608:2016 (International Organization for Standardization, 2016) in the longitudinal direction over the width of the terrain and determine the average displacement PSD.
6. Compare the various displacement PSDs with the Can-can data set.

Aerial images of the Belgian paving test track were collected at three different altitudes (12.2 m, 21.3 m and 30.5 m above ground level) and for three image overlap ratios (50 %, 70 % and 90 %), resulting in nine different data sets of the Belgian paving. The image overlap ratios were the same in the longitudinal and the lateral directions.

2.1. Surveying the terrain

A DJI Mavic 3 Enterprise UAV equipped with the DJI Real-Time Kinematics (RTK) module and DJI D-RTK 2 High-precision GNSS

https://doi.org/10.56884/2I9ZL94A

Table 1
DJI Mavic 3E specifications (DJI, 2024)

Aircraft specifications	
Max flight speed (at sea level, no wind)	15 m/s (Normal Mode)
Max wind speed resistance	12 m/s
Max take-off altitude above sea level	6000 m (without payload)
Max flight time (no wind)	45 min
Max flight distance	32 km
GNSS	GPS + Galileo + BeiDou + GLONASS
Hovering accuracy	Vertical: ±0.1 m (with RTK) Horizontal: ±0.1 m (with RTK)
Camera specifications (wide-angle lens)	
Sensor	4/3 CMOS Effective pixels: 20 MP
Lens	FOV: 84° Format Equivalent: 24 mm Aperture: f/2.8-f/11 Focus: 1 m to ∞
ISO range	100-6400
Shutter speed	Electronic Shutter: 8-1/8000 s Mechanical Shutter: 8-1/2000 s
Max image size	5280 × 3956 pixels
Gimbal	
Stabilization	3-axis (tilt, roll, pan)
Mechanical range	Tilt: -135° to 100° Roll: -45° to 45° Pan: -27° to 27°
RTK module	
RTK Positioning Accuracy	Horizontal: 1 cm + 1 ppm Vertical: 1.5 cm + 1 ppm

mobile base station is used for this investigation. The system specifications are documented in Table 1 (DJI, 2024)

An area route is created with the DJI Pilot 2 application on the DJI RC Pro remote controller. A mapping area is drawn that encompasses the area of interest. In this study, the exact same mapping area was used for all survey flight missions. The appropriate aircraft and camera are selected (in this case the DJI Mavic 3 Enterprise can only collect data with the wide-angle lens). Only orthographic aerial images were collected, based on the findings of our previous study. Figure 1 shows a typical flight route for a survey mission and Fig. 2 shows a typical aerial image collected during a survey.

2.2. Creating point clouds

Once the images are retrieved from the UAV, photogrammetry is used to create a 3D point cloud. RealityCapture (Epic Games, 2025) is used to create the point clouds. Each image captured with the UAV contains information about the UAV position and heading and details on the camera settings in its metadata. Reality Capture uses this information to remove lens distortion and to construct the 3D point cloud using photogrammetry techniques. The user can specify the detail of the point cloud, e.g. by limiting the number of points, etc. For the purpose of this investigation, the highest detail was selected. The point cloud is exported as a *.las object. Figure 3 shows an example of a terrain model in RealityCapture.

2.3. Postprocessing the point cloud

The next step is to clean-up the point cloud before reorganizing it into a regular grid. CloudCompare (2021), a 3D point cloud and mesh processing software (an Open Source Project used under a GNU General Public License) is used during this step. The point cloud is cropped to the area of interest. The segmented point cloud is exported in *.pcd format.

Fig. 1. An example of the flight routes for a mapping area when collecting images

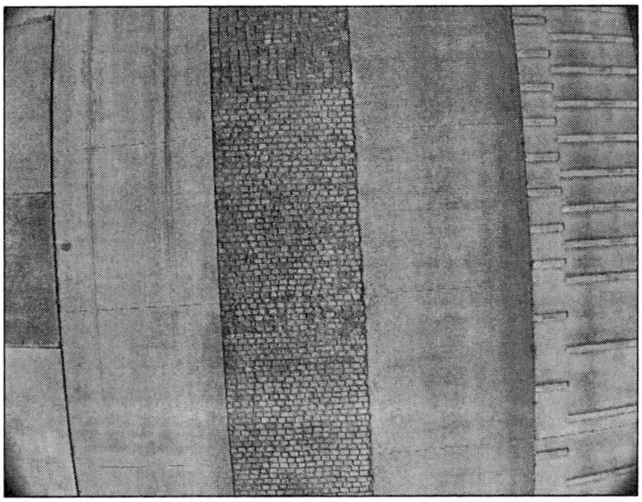

Fig. 2. Orthographic aerial image of the Belgian paving

Fig. 3. Terrain model created in RealityCapture (Epic Games, 2025)

The next step is to remove any unwanted rotations from the point cloud. As can be seen in Fig. 2, the Belgian paving is flanked on all sides by a smooth and level concrete road surface. This road surface was used as the datum plane. Any rotation from the measurement technique is removed with the following steps (note that these steps are only applicable to straight-line tracks and have to be developed further for curved roads):

1. The terrain model is manually cropped to the region of interest. Care is taken to include a small portion of the surrounding concrete surface that will be used as a datum plane.
2. Three points on the corners of the data set are manually picked ($\mathbf{P_1}, \mathbf{P_2}$ and $\mathbf{P_3}$). Note that each point is represented by x, y, z-coordinates.
3. The coordinates of the three points are used to define two vectors ($\vec{v}_1$ and $\vec{v}_2$) with Eq. 1. These two vectors are assumed to lie in the datum plane (the x-y plane in a typical vehicle coordinate system). By determining the cross product of the two vectors, the normal vector ($\vec{u}$) of the plane is determined (see Eq. 2).
4. The rotation axis is calculated by determining the cross product of the normal vector of the plane with the vertical unit vector ($\hat{k}$) (see Eq. 3).
5. The rotation angle about the rotation axis is determined by performing the dot product of the plane normal vector and the vertical unit vector (see Eq. 4)
6. The point cloud is rotated so that its normal vector is parallel to the vertical unit vector (see Eq. 5). The rotation matrix, **R**, is determined with the MATLAB function `axang2rotm` (MathWorks, 2016). The rotation axis ($\vec{e}$) and rotation angle about the axis (θ) are the input arguments to `axang2rotm`.
7. Finally, this point cloud is rotated about the vertical unit vector (which is now parallel to the plane's normal vector) to ensure the x-coordinate of the test track is in the longitudinal direction.

$$\vec{v}_1 = \mathbf{P_2} - \mathbf{P_1} \text{ and } \vec{v}_2 = \mathbf{P_3} - \mathbf{P_1}, \tag{1}$$

$$\vec{u} = \frac{\vec{v}_1 \times \vec{v}_2}{\|\vec{v}_1 \times \vec{v}_2\|} \tag{2}$$

$$\vec{e} = \vec{u} \times \hat{k}, \text{ note } \hat{k} = \langle 0 \quad 0 \quad 1 \rangle \tag{3}$$

$$\theta = \cos^{-1}(\vec{u} \cdot \hat{k}) \tag{4}$$

$$\begin{Bmatrix} x' \\ y' \\ z' \end{Bmatrix} = \mathbf{R} \begin{Bmatrix} x \\ y \\ z \end{Bmatrix} \tag{5}$$

At this point, the point cloud is located in the x-y plane (ground plane) and the test track is oriented along the x-axis. However, the point cloud remains an unorganized point cloud, with points located wherever the stereo-photogrammetry could find corresponding features. For the purposes of calculating the displacement PSD, the point cloud is organized into a regular grid. This is done with the `griddata` MATLAB function (MathWorks, 2016). This function fits a surface to the scattered point cloud data. The surface always passes through the points in the original data set. The nearest neighbor interpolation technique is used.

Various interpolation resolutions are used. The benchmark Can-can data set was recorded with a longitudinal resolution of 10 mm and a lateral resolution of 40 mm. For the purpose of this study, a longitudinal resolution of 10 mm is used for all the UAV-based datasets. Lateral resolutions varying from 10 mm to 65 mm in 5 mm increments are used. The highest resolution was based on the work by Els and Scholtz (2017), who found that resolutions finer than 15 mm x 15 mm for a typical sports-utility-vehicle (SUV) tire were sufficient for ride comfort studies.

After interpolation, outliers are removed with the `filloutliers` MATLAB function (MathWorks, 2016). A center fill method is used in conjunction with the moving mean find method with an 85-element window. Figure 4 shows a portion of the post processed terrain model.

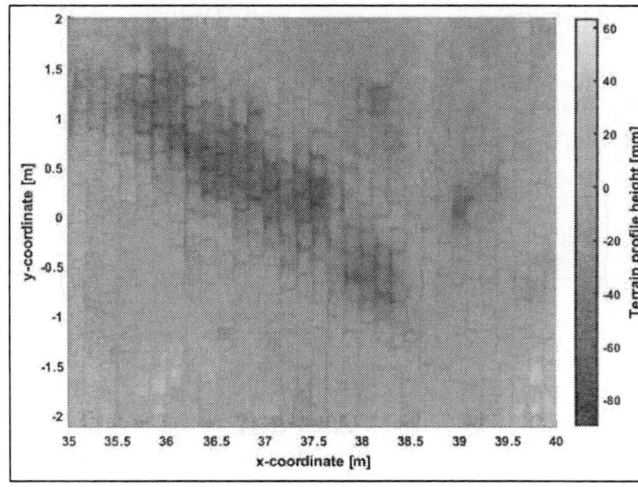

Fig. 4. An example of the post-processed terrain model

2.4. Calculating the displacement PSD

With the points neatly organized into an evenly spaced grid, a displacement PSD of the terrain profile can be calculated along the longitudinal direction. A displacement PSD is calculated for each longitudinal line using the method described in Fig. 5. This results in a set of displacement PSDs over the width of the terrain. The average displacement PSD is then determined, resulting in a single displacement PSD for the terrain. Note that the smooth concrete road surface surrounding the Belgian paving test track is not included in the displacement PSD. The displacement PSD thus only represents the undulating Belgian paving test track.

```
npts        = determine number of points along x-coordinate of
              grid
fs          = spatial sampling frequency of grid

for k = 1 to npts
    fft_x         =    Fast Fourier Transform of line k z-coordinates
    mag_fft_x     =    Single-sided, scaled amplitude of fft_x
    psd_x         =    (mag_fft_x^2)/(2*fs)

    psd_road(:,k)      = store psd_x in column k of an array
end for

psd_mean = mean of psd_road along the y-direction at each x-
           coordinate
```

Fig. 5. Determining the mean displacement PSD of the terrain profile

3. Results

Nine surveys of the Belgian paving were conducted at three different altitudes and three image overlap ratios. The exact same area was used for each survey, covering a total surface area of 776.7 m^2. Each survey mission was used to construct a 3D terrain model. Additionally, three models were created by combining the surveys with the same overlap ratio. These additional models thus combine the surveys conducted at different flight altitudes. It is hoped that these additional terrain models may improve accuracy of the longer wavelengths present in the terrain. Images captured at lower altitudes contain more pixels per area, but may cause errors at longer wavelengths when stitched together. The high-altitude images cover a larger surface area, thus possibly improving the long wavelength road profile accuracy.

Additionally, the effect of lateral distance between displacement PSDs is investigated. Since the point cloud is organized into a regularly spaced grid, a decision has to be made on the longitudinal and lateral distances between points in the grid. Naturally, a shorter distance leads to more points, significantly increasing the file size of the dataset and potentially leading to increased simulation time when used for multibody dynamics simulations. 5 mm lateral increments from 10 mm to 65 mm were investigated in this study. The lower limit was based on the work by Els and Scholtz (2017) and the upper limit was arbitrarily chosen. The benchmark Can-can machine data set was recorded at a 40 mm lateral spacing. Can-can machine data was recorded at 10 mm longitudinal increments, and hence a 10 mm increment was used for the UAV-based datasets too. Investigating the effect of longitudinal resolution is left for a future, more comprehensive study.

Table 2 lists the terrain models, the survey parameters and resulting number of images, points, density and average distance between points. Note that model 5 was used for the interpolation investigation.

The effects of surveying with different overlap ratios and at different altitudes are clearly seen in Table 2.

A higher overlap ratio and a lower altitude result in more images, leading to an increased number of points and consequently, a higher point density. The point density is inversely proportional to the average distance between points. An unexpected result is that the combination of multiple survey altitudes into a single model (see Models 4, 8 and 12) does not lead to significantly lower average distance between points, although the accuracy of the points may be influenced.

The displacement PSDs obtained for the same overlap ratios, but at different surveying altitudes are shown in Fig. 6 to Fig. 8.

Table 2
Parameters of each terrain model

Model No.	Overlap ratio [%]	Altitude [m]	No. of images	No. of points	Point density [points/m²]	Avg. distance between points [mm]	Point cloud file size [MB]
1	50	12.2	26	6 008 470	7 735.9	11.37	211
2	50	21.3	15	2 198 002	2 829.9	18.80	77
3	50	30.5	11	1 097 225	1 412.7	26.60	38
4*	50	12.2, 21.3 & 30.5	52	9 074 003	11 682.8	9.25	319
5	70	12.2	42	9 117 416	11 738.7	9.22	320
6	70	21.3	24	2 799 975	3 605.0	16.65	98
7	70	30.5	17	1 299 978	1673.7	24.44	45
8*	70	12.2, 21.3 & 30.5	83	9 436 278	12 149.2	9.07	331
9	90	12.2	435	15 054 681	19 382.9	7.18	529
10	90	21.3	155	4 238 858	5 457.5	13.54	149
11	90	30.5	81	2 026 109	2608.6	19.58	71
12*	90	12.2, 21.3 & 30.5	671	15 563 702	20 038.2	7.06	547

*Note: Model constructed from images of multiple surveys

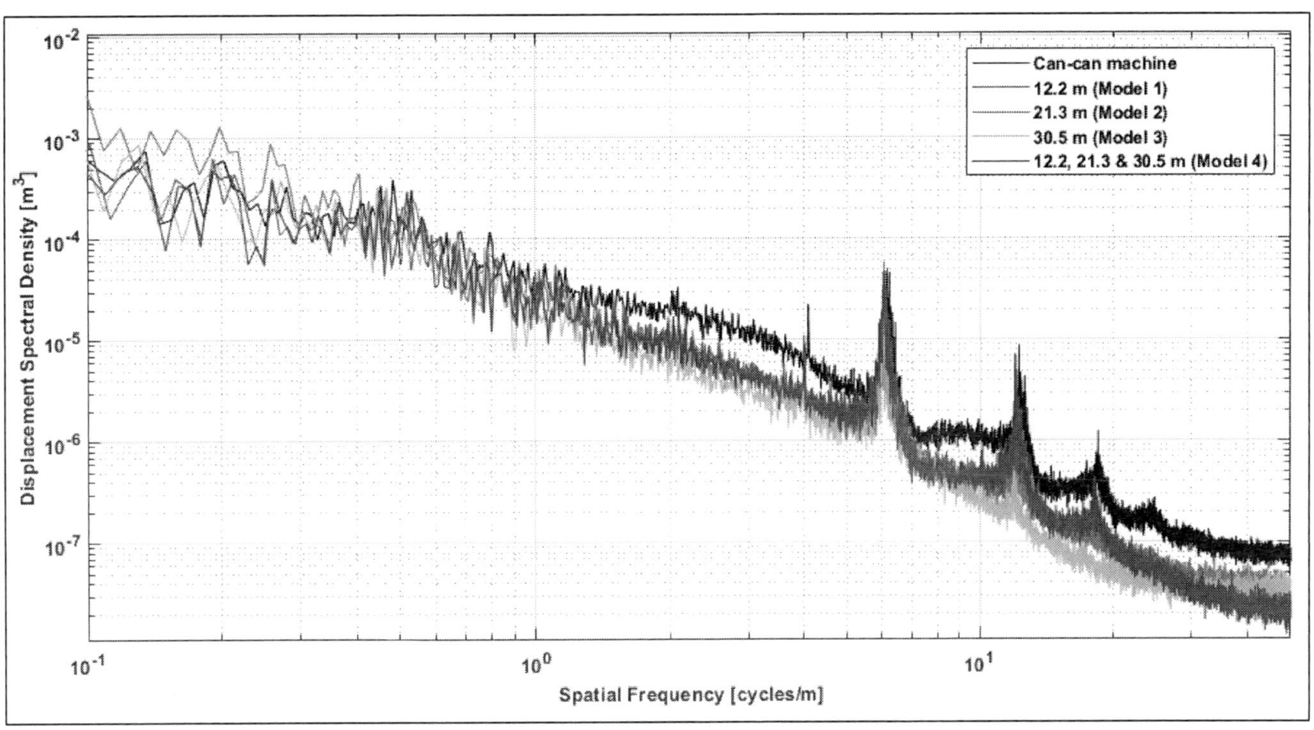

Fig. 6. Displacement PSD of 50 % overlap images at various altitudes

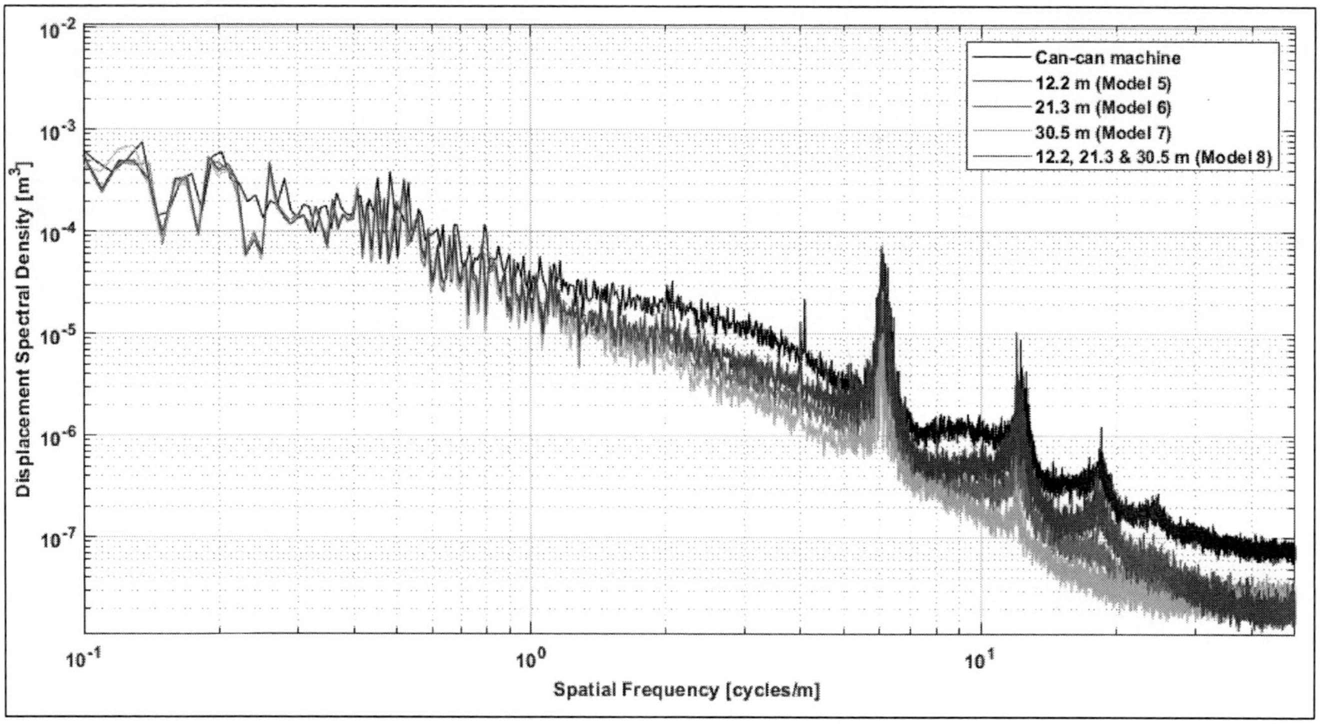

Fig. 7. Displacement PSD of 70 % overlap images at various altitudes

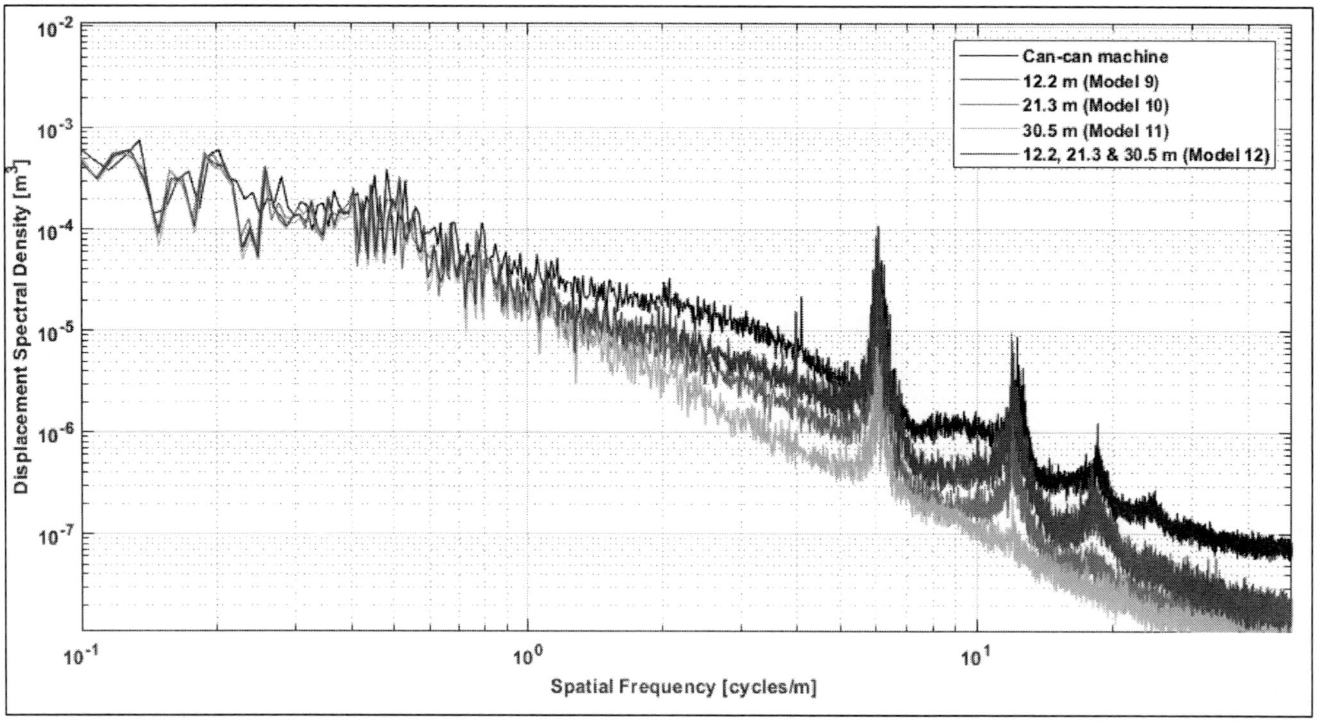

Fig. 8. Displacement PSD of 90 % overlap images at various altitudes

The results in Fig. 6 to Fig. 8 exhibit prominent peaks at spatial frequencies of 6.1, 11.9 and 18.3 cycles/m. These peaks correspond to the cobblestone dimensions and the gaps between cobblestones. Furthermore, the results show that, as expected, surveying altitude has a significant effect on the displacement PSD, especially at higher spatial frequencies. This is because the average distance between points at higher altitudes reduces, leading to a higher average distance between points. This is clearly visible in the diminishing peak at 18.3 cycles/m as the surveying altitude increases. In all three image overlap cases, surveying at the lowest altitude results in a dataset the closest to the Can-can machine benchmark. However, significant discrepancies are seen between all the

UAV-based data sets and the Can-can data. At present, this is not yet fully understood and warrants further investigation.

An interesting observation is that creating a terrain model from surveys conducted at different altitudes (Models 4, 8 and 12) does not yield significant improvements over the model created from the lowest altitude survey. The results indicate that the model created from the lowest altitude survey has almost identical results across the spatial frequencies from 0.1 to 50 cycles/m. This is somewhat surprising; it was anticipated that the combination of images from multiple altitudes will result in different roughness at lower spatial frequency ranges. This does warrant some deeper investigation, as the Belgian paving is only 100 m long, limiting insights into much longer profiles such as a winding road or circuit.

Figure 9 shows the influence of image overlap ratio on the displacement PSD. All displacement PSDs in Fig. 9 were obtained at a survey altitude of 12.2 m. There are small discrepancies between these displacement PSDs, but these are almost insignificant. This is somewhat unexpected, but may once again be explained by the average distance between points listed in Table 2.

The average distance between points for Models 1, 5 and 9 range from 11.37 mm to 9.07 mm. This is a relatively small variation. Furthermore, the results in Fig. 1 were obtained with a longitudinal resolution of 10 mm. The range of image overlap ratios is thus not wide enough to see a significant change in model accuracy. It should be noted that lower overlap ratios were attempted, but insufficient common features could be obtained to build 3D models with RealityCapture.

Finally, the effect of lateral interpolation resolution was investigated. Figure 10 shows the results for the 15 mm, 30 mm, 45 mm and 60 mm (other results were omitted, but followed the same trend as discussed here). The lateral resolution had limited influence on the displacement PSDs at spatial frequencies lower than approximately 15 cycles/m. Higher spatial frequencies are affected, with the terrain model becoming rougher with a coarser grid. 15 cycles/m correspond to a wavelength of 66.7 mm and is thus directly related to the lateral distance between longitudinal lines. Care must be taken when selecting the interpolation grid resolution, but the average distance between points must also be considered. The shortest average distance between points in the twelve models is 7.06 mm. Resolutions finer than this should be avoided.

4. Conclusion and recommendations

This study set out to further investigate the use of a UAV to create 3D terrain models. It follows on previous work that recommended a deeper investigation into the data collection and post-processing methods. The effects of survey altitude, image overlap ratio and lateral interpolation resolution were investigated.

The results show that the 3D terrain models are significantly influenced by the UAV flight altitude during a survey. Flying at a lower altitude results in a denser point cloud. The anticipated benefit of surveying at higher altitudes was not evident in the displacement PSDs. No discernible difference was noted between surveys conducted at 12.2, 21.3 and 30.5 m at spatial frequencies below 1 cycle/m. Combining surveys conducted at different altitudes also made no discernible difference. It can thus be recommended to fly as low as possible.

The image overlap ratio had an effect on the higher spatial frequencies of the displacement PSDs, with some differences visible above 7 cycles/m. However, this effect was not as significant as the survey altitude. It is thus more important that surveys are conducted at the lowest possible altitude. Care must be taken when selecting an image overlap ratio if higher spatial frequencies are of interest (e.g. due to high vehicle speeds, or if noise, vibration and harshness (NVH) are investigated).

Finally, the lateral interpolation resolution was found to influence the higher spatial frequencies of the displacement PSD. The resolution is inversely proportional to the point where displacement PSD start to diverge. Again, care must be taken when selecting an interpolation resolution that is suitable to the intended simulation application.

Recommendations stemming from this investigation are:

- Using the 3D terrain models in a multibody vehicle dynamics simulation.
- Comparing the abovementioned simulation results with experimental measurements on a test vehicle.
- Adapting the technique described in this study to curved roads.

5. Acknowledgements

The support of Gerotek Test Facilities (Armscor Defence Institutes SOC, n.d.) in granting access to the test tracks for the purposes of this study.

6. Declaration of competing interest

The authors declare that they have no known competing financial interests or personal relationships that could have appeared to influence the work reported in this paper.

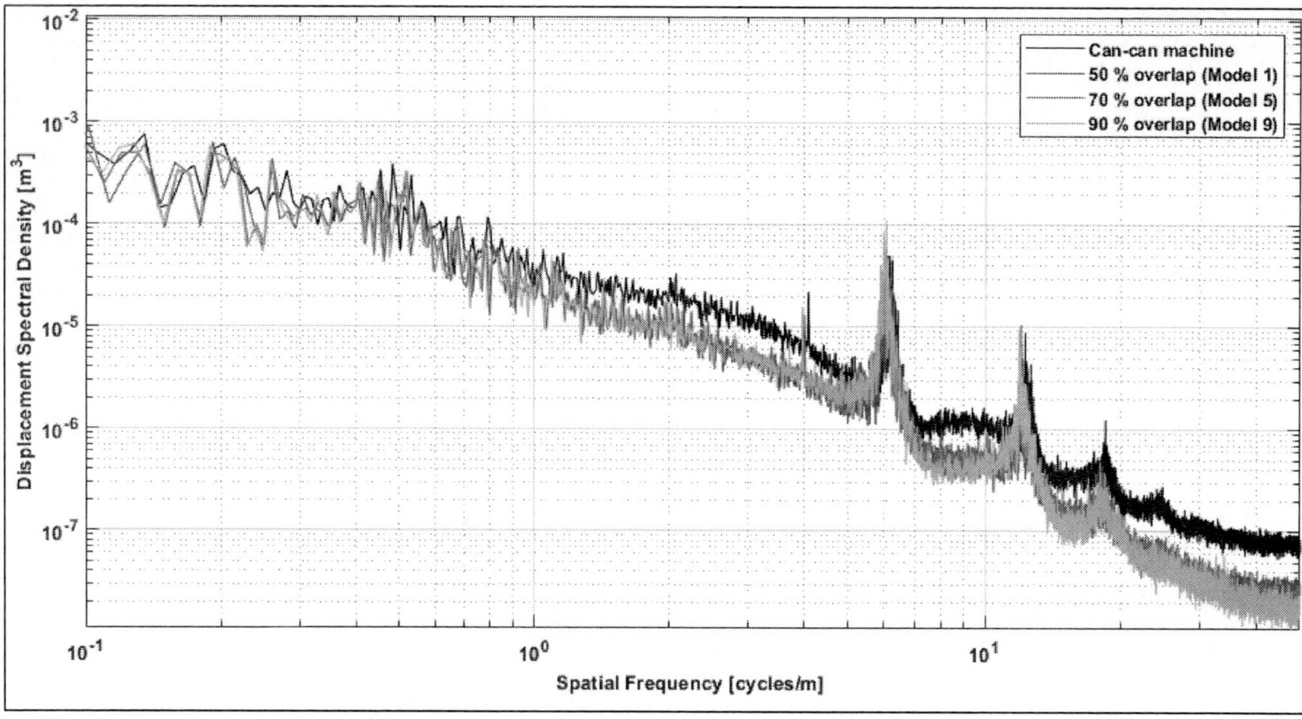

Fig. 9. Displacement PSD of different image overlap ratios at 12.2 m altitude

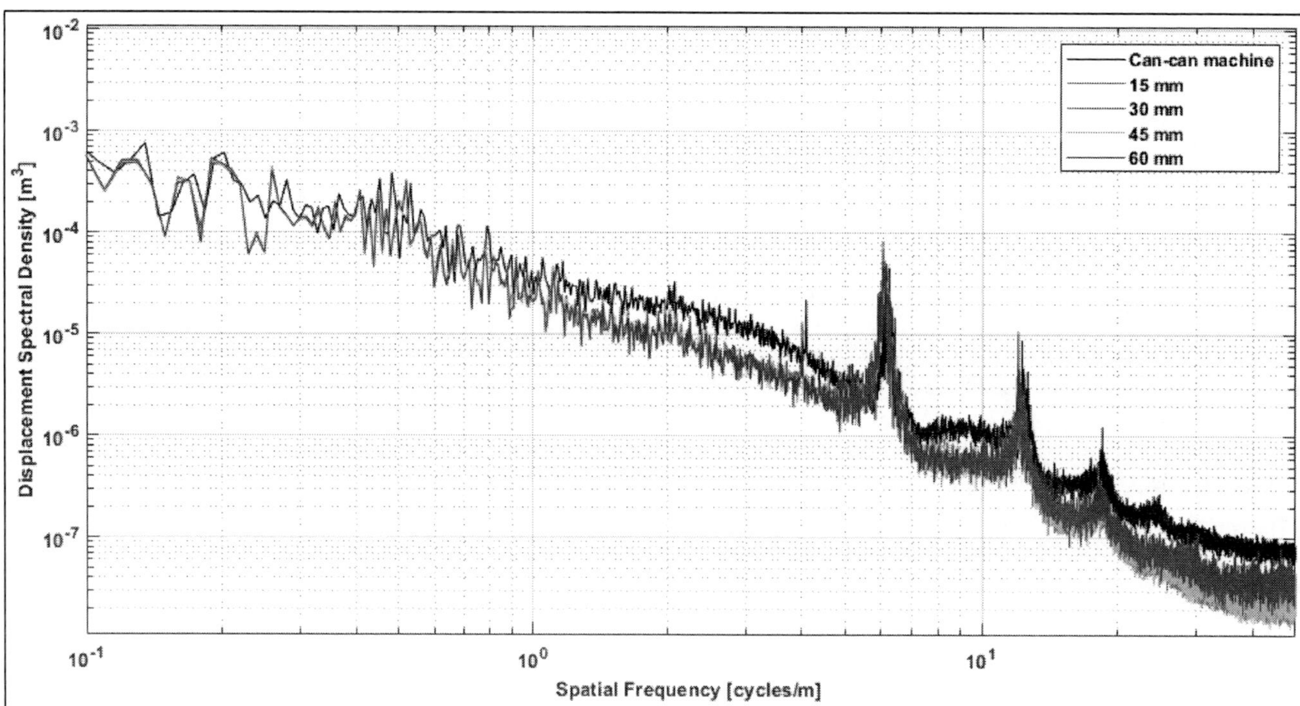

Fig. 10. Displacement PSD of different lateral interpolation distances, 70 % overlap, 12.2 m altitude (Model 5) shown

7. References

Armscor Defence Institutes SOC, n.d. Gerotek Test Facilities [Online]. Available: https://www.armscor.co.za/gerotek/ [Accessed 15 February 2025].

Becker, C. M., Els, P. S., 2014. Profiling of rough terrain. International Journal of Vehicle Design, 64, 240-261.

CloudCompare, 2021. CloudCompare (Version 2.10.3) 3D point cloud and mesh processing software Open Source Project [GPL software] [Online]. Available: https://www.cloudcompare.org/ [Accessed 14 May 2024].

DJI, 2024. DJI Mavic 3 Enterprise Specs [Online]. Available: https://enterprise.dji.com/mavic-3-enterprise/specs [Accessed 8 February 2024].

Els, P. S., Scholtz, O., 2017. Sensitivity of tyre simulation to road profile measurement resolution. Dynamics of Vehicles on Roads and Tracks Vol 1: Proceedings of the 25th International Symposium on Dynamics of Vehicles on Roads and Tracks (IAVSD 2017), 14-18 August 2017, Rockhampton, Queensland, Australia (1st ed.). CRC Press.

Epic Games, 2025. Reality Capture 1.5 Data Sheet [Online]. Available: https://www.capturingreality.com/assets/Documents/RealityCapture1-5_Datasheet.pdf [Accessed 13 May 2025].

Hamersma, H. A., van Aswegen, J. C., Guthrie, A. G., Els, P. S., 2024. UAV-Based Three-Dimensional Rough Terrain Modelling. Proceedings of the 21st International and 12th Asia-Pacific Regional Conference of the ISTVS. Yokohama, Japan, October 28-31, 2024.

International Organization for Standardization, 2016. ISO 8608:2016 - Mechanical vibration - Road surface profiles — Reporting of measured data. Geneva, Switzerland.

MathWorks, 2016. MATLAB and Simulink [Online]. Available: http://www.mathworks.com/ [Accessed 20 July 2016].

6965 | Correlation between optical reflectance and load-bearing capacity of soils

György Pillinger *, Peter Kiss

Institute of Technology. Hungarian University of Agriculture and Life Sciences (MATE). Gödöllő. Hungary

** Corresponding author: pillinger.gyorgy@uni.mate.hu*

ABSTRACT

An important factor in route planning for autonomous off-road vehicles is determining soil condition, particularly load-bearing capacity. Direct (in situ) measurement of this soil property is challenging and its determination through remote sensing technologies is an emerging area of research.

In laboratory tests using prepared soil samples a strong correlation was observed between the soil surface reflectance in the visible range (400-700 nm) and the load-bearing capacity. A bevameter device was used to determine the load-bearing capacity of each sample at various moisture contents. Measurements were taken for eight different soil textures at different moisture levels, starting from the dry state and gradually increasing to full saturation. This approach allowed us to observe how changes in moisture content affect the load-bearing capacity of each soil texture. Reflectance measurements were then performed on these samples using a portable spectrophotometer. The optical reflectance data were compared with the load capacity measurements obtained from the bevameter. This process enabled the development of a predictive equation that incorporates constants representing the soil's reflectance in its dry state. The resulting equation accurately captures the relationship between reflectance and load-bearing capacity under laboratory conditions.

While these results are based on a controlled environment, they offer a valuable foundation for estimating load-bearing capacity in real-world scenarios. Further research, incorporating field data and accounting for environmental variables, will be essential for refining this approach and developing more accurate predictive models for autonomous off-road vehicle navigation.

Keywords
Remote sensing
Optical reflectance
Soil condition
Load-bearing capacity
Soil moisture content
Load-bearing capacity factor

1. Introduction

Off-road autonomous navigation remains a formidable challenge due to the highly unstructured nature of natural terrain. Unlike paved roads with clear markings and reliable surfaces, off-road environments present uneven ground, obscuring vegetation and variable substrates such as mud or sand – any of which can drastically alter a vehicle's traversability. An autonomous vehicle that might easily follow lane lines on asphalt must instead interpret a chaotic landscape with few predictable cues (Wijayathunga et al., 2023). These sudden and constant changes in terrain (e.g., soft soil, loose rock, hidden holes) require continuous adaptation of the navigation strategy. Consequently, autonomous off-road navigation systems face not only perception difficulties but also the physical challenge of moving safely over deformable ground. Ensuring that a robot can maneuver across rugged unstructured terrain without getting stuck or overturned is a critical hurdle identified in recent surveys of the field. The low predictability of ground conditions in such environments continues to limit the reliability of off-road autonomous vehicles (Jiang et al., 2025).

One particularly crucial factor is the load-bearing capacity of the soil – essentially the soil's ability to support vehicle loads without excessive deformation. In off-road conditions, the ground itself becomes a "dynamic" hazard: if the soil cannot bear the weight, wheels may sink in or slip, immobilizing the vehicle. This is not just a theoretical concern it has direct implications for mobility and safety. On the other hand, the higher the soil deformation the higher the energy consumption of the vehicle will be. A vehicle's mobility and the safety of its operation depend strongly on the soil's load-bearing capacity, especially in off-road scenarios where there are no engineered surfaces (Jiang et al., 2025).

Traditionally, evaluating soil load-bearing capacity or trafficability has required direct mechanical testing of the ground – methods such as penetrometer cone index measurements, California Bearing Ratio (CBR) tests, or bevameter pressure–sinkage experiments. While these in-situ mechanical tests provide reliable measures of soil strength, they come with significant limitations. They are point-based, labor-intensive, and time-consuming. So, coverage of large areas is sparse and slow (Li et al., 2022). For example, to assess a field's load bearing capacity, one might have to manually conduct dozens of penetrometer insertions at different spots, each yielding data only for that small location. In fast-changing environments, or over wide off-road expanses, such traditional testing is impractical for real-time mobility planning. By the time a sufficient grid of measurements is made, soil conditions (like moisture) could have changed, or an autonomous vehicle could already have stumbled into an unsafe patch. Moreover, deploying personnel or instruments into difficult terrain for testing defeats the purpose of autonomy and can pose safety risks. The sparsity of in situ strength observations means that vehicles often must operate with incomplete knowledge of the ground ahead. This gap has motivated researchers to seek alternative approaches that can quickly and noninvasively infer soil properties over large areas.

One promising approach is the use of remote sensing techniques to predict soil physical properties related to strength. Optical sensing of the soil's surface reflectance in the visible spectrum (400–700 nm) has attracted interest as a proxy for certain soil conditions. The underlying premise is that many soil properties which affect load-bearing capacity (such as moisture content, texture, and organic matter) also influence the soil's color and optical reflectance. Early experimental work by Bowers and

https://doi.org/10.56884/G5E3K4K9

Hanks, (1965) systematically demonstrated how soil reflectance varies with moisture content and surface conditions, establishing that reflectance decreases significantly as soil becomes wetter. This finding laid the groundwork for using optical sensing to infer soil physical properties in later studies. For instance, wet soil typically appears darker (lower reflectance) than the same soil when dry, because water increases absorbance of light (Whalley and Leeds-Harrison, 1991). Several studies have demonstrated that visible and near-infrared (VIS-NIR) reflectance spectroscopy can be used to estimate key soil physical and chemical properties, including texture, organic carbon, and moisture content (Pinheiro et al., 2017). These findings support the use of reflectance data as a proxy instead of more complex or destructive soil measurements. Decades of soil optics research have quantified these relationships: soil reflectance correlates strongly with various soil parameters. Classic studies demonstrated an inverse exponential relationship between soil moisture and reflectance – as moisture content rises, reflectance drops sharply.

Understanding and monitoring soil moisture is vitally important for off-road mobility. Soil moisture content often acts as the lynchpin variable that determines whether soil will behave like firm ground or a quagmire. Even a normally solid soil can lose most of its bearing capacity when saturated – as soil approaches the liquid limit its shear strength for supporting vehicle loads drops precipitously (Pundir and Garg, 2022). Consequently, a small increase in moisture (for example after rainfall) can turn a passable area into impassable mud. Studies have shown that incorporating real-time soil moisture data substantially improves the accuracy of vehicle mobility predictions.

Stevens et al. (2017) found that using actual measured or modeled soil moisture in a mobility simulation (as opposed to assuming average conditions) could alter predicted travel times and outcomes significantly, especially in wet scenarios. In their analysis soil moisture was identified as an important terrain factor alongside soil type and slope underscoring that knowing moisture content is key to predicting trafficability and accurate soil moisture information is critical for anticipating off-road vehicle performance.

This research is currently ongoing: while the optical reflectance measurements have been completed, the load-bearing capacity tests are still in progress. The present study focuses on the load-bearing characteristics of sandy loam and clay soils. Building on our previous investigations, we apply Eq.s (1) (Nihal, 2022), (2) (Pillinger et al., 2023), and (3) (Pillinger et al., 2024) which were developed and validated in earlier phases of this work.

2. Materials and Methods

This section describes the experimental setup and procedures used to investigate the relationship between soil surface reflectance and load-bearing capacity. It includes the preparation of soil samples with controlled moisture content and initial density, the methodology for measuring mechanical strength using a bevameter, and the protocol for capturing optical reflectance data with a spectrophotometer.

2.1. Soil sample preparation

To investigate the relationship between optical reflectance and load-bearing capacity a diverse set of soil samples was prepared in a controlled laboratory environment. Eight different soil textures were selected to represent a broad range of mechanical and optical characteristics. The parameters of the soil textures investigated are shown in Table 1.

Each soil type was first air-dried and passed through a 2 mm sieve to remove large aggregates and debris. This preprocessing step ensured uniform particle distribution. The dried samples were then stored in sealed containers to maintain consistent initial conditions.

To simulate real-world variations in moisture content each soil texture was prepared at multiple moisture levels. Starting from the oven-dry state deionized water was incrementally added and thoroughly mixed into the soil until saturation was reached. The mixing process was performed manually in small batches to ensure even moisture distribution throughout each sample.

After each hydration step, the soil was allowed to equilibrate in sealed containers for a minimum of 12 hours to stabilize the moisture content. Soil samples were then compacted into cylindrical containers with known dimensions to maintain consistency during load-bearing capacity measurements.

To verify soil moisture content, a moisture analyzer was used in addition to the conventional drying oven method. Moisture measurements were performed using the HE73 moisture analyzer, which combines a precision balance (with 1 mg accuracy) and an integrated heating element. During testing, the soil sample was placed on a tray positioned on the scale, and the device lid was closed to initiate the measurement process. The heating element operated at 105 °C, continuously drying the sample while the device monitored changes in mass over time. The measurement proceeded until the change between successive mass readings fell below a predefined threshold, indicating that the sample had reached a stable dry state. Based on the initial and final masses, the device calculated the soil's moisture content on a dry basis, expressed as a percentage by mass.

Table 1
Investigated soil textures

Soil texture	Clay [$_m$%]	Silt [$_m$%]	Sand [$_m$%]
Sand	2.20	2.10	95.7
Sandy loam	11.9	16.5	71.6
Loamy sand	5.80	12.1	82.1
Silty Clay	44.6	52.1	3.30
Silty loam	26.2	62.2	11.6
Loam	19.0	49.7	31.3
Clay loam	28.5	40.7	30.8
Clay	55.6	31.3	13.1

2.2. Load-bearing capacity measurement using a bevameter

Each soil sample was prepared directly within the measuring bin of the bevameter device to ensure uniformity and consistency during load-bearing capacity testing. In addition to moisture content the initial bulk density of the soil was recognized as a critical factor influencing mechanical behavior. Therefore, the target density was precisely set during the preparation phase and maintained consistently across all samples.

The soil was carefully filled into the bin in controlled layers to promote uniform compaction and achieve maximum homogeneity throughout the sample depth. This meticulous process minimized internal variation and ensured reproducibility of the test results. After the sample was fully compacted and leveled the bevameter was used to apply vertical pressure and measure the resulting deformation. These measurements were used to characterize the load-bearing capacity of the soil under different moisture conditions.

Ground pressure varies significantly across different off-road vehicles, influenced by factors such as vehicle weight, tire or track design. Off-road vehicles like SUVs and trucks typically exert ground pressures ranging from 0.7 to 2.1 bar. All-Terrain Vehicles (ATVs) and Utility Task Vehicles (UTVs) usually operate with ground pressures between 0.2 to 0.6 bar. Tracked vehicles, such as bulldozers and snowmobiles, distribute weight more evenly, resulting in ground pressures as low as 0.07 to 0.35 bar. The maximum ground pressure applied during the measurements was set at 3 bar, as this value aligns with the upper limit typically exerted by existing off-road vehicles. Beyond this threshold, most vehicles do not apply higher ground pressures, making 3 bar a practical and representative maximum for such assessments.

In laboratory measurements conducted within a soil bin, it is essential to consider the potential influence of bottom and sidewall effects, as these phenomena can significantly impact the accuracy of the results. Adequate soil thickness must be ensured to minimize these boundary effects. The relationship between the diameter of the pressure plate (d), the soil thickness (H), and the critical depth (z_0/d) is as follows:

$$\frac{z_0}{d} = \frac{H}{d} - 1 \tag{1}$$

When the soil cone formed ahead of the pressure plate reaches the critical depth, it begins to contact the bottom of the bin. From this point onward, the load cell also detects the influence of the bin bottom. Thus, to examine sinkage up to a relative depth of z_0/d=0.4, the minimum required soil thickness is 21 cm for a pressure plate diameter of 15 cm, and 28 cm for a diameter of 20 cm. If a soil thickness of 30 cm is selected and both pressure plates are used for testing, the critical depth will correspond to 15 cm for the smaller plate and 10 cm for the larger plate. This implies that, with the smaller pressure plate, the soil can be loaded up to half of its total thickness, while with the larger plate, loading is limited to approximately one-third of the soil thickness before bottom effects begin to influence the measurement. Thus, the H/d ratio will be 1.5 for the measurement configuration.

The bulk density of Hungarian soils (dry mass per unit volume) typically averages on the order of around 1.1–1.4 g/cm^3 for topsoils. In general, most undisturbed mineral soils have bulk densities in the range of 1.1 to 1.6 g/cm^3. Within this interval, an approximate nationwide mean ~1.2 g/cm^3 can be expected for topsoil (0–20 cm) when considering all land uses together. This average reflects the mixture of higher-density agricultural soils and lower-density natural soils across the country. Accordingly, a bulk density of 1.4 g/cm^3 was selected for the measurement.

2.3. Optical reflectance measurement setup with the spectrophotometer

Immediately prior to conducting the pressure test the optical reflectance of the soil surface was measured using a portable spectrophotometer. This timing ensured that the reflectance values corresponded precisely to the soil's state during the load-bearing test, particularly regarding moisture and compaction. Reflectance was recorded in the visible wavelength range (400–700 nm).

A Konica Minolta CM-700d spectrophotometer was used to perform reflectance measurements. This device operates by projecting its own internal light source onto the surface and measuring the intensity of the reflected light, eliminating the influence of external illumination. The spectrophotometer records both specular components included (SCI) and specular component excluded (SCE) reflectance values for each measurement. In this study, only the diffuse reflectance (SCE) data were used for analysis. The device supports measurements using interchangeable aperture heads of 3 mm and 8 mm in diameter; all measurements in this experiment were conducted using the 8 mm (MAV: Ø8 mm) configuration. The instrument operates within a reflectance range of 0–175% and a spectral range of 400–700 nm. Reflectance curves obtained from each measurement were used for detailed analysis of the soil surface characteristics, particularly for evaluating color variation in sandy samples. Although the device also computes additional color parameters (e.g., CIE X, Y, Z and L*, a*, b* values), these were not included in the present analysis. To ensure accurate and repeatable measurements the device must be placed in direct contact with the soil surface. thereby preventing interference from ambient light or shadows. While this configuration does not allow for non-contact or remote measurements, it provides high precision in laboratory conditions by minimizing environmental variables that could affect surface color perception. This controlled approach enhances the reliability of spectral data and simplifies the identification of consistent patterns in reflectance behavior across soil types and moisture levels.

To characterize the reflectance properties of soils, it is essential to determine the reflectance number under both dry and saturated conditions. The completely dry state of the soil was defined as the condition achieved after drying the sample at 105 °C for 48 hours. Moisture content levels were then established by calculation, using the known dry mass of the soil and the precise mass of added water. Accordingly, all reported moisture content values are expressed on a dry basis, as weight percentages. To achieve full saturation a separate procedure was applied: soil samples were gradually saturated using the drip method within a ceramic container. Water was added continuously until each droplet was fully absorbed and the soil surface began to exhibit visible reflectance. This condition was considered the fully saturated state for which the reflectance measurements were conducted.

3. Results

This chapter presents the results of the optical reflectance analysis conducted on various soil and load-bearing capacity tests conducted on two soil types under controlled laboratory conditions. The findings include both direct measurements and derived parameters, which together form the basis for evaluating the relationship between surface reflectance and mechanical soil behavior.

3.1. Reflectance-Based Results and Analysis

Based on previous studies the color number (C) is defined as the average reflectance value measured by the spectrophotometer within the wavelength range of 600–700 nm (Eq. 2).

$$C = \frac{1}{11} \cdot \left(\sum_{600}^{700} R_i \right) \tag{2}$$

Table 2 presents the tested soil textures. These include the optical reflectance number recorded at both the completely dry (C_0) and fully saturated states (C_{SAT}) at the saturation moisture content (ω_{SAT}), and the difference between the two reflectance extremes (ΔC_{max}) representing the measurable range of reflectance variation.

Table 2
Optical reflectance properties of investigated soil textures

	Dry	Fully saturated		Reflectance range
Soil texture	C_0 [%]	C_{SAT} [%]	ω_{SAT} [m%]	ΔC_{max} [%]
Sand	17.5	6.9	30.0	10.6
Sandy loam	34.4	13.4	31.9	21.0
Loamy sand	15.0	5.3	34.8	9.70
Silty Clay	18.4	7.1	36.4	11.3
Silty loam	12.2	5.0	36.6	7.20
Loam	20.5	5.5	42.1	15.0
Clay loam	19.7	7.0	47.2	12.7
Clay	15.8	5.5	53.5	10.3

As illustrated in Fig.s 1 and 2, the reflectance number of soils exhibits greater variation at lower moisture contents, while the rate of change in reflectance progressively decreases as the soil approaches its saturation moisture content. At full saturation, reliable reflectance measurements cannot be obtained using a colorimeter, as the presence of free water on the soil surface alters the reflection properties and interferes with the measurement. Consequently, the usable reflectance range depends on the specific soil texture.

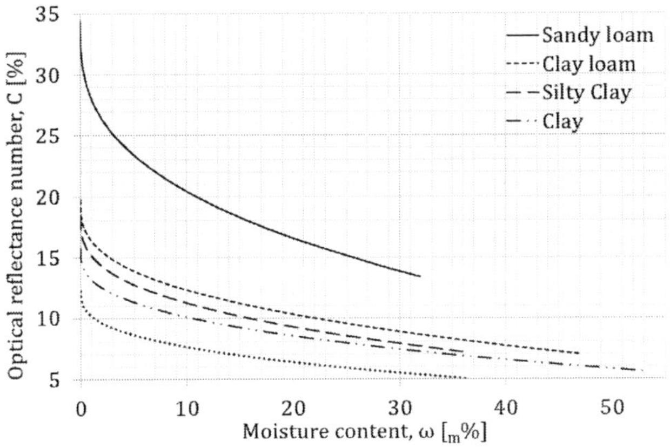

Fig. 1. Optical reflectance number as a function of moisture content for different soil textures

The relationship between soil moisture content and the reflectance number is defined by Eq. 3.

$$\omega = \sqrt[n_C]{\frac{(C_0 - C)}{A}} \tag{3}$$

This formulation describes how the moisture content ω increases as the soil's optical reflectance decreases from its dry-state value (C_0). The difference (C_0–C) quantifies the reduction in reflectance caused by increasing moisture content. Dividing by A serves to normalize this difference, while taking the (n_C) th root introduces a nonlinear relationship that reflects the soil-specific sensitivity of reflectance to moisture. A higher n_C value indicates a more gradual increase in moisture with decreasing reflectance, whereas a lower (n_C) results in a steeper moisture response. Table 3 lists the parameter values for Eq. 3

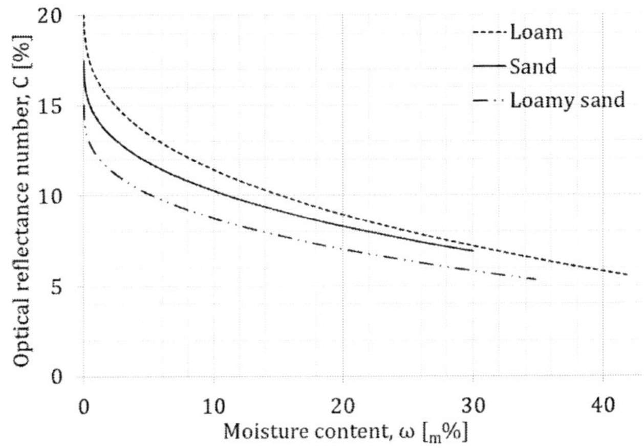

Fig. 2. Optical reflectance number as a function of moisture content for different soil textures

In Fig.s 1 and 2, all soil textures exhibit a clear nonlinear decrease in optical reflectance number (C) as moisture content (ω) increases. This trend confirms that reflectance is highly sensitive to changes in moisture, particularly in the dry to moderately moist range. The initial drop in reflectance is steep for all soils, indicating that even small increases in moisture content can lead to substantial decreases in reflectance. As the soil approaches saturation, the reflectance curve tends to flatten, suggesting that additional water has a diminishing effect on optical behavior.
The specific position and steepness of each curve vary by soil texture:
Sandy and loamy soils generally exhibit higher reflectance values in the dry state and a wider range of reflectance change, indicating greater sensitivity to moisture changes.
Clay-rich and silty soils tend to have lower initial reflectance and more gradual declines, reflecting their naturally darker appearance and finer particle size, which influence light scattering and absorption.

Table 3
Equation 3 constants for various soil textures

Soil texture	Constants	
	A [-]	n_C [-]
Sand	3.2	0.35
Sandy loam	6.3	0.35
Loamy sand	2.8	0.35
Silty Clay	3.2	0.35
Silty loam	2.0	0.35
Loam	4.1	0.35
Clay loam	3.3	0.35
Clay	2.6	0.35

3.2. Evaluation of Load-Bearing Behavior

By analyzing the sinkage–pressure curves shown in Fig. 3, the load-bearing capacity factor (k) can be determined for any given moisture content. The figure illustrates the relationship between relative depth (z/d) and ground pressure (p) for sandy loam soil, measured using a pressure plate of diameter d=20 cm and a bulk density of ρ=1.4 g/cm³. The curves represent different moisture contents ranging from 0 % to 25 % by mass and are shown as interpolated trend lines derived from experimental data points.

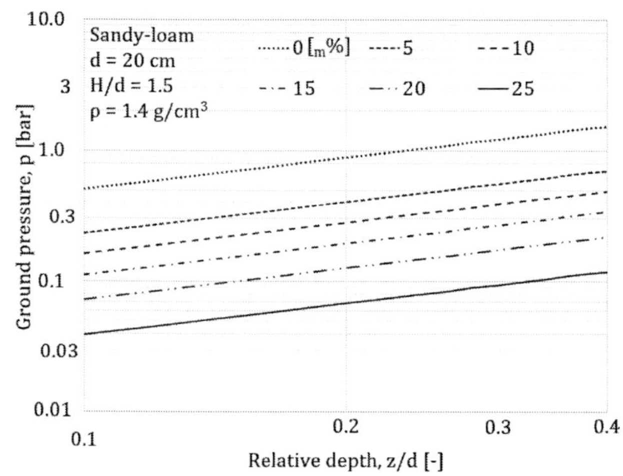

Fig. 3. Measured ground pressure as a function of relative sinkage for different moisture contents

The curves clearly show that ground pressure required to achieve a given normalized depth decreases as moisture content increases. This trend highlights the softening effect of water on soil, as it reduces inter-particle friction and cohesion. The most significant decrease in ground pressure occurs in the low-moisture range (0–10 %), while the effect becomes more gradual at higher water contents.

Further analysis of these curves allows for the calculation of the load-bearing capacity factor (k), which provides a useful single-value summary of soil strength. The calculation of (k) was based on the Saakyan formula:

$$k = p \cdot \left(\frac{d}{z}\right)^{-n} \tag{4}$$

where (p) is the applied ground pressure, (z) is the vertical sinkage, (d) is the pressure plate diameter, and (n) is an empirical exponent that describes the shape of the pressure–sinkage curve. In this study, a value of n=0.8 was used based on curve fitting and literature guidance for granular soils.

Figure 4 illustrates the relationship between the load-bearing capacity factor (k) and the moisture content (ω) for two soil textures: sandy loam and clay. Clay soils retain a higher load-bearing capacity across a broader range of moisture contents compared to sandy loam. While the (k)-values for sandy loam drop sharply with increasing moisture—especially below 15%—clay exhibits a much slower decline, maintaining relatively high strength even as moisture rises to 20–30%.

This difference can be attributed to the fundamental structural properties of the two soil types. Clay particles are significantly smaller and have a much higher surface area, which enhances cohesion and water retention through capillary action. This allows clay to remain relatively stable under loading, even when moisture content increases. In contrast, sandy loam, with its larger particle size and lower cohesive strength, loses mechanical integrity rapidly as water reduces interparticle friction.

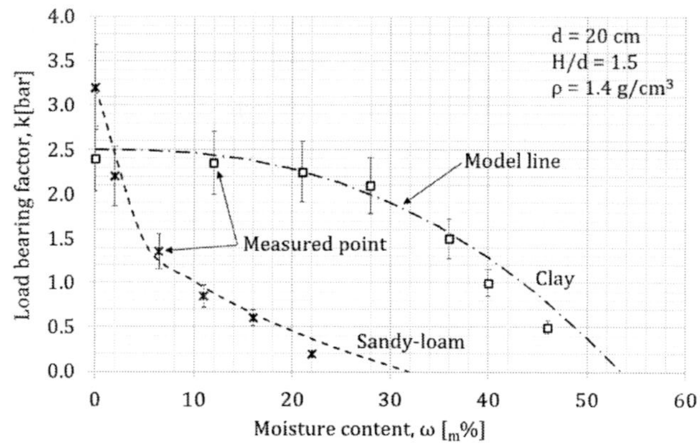

Fig. 4. Load-bearing capacity factor as a function of moisture content

In Fig. 4 measured values are indicated by black crosses for sandy loam and black squares for clay, while the dashed lines represent the fitted model curves for each soil type, as described by Eq. 5.

$$k = k_0 \cdot \left(1 - \left(\frac{\omega}{\omega_{SAT}}\right)^{n_\omega}\right) \tag{5}$$

This model captures the observed nonlinear decline in k with increasing moisture content, where (k_0) is the load bearing capacity factor for dry soil and n_ω is an empirical exponent. As shown, the load-bearing factor decreases sharply in the early stages of wetting, with diminishing effects as the soil approaches saturation. Each measured value includes an error bar representing ±15% uncertainty to account for measurement variability and soil heterogeneity.

3.3. Correlation between Optical Reflectance and Load-Bearing Capacity

Equation 6 establishes a predictive relationship between soil reflectance and its load-bearing response by integrating optical and mechanical parameters. By combining Eq.s 3, 4, and 5, we derive a model that predicts ground pressure p based on the soil's optical reflectance characteristics. This model incorporates constants representing the soil's reflectance and strength properties in its dry state.

$$p = k_0 \cdot \left[1 - \frac{(C_0 - C)^m}{{\omega_{SAT}}^{n_\omega} \cdot A^m}\right] \cdot \left(\frac{z}{d}\right)^n \tag{6}$$

Where, (k_0) represents the load-bearing capacity factor, and (C_0) is the optical reflectance number corresponding to the dry state of the soil. The exponent m is defined as $m=n_\omega/n_C$, linking the moisture and reflectance sensitivities, for the investigated sandy loam m=0.94. The final term, $(z/d)^n$, accounts for the mechanical response of the soil under loading, following the formulation of the Saakyan equation.

It is important to note that Eq. 6 is valid only under the assumption that the soil moisture content is uniform throughout the sample volume and matches the moisture at the surface, which serves as the basis for the reflectance measurement. If moisture gradients exist within the soil, the accuracy of the prediction may be compromised. Additionally, the equation assumes that the initial soil density is uniform throughout the entire soil volume, as variations in compaction can also influence both reflectance and mechanical behavior, leading to deviations from the modeled values.

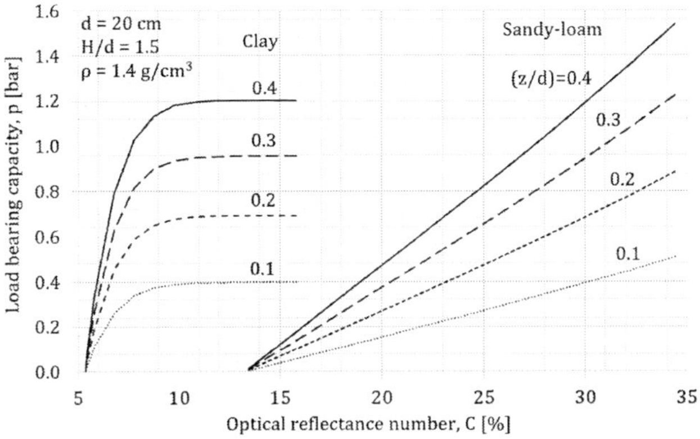

Fig. 5. Load-bearing capacity as a function of the optical reflectance number for different z/d values

Figure 5 presents the modeled relationship between the optical reflectance number (C) and the load-bearing capacity (p) for two soil textures—clay and sandy loam—at various relative depths (z/d=0.1, 0.2, 0.3, 0.4). All simulations assume a constant bulk density of ρ=1.4 g/cm^3 and a pressure plate diameter of d=20 cm, based on Eq. 6.

For sandy loam, the curves show an almost linear increase in load-bearing capacity with increasing reflectance across all (z/d) levels. This reflects the texture's sensitivity to moisture: as reflectance rises (indicating drier conditions), the soil's strength increases proportionally.

In contrast, the clay curves display a distinctly nonlinear trend. Load-bearing capacity increases with reflectance only up to a certain point, after which it plateaus. This behavior is characteristic of clay's higher moisture retention and structural cohesion—beyond a certain dryness (i.e., reflectance threshold), further drying does not significantly enhance mechanical strength.

At every (z/d) level, sandy loam consistently shows higher reflectance sensitivity, while clay demonstrates a more limited response. The flattening of the clay curves also illustrates that for higher depths, clay's load-bearing capacity becomes less dependent on surface reflectance, especially in near-saturated or structurally stable conditions.

Table 4
Equation 6 constants

	Sandy-loam	Clay
A	6.30	2.60
C_0	34.4	15.8
k_0	3.20	2.50
n	0.80	0.80
n_ω	0.33	2.50
m	0.94	7.10
ω_{SAT}	31.9	53.5

Table 4 summarizes the model parameters used in Eq. 6 to describe the relationship between optical reflectance and load-bearing capacity for sandy loam and clay soils. The dry-state reflectance (C_0) is significantly higher for sandy loam (34.4 %) than for clay (15.8 %), consistent with the brighter appearance and coarser texture of sandy soils. The maximum load-bearing capacity factor in the dry state (k_0) is also higher for sandy loam (3.2 bar) compared to clay (2.5 bar), indicating stronger mechanical behavior under dry conditions.

The saturation moisture content (ω_{SAT}) reflects the soils' differing water retention capacities, with sandy loam reaching saturation at 31.9%, while clay saturates at a much higher 53.5%, owing to its finer texture. Both soil types share the same value for the sinkage shape exponent (n=0.8), implying similar assumptions about the vertical deformation profile under loading.

A key distinction is found in the value of the exponent m, which controls the nonlinearity of the reflectance–strength relationship. Sandy loam has a relatively low value (m=0.94), resulting in a near-linear increase in strength with decreasing moisture. In contrast, clay shows a much higher exponent (m=7.1), producing a more abrupt drop in strength as moisture increases — a behavior that aligns with its cohesive nature and sensitivity to water content.

These parameters emphasize the need for texture-specific calibration when applying reflectance-based soil strength models and demonstrate the flexibility of Eq. 6 in capturing contrasting soil behaviors.

4. Conclusions and future work

This study demonstrates that soil surface reflectance, measured in the visible spectrum, exhibits a strong and predictable correlation with load-bearing capacity under controlled laboratory conditions. Its applicability is limited, as soil color becomes less sensitive to moisture near saturation, reducing calculation accuracy.

Using a combination of optical measurements and mechanical tests on various soil textures, including sandy loam and clay, we established that both moisture content and reflectance significantly influence soil strength. The introduction of the reflectance number (C), along with derived empirical models, enabled the estimation of the load-bearing capacity factor (k) and ground pressure (p) as functions of soil reflectance and relative depth.

The development of a composite predictive equation (Eq. 6) integrating optical reflectance and mechanical behavior parameters marks a key outcome of this work. This equation provides a practical means of estimating the bearing capacity of soils based on optical measurements — a major step toward enabling real-time, non-invasive terrain evaluation for autonomous off-road vehicles. The findings also highlight the importance of soil moisture uniformity, as the validity of the reflectance-based predictions depends on consistent moisture distribution throughout the soil profile. For example, sandy and clay soils differ significantly in how they absorb and transmit moisture. In sandy soils, water infiltrates quickly, leading to relatively rapid wetting, whereas in clay soils, moisture penetration occurs much more slowly due to lower permeability. This discrepancy often results in non-uniform moisture distribution throughout the soil profile. Consequently, shortly after rainfall, relying solely on surface measurements may lead to the erroneous conclusion that the soil is fully saturated and exhibits poor load-bearing capacity, even though the deeper layers may still remain relatively dry and structurally sound.

It is also important to emphasize that soils with identical or similar textures can exhibit considerable variability in reflectance.

Therefore, the reference values (C_0) and (k_0) must be determined individually for each soil type. To characterize soil properties in absolute terms, additional parameters should be incorporated. In the later stages of this research, we aim to identify typical exponent values that can be reliably associated with sandy and clay soils, thereby improving the generalizability of the model.

While the current results are based on well-controlled laboratory experiments, they offer a strong theoretical foundation for future field deployment. The next phase of this research will involve validating the models under real-world conditions, where surface irregularities, vegetation, ambient lighting, and vertical moisture gradients may introduce additional complexity. Incorporating remote sensing platforms, such as UAV-mounted spectrometers or multispectral cameras, could further expand the operational potential of the reflectance-based approach.
Future work will also focus on:

- In future stages of the research, we aim to complete the investigation for all the previously mentioned soil textures;
- Refining the model parameters to improve generalizability across soil types;
- Investigating the use of machine learning techniques to enhance predictive accuracy;
- Integrating reflectance-based soil strength estimation into autonomous navigation systems.

Overall, the methodology and models presented here contribute to a growing body of research aimed at improving terrain-awareness and decision-making for autonomous systems operating in complex, off-road environments.

5. Nomenclature

A	Soil reflectance specific constant	[-]
C	Optical reflectance number	[%]
C_0	Reflectance number in the dry state	[%]
C_{SAT}	Reflectance number in the fully saturated state	[%]
ΔC_{max}	Range of reflectance variation	[%]
d	Diameter of the pressure plate	[cm]
k	Load-bearing capacity factor	[bar]
k_0	Load-bearing factor at dry state	[bar]
m	Composite exponent, $m=n_\omega/n_C$	[-]
n	Shape exponent in the pressure–sinkage relationship (Saakyan model)	[-]
n_C	Reflectance sensitivity exponent	[-]
n_ω	Moisture sensitivity exponent	[-]
p	Ground pressure	[bar]
R_i	Reflectance value at a given wavelength	[%]
ω_0	Moisture content on a dry mass basis	[$_m$%]
ω_{SAT}	Moisture content at full saturation	[$_m$%]
z	Vertical sinking	[cm]
z/d	Relative depth	[-]

6. Acknowledgements

Project no. 2022-2.1.1-NL-2022-00012 has been implemented with the support provided by the Ministry of Culture and Innovation of Hungary from the National Research. Development and Innovation Fund. financed under the 2022-2.1.1-NL funding scheme.

7. Declaration of competing interest

The authors declare that they have no known competing financial interests or personal relationships that could have appeared to influence the work reported in this paper.

8. References

Bowers, S. A., Hanks, R. J. (1965). Reflection of radiant energy from soils. *Soil Science*, *100*(2), 130–138. https://doi.org/10.1097/00010694-196508000-00009

Jiang, C., Wei, Y., Chen, Z., Kuang, X. (2025). The prediction method and application of off-road mobility for ground vehicles: A review. *World Electric Vehicle Journal*, *16*(1), 47. https://doi.org/10.3390/wevj16010047

Li, T., Mu, T., Liu, G., Yang, X., Zhu, G., Shang, C. (2022). A method of soil moisture content estimation at various soil organic matter conditions based on soil reflectance. *Remote Sensing*, *14*(10), 2411. https://doi.org/10.3390/rs14102411

Pinheiro, É. F. M., Pereira, M. G., Anjos, L. H. C. (2017). Prediction of soil physical and chemical properties by visible and near-infrared diffuse reflectance spectroscopy in the Central Amazon. *Remote Sensing*, *9*(4), 293. https://doi.org/10.3390/rs9040293

Pillinger, G., Ahmed, E. E. A., Bessenyei, K. (2023). Correlations between moisture content and color spectrum of sandy soils. *Journal of Terramechanics*, *108*, 39–45. https://doi.org/10.1016/j.jterra.2023.05.002

Pillinger, G., Ahmed, E. E. A., Kiss, P. (2024). Spectral determination of soil moisture content based on the dry color of the soil. In *Proceedings of the 21st International and 12th Asia-Pacific Regional Conference of the ISTVS*. https://doi.org/10.56884/GV41C6MW

Pundir, S. K., Garg, R. D. (2022). A comprehensive approach for off-road trafficability evaluation and development of modified equation for estimation of RCI to assess regional soil variation using geospatial technology. *Quaternary Science Advances*, *5*, 100042. https://doi.org/10.1016/j.qsa.2021.100042

Salman, N. D. (2022). *Load bearing capacity of soil as a homogeneous finite half-space* (Doctoral dissertation, Hungarian University of Agriculture and Life Sciences). https://phd.mater.uni-mate.hu/199/2/Thesis_Nihal_DOI.pdf

Stevens, M. T., McKinley, G. B., Vahedifard, F. (2017). Full-featured ground vehicle mobility analysis using different soil moisture sources. *International Journal of Vehicle Performance*, *3*(1), 19–35. https://doi.org/10.1504/IJVP.2017.081262

Whalley, R. W., Leeds-Harrison, B. P. (1991). Estimation of soil moisture status using near infrared reflectance. *Hydrological Processes*, *5*, 321–327. https://doi.org/10.1002/hyp.3360050312

Wijayathunga, L., Rassau, A., Chai, D. (2023). Challenges and solutions for autonomous ground robot scene understanding and navigation in unstructured outdoor environments: A review. *Applied Sciences*, *13*(17), 9877. https://doi.org/10.3390/app13109877

7125 | A Data-driven parameter identification method of characterization of wheel-regolith interaction for lunar surface vehicles

Hiroaki Kawamura [a,*], Tadashi Matsuura [a], Kohei Shintani [a], Nobutaka Tanishima [b], Natsu Fujioka [b], Shinichiro Noda [a]

[a] *Toyota Motor corporation, Toyota, Japan*
[b] *Japan Aerospace Exploration Agency, Tsukuba, Japan*

* *Corresponding author: hiroaki_kawamura@mail.toyota.co.jp*

ABSTRACT

This paper describes a method to clarify the interaction between wheels and regolith particles when a rover runs on the lunar surface. The regolith dust that lifts during movement can adhere to the equipment mounted on the vehicle, potentially causing significant impacts on its functionality. To mitigate the effects of regolith dust on the onboard equipment, it is crucial to understand the behavior of regolith particles that are lifted by the wheels. Additionally, quantitatively understanding the interaction between wheels and regolith particles is important because it affects the wheels' traction and the vehicle's running performance. Many experimental and analytical studies have been published to address this. In the case of numerical analysis, studies using Discrete Element Methods (DEMs) to clarify microscopic interactions have been conducted. To represent the behavior of regolith particles in numerical analysis, it is necessary to identify many parameters that characterize the physical behavior of particles based on test results, which can be time-consuming due to trial and error in deriving these parameters. In this study, a new data-driven set-based searching method, called Differential Evolution Based on Adaptive Sampling (DEBAS), is adopted for the parameter identification of the particles. The proposed method can efficiently find combinations of particle characteristic parameters that match the desired element tests. The parameters obtained through this method will be verified by comparison with several characteristic tests. Furthermore, obtained parameters are validated by comparison with a scaled wheel-regolith simulant interaction test.

Keywords
Regolith
Discrete element method
Differential evolution based on adaptive sampling
Soil characteristics
Vacuum environment

1. Introduction

Solar rocky bodies such as the Moon and Mars are covered with fragmental debris called regolith. There is a concern that the loose regolith will adhere to the surface of the aircraft and to the humans active on the surface of the celestial body, potentially causing various adverse effects. In a rover traveling on a terrain surface, the interaction between the wheels and the surface soil exerts a traction force for driving, making it particularly important to understand the mechanical interaction between the wheels and the soil.

The mechanical analysis of the interaction between wheels and soil has been conducted by researchers such as Bekker (1960) and Wong (2008), focusing on the forces and stress distribution generated by the wheels. Applications of these terramechanics-based mechanical approaches to lunar and planetary exploration rovers include traction control of wheels, analysis of steering characteristics (Ishigami et al., 2007), experimental performance evaluation of wheels, and analysis of wheel performance in low-gravity environments. These methods are very effective for predicting and evaluating driving performance; however, they do not consider the behavior of the regolith that is disturbed by the wheels. On the other hand, there are examples of analysis using the discrete element method (DEM) that approach the interaction between wheels and soil at a more microscopic level. DEM is one of the effective methods for predicting the behavior of regolith particles. To predict the mechanical behavior of particles using DEM, tests are required to identify particle characteristic parameters. Additionally, since identifying these parameters through trial and error can be time-consuming, a method using the Taguchi method for identification has been proposed (Tasnim et al., 2023).

In general, multi-objective optimization methods are effective for parameter identification to achieve desired characteristics. Differential Evolution: DE (Stom and Price, 1997) is a type of metaheuristic algorithm aimed at optimizing functions where the variables take real values. It does not require assumptions of differentiability or convexity for the objective function and can explore global optimal solutions. Other known evolutionary computation algorithms include Genetic Algorithms: GA (Holland, 1975), Particle Swarm Optimization: PSO (Kennedy and Eberhart, 1995), and Covariance Matrix Adaptation Evolution Strategy: CMA-ES (Hansen and Ostermeier, 1996). Although DE has been criticized for lacking theoretical guarantees regarding the optimality and convergence of solutions, as well as being sensitive to hyperparameters, it has the advantage of having fewer hyperparameters and being a simple algorithm. Recently, many DE-based methods have been proposed such as Self-Adaptive DE: jDE (Brest et al., 2006), Adaptive DE with Optional External Archive: JADE (Zhang, 2009), and Success-History Based Parameter Adaptation for DE: SHADE (Tanabe, 2013).

However, these methods are point-based design methods that search for a single global optimal solution, making it difficult to explore a set of feasible solutions (Singer, 2009). On the other hand, a set-based design method has been proposed (Sobek, 1999) that can find feasible solutions not at a point but over a range, consistent with multiple test results. In DEM, even if the same particle characteristic parameters are given, the kinematic behavior may differ due to factors such as particle size and shape. Therefore, it is effective to explore a set of solutions that satisfy the constraints in order to obtain more robust characteristics. This paper proposes a set-based design method based on DE that can explore a set of feasible solutions and applies it to the identification of particle characteristic parameters in DEM. Target

https://doi.org/10.56884/IAO0I4NI

particles were lunar regolith simulants, and experiments under vacuum conditions were conducted for parameter identification. Additionally, using the obtained particle characteristic parameters, the numerical calculation results from a scale model simulating wheels will be compared with experimental results to validate the effectiveness of the proposed method.

2. Proposed formulae

2.1. Differential evolution

Differential evolution (DE) is one of the optimization methods of searching for the optimization problem as $f: \mathbb{R}^n \to \mathbb{R}$ with the design variable $\boldsymbol{x} \in \mathcal{X} = \{\boldsymbol{x} \in \mathbb{R}^n | \underline{x_i} \leq x_i \leq \overline{x_i}, i = 1,2,\dots,n\}$. Figure 1 illustrates the flow of DE. First, each individual solution, which is called vector, is generated randomly in the initialization step. Next, the target vector is selected randomly, and the trial vector is generated by changing target vector in the mutation and crossover process. Finally trial vector and target vector are compared and the superior vector is selected for the next generation.

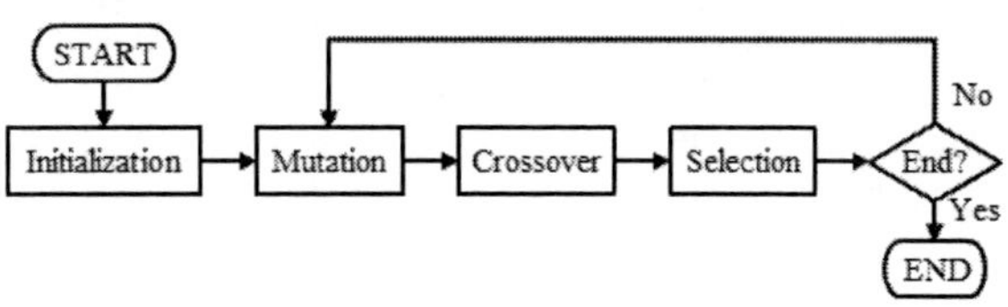

Fig. 1. Flow chart of Differential evolution

2.1.1. Initialization

Initial population $\boldsymbol{P}_0$ at generation $G = 0$ is generated randomly as follows.

$$\boldsymbol{P}_0 = (\boldsymbol{x}_1, \boldsymbol{x}_2, .., \boldsymbol{x}_m) \in \mathbb{R}^{n \times m}, \tag{1}$$

where m and n are size of population and the dimensions of the design variable.

2.1.2. Mutation

For each vector $\boldsymbol{x}_i, i = 1,2,\dots,m$, a trial vector $\boldsymbol{v} \in \mathbb{R}^n$ is generated as follows,

$$\boldsymbol{v} = \boldsymbol{x}_b + F(\boldsymbol{x}_{r_1} - \boldsymbol{x}_{r_2}), \tag{2}$$

where $\boldsymbol{x}_b, \boldsymbol{x}_{r_1} - \boldsymbol{x}_{r_2}$ and F are the base vector and the differential vector and a constant positive scale factor, respectively.

$$\boldsymbol{x}_b \in \boldsymbol{P}_G, \quad b = \{1,2,\dots,m\}, \quad b \neq i, \tag{3}$$

$$\boldsymbol{x}_{r_1} \in \boldsymbol{P}_G, \quad r_1 = \{1,2,\dots,m\}, \quad b \neq \{i.b\}, \tag{4}$$

$$\boldsymbol{x}_{r_2} \in \boldsymbol{P}_G, \quad r_2 = \{1,2,\dots,m\}, \quad b \neq \{i,b,r_1\}. \tag{5}$$

2.1.3. Crossover

Next step, a trial vector $\boldsymbol{u} \in \mathbb{R}^n$ is generated in the crossover process from target vector $\boldsymbol{x}_i$ and mutate vector $\boldsymbol{v}$ as follows,

$$u_i = \begin{cases} v_i & \text{if } (r_i \leq C_r) \vee (i = R) \\ x_i & \text{otherwise} \end{cases}, \tag{6}$$

where $C_r \in [0,1]$ and R are given crossover ratio and one arbitrarily chosen index, and $r_i \sim U(0,1)$ is uniform random number.

Finally, by comparing the trial vector $\boldsymbol{u}$ generated by equation (2) with the target vector $\boldsymbol{x}$, better solutions are selected as candidates for a new solution population $\boldsymbol{P}_{new}$ using the following equation.

$$\boldsymbol{x}_{i,new} = \begin{cases} \boldsymbol{u}_i & \text{if } f(\boldsymbol{u}_i) \leq f(\boldsymbol{x}_i) \\ \boldsymbol{x}_i & \text{otherwise} \end{cases}. \tag{7}$$

2.2. Differential evolution based adaptive sampling (DEBAS)

This section explains the details of the proposed method for searching not only for optimal solutions but also for the feasible range of solutions, which is called Differential Evolution-based Adaptive Sampling (DEBAS). Figure 2 shows the flow of the proposed method.

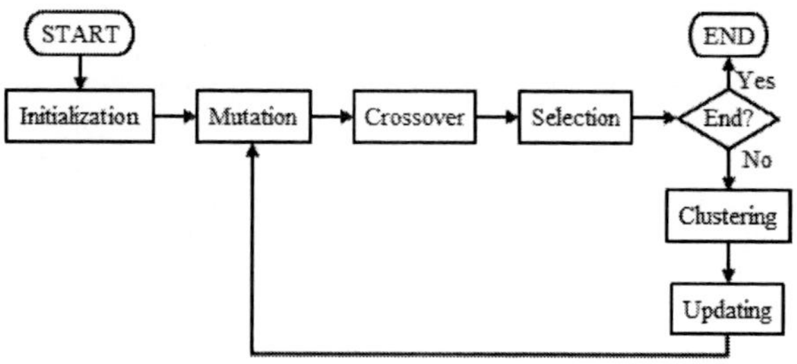

Fig. 2. Flow chart of DEBAS

Using the proposed method, it is possible to explore the feasible region $\boldsymbol{Q}$, where l target responses $\boldsymbol{g}(\boldsymbol{x}) = \{g_1(\boldsymbol{x}), g_2(\boldsymbol{x}),\dots, g_l(\boldsymbol{x})\}$ are satisfied, given the n-dimensional design variable $\boldsymbol{x} \in \mathcal{X} = \{\boldsymbol{x} \in \mathbb{R}^n | \underline{x_i} \leq x_i \leq \overline{x_i}, i = 1,2,\dots,n\}$ within the provided design variable space. In DEBAS, a single cost function is defined using multiple constraint functions $g_i(\boldsymbol{x}) \leq 0. i \in \{1,2,\dots,l\}$. The maximum value $\max_i g_i(\boldsymbol{x})$ of a constraint function $g_i(\boldsymbol{x})$ is approximated using the Kreisselmeier-Steinhauser function (Kreisselmeier and Steinhauser, 1979) as follows,

$$g_{KS}(\boldsymbol{x}) = \frac{1}{\alpha} \ln\left(\sum_{i=1}^{l} \exp\left(\alpha H(g_i(\boldsymbol{x})) g_i(\boldsymbol{x})\right)\right) - \frac{\ln(l)}{\alpha}, \tag{8}$$

where $\alpha > 0$ is a parameter that adjusts the approximation of the max operation. $H(\cdot)$ represents Heaviside function, with $H(0) = 0$. In the above equation, $\boldsymbol{x}$ that satisfies $g_{KS}(\boldsymbol{x}) = 0$ is a design variable that exists within the feasible region $\boldsymbol{Q} = \{\boldsymbol{x} \in \mathcal{X} | g_i(\boldsymbol{x}) \leq 0, i \in \{1,2,\dots,l\}\}$. Based on the above equation, the objective function is given by following expression.

$$J_0(\boldsymbol{x}) = g_{KS}(\boldsymbol{x}) - \frac{1}{\Pr(\boldsymbol{x})}(1 - H(g_{KS}(\boldsymbol{x}))), \tag{9}$$

where $\Pr(\boldsymbol{x})$ is the probability density from the distribution of discovered feasible solutions.

In the proposed method, the set of discovered feasible solutions is clustered using the following approach. The set of feasible solutions $\boldsymbol{Q}_k$ in generation k is divided into clustered sets $\boldsymbol{Q}_{i,k} \subseteq \boldsymbol{Q}_k$, where $i = \{1,2,\dots,K\}$, and the set of feasible solutions that do not belong to any cluster, $\boldsymbol{Q}_{0,k} = \boldsymbol{Q}_k \backslash \cup_{i=1}^{K} \boldsymbol{Q}_{i,k}$. K

represents the number of clusters and is initially set to 0. When the number of feasible solutions $|Q_{0,k}|$ increases and reaches the cluster size limitation N_{CS}, a new cluster $\boldsymbol{Q}_{K+1,k} \coloneqq \boldsymbol{Q}_{0,k}$ is generated. N_{CS} is given as a hyperparameter.

The probability density of feasible solutions is given as follows,

$$\Pr(\boldsymbol{x}) = \sum_{j=0}^{K} \pi_j \Pr(\boldsymbol{x}|\boldsymbol{\theta}_j), \tag{10}$$

where $\Pr(\boldsymbol{x}|\boldsymbol{\theta}_j)$ represents the probability function of j-th feasible solution cluster with a parameter $\boldsymbol{\theta}_j$, and π_j represents weight function of j-th cluster as follows,

$$\pi_j = \frac{|\boldsymbol{Q}_{j,gen}|}{|\boldsymbol{Q}_{gen}|}, \quad j \in \{0,1,2,\dots,K\}. \tag{11}$$

Any probability distribution function will be used for $\Pr(\boldsymbol{x})$, such as Gaussian or Poisson distribution. In this paper, it is defined as a Gaussian mixture model with N_{gc} Gaussian distributions, and the parameter $\boldsymbol{\theta} = \{\boldsymbol{\theta}_0, \boldsymbol{\theta}_1, \dots, \boldsymbol{\theta}_K\}$ is estimated by the EM algorithm.

In the proposed method, scale factor is given as follows instead of Eq. 2,

$$F_{i,k} = \underset{F_{i,k}}{\operatorname{argmax}} \left(\min_{\boldsymbol{x}_k \in \boldsymbol{P}_k} \|\boldsymbol{v}(\boldsymbol{F}) - \boldsymbol{x}_k\| \right), \tag{12}$$

where $\boldsymbol{P}_k$ represents the whole set of target vector.

3. Parameter calibration tests

3.1. Dynamic rest angle test

In order to identify DEM parameters, a dynamic rest angle test (Chatham et al., 2019) was carried out. Figure 3 shows a schematic configuration of the test. The regolith simulant LHS-1 was adopted as the sample for the test. The test samples are baked before the test to eliminate the effect of cohesion caused by moisture. After baking, the sample was placed in a drum, which was then depressurized to less than 5 Pa using a vacuum pump. Two measurements were focused on in this test, as shown in Fig. 4. The first is the avalanche angle θ_a, which is measured when the total potential energy is maximized before the sample avalanches. The second is the rest angle θ_r, which is measured when the total potential energy is minimized after the sample has avalanched. Table 1 shows the test condition of the dynamic rest angle test.

The results of the avalanche angle are shown in Fig. 5. For reference, the test results conducted under air conditions are also presented in the figure. When the drum's rotation rate is low, the avalanche angle in a vacuum is not different from that in the air. However, when the drum's rotation rate exceeds 10 rpm, the avalanche angle in the vacuum becomes larger compared to that in the air. It is inferred that the avalanche angle decreases with the increase in rotation rate due to the increased fluidity caused by the air being drawn into the sample within the drum. It is inferred that the shear stress at the contact area of the wheel during rotation under vacuum conditions will be greater than that in atmospheric conditions, leading to an increase in traction force.

The results of the rest angle are shown in Fig. 6. The rest angle shows a difference between air and vacuum when the drum's rotation rate exceeds 10 rpm.

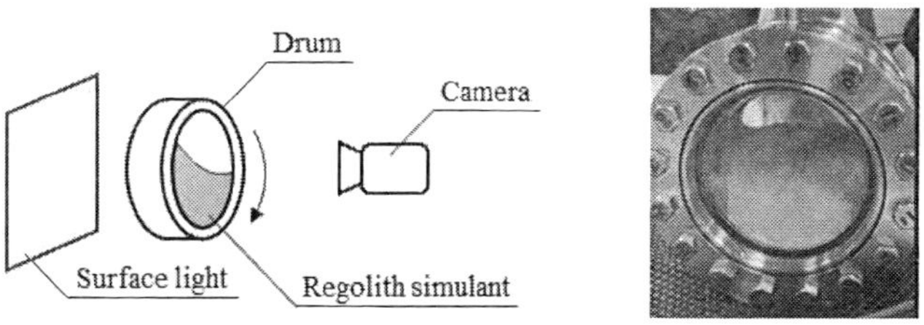

Fig. 3. Configuration of the dynamic rest angle test and a photo of the drum

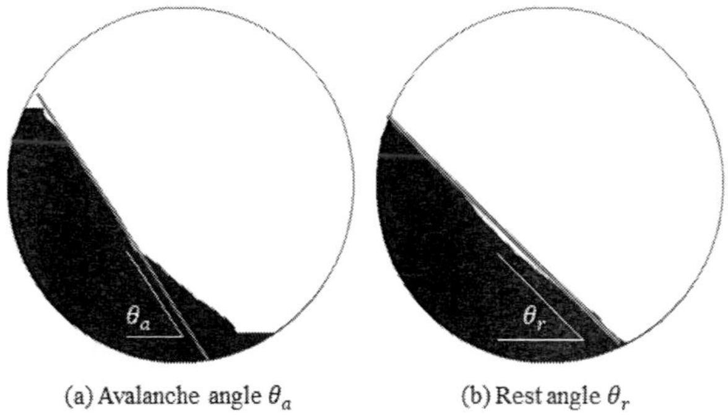

Fig. 4. Captured image of LHS-1 simulant under vacuum environment

Table 1
Test conditions of dynamic rest angle test.

Sample	Regolith simulant LHS-1
Gas in the drum	Air, Medium vacuum (≤ 5 [Pa])
Rotation rate	2, 5, 10, 15, 20 [rpm]

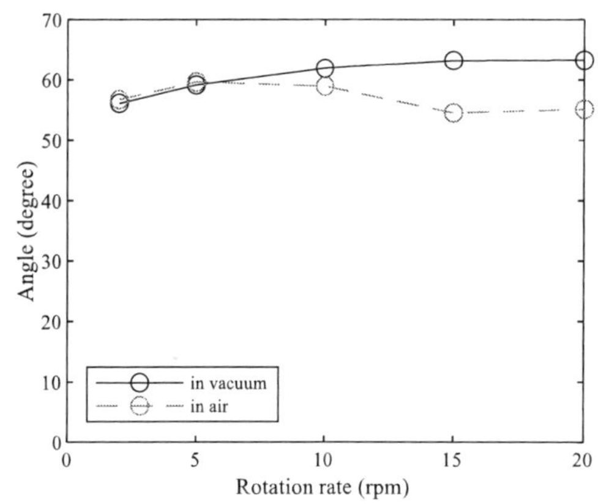

Fig. 5. Relationship between Avalanche angle θ_a and rotation rate

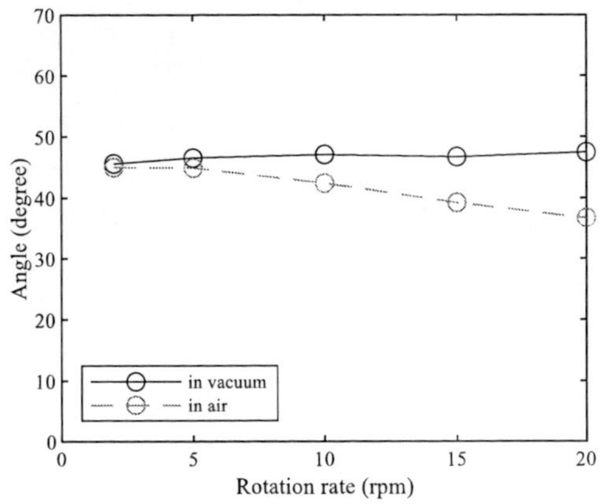

Fig. 6. Relationship between rest angle θ_r and rotation rate

3.2. Uniaxial compression test

In order to evaluate the cohesion under compression conditions, a uniaxial compression test (Kuentz and Schirg, 2013) was carried out. Figure 7 denotes a procedure of uniaxial compression test. In this test, we can obtain unconfirmed yield stress of the sample.

Figure 8 shows the setup of the test implementation. Figure 9 shows a photo at the unconfined compressing process in the test. Table 2 shows the test conditions of the uniaxial compression test.

The test results of unconfined uniaxial compression in air are shown in Fig. 10. The tests were conducted three times ($n = 1,2,3$) under the same conditions. There is some scatter in the test results. Samples were manually placed in the measurement cup. The variance of the tests may be attributed to the air present in the measurement cup.

The test results in a vacuum are shown in Fig. 11. The unconfined yield stress is approximately 1.0 kPa under these conditions. In a vacuum environment, the unconfined yield stress is lower than in an air environment.

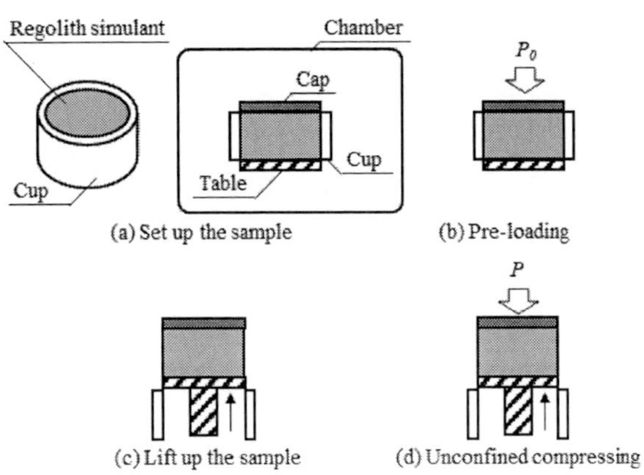

Fig. 7. Procedure of uniaxial compression test

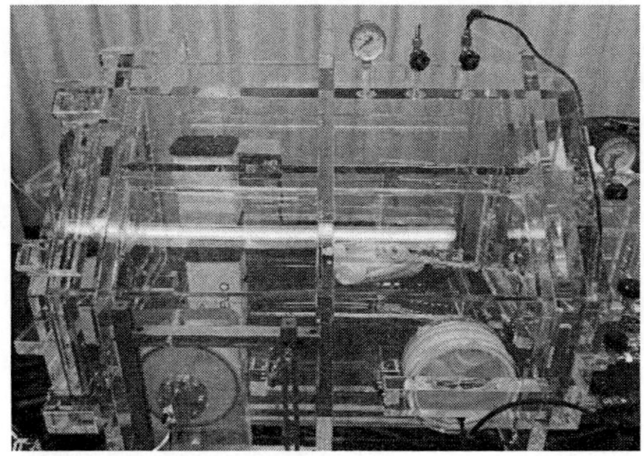

Fig. 8. uniaxial compression test set-up

Fig. 9. A photo at unconfined compression

Table 2
Test conditions of dynamic rest angle test.

Sample	Regolith simulant LHS-1
Gas in the Chamber	Air, Medium vacuum (≤ 5 [Pa])
Pre-pressure	20 [kPa]
Duration of pre-loading	5 minutes
Compression rate (pre-load)	0.2 mm /sec.
Compression rate (unconfined)	1.0 mm /sec.

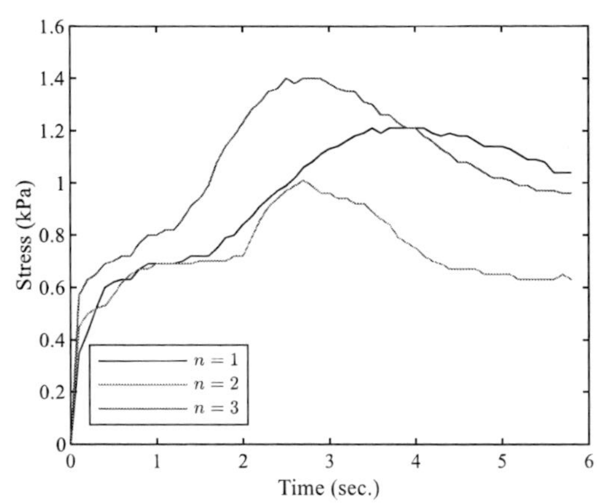

Fig. 10. Test result of unconfined uniaxial compression in air

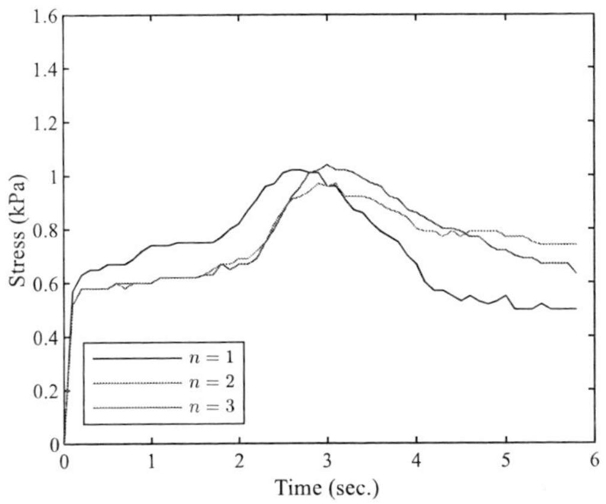

Fig. 11. Test result of unconfined uniaxial compression in vacuum

4. Parameter identification by DEBAS

Table 3 shows design variables for parameter identification. These variables are defined to represent the mechanical interactions between particles or between particles and geometry, and they are used in EDEM®, which is one of the commercial DEM software. Table 4 shows constraint functions. This paper focuses on the avalanche angle and rest angle in dynamic rest angle tests conducted at two different drum's rotational rates, as well as the unconfined yield stress in uniaxial compression tests. Each test was conducted in vacuum described in Section 3. Table 5 shows the specifications of the computer used for the numerical simulations.

Figure 12 shows an example using DEM that simulates the dynamic rest angle test. The particles are represented in two colors for the purpose of verifying the simulation results, but there is no difference in the properties used in the calculations. Figure 13 shows an example of a DEM model for the uniaxial compression test. In order to save the calculation time, the duration of pre-loading was set to 2 seconds, which is different from that used in the experiments. Table 6 shows invariable parameters for the DEM model. Particle diameter is set to a fixed value, which differs from the actual distribution. This is due to the constraints of computational cost. Other parameters are referenced from prior research (Tasnim et al., 2023).

Figure 14 shows the results of the parameter identification. Red markers indicate feasible combinations of design variables for all constraint functions, while black markers indicate unfeasible variables. From the figure, it can be observed that the influence of the design variable x_5 on the test results of this study is smaller compared to the other design variables. There are many combinations of design variables that satisfy all constraint functions. This demonstrates the effectiveness of the proposed clustering method. In this research, approximately 24,000 combinations are evaluated in total, and 1,600 solutions are identified as feasible. The process took about four weeks in total. As illustrated in the color transition sections of the figure, certain combinations of design variables may fail to satisfy the constraints due to minor variations in the variables. Conversely, combinations of design variables that exhibit robustness against such variations can also be identified. It is possible to obtain parameters that satisfy all given constraints and are robust to variations in the variables. Finally, the identified design variables are obtained as shown in Table 7. The identified variables have the minimum Euclidean distance to the goal in the problem domain $\boldsymbol{g}(\boldsymbol{x})$.

Table 3
Design variables.

	Category	Design variable	LB	UB
x_1	Particle to particle	Restitution	0.01	1.0
x_2		Static friction	0.0	1.0
x_3		Rolling friction	0.0	1.0
x_4		Surface energy	0.0	1.0
x_5	Particle to geometry	Restitution	0.01	1.0
x_6		Static friction	0.0	1.0
x_7		Rolling friction	0.0	1.0
x_8		Surface energy	0.0	1.0

Table 4
Inequality constraint functions.

	Cost functions	LB	Goal	UB
g_1	Avalanche angle at 5 rpm	42.0	59.0	60.0
g_2	Rest angle at 5 rpm	45.0	46.0	47.0
g_3	Avalanche angle at 10 rpm	44.0	63.5	64.5
g_4	Rest angle at 10 rpm	45.5	46.5	47.5
g_5	Unconfined yield stress	0.9	1.0	1.1

Table 5
Computer Specifications

Component	Specification
CPU	Intel ® Core ™ i9-14900 @ 2.0GHz
RAM	128GB DDR5-4800 UDIMM
GPU	Nvidia RTX 4080
Storage	2TB SSD
Operating system	Windows 11 Enterprise

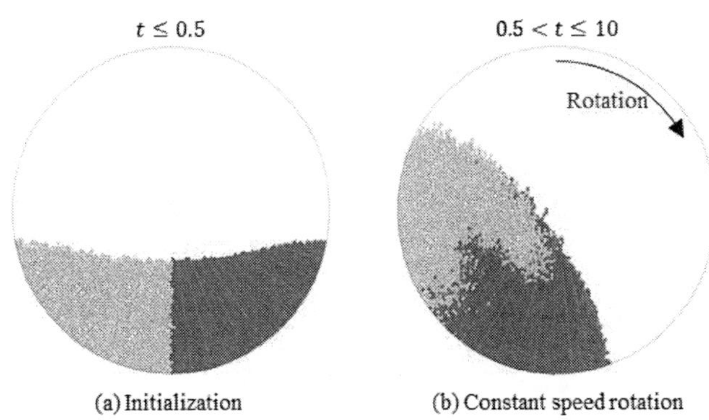

Fig. 12. An example of DEM Simulation for the dynamic rest angle test.

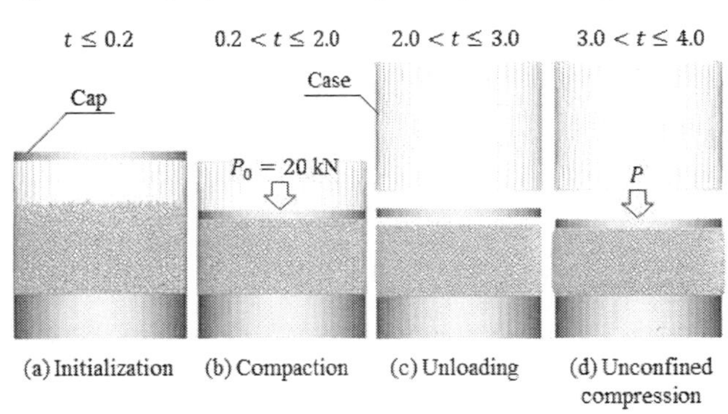

Fig. 13. An example of DEM model for the uniaxial compression test

Table 6
Input parameters for the DEM model

Parameters	Values
Particle diameter	1.0 mm
Particle solid density	2700 kg/m^3
Shear modulus	13.8 MPa
Poisson's Ratio	0.3

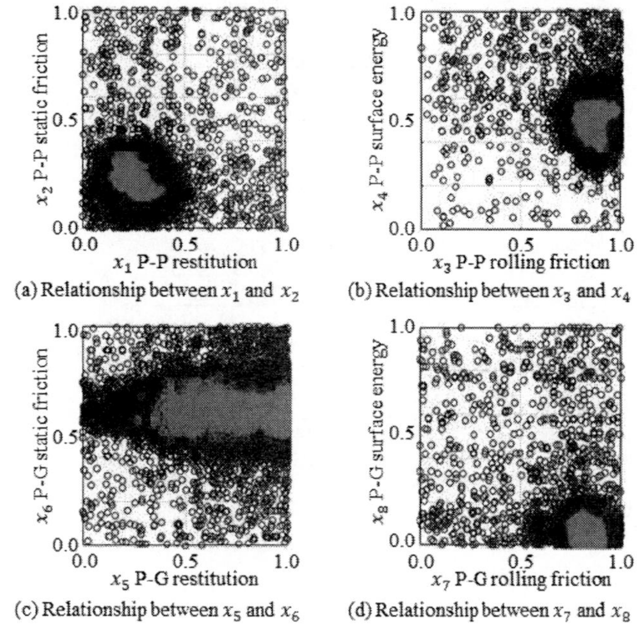

Fig. 14. Distributions of feasible solutions by proposed method

Table 7
Identified design variables

	Category	Design variable	value
x_1	Particle to particle	Restitution	0.163
x_2		Static friction	0.048
x_3		Rolling friction	0.750
x_4		Surface energy	0.458
x_5	Particle to geometry	Restitution	0.996
x_6		Static friction	0.468
x_7		Rolling friction	0.816
x_8		Surface energy	0.170

5. Validation of identified parameters

To validate the effectiveness of the obtained mechanical interaction parameters, a scaled wheel test was conducted and compared with the numerical analysis results.

5.1. Scaled wheel test

Figure 15 shows the schematics of the scaled wheel test equipment. Figure 16 shows a photo of the test equipment. The metal wheel and axle are rotated by a motor through a gearbox. The fixture supporting the axle is supported by the frame via two cylinders, allowing for free vertical movement. The contact pressure between the wheel and the soil bed is adjusted by a counterweight. Figure 17 shows the setup of the particle image velocimetry (PIV) system. The behavior of the regolith simulant that is kicked up by the wheel is captured using a high-speed camera. The field of view of the high-speed camera is depicted as Part A in Fig. 15. Table 8 shows the test conditions. The tests were conducted according to the following procedure:

1. Set up a soil bed covered with baked regolith simulant under the wheel.
2. Restrain the vertical movement of the wheel such that the lowest point of the wheel is 10 mm above the soil bed.
3. Maintain the absolute pressure inside the chamber to be below 5 Pa.
4. Rotate the wheel using a motor, with the motor torque controlled to achieve the target angular velocity of the wheel.
5. Release the restraint on the vertical direction of the wheel and allow it to fall freely.

Table 8
Conditions of the scaled wheel test

Soil	Regolith simulant LHS-1
Degree of vacuum	≤ 5 Pa
Diameter of wheel	130 mm
Width of wheel	50 mm
Mass of wheel (with axis support)	10 kg
Rotation rate of wheel	200 rpm

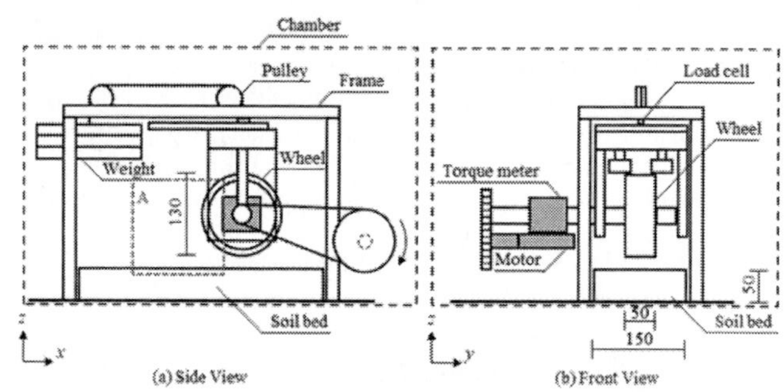

Fig. 15. Schematics of the scaled wheel test equipment.

Fig. 16. Test settings of scaled wheel model in a vacuum chamber.

Fig. 17. Setup of PIV systems, high speed camera and LED light.

Figures 18 and 19 show the results of the PIV analysis. Figure 20 shows the time history of the sinkage of the wheel and the wheel torque. The timestamps indicate the elapsed time since the bottom of the wheel touched the simulant. Red arrows indicate the velocity vectors of the particles. Immediately after the wheel lands, a large amount of the simulant is kicked backward behind the wheel. Meanwhile, a certain amount of the simulant is observed to be lifted upward along with the rotation of the wheel. The amount of the simulant that is lifted upward along the wheel is highest immediately after landing, as shown in Fig. 18; however, similar behavior was observed while the wheel is rotating, as indicated in Fig. 19. When the elapsed time was 3.55 seconds, as shown in Fig. 19, the number of kicked backward particles appeared to be less than the number of lifted-up particles.

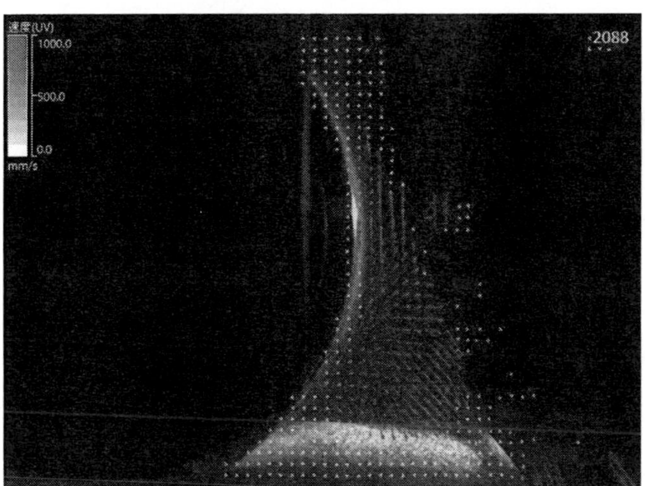

Fig. 18. Particle velocity distribution at time 0.25 sec.

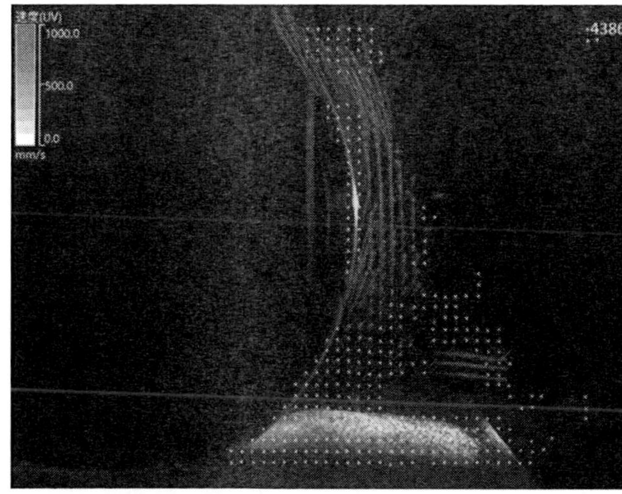

Fig. 19. Particle velocity distribution at time 3.55 sec

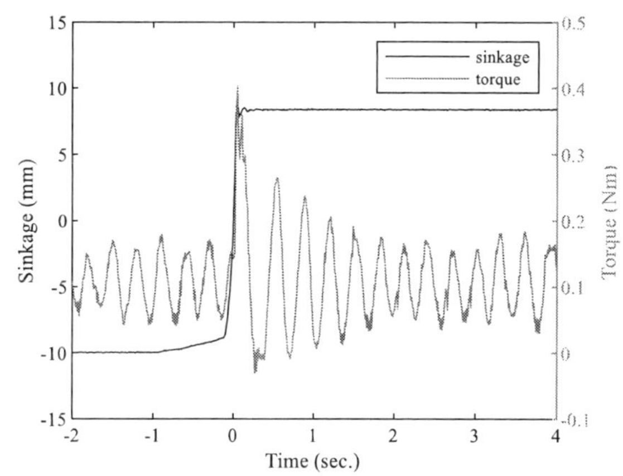

Fig. 20. Time history of wheel sinkage and wheel torque

5.2. Numerical example

A numerical analysis simulating the wheel-scale test was conducted. Figure 21 shows the numerical analysis model. The model is a coupled analysis model that integrates DEM analysis and multibody dynamics (MBD) simulation. The parameters for the DEM were set to the same values as those shown in Tables 6 and 7. The total number of particles is approximately 2.9 million. The timestep was set to 5×10^{-6} seconds. Additionally, the MBD model is designed such that a constant angular velocity is applied to the wheel, allowing it to fall freely under the influence of gravity. In Fig. 21, Part A and Part B are connected by a translational joint along the vertical axis. Part A is fixed to the ground. In this model, the frictional resistance at the mechanical parts is not taken into account. The computational time was 1.2 hours using a computer whose specifications are shown in Table 5.

Figure 22 shows the particle behavior results. The arrows in the figure indicate the velocity vectors of the particles that model the regolith simulant. It can be observed from the figure that the scattering of particles is most significant between $t = 0.15$ and $t = 0.3$, and the amount of scattering decreases after $t \geq 0.5$. between $0.3 \leq t \leq 0.4$, the simulant particles are seen scattering while remaining linear. This is due to the clumps formed at the bottom of the wheel as a result of compression when the wheel contacts the simulant particles, and they are scattered due to the centrifugal force from the wheel's rotation, peeling away from the wheel surface. The numerical results matched the experimental results qualitatively. The quantitative validation of the numerical results remains a future challenge. Additionally, in this paper, one feasible solution was selected from among many shown in Figure 14 and applied to the calculations of the wheel-scale test. Further investigation into the sensitivity of these parameters is also necessary in future research.

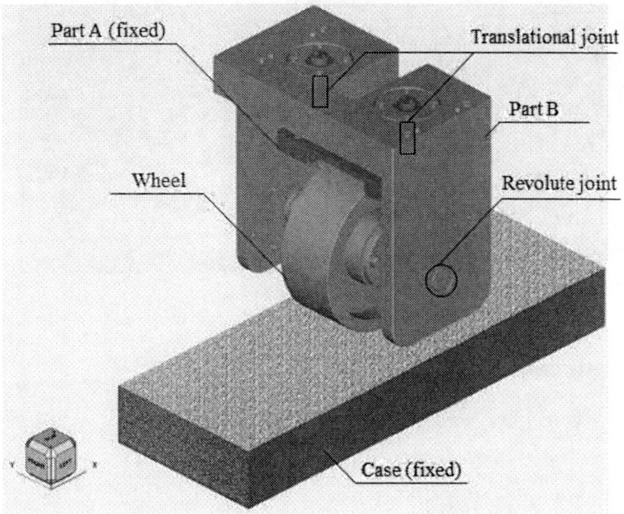

Fig. 21. A DEM simulation model coupled with multibody dynamics

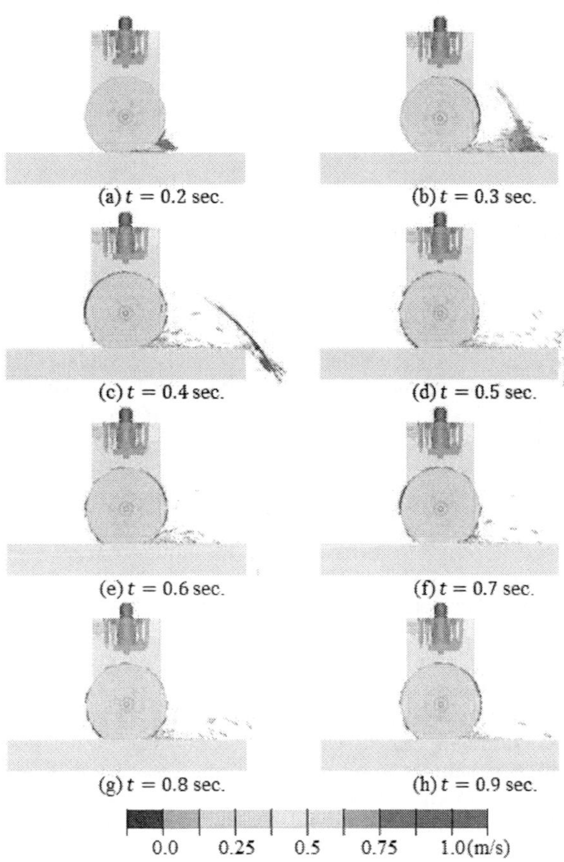

Fig. 22. Particle behavior in numerical simulation.

Figure 23 presents the numerical results of particle behavior under a gravitational acceleration of 1.62 m/s^2. These results indicate that by reducing the gravitational acceleration, a larger quantity of particles is lifted, resulting in a greater height of ascent. This can be attributed to two factors: first, the resistance of the soil is relatively lower when the gravitationally compacted regolith particles come into contact with the wheel; and second, the drop velocity at the point of wheel contact is $\sqrt{1/6}$ times that of the ground test, allowing the regolith simulant to be kicked up for a longer duration.

Figure 24 presents a comparison of the trajectories of simulant particles at $t \geq 0.5$ seconds. The figure shows the positions of the particles at intervals of $\Delta t = 0.01$ seconds, with colors indicating the magnitude of the particles' velocity vectors. The circular markers in the figure represent particles that are adhered to the wheel surface. It should be noted that the computational model does not account for the space where $Z \geq 0.2$ m, resulting in the absence of particle depiction in that region. Additionally, particles that are moving away from the wheel have velocities below the surface speed of the wheel, which is 1.36 m/s. Furthermore, under conditions of lower gravitational acceleration, the projection angle of the particles is smaller compared to conditions of higher gravitational acceleration. This is presumed to be due to the reduced initial adhesion force between the particles and the wheel, resulting from lower contact pressure.

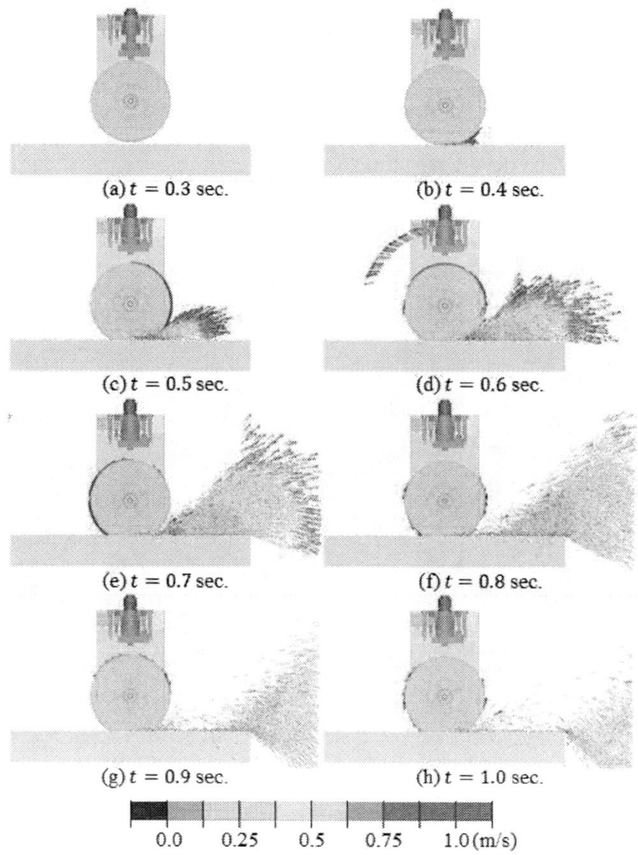

Fig. 23. Particle behavior under 1/6 G condition.

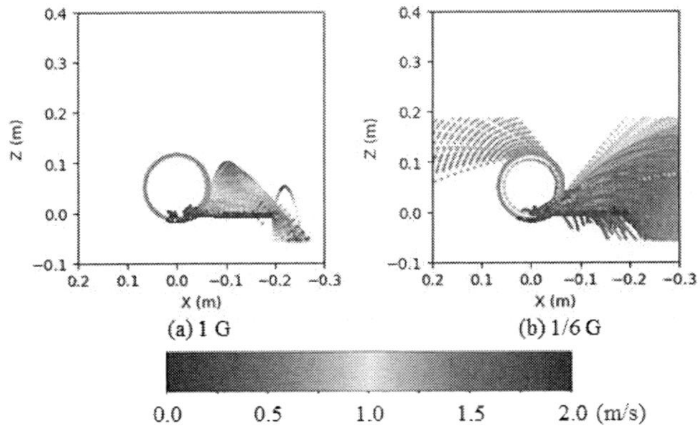

Fig. 24. Comparison of Particle behavior during t ≥ 0.5 seconds

6. Conclusions and future work

This paper proposes a new set-based design method, DEBAS, which improves upon DE and applies it to the identification of particle parameters in DEM. To eliminate interactions caused by air, tests necessary for parameter identification were conducted under vacuum conditions. Clear differences were observed between the test results under vacuum conditions and those conducted in air for both dynamic rest angle tests and uniaxial compression tests. Using the parameters identified by the proposed method, tests were conducted to confirm the mechanical interactions between the wheel and the regolith simulant, and comparisons were made with the numerical calculation results. The comparison results demonstrated the effectiveness of parameter identification using the proposed method. In this study, all DEM parameters except for contact were fixed values. The influence of particle size and particle shape on uplift behavior is considered significant. Additionally, the mechanical interactions under vacuum conditions raise concerns about the effects of frictional charging on the regolith simulant. Further investigation into these effects is necessary.

References

Bekker, M. G., 1960. Off-The-road locomotion, The university of Michigan press, Ann Arbor, MI.

Brest, J., Greiner, S., Boskovic, B., Mernik, M., Zumer, V., 2006. Self-Adapting Control Parameters in Differential Evolution: A Comparative Study on Numerical Benchmark Problems, IEEE Transactions on Evolutionary Computation, vol. 10, no. 6, pp. 646-657.

Chatham, C. A., Long, T. E., Williams, C. B., 2019. Powder bed fusion of poly (phenylene sulfide) at bed temperatures significantly below melting, Additive Manufacturing, 28, 506-516.

Hansen, N., Ostermeier, A., 1995. Adapting arbitrary normal mutation distributions in evolution strategies; The covariance matric adaptation, Proceedings of IEEE international conference on evolutionary computation, IEEE.

Holland, J. H., 1975. Adaptation in natural and artificial systems: an introductory analysis with applications to biology, control, and artificial intelligence, The university of Michigan press, Ann Arbor, MI.

Ishigami, G., Miwa, A., Nagatani, K., Yoshida, K., 2007. Terramechanics-based model for steering maneuver of planetary exploration rovers on loose soil, J. of Field Robotics, 24, 3, 233-250.

Kennedy, J., Eberhart, R., 1995. Particle swarm optimization, Proceedings of ICNN'95 - International Conference on Neural Networks, Perth, WA, Australia, 4, 1942-1948.

Kreisselmeier, G., Steinhauser, R., 1979. Systematic control design by optimizing a vector performance index. IFAC Proceedings Volumes, 12, 7, 113–117.

Kuentz, M., Schirg, P., 2012. Powder flow in an automated uniaxial tester and an annular shear cell: A study of pharmaceutical excipients and analytical data comparison. Drug development and industrial pharmacy. 39.

Singer, D. J., Doerry, N., Buckley, M. E., 2009. What Is Set-Based Design?, Naval Engineers Journal, 121, 4, 31-43.

Sobek, D., Ward, A.C., Liker, J., 1999. Toyota's Principles of Set-Based Concurrent Engineering, Sloan Management Review. 40.

Stom, R. and Price, K., 1997. Differential evolution: A simple and efficient heuristic for global optimization over continuous spaces, J. Global Optimization, 11, 4, 341-359.

Tanabe, R., Fukunaga, A., 2013. Success-history based parameter adaptation for Differential Evolution, 2013 IEEE Congress on Evolutionary Computation, Cancun, Mexico, pp. 71-78

Tasnim, Z Chen, Q., Badal, J., Wang, L., 2023. Characterization and discrete element modeling of LHS-1 Lunar highlands simulant, Geotechnical Special Publication, 2023-March, GSP 342, 74-84.

Wong, J., 2022. Theory of ground vehicles, fourth ed., Wiley.

Zhang, J., Sanderson, A. C., 2009. JADE: Adaptive Differential Evolution With Optional External Archive, IEEE Transactions on Evolutionary Computation, vol. 13, no. 5, pp. 945-958.

7222 | A simulation framework for autonomous lunar construction work

Mattias Linde [a], Daniel Lindmark [a], Sandra Ålstig [a], Martin Servin [a,b,*]

[a] *Algoryx Simulation AB, Umeå, Sweden*

[b] *Umeå University, Umeå, Sweden*

* *Corresponding author: martin.servin@algoryx.com*

ABSTRACT

We present a simulation framework for lunar construction work involving multiple autonomous machines. The framework supports modelling of construction scenarios and autonomy solutions, execution of the scenarios in simulation, and analysis of work time and energy consumption throughout the construction project.

The simulations are based on physics-based models for contacting multibody dynamics and deformable terrain, including vehicle-soil interaction forces and soil flow in real time. A behaviour tree manages the operational logic and error handling, which enables the representation of complex behaviours through a discrete set of simpler tasks in a modular hierarchical structure. High-level decision-making is separated from lower-level control algorithms, with the two connected via ROS2. Excavation movements are controlled through inverse kinematics and tracking controllers.

The framework is tested and demonstrated on two different lunar construction scenarios that involve an excavator and dump truck with actively controlled articulated crawlers.

Keywords
Lunar construction
Excavation
Physics-based simulation
Automation
Behaviour trees

1. Introduction

Planning construction work on the Moon, such as the Artemis project lunar base, is a challenging task. Many aspects remain unknown, including local ground conditions, how the machines should be designed and operated to accomplish the task within the stipulated time and adhere to limited energy budgets. The machines should possess a high degree of autonomy due to the significant delays associated with remote control from Earth. Additionally, it is important that multiple machines can work in a well-coordinated manner, avoiding idle times and utilizing the equipment as efficiently as the energy supply allows for.

To succeed in this and similar tasks, simulation-based tools are crucial for studying how different configurations of the lunar conditions, machine design, and control system affect the time and energy consumption in a given construction project. These tools are also essential for detecting weaknesses in the entire approach and reducing the risk of severe failures. Additionally, such tools can also be used for generating large amounts of synthetic data for training deep learning models. To effectively iterate over numerous scenarios, each involving long sequences of machine operations, the simulations must be highly computationally efficient while remaining sufficiently realistic in terms of forces, power consumption, material flow, and vehicle behaviour. Furthermore, it must be easy to define and test different work plans, autonomous functions, and combinations of sensors.

To this end, we develop a framework for simulating autonomous ground construction tasks in a lunar environment. The framework leverages physics-based simulations of autonomously controlled construction machines interacting with dynamic terrain, utilizing the real-time AGX Dynamics physics engine (Algoryx, 2025; Servin 2021). The systems' autonomous control is modeled using behaviour trees (Colledanchise, 2018, Iovino 2022), which enables the representation of complex behaviours (tasks) through a discrete set of simpler tasks organized in a hierarchical and modular structure. High-level decision making is performed in the behaviour tree, whereas low-level control algorithms are handled by skills on the different machines. This approach makes it easy to replace and introduce new capabilities and new machines. Communication between the machine's controller and the current state of the lunar environment occurs via ROS2 (Macenski, 2022), simplifying the replacement of virtual components with their physical counterparts.

Our main contribution in the present work is a description of said framework and results from testing it on two scenarios, namely extracting lunar regolith and excavating a predefined ground structure. The tests involve one excavator and one dump truck, but the framework is inherently designed to support many different machines. The work illustrates that the framework supports analysis of how the construction performance and energy use depend on different environmental factors.

1.1. Related work

There are few prior publications on simulation frameworks for autonomous lunar construction. A recent work is the framework introduced in (Batagoda, 2024). It supports synthetic sensors, vehicle dynamics, and terramechanics using the Chrono physics engine. The terramechanics models include the Soil Contact Model (SCM) for wheel-terrain interaction, and continuum (SPH) and discrete (DEM) soil models for general use, including excavation and bulldozing. The SynChrono module give support for multi-agent systems, e.g., sharing of state information with synchronised but distributed simulation across computing nodes. The

https://doi.org/10.56884/PQZP98ES

simulations involving ground construction work are too computationally intense for use in applications with real-time or higher performance. Case studies in the paper focus on generation and analysis of synthetic images using physically-based rendering for realistically capturing lunar lighting conditions.

In (Thangavelautham, 2017), autonomous multirobot excavation for lunar applications was controlled using artificial neural tissue (ANT), trained using multi-agent grid world simulations and evaluated using a higher-fidelity 3D simulation based on the tool Digital Spaces with the fundamental earthmoving equation (FEE) for bucket-soil interaction (El Samid 2008). Modeling or analysis of power transmission and consumption is not addressed. In simulation, the ANT controller was found to achieve 30% better fitness (target grid cells being excavated) than with a conventional neural network.

From NASA, the Digital Lunar Exploration Sites Unreal Simulation Tool (DUST) (Bingham, 2023) makes lunar terrain data and several computational tools available in the Unreal Engine 3D environment to support the Artemis mission. NASA-Ames and Open Robotics have developed a Gazebo-based lunar rover simulator. Wheel-terrain interaction is accounted for by empirical drawbar-pull and slip models, and terrain deformations are not supported.

OmniLRS is a simulation framework under development that uses IsaacSim from NVIDIA for photorealistic simulation of planetary rovers on lunar terrain (Richard, 2024) with a data-driven wheel-terrain interaction model (Kamohara, 2024). As far as we are aware, the framework does not support excavation or bulldozing. Perception models trained on synthetic data achieved a performance 5% short of a model trained on real-world data and 14% better when finetuned with real data.

In (Sutoh, 2024), a tracked excavation rover platform was studied in simulations based on the Vortex physics engine from CM Labs and by field experiments. No conclusions were reported.

Teleoperation tasks for lunar surface construction, in particular excavation, were studied in virtual reality using the Unity Engine in (Seo, 2024) to examine the effect of time delay. The outcomes indicate that time delays significantly degrade task performance, and the operators modify their control strategies to cope with the time delay conditions.

None of the related work offers the combined functionality of real-time physics-based simulation of lunar construction work and representation of complex behaviour of autonomous machines for executing the construction work plan.

2. Method

This section introduces the underlying methods for physics-based simulation, behaviour tree-based task planning for multiple mobile machines, and ROS for interconnecting them.

2.1. Physics-based simulation

For modeling and simulation, we rely on the physics engine AGX Dynamics (Algoryx, 2025). It is based on the framework for contacting multibody dynamics introduced in (Lacoursière, 2007), extended to the discrete elements in (Servin, 2014), and to dynamic deformable terrain in (Servin, 2021). Specifically, it uses maximal coordinate representation in terms of rigid bodies and kinematic constraints for joints, motors, and frictional contacts. This has been used in previous work to develop machine learning models for autonomous control of earthmoving equipment (Backman, 2021; Aoshima, 2024) and active suspension system for rough terrain navigation (Wiberg, 2021) with successful transfer of controllers trained in simulation to reality (Wiberg, 2024).

The time-stepper and solver are optimized particularly for fixed timestep real-time simulation of multibody systems with non-ideal constraints and non-smooth dynamics (Lacoursière, 2011). The hybrid direct-iterative solver supports fast and stable simulations at high accuracy for mathematically stiff and ill-conditioned systems, such as robots and vehicles, and scalability to large-scale dynamic contact networks, such as for granular systems, at the price of lower accuracy (Servin, 2014).

For deformable terrain, AGX uses the multiscale model described in (Servin, 2021). It has been demonstrated to produce digging forces and soil displacements with an accuracy of 75-90% to that of a resolved discrete element method (DEM), coupled with multibody dynamics, and with field tests involving full-scale construction equipment (Aoshima, 2024). The model can be regarded both as a reduced-order model of DEM and as a multibody dynamics generalization of the FEE. The idea is to dynamically resolve only the part of the terrain inside a well localized region of shear failure. The terrain is assigned a set of bulk mechanical parameters for its physical behaviour in nominal bank state. When a digging tool contacts the terrain surface, a zone of active soil displacement is predicted. Only inside the active zone is the soil resolved into particles. These are modeled using DEM with specific mass density and contact parameters that ensure a bulk mechanical behaviour consistent with the assigned bulk parameters through a pre-calibration procedure. See, Ref. (Servin 2021) for details. The reaction force on the bucket consists of frictional-cohesive contacts with an aggregate body, which inherits the physical properties of the soil in the active zone, and a penetration resistance constraint, which is a function of the bucket geometry, soil parameters, and the soil pressure at the cutting edge.

The particle size varies dynamically in the multiscale model by mass exchange between particles and with the terrain. This provides a balanced tradeoff between fine resolution, close to the bucket cutting edge and the terrain surface, and coarser elsewhere. To avoid the DEM simulation being a computational bottleneck, large particles are favored by the algorithm. With the multiscale approach, the errors in digging forces remain smaller than if the terrain were composed of DEM particles. However, large particles misrepresent the real soil flow and distribution inside the bucket. Specifically, the volume fraction is overestimated at the soil-bucket interface. Consequently, with large particles the bucket filling is typically underestimated unless one compensates by assigning the particles a higher mass density by a factor proportional to the particle cross-section area over the bucket's internal surface area.

The terrain model can also be assigned to dynamic bodies, such as the load-carrying truck bed of a dump truck. Terrain particles that are dumped and settling to rest, are converted to a local terrain that co-moves with the dump truck bed and adds to its mass and inertia tensor. When the truck bed is raised and tilted to a critical angle, the terrain starts avalanching and is resolved into particles until they come to rest again.

2.2. Behaviour trees

To achieve modular, scalable, and explainable machine behaviours, behaviour trees (BT) are utilized. They are widely used in artificial intelligence applications for representing task switching policies. See (Iovino, 2022) and (Colledanchise, 2018) for an in-depth treatment. A BT is a rooted tree structure where the nodes are connected with parent-child relationship. When the tree is executed, or ticked, traversal begins at the root and proceeds according to the types of nodes and the status of the nodes. Each node returns one of the following status values, *Success, Failure, Running* and the status values affect the traversal for the common node types in the following way:

Sequence A sequence node ticks all its children in order. If any child returns *Failure* or *Running*, the sequence halts and returns the same status. The sequence returns *Success* only if all children succeed.

Selector A selector node, also called fallback node, ticks its children in order until one returns *Running* or *Success* and the selector will return the same status. If all children fail, then the selector also returns *Failure*.

Parallel A parallel node ticks all children concurrently. A policy (e.g. succeed on one, succeed on all, succeed on specific child) determines how the children's statuses are aggregated into the parallel nodes return status.

Decorator A decorator is a node with a single child that can change the child status, e.g. an Inverter swaps *Success* and *Failure.*

Condition A condition is a leaf node that checks a condition and returns either *Success* or *Failure.*

Task A task or action is a leaf node that represents an action. It returns *Running* while the action is ongoing, *Success* when the action is completed, and *Failure* if the action could not be performed.

2.3. Robot Operating System

Robot Operating System (ROS) is a middleware framework for robot software development (Quigley, 2009). It provides tools, libraries, and conventions for creating complex and robust robot behaviour across different robotic platforms. With standardised interfaces, ROS encourages modular development with packages for specific robotic functionalities and services designed for heterogeneous computer clusters, including hardware abstraction and low-level device control. With ROS2, it also functions as a middleware for communication (Macenski, 2022). The message passing interface enables independent processes, nodes, to communicate via publications and subscriptions.

3. Simulation framework

This section begins with a general overview of the autonomous construction framework, followed by a detailed explanation of how its components are used to control machines for performing autonomous tasks.

The autonomous simulation framework is implemented in Python and consists of two main components: *a high-level planner* and a *simulator.* These components run as separate processes and communicate using ROS2. This separation introduces a key feature: the planner does not have direct access to the ground truth of the world state, as that information resides within the simulator. Instead, the planner must rely on observations being sent from the machines to update its own world state. Although there is no technical barrier preventing the components from being in the same process, maintaining them as separate processes offers several advantages. It enhances modularity, facilitates the replacement of individual parts and supports the development of more robust autonomous systems where perfect information cannot be assumed. Furthermore, this architecture simplifies future extensions, such as introduction of extra communication delays or disturbances. An overview illustration of the framework is found in Fig. 1.

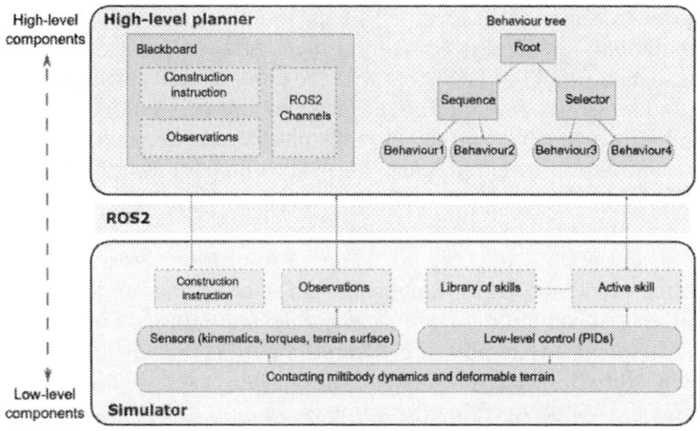

Fig. 1. Overview of the framework with its two main components, the high-level planner and the simulator interconnected with ROS.

3.1. High-level planner

The tasks to be performed by the planner are defined using a behaviour tree. This tree is composed of PyTrees' built-in *composites* and *decorators*, along with a set of predefined framework behaviours, and when necessary, user-defined behaviours. Tree nodes can share data via a common blackboard and may add information for future node ticks. The framework makes use of the *memory* feature for sequence nodes – where the status *running* results in the sequence being restarted from the running node on the next tick instead of from the beginning.

Users define the behaviour tree using a structured text format. Each non-empty line includes an *optional node name*, a *node specification* and an *optional machine context.* Indentation levels are used to indicate parent-child relationship between nodes. Comments can also be added using a simple annotation format. The node specification can either reference a class name or a Python method that returns a node or subtree. This allows for modular reuse of subtrees across different machines, minimizing code duplication. The machine context, specified as `-> machinename`, indicates which machine a subtree should interact with. This enables the reuse of identical node types multiple times within a tree, each instance operating on a different machine. An example of textual representation of a behaviour tree is illustrated in Fig. 2 with the corresponding tree representation in Fig. 3.

```
Root : Sequence
   Parallel : ParallelSuccessOnFirst
      # Scenario completed ? -> Tree Success
      Scenario : Sequence
            ...
      # Decorator between machine subtrees
      # and parallel node are omitted .
      Machine1 : Sequence -> excavator1
            ...
      Machine2 : Sequence -> dumptruck1
            ...
```

Fig. 2. Example of a textual representation of a behaviour tree. Dots are used to indicate omitted nodes or subtrees to keep the example short.

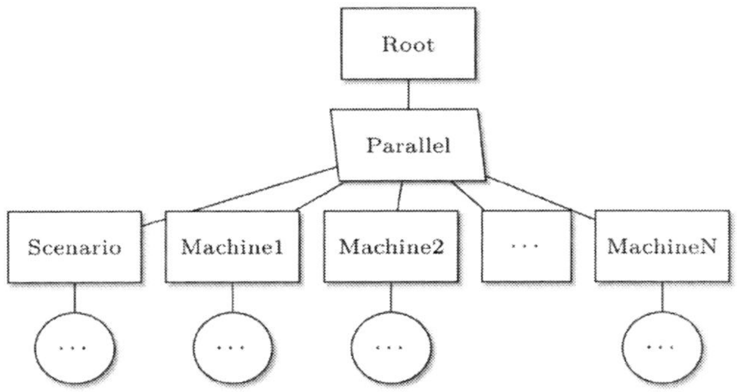

Fig. 3. Example structure on a high level for a behaviour tree with multiple machines.

3.2. Simulator

The simulator component features a dynamic terrain and one or more machines. Each machine is equipped with a set of *skills* - capabilities that can be activated by the planner. Skills may also accept optional input data required for their execution. These skills correspond to specific actions that machines can perform through their low-level control systems.

Communication between the planner and simulator occurs over multiple ROS2 topics. Each node in the behaviour tree can have an associated skill on the machine, and the communication topic incorporates the machine context. For example, a topic could use `/{machine context}/target/drive` as name template. This naming convention allows generic nodes within the planner's behaviour tree to interact with specific machines, supporting modular and scalable control logic. The same machine-context naming scheme is used for the planner to subscribe to telemetry and evaluation data from machines. These topics provide metrics such as execution time, energy consumption and skill status. Collecting this data enabled the assessment of a work plan's efficiency and overall system performance.

Case study: *Excavating terrain to match a target profile.* To execute an excavation task using the autonomous framework, the process begins by selecting which machines will participate in the operation. Afterwards, a logical plan must be formulated outlining how the task should be executed and then translated into a behaviour tree structure. With the tree setup, the next step involves preparing an initial configuration, currently done primarily using JSON files, which ensure that both the simulator and planner share a consistent initial view of the scenario configuration. Once this setup is complete, both components can be launched, allowing the machines to begin autonomous operation.

Assuming a similar structure as depicted in Fig. 3 and that a single excavator and one dump truck are involved, the behaviour tree structure offers a highly intuitive way to model and reason about the roles and actions of each machine. The use of a *parallel* node within the tree architecture allows each machine's subtree to operate independently. This modularity simplifies the addition of new tasks or machines without devolving into the complexity associated with traditional state machines.

A logical breakdown of the excavators' tasks is: if the bucket is empty, plan where digging should be performed and plan a drive trajectory towards the location, drive, and dig. If the bucket is non-empty, then a dig cycle has been completed, and then a dump cycle should be performed. If the bucket has material, plan where the material should be offloaded and plan a drive trajectory to the location, drive, wait for a dump truck to be in position, offload material. This workflow is illustrated in Fig. 4.

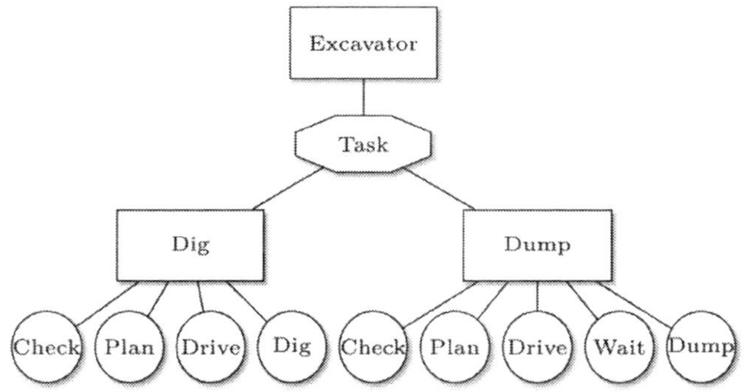

Fig. 4. Subtree for an excavator.

The tasks for a dump truck can be divided and expressed in a similar way with two subtrees: Plan where to accept material and plan a route to the loading location if the truck bed is not full, then drive to the location. If the truck bed is full, plan where to offload material and plan a route to the target, wait until no excavator is offloading material, drive, and empty the truck bed.

The nodes shown in the subtrees in Fig. 4 and 5 map to functionality implemented in the framework. Nodes responsible for various types of planning do not activate machine skills directly. Instead, they carry out computation and store the result on the blackboard, allowing subsequent nodes to make use of that information. For instance, a drive node retrieves the precomputed route information, activates a driving skill on a machine, and passes along waypoint data required for execution.

To maintain execution integrity, especially in scenarios where certain tasks may fail and require re-execution, *FailureIsRunning* decorators are placed between a machine's subtree and the parallel node. These decorators prevent PyTrees from reinitializing all subtrees in response to a failure, thereby supporting selective re-planning and robust recovery mechanisms.

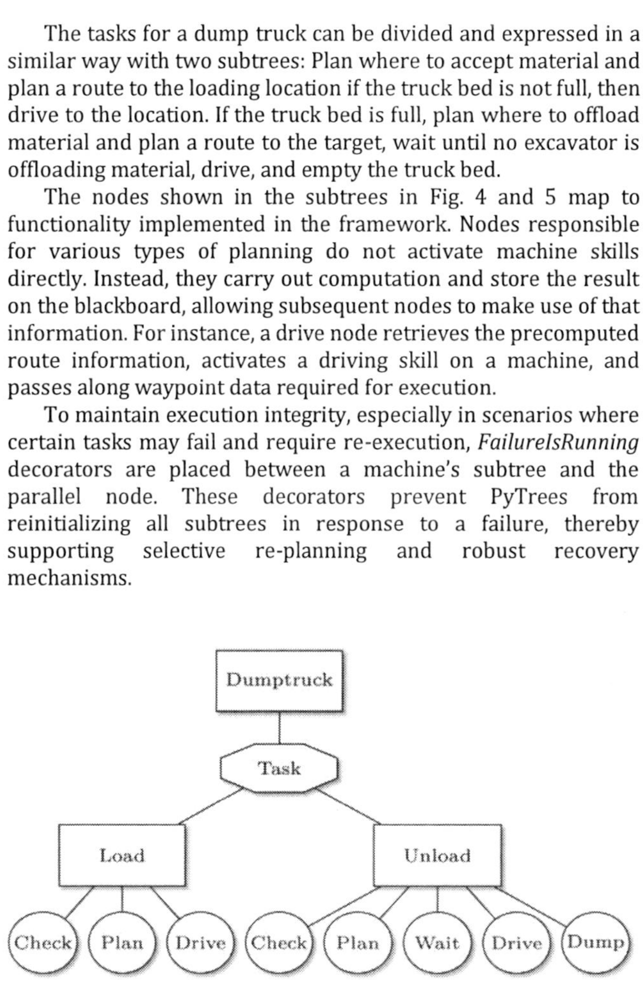

Fig. 5. Subtree for a dump truck.

The final portion of the behaviour tree addresses scenario completion. Determining when a scenario should be considered complete depends on how tasks, primarily the digging, are organized and carried out. A common strategy involves dividing the construction site into a local grid of cubic cells, with the current cell index stored on the blackboard. During dig planning, the target profile is compared against the current world state for the active grid cell. If the current terrain matches the target within a predefined tolerance, the cell index is increased and planning restarted. If material needs to be removed, a dig trajectory is planned for the current active cell, taking the target profile into account. The scenario is considered complete when the final grid cell has been processed. This condition can be verified by a behaviour tree node that reads from the blackboard and checks if all grid cells have been successfully handled.

3.3. Motion planning and control

When the excavator should perform a certain task, for example digging out a section of the terrain, local joint-space values are needed that result in the end-effector performing the desired motion in task space. To compute these joint values, the Inverse Kinematics (IK) module in AGX Dynamics is used. The excavator model in the simulation contains all the information that is needed to do IK computations without specifying additional link lengths or joint positions. The attachment frames for the kinematic constraints already hold all the needed information, including possible joint ranges. Therefore, redesign of the machine can be done during prototyping without maintaining redundant values that risk becoming out of sync. The types of constraints that are supported are Hinge (rotational), Prismatic (sliding) and Lock (fixed).

The API for computing Inverse Kinematics in `agxModel` accepts an `AffineMatrix4x4` that contains the desired end-effector transform. That transform might not be reachable due to being e.g outside of the working range, or the desired orientation might not be achievable due to the degrees of freedom available in the mechanical structure.

To simplify the usage of IK computations, the `LunarExcavator` class has a method, `calculate_ik`, that accepts a world position and an angle versus the ground for the digging direction. This is to avoid requesting rotations in directions the excavator bucket can not physically rotate. Despite this, it is possible to request end-effector transforms which can not be realised by the machine.

When a transform can not be reached with sufficient precision, an error code is returned as well as the best joint values the IK-solver could find. These joint values can be used in a forward computation to see where the end-effector could go. Comparing the reachable transform with the requested transform is something a control algorithm can do as a step to decide how to proceed.

4. Numerical experiments

The framework was tested for its ability to support the analysis of autonomy solutions for lunar ground construction with multiple machines. This was done with numerical experiments covering two scenarios, one where lunar material is loaded and transported from an excavation area to a dump site, and a second one where the goal is to create a ground construction of a given shape. The simulations run approximately in real time with a time-step of 10 ms with the behaviour tree accounting for about 0.3 ms of the computing time per time-step. Simulations were run on an AMD Ryzen 9 5900X desktop processor with 12 CPU cores and 3.7 GHz clock speed (GPU does not accelerate these simulations). Video material from the simulations can be found at https://www.algoryx.se/papers/lunar-construction.

4.1. Construction machines and lunar environment

Both scenarios involve an excavator and a dumper. They are both equipped with crawlers with actively controlled sub-crawlers that promote high terrain mobility. When turning, the sub-crawlers are lifted to reduce turning resistance. While excavating and driving straight, the sub-crawlers are lowered to reduce sinkage and actively controlled to smoothly follow the terrain surface. The speed of the crawlers are controlled individually via a rotational motor for each sprocket. The motion of the sub-crawlers relative to the main crawler is also controlled via rotational motors.

Fig. 6. 3D models of the excavator and the dump truck. Actuated joints are marked in green. The machine width is 2.5 m.

The machines are 3.3 m long and 2.5 m wide. The excavator has 3 ton mass, and the empty dump truck weighs 2.0 tons. The actuators can deliver torque up to a maximum value but the required torques never reached these limits in the experiments. The target drive speed of the machines was 0.3 m/s when carrying a load and 0.35 m/s when driving empty. The crawlers were modelled using the AGX Tracks module. The number of track elements is about 50 and 20, for the main crawlers and sub-crawlers, respectively. To support numerical stability at integration with 10 ms time-steps, strong numerical viscous damping is enforced in the track element interconnection joints. That results in excessive power consumption when driving the crawlers. Therefore, the computed power was normalised by subtracting the dissipation observed when driving the tracks with no resistance from the terrain (having the vehicle elevated above ground) and introducing an efficiency factor (constant).

The lunar surface has an initial shape represented by an artificially generated elevation map. The terrain is given uniform soil properties, specifically the regolith was assigned a bank state internal friction angle of 0.80 rad, cohesion 900 Pa, dilatancy angle 0.23 rad, mass density of 1580 kg/m3 at packing density 0.66, and compression index of 0.11.

4.2. Scenario 1 - Excavate, haul, and dump lunar regolith

The construction instruction in this scenario was to repeatedly excavate and move soil from a dig area to a dump area with boundaries specified by coordinates. The excavator and dump truck have behaviour trees as described in the previous section.

Two variations of the scenario were made, and we refer to them as *flat* and *sloped*. They differ by the location of the dump area. In the flat case, the dump area is located in the same region of rather flat terrain as the dig area. See Fig. 7 for an illustration. In the sloped case, the dig and dump areas are separated by a slope that the excavator must cross each digging cycle to reach the dump truck at the lower region. The distance to reach the dump truck from the dig area is, on the other hand, 20% shorter on average in the sloped case.

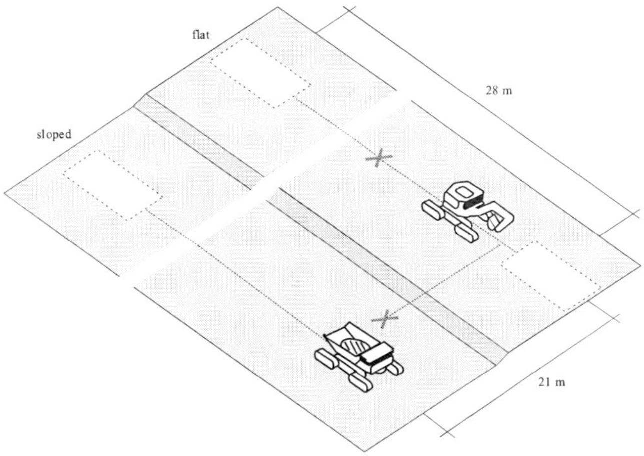

Fig. 7. Illustration of Scenario 1 - Excavate lunar regolith from a dig area (right dashed rectangle), offloaded on a dump truck that carries the material to a dump area (left dashed rectangle) where it is offloaded. Two dump areas are considered, separated by a slope. The scales are not accurate to the real scene.

The dump truck will start the dumping process when its target load is reached. When driving to the dumping position, an extra waypoint will be created where the dump truck will get the correct dumping orientation. It will then drive straight to the final position. This is to avoid unwanted interaction between the pile and the dump truck. When dumping, the dump bed will rest in a raised position for a couple of seconds. The dump truck will then move forward with the bed still raised before lowering the bed and driving towards a loading position.

The construction work was simulated over 30 digging cycles, and virtual sensor data was collected, including time series for vehicle pose, kinematics, actuator torques, and loaded and dumped soil mass. Power consumption was computed by integrating the (positive) work exerted by the actuators. Image sequences from the two simulations can be found in Fig. 8 and Fig. 9.

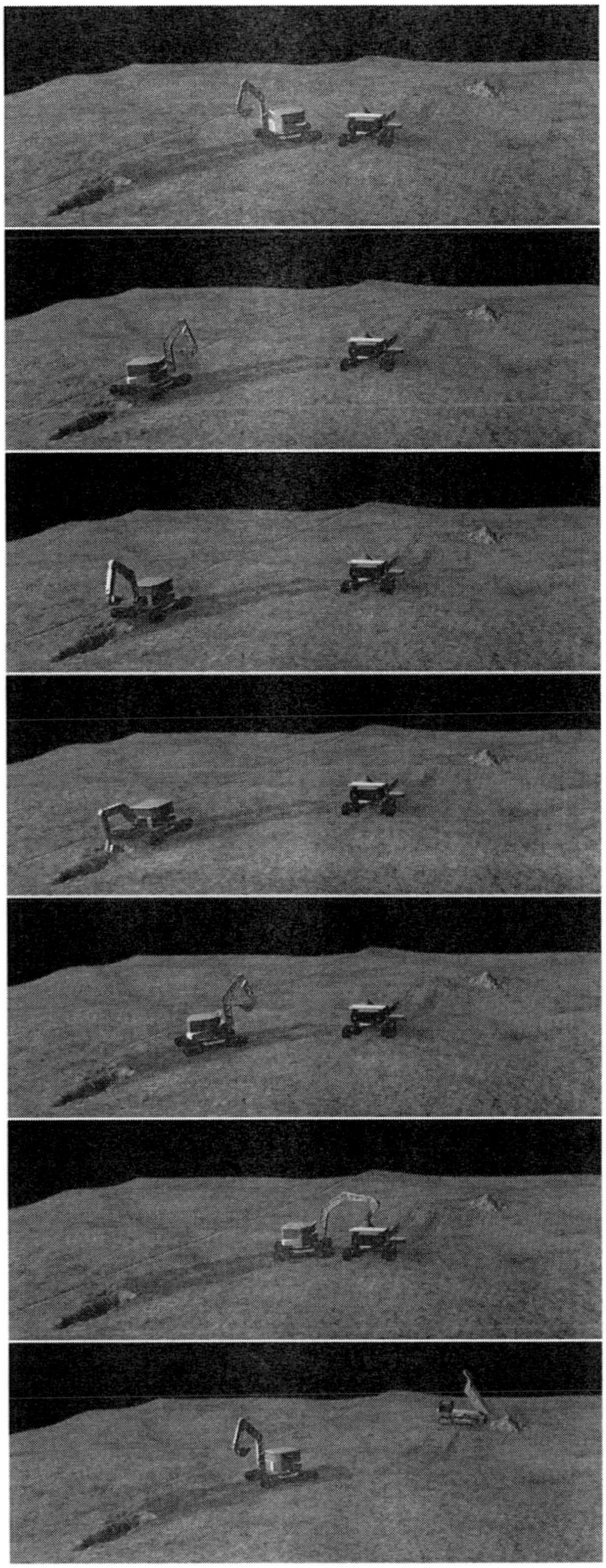

Fig. 8. Image sequence from one excavation cycle in the scenario `Excavation-flat`. The sequence ends with the crawler dumping its load at a dump area.

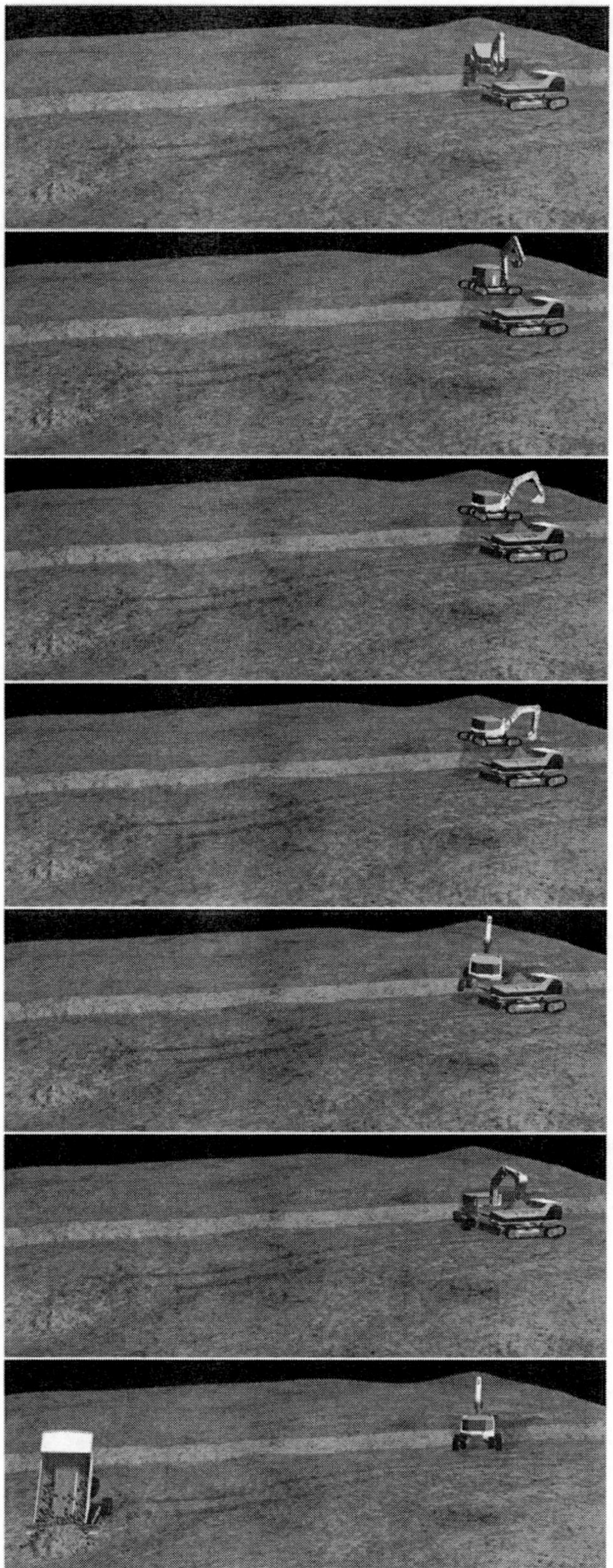

Fig. 9. Image sequence from one excavations cycle in the scenario `Excavation-slope`. *The excavator must cross a small slope between the dump truck and the dig area. The sequence ends with the crawler dumping its load at a dump area.*

A summary of the results is found in Table. 1. The average loaded mass was 8% higher in the flat case. This is attributed to the fact that it is more difficult to track the excavator's planned path across the slope than over the flat terrain. Consequently, the excavator sometimes ends up in a skewed pose where it is harder to fill the bucket. In both cases, there is material spilling (6-7%) from the bucket. This occurs mainly when turning the chassis, apparently causing too large peaks in acceleration. The average cycle times are very similar but the power consumption is 14% higher for the sloped case due to the slope traversal.

Table 1
Mean and standard deviation for loaded and spilt mass, duration, and work over 30 cycles in Scenario 1 on flat and sloped terrain.

Case	Mass [kg]	Spill [kg]	Time [s]	Work [kJ]
Flat	69±26	5±7	117±13	66±16
Sloped	64±24	4±5	118±15	74±24

The evolution of the excavated and dumped mass, working time, and work per cycle are displayed in Fig. 10, 11, and 12, respectively. The moments when the excavator switches to a new cell are marked with dashed lines. We observe that the mass (and power consumption) drops at the switching events. This indicates that there is room for improving the planning to maximize the equipment utilization. The working time for different tasks is nearly constant, with the exception of when the excavator waits for the dump truck to return. The power consumption is dominated by running the crawlers although it was rescaled for the excessive dissipation mechanism for numerical stabilization. As can be more clearly seen in Fig. 13, the arm actuator dominates the power consumption during digging and varies a lot over the cycles, which can be understood by the fact that the planned bucket trajectory adapts to the shape of the terrain, which evolves over time. We hypothesize that the greater power consumption in the sloped case is due to the fact that more uneven terrain leads to more situations where it is difficult, or even impossible, not to slip significantly with the large tracks.

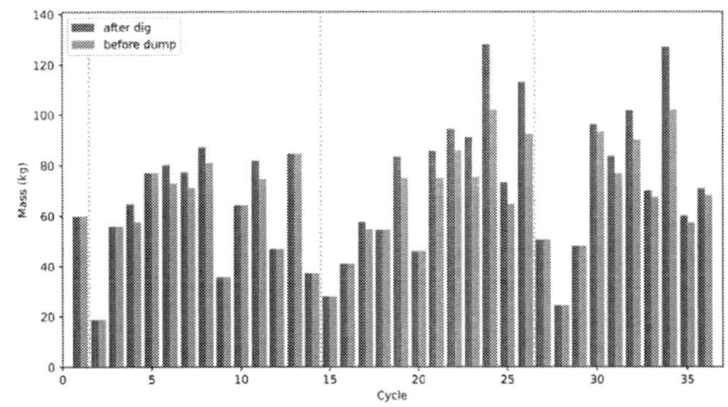

(a) flat

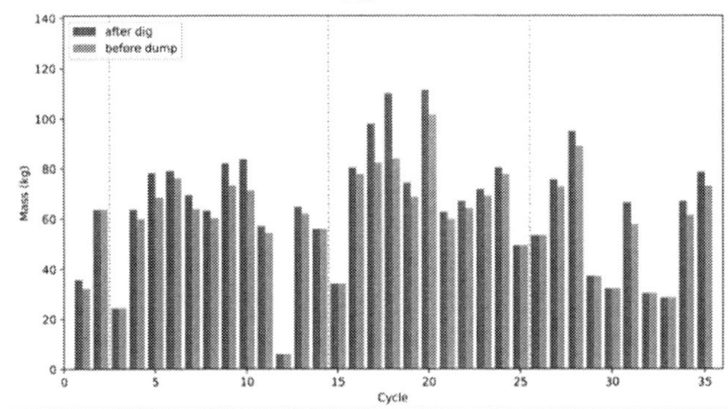

(b) sloped

Fig. 10. Evolution of the excavated and dumped mass over 30 cycles for the two cases of Scenario 1.

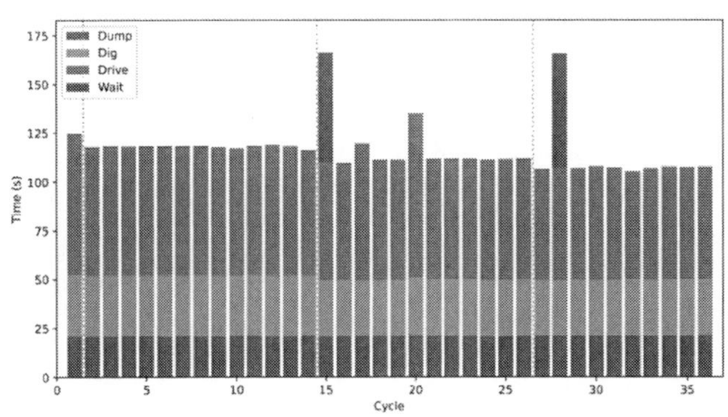

(a) flat

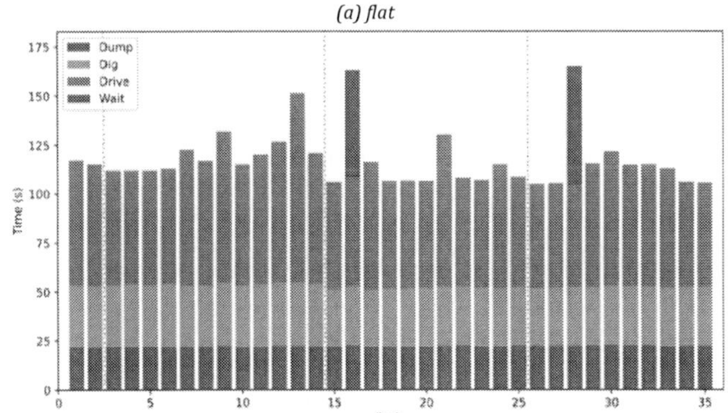

(b) sloped

Fig. 11. Evolution of the working time over 30 cycles for the two cases of Scenario 1.

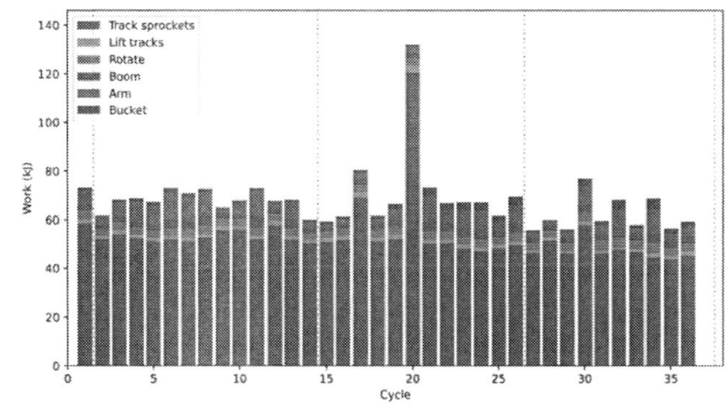

(a) flat

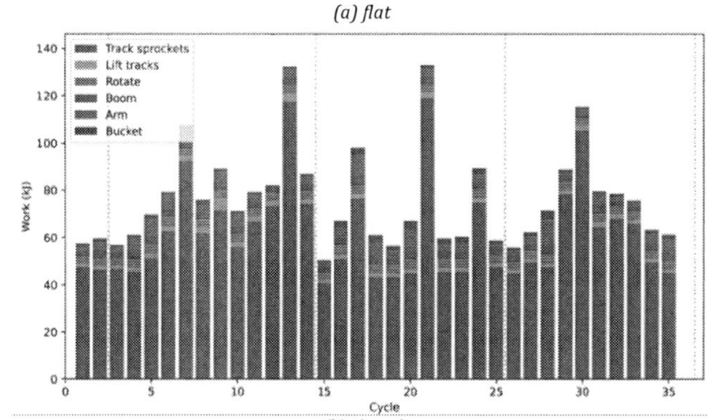

(b) sloped

Fig. 12. Evolution of the work over 30 cycles for the two cases of Scenario 1.

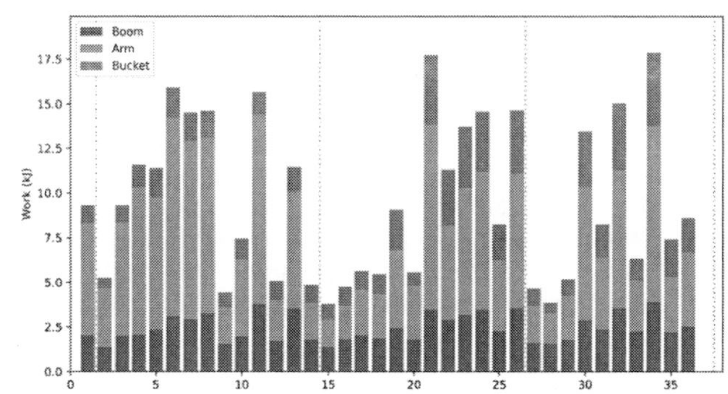

(a) flat

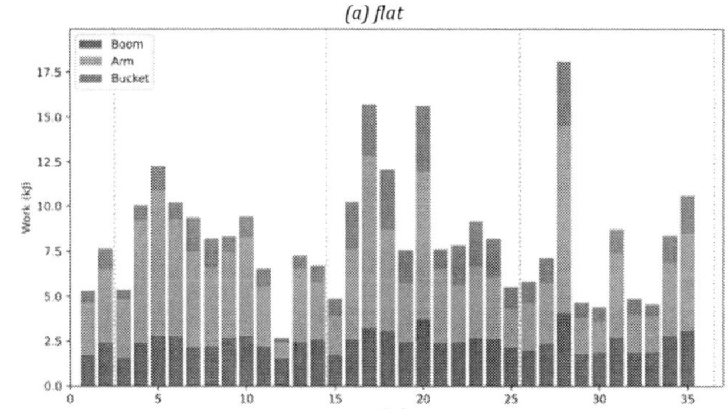

(b) sloped

Figure 13. Evolution of the work for the excavation process only over 30 cycles for the two cases of Scenario 1.

4.3. Scenario 2 - Creating a predefined ground construction for habitat modules

In this scenario, an excavator and a dump truck collaborate to shape the terrain towards a specified 3D profile where habitat modules are to be placed and partially covered with regolith. The dump truck must adapt the location for parking to both the current dig location and the already excavated area in order to be easily accessible to the cycling excavator. See Fig. 14 for an illustration.

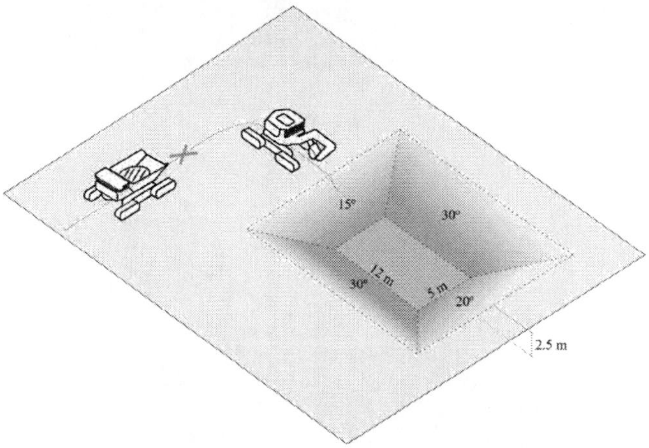

Fig. 14. Illustration of Scenario 2 - Creating a predefined ground construction for habitat modules by repeated excavation and final grading using the dump truck blade.

The beginning of the scenario is displayed in Fig. 15 and the end in Fig. 16. When a cell is excavated, a new bucket tip trajectory is planned for each dig cycle. The positions are planned with the excavation depth down from the current terrain surface, but no deeper than the target surface or maximum digging depth. The target angle for the bucket is adapted to the target slope of the terrain. Excavating the entire structure would require roughly 10,000 work cycles which would amount to about a week of serial computing time. In our tests, we jumped to selected intermediate states and verified the operational performance over a sequence of cycles.

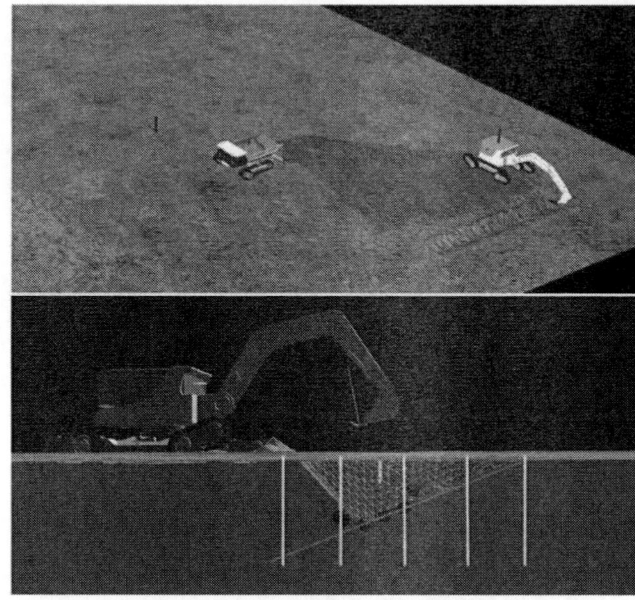

Fig. 15. Beginning of excavating the ground construction for habitat modules. The lower image shows a cross-section with the target surface indicated in red and the estimated cell boundaries in yellow.

After the target 3D profile has been excavated, it is graded. When grading is used, the heights are different during the dig trajectory. If, on the other hand, the target profile is flat and trench digging is used, then a flat dig trajectory is used. For both cases, the shovel first reaches for a point slightly before and above the trajectory start, which improves the ability to get to the trajectory start with the specified angle.

Fig. 16. End of excavating the ground construction for habitat modules.

During excavation, the terrain will avalanche, both sideways and also from the next row of cells that are closer to the excavator. Material from cells which have not been excavated yet will have material removed in the process of getting the current cell to its desired shape.

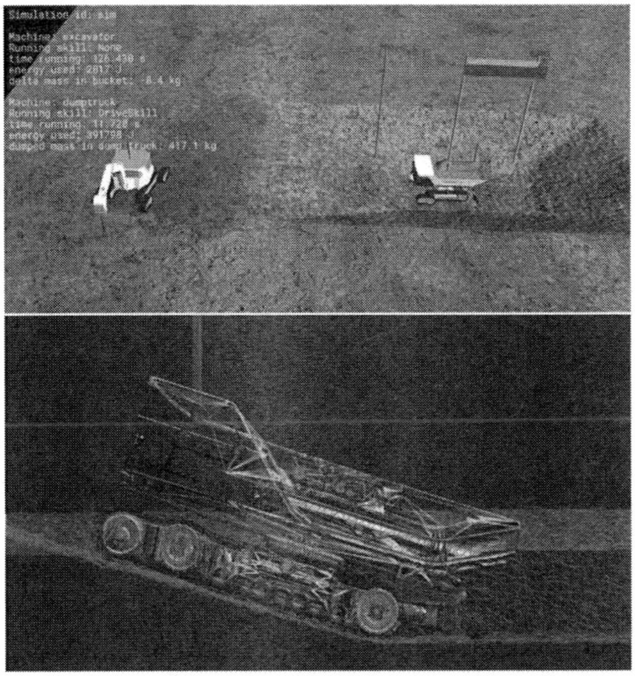

Fig. 17. After excavating the ground construction for habitat modules, the excavator leaves room for the dump truck to perform final grading of the target surface. The lower image is a cross-section of the grading with the actively controlled blade highlighted in blue.

The dump truck operates as in the previous scenario until the excavator has finished grading the target shape. The dump truck is then performing final grading of the bottom of the shape using its blade, see Fig. 17. The leveling/start_position, leveling/end_position, and leveling/offset variables from the last work setting provided are used to plan out a series of runs across the bottom of the shape. These runs are then performed with the shovel lowered to level the terrain. The blade position is dynamically adjusted using a PID controller with parameters provided by the machine manufacturer.

5. Conclusions and future work

We have found that the simulation framework serves the purpose of analysing how a system for autonomous lunar construction performs under different circumstances, which can include differences in the lunar environment, machine design and control algorithms, and more. Future work should include more elaborate models for sensors and lunar soil, and explore the framework's scalability for a large number of machines. The normalization procedure for compensating for numerical damping in the tracks should be experimentally validated to ensure that the energy computations are reliable. Automated generation and improvement of construction instructions using artificial intelligence is also interesting to evaluate.

6. Acknowledgements

The research was commissioned from Komatsu Ltd, which is commissioned for the Space construction innovation project as part of the Stardust Program in Japan, and supported in part by the Swedish National Space Agency through Rymdtillämpningsprogrammet dnr 2024-00310 (AILUR).

7. Declaration of competing interest

The authors are affiliated with Algoryx Simulation AB, which provides simulation software and services on a commercial basis, some involved in this research paper.

8. References

Aoshima, K., Fälldin, A., Wadbro, E., Servin, M., 2024. World modeling for autonomous wheel loaders. Automation. 5(3):259-281.

Aoshima K., Servin M., 2024. Examining the simulation-to-reality gap of a wheel loader digging in deformable terrain. Multibody System Dynamic. doi: 10.1007/s11044-024-10005-5

Backman, S., Lindmark, D., Bodin, K., Servin, M., Mörk, J., Löfgren, H., 2021. Continuous control of an underground loader using deep reinforcement learning. Machines 9(10):216.

Batagoda, N.M., Chen, B.-H., Zhang, H., Serban, R., Negrut, D., 2024. A Physics-based sensor simulation environment for lunar ground operations. arXiv:2410.04371

Bingham, L., Kincaid, J., Weno, B., Davis, N., Paddock, E., Foreman, C., 2023. Digital Lunar Exploration Sites Unreal Simulation Tool (DUST). IEEE Aerospace Conference. 2023, pp. 1–12.

Colledanchise, M., Ögren, P., 2018. Behavior Trees in Robotics and AI. CRC Press.

El Samid, A., Thangavelautham, J., Richard, J., Boucher, D., D'Eleuterio, B.M.T., 2008. Infrastructure robotics: A technology enabler for lunar in-situ resource utilization, Habitat construction and maintenance. Proceedings of International Astronautic Conference. 2008, pp. 2045–2058.

Iovino, M., Scukins, E., Styrud, J., Ögren, P., Smith, C., 2022. A survey of Behavior Trees in robotics and AI. Robotics and Autonomous Systems 154, p. 104096.

Kamohara, J., Ares, V., Hurrell, J., Takehana, K., Richard, A. Santra, S., Uno, K., Rohmer, E. Yoshida, K., 2024. Modeling of Terrain Deformation by a Grouser Wheel for Lunar Rover Simulation. arXiv:2408.13468.

Lacoursière, C., 2007. Ghosts and machines: regularized variational methods for interactive simulations of multibodies with dry frictional contacts. PhD thesis, Umeå University, Umeå, SWEDEN.

Lacoursiere, C., Linde, M., 2011. Spook: a variational time-stepping scheme for rigid multibody systems subject to dry frictional contacts. UMINF report, 11, 36.

Macenski, S., Foote, T., Gerkey, B., Lalancette, C., Woodall, W., 2022. Robot Operating System 2: Design, architecture, and uses in the wild. Science Robotics 7(66):eabm6074.

Quigley, M., Gerkey, B., Conley, K., Faust, J., Foote, T., Leibs, J., Berger, E., Wheeler, R., Ng, A., 2009. ROS: an open-source Robot Operating System. ICRA workshop on open source software. Vol. 3. 3.2. Kobe. 2009, p. 5.

Richard, A., Kamohara, J., Uno, K., Santra, S., van der Meer, D., Olivares-Mendez, M., Yoshida, K., 2024. OmniLRS: A photorealistic simulator for lunar robotics. IEEE International Conference on Robotics and Automation (ICRA). IEEE. 2024, pp. 16901–16907.

Seo, M., Gupta, S., Ham, Y., 2024. Exploratory study on time-delayed excavator teleoperation in virtual lunar construction simulation: Task performance and operator behavior. Automation in Construction 168, p. 105871.

Servin, M, Berglund, T., Nystedt, S., 2021. A multiscale model of terrain dynamics for real-time earthmoving simulation. Advanced Modeling and Simulation in Engineering Sciences 8(1):1–35.

Servin, M., Wang, D., Lacoursière, C., Bodin, K., 2014. Examining the smooth and nonsmooth discrete element approaches to granular matter. International Journal for Numerical Methods in Engineering 97(12): 878–902.

Thangavelautham, J., Law, K., Fu, T., El Samid, N.A., Smith, A.D.S., D'Eleuterio, G.M.T., 2017. Autonomous multirobot excavation for lunar applications. Robotica 35(12):2330–2362.

Wiberg V., Wallin, E., Servin, M., Nordfjell, T., 2021. Control of rough terrain vehicles using deep reinforcement learning. IEEE Robotics and Automation Letters, 7(1):390-397.

Wiberg, V., Wallin, E., Fälldin, A., Semberg, T., Rossander, M., Wadbro, E., Servin, M., 2024. Sim-to-real transfer of active suspension control using deep reinforcement learning. Robotics and Autonomous Systems 179:104731.

7265 | Evaluation of a new friction measurement device on arctic terrain

Sally Shoop [a,*], Bruce Elder [a], Steven Meurer [b], Owen Parker [a], Nathan Kamprath [c], Don Halliday [d]

[a] US Army Corps of Engineers, Engineer Research & Development Center - Cold Regions Research and Engineering Laboratory (CRREL) , Hanover, NH
[b] US Army, Arctic Regions Test Center (ARTC), Fort Greeley, AK
[c] US Army, Ground Vehicle Systems Center (GVSC), Warren, MI
[d] Halliday Technologies, Delaware, OH

* Corresponding author: sally.a.shoop@erdc.dren.mil

ABSTRACT

The RT3 Simplified Curve Unit (SCU) friction tester, by Halliday Technologies, measures surface friction by operating the test wheel at a small side-slip angle. It can be operated at highway speeds and can be used with either aircraft or truck tires and was specifically designed for robust operations in cold environments in the far north. The US Army Ground Vehicle Systems Center (GVSC), Ground Vehicle Power and Mobility, Tires Track & Suspension Team purchased the Halliday RT3 SCU (SCU) friction tester for use at the Arctic Regions Test Center (ARTC) in Alaska. Before sending it to Alaska, the SCU was evaluated for comparison to standard friction test techniques and cold room performance at the Cold Regions Research and Engineering Laboratory in Hanover, NH. On showing good performance at temperatures down to -45 °C (-50 °F), and excellent correlation with standard friction test techniques over a wide range of friction surfaces, it was shipped to Alaska and successfully used on snow and ice test surfaces during March and April of 2025.

This paper presents how the SCU tester measures and reports friction values, test results for the SCU unit and other friction measurement techniques on a wide variety of on- and off-road surfaces, the results of cold room testing, and the operational and exploratory measurements on snow and ice fields in Alaska. A discussion of the measurement technique and results is presented along with conclusions for optimizing use on arctic terrain surface characterization and monitoring for vehicle and tire performance testing.

Keywords
Cold regions
Arctic
Snow
Ice
Friction
Tire
Monitoring
Tire testing
Characterization
Winter

1. Introduction

The interface friction between tires and terrain is a fundamental detail for evaluation of off-road cold regions mobility yet it is complex and varies based on how it is measured and reported. Interface friction is dependent upon both the adhesive and asperities of the terrain surface material as well as the tire construction, compound, tread pattern, tread depth, and winter tire unique enhancement features such as siping or studding. The measurement of terrain friction is of interest for many different purposes including agriculture and forestry, roadway pavement for snow removal and maintenance (i.e. plowing and deicing) runway friction for safe aircraft performance and maintenance scheduling, tire and vehicle performance, and military operations for a wide variety of vehicles and surfaces, among others. Thus, the best measurement techniques are usually those that mimic the purpose of the measurement most closely; for example, to quantify runway friction for aircraft performance it makes sense to use a tire similar to an aircraft tire, operating in braking slip traveling in a straight line at high speed to estimate the friction required to land safely, and a wide variety of runway friction measurement techniques have been developed with this in mind. However, for off-road vehicle terrain characterization it would be more appropriate to use a vehicle tire operating at slower speed and include maneuvers more typical of off-road vehicles, such as turning.

The objective of this project was to explore the use of a new friction measurement device designed for use in cold and arctic operations and develop guidance for using the equipment for monitoring and characterization of test areas for winter vehicle mobility testing, namely for tire and track development, in arctic environments.

2. Background

U.S. military ground forces require improved mobility and safety for challenging winter conditions to enhance Warfighter readiness and driving confidence for operations in austere, unsparing, and hazardous winter weather, winter temperatures below -18°C (0°F) are not uncommon and frostbite can occur within minutes. The Joint Force has repeatedly highlighted in numerous strategy and policy documents the need for improved arctic mobility, including in; *The White House's National Strategy for the Arctic Region (2022), The DOD Arctic Strategy (2024), Army Chief of Staff Paper #3 Regaining Arctic Dominance (2021), The Army's Climate Strategy for FYs 2023-2027 (2022), The Department of Defense's National Defense Strategy (2022), and the Arctic Requirements Oversight Council-Equipping the Arctic (2023).* Additionally, the DOD has known capability gaps with tire and track components that are not designed for extreme negative temperatures or ice/snow road conditions commonly experienced in cold/arctic regions during winter months. Improved winter mobility developmental and validation testing of tires and track components will allow for US and Allied Warfighters to more safely respond or engage during winter in; Europe, throughout the state of Alaska, South Korea near the Demilitarized Zone (DMZ), Northern Japan, and even Northern CONUS, among other current and future cold operational areas.

There are numerous technologies used to measure friction for vehicle related surface characterization. These techniques often use an instrumented wheel, but also include spinning disks, pendulum testers, accelerometers, etc. Of those friction test methods, an instrumented wheel slipping on the surface mimics

https://doi.org/10.56884/3PKE9TEN

the conditions of a vehicle tire slipping on a surface. Since a large market for friction testing is for runway characterization for safe aircraft landing, many of the friction test tires are built like an aircraft tire, and to simplify things they are often operated at a fixed longitudinal braking slip value, therefore friction is reported at that value. Shoop et al. (1994) compared three instrumented wheel friction measurement techniques for use on winter surfaces and also discussed the different methods and nuances of reporting a single friction value rather than a full mu-slip curve indicating maximal then declining friction as tire traction is lost and slip relative to the road surface increases A graphic comparing the traction, or friction curves for a range of slip velocity (also called Differential Interface Velocity, DIV, in this graphic) are shown in Fig. 1. A similar curve for lateral friction/traction over a range of tire slip angles between travel direction of motion and the wheel rotational plane is shown in Fig. 2 (Coutermarsh and Shoop, 2009). Based on the vastly different friction values or curve shapes for the full range of winter surfaces, you can see how reporting a single value of traction or friction, either longitudinal or lateral, could miss capturing the full impact of how tire forces (longitudinal or lateral) vary with slip or slip angle, which is important information for fully understanding, interpreting or modeling tire or vehicle performance for off-road surfaces. Generally, on-road vehicles operate at low slip or slip angle on high friction surfaces; whereas off-road vehicles often operate on low friction surfaces and at higher slip and slip angle, especially winter surfaces. In general, Shoop et al. (1994) found the SAAB friction tester, which operates at a set longitudinal slip value of 12%, to have friction values comparable to the CRREL Instrumented Vehicle when fiction values at the same slip were compared.

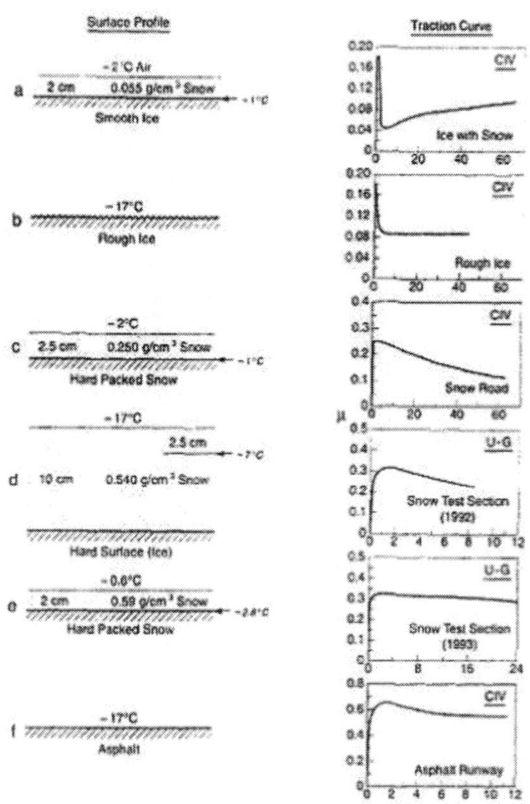

Fig. 1. Longitudinal friction vs slip curves for different winter surfaces (from Shoop et al., 1994)

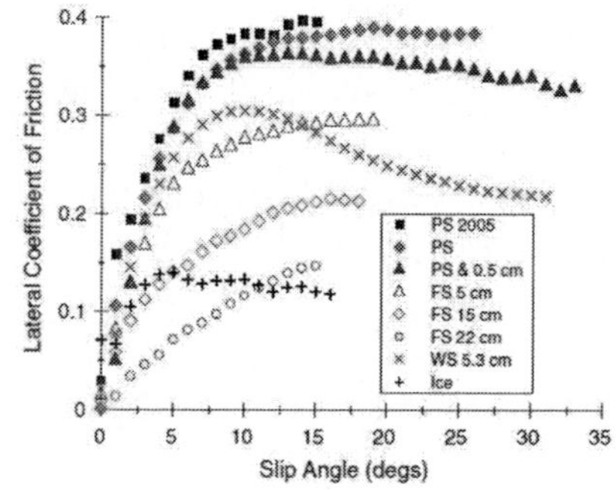

Fig. 2. Lateral friction vs wheel slip angle for different winter surfaces (from Coutermarsh and Shoop, 2009). PS = Packed Snow; WS = Wet Snow, FS=Fresh Snow

Recently, a portable friction test capability that could be easily transported to various remote regions test sites was desired. For this, a study by Menke et al. (2021) was completed to evaluate portable pavement friction testers for use in the cold and on cold regions surfaces. Menke first tested the instruments on different friction surfaces in cold rooms and evaluated the stability of the results with temperature and time under cold temperatures. They determined that the portable walk behind Micro-Grip Tester (μGT) was the most robust for cold operations and gave the most consistent results. The μGT was then taken for winter field testing and performed well, with results comparable to the larger SAAB Friction tester (Menke et al., 2021). Similar to the SAAB Friction Tester, the μGT records a continuous readout of a single "friction" value across the measurement surface and uses a scaled aircraft tire for the measurement tire, operating at a 12% longitudinal slip. Friction is continuously measured across the surface traversed, and then generally reduced to an average value for the full surface, or the portion of the surface of interest. The μGT friction tester is commonly used across the US, and we have a high level of confidence in its use on cold terrain, so it was included in this study for comparison.

Over the past few years, however, a new effort to evaluate and develop winter tires for larger military vehicles required an easy and rapid way to reliably measure friction on a large test surface to characterize and monitor the repeatability of the test surface for tire performance comparisons under arctic conditions. For this purpose, a newer technology that used a full-scale tire as the standard measurement tire and operates at a small lateral slip angle was judged to be a good way to characterize the surface specific to interactions for ground vehicle tire performance. Similar equipment was cold proven through year-round use at airports in Alaska but was further streamlined to reduce components not needed for our purposes to make it even more robust for use in extreme temperatures. This RT3 Simplified Curve Unit (SCU) by Halliday Technologies was then evaluated for suitability and operational use at Arctic test areas.

3. Methodology

3.1. The RT3 SCU friction tester

The RT3 SCU friction tester is a trailer attached to the vehicle through a tow hitch with two stabilizer bars. Two test wheels are utilized with inward slip angles for balance and to measure the lateral forces on the wheels. Some of the specifications of the tester are listed below. For our measurements, the slip angle was set at 1.75 degrees, the vertical load at 140 kg (310 lb) using the ASTM 2493 SRT radial tire at 124 kPa (18 psi) inflation pressure. The advantages of this equipment include adjustable slip angle, tire size and test speed, along with the suitable load cell and operational temperature ranges. The original RT3 design has been used successfully by the US Air Force in Alaska for several years, and the new SCU design with minimal moving hardware and hydraulics, would be even more beneficial for minimizing operation or measurement issues at extremely cold temperatures. In addition, the SCU was configured in two ways, either centered behind the vehicle with the test tires inboard of the tow vehicle tire paths, or offset to straddle the tow vehicle tire path. The reason for this becomes apparent in the results section. A picture and illustration of the SCU showing key features is shown in Fig. 3, and a brief overview of the specifications are listed below.

- slip angle is variable from 0 to 7.5 degrees
- SCU can operate with alternate tire sizes, but because of the requirement of testing on low friction (arctic) surfaces for vehicle performance was decided to use the treaded ASTM F2493 Standard Reference Test Tire (SRTT) – a modern P225/60R16 97S radial test tire made by BF Goodrich (Michelin)
- test tire size is variable, with options for plate adaptors to fit different wheels (estimated maximum wheel radius is 43.3 cm with 2.5 cm clearance, and a width of 27cm)
- temperature range is specified from -46 °C to 43 °C (-50 °F to 110 °F)
- lateral load cell ranges from 0 to 2669 N (600 lb)
- test speed ranges from 10 to 130 kph (5 to 80 mph)
- the user interface allows the customer to choose between different test tires and units.
- the display samples data at 100 Hz, but data was saved through Bluetooth to an android device at 10 Hz. The saved data can be stored on the cloud, presented real-time in google maps or street maps, or displayed in a variety of different formats (we mostly used Excel for our analysis).
- the data acquisition converts voltage to a "true" friction value or to the Halliday Friction Number (HFN), which is a percentage of the friction and is easily interpreted (similar to Mu), or as a force.

The values reported in this paper were the HFN values as those correlated most closely with expected results and that of the μGT.

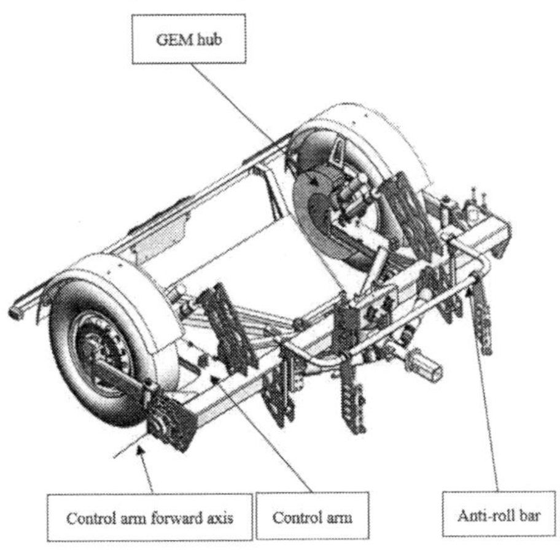

Fig. 3. The RT3 SCU friction tester indicating some of the major features.

3.2. Summer field experiments at CRREL (Hanover, NH)

The experiments completed during the summer of 2024 were primarily to gain a thorough understanding of the operational use of the SCU and to tune the adjustments on configurable options for best use on a wide variety of surfaces. The summer experiments focused on a comparison of the SCU and the μGT operation and results for a wide range of surface types. The major differences between the SCU and the μGT are in Table 1 below:

Table 1
Comparison between SCU and μGT major parameters.

	RT3 SCU	μGT
Test tire	Radial Truck Tire	Scaled aircraft tire with longitudinal grooves
Tire slip used	1.75° slip angle	12% longitudinal slip
Operation speed	8 to 80 kph	Walking speed (apprx. 8 kph)

The experiments were designed to answer the following questions:

i. How well does the SCU work on a wide range of surfaces, deformable and non-deformable
ii. Comparison of SCU to μGT friction values
iii. Repeatability of values for a variety of surface types
iv. Guidelines for standardized operations

3.3. Cold room experiments

In preparation to deploy the SCU to Alaska for arctic operations, the effect of cold temperatures on the SCU was measured using CRREL's Cold Climatic Chamber. A range of temperatures from approximately 24 °C (75 °F) to -46 °C (-50 °F) were tested. Note that -50 °F is the specified low temperature operational limit for most Army Tactical Vehicles. Starting with room temperature (24 °C), the load cells were checked by raising the wheels off the ground at the axle, spinning the wheels to make sure they could operate, and recording the reading. The general condition of the equipment was also observed. The temperature was then decreased in increments, allowing the SCU to equilibrate at each temperature for at least 4 hours before testing and moving to the next temperature.

3.4. Winter SCU testing at the Arctic Regions Test Center, Fort Greeley, AK

Lastly, the testing of the SCU in Alaska was completed so the personnel at ARTC could gain experience with using the test unit and collaborate with CRREL on shakedown testing and explore nuances to develop best practices for operational use. Experimental testing included:

1. Measurements on a full range of cold regions surfaces (snow, ice, deformable, non-deformable)
2. Effects of speed on friction measurements on snow and ice
3. Responsiveness of the measurement technique when transitioning between surfaces and the ability to capture variations in the surfaces
4. The effect of turning maneuvers on the measured friction values. (requires further experimental investigation and Test Operations Procedure (TOP) development)

4. Results

4.1. Comparison of SCU and µGT on a wide variety of surface types (summer in Hanover, NH)

Prior to deploying the friction tester to Alaska, it was sent to the Cold Regions Research and Engineering Laboratory (CRREL) in Hanover, NH for shake down testing and generating guidelines for operational use. After assembly and discussions on settings and use, the SCU was assessed for arctic use by testing on both rigid and deformable surfaces with a wide range of friction characteristics to determine the limitations of use on off-road surfaces. Fig. 4 shows some of the surfaces tested: packed gravel parking lot, grass playing fields, wetted HDPE (for low friction), deformable soils, and many others. For each surface, the portable µGT was also used so we could determine the relationship between the measurements of the two pieces of equipment. Table 1 gives the data collected for each of the surfaces. The friction values for these surfaces ranged from 0.26 to 1.01. The experiments on this wide range of surfaces in a controlled environment aimed to find the limits of the equipment and help develop guidance for its operational use.

Fig. 4. Select test surfaces illustrating the variety of surfaces tested by both the SCU and the µGT at CRREL, summer of 2024: SCU on packed gravel parking lot (upper left); µGT on grass (upper right), SCU on wet and soapy HDPE to simulate low friction surfaces (lower left), and SCU in offset-mounting configuration on deformable soils (lower right).

Table 2
Summary of results from the µGT and SCU testing at CRREL, color coded by terrain surface type.

Date	Time	Air Temp (C)	Surface	Surface condition	MicroGT	SCU HFN
6/28/2024	12:03:00 PM	21	Asphalt	Dry	1.00	0.83
6/28/2024	12:08:00 PM	21	Asphalt	Dry	0.99	0.82
6/28/2024	12:11:00 PM	21	Asphalt	Dry	0.99	0.81
7/1/2024	13:42:00 PM	26	Gravel	Dry	0.50	0.43
7/1/2024	13:42:00 PM	26	Gravel	Dry	0.48	0.48
7/1/2024	13:42:00 PM	26	Gravel	Dry	0.50	0.43
7/1/2024	13:42:00 PM	26	Grass	Dry	0.69	0.34
7/1/2024	13:42:00 PM	26	Grass	Dry	0.81	0.34
7/2/2024	7:51:00 AM	16	Grass	Wet	0.70	0.34
7/2/2024	7:51:00 AM	16	Grass	Wet	0.71	0.35
7/8/2024	14:26:00 PM	27	Epoxy	Dry	0.83	0.61
7/8/2024	14:30:00 PM	27	Epoxy	Dry	0.80	0.61
7/8/2024	14:32:00 PM	27	Epoxy	Dry	0.79	0.60
7/9/2024	11:09:00 AM	28	Concrete	Dry	0.71	0.65
7/9/2024	11:11:00 AM	28	Concrete	Dry	0.73	0.65
7/9/2024	11:13:00 AM	28	Concrete	Dry	0.72	0.65
7/9/2024	11:36:00 AM	29	Asphalt	Dry	1.01	0.8
7/9/2024	11:40:00 AM	29	Asphalt	Dry	1.01	0.8
7/9/2024	11:44:00 AM	29	Asphalt	Dry	1.01	0.79
7/11/2024	7:55:00 AM	25	Asphalt	Damp	0.91	0.81
7/11/2024	7:58:00 AM	25	Asphalt	Damp	0.9	0.82
7/11/2024	8:01:00 AM	25	Asphalt	Damp	0.89	0.81
7/15/2024	9:00:00 AM	25	Tarp	Dry	0.71	0.7
7/15/2024	9:00:00 AM	25	Tarp	Dry	0.76	0.7
7/15/2024	9:00:00 AM	25	Tarp	Dry	0.75	0.71
7/15/2024	9:10:00 AM	25	Plastic	Dry	0.78	0.63
7/15/2024	9:10:00 AM	25	Plactic	Dry	0.77	0.63
7/15/2024	9:10:00 AM	25	Plactic	Dry	0.79	0.62
7/15/2024	9:30:00 AM	25	Trap	Wet	0.65	0.66
7/15/2024	9:30:00 AM	25	Tarp	Wet	0.66	0.65
	9:30:00 AM	25	Tarp	Wet	0.68	0.65
7/15/2024	9:40:00 AM	25	Plastic	Wet	0.64	0.62
7/15/2024	9:40:00 AM	25	Plastic	Wet	0.66	0.62
7/15/2024	9:40:00 AM	25	Plastic	Wet	0.67	0.6
7/29/2024	9:00:00 AM	22	HDPE	Dry	0.65	0.52
7/29/2024	9:00:00 AM	22	HDPE	Dry	0.7	0.51
7/29/2024	9:00:00 AM	22	HDPE	Dry	0.65	0.5
7/29/2024	9:30:00 AM	22	HDPE	Wet	0.45	0.41
7/29/2024	9:30:00 AM	22	HDPE	Wet	0.45	0.43
7/29/2024	9:30:00 AM	22	HDPE	Wet	0.5	0.44
7/29/2024	10:00:00 AM	22	HDPE	Soap	0.27	0.4
7/29/2024	10:00:00 AM	22	HDPE	Soap	0.26	0.38
7/29/2024	10:00:00 AM	22	HDPE	Soap	0.27	0.39

The friction measurement of the SCU and the µGT for all the surfaces in Table 2 was used to develop a correlation between the two instruments. This is plotted in Fig. 5. The plot on the top includes the data and shows a strong linear correlation between the SCU and µGT, although it is not a 1:1 correlation.

The plot of the bottom removes some of the more unusual surfaces measured, specifically the testing on the grass and the soapy water on HDPE, and this improves the correlation slightly. Either plot shows good correlations between the two instruments, with correlation coefficients, R^2 greater than 0.97 for both analysis methods.

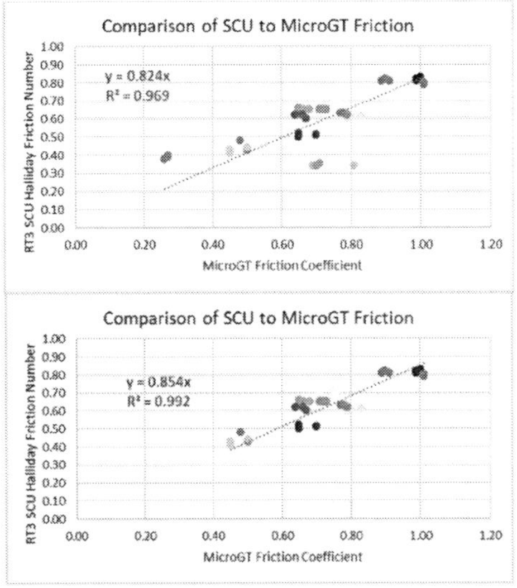

Fig. 5. Plots of RT3 SCU friction vs μGT friction showing strong liner relationship between the two measurements for all surfaces (top). With the more irregular surfaces of grass and soapy plastic removed, the correlation is even stronger (bottom). Both correlations show R^2 values above 0.97 and slopes of 0.824 and 0.854 respectively, using an intercept of 0.0.

4.2. Cold Room Testing of the SCU

The cold room tests were completed over the course of 2 days starting at room temperature and dropping by increments down to -46 °C (-50 °F). The SCU was held at the test temperature for at least 4 hours before observing the operational condition. To check this, the SCU was raised on jacks and the tires were rotated. The load cell readings were checked to make sure they remained constant at "zero" (i.e. the wheels lifted off the surface, therefore no load) at each temperature. The SCU functioned satisfactorily at the full range of temperatures although some small leakage from the hydraulics was noticed after the SCU was brought back to room temperatures after the -45 °C (-50 °F) test. A picture of the SCU in the cold room is shown in Fig. 6.

Fig. 6. Checking the load cell readings during the cold room testing from 21 to -46°C (70 to -50°F).

4.3. Winter SCU tests at the Arctic Regions Test Center

At the end of the summer testing, the SCU was shipped to the Arctic Regions Test Center at Fort Greeley, Alaska for winter testing. Due to a busy test schedule, the SCU wasn't able to be used until March of 2025, which is fortuitous as the snow and ice test surfaces are often at their full depth and most interesting in early March. At this time of the year, the more routine testing is nearly done, allowing more time to accommodate more exploratory testing and experiments.

Personnel at ARTC conducted several experiments during March and April 2025. These included testing on their groomed snow and ice surfaces routinely used for testing, plus exploring the impacts of speed, slopes and directional impacts through a zigzag test in a slalom-like pattern. In addition, testing was completed on snow packed roads with mixed surfaces of gravel on snow and asphalt, to explore the resolution of data on mixed surfaces. Lastly, the effects of temperature on friction on snow or ice was tested during the early days of the spring transition when temperature fluctuates widely.

4.3.1. Winter surface test measurements

A summary of the surfaces tested, and the average Halliday Friction Number (HFN) are given in Table 3 below. Pictures of the various test surfaces are given in Fig. 7. The friction values on the surfaces ranged from an average of 0.05 on ice to 0.9 on dry asphalt, so covered a wide range. The values measured on the surfaces align with those expected based on experience, previous testing, and those reported in the literature.

Table 3
Summary of conditions tested at Arctic Regions Test Center, March and April 2025.

Surface	Test Speeds (kph)	Halliday Friction Number
smooth ice	8, 16, 32, zig zags	0.05 - 0.16
packed snow with fresh dusting	16 to 24	0.21
dry asphalt (rd) & packed snow (sn)	16 to 32	0.2 - 0.4 (sn) 0.9 (rd)
groomed snow test area	24, 40, 56, 72, zig zags at 24	0.17 - 0.22
smooth ice test area	28	0.05 - 0.16
Packed snow road w/ gravel	16 to 31	Avg = 0.25
5 to 7 cm heavy, wet snow	26 to 40	0.14
Asphalt (rd) to snow (sn) transition	27 to 35	0.9 (rd); 0.21 (sn)

Fig. 7. Selection of pictures of test surfaces illustrating the variety of cold regions surfaces measured. Test surfaces on snow packed roads on Fort Greeley with small area where gravel was spread on test area access road (second from top), packed snow and areas of dry asphalt on Robin Rd (top); and packed snow along the return path of the test sections with a dusting of fresh snow, clearly showing the wheel paths of the test wheels and tow vehicle (second to bottom) and 5 to 8 cm of heavy wet snow (bottom).

4.3.2. Speed Effects

The two standard test courses of packed snow and smooth ice (Fig. 8) were used to study the effect of speeds. The pack snow surfaces are created by packing and grooming the snow and checked with a snow penetrometer called a CTI to ensure that the compaction stays within the range specified by ASTM. The ice test section is created by spreading water on the surface and brushing the surface to remove falling snow as needed. Commercial tire testing is done at specific standards according to ASTM, but for our testing purposes we did not need to test only at standard surfaces and wanted to explore the effect of other surface types as well.

Testing was performed at speeds ranging from 24 to 80 kph (15 to 50 mph) on the snow test area (snow field); and from 13 to 39 kph (8 to 24 mph) on the ice field where you need to drive slower for safety reasons. The data is plotted in Fig. 9 for both surfaces. Speed did not have a measurable effect on the friction values for either surface, with the slope of friction vs speed for the snow being zero to three decimal places and the slope for ice being only slightly more than zero, at 0.001 HFN/kph. These data sets also illustrate the small variability of these highly controlled groomed surfaces, and document whatever variability is present, which is one of the desired purposes for using this equipment.

Fig. 8. Two of the standard test surfaces used for cold regions vehicle and tire performance testing are a groomed snow surface (top) and a smooth ice surface (bottom).

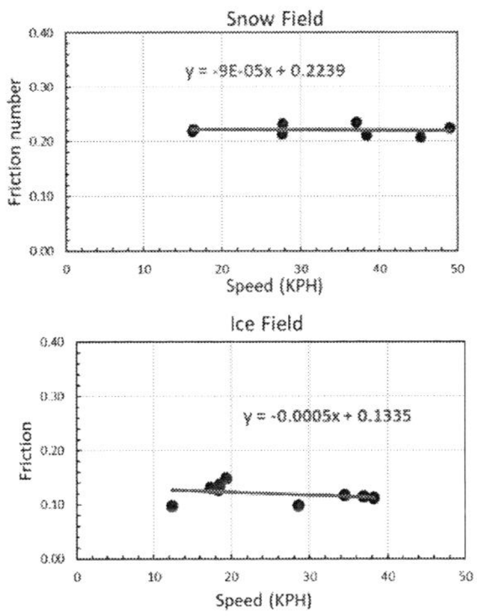

Fig. 9. Effect of speed on friction measurement on the groomed snow field (left) and ice (right). The average effect on speed is zero. The surface variability can be seen in both plots.

4.3.3. Responsiveness to surface transitions

The responsiveness of the RT3 SCU to rapid changes in surface friction was checked by driving it over surfaces of rapidly varying friction. The RT3 SCU was indeed extremely responsive to surface conditions and there was no noticeable delay in the change in friction values. We did note that if the test is not intended to measure average values over a uniform surface it will be important to review and post process the raw data set (as shown in the figures below).

A test checking response time for transitions between vastly different friction surfaces, from asphalt to snow, is shown in Fig. 10. Similarly, the road test with the alternating packed snow and dry asphalt areas also clearly shows the friction variability in Fig. 11.

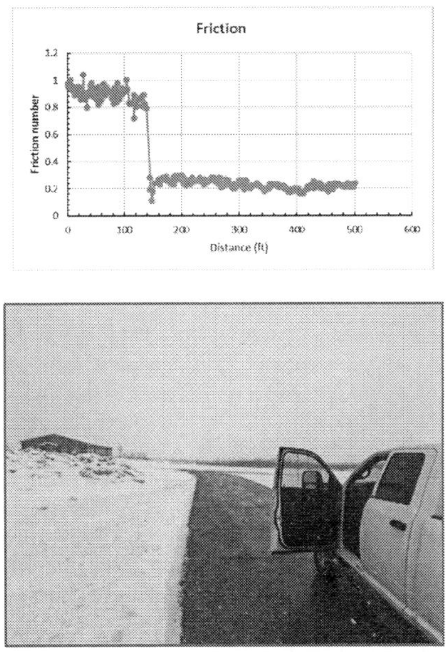

Fig. 10. Testing the responsiveness of the friction measurements to rapid and extreme changes in friction going from asphalt to snow on the road. The graph of friction with distance (top) illustrates that the SCU very quickly captures the different surfaces driven (bottom).

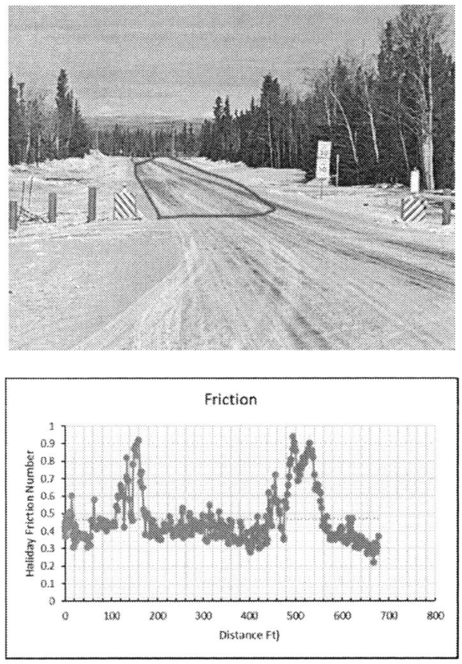

Fig. 11. The responsiveness to rapid and extreme changes in friction across a road surface with both packed snow and dry asphalt. The area of the test (top) and graph of friction with distance (bottom) illustrate that the SCU very quickly captures the variation in surface friction.

4.3.4. Deformable surfaces

When measuring the friction on deformable materials, the forces resisting the movement of the tire through the material is inherently included in the measurement. This force is called the motion resistance and is generally measured separately using a free rolling wheel. Since that was not possible with the friction tester without reconfiguring it, we could not separate the frictional forces from the resisting forces, thus the friction values include the sum of all forces and are much lower than what would be measured on the same type of material in a non-deformable state. We did two measurements on highly deformable materials: the deformable soil shown in Fig. 4 (lower right), and the fresh wet snow shown in Fig. 6 (bottom). The deformable soil resulted in approximately 2.5 to 5 cm (1 to 1.5 inches) of sinkage of the test tires, and an average friction value of less than 0.1. The soil was tilled so it was free to compact with the passage of the test wheels. This same soil in a less deformable (more compacted) state would usually have a friction value more on the order of 0.4 to 0.5, so we can assume that the low measurement is due to the resisting forces on the tire. Similarly, for the deformable snow we measured a friction value of 0.14 whereas the friction of snow is usually closer to 0.3 or more when there are not forces resisting the tire movement during the tire-snow deformation. The offset configuration shown in Fig. 4 (lower right) works best for deformable materials because of the increased distance reducing the chance of interference between the test tires and the tow vehicle wheel paths. The potential for terrain deformation interaction between the test tires and tow vehicle ruts when using the central mounting configuration is noticeable in the test on deformable snow in Fig. 7 (bottom).

4.3.5. Impact of turning and maneuvers while testing

While the SCU is designed to be operated both straight and in curves, the details of the interaction between the surface on low friction surfaces while driving on a curve or other maneuvers was explored. Two types of maneuvers were performed on the standard test surfaces: a gradual curve and a more rapid maneuver performing a zigzag or slalom-type of maneuver. These were meant to explore the limits of the device and to help in developing the TOP measurement protocols for routine testing on the snow and ice test surfaces. The zigzag test GPS trace and resulting friction measurement with distance are shown in Fig. 12. Because of the small radius turning and low friction surfaces, these tests clearly show the need to consider that the surface friction along with the rate and degree of steering input does not result in the SCU sliding sideways on the test surface. As such, the guidance for routine friction monitoring on arctic surfaces is to perform straight-line testing, at least for now until further experimental investigation is completed for better understanding

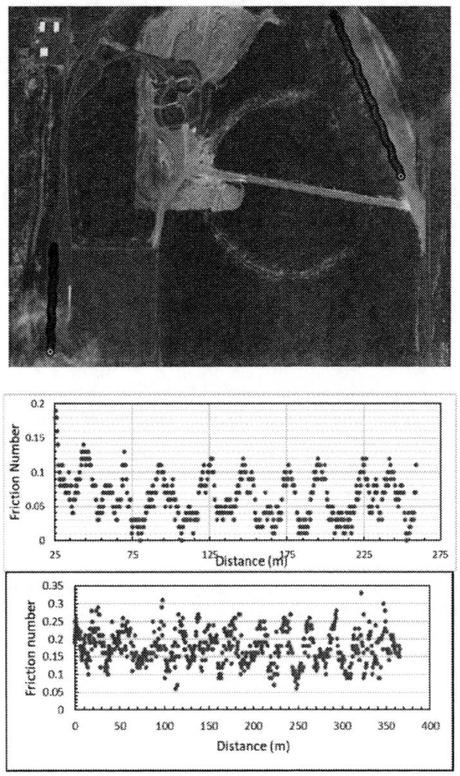

Fig. 12. Friction on the ice surface varied from almost 0.0 to 0.14 during a slalom or zigzag maneuver. Image of maneuver (top, on summer google earth imagery), ice friction data (center), and snow friction data (bottom).

In the future, we need to explore the use of the SCU on curves for winter surfaces as this could be a powerful tool for tire testing. There is a question as to the efficacy of cornering data relative to straight-line friction data for all Continuous Friction Measurement Equipment (CFME). The understanding of these differences is important when analyzing friction data. For the next phase of ASTM standard submission, a sequence of cornering tests on wet asphalt will be run at 32 kph (20 mph) for 45.7 to 152.4 m (150 to 500 ft) radii curves. These tests will be run at the Transportation Research Center (TRC) in October this year. Because of the variability on road surfaces, we have determined that the only way to understand these friction result differences between cornering and straight-line testing is to compare the friction results generated by testing over the same surface when the CFME is being driven in both a cornering and straight-line mode. That is; after running a constant radius test, a second test should be run in a straight tangential line directly over a section of the same path. The small straight section line data is then compared to the cornering data at this coincident point.

When the surface friction is very low, as with an ice surface, this test may have to change either the radius and/or the speed because for the RT3 CFME to read friction correctly, it is necessary that the tow vehicle track over the surface without sliding. Therefore, next winter it would be useful to explore any variation of measured friction when the SCU is driven in a curve and when driven in a straight line.

5. Operational Guidelines

The equipment operation was user friendly after initial training was provided by the manufacturer No problems were experienced during operations at temperatures of -32 °C (-25 °F). For temperatures closer to -46 °C (-50 °F) recommend storage of the tester in a heated garage to minimize the occurrence of hydraulic dampers leakage. At any temperature, if the SCU is stored in a garage it should be set outside for an appropriate interval to equilibrate to the ambient outdoor temperature before being used for test measurements. Other considerations for optimizing use are:

- Wait until at steady-state nominal test speed to initiate the start of test measurements
- In order to most effectively use the standard quick friction calculation features, consider choosing the test distance estimate to be equal to or greater than the distance of the traverse. If driving for less distance than entered into the software, you can easily reanalyze the data in Excel or other software to obtain the desired values, but it would entail an extra step in data analysis, which could be time consuming when acquiring large amounts of data.
- Measurements on deformable surfaces will include effects of resistance forces on both the front and the sides of the tires since the tire is at an angle.
- Towing the SCU centered between the tow vehicle tire paths is the preferred operational method for the standard packed snow and ice test surfaces; however, the offset configuration is beneficial for measurements on deformable surfaces.
- 32 kph (20 mph) is good target test speed for efficiency and safety on most of the surfaces however a slower speed should be chosen for the more slippery ice surface.
- Although the RT3 SCU is designed for use on straight roads or on curves of up to 45 m (150 ft) radius on high friction surfaces, the ARTC test measurements will most likely be taken in a straight line. For low friction surfaces the turning radius would need to be greater than 45 m (150 ft) radius and should be studied in future experiments as it could be very useful for characterizing dynamic performance of different tires on arctic surfaces in other-than-straight-line maneuvers.
- Data acquisition doesn't start until the speed is at 8 kph (5 mph) or above.

6. Summary and Conclusions

Recent interest in developing winter tires for military vehicles has revealed the need to have a rapid and reliable friction testing method for use on winter surfaces to characterize and monitor the winter test surfaces in the arctic. The desired capabilities resulted in the purchase of a RT3 SCU friction tester by Halliday Technologies, which uses an innovative tire slip angle design and a radial tire to more accurately represent the surface friction between a vehicle and arctic surfaces for military vehicle maneuvers. Tire mechanics theory supports the use of a test method engaging side slip as more representative of the conditions for ground vehicles operating on winter surfaces. However, the value of the slip angle has a strong impact on the friction value, especially for snow and other off-road surfaces. The SCU friction tester has a side friction angle that is adjustable but was set at a small value of 1.75 degrees for these tests. This would represent incipient side slip or yaw of a vehicle, generating enough side force for adequate vehicle control in a straight ahead or turning maneuver.

The new friction tester was sent to CRREL in Hanover, NH for evaluation during the summer of 2024 prior to being deployed to ARTC in Alaska. At CRREL, the SCU was evaluated on a variety of surfaces of different friction and deformability, as well as cold room testing down to -46 °C (-50 °F) to determine limits and guidelines for operational use. The RT3 tester performed well in all conditions and had excellent correlation with the µGT friction measurement device over a broad range of surfaces. Slight upgrades to the hydraulics for extreme temperatures may prove beneficial. In Alaska, the friction tester was used on snow, ice and mixed surfaces to assess the SCU operation and use, and explore impacts of speed and variable surfaces. The SCU friction tester was found to be both effective and easy to use in a wide variety of conditions including variable surfaces and temperatures.

The use of this friction tester was proven to be capable, efficient, and valuable for characterization of the surface conditions of standard test courses that are used for both military and commercial testing for tire and vehicle performance in Arctic and winter conditions.

7. References

Arctic Requirements Oversight Council, 2023. Equipping the Arctic.

ASTM (In review). Standard Test Method for Measuring the Friction of Pavements and Other Trafficked Surfaces Using the Technique of Continuous Reading from an RT3® Curve Continuous Friction Measurement Equipment (CFME) Fixed Side-Slip Angle Device. ASTM Subcommittee E17.21 on Field Methods for Measuring Tire Pavement Friction.

ASTM 2493-20. Standard Specification for P225/60R16 97S Radial Standard Reference Test Tire

Coutermarsh, B. Shoop, S., 2009. Tire Slip-Angle Force Measurements on Winter Surfaces. J. Terramech. Special Issue on Snow Mobility. 46(4), p. 158-163.

Menke, A., Shoop, S., Elder, B., 2021. Operation of Pavement Friction Testing Equipment in Cold Environments. Proc. Regional Conf. on Permafrost and Int'l. Conf. on Cold Regions Engr., 24-29 Oct, Boulder, CO. Also ERDC/CRREL MP-22-7. Hanover, NH: U.S. Army Engineer Research and Development Center. (Feb 2022) http://dx.doi.org/10.21079/11681/43380

Society of Automotive Engineers SAE J2014, 2022. Pneumatic Tire/Wheel/Runflat Assembly Qualifications for Military Tactical Wheeled Vehicles.

Shoop, S., Young, B., Alger, R., Davis, J., 1994. Effect of Test Method on Winter Traction Measurements. J. Terramech. 31(3), p. 153-161.

Shoop, S.A., et al. 2021. Improving Winter Traction for Vehicles in Northern Operations", In Proceedings of the Ground Vehicle Systems Engineering and Technology Symposium (GVSETS), NDIA, Novi, MI, Aug. 10-12, 2021.

Shoop, S., Garibay, E., Parker, M., Olivier, J., 2024. Understanding mobility in the arctic. Ground Vehicle Systems Engineering and Technology Symposium (GVSETS). August 13-15, 2024. Novi, Michigan.

The White House, 2022. National Strategy for the Arctic Region. https://www.whitehouse.gov/wp-content/uploads/2022/10/National-Strategy-for-the-Arctic-Region.pdf

US Department of the Army, 2021. Regaining Arctic Dominance – The US Army in the Arctic (Chief of Staff Paper #3), Jan. 19.

US Department of the Army, 2022. Army Climate Strategy Implementation Plan Fiscal Years 2023-2027.

US Department of the Army, 2023. Military Performance Specification MIL-PRF-32759. Dec: Tire, Winter Weather - High Mobility Multipurpose Wheeled Vehicle (HMMWV).

US Department of Defense, 2024. Arctic Strategy.

US Department of Defense, 2022. National Defense Strategy.

US Office of the President, 2021. Executive Order 14008. Tackling the Climate Crisis at Home and Abroad. 2 January.

US Office of the Under Secretary of Defense for Policy, 2019. Report to Congress: Department of Defense Arctic Strategy, 18 p.

Witte C., Welling, O., Parker, M., Kamprath, N., 2023. Modifying military tires for improved winter traction. Ground vehicle systems engineering and technology symposium. August 15-17, Novi, Michigan.

7559 | Dynamic contact patch estimation from 3D contact patch measurements inside a moving tyre

Amal Patel, P. Schalk Els *

Department of Mechanical and Aeronautical Engineering, University of Pretoria, Pretoria, South Africa

** Corresponding author: schalk.els@up.ac.za*

ABSTRACT

Vehicle mobility is inherently related to tyre-terrain interaction. Slip conditions arising from tyre forces generated due to the terrain and vehicles state is a valuable indication of a vehicle's stability. Measurement and analysis of the tyre contact patch is complex, yet abundant information can be derived from contact patch measurements.

Literature indicates that analysis of the contact patch is difficult, and indirect methods are used to determine tyre states under ideal conditions of paved roads. Few researchers attempt to directly analyse the contact patch and extract key tyre states: slip angle, pressure distribution, terrain deformation, contact patch velocity and dynamic rolling radius.

This study uses the T2Cam apparatus, a mechanism that fits into a tyre and provides the capability to attach cameras inside the wheel that remain stationary despite the tyre rotating. The use of T2Cam and an RGBD camera provides the capability of a non-contact method using DIC (Digital Image Correlation) to measure tyre contact patch deformation and relate that to key tyre states. The purpose of this study is to develop algorithms to extract key states from contact patch deformation of the inside of the tyre carcass under varying tyre conditions.

Preliminary findings show tyre deformation of the contact patch is measurable and accurate to be used as an input in an algorithm to calculate slip angle, contact patch velocity and dynamic rolling radius. Overall, determining these factors will enhance the understanding and visualisation of the dynamic contact patch and prediction of vehicle stability on many types of terrain.

Keywords
Digital image correlation
Tyre contact patch
Slip angle
Pressure distribution
Dynamic rolling radius
Contact patch velocity
Intelligent tyres

1. Introduction

A vital aspect of vehicle dynamics lies in the tyre-terrain interaction. The mechanism through which this interaction takes place is the tyre contact patch and forms a basis on how tyres generate forces on which the vehicle demands. When generating these forces, the tyre deforms in different ways under different conditions: stationary, rolling, cornering, braking, and accelerating. This deformation that the tyre experiences, both in the tyre sidewall and the contact patch, becomes a factor of interest due to its relationship with tyre force generation. Therefore, understanding the tyre contact patch would be extremely valuable in not only confirming the validity of current research but also in investigating and relating the vehicle state to the information derived from the contact patch.

Key tyre states, slip angle and longitudinal slip, are physically developed in the contact patch region and is linked to deformation behaviour in this region, especially due to terrain.

The concept of intelligent tyres is newly formed where it is thought that the combination of embedding sensors into the tyres will inform valuable data and information from them. Intelligent tyres are an important concept to consider when investigating the tyre-terrain interaction.

Literature exists whereby key vehicle states are measured indirectly or contact patch measurements are static. There are limitations in the implementations that leverage the non-contact 3D measurements whereby systems are fixed to external test rigs or move externally with the vehicle. The Vehicle Dynamics Group at the University of Pretoria leverages an apparatus designed by Guthrie (2016) to measure internal deformation of a moving tyre. This system has been the subject of many proceeding works, whereby Kabutz and Els (2024) are able to derive the outside contact patch from the inner deformation 3D measurement.

This work focuses on building upon Kabutz and Els (2024) to dynamically measure the contact patch area and shape, as well as explore the measurement of slip angle and longitudinal slip. Consequently, this work aims to use digital image correlation to measure the tyre deformation directly in a non-contact fashion to provide a custom hardware and real-time capable software infrastructure. The current software pipeline is capable of:

1. Close depth 3D point tracking
2. Tyre contact patch deformation

Ultimately, this research will aim to investigate the use of computer vision to dynamically measure the contact patch from the inside of the tyre. This investigation focuses on specifically finding the shape of the outside contact patch such it can be studied under the influence of varying slip angle.

2. Related Work

Currently, there are researchers that statically and dynamically measure the contact patch. The static measure of the contact patch does provide actionable information, however, the dynamic measurement of the contact patch provides valuable information in terms of the deformation of the contact patch under various conditions, such as speed, slip angle and terrain.

Static measurements of the contact patch have been attempted and reveals viability in contact patch measurements with varying change in tyre parameters. Optical methods using cameras placed under glass to measure contact area have already been attempted. (Derafshpour et al., 2019) Additionally, Liang et al. (2021)

https://doi.org/10.56884/3C89EZ94

investigates the pressure distribution estimation using frustration of total internal light reflection (FTIR) methods through a glass plate and validates the contact patch shape under differing conditions. Tyre tread depth measurement using stereovision from the outside of the tyre had been investigated with sub-millimetre accuracies below 0.2mm (Wang et al., 2019). Furthermore, contact area deformation on noise, grip and rolling resistance were investigated (Wang et al., 2021) in which firm correlations were made. Slip angle measurement comparison from sidewall deformation using digital image correlation was investigated and provided positive results (Gao et al., 2024).

This reveals the fact that exploration should take place on experimental setups that are not stationary but move with the vehicle and is able to continuously analyse the tyre contact patch.

3. Methodology / System Design

To investigate the deformation of the tyre, a full pipeline must be designed and deployed. Two approaches were explored to measure deformation in the contact patch.

The first approach explores measuring the inner deformation of a tyre by comparing the Red, Blue, Green, Depth (RGBD) image of the deformed tyre to a reference 3D geometry that resembles the tyre in its undeformed state. Furthermore, the rubber-incompressibility assumption was used to project the inner deformation to the outside of the tyre to extract outer contact patch deformation.

The second approach explores the tracking of 3D points on the surface of the inner tyre to extract point-wise displacements and velocities over consecutive RGBD image frames. This method uses the comparison with previous frames instead of a 3D reference geometry, while still providing dense deformation data.

Both approaches capture the deformation behaviour in the tyre but differ in assumptions made, comparisons made to as well as in the outputs yielded. The following sections detail the design and implementation of each approach as well as the experimental setup.

3.1. Physical Apparatus and Hardware

This experimental setup consists of using the T2Cam apparatus design by Guthrie (2016), in Fig. 1, whereby the hardware is upgraded to Intel RealSense D405 Depth Camera as its stereovision solution as well as the Nvidia Jetson Orin Nano for compute hardware.

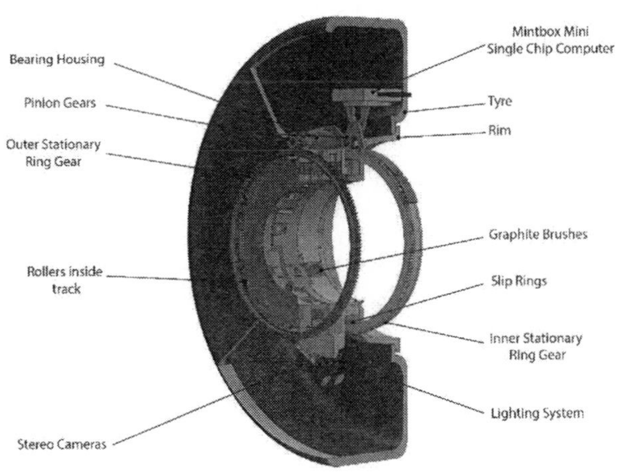

Fig. 1. T2Cam Apparatus for Camera Stabilisation

The Intel RealSense D405 delivers synchronised RGBD images at various resolutions and frames per second (fps). The image resolution of 848 x 480 at 30 fps was used to reduce compute cost while retaining high detail at close depths. Additionally, to improve depth accuracy, a speckle pattern on the inside of the tyre was used such that RGB images had high contrast and interest points. The Nvidia Jetson Orin Nano allows for the deployment of real-time capable algorithms by leveraging the edge-computing capabilities of the CUDA cores as well as it supports of many accelerated python libraries.

The mechanism shown in Fig. 1 is designed to mount peripherals inside the tyre and keep it stationary relative to the tyre moving as well as to stabilise Intel RealSense D405 Depth Camera. The stabilisation of this RGBD camera allows for it to be constantly pointing in the direction of the inner surface of the contact patch and thus is the mechanism by which the algorithms were built upon.

3.2. Digital Image Correlation and 3D Data Processing

This work leverages digital image correlation and other computer vision techniques through OpenCV (Bradski, 2000). Additionally, Open3D serves as the 3D data processing library to analyse and measure the contact patch (Zhou et al., 2018) as well as supports Intel RealSense Depth Cameras.

3.3. Optical Flow

Optical flow involves the process of estimating the motion of pixels through two different frames. This algorithm allows for subpixel tracking whereby the displacement in the image plane is estimated. OpenCV implements CUDA accelerated optical flow methods whereby a dense optical flow algorithm was chosen (Bradski, 2000). The Farnebäck dense optical flow method was implemented due to the algorithm modelling local neighbourhoods with polynomial models whereby one can leverage to the continuity of the inner tyre surface geometry (Farnebäck, 2003). Additionally, Farnebäck is more lightweight in terms of computation and more robust to large movements between consecutive frames

3.4. 3D Scanning

Structure from Motion (SfM) methods were used to acquire 3D models, in Fig. 2, depicting the undeformed inner and outer surface geometry of the tyre.

Fig. 2. Reality Capture Scans of Undeformed Outer Tyre (top) and Inner Tyre (bottom)

The Reality Capture software pipeline from Epic Games was used and allows for the comparison of the inner deformed tyre surface from the RGBD images with the Reality Capture models in Fig. 2. Therefore, a correspondence between the inner deformed RGBD image and the undeformed 3D inner tyre model was generated. Additionally, the relationship between surface points on the inner undeformed 3D model and the outer undeformed 3D model can be related using the proposed solution from Kabutz and Els (2024). This was done by extending the discretized 2D cross sectional slicing and projection method to 3D ray casting approach using surface normals inherent to the 3D model geometries.

3.5. RGBD and Point Cloud Processing

Each frame captured by the Intel RealSense Depth Camera can be converted into a point cloud and is the data structure that is primarily used to determine deformation in the contact patch. The point cloud, shown in Fig. 3, is an example of the data captured using the T2Cam apparatus and is positioned in such a way that the middle point in the point cloud is central to the static contact patch under vertical load only.

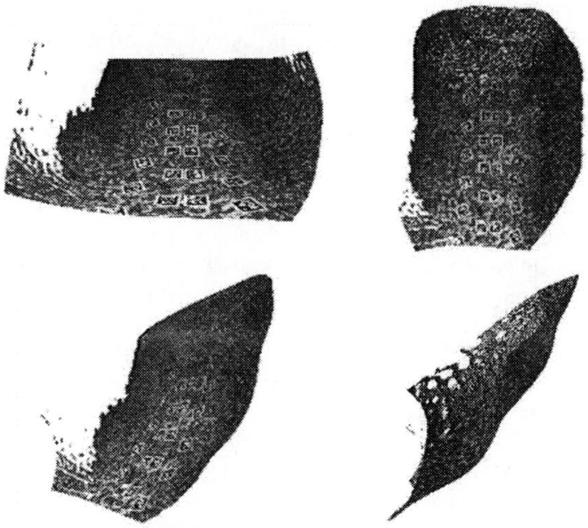

Fig. 3. RGBD Image from T2Cam Apparatus with point cloud representation with front view (top-left), top view (top-right), diagonal view (bottom-left) and side-on view (bottom-right)

The Open3D library is leveraged to in terms of its point cloud registration algorithms to correctly position the T2Cam RGBD image relative to the full inner tyre 3D model.

3.6. Inner-to-outer surface estimation (Approach 1)

The first approach to contact patch deformation is implemented through the following process:

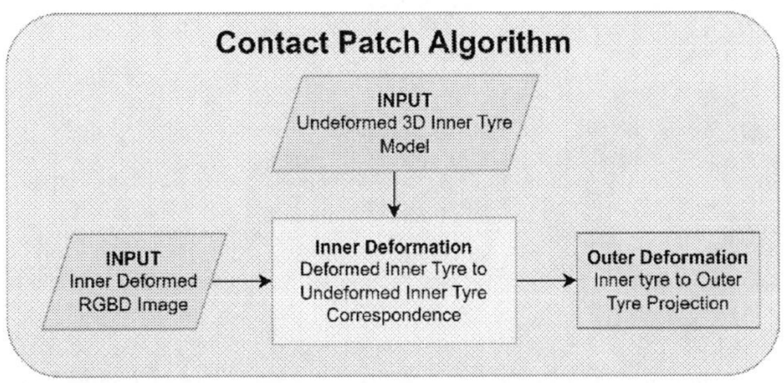

Fig. 4. Flow Diagram of Summarised Contact Patch Algorithm

Figure 4 reveals the correspondence between the undeformed inner tyre model and the deformed RGBD image is calculated first. This uses a nearest neighbour search (NNS) approach due to the density of points in both the Reality Capture model and RGBD image. Moreover, the inner undeformed model was used as the source for rough global alignment using April tags (Olson, 2011) and secondly iterative closest point (ICP) registration to refine a more local alignment, such that the NNS approach is valid.

Additionally, the correspondence between the undeformed inner tyre surface and outer tyre surface generated by Reality Capture was processed and represented as the vector distance between these surfaces. Lastly, this vector distance is applied to the inner deformed RGBD image to estimate the outer deformation of each point in that image.

3.7. 3D Point-Velocity Tracking (Approach 2)

The second approach to contact patch deformation is to temporally track points on the inner tyre surface and the given process in summarised in Fig. 5:

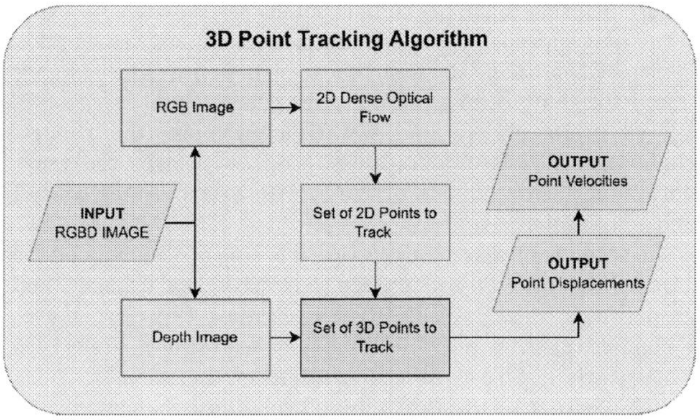

Fig. 5. Flow Diagram of Summarised 3D Point-Velocity Tracking Algorithm

This approach uses 2D dense optical flow on all the RGB images from the Intel RealSense Depth Camera, such that image displacements can be tracked with subpixel accuracy for each pixel in the first image. The depth map and camera intrinsics are used to take the 2D point tracking and extend it to a metrically scaled 3D space. This pipeline temporally extends this process to get point displacements and velocities after every frame. This ultimately leads to 3D full-field velocities and displacements, providing motion estimates for each tracked point in the scene.

4. Experiments and Results

To evaluate the performance of the two approaches undertaken, test data was captured and applied such that it was able to showcase the visual deformation of the tyre contact patch.

4.1. Test Data from Experiments

The test data consists of a series of 300 RGBD frames as a dataset whereby the Intel RealSense Depth Camera was mounted with the T2Cam apparatus, while **the tyre was rocking back and forth repeatedly**, with the RGBD camera being stationary relative to the tyre. The Trelleborg Agricultural Tyre was used unpressurized and not beaded, such that maximum deformation can be visualised.

4.2. Test Case: Tyre rocking back and forth with stationary camera relative to tyre: Approach 1

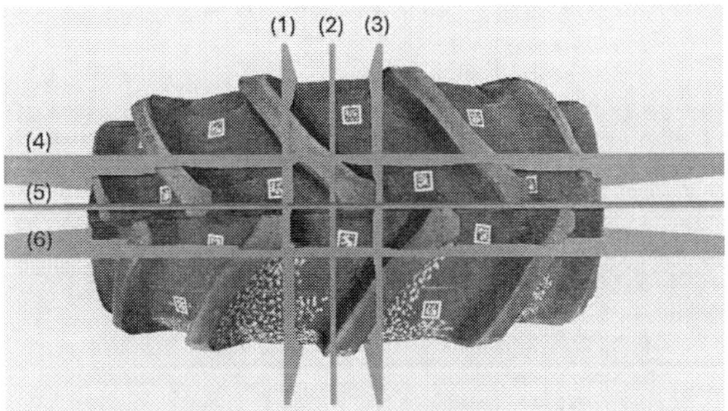

Fig. 6. Outer Undeformed Model showing the longitudinal and lateral cross-sections to be evaluated (pink) at (1) x = -0.04m, (2) x = 0.0m, (3) x = 0.04m, (4) z = 0.09m, (5) z = 0.13m, (6) z = 0.17m

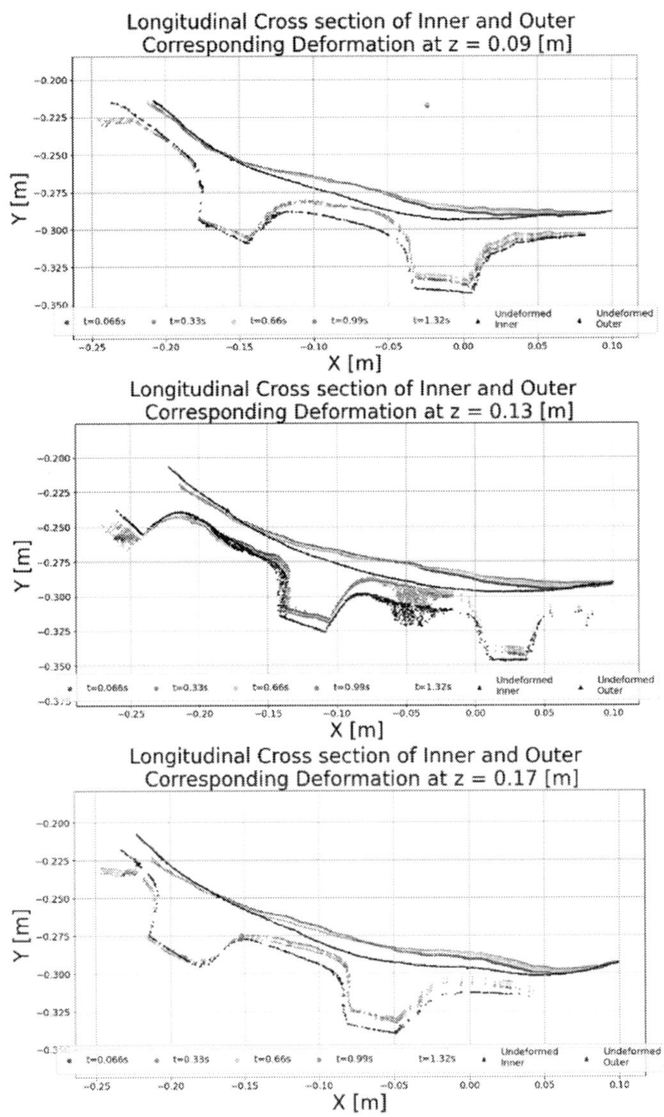

Fig. 7. Deformation in the Contact Patch over time compared to the undeformed tyre (black) measured with longitudinal cross sections at z = 0.09, 0.13, 0.17m

In order to evaluate the deformation, cross sections of the tyre were taken in the longitudinal and lateral directions. The pink highlighted planes in Fig. 6 refer to the cross-sections evaluated for:

- Undeformed Inner and Outer 3D Tyre Surface Model
- Five Deformed Inner and Outer Tyre Surface from RGBD Images at t=0.066s intervals

This allows for the comparison between deformed cross sections temporally as well as comparison to the undeformed tyre cross section at each time step.

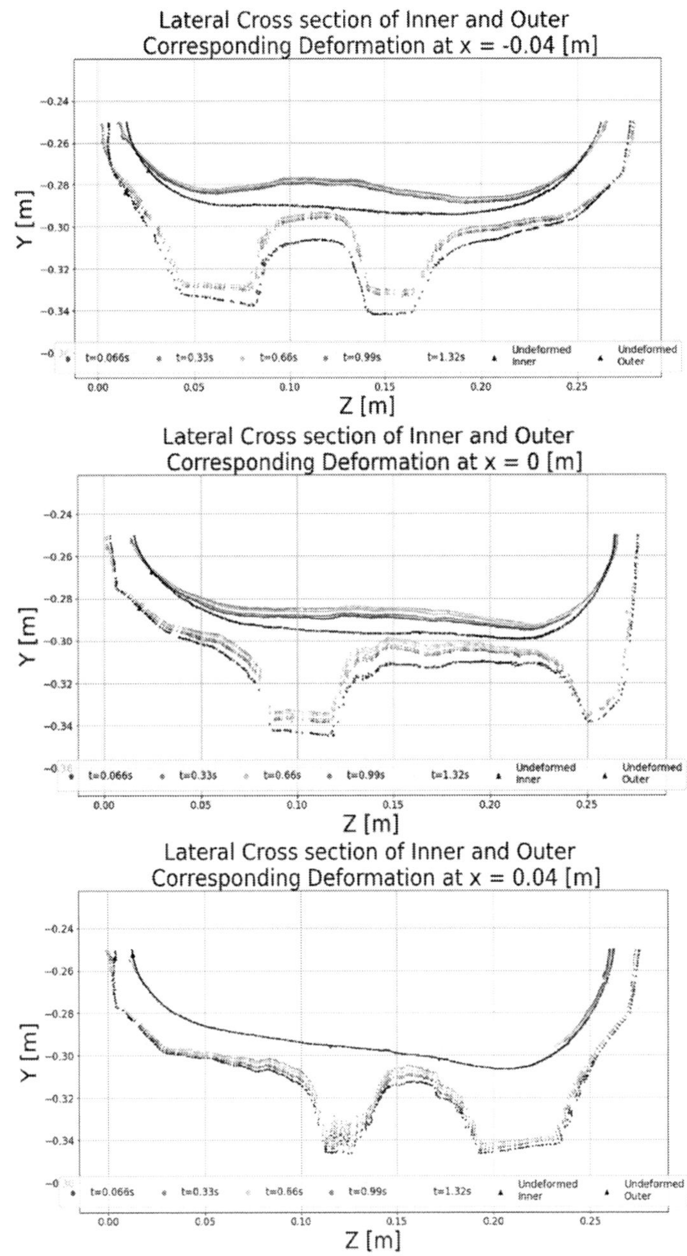

Fig. 8. Deformation in the Contact Patch over time compared to the undeformed tyre (black) measured with lateral cross sections at z = -0.04, 0, 0.04m

Figure 7 and 8 illustrates the temporal deformation in relation to the undeformed tyre surface at that specific cross section. The vertical load on the uninflated tyre, being under its own weight, results in high displacements from the undeformed surface to deformed surfaces. Additionally, the maximum deformation in this case is seen to be centred to the vertical load placed on the tyre. Therefore, as time increases, the tyre rocks forward past t = 0.99s and the vertical load matches the change in which the tyre contacts the ground. Figure 8, emphasis how the deformation increases as the longitudinal z axis approaches the centre of the tyre. Furthermore, Fig. 7 and 8 confirms the non-linearity in the surface deformation as the sidewall has minimal displacement relative to the undeformed surface. Lastly, the tyre lug behaviour is captured on the outer surface estimation, that matches how the vertical load changes across the tyre surface.

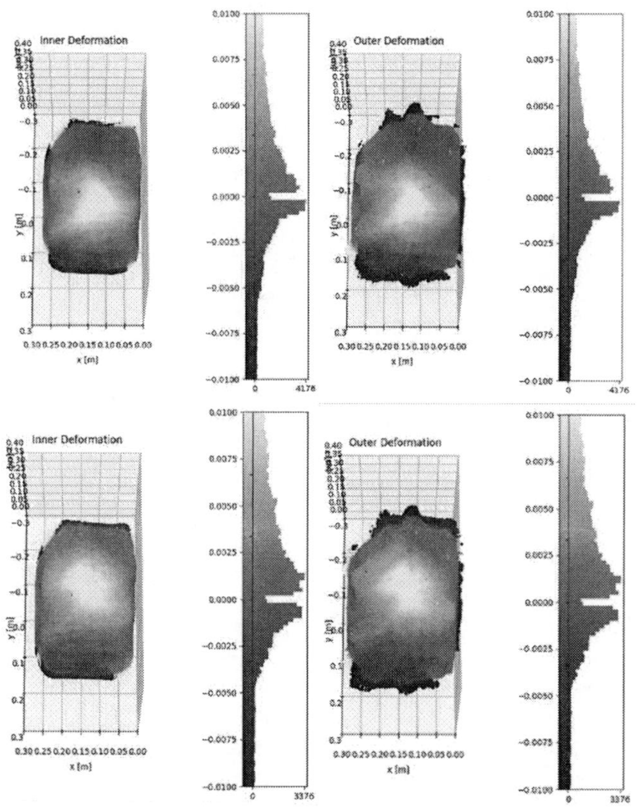

Fig. 9. 3D Deformation Maps with Histogram of Signed Distances: Inner Deformation at t = 0.066s (top-left), Outer Deformation at t = 0.066s (top-right), Inner Deformation at t = 1.32s (bottom-left), Outer Deformation at t = 1.32s (bottom-right)

Figure 9 illustrates the point cloud representation of the deformation with accompanying histogram of signed distances. This figure aims to show the extent of the rocking motion applied to the tyre where at t = 0.066s, the highest deformation and, thus, the main contact area, indicated by the yellow region, resides lower relative to the centre. However, while at t = 1.32s, the contact area shifts upwards. Additionally, the histogram of signed distances, reveals a valley at zero, showing the limitations of NNS where point cloud representations of surfaces can yield larger distances than in reality. This confirms that the large tyre lug geometry can be determined from the inside tyre surface only.

Finally, Fig. 7, 8 and 9 visualize the results from different perspectives, but consistently highlight common trends observed in established tyre literature, whereby the contact patch deformation exhibits non-linearity under dynamic and temporal tyre loading conditions.

4.3. Test Case 3D Displacement and Velocities of inner tyre surface while rocking back and forth: Approach 2

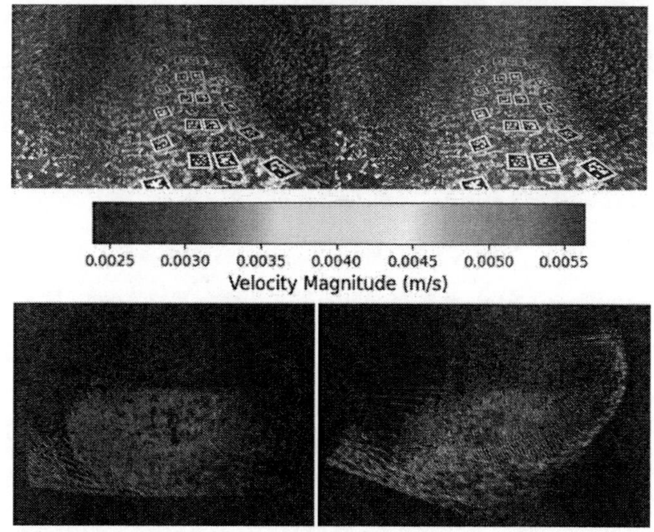

Fig. 10. 3D Point Tracking across two consecutive frames at t=0.462s to t=0.528s. RGB images of the two consecutive frames (top). Jet colormap corresponding to point velocity magnitudes (middle) and the associated metrically scaled 3D space with velocity vectors at different views (bottom)

Figure 10 shows the 3D tracking between two consecutive RGBD frames. This approach only tracks the change in surface deformation as the set of initial points to track consist of the deformed surface. Therefore, the point-velocities in red reveals a central shape in the point cloud surface that lines up with the surface normal of the ground on which the tyre is rocked. This approach is robust to small changes in the surface deformation whereby the top-left and top-right frames look indistinguishable to the naked eye. Ultimately, we are able to capture contact patch behaviour from successive frames by which its shape is contrasted well with respect to the surface geometry.

4.4. Limitations and Sources of Error

The results obtained are limited by means of the validation of the deformation of the outer surface due to it being occluded during testing. Additionally, the method by which deformation is measured has many sources of errors, starting with the depth estimation of the RGBD camera, the reconstruction of the undeformed model in Reality Capture, the point cloud registration techniques, as well as the optical flow estimation. All of the above yields a building of errors that compound in such a pipeline used in this work.

5. Conclusions and Future Work

Ultimately, the deformation in the contact patch was measured using two approaches in which, multiple outputs can be further derived. The first approach validated the behaviour of the vertical deformation in the contact patch under a rocking motion whereby lateral and longitudinal cross sections temporally deform according to the rocking motion of the tyre. The second approach validated the behaviour of not only the vertical deformations that occur from the tests but the longitudinal and lateral deformations as the tyre rolls back and forth.

Future work will focus on combining the two approaches in one, such that the point-wise vertical, lateral and longitudinal point-displacements, velocities and deformation can be projected to the outer surface. Additionally, contact patch velocity under of a moving tyre should be derived as well as slip angle which is inherent to lateral and longitudinal velocities in the contact patch of the tyre.

6. Acknowledgements

This work is funded by the DSI through the CSIR as part of the South African Government's commitment to promote and support the Collaborative Program in Foundational Digital Capacity Development. Special gratitude is expressed to the Vehicle Dynamics Group at the University of Pretoria.

7. Declaration of competing interest

The authors declare that they have no known competing financial interests or personal relationships that could have appeared to influence the work reported in this paper.

8. Declaration of generative AI in scientific writing

During the preparation of this work the authors used ChatGPT in order to improve paper readability. After using this tool/service, the authors reviewed and edited the content as needed and takes full responsibility for the content of the publication.

9. References

Bradski, G., 2000. The OpenCV library. Dr. Dobb's Journal of Software Tools.

Derafshpour, S., Valizadeh, M., Mardani, A., Tamaddoni Saray, M., 2019. A novel system developed based on image processing techniques for dynamical measurement of tire-surface contact area. Measurement, 139, 270-276.

Farnebäck, G., 2003. Two-Frame Motion Estimation Based on Polynomial Expansion. In: Bigun, J., Gustavsson T., eds. Image Analysis, 2003// Berlin, Heidelberg. Springer Berlin Heidelberg, 363-370.

Gao, X., Zhang, Z., Zhang, H. & Fan, W., 2024. Analyzing the interaction mechanism between tire side-slip angle and sidewall flexible deformation using enhanced digital image correlation. Nondestructive Testing and Evaluation, 1-24.

Gutherie, A.G., 2016. 3D Computer vision contact patch measurements inside off-road vehicle tyres, University of Pretoria (South Africa).

Kabutz, T.B. ELS, P.S., 2024. A novel measurement method to aid development of soft soil tyre-terrain interaction models. Vehicle Dynamics Group: University of Pretoria.

Liang, C., Zhu, D., Wang, G., Shan, M., 2021. Experimental Study on Tire-Road Dynamic Contact Pressure Distribution Using FTIR Imaging. International Journal of Automotive Technology, 22, 1305-1317.

Olson, E., 2011. AprilTag: A robust and flexible visual fiducial system. 2011 IEEE International Conference on Robotics and Automation, 9-13 May 2011 2011. 3400-3407.

Wang, G., Yu, K., Liang, C., Mei, Y. 2021. Influence of Contact Area Deformation Distribution of Tire on Tire Noise, Rolling Resistance and Dry Grip Performance. International Journal of Automotive Technology, 22, 231-242.

Wang, X.-B., Li, A.-J., Ci, Q.-P., Shi, M., Jing, T.-L., Zhao, W.-Z., 2019. The study on tire tread depth measurement method based on machine vision. Advances in Mechanical Engineering, 11, 1687814019837828.

Zhou, Q.-Y., Park, J. & Koltun, V., 2018. Open3D: A modern library for 3D data processing. arXiv preprint arXiv:1801.09847.

8140 | Numerical modeling of plate sinkage and in situ Bevameter shear tests using FEM and the density-dependent NorSand critical state constitutive soil model

Ray Kruger [a], Petrus Arnoldus Crous [b], Yuderka Trinidad González [c,*], Mehari Tekeste [d]

[a] *Department of Mechanical Engineering, Iowa State University, Ames, United States*

[b] *Department of Civil Engineering, University of Pretoria, Pretoria, South Africa*

[c] *Department. of Civil Engineering, Iowa State University, Ames, United States*

[d] *Department of Agricultural Engineering, Iowa State University, Ames, United States*

* *Corresponding author: yuderkat@iastate.edu*

ABSTRACT

The Bekker-Wong terramechanics model remains a cornerstone for predicting wheel–soil interactions, relying on Bevameter tests to characterize soil behavior via plate sinkage and in situ shear tests. However, accurately modeling these tests, and the underlying physics, remains challenging. Physics-based methods like the Discrete Element Method (DEM) lack key critical state soil mechanics features such as density dependence, yield functions, and pore pressure effects. This study employs an implicit, large-strain Finite Element Method (FEM) formulation and the NorSand critical state model to simulate plate sinkage and shear testing in partially saturated Cullinan sand. The advanced NorSand model captures soil behavior through evolving yield surfaces and stress–dilatancy relationships linked to effective stress, void ratio, and density-dependent soil strength. Model parameters were calibrated using 18 CU and CD triaxial tests via a novel, automated Bayesian optimization scheme, with calibration accuracy enhanced by empirical meta-models for the shear modulus *G* and hardening parameter *H*. Validation against field Bevameter experiments was conducted using a forward-only modeling approach. The model shows good agreement with in situ shear tests (6–15% relative error in the pre-failure region) and captures peak shear strength and post-peak softening trends across a range of densities. Plate sinkage predictions match early-stage stiffness (16% error at intermediate density) but diverge at large settlements due to numerical instability. This study highlights how physics-based methods, grounded in critical state soil mechanics, can bridge the gap between laboratory calibration and reliable field predictions.

Keywords
Plate Sinkage
In Situ Shear Tests
Bevameter
Critical State Soil Mechanics (CSSM)
Finite Element Method (FEM)
NorSand Constitutive model

1. Introduction

The Bekker–Wong model is an established framework and cornerstone of terramechanics for predicting vehicle traction on soft soil. This model relies on empirical parameters (e.g., k_c, k_ϕ, n, c, ϕ, K) derived from in-situ plate sinkage and in situ Bevameter shear tests. However, the Bekker Bevameter-based approach to soil parameter estimation is subject to variability due to experimental test setup factors (He et al., 2020; Apfelbeck et al., 2011). Estimating shear strength parameters such as cohesion (c) and friction angle (ϕ) from Bevameter in situ shear tests is particularly challenging. For example, it has been shown that the Mohr–Coulomb total stress friction angle for the same soil ranged from 16.5° to 37.6° depending on shear mechanism, tool shape, and estimation methodology, challenging the reliability of Bevameter-based soil characterization and, consequently, the Bekker–Wong soil-wheel model that depends on Bevameter tests (Kruger et al., 2023). Similarly, plate sinkage Bevameter tests fit empirical models that do not represent truly unique intrinsic soil parameters, but rather reflect system response variables (He et al., 2019). Furthermore, important factors that influence in situ shear test, such as bulldozing force and lateral grouser drag, remain difficult to quantify and measure experimentally (Kruger et al., 2023). The empirical plate sinkage and shear test models are also unsuitable for generalizing to new conditions; any alterations in soil state, moisture content, tool geometry, or boundary conditions necessitate recalibration. Significant epistemic model uncertainty also exists in mapping vertical flat plate stresses directly to round or flexible wheel geometries, especially in the case of smaller wheels, further impeding predictive efforts with the empirical Bekker-Wong approach (Meirion-Griffith and Spenko, 2011). These issues have widespread implications for planetary exploration, agriculture, and military vehicle mobility applications.'

Physics-based soil models, especially those grounded in critical state soil mechanics principles, represent a promising avenue for replacing empirical models in terramechanics applications. These models provide insights into the macroscopic behavior of the material rather than system-level responses, and rely on truly invariant soil parameters, such as the critical state friction ratio, that are not functions of geometry, stress path, or soil state (Wood, 1991). To model soil behavior under general loading, varying soil states, and coupled hydro-mechanical effects, constitutive models governing macroscopic soil behavior are required (Karstunen and Amavasai, 2017). However, applying these emerging physics-based constitutive models directly to full tire–soil interaction introduces additional complexity due to the involvement of complex sub-systems, such as flexible tires. Therefore, it is proposed that these numerical methods first be validated against Bevameter in situ plate sinkage and shear tests before extending them to full wheel–soil interaction modeling.

Several studies have utilized physics-based soil models to simulate plate sinkage tests, but no published work has simulated Bevameter in situ translational or rotational shear tests. On the other hand, simulations of plate sinkage tests have utilized a range of numerical approaches, including Finite Element Methods (FEM), Discrete Element Methods (DEM), and more recently, emerging frameworks such as Smoothed Particle Hydrodynamics (SPH) (Hemmat et al., 2012; Xia et al., 2014)

Although the DEM approach handles large deformations and discontinuities well, its explicit particle–contact formulation requires very small time steps and a large number of particles, resulting in high computational cost. It lacks mechanisms to

https://doi.org/10.56884/RD0R8NHU

strictly enforce macroscopic material yield surfaces and lacks incorporation of critical state soil mechanics (CSSM) concepts (Rojek, 2018; Jiang et al., 2025). Tracking continuum variables like void ratio and density is challenging, particularly due to the typical upscaling of soil particles (Zeraati-Shamsabadi and Sadrekarimi, 2025). Standard DEM, without coupled flow models, neglects effective stress and pore-pressure effects, key factors in real soil behavior (Liu et al., 2015). DEM contact models also struggle to reproduce macroscopic responses such as damage, post-peak softening, and shear band orientation (Xiong and Wang, 2023), and fail to capture how dilatancy and shear stiffness evolve with mean stress and strain (Aboul Hosn et al., 2017), making them less suitable for continuum-scale stress–strain soil modeling. To the best of the author's knowledge, no DEM formulation currently includes void ratio-dependent effects on shear strength and mobilized friction angle, which are central concerns in this study.

SPH and Physics-Informed Neural Networks (PINNs) are promising emerging methods in geomechanics and terramechanics. SPH's mesh-free nature makes it adept at handling large deformations; however, it lacks robust formulations for enforcing yield surfaces, has difficulty following macroscopic stress-strain paths, and lacks critical state principles (Islam et al., 2020). PINNs, while powerful for solving general PDE-constrained problems, remain early stage of development for terramechanics and geotechnical soil modeling, with experimental studies simulating only basic soil behaviors (Lu and Mei, 2022). PINNs application to specific tests like the Bevameter tests remains unexplored. They also require a significant amount of data for training and calibration (Monaco and Apiletti, 2023). Due to these limitations and lack of robust formulations for complex soil behaviors in the above-mentioned physics-based approaches were not adopted in the present study.

Among continuum-based FEM approaches, a variety of constitutive formulations exist. The Mohr–Coulomb model is widely used due to its simplicity, but it assumes linear elasticity followed by perfect plasticity, neglecting nonlinear strain hardening/softening behavior (Yu, 2007). Mohr–Coulomb parameters are also not unique, intrinsic fundamental soil properties (Craig, 2004). Similarly, the Drucker–Prager model is also a very simple, elastic-perfectly plastic model. The cap hardening variant introduces some nonlinearity in compressive regimes but still lacks a critical state framework, and has a fixed yield surface with no dilation capability or density dependence, and requires different model parameters for different soil states (Sinha et al., 2010). Cam-Clay and Modified Cam-Clay models incorporate more advanced critical state soil mechanics concepts and better represent volume change effects on shear strength. However, they assume a unique relationship between void ratio e and mean effective stress p', which is not valid for sandy soils (Shuttle and Jefferies, 2010). Sandy soils exhibit an infinite number of Normal Consolidation Lines (NCLs) depending on initial density, and this behavior cannot be captured by the fixed Cam-Clay yield surface.

To overcome these limitations, this study adopts the NorSand model, implemented within a large strain FEM solver framework (Jefferies, 1993). NorSand expands and generalizes Cam-Clay by introducing a state parameter $\psi = e - e_c$, representing the distance between the current void ratio and the critical state void ratio (Jefferies, 1993; Shuttle and Jefferies, 2010). This state-dependent formulation allows the yield surface and shear strength in NorSand to evolve with loading, capturing both contractive and dilative behavior across a wide density range. The evolving yield surface is governed by a density-dependent mobilized friction ratio η_i, which adapts to the current soil state, unlike the fixed critical state friction ratio M_{tc} in traditional models (Shuttle and Jefferies, 2010). This enables NorSand to represent behavior from loose, liquefiable states to dense, dilative conditions. It also explicitly models dilatancy and tracks the evolving void ratio, a capability absent in many traditional constitutive FEM models but essential for simulating realistic sandy soil response under Bevameter loading. Hence, the Norsand model is selected for this study and solved using a large strain FEM approach. This enables mechanistically accurate predictions of plate sinkage and Bevameter in-situ shear tests across varying soil densities and mobilized friction ratios, advancing terramechanics modeling beyond traditional empirical curve fitting methods toward a more rigorous critical state soil mechanics approach.

1.1. The NorSand Constitutive Soil Model Formulation

The NorSand constitutive soil model consists of a system of simultaneous algebraic and nonlinear partial differential equations that require solving using numerical integration schemes; no closed form solutions exist (Jefferies and Been, 2016). This study utilizes NorSand-M version 6 (Shuttle and Jefferies, 2010), building on the original formulation by Jefferies (1993). The governing equations of the NorSand constitutive model are indicated in Table 1.

The soil's evolving state parameter vector (Equation 1) is tracked at each FEM integration point throughout the FEM loading increments. All NorSand parameters are rooted in fundamental soil mechanics and carries deep physical meaning and significance (Shuttle and Jefferies, 2010).

$$x_i = \{e, e_c, p', p_{\mathrm{i}}', q, \sigma_{\mathrm{ij}}, \varepsilon^{\mathrm{p}}{}_{\mathrm{ij}}, \dot{\varepsilon}_q^p, \dot{\varepsilon}_v^p, \xi_{3,tc}, \xi_{3,te}, \dot{\varepsilon_3}, \dot{\varepsilon_1} \\ \psi, \psi_{\mathrm{i}}, \eta, \eta_c, \theta, M, M_{tc}, M_{\mathrm{i}}, M_{\mathrm{i},tc}, M_{\mathrm{i},te}, H, N, \chi, \chi_{tc} \\ K, \lambda, \Gamma, G, \nu, D^{\mathrm{p}}, D^{\mathrm{p}}{}_{tc}, D^{\mathrm{p}}{}_{te}, a, b, c\} \tag{1}$$

In contrast to the NorSand constitutive model summarized in Table 1, the traditional Bekker terramechanics model, commonly used to model Bevameter plate sinkage and in situ shear test, relies on a fixed set of empirical coefficients that do not evolve with changing stress, strain, void ratio, or loading history:

$$\text{Bekker Parameters} = \{k_c, k_\phi, n, c, \phi, K\} \tag{2}$$

The Bekker modeling approach utilizes empirical coefficients unconnected to spatially resolved fields, limiting its ability to capture soil deformation or determine intrinsic soil parameters. Parameters like cohesion c and friction angle ϕ are not soil constants but response variables dependent on density, moisture, pore pressure, and loading history. This variability is amplified in Bevameter shear tests due to experimental factors (Kruger et al., 2023). NorSand, however, is grounded in critical-state soil mechanics, employing a fundamental critical state friction ratio M_{tc} (constant for a given soil) while allowing the mobilized friction ratio η_i to evolve with state parameter ψ. As soil densifies, η_i increases, capturing density-dependent shear strength without requiring ad hoc empirical coefficients.

In the state vector (Equation 1), some parameters remain fixed while others evolve dynamically (indicated by subscript i). Internal variables like the image condition friction ratio M_i are not directly accessible, but are crucial for yield surface evolution. Key NorSand components include:

1. **Internal State Variables:** State parameter $\psi = e - e_c$ quantifies soil density relative to critical state, with ψ_i adjusting for stress and M_i modifying friction ratio for state-dependent strength.
2. **Critical State:** The CSL, $e_c = \Gamma - \lambda \ln(p')$, connects void ratio to mean effective stress, while the critical state friction ratio M_{tc} varies with Lode angle θ in three-dimensional stress states.
3. **Yield Surface:** Evolves dynamically with the mobilized friction ratio $\eta = q/p'$, which depends on the state parameter ψ_i through the image condition M_i. Its maximum extent is bounded by an internal cap defined by χ and ψ_i, enabling state-dependent control of shear strength.
4. **Hardening Rule:** Modulus H governs yield surface rate of evolution with plastic strain rate, enabling stress-state dependent hardening/softening.
5. **Stress–Dilatancy Relation:** Shear and volumetric strain rates are linked via the stress ratio and state parameter, with N and χ_{tc} controlling state-dependent dilation and contraction.
6. **Elasticity:** Shear (G) and bulk (K) moduli relate through Poisson's ratio ν.

A nonlinear, large-strain, quasi-static implicit (i.e. error controlled) FEM formulation solves the NorSand constitutive equations for arbitrary boundary problems. The domain is discretized with constitutive behavior defined at integration points. Newton's method minimizes residuals between internal and external forces using a consistent tangent matrix assembled from partial derivatives of NorSand variables. A stress return mapping algorithm enforces yield surface constraints while an incremental load stepping approach with large strain theory handles geometric nonlinearity. Line search techniques, with additional strain energy-based residuals, enhance convergence robustness. NorSand is highly nonlinear, with evolving yield surfaces, softening, and hardening, making FE convergence challenging using standard FEM solvers.

Table 1
Summary of the NorSand governing equations and associated state parameters (adapted from Shuttle and Jefferies (2010)).

Equation Description	Governing Equations	Internal Variables + Parameters
Internal State Variables	$\psi_i = \psi + \lambda \ln(p'_i/p')$ where $\psi = e - e_c$ $M_i = M\left(1 - \frac{N\chi\lvert\psi\rvert}{M_{tc}}\right)$	$\psi_i, \psi, p'_i, p', e, e_c, M_i, M, M_{tc}, N, \chi, \psi$
Critical State Relationship in e-p and q-p Space	$e_c = \Gamma - \lambda \ln(p')$ *and* $\eta_c = M$ where $M = M_{tc} - \frac{M_{tc}^2}{3+M_{tc}} \cos\left(\frac{3\theta}{2} + \frac{\pi}{4}\right)$	$e_c, \Gamma, \lambda, p', \eta_c, M, M_{tc}, \theta$
Yield Surface Shape & Internal Cap Shape	$\frac{\eta}{M_i} = 1 - \ln\left(\frac{p'}{p'_i}\right)$ *with* $\left(\frac{p'_i}{p'}\right)_{max} = \exp\left(-\frac{\chi_{tc}\psi_i}{M_{i,tc}}\right)$	$\eta, M_i, p', p'_i, \chi_{tc}, \psi_i, M_{i,tc}$
Plastic Hardening Rule (on outer yield surface)	$\frac{\dot{p}'_{m,i}}{p'_{m,i}} = H \frac{M_i}{M_{i,tc}} \left(\frac{p'}{p'_i}\right)^2 \left[\left(\frac{p'_i}{p'}\right)_{max} - \frac{p'_i}{p'}\right] \dot{\epsilon}_q^p$	$p'_{m,i}, \dot{p}'_{m,i}, H, M_i, M_{i,tc}, p', p'_i, \dot{\epsilon}_q^p$
Stress-Dilatancy & Strain Rate Relationships	$D^p = \frac{\dot{\epsilon}_q^p}{\lvert\dot{\epsilon}_q^p\rvert M_i} - \eta$, and $D_{tc}^p = D^p M_{i,tc}/M_i$ and $D_{te}^p = D^p M_{i,te}/M_i$ $\xi_{3,tc} = \frac{2D_{tc}^p - 3}{6 + 2D_{tc}^p}$ *and* $\xi_{3,te} = \frac{2D_{tc}^p - 6}{3 + 2D_{te}^p}$ $\frac{\dot{\varepsilon}_3}{\dot{\varepsilon}_1} = \xi_{3,tc} - (\xi_{3,tc} - \xi_{3,te})\cos\left(\frac{3\theta + 90}{2}\right)$ Define $a = \frac{\sin\theta + \sqrt{3}\cos\theta}{3}$ and $b = -\frac{2\sin\theta}{3}$ and $c = \frac{\sin\theta - \sqrt{3}\cos\theta}{3}$ $\frac{\dot{\varepsilon}_2}{\dot{\varepsilon}_1} = (aD^p - 1 + \frac{\dot{\varepsilon}_3}{\dot{\varepsilon}_1}(cD^p - 1))/(1 - bD^p)$	$D^p, \dot{\varepsilon}_q^p, \dot{\varepsilon}_v^p, \eta, D_{te}^p, M_i, M_{tc}, M_{i,tc}, M_{i,te}$ $\xi_{3,tc}, \xi_{3,te}, \dot{\varepsilon}_3, \dot{\varepsilon}_1, \theta, a, b, c$
Elasticity	G *with* $K = \frac{2(1+\nu)}{3(1-2\nu)} G$	G, K, ν

2. Methodology

2.1. Soil Characteristics

The soil used in this study is a naturally occurring fine quartz sand known as Cullinan sand, with a mean grain size of 149μm. It is sourced from a sand quarry located 50km east of Pretoria, South Africa, and serves as a standardized medium for geotechnical, terramechanics, and centrifuge testing at the University of Pretoria, including in both indoor and large outdoor soil bins. Therefore, it has been extensively used in prior research (Archer and Heymann, 2015; Ganal et al., 2022; Murison and Heymann, 2024; Kruger et al., 2023; Crous and Jacobsz, 2024a; Crous and Jacobsz, 2024b).

Lunar regolith, as encountered by Moon rovers, is also a fine-grained, angular, cohesionless material. Although lunar soils span a broad particle sizes, several simulants have been developed to approximate their mechanical behavior. One widely used example is JSC-1, modeled after Apollo lunar sample 14163, has a mean particle size of d_{50} = 117μm (McKay et al., 1994). This closely matches the d_{50} = 149μm of Cullinan sand, which, along with its angular particle shape, makes Cullinan sand a suitable physical analogue for lunar vehicle mobility studies.

The particle size distribution and physical properties of Cullinan Sand are presented in Appendix A. Field tests were conducted under partially saturated conditions to better simulate terrestrial vehicle environments. The dry-basis moisture content (MC) was maintained at 6.5 ± 0.5%, corresponding to the optimal Modified Proctor moisture content. The resulting degrees of saturation ranged from 20.2% to 35.4%, depending on density. See Kruger et al. (2023) for further details.

2.2. Laboratory Experiments

To calibrate the NorSand constitutive FEM soil model, a series of consolidated undrained (CU) and consolidated drained (CD) triaxial tests were performed on Cullinan sand specifically with the calibration of the NorSand model in mind. According to Shuttle and Jefferies (2010), at least two CD and two CU tests are required for the calibration of the NorSand constitutive model. The goal is to calibrate the model across a wide range of density states and then interpolate material behavior in field conditions using the observed laboratory training data, as recommended by Shuttle and Jefferies (2010).

CU laboratory triaxial tests are conducted to facilitate comprehensive calibration of the NorSand model under coupled hydro-mechanical conditions, capturing pore fluid pressure effects in general settings. Additionally, CU tests on loose samples more readily reach the critical state, whereas CD tests often encounter difficulties due to shear band formation. However, for field conditions, CD tests are a better analog, as Cullinan sand's high permeability allows rapid pore pressure dissipation. This is supported by prior results by Kruger et al. (2023), which show no strain rate sensitivity - indicating negligible pore pressure buildup. Thus, fully drained conditions can be assumed in the field, despite partial saturation.

In addition to the minimum of two CU and two CD tests for NorSand calibration recommended by Shuttle and Jefferies (2010), additional tests were conducted under various soil conditions, including a range of initial void ratios and confining pressures for both CU and CD tests for a total of 18 triaxial tests. These additional tests were performed to ensure the model was trained on a diverse set of soil states and scenarios in the laboratory. This enhances the model's robustness, enabling it to interpolate and generalize effectively across a wide spectrum of potential field conditions.

To measure small-strain elasticity, bender element shear wave transducers were embedded in the triaxial test specimens and activated during triaxial testing at various confining pressures and void ratios to determine the elastic shear modulus as a function of void ratio and mean effective pressure.

2.3. Field Experiments

For field tests, Bevameter in situ plate sinkage data and in situ translational shear test results were obtained from the study conducted by Kruger et al. (2023) on the Cullinan sand. The test apparatus used for these experiments is shown in Figure 1a), while Figures 1b) to e) illustrate examples of the two specific tests investigated in this study: (1) plate sinkage and (2) in-situ translational shear. Although only one example of each test is shown, multiple tests were conducted with soils prepared at different densities, as indicated in Table 2, to investigate their effect on soil strength. For more details on the experimental procedures and methods used to achieve different soil states, refer to Kruger et al. (2023). Note that the Bevameter apparatus utilizes a custom numerically optimized 4-DOF load cell to directly measure soil reaction forces in the different DOF simultaneously.

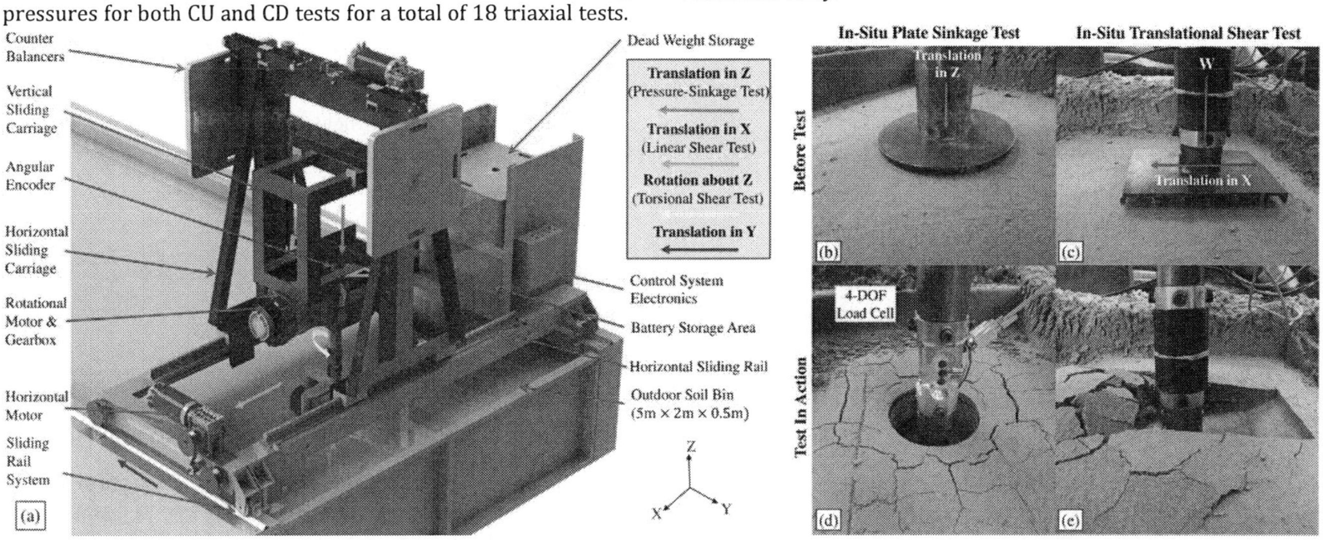

Fig. 1. In situ Bevameter testing. (a) CAD model of the apparatus in an outdoor soil bin. (b) Plate-sinkage test setup. (c) Translational shear test setup. (d, e) Tests in action showing soil deformation. Adapted from Kruger et al. (2023).

Table 2
Summary of measured dry densities, void ratios, and Critical State Soil Mechanics (CSSM) state parameters for different soil density states.

State	Designation	Dry Density	Void Ratio e	Mod. Proc.	RD[1]	State Parameter[2] ψ_0	CSSM Classification[3]
1	'Loose'	1390 kg/m^3	0.908	77.4 %	−29 %	0.030	*Loose*
2	'Medium'	1511 kg/m^3	0.755	84.7 %	56.4 %	−0.123	*Dense*
3	'Dense'	1538 kg/m^3	0.723	86.2 %	73.1 %	−0.153	*Dense*

Note: [1]Based on ASTM D4254-91-1A and D4253-92-1B. [2]Evaluated at $p' = 1$ kPa, $\gamma = 0.877$, and $\lambda = 0.021$. [3]Density classification based on if the soil state lies above (loose) or below (dense) the Critical State Line in the CSSM framework.

An important aspect of these tests was the measurement of in situ bulk and dry density, which is crucial for determining the CSSM state parameter ψ_0, which is to be used in the NorSand model for soil strength predictions. The in situ bulk density was measured using the photogrammetry method described by Berney et al. (2018). In this process, a 15 cm diameter hole with a depth of 10 cm is excavated, followed by a 3D scan to accurately measure the excavated volume. This method provides precise in situ density measurements, similar in principle to the sand cone method.

2.4. Setup of the Finite Element Model Boundary Conditions

Figure 2 illustrates the FEM setup and boundary conditions for the CU/CD triaxial and oedometer test simulations. Both models adopt an axisymmetric formulation with full 3D stress evaluation, implemented using a single 8-noded quadratic quadrilateral element.

For brevity, field simulation mesh setups are omitted; deformed meshes are shown in Figures 10 and 12. The plate sinkage model is axisymmetric, with fixed right and bottom boundaries, a roller on the left (axis of symmetry), and a 0.5m deep, 0.75m wide soil bin. A 150mm diameter plate is loaded via vertical Dirichlet displacement. The in-situ shear test uses a 2D plane strain model with fixed boundaries except the top; the soil bin is 0.5m deep and 2m wide, matching the experimental setup. The shear plate is modeled as a rigid 200mm × 10mm elastic steel body. Gravity loading initializes geostatic stress. Both models use 8-noded quadratic quadrilateral elements for high order accuracy and mid-side flexure.

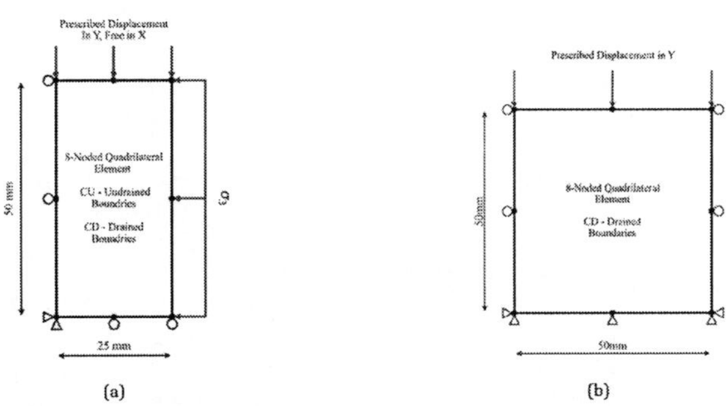

Fig. 2. FEM mesh and boundary conditions for (a) CU and CD triaxial test simulations, and (b) oedometer test simulations, using an axisymmetric 8-noded quadrilateral element.

2.5. NorSand Constitutive Model Calibration

2.5.1. The Calibration Approach

The NorSand constitutive model was calibrated using an extensive laboratory dataset comprising 18 triaxial tests covering diverse density states, confining pressures, and pore pressures. Following recommendations by Shuttle and Jefferies (2010), the model was calibrated across a wide range of soil states to enable interpolation of material behavior in field conditions rather than extrapolation. Field tests, such as the Bevameter, validated these interpolated predictions.

A meta-model approach was employed to improve calibration accuracy. Extending the meta-modelling strategy proposed by Shuttle and Jefferies (2010), initially focused only on the hardening parameter, this study introduced empirical meta-modeling of the shear modulus to enhance predictive accuracy. Furthermore, rigorous separation between training and validation datasets was maintained, complemented by fully automated calibration algorithms to eliminate human bias, ensuring reproducibility and robustness.

A novel data-driven algorithm was developed to automatically estimate the hardening parameter H, systematically determining the intercept H_0 and gradient H_y, thereby replacing manual tuning methods with an unbiased and efficient optimization procedure. Additionally, a new empirical meta-model was introduced for estimating the elastic small-strain shear modulus G_0 based on the soil state parameter ψ, enabling practical in situ modulus predictions from density measurements alone, thus avoiding expensive in situ testing.

The remaining NorSand parameters, including critical state friction ratio M_{tc}, slope λ and intercept Γ of the critical state line, and dilatancy-related parameters N and χ, were calibrated using standard methods recommended by Shuttle and Jefferies (2010). Each parameter retains clear physical significance and was determined using established experimental techniques.

2.5.2. Elasticity - Estimating Stiffness Parameters: G *and* ν

The small-strain shear modulus G_0 was determined from laboratory bender element tests, following the approach recommended by Shuttle and Jefferies (2010). In these tests, piezoelectric bender elements embedded in consolidated triaxial specimens generated shear waves; shear wave velocity V_s was then calculated from the measured arrival times. G_0 was computed from V_s as:

$$G_0 = \rho_{bulk} \times V_s^2 \tag{3}$$

where ρ_{bulk} is the bulk soil density. Tests were performed at various densities and confining pressures. The following generic form, commonly attributed to the Bolton-type model Seed and Idriss (1970), was fitted to the data, allowing G_0 to be modeled as:

$$G_0(p', e) = A \times \left(\frac{1}{1+e}\right)^n p'^{m} \tag{4}$$

where parameters A and m were fitted to the experimental shear wave data (see Figure A2 in appendix), and the void ratio exponent $n = 2.17$ was adopted from literature for fine sand (Ha Giang et al., 2017).

Since the small-strain shear modulus G_0 is well known to overestimate soil stiffness at strains greater than 10^{-5}, strain-dependent stiffness degradation was applied using the Hardin and Drnevich (1972) hyperbolic type model:

$$\frac{G}{G_0} = \frac{1}{1 + \left(\frac{\gamma}{\gamma_{ref}}\right)^a} \tag{5}$$

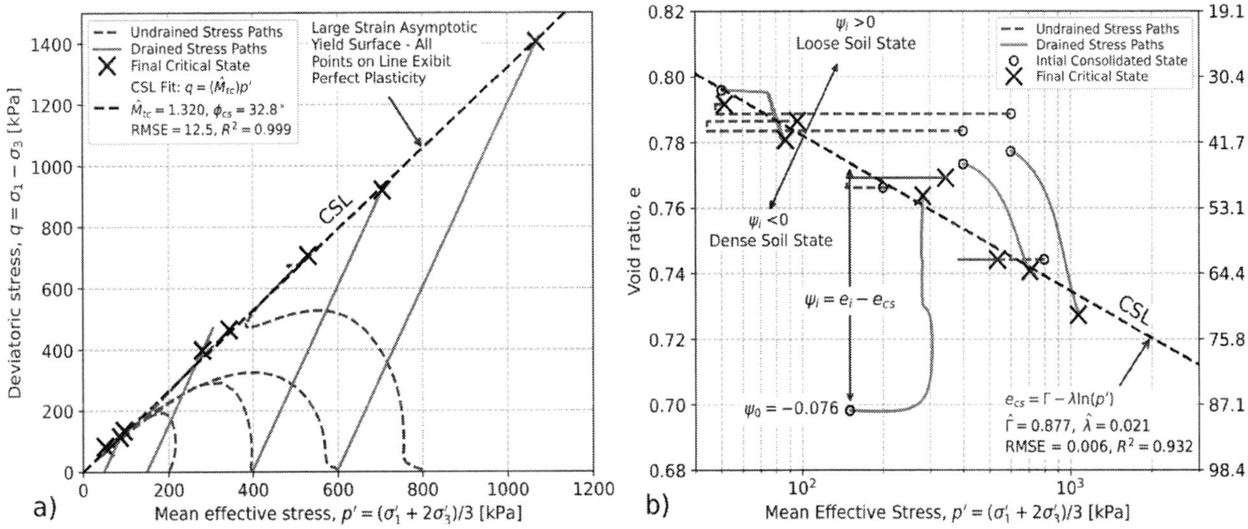

Fig. 3. Estimation of NorSand Critical State Line (CSL) parameters from triaxial tests on Cullinan Sand: (a) Shear strength behavior in q–p' space used to estimate M_{tc} from CD and CU tests; (b) Volumetric behavior in e–log(p') space used to estimate λ and Γ

Where, G is the shear modulus at a given strain γ, G_0 is the small-strain modulus, and γ_{ref} is the reference strain where significant softening begins. Both γ_{ref} and the exponent a are material dependent and typically determined from experiments. However, due to scatter in the triaxial data, direct data-driven calibration was not feasible. Therefore, literature values were adopted from studies on similar fine quartz sand (d_{50} = 160 μm) that utilized high-precision resonant column instrumentation (Senetakis et al., 2013).

Shuttle and Jefferies (2010) does not specify the strain level at which the shear modulus G should be evaluated for NorSand. For practical constitutive modeling, it is proposed to evaluate Equation 5 at a shear strain of γ = 0.1%, which corresponds to the typical initial strain onset in triaxial tests. For the specific sand used in this study, Equation 5 yields a modulus reduction ratio of G/G_0 = 0.53 at 0.1% strain.

This approach of measuring $G_0(p',e)$ using bender elements in the lab at different pressures and void ratios and reducing by G/G_0 allows practical estimation of the field large-strain shear modulus using easily measurable in-situ soil density and stress conditions without the need for expensive in situ geophysics shear wave experiments. An elastic Poisson's ratio of 0.2 was assumed, consistent with typical fine sand values (Shuttle and Jefferies, 2010).

2.5.3. Plasticity - Estimating the CSL Parameters: M_{tc}, λ, Γ

The estimation of the Critical State Line (CSL) parameters follows well-established procedures. The stress paths and critical state stress points in the $q - p'$ space are indicated in Figure 3a, with the corresponding volumetric behavior in the $e - \log(p')$ space illustrated in Figure 3b. The CSL represents the long-term limit of plastic deformation, where the soil undergoes continuous shearing at constant volume and stress, marking the onset of perfectly plastic flow.

The experimental data validate the assumptions of the critical state soil mechanics framework, with agreement between the CSL fits and experimental test data. In the $q - p'$ space, both drained and undrained triaxial tests converge clearly to the critical state line, yielding R^2 values exceeding 0.999. In the $e - \log(p')$ space, volumetric behavior also follows the theoretical expectations, albeit with slightly greater scatter. Nonetheless, the trend is robust, with R^2 values above 0.93 and a root mean square error (RMSE) of 0.006. The key takeaway is that all the specimens consistently approach the CSL at large strains, regardless of their initial density state or the level of confining pressure.

It should be noted that the Critical State Line (CSL) parameters were primarily derived from loosely prepared, contractive specimens. Loose soil specimens more reliably approach the critical state without the onset of strain localization or shear banding typical of dense samples (Shuttle and Jefferies, 2010). Dense samples are also expected to approach the critical state at large strain, provided that tests are not stopped prematurely by experimental limitations such as shear banding.

Unlike most classical critical state constitutive models, like cam-clay, where M_{tc} defines the boundary of the yield surface, the NorSand model features a associative but state-dependent, evolving yield surface that adjusts dynamically with the soil's volumetric behavior and state parameters. The actual peak mobilized shear strength (or yield surface) is governed by the state-dependent shear stress ratio η_L, which is a nonlinear function of both M_{tc} and the state parameter ψ. This distinction becomes particularly significant in dense specimens, where shear strength increases due to dilative behavior. However, at very large strains, even dense samples gradually return to the CSL and asymptotically tend to M_{tc}. This coupling allows the NorSand model to capture density-dependent yield strength, a well-documented phenomenon in sandy soils.

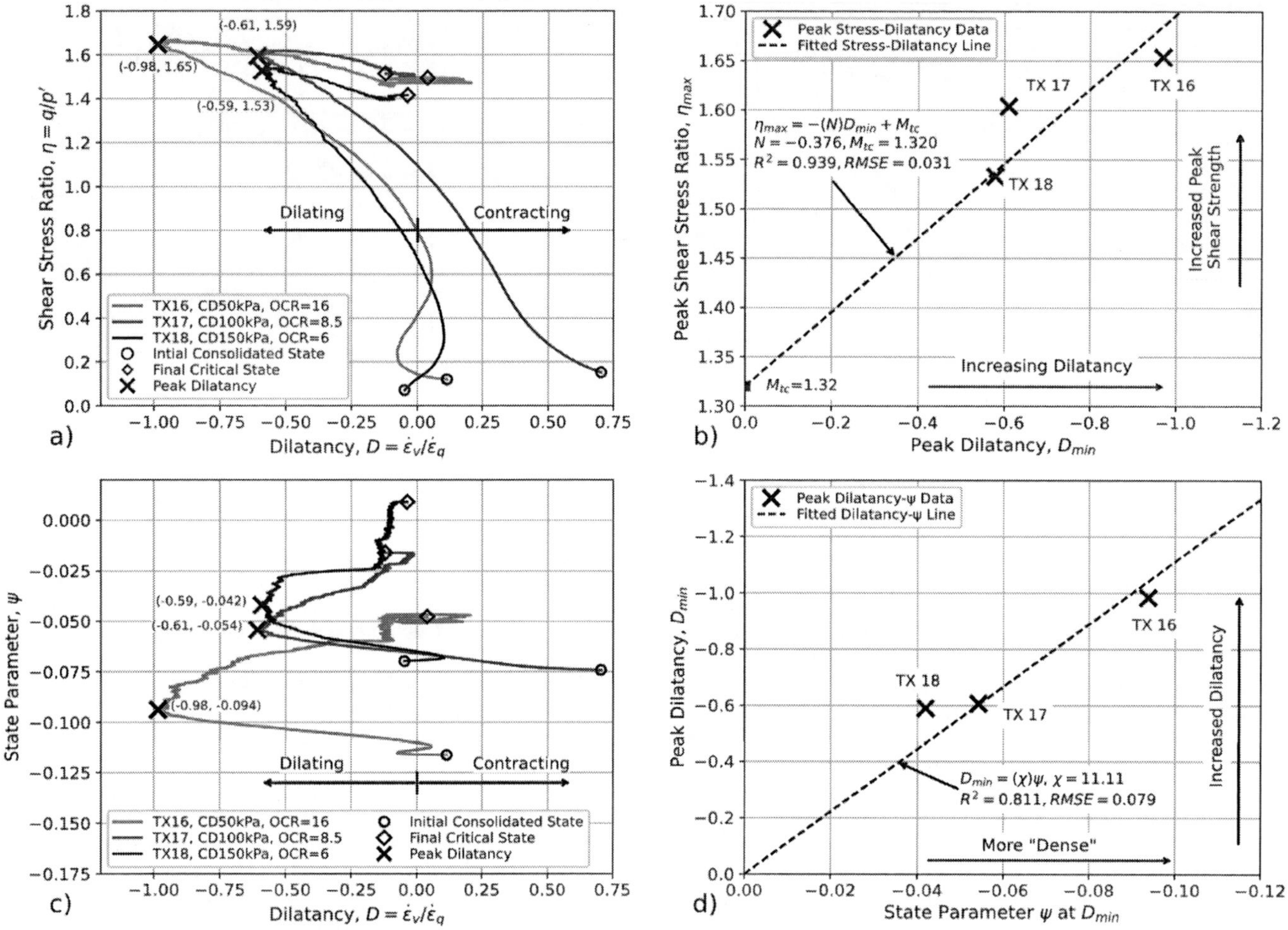

Fig. 4. Estimation of NorSand stress–dilatancy coupling parameters from dense CD triaxial tests: (a) Stress–dilatancy evolution during shearing; (b) Peak stress ratio η_{max} vs. peak dilatancy D_{min}, used to estimate the stress–dilatancy parameter N; (c) Evolution of dilatancy with state parameter ψ; (d) Peak dilatancy D_{min} vs. ψ, used to estimate the state–dilatancy coupling parameter χ.

2.5.4. Plasticity - Estimating the Volumetric Expansions and Shear Strength Coupling Parameters: χ & N

Figure 4 illustrates the estimation of NorSand's plasticity coupling parameters N (stress–dilatancy) and χ (state–dilatancy) using dense CD triaxial test data. In subplot (a), the stress ratio increases with negative dilatancy, indicating a clear peak followed by contraction toward the critical state. Subplot (b) shows a strong linear relationship between the peak dilatancy D_{min} and the peak mobilized stress ratio η_{max}, from which the stress–dilatancy coupling parameter N is obtained. Subplot (c) presents the evolution of the state parameter vs dilation ψ, while subplot (d) shows that more negative state parameters, corresponding to denser specimens, lead to greater peak dilatancy. This provides strong support for the use of the NorSand model, which effectively captures the coupled influence of dilatancy and density on peak strength and mobilized friction ratio, capabilities absent in conventional constitutive FEM soil models.

Figure 5 further supports the rationale for adopting the NorSand model, with panel (a) showcasing how the NorSand model uses parameters χ and N to predict the peak mobilized friction ratio for dense soil samples, validated against experimental data. Figure 5 (b) depicts how the NorSand yield surface develops and enhances the peak friction ratio.

2.5.5. Plasticity - Estimating the Hardening Rate Parameters: H, H_0 and H_y

The plasticity parameters H, H_0, and H_y govern the plastic hardening behavior in the NorSand model, more specifically, how quickly shear strength increases with increasing deviatoric strain until the critical state is reached. According to Shuttle and Jefferies (2010), the hardening modulus H must be calibrated for each individual triaxial test by fitting the model's stress–strain response to experimental deviatoric stress versus axial strain curves. This yields a unique optimal H value for each test, depending on its initial densities and mean stress state.

A linear meta-model is then fitted to relate these optimal H values to the initial state parameter ψ_0, from which the hardening intercept H_0 and hardening gradient H_y are determined. This results in a generalized hardening law that allows H to be estimated directly from the current density state parameter. The model is trained across a wide range of densities in the laboratory, enabling it to infer the appropriate hardening rate under in situ conditions based on experimental measured field density. I.e, a density-dependent hardening meta-model.

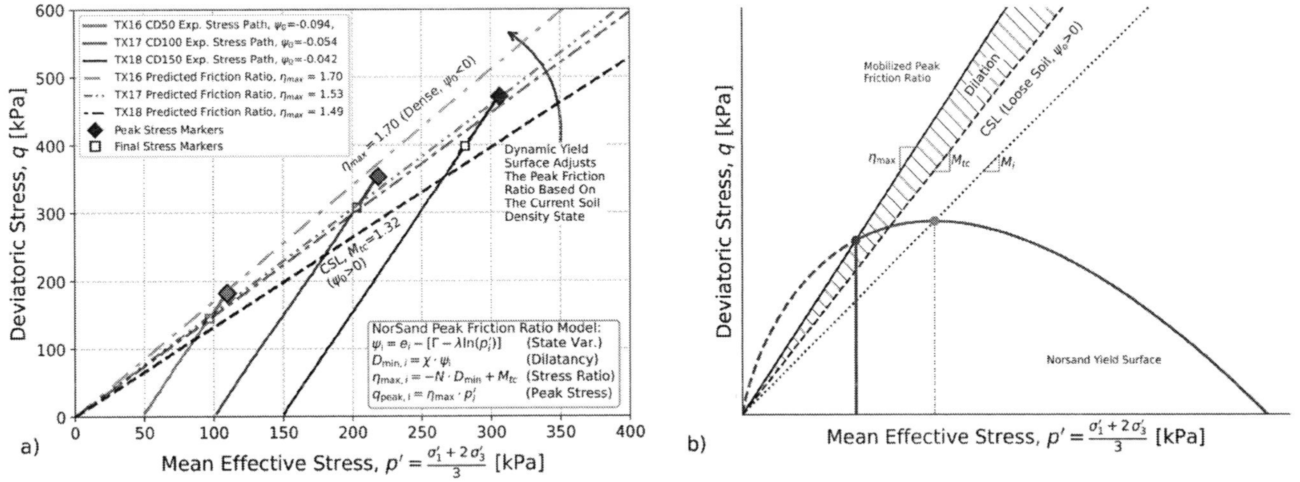

Fig. 5. a) Illustration of how NorSand dynamically adjusts the yield surface and predicts the peak mobilized friction ratio (ηmax) using parameters χ and N estimated in Figure 4, compared with experimental data for Cullinan Sand. b) Corresponding theoretical representation of the NorSand yield surface.

Shuttle and Jefferies (2010) does not provide a specific formal methodology or algorithm for performing this optimization. Similarly, the setup of the optimization objective function remains unclear. Prior studies that have applied the NorSand model used manual or 'hand' tuning, where nonlinear shear stress vs. axial strain curves of the model were visually fitted to the experimental data for each individual triaxial test. Given the large number of triaxial tests (18) in the NorSand model training dataset, manually optimizing by hand tuning is neither practical nor guaranteed to yield the theoretically optimal solution or unbiased parameter estimates.

This study improves on the calibration procedure of Shuttle and Jefferies (2010) for H by employing a formal and fully automated algorithm for estimating the hardening parameter H for each triaxial test. Subsequently, the dependent hardening intercept H_o and hardening gradient H_y. Shuttle and Jefferies (2010) used of Plaxis FEM software and Microsoft Excel, which makes automating FEM solver-inside optimization loop difficult. With custom FEM code and Rockscience RS2 software with a Python API it is trivial to run automated algorithms for fully automated nonlinear and (optionally also probabilistic) parameter estimation and calibration. The nonlinear optimization method employed is Bayesian Optimization (BO), a gradient-free, probabilistic approach that efficiently calibrates parameters of expensive black-box functions, such as the finite element formulation of the NorSand model. BO utilizes a surrogate model, commonly a Gaussian Process or Radial Basis Function (RBF) network, fitted via Bayesian inference to approximate the response surface. Parameters are selected adaptively through an acquisition function (e.g., expected improvement), effectively balancing exploration and exploitation. This makes BO particularly suited for calibrating NorSand parameters, where forward finite element simulations are computationally intensive, gradients are unavailable, and data can be noisy. BO typically achieves faster convergence and requires fewer simulations compared to traditional optimization techniques.

Most NorSand studies assume a fixed hardening modulus H_0, typically adopted from literature or manually adjusted. In contrast, we implement a Bayesian optimization framework using a surrogate based RBF model with the full nonlinear FEM simulation embedded in the loop. At each iteration, ψ_0 and H_0 are updated to minimize the error between simulated and experimental stress–strain responses from CU and CD tests. As shown in Figure 6, the results reveal a weak negative correlation, with H increasing as ψ_0 decreases. Despite the scatter, a trendline was fitted and used in this study. Nevertheless, the variability suggests that a constant H value in the range of 1000–2000 may be reasonable for practical applications.

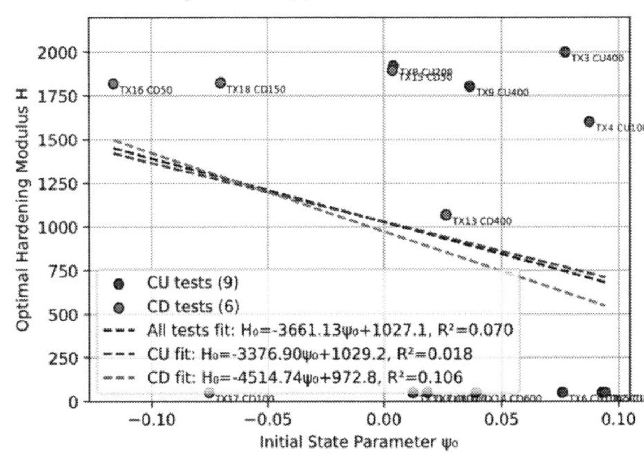

Fig. 6. Data-driven calibration of NorSand FEM cap hardening behavior - Linear trend of optimized H values against initial state parameter ψ0 across 18 tests to estimate global hardening parameters Hy and H0.

3. Results

3.1. Final Parameters Estimates

Tables 3 and 4 summarize the calibrated NorSand parameters and corresponding empirical metamodels. Table 3 groups the parameters by constitutive category: elasticity, critical state, plasticity, and initial state. Key parameters such as the shear modulus G, hardening parameter H, and state parameter ψ_0 depend on calibrated meta-models. Table 4 presents the functional forms and coefficients of these meta-models, with elasticity modeled as a function of effective stress p' and void ratio e, and plasticity as a function of the state variable ψ_0, enhancing model accuracy and generalization.

3.2. NorSand FEM Predictions of Triaxial Tests (Seen/Optimized Training Data).

The FEM simulation results from the CU triaxial calibration tests are indicated in Figure 7. Figure 7 presents the comparison between NorSand FEM predictions and experimental results for drained triaxial tests on both loose (left column) and dense (right column) specimens. The plots show deviatoric stress, volumetric strain, stress paths in p–q space, and void ratio evolution as functions of mean stress. Note that the data and predictions are from the training set used to optimize the NorSand parameters (specifically the hardening rate) using formal algorithms, namely Bayesian optimization. This formal inverse modeling approach eliminates the need for the manual 'hand-tuning' and associated human bias.

For both material states, the model captures the general trends of stress-strain response, including the strain hardening behavior in loose samples and the softening followed by plateau behavior in dense samples. The FEM predictions align closely with the experimental curves in the early and peak strain stages, with some deviation noted in post-peak volumetric response. The agreement between model predictions and test data demonstrates the effectiveness of the calibrated NorSand parameters in reproducing drained behavior across a range of initial densities.

In loose specimens, the model accurately predicts contractive volumetric strains and a monotonic increase in deviator stress, consistent with critical state soil mechanics principles. In dense specimens, the onset of dilatancy and subsequent reduction in stiffness are well captured, indicating appropriate sensitivity of the model to the state parameter ψ_0.

The correspondence in p–q and e–p' space confirms the model's ability to follow the evolving critical state line (CSL), supporting its applicability to simulating strain localization and phase transformation behavior under drained conditions.

Table 3
Final calibrated NorSand FEM soil model parameters.

Parameter	Description	Value	Source
Elasticity			
G	Shear modulus [MPa]	Meta Model: $G(e_0, p_0')$	Data Driven Calibration (Table 4)
ν	Poisson's ratio	0.2	Literature (Shuttle and Jefferies, 2010)
CSL (Volumetric Behavior)			
Γ	Altitude of CSL at 1 kPa	0.877	Data Driven Calibration (Figure 3)
λ	Slope of CSL (Semi-Log scale)	0.0217	Data Driven Calibration (Figure 3)
Plasticity			
M_{tc}	Critical State Friction Ratio	1.32	Data Driven Calibration (Figure 3)
H	Strain Rate Hardening Parameter	Meta Model: $H(\psi_0)$	Data Driven Calibration (Table 4)
N	Stress–Dilatancy Coupling Parameter	0.376	Data Driven Calibration (Figure 4 & 5)
χ_{tc}	State–Dilatancy Coupling Parameter	11.11	Data Driven Calibration (Figure 4 & 5)
Initial State Parameters (Evolving)			
ψ_0	Density State Parameter	Test Dependent $\psi_0(e_0, p_0')$	Measured Experimentally (Table 2)
p'	Mean Effective Stress	Test Dependent	Controlled for Triaxial & Gravity loading for field

Table 4
Additional empirical meta-models for improving NorSand accuracy.

Parameter	Meta-Model Formula	Parameter Values	Source
Elasticity			
G_0	$G_0(e,p') = A\left(\frac{1}{1+e}\right)^n p'^m$	$\hat{A} = 17.5, \hat{m} = 0.51,$ $\hat{n} = 2.17$	Data-driven calibration for parameters A and m (Figure A2) Parameter n from Literature (Ha Giang et al., 2017)
G	$G(G_0) = \frac{G_0}{1+\left(\frac{\gamma}{\gamma_{ref}}\right)^a}$	$\hat{\gamma}_{ref} = 0.0075\%,$ $\hat{a} = 1.02$	Literature (Senetakis et al., 2013)
Plasticity			
H	$H(\psi_0) = H_0 + H_y(\psi_0)$	$\hat{H}_0 = 1027, \hat{H}_y = -3376$	Data-driven calibration (Figure 6)

3.3. NorSand FEM Model Predictions of Bevameter Plate sinkage and in-situ shear tests (Forward model validation on Unseen Field Data)

Figure 9 presents the comparison between NorSand FEM predictions and experimental results for in-situ plate sinkage tests conducted using a 150mm diameter plate across three distinct density states. The vertical axis shows the average vertical stress on the plate, while the horizontal axis denotes the measured settlement. The NorSand model simulations were carried out using parameters derived from previously calibrated meta-models based on triaxial tests, with no additional tuning. For all three soil density cases, the NorSand FEM predictions capture the overall trend of experimental data, with a clear increase in vertical stress due to the effect of soil density (or state parameter), with deviations primarily emerging at larger displacements, around 12mm.

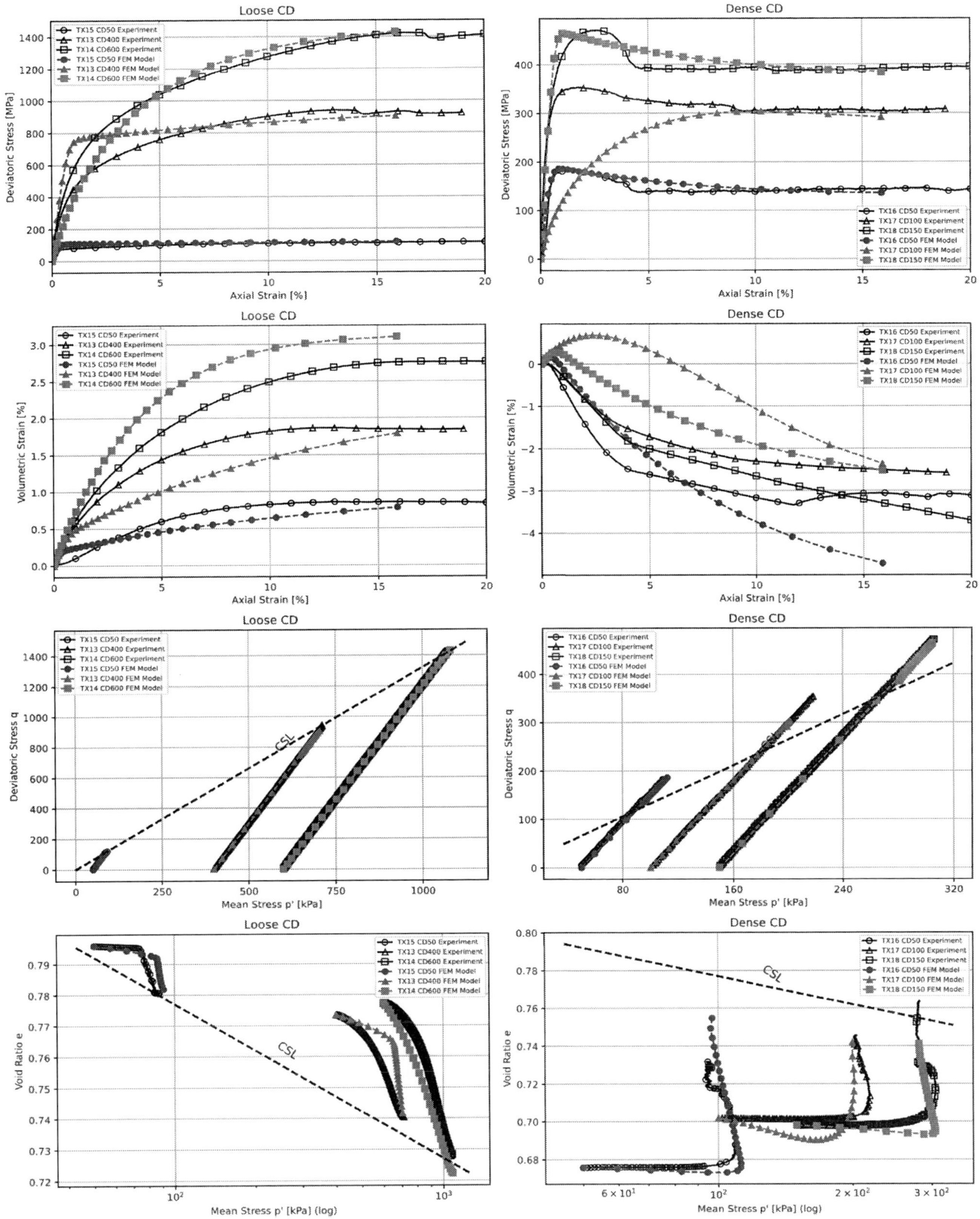

Fig. 7. NorSand FEM predictions of Drained triaxial test vs. experimental data, left is loose right column is dense.

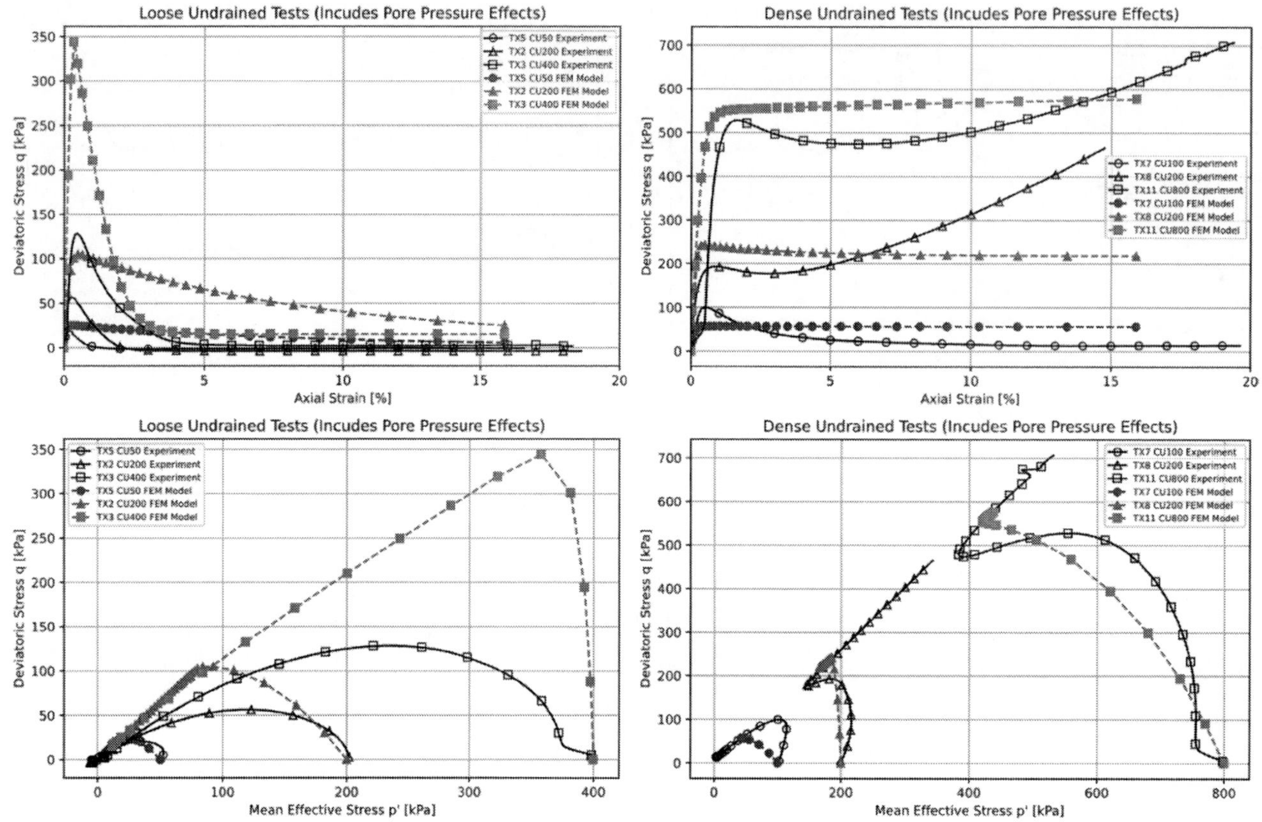

Fig. 8. NorSand FEM predictions of hydro-mechanically coupled undrained triaxial test vs. experimental data: Left column is loose, right column is dense.

The close agreement in the early loading stages in Figure 9 highlights the accuracy of the elasticity meta-model $G(e,p')$, defined as a function of normal stress and void ratio and calibrated using bender element shear wave data. Divergence beyond approximately 13mm of settlement corresponds to numerical instability in the implicit FEM solver, attributed to strain localization, softening, and mesh distortion — typical challenges in large-deformation plasticity. This is a limitation of the numerical method rather than the governing constitutive model and could be addressed using arclength solvers or mesh-free PDE solvers. Additional error could also be attributed to the continuum FEM approach not being able to model the "punching" shear. However, the simulations successfully reproduce density-dependent behavior: looser states respond more compliantly, while denser states mobilize higher resistance at smaller settlements. This response, controlled by the evolving state parameter ψ_0 and the hardening modulus $H(\psi)$, is a feature of the NorSand model not captured by conventional models like Mohr–Coulomb or Drucker–Prager. This field validation demonstrates NorSand's ability to generalize to field conditions using parameters calibrated solely from laboratory triaxial tests across a wide range of densities and stress states, and interpolating for the field states.

Figure 10 shows the simulated displacement field at the final stage of the NorSand FEM plate sinkage test for Density State 3, corresponding to a settlement of 40mm. The contours reveal a localized plastic zone beneath the circular plate, extending downward and outward in a bulb-like pattern, with deformation concentrated under the plate center and spreading outward in a fan-like shape. Soil adjacent to the plate is displaced upward, as indicated by the red zone. This reflects shear mobilization and stress arching typical of dense granular media under circular footings. The continuum-based FEM solution shows gradual downward bending near the boundaries unlike the experimentally observed punching failure in that region (Figure 1). This is due to tensile stresses in those elements, which NorSand does not model to incorporate damage or damage evolution. For improved realism in such regions, damage evolution modeling, e.g., via phase-field methods or mesh-free PDE solvers that accommodate discontinuities, could be incorporated into the FEM formulation to simulate material weakening under tension. Nonetheless, the highly localized nature of the deformation beneath the footing confirms the adequacy of the NorSand model in capturing axisymmetric failure mechanisms without requiring additional localization rules.

Figure 11 compares the NorSand FEM prediction against experimental results for in-situ shear tests across three different initial densities. The shear stress–displacement response for the dense condition ($e = 0.723$, $\psi_0 = -0.153$) shows strong agreement between the model and measured data, including peak stress and softening behavior. The model accurately predicts the initial stiffness, peak shear resistance, and residual strength for the dense case.

The high-fidelity match between simulation and experiment highlights the NorSand model's capacity to simulate soil shear behavior very well, including peak shear strength mobilization and post-peak softening. This alignment reinforces the importance of the state parameter ψ_0 and its control over dilatancy and hardening/softening response. The transition from peak to residual stress reflects the critical state framework, validating the model's ability to replicate real field shear deformation

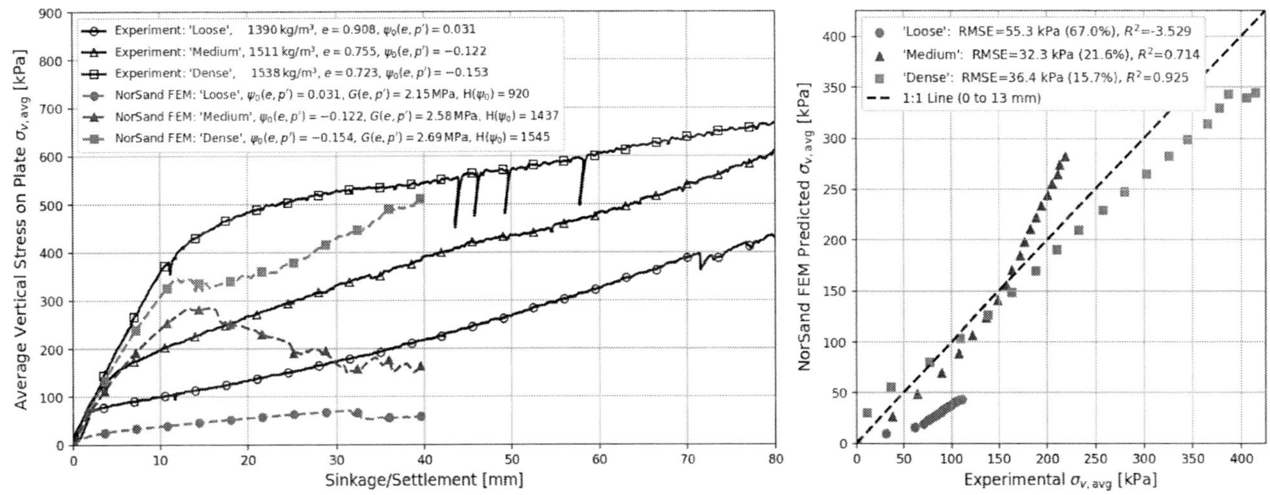

Fig. 9. NorSand FEM prediction of in-situ plate sinkage vs. experimental data. 150mm diameter plate. Experimental data from Kruger et al. (2023).

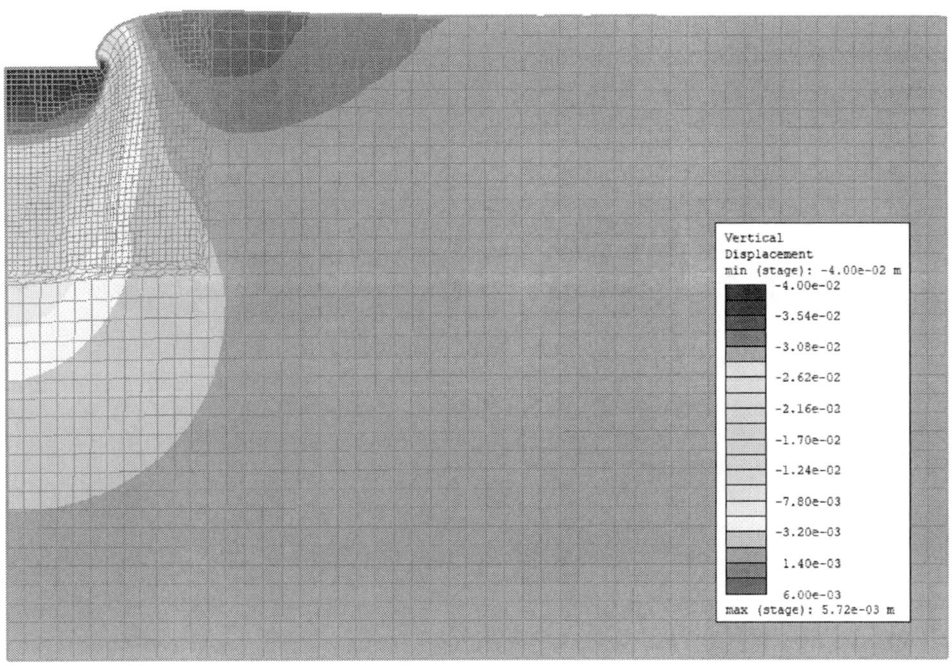

Fig. 10. NorSand FEM predicted displacement field for in-situ plate sinkage at 40mm displacement, 'Dense', e= 0.723, ψ_0= -0.153.

Figure 12 presents the vertical displacement contours at the end of Stage 1 of the Bevameter in-situ shear test simulation, where vertical stress (σ_v = 93kPa) has been applied. The settlement beneath the shear plate is visibly localized, with downward movement concentrated below the plate center and gradually dissipating radially. The observed displacement field is characteristic of dense sands, where compressive settlement occurs without widespread displacement. The stress is distributed effectively due to increased stiffness at lower void ratios, validating the NorSand model's ability to simulate realistic vertical deformation fields.

Figure 13 shows the predicted lateral stress during Stage 2 of the Bevameter shear test, where lateral displacement is applied under constant vertical load. The shear band initiates beneath the plate and propagates asymmetrically along the loading direction, with displacement intensity decreasing away from the shearing zone.

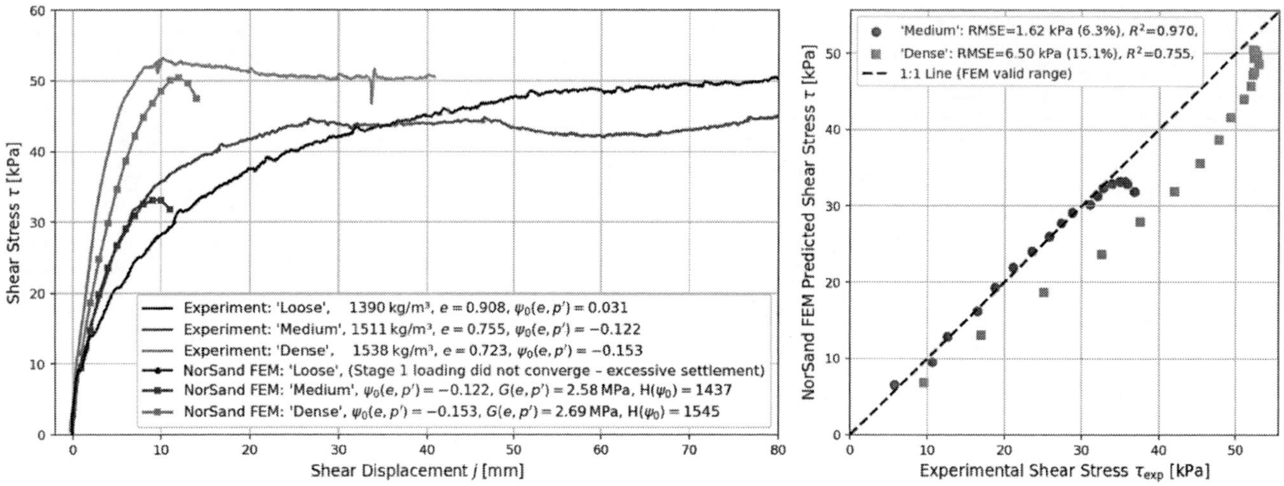

Fig. 11. NorSand FEM prediction of Bevameter in-situ Shear test vs. Experimental Data (σ_v=93kPa). Experimental data from Kruger et al. (2023).

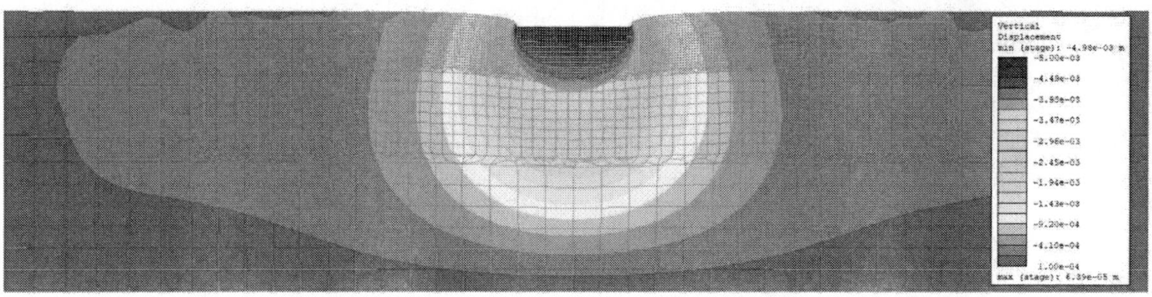

Fig. 12. Stage one (1): NorSand FEM prediction of Bevameter in-situ shear test, settlement after applying vertical load ('Dense' state, σ_v=93kPa).

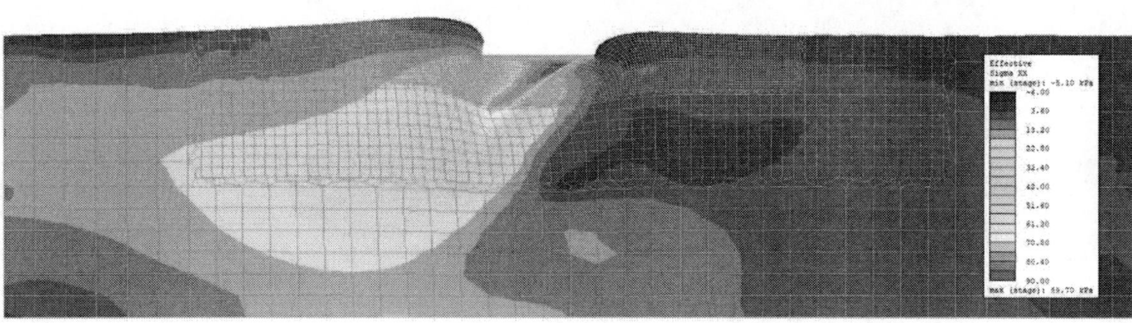

Fig. 13. Stage (2): NorSand FEM prediction of Bevameter in-situ shear test, lateral displacement under constant vertical stress, with resulting lateral stress field σ_{xx} ('Dense' state, σ_v = 93kPa, 15mm displacement).

The asymmetric displacement pattern illustrates the formation of a shear zone consistent with dense granular response under high confinement. The NorSand model effectively captures the coupling between vertical and lateral loading stages, where shear localization develops within the preloaded zone. The results align with field-observed failure mechanisms, reinforcing the model's suitability for simulating complex, staged in-situ loading.

4. Conclusion and future work

This study presents a physics-based finite element modeling framework for simulating in situ plate sinkage and shear testing using the NorSand critical state soil model. Unlike traditional empirical terramechanics approaches, such as the Bekker–Wong formulation, this work adopts a first-principles methodology grounded in critical state soil mechanics, enabling the simulation of density-dependent plasticity, volumetric coupling, and evolving yield surfaces. The finite element implementation incorporates large-strain plasticity with adaptive load-stepping, allowing stable simulation of soil deformation up to substantial strains before the onset of numerical instability. A novel, fully automated data-driven calibration scheme is introduced for estimating NorSand parameters from laboratory triaxial data using Bayesian optimization. Calibration accuracy is further enhanced by the introduction of empirical meta-models for the shear modulus G and the hardening parameter H. The model is validated against full-scale Bevameter field experiments, marking the first known application of the NorSand model in FEM-based simulations of tool–soil interaction. Critically, to our knowledge, this work represents the first rigorous attempt to bridge the long-standing gap between laboratory-calibrated parameters, especially ones rooted in truly invariant soil properties such as the critical state friction ratio, and field-scale Bevameter predictions within a density-dependent framework. In the pre-failure region, before the onset of numerical instability, the model achieves 6–15% relative error in *in situ* shear tests across multiple density states, and approximately 16% error in plate sinkage prediction at intermediate density, all without fine-tuning or backfitting to field data. These results demonstrate the feasibility and accuracy of a forward-only modeling approach, driven purely by measurable in-situ density state parameters and supported by in-depth laboratory characterization across a wide range of soil states and conditions (18 triaxial tests).

Future work will explore inverse modeling with Bayesian optimization to estimate NorSand in-situ state parameters (e.g., ψ_0, H) directly from Bevameter tests, further enhancing accuracy and allowing in situ soil parameter estimation. Future work will also investigate hybrid particle–mesh integration schemes, such as the Material Point Method (MPM), to solve NorSand's governing PDEs and extend applicability to problems involving severe mesh distortion, granular flow, and post-failure behavior. Integrating NorSand into an MPM framework could enable simulation of extremely large strains and complex contact interactions, as required for wheel–soil modeling, while preserving the critical state soil mechanics rigor demonstrated in this study. This would expand the range of terramechanics problems addressed, including wheel–soil interaction and post-failure loading, with applications requiring high-fidelity, physics-based simulation grounded in rigorous soil mechanics and truly invariant soil properties such as the critical state friction ratio. Such a framework could enable near real-time mobility predictions, solving in seconds, rather than minutes or hours, while providing enhanced physical rigor, and substantially greater generalizability across varied geometries, boundary conditions, and soil states, compared to empirical models. Key applications include autonomous planetary rover navigation, precision agriculture, and military vehicle mobility prediction.

5. References

Apfelbeck, M., Kuß, S., Rebele, B., Schäfer, B., 2011. A systematic approach to reliably characterize soils based on Bevameter testing. Journal of Terramech. 48(5):360– 371, 10. ISSN 00224898. doi: 10.1016/j.jterra.2011.04.001. URL https://linkinghub.elsevier.com/retrieve/pii/S0022489811000206.

Archer, A., Heymann, G., 2015. Using small-strain stiffness to predict the load-settlement behaviour of shallow foundations on sand. Journal of the South African Institution of Civil Engineering, 57 (2):28–35. ISSN 23098775. doi: 10.17159/2309-8775/2015/v57n2a4.

Berney, E.S., Ganesh, N.B., Pratt, T.C., 2018. A Photogrammetric Method for Obtaining Soil Density. Technical Report June, Washington.

Craig, R.F., 2004. Craig's Soil Mechanics. CRC Press. ISBN 9780429209307. doi: 10.4324/ 9780203494103. URL https://www.taylorfrancis.com/books/9780203494103.

Crous, P.A., Jacobsz, S.W., 2024a. Centrifuge and Numerical Modelling of a Slope Failure Triggered by a Loss of Confinement. Journal of Geotechnical and Geoenvironmental Engineering. Journal of Geotechnical and Geoenvironmental Engineering (under review).

Crous, P.A., Jacobsz, S.W., 2024b. Insights on Slope Stability from Centrifuge Modelling. Journal of Geotechnical and Geoenvironmental Engineering. Journal of Geotechnical and Geoenvironmental Engineering (under review).

Ganal, A., Reul, O., Jacobsz, S.W., 2022. Centrifuge tests on foundations under alternating loads in overconsolidated clay. In 10th International Conference on Physical Modelling in Geotechnics. ISBN 9788995219775. doi: 10.1680/geot.1974.24.3.451.

Ha Giang, P.H., Van Impe, P.O., Van Impe, W.F., Menge, P., Haegeman, W., 2017. Small-strain shear modulus of calcareous sand and its dependence on particle characteristics and gradation. Soil Dynamics and Earthquake Engineering, 100:371–379, 9. ISSN 02677261. doi: 10.1016/j.soildyn.2017.06.016. URL https://linkinghub.elsevier.com/retrieve/pii/ S0267726116306157.

Hardin, B.O., Drnevich, V.P., 1972. Shear Modulus and Damping in Soils: Design Equations and Curves. Journal of the Soil Mechanics and Foundations Division, 98(7):667–692, 7. ISSN 0044-7994. doi: 10.1061/JSFEAQ.0001760. URL https://ascelibrary.org/doi/10.1061/JSFEAQ. 0001760.

He, R., Sandu, C., Mousavi, H., Shenvi, M.N., Braun, K., Kruger, R., Els, P.S., 2020. Updated standards of the international society for terrain-vehicle systems. Journal of Terramech. 91:185–231. ISSN 00224898. doi: 10.1016/j.jterra.2020.06.007.

He, R., Sandu, C., Khan, A.K., Guthrie, A.G., Els, P.S., Hamersma, H.A., 2019. Review of terramechanics models and their applicability to real-time applications. Journal of Terramech. 81: 3–22. ISSN 00224898. doi: 10.1016/j.jterra.2018.04.003.

Hemmatj, A., Nankali, N., Aghilinategh, N., 2012. Simulating stress–sinkage under a plate sinkage test using a viscoelastic 2D axisymmetric finite element soil model. Soil and Tillage Research, 118:107–116, 1. ISSN 01671987. doi: 10.1016/j.still.2011.10.016. URL https://linkinghub.elsevier.com/retrieve/pii/S0167198711001826.

Hosn, R.A., Sibille, L., Benahmed, N., Chareyre, B., 2017. Discrete numerical modeling of loose soil with spherical particles and interparticle rolling friction. Granular Matter, 19(1):4, 2. ISSN 1434-5021. doi: 10.1007/s10035-016-0687-0. URL http://link.springer.com/10.1007/s10035-016-0687-0.

Islam, M.R., Rahman, Md.-A., Hayano, K., 2020. Application of smoothed particle hydrodynamics (SPH) for simulating various geotechnical problems. SN Applied Sciences, 2(4):687, 4. ISSN 2523-3963. doi: 10.1007/s42452-020-2379-y. URL http://link.springer.com/10.1007/s42452-020-2379-y.

Jefferies, M., Been, K., 2016. Soil Liquefaction. CRC Press, Boca Raton, USA, second edition, 9. ISBN 9780203301968.

Jefferies., M.G., 1993. Nor-Sand: A simple critical state model for sand. Geotechnique, 43(1):91–103. ISSN 17517656. doi: 10.1680/geot.1993.43.1.91.

Jiang, M., Barreto, D., Ding, Z., Yang, K., 2025. Discrete Element Study of Particle Size Distribution Shape Governing Critical State Behavior of Granular Material. Fractal and Fractional, 9(1):26, 1. ISSN 2504-3110. doi:10.3390/fractalfract9010026. URL https://www.mdpi.com/2504-3110/9/1/26.

Karstunen, M., Amavasai, A., 2017. BEST SOIL: Soft soil modelling and parameter determination. Technical Report November. URL https://research.chalmers.se/en/publication/ ?id=522789.

Kruger, R., Els, P.S., Hamersma, H.A., 2023. Experimental investigation of factors affecting the characterisation of soil strength properties using a Bevameter in-situ plate sinkage and shear test apparatus. Journal of Terramech. 109:45–62. ISSN 00224898. doi: 10.1016/j.jterra. 2023.06.002.

Liu, G., Rong, G., Peng, J., Zhou, C., 2015. Numerical simulation on undrained triaxial behavior of saturated soil by a fluid coupled-DEM model. Engineering Geology, 193:256– 266, 7. ISSN 00137952. doi: 10.1016/j.enggeo.2015.04.019. URL https://linkinghub.elsevier.com/retrieve/pii/S0013795215001350.

Lu, Y., Mei, G., 2022. A Deep Learning Approach for Predicting Two-Dimensional Soil Consolidation Using Physics-Informed Neural Networks (PINN). Mathematics, 10(16):2949, 8. ISSN 2227-7390. doi:

10.3390/math10162949. URL https://www.mdpi.com/2227-7390/10/16/2949.

McKay, D.S., Carter, J.L., Boles, W.B., Allen, C.C., Allton, H.H., 1994. New lunar soil simulant. Proceedings of the 4th International Conference on Engineering, Construction and Operations in Space, pages 857–866.

Meirion-Griffith, G., Spenko, M., 2011. A modified pressure-sinkage model for small, rigid wheels on deformable terrains. Journal of Terramech. 48(2):149–155. ISSN 00224898. doi: 10.1016/j.jterra.2011.01.001. URL http://dx.doi.org/10.1016/j.jterra.2011.01.001.

Monaco, S. Apiletti, D., 2023. Training physics-informed neural networks: One learning to rule them all? Results in Engineering, 18:101023, 6. ISSN 25901230. doi: 10.1016/j.rineng. 2023.101023. URL https://linkinghub.elsevier.com/retrieve/pii/S2590123023001500.

Murison, R.A., Heymann, G., 2024. Using shear modulus to predict the bearing capacity of strip foundations on sand. E3S Web of Conferences, 544:1–8. ISSN 22671242. doi: 10.1051/e3sconf/202454407005.

Rojek, J., 2018. Contact Modeling in the Discrete Element Method. pages 177–228. doi: 10. 1007/978-3-319-90155-8{_}4. URL http://link.springer.com/10.1007/978-3-319-90155-8_4.

Seed, H.B., Idriss, I.M., 1970. Soil Moduli and Damping Factors for Dynamic Response Analyses. Technical report. URL https://ntrl.ntis.gov/NTRL/dashboard/searchResults/titleDetail/PB197869.xhtml.

Senetakis, K., Anastasiadis, A., Pitilakis, K., 2013. Normalized shear modulus reduction and damping ratio curves of quartz sand and rhyolitic crushed rock. Soils and Foundations, 53(6):879–893. ISSN 00380806. doi: 10.1016/j.sandf.2013.10.007. URL http://dx.doi.org/10.1016/j.sandf.2013.10.007.

Shuttle, D., Jefferies, M., 2010. Norsand: Description, Calibration, Validation and Applications.

Sinha, T., Curtis, J.S., Hancock, B.C., Wassgren, C., 2010. A study on the sensitivity of Drucker–Prager Cap model parameters during the decompression phase of powder compaction simulations. Powder Technology, 198(3):315–324, 3. ISSN 00325910. doi: 10.1016/j.powtec. 2009.10.025. URL https://linkinghub.elsevier.com/retrieve/pii/S0032591009005889.

USDA, 2017. Soil Survey Manual, U.S. Department of Agriculture Handbook. volume 18, chapter 3. doi: 10.2307/1233734.

Wood, D.M., 1991. Soil Behaviour and Critical State Soil Mechanics. Cambridge University Press, 4. ISBN 9780521332491. doi: 10.1017/CBO9781139878272. URL https://www.cambridge.org/core/product/identifier/9781139878272/type/book.

Xia, T., He, G., Hu, P., Li, X.L, 2014. Numerical Simulation of Rigid Cylindric Plate's Sinkage on Soil Based on SPH/FEM Method. Advanced Materials Research, 1049-1050:483–486, 10. ISSN 1662-8985. doi: 10.4028/www.scientific.net/AMR.1049-1050.483. URL https://www.scientific.net/AMR.1049-1050.483.

Xiong, X.B., Wang, X.J., 2023. A cohesive strengthening/damage-plasticity model and its application for DEM numerical investigation in hardening/softening of rock and soil. Computers and Geotechnics, 159:105500, 7. ISSN 0266352X. doi: 10.1016/j.compgeo.2023.105500. URL https://linkinghub.elsevier.com/retrieve/pii/S0266352X23002574.

Yu, H.-S., 2007. Plasticity & Geotechnics. ISBN 9780387335971.

Zeraati-Shamsabadi, M., Sadrekarimi, A., 2025. A DEM study on the effects of specimen and particle sizes on direct simple shear tests. Granular Matter, 27(2):36, 4. ISSN 1434-5021. doi:10.1007/s10035-025-01513-y. URL https://link.springer.com/10.1007/s10035-025-01513-y.

Appendix

Table A1
Additional Cullinan sand soil properties from Kruger et al. (2023).

Soil property	Value	Test
Specific gravity [kg/m³]	2651	SANS 5845:2006
Max dry density ρ_{max} [kg/m³]	1580.7	ASTM D4253-93-1B
Min dry density ρ_{min} [kg/m³]	1429.6	ASTM D4254-91-1A
Mod. Proctor max dry density ρ_{max} [kg/m³]	1784	SANS 3001-GR30:2015
Optimum moisture content [%]	6.5	ASTM D4643-93
D_{50} [μm]	149	ASTM D6913-17
D_{60} [μm]	181.5	SANS 3001-GR30:2015
D_{10} [μm]	73.5	ASTM D2487-06
Coefficient of uniformity C_u	2.46	ASTM D2487-06
Uniformity classification	Uniformly graded	USDA Soil Classification
Sand [%]	96.2	
Silt [%]	2.8	
Clay [%]	<1	
Textural classification	Fine sand	
Particle shape	Angular	Note: Particle shape from electron microscope imaging
Effective friction angle ϕ' [Deg]	36.1	ASTM 3080-72
Effective cohesion c [kPa]	0	Direct shear box test
Target moisture content [%]	6.5	ASTM D4643-93
Degree of saturation at max density [%]	35.4	
Degree of saturation at min density [%]	20.2	

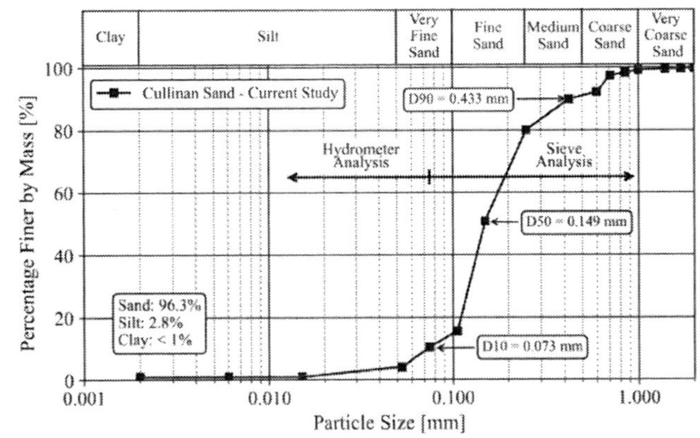

Fig. A1. Cumulative particle size distribution of Cullinan sand, showing USDA classification boundaries (USDA, 2017). Adapted from Kruger et al. (2023).

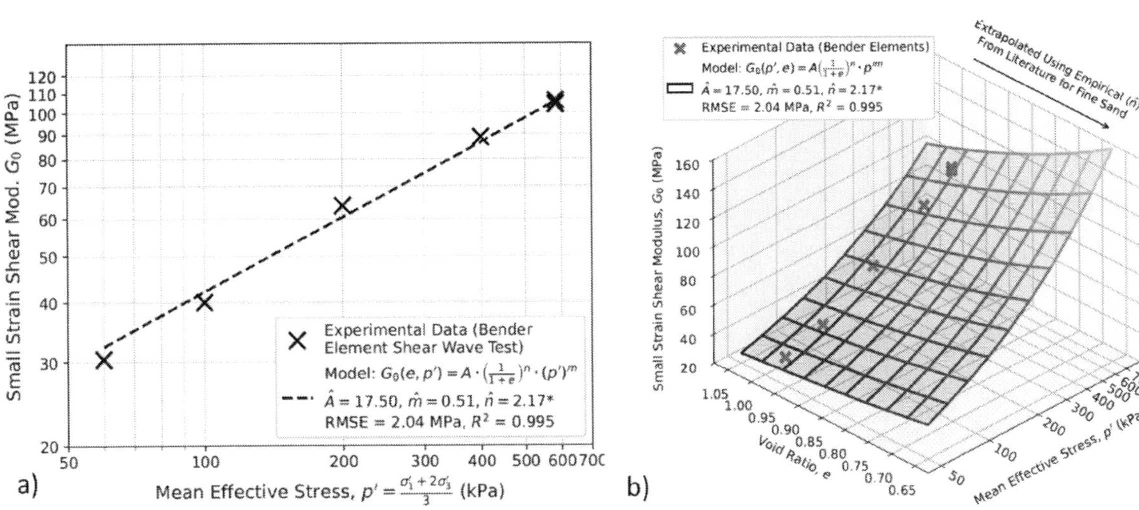

Fig. A2. Additional calibration results of the small-strain shear modulus G_o using bender elements: (a) Model fit to mean stress data; (b) Surface extrapolated over void ratio direction using empirical formulation and constants from Ha Giang et al. (2017).

8342 | The effect of snow subduction on vehicle mobility

Mark Bodie *, Michael Parker

US Army Engineer Research and Development Center
Cold Regions Research and Development Center, Hanover, NH

** Corresponding author: mark.o.bodie@erdc.dren.mil*

ABSTRACT

Winter surfaces are characterized by low strength, which can result in deep vehicle sinkage and high vehicle motion resistances. If the vehicle's sinkage is greater than the vehicle's ground clearance, snowplowing forces occur, further increasing the vehicles motion resistance. The increased motion resistance can lead to vehicle immobility if the vehicle's available traction is exceeded. In this work we investigate the effect of snow subduction on vehicle mobility. When snowplowing occurs a portion of the plowed snow can be subducted under the vehicle's belly. The subducted snow is compressed resulting in both an increase in motion resistance and a normal load reduction under the vehicle's track or tire. We developed an algorithm which includes both snow subduction effects on the vehicle's mobility. In the study we found that the snow subducted under the vehicle has a substantial effect on the vehicle's mobility. We simulated a 82.74kPa ground pressure vehicle with 0.1m snow subduction zone. When snow subduction was included in the analysis, we found that vehicle immobility occurred at a snow depth of 0.59m as compared to a depth of 0.7m when subduction was not included. The large variation in snow depth at vehicle immobility, underlines the importance of including snow subduction in vehicle mobility and route planning analyses. In future work we will improve the accuracy of the algorithm by incorporating snow stratigraphy. The current algorithm uses bulk density to determine snow strength, in future analyses we will incorporate snow layer density and grain size to improve the algorithms accuracy.

Keywords
Snowplowing
Motion resistance
Traction
Mobility
Sinkage

1. Introduction

Winter surfaces are characterized by low strength, which can result in deep vehicle sinkage and high vehicle motion resistances. If the vehicle's sinkage is greater than the vehicle's ground clearance, snowplowing forces occur which greatly increase the vehicles motion resistance. These higher motion resistances, if greater than available traction, lead to vehicle immobility. The ability to accurately predict a vehicles motion resistance and available traction are important for mission planning, navigation, and vehicle assessment.

In this work, we develop an algorithm to predict the effect of snow subduction on vehicle mobility. During a snowplowing event, vehicle sinkage exceeds the vehicle's ground clearance with a portion of the plowed snow subducted under the vehicle's belly. The subducted snow is compressed, resulting in both an increase in motion resistance and a reduction in normal load under the vehicle's drive element (track or tire).

2. Vehicle Forces

Vehicle immobility occurs when the vehicle's motion resistance exceeds either the drive elements available traction or the powertrains thrust capability,

$$MR > \min\,(PT, T_{avail}) \tag{1}$$

Where MR is the vehicle's total motion resistance, PT is the powertrain thrust, and T_{avail} is the traction limit between the drive element and the surface.

In this work we are concerned with low strength winter surfaces and assume that the vehicle's powertrain can exceed the traction limit which reduces Eq. 1 to,

$$MR > T_{avail} \tag{2}$$

2.1. Available Traction

We use an existing relationship to calculate the vehicle's available traction. A vehicle's available traction is a function of 1) the vehicle's ground pressure, 2) interaction between the vehicle's drive element and the surface, and 3) the shear strength of the surface. Richmond et al. (1995) provides the following equation describing a vehicle's available traction over undisturbed snow on top of a strong substrate,

$$T_{avail} = 0.851N^{0.823}A \tag{3}$$

Where N is the ground pressure of the drive element (kPa), A is the area of the drive element (m^2), and T_{avail} is the available traction of the drive element (kN). The total vehicle available traction, T_{avail}, is the sum of the available traction for all the vehicle's drive elements. If we assumed that the drive element ensures the surface shear strength failure, Eq. 3 defines the surface shear strength as a function of ground pressure,

$$\sigma_s = {}^{T_{avail}}\!/_{A} = 0.851N^{0.823} \tag{4}$$

2.2. Motion resistance

We determine the vehicle's deep snow motion resistance using the categories define in Welling et al. (2022): 1) drive element motion resistance, 2) snowplowing force, 3) snow subduction compression force, and 4) snow subduction frictional force. The

https://doi.org/10.56884/7AEYKXDK

drive element compression defines the work required to compress the snow under the drive element. The snowplowing force describes the force required to push the snow in front of the vehicle. The snow subduction compression force is similar to the drive element compression term, with motion resistance resulting from the work required to compress the subducted snow. The final major motion resistance factor is the frictional force between the compressed subducted snow and the vehicle belly.

2.2.1. Drive Element Motion Resistance

To determine the drive element motion resistance, we use the method described in Shoop et al (2020). In this method, the drive element work is equated to the snow compression work or,

$$W_e = MR_e x \tag{5}$$

where W_e is the vehicle work associated with snow compression under the drive element, MR_e is the vehicle's motion resistance associated with the drive element snow compression, and x is the distance traveled.

The work associated with the snow compaction, W_s, is given by,

$$W_s = \int_0^{Z_f} 2F_s(z)\, dz \tag{6}$$

Where $F_s(z)$ is the force required to compress the snow, a function of sinkage z, and Z_f is the final snow sinkage. Rewriting Eq. 6 in terms of snow compression stress results in,

$$W_s = 2bx \int_0^{Z_f} \sigma(z)\, dz \tag{7}$$

Where b is the drive element width, the factor 2 represents the two ruts created by the vehicle, and $\sigma(z)$ is the snow compressive stress as a function of sinkage height. The vehicle's motion resistance associated with the snow compression under the drive element is given by equating Eq. 7 and 5,

$$MR_e = 2b \int_0^{Z_f} \sigma(z) dz \tag{8}$$

Next, we use the Wong, Graber, and Preston-Thomas equation (Wong, J.Y. et al. (1984)) to describe the snow pressure sinkage characteristics and substitute into Eq. 8,

$$MR_e = 2b \int_0^{Z_f} P_w \left[-ln\left(1 - \frac{z}{z_w}\right)\right] dz \tag{9}$$

Where z_w and P_w are regression constants. Equation 9 is then integrated,

$$MR_e = 2bz_w P_w \left[\left(1 - \frac{z}{z_w}\right) ln\left(1 - \frac{z}{z_w}\right) + \frac{z}{z_w}\right] \tag{10}$$

Shoop et al (2020) provides values for z_w and P_w using measured vehicle sinkages for: 1) snow depths ranging from 0.035 to 0.64m, 2) undisturbed snow densities of 0.090 to 0.410 g/cm^3, and 3) vehicle ground pressures from 13.3 to 413.7kPa. The regression terms z_w and P_w are defined as,

$$P_w = 48.36\rho_o \tag{11}$$

$$z_w = h\left[1 - \frac{\rho_o}{0.416604\left[1 + {0.614817}/{b}\right]}\right] \tag{12}$$

Where ρ_o is the undisturbed snow density, h is the snow depth, and b is the drive element width. The drive element sinkage is given as

$$z = z_w\left[1 - e^{\left({P_c}/{P_w}\right)}\right] \tag{13}$$

Where P_c is the drive element's ground pressure.

2.2.2. Snowplow Force

We take the vehicle snowplowing force directly from Olson et al. (1997).

$$MR_p = A_p \frac{11.1kPa\; g/cc}{0.5g/cc - \rho_o} \tag{14}$$

Where A_p is the vehicle's plowing area and MR_p is the motion resistance associate with snowplowing.

2.2.3. Snow Subduction Compression Motion Resistance

The motion resistance due to snow subduction beneath the vehicle belly is governed by the same mechanical principle as that acting on the drive elements. This resistance is assessed by equating the work required for vehicle movement to the work involved in compressing the snow. The snow subduction compression motion resistance is determined using Eq. 10, by noting that the 2b factor is replaced with the vehicle width between the drive elements and the sinkage term, z_s, is the sinkage of the vehicle belly.

$$MR_s = w_S z_{ws} P_w \left[\left(1 - \frac{z_s}{z_{ws}}\right) ln\left(1 - \frac{z_s}{z_{ws}}\right) + \frac{z_s}{z_w}\right] \tag{15}$$

$$z_{ws} = h_s\left[1 - \frac{\rho_o}{0.416604\left[1 + {0.614817}/{w_s}\right]}\right] \tag{16}$$

Where w_S is the vehicle width between the drive elements, z_s is the depth of the subduction zone, h_s is the depth of the snow under the vehicle belly, and z_{ws} is the regression term using the vehicle width as the size component.

2.2.4. Belly Friction Motion Resistance

Equation 4 defines the surface's shear strength as a function of the vehicle's ground pressure. This equation can be used to determine the shear strength of the subducted snow under the vehicle's belly. The normal load is the load that the vehicle belly applies to the subducted snow. The belly friction motion resistance is determined by multiplying the surface stress by the vehicle's belly area.

$$MR_b = 0.851{N_b}^{0.823} A_b \tag{17}$$

Where MR_b is the motion resistance force associated with shearing the snow under the vehicle belly, N_b is the normal load applied by the belly on the subducted snow, and A_b is the vehicle belly area.

3. Snow Subduction Algorithm

The vehicle's motion resistances are described by Eq.s 10, 14, 15, and 17, while the vehicle's available traction is provided by Eq. 3. To solve the listed equations, we must first determine the vehicle sinkage. Here we note that the vehicle sinkage is a function not only of the drive elements ground pressure but also by the snow subducted under the vehicle. The subducted snow is compressed and supports a portion of the vehicles weight which reduces the normal load applied by the drive element. This reduction in drive element normal load reduces the vehicle's available traction as defined by Eq. 3.

We determine the vehicle sinkage by balancing the vehicle's weight with the normal forces produced by the snow compressed under the drive elements and vehicle's belly.

$$W_v = A_t P_c(z) + A_b P_b(z_s) \tag{18}$$

Where W_v is the vehicle weight, $P_c(z)$ and $P_b(z)$ are the snow compressive stresses under the driven elements and vehicle belly, and A_t is the total area of the drive elements. We assume that all contact areas are drive elements.

We solve Eq. 13 in terms of compressive stress and then apply to the drive elements and the subducted snow,

$$P_c = P_w\left[-\ln\left(1 - \frac{z}{z_w}\right)\right] \tag{19}$$

$$P_b = P_w\left[-\ln\left(1 - \frac{z_s}{z_{ws}}\right)\right] \tag{20}$$

We use an iterative method to balance the vehicle's weight between the drive elements and vehicle belly by modifying the vehicle sinkage. We initially calculate the vehicle sinkage with consideration to only the drive elements. This represents the maximum vehicle sinkage, and if the vehicle's ground clearance is not exceeded, represents the vehicle sinkage. If the vehicle sinkage exceeds the vehicle's ground clearance, a portion of the vehicle's weight is supported by the compression subducted snow, reducing the vehicle sinkage. We modify the vehicle sinkage until Eq. 17 is satisfied.

After determining the vehicle sinkage, we solve Eq. 3 to determine the vehicle's available traction. Next, the four motion resistances described by Eq.s 10, 14, 15, and 17 are summed to calculate the vehicle's total motion resistance. Vehicle immobility occurs when the vehicle's total motion resistance exceeds the vehicle's gross traction.

3.1. Snow Subduction Algorithm Sinkage Calculation

The following provides a numerical example of the algorithm described above. We start with a conceptional tracked vehicle (82.74kPa ground pressure, 0.36m ground clearance, 0.1m subduction zone) transversing over a snowpack (0.75m depth 0.25g/cc density) as shown in Fig. 1.

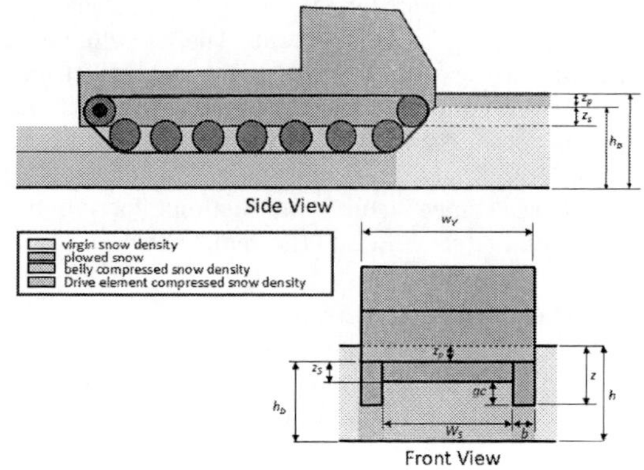

Fig. 1. Conceptual Track Vehicle with Rut

The vehicle parameters are: 1) drive element width, b, 2) subduction sinkage, z_s, 3) vehicle belly width, w_s, 4) plow depth, z_p 5) vehicle width, w_v, 6) vehicle sinkage, z, 7) snowpack depth, h, and 8) ground clearance, gc. We recognize that the subduction sinkage, zs, and plow sinkage, z_p, are a function of the vehicle's over all sinkage and ground clearance. The plow sinkage is,

$$z_p = z - (gc + z_{sm}) \tag{21}$$

Where z_{sm} is the maximum vehicle's maximum subduction height. If Eq. 21 is greater than zero the vehicle is plowing snow and the area used in Eq. 13, Ap is,

$$A_p = z_p w_v \tag{22}$$

The subduction height is given by,

$$z_s = z - gc \tag{23}$$

The subduction height, z_s, is limited to the maximum subduction depth z_{sm} and positive values.

Initially, we assume that the vehicle is fully supported by the drive elements and use Eq. 13 to calculate the vehicle sinkage. Because we assume that all the vehicle's weight is supported by the drive elements and none by the subducted snow, this initial calculated vehicle sinkage represents a maximum vehicle sinkage. If the initial calculated vehicle sinkage exceeds the vehicle's ground clearance, we know that snow is subducted under the vehicle and will support a portion of the vehicle's weight. The weight supported by the subducted snow will reduce the normal load on the drive elements by an equal amount. We use Eq.s 17-19 to determine the snow strength under the drive elements and vehicle belly. We then vary the vehicle sinkage until the supportive snow strength equals the vehicle weight. Figure 2 shows the calculated snow strength as a function of vehicle sinkage. The final calculated vehicle sinkage is determined at the point where the vehicle weight and snow strength curves cross. In this case vehicle sinkage, z, equals 0.586m, the subducted depth, zs, is 0.1m, and the plow depth, z_p, is 0.130m. The snow depth under the vehicle belly is 0.619m. The snow compressive strength under the drive element is 54.398kPa, calculated from Eq. 18, and under the belly is 4.007kPa, calculated from Eq. 19.

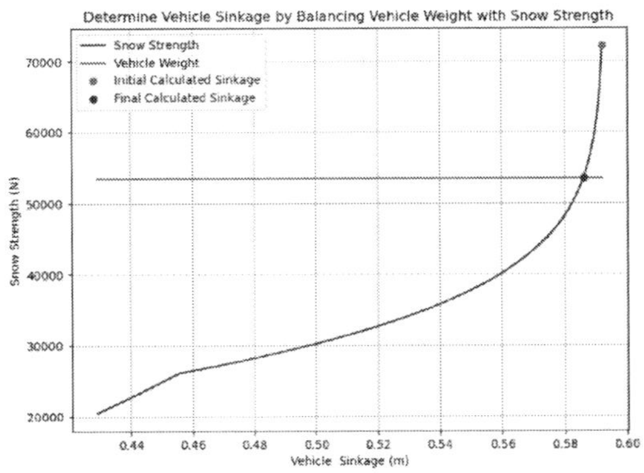

Fig. 2. Sinkage Determination

3.2. Snow Subduction Algorithm Immobility Calculation

Vehicle immobility occurs when the motion resistance of the vehicle is greater than the vehicle's available traction. In the following analysis we determine the snow depth at which the conceptual vehicle described above becomes immobile. We determine snow depth at vehicle immobility when the vehicle has a snow subduction zone and when the vehicle is without a subduction zone. In the latter case, when the vehicle's sinkage exceeds the vehicle's ground clearance all excess snow will be plowed. In the former case the vehicle will contain a 0.1m subduction zone. Sinkage depths that exceed the subduction zone will be plowed. The vehicle's motion resistance and gross traction will be evaluated for snow depths from 0.2 to 0.75m at increments of 0.01m with a snow density of 0.25g/cc.

In Fig. 3, the vehicle with a snow subduction zone of 0.1m became immobile at a snow depth of 0.576m, the depth where the vehicles motion resistance equals the vehicles available traction. From the figure, when the vehicle's ground clearance is exceeded, the vehicle's motion resistance increases rapidly, and the available traction is reduced. The motion resistance is increased due to the subducted snow's compression, snow plowing, and belly frictional force. The gross traction is reduced by the reduction in the normal load applied to the drive element. In Fig. 4 the vehicle without a snow subduction zone became immobile at a snow depth of 0.68m. As with the subduction case, the motion resistance of the vehicle increases rapidly when the vehicles ground clearance is exceeded. This motion resistance increase is due to the snow plowing force. Unlike in the subduction case the available traction is not reduced. The net result is that the vehicle immobility occurs at a snow depth 0.1m greater than in the subduction case.

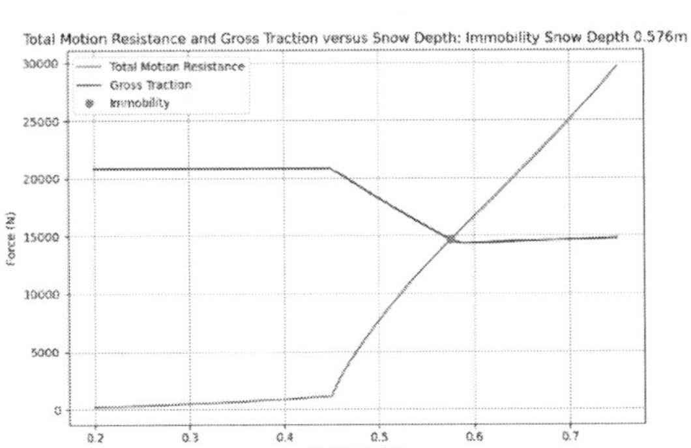

Fig. 3. Immobility With 0.1m Subduction Zone

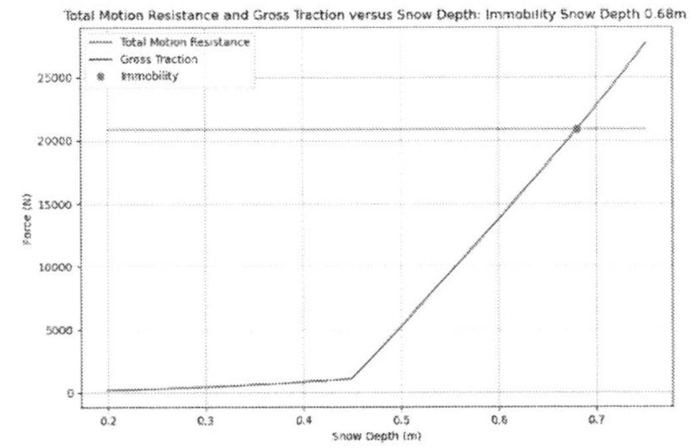

Fig. 4. Immobility Without Subduction Zone

4. Conclusions and Recommendations

In this work we investigated the effect of snow subduction on vehicle immobility. We identified that snow subduction both increases vehicle motion resistance and reduces available traction. The subducted snow supports a portion of the vehicle's weight, reducing the drive elements normal load and available traction. In a conceptual vehicle simulation, when snow subduction was included in the analysis, vehicle immobility occurred at a snow depth of 0.1m lower than when snow subduction was not included. This snow depth discrepancy highlights the importance of adding snow subduction effects in mission planning, navigation, and vehicle assessment software.

In future work, we look to validate the snow subduction algorithm by measuring the motion resistance and available traction for vehicles in deep snow conditions were snowplowing occurs. We look to measure the depth and density of the ruts formed by the drive elements and vehicle belly. Finally, we will improve the accuracy of the algorithm by incorporating snow stratigraphy. The current algorithm uses bulk density to determine snow strength, in future analyses we will incorporate snow layer density and grain size to improve the algorithm's sinkage and motion resistance accuracy.

5. References

Olson, W.W., Kempainen, A., Milacic, D., Osborne, M., 1997. Dynamic modeling of forces on snowplow equipped trucks, SAE Technical Papers (1997) doi:10.4271/973193.

Richmond, P.W., Shoop, S.A., Blaisdell, G.L., 1995. Cold regions mobility models. CR 95-1. US Army Corps of Engineers, Cold Regions Research & Engineering Laboratory, Hanover, NH.

Shoop, S.A., Bodie, M.O., Frankenstein, S., Bigl, M.F., 2020. Recent improvements to snow mobility algorithms, Cold Regions Research and Engineering Laboratory Technical Report, MP-20-9, 2020. doi:10.21079/11681/37821.

Welling, O., Shoop, S., Letcher T., Elder, B., and Bodie, M., 2022. Estimation of Plowing Forces on Vehicle Driving Through Deep Snow. Journal of Terramech. 104: 25-29.

Wong, J.Y., Garber, M., Preston-Thomas, J., 1984. Theoretical prediction and experimental substantiation of the ground pressure distribution and tractive performance of tracked vehicles. Proc. Inst. Mech. Engrs. 198 (D15): 265-285.

8352 | Assessing the strength of river sediments for vehicle fording

Sally A. Shoop [a,*], John B. Eylander [b], Theodore W. Letcher [a], Kathleen A. Staebell [b]

[a] *ERDC-Cold Regions Research and Engineering Laboratory, Hanover, NH*

[b] *ERDC-Coastal and Hydraulics Laboratory, Vicksburg, MS*

* *Corresponding author: sally.a.shoop@erdc.dren.mil*

ABSTRACT

Vehicle fording can be difficult to assess because of numerous requirements to ensure safe crossing based on the vehicle and the changing river characteristics. While some of these requirements are well-known, others are so complex and variable that they remain ill-defined. This paper reviews river crossing guidance with emphasis on the river and riverbed characteristics. While seasonal changes in flow rate are somewhat predictable with hydrologic models, they rely on many assumptions about the river bathymetry and discharge. The riverbank conditions can also change drastically with season, affecting ingress and egress. Lastly, the competence of the river bottom sediments (strength and stability) is extremely difficult to estimate or measure, so this was a primary focus for this research.

Recent progress in fluvial geomorphology and a new field called river morphodynamics may help in determining the best places to cross a river based on river type, processes, and adjacent landforms. Equations from fluvial geomorphology and geotechnical engineering are used to relate river flow conditions to bottom sediment type and strength. Exploratory field measurements of river bottom strength were performed. A methodology to evaluate the strength of the river bottom sediments was developed and results using a modified static cone penetrometer were analyzed.

This research represents a major step forward in understanding and predicting suitable river bottoms for fording, presents a methodology to measure strength in river sediments, and highlights areas of future research needs. Applicable fording guidance, geomorphology, and recent advancements in terramechanics as related to river crossings are summarized to provide context.

Keywords
River Crossing
Fluvial Geomorphology
Gravel
Strength
Low Water Crossing
River Sediments
River Bottom
Grain Size
Wet Gap Crossing
Measurement
Vehicle Fording

1. Introduction

River crossing by fording can be expeditious when the water level is low, water velocity is slow, the river bottom is competent, and the vehicle is capable. Fording is also sometimes called "low water crossing" or "hasty crossing". Fording can be an advantageous means to bypass otherwise long or arduous journeys around river or stream obstacles. Potential fording sites are of interest to agriculture and forestry for access to adjacent work sites; for off-road vehicle enthusiasts in search of expedient and challenging terrain crossings; and for the military as part of achieving tactical cross country mobility objectives. For military operations in temperate environments, low water crossing (LWC) is a common occurrence in cross country mobility, but it is difficult to train for, primarily because environmental restrictions around riparian zones limit training opportunities.

Recently, predicting fording sites has been revisited by Fishell et al (2021) and Rybansky et al. (2018 and 2024) who have studied river fording from a geospatial viewpoint. Still, in these and other studies the state of the river bottom is alluded to as needing to be "strong," but is not specifically quantified or predicted so it remains largely unknown or unpredictable.

This paper looks at some of the existing guidelines for fording, discusses recent work reported in the literature, and emphasizes the need for new methods to quantify, qualify, and predict the strength characteristics of the river bottom. With this aim in mind, we place a heavy focus on recent advancements in the fields of fluvial geomorphology and morphodynamics, and on how these concepts influence river shape and riverbed particle grain size, and how they can be utilized together with geotechnical engineering knowledge to make significant advancements in developing new methods to predict the characteristics of the river bottom. Fluvial morphology concepts were considered during an exploratory field experiment where the strength of the river bottom was measured. The findings clarify the need for specific research in the many aspects of quantifying the nature and physical processes involved in this complex environment.

1.1 Scientific Purpose:

The broader scientific purpose of the work was to initiate study of the physical processes controlling the soils in and around streams and rivers, and integrating the latest state-of-the art to benefit the prediction of fording sites. Specifically, we aimed to:

- Research how water redistributes soil material, impacting the soil strength adjacent to and within the stream bed.
- Understand streamflow and stream geophysical. processes to support forecasting and crossing decisions
- Describe, quantify, and predict gravel content of streams and how it increases streambed strength and facilitates fording opportunities.
- Develop measurement methods to quantify streambed strength for vehicle fording.

2. Background

2.1. Guidelines for fording a stream or river

Documented guidance on how to ford a river or stream is minimal, but the common-sense approach taken by the off-road vehicle recreational community is emphasized by some of the

https://doi.org/10.56884/JF8WN0WV

suggestions below. Since many rivers are too wide, deep, or fast moving for vehicles to cross, this guidance is most applicable to low water crossings (LWC) of small rivers and streams, or drainage systems. While these and other guidelines tend to be general, they are helpful for understanding the issues involved and were assembled from a wide variety of documents aggregated from the USDA Forest Service; crossing management practices from the States of Idaho, Michigan, Virginia and Oklahoma; fording guidance from Iceland recreational vehicles (Campervan Reykjavik, 2023); and a remote fording site analysis study by Coutermarsh and Dwinal (2003). The guidance gleaned is grouped into those related to a) the capabilities of the vehicle and vehicle operators, b) the river conditions at the time of crossing, and c) advice regarding the river bottom and adjacent terrain.

Vehicle operation guidance:

- You must have a 4x4 (or better) with good traction and high ground clearance.
- The vehicle must be capable of being in water without severely damaging it. A snorkel is critical for air intake so you don't ruin your engine. Use coverings (radiator blind, car bra, or even a tarp) to keep water from flowing into the radiator and vehicle electronics compartments.
- During crossing, drive slowly at a constant speed to avoid waves, spinning the wheels, and changing gears.
- If you stall due to the engine taking in water do not restart the engine (causes additional damage) – you will need to be towed.
- After crossing, stop to let water drain from the vehicle and check that electronic areas have drained (Campervan Reykjavik, 2023).

River flow velocity and depth:

- Avoid turbulent areas.
- Look for slow, wider crossings rather than fast moving water.
- Areas upstream of riffle (shallow fast-moving) sections can make good crossings, typically having a solid stream bottom and shallow depth.
- If you can't easily see the river bottom to inspect for drop-offs and sharp rocks, you may need to walk the stream/river to inspect it (secure the person, traverse with a rope, use walking or probing poles, use flotation devices and other safety measures).
- If you are unstable while walking, retreat. If you cannot walk across the river safely you should not drive your vehicle through it, and instead travel around it and/or look for an alternate crossing site.
- Flow rates can be measured by timing a float over a set distance. The float should be thrown to the fastest part of the river, commonly, but not necessarily, in the middle. High flow rates can push the side of the vehicle moving it downstream or catch the underbody of the vehicle and flip it.
- For glacial rivers and streams, water levels are lower in early morning before the sun and warm temperatures cause glacier melting and increased river flow (Campervan Reykjavik, 2023).
- River flow rates can be measured, estimated based on stream gradient and roughness, or predicted from weather and topography. This is discussed briefly later on in this paper.

River bottoms and banks:

- Look for gradual banks (20% slope or less is recommended, unless you are reinforcing it).
- Look for a stable river bottom of coarse-grained soil, firm sandy soil, coarse gravel, rock or bedrock (numerous state and USDA references). Angular or semi-angular rock of at least 2.5 cm diameter are best (noting that river gravels are often well rounded).
- Clay, weak, and any soft or fine soils on the riverbed will need engineering improvements, especially for multiple crossing. Improving these soils may require removing the soft soils and replace them with geosynthetics and/or gravel.
- Avoid areas with marshes near the entry or exit as they could impact mobility during ingress, egress or further movement away from the river (Coutermarsh and Dwinal, 2003)
- Areas near existing or destroyed bridges tend to have steep banks (Coutermarsh and Dwinal, 2003)
- Near perpendicular entry into the water can minimize bank damage during ingress and egress, facilitating multiple crossings.

Summarizing these guidelines, the best river conditions for vehicle fording occur in a small subset of river types that include coarse-grained bottom sediments, shallow water, low velocity flow, and gradual river banks.

2.2. Fording for military operations

For the military, a ford site is considered a critical point in route reconnaissance and the engineering assessment of a ford site follows a formalized procedure to document the river geometry, flow conditions, bottom type, and seasonality. Yet, the expedient analysis of the specific site is often left to the soldiers driving the vehicle. Military vehicles are usually more competent with respect to fording, having excellent traction and maneuverability and stated performance requirements for fording. As an example, Rybansky et al. (2018) gives military vehicle specifications in Table 1 below. In the US, many of the military vehicles were up armored over the years, significantly increasing their weight resulting in decreased mobility putting them at serious disadvantage for crossing over a weak river bottom. For the more agile vehicles, there are add-on fording kits, and some vehicles are specifically designed as amphibious, having swimming capability.

Table 1
Fording depths for military vehicles for the Czech Republic from Rybansky (2018), in meters.

Units/ Equipment	Permissible depth of watering at flow speed		
	Up to 1 ms^{-1}	Up to 2 ms^{-1}	over 2 ms^{-1}
Motorized infantry units – by foot 1.00	1.00	0.80	0.60
Automobiles: passenger GAZ-69 A	0.60	0.50	0.40
truck 1.5t up to 2t	0.60	0.50	0.40
truck 3t up to 3.5t	0.80	0.70	0.60
truck 5t	0.90	0.80	0.70
Armored personnel carriers	0.70	0.60	0.50
Truck-drawn artillery can pass the same depth as trucks			
Artillery drawn by full tracks:	1.00	0.90	0.80
Tanks medium	1.20	1.10	1.00
heavy	1.50	1.40	1.30

Ekegren and Bailey (2017) expounded on the need for better reconnaissance and training methods for fording and echo the need for high-resolution real-time intelligence suggested by

Coutermarsh and Dwinal (2003). Platoons need to be able to cross at suitable crossing sites, but those sites also must fit the overall tactical scenario (Ekegren and Bailey, 2017).

2.3. Predicting weather-based mobility and river flow characteristics

Advances in geospatial data analysis and availability, weather predictions, and computational resources have enabled global predictions of soil moisture at high-spatial and temporal resolution within the easy-to-use GeoWATCH software. GeoWATCH uses high resolution geospatial data with high temporal resolution Air Force weather data to predict real-time, historical, or future soil moisture, which is then used for predicting soil strength and cross-country mobility (Eylander et al., 2021). Similarly, the availability of global weather data enables high-resolution streamflow prediction using the Steamflow Prediction Tool (Snow et al., 2017) and Autoroute (Follum et al., 2020), which predict stream width, depth and flow rates based on weather and terrain inputs.

We propose using the real-time hydrology for stream width, depth, and flow rates for vehicle fording calculations; and then merging these with cross-country mobility models within GeoWATCH to optimize wet-gap crossing predictions. Similarly, Fishchell (2021) proposed a prototype fording modeling capability using VEHDYN on a range of river geometries for specific soil strengths to develop tables that could be merged with hydrologic models. Recently, Rybansky et al. (2018, 2024) developed a framework for geography-based hydrologic modeling to calculate flow velocity based on drainage area, topography, roughness, vegetation, etc. Their framework includes the impact of water storage in depressions, which can be critical for vehicle mobility. While these studies provide crucial insight on the geospatial weather-based aspects of hydrologic modeling, all of these models are missing information on how to predict, infer or otherwise measure the ill-defined nature of the riverbed or river sediment strength, which is critical to locating or predicting suitable fording sites. Therefore, the remainder of this paper investigates how to obtain or estimate the nature of the river sediments and their strength for fording.

3. What can fluvial geomorphology tell us about the condition of the river bottom?

3.1 Basics of fluvial geomorphology

On a scale of regional geomorphology or landforms, river classes can be related to flow characteristics and geometry. There are numerous river classification methods depending on the purpose. Recent reviews of river classification schemes are given in Buffington and Montgomery (2022) and Kondolf and Lisle (2016). To illustrate the wide ranges of classification schemes based on applications and scale, Kondolf et al. (2016) evaluated the variables and scales of 21 such classification methods. These variables included geographical unit, valley or valley bottom morphology, channel pattern/sinuosity, channel morphology, gradient, stream power, channel bottom grain size, sediment load, morphodynamic units, morphodynamic adjustments, and spatial scale category (ranging from 10s of meters, to 1000 km or more). Dominant characteristics of gravel rivers include pattern, profile and cross section, and nature of transport and deposition. An example classification scheme of gravel river types is shown in Fig. 1, based on characteristic river pattern, types of islands or bars, and the lateral activity or movement (from Galay et al. 1998). In this scheme, gravel supply, lateral instability, valley slope and ratio of bedload to total load increase from left to right.

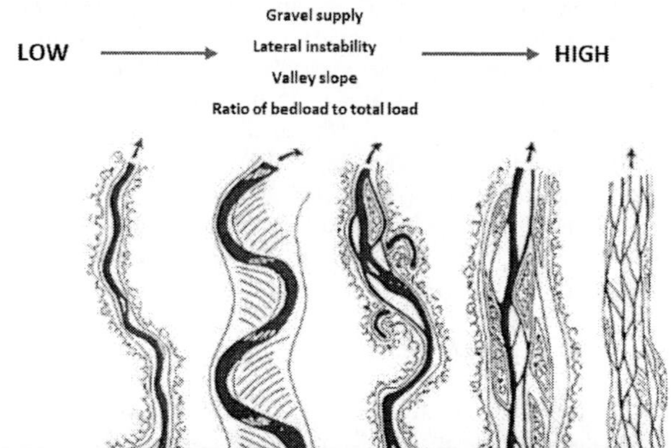

TYPE	IRREGULARLY SINOUS	MEANDERING	WANDERING	ANASTOMOSE D	BRAIDED
PATTERN	straight, sinuous irregular	irregular or regular meanders	irregular	irregular	
ISLANDS	none, occasional	occasional	frequent: regular or irregular to split	split to anastomosed	(none)
BARS	none, slide bars diagonal	diagonal, point bars	diagonal, point mid-channel	split to anastomosed	diamond mid-channel (braided)
LATERAL ACTIVITY	none, limited	downstream progression	irregular, avulsion	irregular, avulsion	avulsion

Fig. 1. Gravel river types noting characteristics of pattern, islands and bars and lateral activity with changes in gravel supply, lateral instability, valley slope and ratio of bedload to total load (based on Galay et al., 1998, with permissions from Water Resources Publication, LLC).

River morphodynamics

River morphodynamics is a relatively new field of study that recognizes riverbed evolution based on the physical processes between the water and the geomorphic environment. Morphodynamics uses fluid mechanics and applied mathematical principles to describe riverbed changes such as scour, bank erosion, river incision and planimetric changes (Crosato, 2007) and can help us understand how riverbeds form and change in response to weather and other processes. While fishery scientists are looking at these processes for understanding or developing fish habitat, we are interested in similar river processes for locating where the river bottom has characteristics that would indicate that it is strong or stable enough for vehicle fording. Church and Ferguson (2015) present an overview of the field and the need to study morphodynamics.

For fording we are concerned with the nature of both the river flow and the river bottom on scales on the order of 10s of meters. Hydrologists define a section of river with similar characteristics as a "reach", thus, a process-oriented classification scheme for river reaches is useful. Buffington and Montgomery (2022) summarize such as scheme in the graphic shown in Fig. 2. Based on this diagram the stream morphologies likely to be most suitable for fording (i.e. most likely to have coarse grained and gravel sediments and suitable slopes) are step pool, plane bed, pool ripple, and dune riffle river forms. Photographs and illustrations of sediment distributions for these river reach types are shown in Fig. 3 from Montgomery and Buffington (1997)

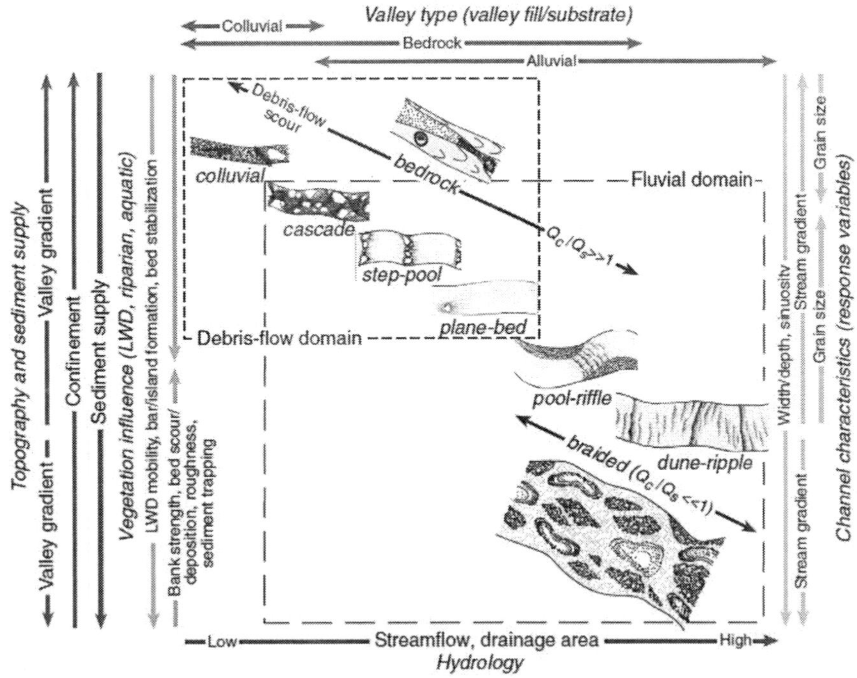

Fig. 2. River reach scale morphology based on basin conditions basin conditions (left) channel characteristics (right), valley type (top) and hydrology (bottom); illustrated with the channel types of Montgomery and Buffington (1997) (center). Reprinted from Treatise on Geomorphology 2nd Ed. Vol 6. Buffington and Montgomery, Geomorphic classification of Rivers: an updated review, Page 1150 (2022) with permission from Elsevier.

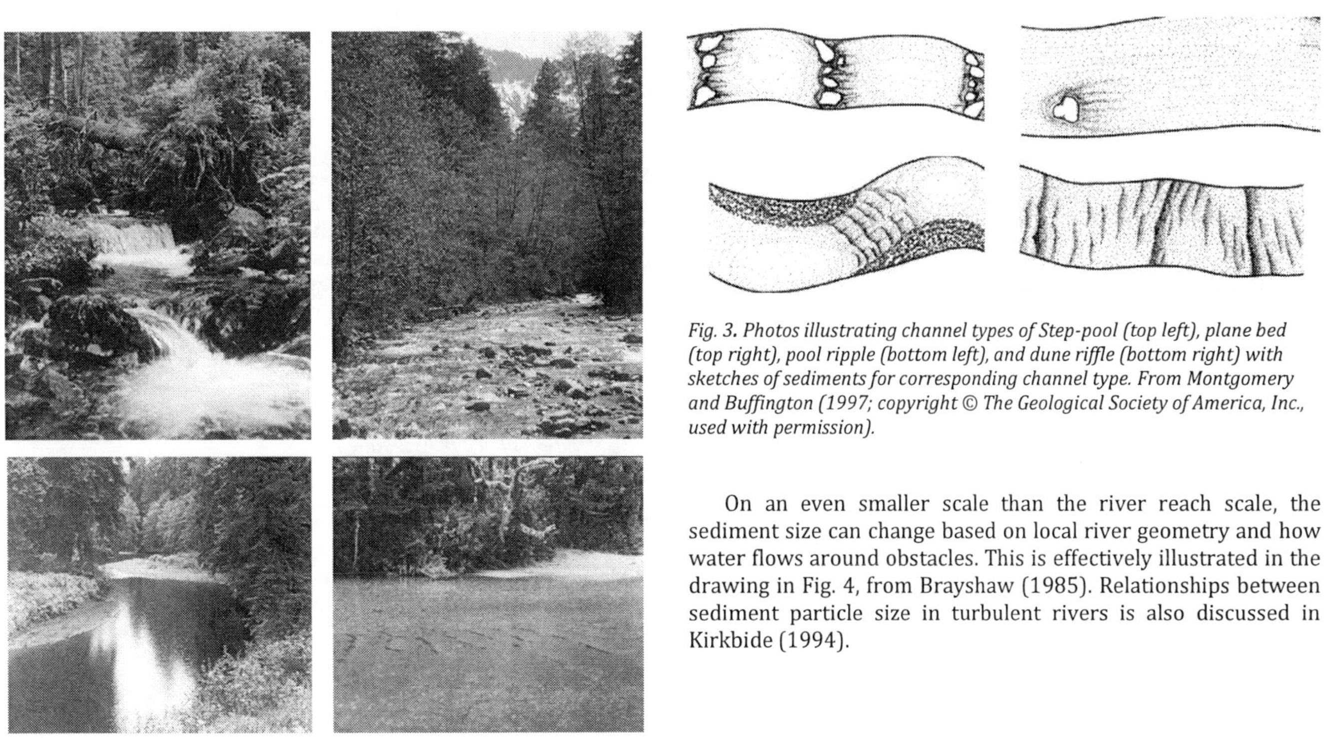

Fig. 3. Photos illustrating channel types of Step-pool (top left), plane bed (top right), pool ripple (bottom left), and dune riffle (bottom right) with sketches of sediments for corresponding channel type. From Montgomery and Buffington (1997; copyright © The Geological Society of America, Inc., used with permission).

On an even smaller scale than the river reach scale, the sediment size can change based on local river geometry and how water flows around obstacles. This is effectively illustrated in the drawing in Fig. 4, from Brayshaw (1985). Relationships between sediment particle size in turbulent rivers is also discussed in Kirkbide (1994).

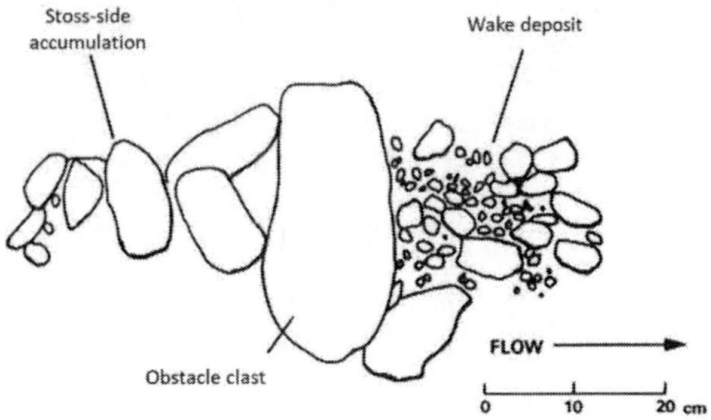

Fig. 4. Sketch of how riverbed grain sizes distribute with flow around an obstacle or boulder. Figure from Brayshaw (1985; copyright © The Geological Society of America, used with permission).

3.2 Predicting Grain Size of Bottom Sediments.

As noted in the introduction, coarse grained river bottoms (coarse sands, gravels and cobbles) are considered better for fording than fine grained soils, but no further guidelines or details are provided. In addition to using fluvial geomorphology and morphodynamics principles to infer riverbed sediment type, we can also use river hydraulics to calculate grain size. Note, however that the flow conditions causing deposition are not necessarily the same flow conditions occurring during fording operations.

As one would expect, fine sediments are found in slow and calm river areas, while coarse sediments are found where flow velocity is higher. To facilitate comparison among grain size distributions, statistics are commonly generated to represent the percentage of grains for different sizes. For example, the median particle diameter (*D*50) is often used in hydrology, geomorphology and engineering as a measure of the dominant tendency of the grain-size distribution. Other grain size indicators (e.g. D15, D84 are common) and ratios or logarithms of such are used to indicate the dispersion, skewness or kurtosis (peakedness) of the grain size distribution.

How can we use hydrology and fluvial geomorphology to predict grain size?

Grain size can be estimated or calculated based on the flow rates of the river and numerous relationships exist relating particle size to river flow and topographic parameters. Bacon and Sion (2024) recently studied fluvial geomorphology specifically for military gap crossing operations, so we will use their approach. They chose the method of Jha et. al., (2022a and 2022b) to relate Total Stream Power, TSP (W/m) as the rate of energy per unit of downstream length, to median grains size (D50, mm):

$$TSP/w = 0.1[(SG-1)gD50)]^{3/2} \qquad (1)$$

Where: TSP = Total Stream Power (W/m)
w = the bank-full channel width (m)
SG = the sediment specific gravity (approximately 2.65 kg/m^3)
g = the gravitational constant (9.81 m/s^2)
D50 = grain size of the 50th percentile of the stream bed (mm)

Bacon and Sion then used the data of van den Berg (1995) and Leopold and Wolman (1957) to develop a simplified empirical equation and compared their results to the observed field conditions reported in Sion et al (2022) to derive the simplified equation below:

$$D50 = 1678.94\,(S)^{0.72} \qquad (2)$$

Where S is the valley slope (an overall slope of the valley on a geomorphology scale, rather than stream gradient, which is at the hydrology "stream reach" scale).

Lastly, while coarse grain river bottoms are generally created by high flow conditions, which are worse for fording, the coarser grains will persist into times with low flow when fording is feasible. Additionally, in northern latitudes, rivers in glaciated terrain usually have source materials of glacial till and outwash, so often have an abundance of non-cohesive and coarse-grained soils and gravels to supply sediments for riverbeds.

4. Literature related to estimating river sediment quality and strength

4.1 Character of gravel river deposits

In fluvial geomorphology, the sediments of gravel rivers are often studied for their qualities related to habitat for aquatic species and, more specifically, for spawning grounds for trout, salmon, or other fish. We can use the methods of these studies to help understand the nature of the gravel river bottoms for fording as well. In fishery science riverbed studies, the structure or framework of the sediment is determined by the amount of gravel and the fines make up the matrix. If there are too many fines in the matrix then the larger, gravel-size particles are not touching and thus provide little support. Kondolf and Lisle (2016) provide an excellent overview of the different methods of measuring riverbed sediment size with emphasis on gravel rivers, and the work of Kondolf (2000) is specific to assessing the framework/structure size and matrix size in gravels for salmon spawning.

From the geotechnical literature, Vallejo (2001) looked at mixtures of fine and coarse-grained sediments and found that if the coarse grain materials make up 70% or more of the material, then the mixture structure is totally supported by the coarse-grained material. Conversely, if the coarse grains made up less than 40% of the mixture, the strength is completely controlled by the fine grains of the mixture.

4.2 Strength of gravelly sediments

The strength of gravels (or coarse-grained soils) is more difficult to obtain than for fine-grained soils since it requires a sample size that is significantly larger than the particles. Thus, for samples that include gravel or cobbles, a large test cell or large in-situ equipment is needed, which quickly becomes cost prohibitive even if you are not working under water. Thus, estimates of gravel strength and gravel strength predictions are relatively uncommon. However, Chang and Cheng (2014) provide excellent data and analysis that can be used for this. They investigated the shear strength for gravel deposits in Taiwan. Shear strength is the controlling bearing capacity strength measure for sediments failing in a general shear mode, which is applicable for a vehicle fording condition. They use the equation by Baker (2004) based on triaxial test data for non-linear failure criteria shown in Eq. 3:

$$\tau = Pa\, A(\sigma/Pa + T) \quad (3)$$

Where: τ = Shear stress
σ = normal stress
Pa = Atmospheric pressure
A = dimensionless parameter related to the magnitude of the shear strength
T = shift parameter on the normal stress axis

They measured the gravel matrix strength, qM, using a drop hammer, calculated percent gravel content and D50 from images, and correlated these to the shear strength parameter A yielding the relationship below.

$$A = -0.84 + 0.45\, qM - 0.0056 D50 + 0.14 CG \quad (4)$$

Where: qM = matrix strength
D50 = median grain size
CG = Gravel content (%)

This equation indicates that the matrix strength has the largest effect on overall gravel shear strength. They simplified Eq. 4 by removing the less significant variables, yielding an equation relating the shear strength parameter (shear strength index, A) to matrix strength only, with an R^2 value of 0.84:

$$A = -0.0575 + 0.457\, qM \quad (5)$$

Significantly, these findings illustrate that the strength of gravel bed deposits is primarily related to the strength of the fine-grained matrix material. This lends credibility in applying soil measurement techniques designed for use in fine-grained material, such as the cone penetrometer, to representing the overall river bottom strength.

4.3 Measuring the strength of submerged soils

There are not many examples in the literature regarding measuring the strength of submerged soils, but applications for the US Navy and the oil industry for sub-sea platforms, and the mining industry for manganese mining of the ocean floor could be useful. Hirst et al (1972) found good comparisons between measured vane shear strength and penetration resistance using a static cone (continuous push) penetrometer in ocean sediments for the US Navy, and found the penetrometer to be a valuable tool for submerged soil strength assessment.

A few terramechanics studies related to vehicle operations on submerged soils have been done in recent years. Aside from Rybansky's important work previously mentioned, Yamada et al., (2021) performed large scale laboratory studies of cone penetration testing on submerged and non-submerged soils, in addition to field tests of a tracked vehicle in a submerged soil configuration, primarily for disaster recovery. Wu et al., (2022) used a model of a vehicle track to study traction in submerged soils for unmanned vehicles in soft river soils. Cho, et al., (2019) designed a pilot seabed vehicle for mining manganese on the seafloor. Interestingly, each of these studies looked at vehicle behavior on fine-grained submerged soils. Additionally, they each noted that the submerged soils developed layering of fragile soils and more competent materials. This layering is consistent with field measurements collected as part of this study.

5. Exploratory field experiments: Measurements on the upper Mississippi River, Sept 2023

5.1 Field work objectives

To answer the question of how to measure the riverbed strength an exploratory field experiment was performed. The objectives of the field observations and measurements were to:

1. Explore how to measure soil strength in and around rivers and streams to support mobility decisions for wet gap crossing, with specific interest in fording sites.
2. Develop measurement equipment, techniques and field protocols for assessment of riverbed strength for fording
3. Facilitate collaboration between the researchers in the fields of meteorology, hydrology and terramechanics; with the goal of linking the science and technology from these fields with geospatial software capabilities to support of gap crossing decision making.

5.2 Field equipment

A primarily focus of the field experiments was to evaluate how we could measure the strength of the river bottom in a variety of flow conditions and bottom types (gravel, silt, bars), the ease of the measurements, and develop any adaptations needed for taking measurements below water – all with the purpose for measuring the strength of river bottoms. Because of the unknown nature of the river bottom, a range of equipment was tested to determine what worked best. Equipment used during the field exploration included waders, wading poles, measuring tapes, a field capable and GIS/GPS enabled computer with preloaded data layers (topography, USCS and USDA soil types, base-maps, and satellite imagery), a dynamic cone penetrometer, three types of trafficability cone penetrometers with two cone sizes, a selection of custom-made extension rods, etc. Fig. 5 shows the various static, or trafficability, cone penetrometers, cones, rod extensions and adaptors.

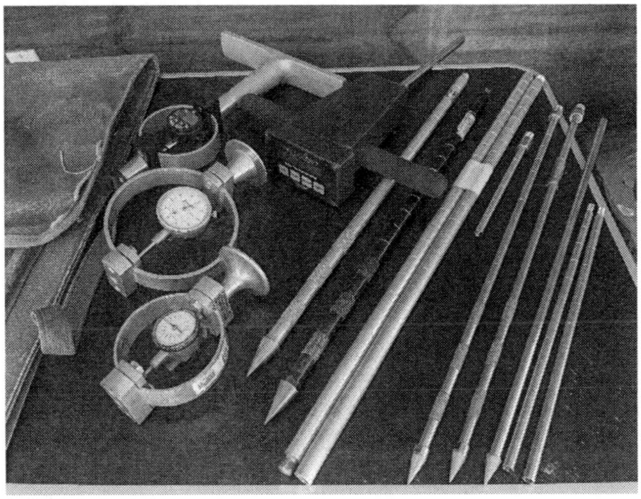

Fig. 5. Static/trafficability cone penetrometer parts used for measuring the strength of river sediments: Force measuring equipment included the Humbolt proving ring with electronic gauge (upper left), 2 proving rings with dial gauges, and the FieldScout penetrometer with electronic load cell with optical depth sensor and digital readout (purple, upper center); along with two size of penetrometer cones and various length extension roads and rod or thread size adaptors (center and right side).

5.3 Site Selection

A reconnaissance trip to Minnesota National Guard (MN NG) Camp Ripley was taken in May 2023 to look for suitable test sites and riverbed conditions for wet gap crossing operations, and especially for vehicle fording. During the initial visit we met with soil scientists and base environmental and facility personnel familiar with the terrain, soils and river conditions. Camp Ripley is in northern Minnesota and lays adjacent to the upper Mississippi River. Portions of the river are used to train military bridging units. The base provides several sites along the Mississippi and Crow Wing rivers, and even some pond sites that are potentially suitable fording and river bottom assessment. Study sites were chosen to explore the character and strength of river bottoms and banks in a variety of conditions.

The exploratory measurements were scheduled to occur at low water levels during September 2023 to enable improved access to the river bottom and adjacent banks. Sites were chosen to include a) previously used bridging exercise site, b) low bank relief and relatively low river conditions with riffles, c) higher current but low water level riffle areas, d) low velocity environment with silty soils, e) adjacent bank and bank terraces.

The characteristics of each site are described briefly below, and the results are discussed in detail the following section.

Site 1, Picnic area 3 (P3) has easy river access to a gravel riverbed with riffles and islands but has a steep bank. This site was used to develop river scouting and measurement techniques.

Site 2, WL4 is 300 m downstream of P3 and has gently sloping ingress with low, terraced riverbanks and gravel riverbed. Measurements at this site consisted of a riverbed traverse, a bank traverse, and riverbed terrace measurements.

Site 3, W2E is upstream of the bridge crossing exercise location on the East side of river (near the bridging site egress). At this site we performed a traverse upstream and downstream of a shallow gravel bar.

Site 4, SD 1 to 4 is a low-flow velocity, silted area upstream of the Sylvan Dam on the Crow Wing River. Measurements were taken from a dock.

5.4 Analysis and Results

Observations from site 1 (P3) were used to develop measurement technique and protocols. The three following sites provided measurements in a variety of river and riverbank conditions. The analysis and results for each site are presented below.

5.4.1 Picnic area 3 (P3)

Picnic area 3 was close to the road and the post cantonment area and had easy access via stairs to the river. We scouted the river and waded out halfway to the nearest bar island to sample a transect back towards the shore (Fig. 6). The river bottom was mostly gravel with hard packed sand and larger rocks. We found that the deepest water was in a fast-moving channel close to the shore and was about 1 m deep. Overall, the river bottom condition would be suitable for fording, but it was difficult to measure the soil strength with the cone-penetrometers due the water depth and rocks, so modifying the equipment and developing the measurement protocol was the focus for this site.

Fig. 6. Scouting the upper Mississippi river at Picnic site 3 to determine how to quantify the river bottom sediment strength for vehicle fording.

Since this was our first time using the different types of equipment in water, the primary result was experimenting with and determining how to configure the equipment for use in a river, and in developing guidance for collecting the data. For example, we experimented with how to operate each piece in different water depths, how far we could go into the river safely, and what type of measurement scheme worked best. To this end, while the tallest of our field team members could entirely cross the river, several people are needed to take measurements adequately and safely, as discussed more later.

Some of the instruments needed adaptations for use in rivers. Specifically, the FieldScout optical depth sensor reflected off the water surface and thus the depth readings would exceed their maximum range before the cone even reached the river bottom. Conversely, the Dynamic Cone Penetrometer is designed to be used for deeper penetration but produced erratic measurements due to the impact load causing the probe to rapidly bounce or jump past cobbles or measured the strength of individual rocks rather than the overall river bottom sediments. The abundance of rocks made taking measurements in many locations difficult once the probe hit a large cobble or boulder, but a probing technique with a bit of finesse in pushing past cobbles was eventually discovered. In addition, the flow rates in portions of this part of the river made it difficult to stand or work while keeping electronics out of the water. Fig. 7 shows measuring water depth at each location to map river bathymetry, and Fig. 8 shows fitting the penetrometers with extension rods.

Fig. 7. Measuring water depth using wading poles by subtracting off the length of pole above the water at site P3.

Fig. 8. Setting up the penetrometers with custom-made extension rods of various lengths at site P3: the Humbolt penetrometer with 20mm (0.75 inch) diameter cone (left) and the FieldScout penetrometer with 12.5mm (0.5 inch) diameter cone (right).

Comments on using the penetrometers

The primary result of our work at this site was to determine the best equipment configuration and technique for penetrometer use. In some areas, it was difficult to get a good penetrometer reading on the river bottom because of numerous large cobbles and even boulders. The optimal penetration technique was to push slowly through the gravel riverbed, gently moving the penetrometer tip past rocks. If the river bottom could not be penetrated to at least 15 cm, the maximum depth was recorded, and the measurement was retried at a nearby location. Notably, this technique resulted in sampling locations which strategically avoided large rocks, and consequently, the soil strengths recorded are more representative of the matrix of finer grained sediments comprising the river bottom. Recall however, that the matrix strength can be a good representation of gravel bed strength as noted in the previous discussion (section 4.1 and 4.2 on gravelly sediments).

We found that the best penetrometers were the ones that were easiest to set up and read. Our favorite penetrometer for these measurements was the Humbolt penetrometer which has a digital readout from a proving ring. We configured the Humbolt penetrometer with the larger cone size and 60-cm length extension rods, but we speculate that it could be useful with a smaller cone as well. The fully manual trafficability cone penetrometer with a proving ring and dial gauge was used with a small cone and narrow shaft. The dial gauge was harder to read than the digital readout on the Humbolt. As mentioned previously, the FieldScout cone penetrometer had issues with the optical depth sensor, but those were mitigated by waiting to initialize the "zero" depth until the penetrometer tip was just above river bottom. However, this mitigation technique required an additional correction for water depth as well as initialization height above river sediments, adding an extra step and potential confusion in data post-processing. Lastly, using the DCP in this environment was ineffective as the dynamic nature of the device did not facilitate pushing through the gravel matrix gradually, instead generating large and erratic jumps in the penetration depth as it slide past or broke rocks.

River bottom assessment protocol

We settled on a measurement method that included using the two most useful penetrometers. One person would measure the water depth and operate the penetrometer calling out each depth or penetrometer reading as it was pushed into the river bottom; the second person took the GPS waypoints and read the penetrometer gauge (for the Humbolt) or the depths (for the FieldScout), and the third person took field notes and pictures. The person taking notes and pictures could then be standing in shallow water or on more stable ground. For our field program we had four people so we could switch duties often and optimize these techniques. We initially attempted to take measurements at specific distances along a traverse, but using a measuring tape or pacing the distance was not possible in rapidly moving water on the uneven, rocky and slippery bottom, so we estimated measuring locations at approximately every 3m (10 ft) in a line roughly perpendicular to the shore. Additional measurements were also taken if the bottom conditions changed and other in areas of interest.

The bullet points below outline recommendations for measuring river bottom strength based on lessons learned at this site.

- Scout the river to determine the quickly flowing areas and note any large rocks, boulders, and holes.
- Choose your traverse path and start at the far shore or mid-river, whichever is most suitable for your purpose and for safe operations.
- Work your way back to the entry location taking measurements at intervals (we used 3-m intervals).
    - Measure and record the water depth (using a wading pole with depth markings is a good way to do that).
    - Take a GPS way point.
    - Use a cone penetrometer with suitable extension rods (we used 30 or 60 cm extensions) initializing a "zero" reading so you don't exceed the maximum depth allowed in preprogrammed electronics. Starting at the top of the river bottom surface, push the penetrometer at a steady rate and record penetration force every 2.5 cm (1 inch). Record the cone size and extension rod length.

- Strive to measure the river bottom strength to at least 15 cm depth, but more is better. If you immediately hit a rock, choose a nearby alternate location.
- Rough estimates of stream velocity can be made using a float by timing its movement across a known distance.

5.4.2 Site WL4

The WL4 site was approximately 300m downstream of the picnic area. This section of river was a continuation of the wide braided channels, and the approaching shoreline was characterized as a gradual, tiered slope with bushes and reeds. Along the shoreline, the soil was generally thick organic muck. The river bottom itself was similar to Picnic area 3, with a rocky gravel bottom. The site had easy access to the water and a wide area to explore the river, with a gradual and terraced bank where we could look at the soil conditions at different terrace levels (Fig. 9). While the river measurements were made at approximately 3 m intervals, the bank measurements were made every 1 m and the terrace measurements were taken in clusters at three terrace levels.

Results from the river measurements are shown in Fig. 10. In many cases, we were not able to probe deeper than 15 cm into the riverbed, indicating that there was a significant "armor" surface layer of gravels that had been washed of its fine matrix. This armor is common in gravel rivers and is illustrated in the sketch and photo in Fig. 10 (bottom). The measurement locations near the shore (shallow water depth) had more fine soils and were softer.

The riverbank data will be used in future studies with other soils measurements nearby to help understand relationships between the riverbed sediments and the adjacent banks and terraces in order to ultimately merge the fording predictions with landward cross-country mobility models.

Fig. 9. Traverse waypoints along the river, banks and terraces at site WL4 (top). Overview of the site showing low slope and marshy riverbanks (bottom left), and bank moisture and strength measurements (bottom right).

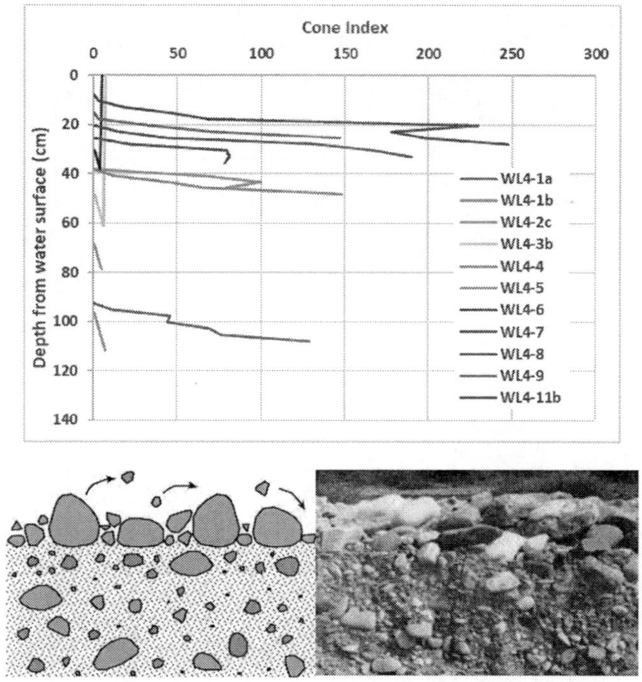

Fig. 10. Riverbed penetration profiles at site WL4 using the water surface as the datum (top) from waypoint 1a near river center to waypoint 11b near shore. Note that many of the measurements were not able to penetrate a minimum 15 cm into the riverbed (shown as "aborted" shallow depth profiles). Measurements in the shallower water (of 40cm or less) close to shore had more fines in the gravel matrix and could be penetrated deeper. Illustration and photo of a gravel bar showing "armoring" common on gravel riverbeds (public domain illustration from Pitlick et al. 2009 and public domain photo from USDA-FS 2008).

5.4.3 Straight section of river with higher flow velocity: W2E

Site W2E was near the bridging exercise site. The egress side of the river crossing was less disturbed than the west side of the river where the bridging staging area was located. The east side of the river was also easily accessible, and we were able to travel upstream to a more natural site. Two traverses were completed at this site, one representing the overall river conditions (waypoints A through H), and a second traverse downstream of a gravel bar (waypoints I through N), as shown in Fig. 11. The results for both traverses are given in Fig. 12.

At both sites, we took measurements in a traverse starting in the main flow of the river. It is evident from the spacing of the sampling locations that we were improving our site assessment technique in measuring the strength in a traverse across the river, starting by wading as far as possible and then gradually working our way back to shore. Initial plans for testing the full width of the river by taking measurement from a boat were later deemed unnecessary. Again, while the tallest member of the team could cross the entire river width, it was difficult to record information if the other members of the field team were not nearby. In addition, while it is possible to modify the equipment for deeper readings, a depth limit of 1 to 1.5 m was beyond the fording depth for most vehicles (as noted in Table 1) so was sufficient for our purposes.

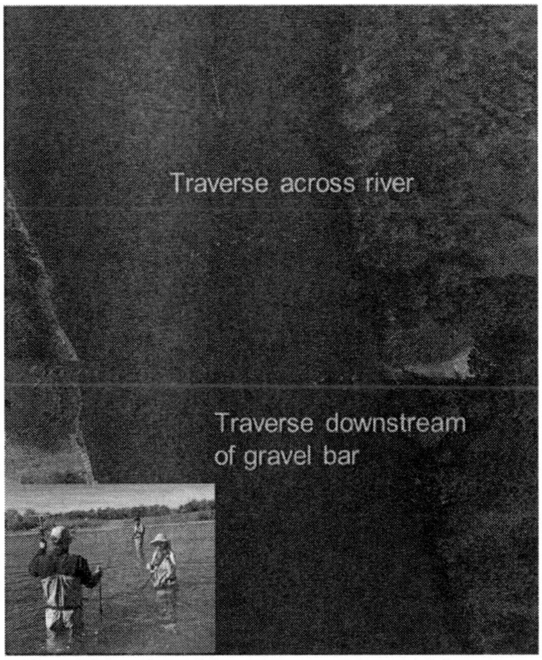

Fig. 11. Waypoint locations of the traverses upstream and downstream of the gravel bar at the W2E site with insert photo of downstream site. Note the bridging ingress site on the far shore and the silty water at this location.

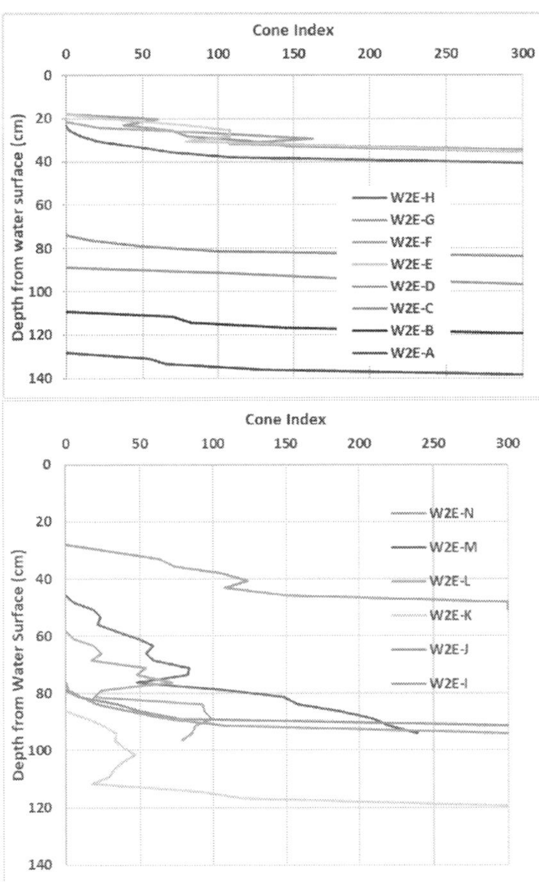

Fig. 12. Penetrometer results for WE2 site upstream (point A in the river center to H near shore, top) and downstream of the gravel bar (point I near river center to N near shore, bottom) showing weaker sediments downstream of the bar. Several profiles show an intermediate weak layer.

5.4.4 *Low velocity area upstream of Sylvan Dam*

After taking measurements at the bridging exercise site, we drove north to scout areas along the Crow Wing River upstream of the Sylvan Dam to measure a low velocity flow area with finer grained soils. A boat was initially used to scout for suitable areas, but problems with the engine along with a heavily vegetated shoreline resulted in restrictive site access. However, there was a picnic area with a dock just upstream from the dam that we used to take measurements from the shore to the end of the dock to obtain low flow conditions without using our waders, which were already dried and packed for shipping. A photo and aerial view of the site along with the measurement results from both the Humbolt and the FieldScout penetrometers are given in Fig. 13.

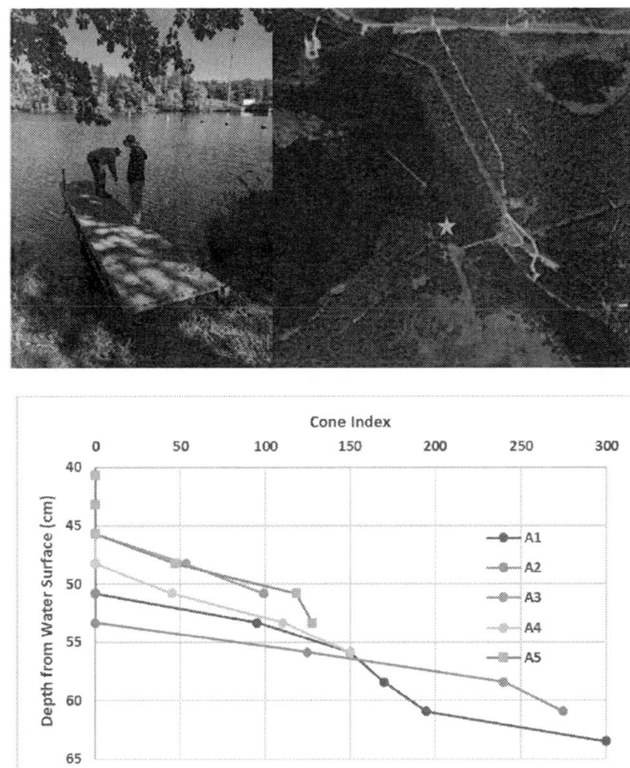

Fig. 13. Field site SD in silty bottom, upstream of dam (top) with red arrow showing the flow direction. Cone index values from the both the Humbolt and the FieldScout penetrometers yielding similar strength gradients of the bottom sediments (bottom) and being much weaker than the gravel beds of the faster flowing Mississippi River in Fig.s 10 and 12.

6. Discussion and Future Needs

Information on river geometry and classification may be useful for incorporating a rudimentary knowledge of fluvial geomorphology to estimate effects of riverbed turns, obstacles and riffles on grainsize distribution to locate gravelly sites or to avoid suspected fine grained or soft areas. Hydrologic and fluvial morphology models are often developed for a "reach" in a river or stream system, but even smaller scale site estimates of riverbed suitability could also be used for detailed, high-resolution site analysis.

Using the process relationships from fluvial geomorphology, morphodynamics, hydrology and hydraulics we can estimate potential fording locations from locations likely to have coarse-grained sediments and also predict suitable streamflow depth and velocity using real-time weather-based hydrologic models. We propose merging these layers within the geospatial framework of GeoWATCH and validate the results in future field explorations. The geospatially derived target areas for likely fording sites can be used to down select sites for further analysis, and for overall operational planning.

Lastly, level terrain is estimated as being trafficable for specific vehicles if the measured or predicted Cone Index is greater than the soil strength required for that vehicle (also known as the Vehicle Cone Index, VCI). In other words, CI must be greater than VCI for the critical later, which is the top 15 cm for coarse-grained soils (US Army 1994). *However*, these generalities were developed for dry or moist coarse-grained soils. Submerged coarse-grained soils may behave quite differently under vehicle loading so the deformation mechanics should be examined, and the strength measurement validated specifically for suitability for estimating vehicle fording performance.

Concluding, while we have made a large step forward, the task of predicting the river bottom conditions and relating this to vehicle fording potential is perhaps the task with the greatest uncertainty, yet the highest payback for advancing our state-of-knowledge.

Some future development and validation needs include:

- Predicting grain size based on river flow rates, and then predicting strength of river bottom based on grain size. Validate with measurements over a range of regional, seasonal and geomorphic conditions
- Validate grain size distribution based on local obstructions and small-scale river morphology (flow around obstacles, scouring)
- Develop and standardize methods for measuring river sediment grain size and strength
- Develop methods to measure the strength below the surface armored layer
- Explore the mechanisms of river sediment response to vehicle loading, and especially multiple vehicle crossings
- Develop levels of crossing site potential based on flow, grainsize, and river-bottom strength
- Bring seasonal river and terrain conditions into the predictions
- Merge fording with cross-country mobility modeling capabilities
- Expand collaboration with different scientific fields and NATO countries to utilize breadth of knowledge to the highest extent possible
- Demonstrate physical fording experiments in various river and seasonal conditions.

7. Summary and Conclusions

The best fording sites are shallow, slow moving river locations with a strong, coarse-grained river bottom. Principles of fluvial geomorphology, which include relationships between streamflow velocity and grain size of bed sediments, can be used to help locate potential gravel sites (formed during high-flow periods) and hydrologic models with real-time weather input can estimate when river flow is appropriate for fording. Algorithms to location suitable fording site locations can be implemented within a geospatial framework such as GeoWATCH and merged with weather-based cross-country mobility models. Still, there remains the question of how to measure and quantify the strength of the riverbed to assess and validate the suitability for fording. To this end, we performed an exploratory field project to answer basic questions on how to measure riverbed strength for fording, and to initiate a scientific study bringing together meteorology, hydraulics and hydrology, geomorphology, geotechnical engineering and terramechanics to develop methods to advance the state-of-the-knowledge.

River bottom conditions were evaluated using cone penetrometers modified with extensions and using measurement techniques suitable for taking measurements in rivers. Several types of river bottoms were studied, and these exploratory measurements were used to develop protocols for measurement and data analysis. The field exploration and analysis were presented within the context of literature from many fields of study.

Lastly, the knowledge both assembled and developed was considered in generating capability gaps and future research needs for advancing the state-of-the-art in measuring, estimating and quantifying suitability of river fording sites to include the riverbed strength.

8. References

Bacon, S., Sion, B., 2024. Conceptual framework of soil landform relations in fluvial-geomorphic settings to support military gap crossing operations. Desert Research Institute (DRI) Technical Report, Task CHL 1A, Contract W912_HZ-23-C-0008. Reno, NV

Baker, R., 2004. Nonlinear Mohr envelopes based on triaxial data. J. Geotech. & Geoenviron. Engineering. 498–506.

Brayshaw, A.C., 1985, Bed microtopography and entrainment thresholds in gravel-bed rivers: GSA Bulletin, v. 96, no. 2, p. 218–223, https://doi.org/10.1130/0016-7606(1985)96

Brayshaw, A., 1984. Characteristics and origin of bluster bedforms in coarse-grained alluvial channels, p 77-85. In Sedimentology of Gravels and Conglomerates. Koster and Steel (eds) Canadian Society of Petroleum Geologists, Calgary, Alberta, Canada, Nov 1984. 441 p.

Buffington, J., Montgomery, D., 2022. Geomorphic classification of Rivers: an updated review. In Treatise on Geomorphology 2nd Ed. Vol 6. Elsevier. p. 1143- 1190.

Campervan Reykjavik. 2023. A guide to fording rivers. https://www.campervanreykjavik.com/post/crossing-rivers-iceland-camper

Chang, K.-T., Cheng, M.-C.. 2014. Estimation of the shear strength of gravel deposits based on field investigated geological factors. Engineering Geology 171 (2014) 70–80.

Cho, S., Park, S., Oh, J., Min, C., Kim, H., Hong, S., Jang, J., Lee, T.. 2019. Design optimization of deep-seabed pilot miner systems with coupled relations between constraints. J. Terramechanics 83 (2019) 25-34.

Church, M., 2006. Bed material transport and the morphology of alluvial rivers. Annual Review of Earth and Planetary Sciences 34, 325–354.

Church, M., Ferguson, R., 2015. Morphodynamics: Rivers beyond steady state. Water Resources Research. P. 1883-1897. DOI: 10.1002/2014WR016862

Coutermarsh, B., Dwinel, B., 2003. Remote determination of bridging/fording sites. ERDC-CRREL TR-03-8. Hanover, NH USA.

Crosato, A., 2007. River morphodynamics: Brief introduction. Lecture notes. Available On line or from the author.

Ekegren, M., Bailey, B., 2017. The first two meters: The importance of fording in reconnaissance. ERDC White Paper.

Eylander, J., Bieszczad, J., Ueckermann, M., Peters, J., Brooks, C., Audette, W., Ekegren, M., 2023. Geospatial Weather Affected Terrain Conditions and Hazards (GeoWATCH) Description and Evaluation. Environmental Modeling and Software, 160. https://doi.org/10.1016/j.envsoft.2022.105606.

Fischell, J., Holden, G., Stevens, M., Jackson, R., Webb, B., Ewing, M., Fernandeza, J., Monroe, J., McKinley, G., 2021. Prototype for a worldwide gap crossing tool. Proceedings of the ISTVS 20th International Conference, Montreal, Quebec, Canada, September 27-29, 2021

Follum, M., Vera, R., Tavakoly, A., Gutenson, J., 2020. Improved accuracy and efficiency of flood inundation mapping of low-, medium-, and high-flow events using the AutoRoute model. Natural Hazards and Earth System Sciences, 20(2), pp.625-641.

Galay, V.J., Rood, K.M., Miller, S., 1998. Human interference with braided gravel-bed rivers. In Gravel-bed rivers in the environment. Edited by PC Klingeman, RL Beschta, PD Komar, and JB Bradley. Water Resources Publications, LLC, Highlands Ranch, Colo: 471-512.

Hirst, T., Richards, A., Inderbitzen, A., 1972. A static cone penetrometer for ocean sediments. ASTM Technical Publication STP38794S. ASTM International. West Conshohocken, Pennsylvania.

Jha, M.K., Asamen, D.M., Allen, P.M., Arnold, J.G., White, M.J., 2022a. Assessing streambed stability using D50-based stream power across contiguous U.S. Water 14, 3646. 10.3390/w14223646.

Jha, M.K., Asamen, D.M., Allen, P.M., Arnold, J.G., White, M.J., Bieger, K., 2022b. Streambed median grain size (D50) across the Contiguous U.S. Water 14, 3378, 10.3390/w14213378.

Kirkbride, A., 1994. Turbulence structure in straight gravel-bed channels. Doctoral Thesis. University of Sheffield, UK.

Kondolf, G., 2000.Assessing salmonid spawning gravels. Transactions of the American Fisheries Society 129: 262–281.

Kondolf, G., Lisle, T., 2016. Measuring bed sediment. Chapter 13 In Tools in Fluvial Geomorphology (Kondolf and Peigay, eds.). John Wiley and Sons Inc., England, 688 p.

Kondolf, G., Piégay, H., Schmitt, L., Montgomery, D., 2016. Geomorphic classification of rivers and streams. Chapter 7 in Tools in Fluvial Geomorphology (Kondolf and Peigay, eds.). John Wiley and Sons Inc., England, 688.

Leopold, L., Wolman, M., 1957. River channel pattern: braided, meandering and straight. Physiographic and hydraulic studies of rivers. USGS Professional Paper 282B, 85p.

Montgomery, D.R., Buffington, J.M., 1997, Channel-reach morphology in mountain drainage basins: GSA Bulletin, v. 109, no. 5, p. 596–611, https://doi.org/10.1130/0016-7606(1997)109<0596:CRMIMD>2.3.CO;2

Pitlick, J., Yantao, C., Wilcock, P., 2009. Manual for Computing Bed Load Transport Using BAGS (Bedload Assessment for Gravel-bed Streams) Software. USDA-FS General Technical Report RMRS-GTR-223. Rocky Mountain Research Station, Fort Collins, CO.

Rybansky, M., Dohnal, F., Hubacek, M., Capek, J., Kovarik, V., 2024. The impact of changes in the river regime on the mobility of off-road vehicles. Proc of the 21International and 12th Asian-Pacific Regions Conference of the ISTVS, Yokohama, Japan, Oct 28-31, 2024.

Rybansky, M., Dohnal, F., Hoskova-Mayerova, S., Svatonova, H., 2018. The impact of drainage on terrain UGV movement. In: 9th IGRSM International Conference and Exhibition on Geospatial & Remote Sensing (IGRSM 2018) 24-25 April 2018, Kuala Lumpur, Malaysia. Kuala Lumpur, Malaysia: IOP Conference Series: Earth and Environmental Science, 2018, p. 1-10. DOI: 10.1088/1755-1315/169/1/012027.

Sion, B.D., Shoop, S.A., McDonal, E.V., 2022. Evaluation of in-situ relationships between variable soil moisture and soil strength using a plot-scale experimental design. Journal of Terramechanics 103, 33–51.

Snow, A., Christensen, S., Swain, N., Nelson, E., Ames, D., Jones, N., Ding, D., Noman, N., David, C., Pappenberger, F., Zsoter, E., 2017. A High-resolution national-scale hydrologic forecast system from a global ensemble land surface model: Journal of the American Water Resources Association (JAWRA), 52(4), pp. 950-964.

US Army, 1994. Field Manual FM 5-430-00: Planning and design of roads, airfields, and heliports in the theater of operations. Headquarters Department of the Army. Washington, DC, 29 September 1994.

USDA-Forest Service, 2008. Stream Simulation: An Ecological Approach to Providing Passage for Aquatic Organisms at Road-Stream Crossings U.S. Department of Agriculture, National Technology and Development Program, San Dimas, CA.

Vallejo, L.E., 2001. Interpretation of the limits in shear strength in binary granular mixtures. Can. Geotech. J. 38, 1097–1104.

van den Berg, J.H., 1995. Prediction of alluvial channel pattern of perennial rivers. Geomorphology 12, 259–279.

Wu, J., Liu, B., Shen, Y., Feng, Z., Jiao, Q., 2022. Piecewise model of track sinkage based on underwater cohesive soil. J. Terramechanics 103 (2022) 1-10.

Yamada, M., Yamauchi, G., Hashimoto, T., 2021. Fundamental study on underwater trafficability for tracked vehicle. J. Terramechanics 98 (2021) 42-49.

8964 | Application of LSTM networks to prediction of a wheel traveling phenomena

Haruki Endo *, Kai Xing, Shingo Ozaki

Yokohama National University, Yokohama, Kanagawa, Japan

** Corresponding author: endo-haruki-kb@ynu.jp*

ABSTRACT

Lunar exploration, which began in the 1960s, is still ongoing and various space exploration programs are planned. In these space exploration programs, it is important to analyze and predict the trafficability of exploration rovers on the lunar and Martian surfaces, and to develop and operate on these ground conditions. Therefore, in this study, we focused on the real-time analysis in operational phase of exploration rovers. We applied a Long Short-term Memory (LSTM) network, an artificial regression neural network architecture used in the field of deep learning, to single-wheel traveling phenomena with the aim of applying it to the real-time prediction in future exploration rover operations. Specifically, numerical simulations were carried out using an extended terramechanics model that takes terrain surface deformation into account, and training and prediction are carried out on several LSTM models. The results show the possibility of real-time prediction of trafficability of a wheel on soft ground due to the characteristics of LSTM networks that can handle time-series data. Furthermore, it is suggested that the proposed method may be applicable to the prediction of precursor phenomena of wheel stuck.

Keywords
Wheel
Trafficability
Simulation
Deep learning
LSTM network

1. Introduction

The principles of terramechanics, an interdisciplinary discipline dealing with machine-ground interaction, are used in various extreme environments, such as mines, disaster sites and even extraterrestrial solid bodies. In particular, the role of tera-mechanics-based rover traveling simulation is becoming increasingly important for exploration projects on extraterrestrial solid objects, as it is difficult to conduct actual tests in advance in the local environment. In fact, as reported in the roadmaps published by several space agencies, exploration rover travel has been identified as a key mission in future space exploration projects (ISECG Technology Working Group, 2019; Kawakatsu et al., 2023; NASA technology taxonomy, 2024).

As mentioned above, numerical simulation techniques play an important role in the development of rovers. Based on pioneering work by Bekker, Wong and Reece in the field of teramecanics (Bekker, 1969; Wong and Reece, 1967a, 1967b; Wong, 2008), semi-empirical terramechanics models have been applied to evaluate the traveling performance of off-road vehicles across a wide range of scales, from small robots to large dump trucks. To date, in addition to the development of formula for stress evaluation, various extended models have been proposed to accommodate behaviors such as braking and lateral slip (e.g., Ishigami et al., 2007, 2009; Haibo et al., 2013; Junlong et al., 2017). Furthermore, by implementing it in a multibody dynamics model, efforts have also been made to analyze the overall behavior of various vehicle models, including exploration rovers (Arvidson et al., 2017; shigami et al., 2007, 2009; Zhou et al., 2014). In recent years, leveraging its advantage of generating driving data under various conditions within a short period of time, terramechanics-based simulation has also been utilized for preparing training datasets for the development of surrogate models based on machine learning (Fediukov et al., 2024).

In planetary rover operations, becoming immobilized—commonly referred to as 'stuck'—is a critical failure mode. Indeed, several exploration missions have encountered such incidents. For example, NASA's Mars Exploration Rover *Spirit* successfully operated for nearly five years; however, its mission ultimately ended when one of its wheels became embedded in soft soil, leaving the rover unable to free itself (Kerr, 2009). However, conventional terramechanics models are incapable of accurately evaluating the deformation of the terrain surface caused by the interaction between the wheel and the ground. Moreover, since these models are fundamentally based on static equilibrium, they cannot adequately capture transient response behaviors. Consequently, it has been challenging to precisely assess key indicators such as drawbar-pull and sinkage, which are closely related to the wheel getting stuck.

Under such circumstances, models that account for terrain surface deformation resulting from wheel–ground interaction have been proposed. Among them, Suzuki et al. proposed an extended terramechanics model (xTerramechanics model), which incorporates a terrain surface deformation mechanism based on the principles of cellular automata (Suzuki et al., 2022). The xTerramechanics model has been shown to not only accurately represent the stuck phenomenon but also adapt to a wide range of driving conditions (Watanabe et al., 2023, Kita et al., 2024; Shimizu and Ozaki, 2024). However, many unknown factors remain in the lunar surface environment, and the utilization of real-time simulation technology is considered ideal for achieving more advanced operations to avoid the stuck phenomenon.

In recent years, artificial intelligence has been applied to real-time prediction and control of vehicle behavior, with ongoing

https://doi.org/10.56884/DUMF7XJZ

efforts to extend its use to off-road vehicles. In this study, aiming at real-time prediction of future exploratory rover behavior, we attempt to apply a Long Short-Term Memory (LSTM) network (Ralf C et al., 2019) to single-wheel travel phenomena. LSTM is a type of recurrent neural network that is capable of learning long-term dependencies (Chung et al., 2014; Hochreiter and Schmidhuber, 1997; Tang et al., 2017). Therefore, based on sequential data of various dynamic quantities (physical values measured by sensors, etc.) during wheel locomotion, it may be possible to predict signs of the stuck phenomenon. It should be noted that constructing an LSTM network requires a large amount of training and validation data. In this study, we utilized numerical simulation data based on the xTerramechanics model described earlier. Furthermore, in the training and prediction processes of the LSTM network, we employed three types of LSTM models—one-to-one, many-to-many, and many-to-more-many—and examined the relationship between model complexity and prediction accuracy and characteristics.

2. Theory

In this study, the interaction between the wheel and soil is analyzed using the xTerramechanics model. The xTerramechanics model is a combination of the conventional terramechanics model, Bekker-Wong-Reece model (Wong and Reece, 1967a) (BWR model) and terrain surface deformation model (Suzuki et al., 2023; Watanabe et al., 2023). The terrain surface deformation model considers soil movement due to the interaction between soil and an object. This section briefly describes each of these theories. In addition, LSTM, a type of recurrent neural network specialized for processing serial data, is also reviewed.

2.1. BWR model

BWR model is widely used to evaluate the traveling performance of off-road vehicles. The vertical stress $\sigma(\theta)$ acting on a wheel surface is defined by the contact angle θ as in Eq. 1.

$$\sigma(\theta) = \left(ck_c' + \rho g l k_\phi'\right)\left(\frac{\lambda(\theta)}{l}\right)^n \tag{1}$$

where c is the soil cohesion [Pa], k_c' is the sinkage modulus influenced by cohesion [-], ρ is the bulk density [kg/m^3], g is the acceleration of gravity [m/s^2], l is the wheel contact length (the length of the arc where the wheel is in contact with the soil) [m], k_ϕ' is the sinkage modulus influenced by friction angle [-], and n is the sinkage exponent [-]. The amount of sinkage $\lambda(\theta)$ at a contact angle θ [rad] is obtained as follows.

$$\lambda(\theta) = \begin{cases} \frac{D}{2}\left(\cos\theta - \cos\theta_f\right) \ (\theta_m < \theta < \theta_f) \\ \frac{D}{2}\left\{\cos\left[\theta_f - \frac{\theta-\theta_f}{\theta_m-\theta_r}(\theta_f-\theta_m)\right] - \cos\theta_f\right\} (\theta_r < \theta < \theta_m) \end{cases} \tag{2}$$

where D is the wheel diameter [m], θ_f is the entry angle [rad], and θ_r is the exit angle [rad]. Also, θ_m is the maximum stress angle [rad], which is the angle at which the wheel receives the maximum normal stress from the soil, and is expressed by the following equation:

$$\theta_m = (a_0 + a_1 s)\theta_f \tag{3}$$

where a_0 and a_1 are the maximum stress angle constants [-], which are parameters depending on the wheel geometry and soil. s is the slip ratio [-] and is defined as

$$s = 1 - \frac{v}{(D/2)\omega} \tag{4}$$

where ω is the angular velocity of the wheel [rad/s] and v is the translational velocity [m/s]. The shear stress τ_x occurring in the longitudinal tangential direction of the wheel surface is given by the following equation using the normal stress in Eq. 1.

$$\tau_x = (c + \sigma(\theta)\tan\phi)\left(1 - e^{-\frac{j_x(\theta)}{k_x}}\right) \tag{5}$$

where ϕ is the internal friction angle [rad] and k_x is the shear deformation modulus along x-axial [-]. $j_x(\theta)$ is the amount of sliding between the wheel and the soil [m], which is obtained by integrating the wheel-soil sliding velocity over time and is expressed by the following equation:

$$j_x(\theta) = \frac{D}{2}\left[\theta_f - \theta - (1-s)\left(\sin\theta_f - \sin\theta\right)\right] \tag{6}$$

From the above, the drawbar-pull F_x and the vertical reaction force F_z can be evaluated as follows:

$$F_x = \frac{D}{2}B\int_{\theta_r}^{\theta_f}\left[\tau_x(\theta)\cos\theta - \sigma(\theta)\sin\theta\right]d\theta \tag{7}$$

$$F_z = \frac{D}{2}B\int_{\theta_r}^{\theta_f}\left[\tau_x(\theta)\sin\theta + \sigma(\theta)\cos\theta\right]d\theta \tag{8}$$

where B is the wheel width [m].

2.2. Terrain surface deformation model

The terrain surface deformation model adopts cellular automaton that discretizes the ground surface into cells and generates terrain surface deformation based on information such as the height of each cell and its contact with the wheel. There are three types of terrain surface deformation actions: rolling action, moving action, and compressive action. In addition, changes in soil density due to wheel compaction are also considered.

The rolling action is the action of sand rolling by gravity when the slope of the formed road surface is greater than the angle of repose (Nishimori and Ouchi, 1993; Suzuki and Ozaki, 2020; Werner, 1995), and is expressed by the following equation:

$$dh_{i,j}^{roll} = -rh_{i,j} + \frac{r}{6}H_{cross} + \frac{r}{12}H_{diag} \tag{9}$$

where the subscripts i and j are the cell numbers corresponding to the x- and y-directions, respectively. $h_{i,j}$ [m] is the cell surface height, and r is the outflow ratio [-]. Also, H_{cross} is the sand inflow term from orthogonal cells, and H_{diag} is the inflow term from diagonal cells, which are given as follows:

$$\begin{cases} H_{cross} = h_{i+1,j} + h_{i-1,j} + h_{i,j+1} + h_{i,j-1} \\ H_{diag} = h_{i+1,j+1} + h_{i+1,j-1} + h_{i-1,j-1} + h_{i-1,j+1} \end{cases} \tag{10}$$

The moving action is caused by the interaction between the object and the ground. This action is represented as

$$\begin{cases} dh_{i,j}^{rem} = -S_{i,j} \\ dh_{i+di,j+dj}^{rem} = R_1 S_{i,j} \end{cases} \tag{11}$$

Eq.11 indicates that the terrain surface height decreases by a specific height $S_{i,j}$ [m], and that soil with a height equal to $S_{i,j}$ multiplied by a specific ratio R_1 [-] moves to the adjacent cell. Note that the specific ratio R_1 is a parameter that takes into account the compaction or expansion of the soil as it moves. The direction of soil movement is in the opposite direction of the stresses that the wheel receives from the soil. In addition, the amount of movement $S_{i,j}$ is given as

$$S_{i,j} = R_2\delta_{i,j} \tag{12}$$

where $\delta_{i,j}$ is the amount of wheel penetration into the soil [m] during each numerical calculation step. The amount of penetration multiplied by a specific ratio R_2 [-] becomes the amount of movement $S_{i,j}$.

The compressive action corresponds to the terrain surface compressing as the wheels pass over it, and is expressed by the following equation:

$$dh_{i,j}^{comp} = -R_3\delta_{i,j} \tag{13}$$

As with the moving actions, the soil is reduced in height by a specific ratio R_3 [-] times the penetration $\delta_{i,j}$ [m], but no movement into other cells takes place. The compressive action is more pronounced in loose soils.

The deformation of the terrain surface is determined by applying the deformations evaluated by the above three actions to all cells as follows:

$$h'_{i,j} = h_{i,j} + dh_{i,j}^{roll} + dh_{i,j}^{rem} + dh_{i,j}^{comp} \tag{14}$$

2.3. Long short-Term Memory networks

The traveling performance of wheel is represented by time-series data. Time-series data points arranged chronologically enable the prediction of future behavior by observing past trends. However, traditional regression models are difficult to predict accurately because of the nonlinear and time-evolving nature of the wheel traveling motion. On the other hand, unlike convolutional neural networks, which process inputs independently without retaining memory, recurrent neural networks and LSTMs are designed to handle sequences with interdependencies. They maintain an internal memory of past inputs, enabling them to make informed predictions about future events based on temporal context (Chung et al., 2014; Hochreiter and Schmidhuber, 1997; Tang et al., 2017).

Therefore, we applied these advanced neural network architectures to model the temporal dynamics of the wheel traveling. Recurrent neural networks were initially used to understand the sequential nature of data. However, because of the vanishing gradient problem, LSTMs have been employed to capture long-term dependences more effectively (Chung et al., 2014). LSTMs, with their unique gating mechanisms, allow the network to learn which information to retain or forget, making them ideal for analyzing time-dependent wheel traveling behavior. In this study, LSTM networks were trained on extensive terramechanics-based simulation data to predict posterior traveling events and perform multivariate analysis.

3. Training data prepared by terramechanics-based simulation

3.1. Outline of data processing

The Simscape® environment was used to simulate the dynamic system, and the xTerramechanics model was implemented using the MATLAB®-based Simscape language. The ode4 solver was employed to solve the equations of motion, and the time increment was set to t=0.002 s. The acceleration of gravity was g = 9.8 m/s^2.

In this study, the analytical data obtained under the conditions described above were used to generate training data for the LSTM network. The analysis conditions and the data preprocessing methods are presented below.

1. In the Simscape model, an additional block was implemented to apply various patterns of traction load (towing load) P_x, and the simulation results during traction driving were saved as MATLAB data.
2. From the generated MATLAB data, time-series data of trafficability indicators such as sinkage, slip ratio, and forces/torques were extracted, saved as CSV files, and organized.
3. Using the LSTM model developed in Python, the time-series data stored in the CSV files were loaded for further processing.

In this study, simulation data were generated for single-wheel straight-line motion under conditions that emulate traveling on uphill, downhill, and uneven terrain. Specifically, various patterns of traction load P_x[N] were applied to a wheel operating under a free-propeller traveling condition, in which angular velocity is the sole input. For example, uphill travel was simulated by applying a traction load P_x [N] equivalent to the load acting on the wheel during actual uphill motion.

3.2. Parameters

The wheel and soil conditions used for training are described below. In this study, the soil parameters of Toyoura sand were employed (Shimizu and Ozaki, 2024). Table 1 shows the wheel specifications used in the analysis. Furthermore, Table 2 shows the basic parameters of the BWR model used for stress evaluation. The parameters related to terrain surface deformation are shown in Table 3, where these parameters were set based on previous studies (Suzuki et al., 2023). Because the bulk density used in this analysis assumed a moderately dense condition, compaction and bulk density change were ignored for the terrain surface deformation. The ground surface was discretized into cells with dx=dy= 5 mm on a side.

In this study, the objective is to predict the posterior behavior of a single wheel during straight-line traveling using a trained LSTM network, based on time-series data obtained up to the current moment. Accordingly, the time-series data used for training consisted of variables that could potentially be obtained in real time during actual vehicle operation, including velocity, slip ratio, sinkage, torque, and drawbar-pull.

Table 1
Specification of rigid wheel.

r	: Wheel radius	0.17	[m]
B	: Wheel width	0.09	[m]
M	: Wheel mass	15.7	[kg]
ω	: Angular velocity	7	[deg/s]

Table 2
Parameters of BWR model.

ρ	: Bulk density	1490	[kg/m^3]
k'_c	: Sinkage modulus influenced by cohesion	0	[-]
k'_ϕ	: Sinkage modulus influenced by friction	55	[-]
n	:Sinkage exponent	1	[-]
k_x	: Shear deformation modulus	0.015	[-]
κ	: Sinkage ratio	0.05	[-]
a_0	: Maximum stress angle constant	0.2	[-]
a_1	: Maximum stress angle constant	0.3	[-]
c	: Soil cohesion	0	[kPa]
ϕ	: Internal friction angle	32	[deg]

Table 3
Parameters of terrain surface deformation model.

r	: Rolling ratio	0.2	[-]
R_1	: Expansion and compression ratio	0.99	[-]
R_2	: Moving ratio	0.2	[-]
$k_{\max\_d}$	: Number of rolling repetitions	6000	[count/s]
d_{act}	: Interval of moving action	0.0005	[m]

LSTM networks have several hyperparameters that significantly influence model performance and behavior. Optimizing these hyperparameters plays a critical role in improving model accuracy. In what follows, we describe the hyperparameters used in this study. Three types of LSTM models were employed: one-to-one, many-to-many, and many-to-more-many.

The one-to-one model generates a single output from one input feature over a specified number of past time steps. In this structure, inter-feature relationships are not considered, and each feature is learned and predicted independently. The many-to-many model generates multiple outputs—equal in number to the input features—based on several features over a sequence of past time steps. This model takes into account the interrelationships among features for learning and prediction. The many-to-more many model generates a greater number of outputs than input features, also considering inter-feature relationships for learning and prediction. In this paper, we focus on a representative many-to-many model.

Table 4 summarizes the hyperparameters used in the many-to-many model. Note that the variable *time_steps* refers to the number of steps in the sequence, not the actual time interval. For instance, time_steps=3 means that the output is generated based on data from the past three time steps for each feature. This is because LSTM treats time as a feature dimension. In this study, the model was primarily trained to predict the subsequent single time step. One epoch refers to one complete iteration through the entire training dataset. Thus, $epoch = 100$ means that the model trains on the entire dataset 100 times. During each epoch, the model updates its weights based on the training data.

The batch size refers to the number of samples passed to the model at once. A larger batch size typically reduces training time per epoch. While batch sizes of 32, 64, or 128 are commonly used, we used $batch\_size = 64$. The learning rate determines the step size during the gradient descent update of weights. It controls the speed of convergence and is generally set between 0.1 and 0.0001. Here, we used $learning\_rate = 0.001$. The dropout rate defines the probability of randomly disabling LSTM units during training, preventing them from participating in weight updates. This helps avoid overfitting by reducing reliance on specific units. Dropout rates are typically set between 0.2 and 0.5; we used $dropout = 0.2$. The hidden layer size specifies the number of LSTM units, which corresponds to the dimensionality of the output vector at each time step. We set $hidden\_layer\_size1 = 100$, meaning the model outputs a 100-dimensional vector at each time step. When two hidden layers are used, the output of the first layer becomes the input to the second, allowing the model to learn more complex patterns. Hidden layer sizes are typically in the range of 32 to 512.

In addition, this study employs a technique called K-fold cross-validation for model training. When $n\_splits = 5$, the entire dataset is divided into five subsets. For each fold, one subset is used as the validation set and the remaining four as training data. The model is trained independently for each fold. The validation-to-training data ratio depends on the value of n_splits, which is 1:4 in our case. The results from each fold are stored, and the average of predictions from all folds is used during inference.

Table 4
Hyperparameters used for training deep-learning models.

time_steps	: Length of time series to input	3
epochs	: Number of study repetitions	100
batch_size	: Sample size of data	64
learning_rate	: Step size for updating weights	0.001
n_spilts	: Number of Folds (divisions)	5
hidden_layer_size1	: Size of hidden layer	100
hidden_layer_size2	: Size of hidden layer	50
dropout	: dropout ratio	0.2

3.3. Example of training data

In this study, a total of 190 sets of single-wheel traveling data were prepared for training the LSTM model. Here, we present a subset of the five time-series features used for training, along with the corresponding time-series data of the applied traction load.

Figure 1 shows an example of the training data. The figure illustrates the behavior of each time-series feature when an overshoot-type traction load was applied. In some cases, the data remain constant after a certain time point, which corresponds to the wheel becoming stuck. This was done to ensure consistent data length across all conditions. In addition to the example shown in Figure 1, other data sets were generated using various traction load patterns, including ramp and sinusoidal waveforms.

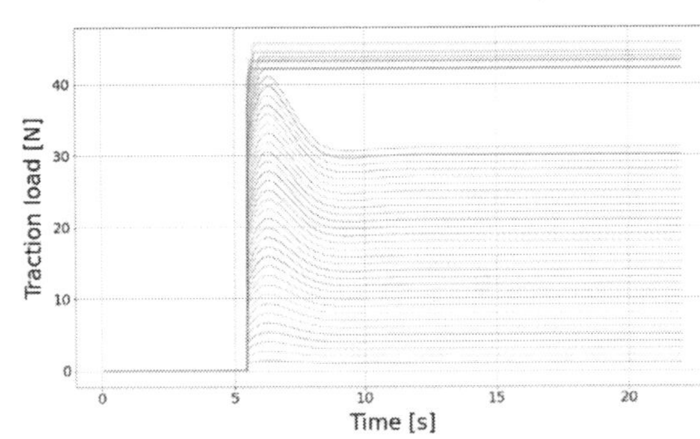

(a) Traction load

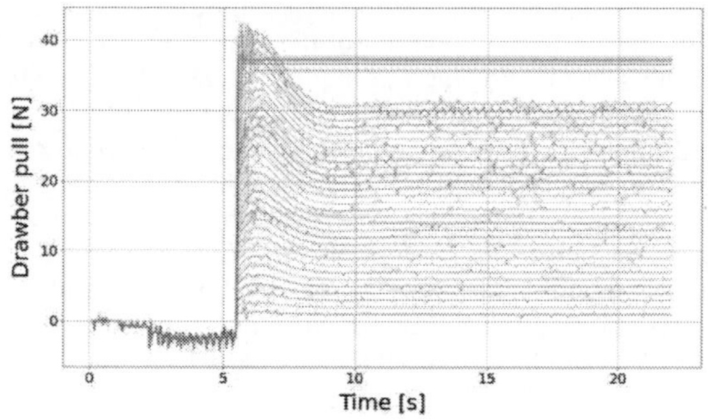

(b) Drawbar-pull

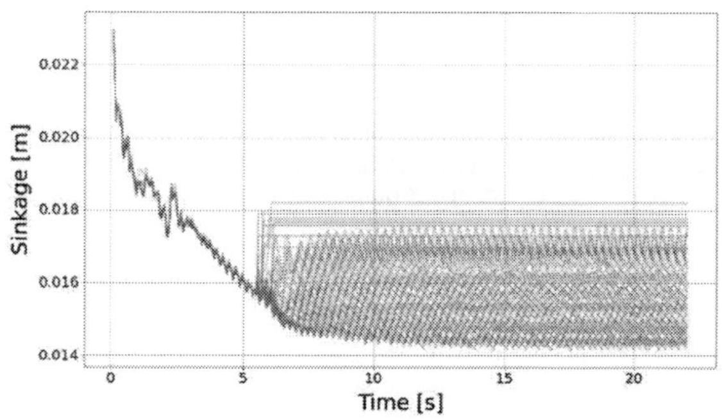

(c) Sinkage

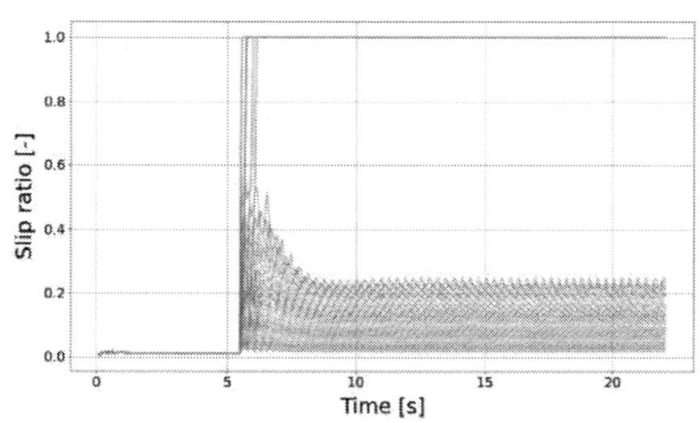

(d) Slip ratio

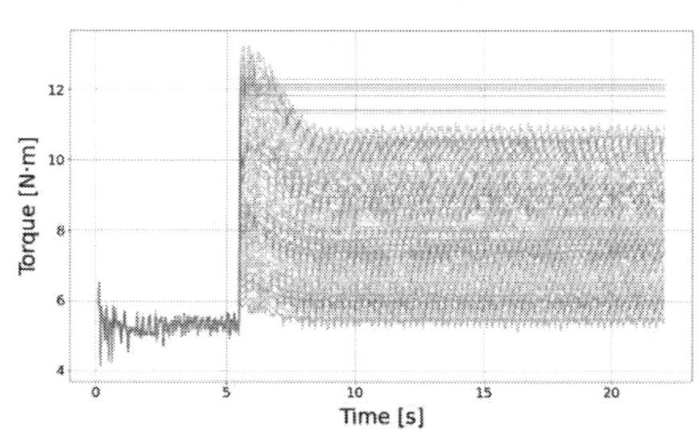

(e) Torque

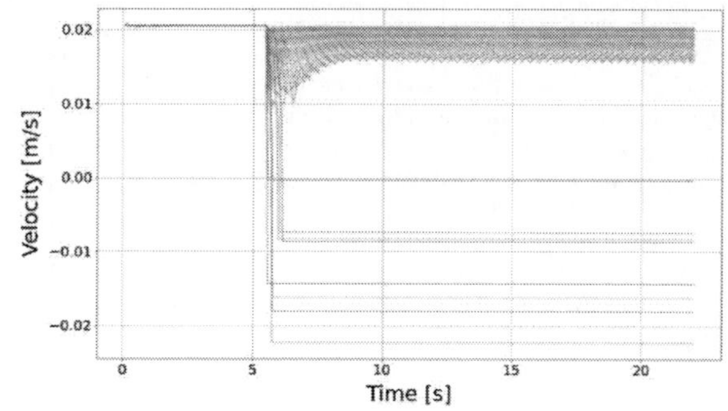

(f) Velocity

Fig. 1. Example of time-series training data.

4. Analysis Results

4.1. Loss Functions

In this study, three evaluation metrics were adopted to assess the training performance: Mean Absolute Error (MAE), Mean Squared Error (MSE), and R-squared (R^2).

Figure 2 presents the training results for slip ratio using the many-to-many LSTM model. In the LSTM model employed in this study, each fold was trained for 100 epochs, resulting in a total of 500 epochs per feature. As shown in the figure, the MAE, MSE, and R^2 values vary across folds. This variation arises from the differing characteristics (or tendencies) of the data assigned to each fold. In terms of overall loss function evaluation, all features show convergence within a relatively small number of epochs. Furthermore, the final values of the loss metrics are very close to 0 for MAE/MSE and close to 1 for R^2, indicating that the model was effectively trained.

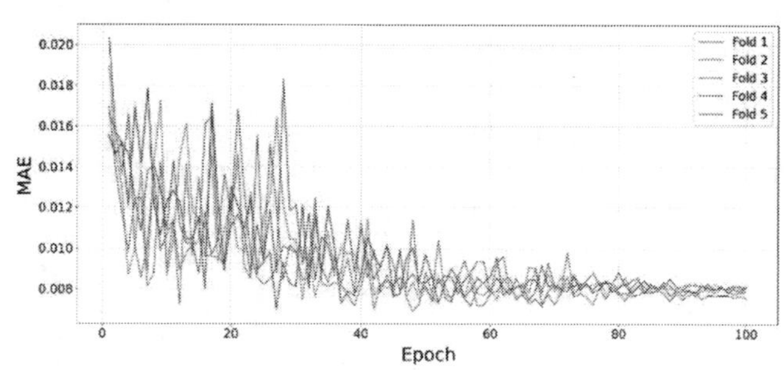

(a) MAE

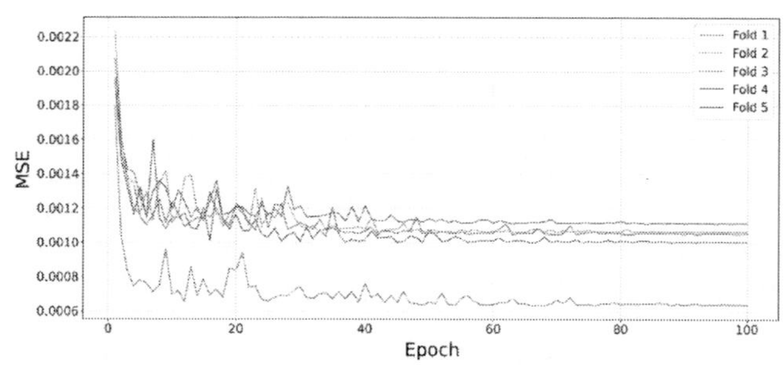

(b) MSE

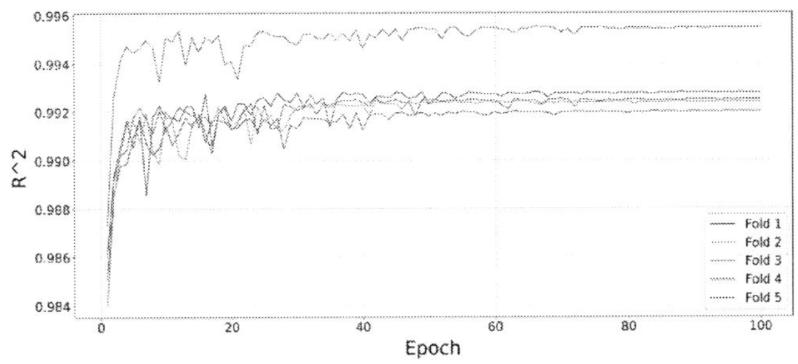

(c) R^2

Fig. 2. Variation of loss functions with epoch (slip ratio).

4.2. Predictions

This section presents the prediction results obtained using the LSTM network trained in Section 4.1. Figure 3 shows the predicted time-series variations of representative mobility metrics: drawbar-pull, slip ratio, and sinkage. The terrain conditions used for prediction correspond to uphill traveling scenarios based on terramechanics-based physical simulations with slopes of 5 degrees and 16 degrees. As seen in the figure, the trained LSTM network successfully performs the prediction for all evaluated trafficability indicators. Notably, the model trained using data simulating uphill travel was able to accurately predict actual uphill performance. Moreover, the one-step-ahead (posterior) predictions from the current time point matched the results of the physical simulations. In particular, in cases where the physical simulation results in wheel stuck conditions, the model demonstrated the ability to capture the transient responses of each indicator and predict the occurrence of a stuck event in advance.

Figure 4 presents the R^2 values for the predicted values compared to the terramechanics-based physical simulations. As shown, R^2 is close to 1 for all five trafficability indicators, indicating strong predictive performance. Additionally, results from other loss metrics further support the high prediction accuracy of the LSTM network.

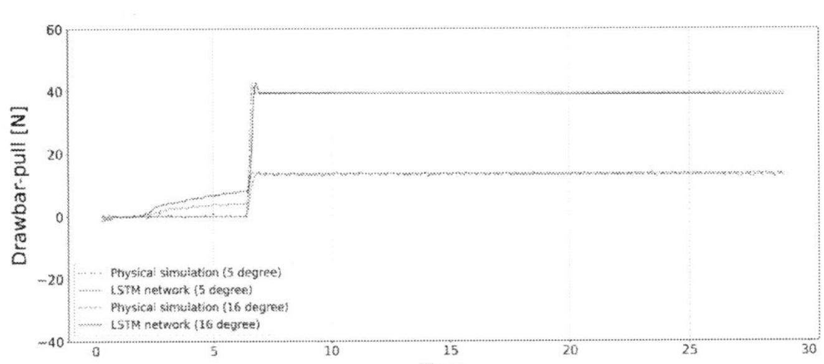

(a) Drawbar-pull

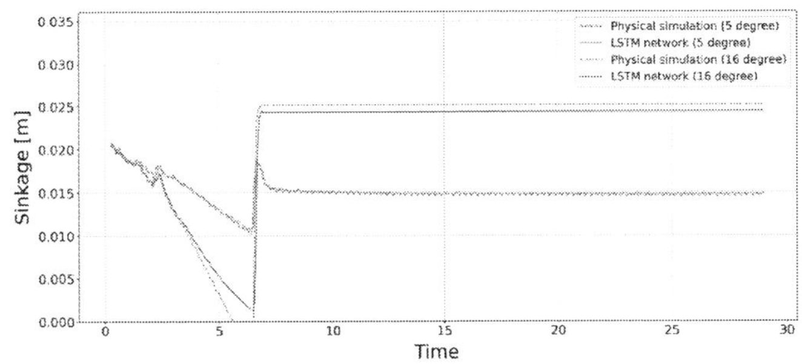

(b) Sinkage

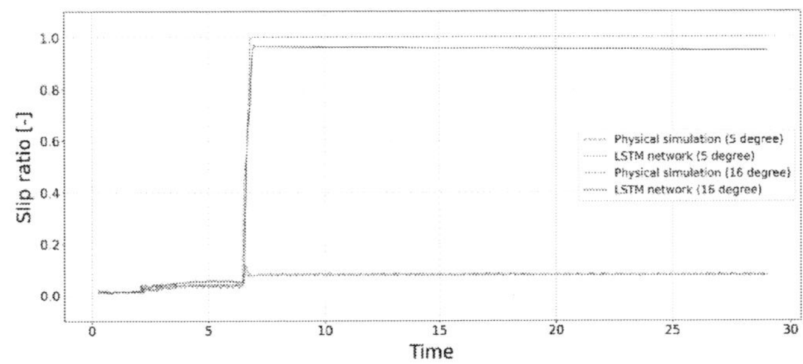

(c) Slip ratio

Fig. 3. Prediction of climbing 5 degrees and 16 degrees (Toyoura sand).

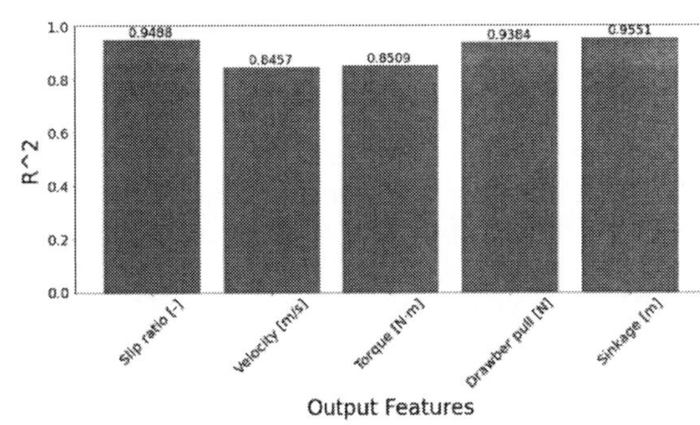

(a) 5 degrees

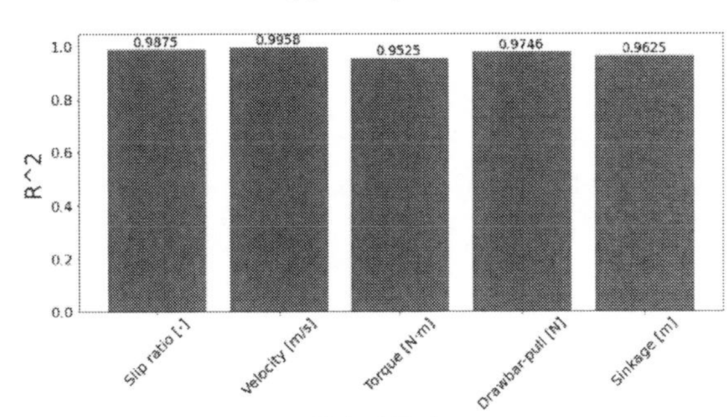

(b) 16 degrees

Fig. 4. Loss function R^2 for slope climbing of 5 degrees (Toyoura sand).

5. Conclusions and future work

In this study, as a preliminary step toward establish a real-time prediction system for rover traveling behavior using an LSTM model, we focused on predicting the single wheel traveling. All the soil parameters used in the training and validation datasets were based on Toyoura sand. To generate a large dataset, we simulated various traveling conditions such as uphill, downhill, and uneven terrain by applying different traction load patterns to a wheel traveling on a flat surface. The validity of this approach was confirmed by comparing the results with actual simulations of uphill travel.

Based on these simulations, we applied three types of LSTM models—one-to-one, many-to-many, and many-to-more many—to three validation datasets obtained from actual uphill driving on Toyoura sand. We then evaluated the differences in prediction accuracy and model characteristics. However, only a portion of the results was presented in this paper. The evaluation metrics used for the models were three loss functions: Mean Absolute Error (MAE), Mean Squared Error (MSE), and R-squared (R^2). As a result, all three models demonstrated sufficient prediction accuracy and were able to accurately predict the posterior behavior in actual uphill traveling. Among the models, the prediction accuracy followed the order: one-to-one > many-to-many > many-to-more many. This trend is considered to be due to the relatively simple nature of the dataset, where simpler models achieved better performance. However, in more complex situations, such as when a wheel traverses a rut and parameters change rapidly or behave non-linearly, the many-to-many model outperformed the one-to-one model in terms of prediction accuracy.

In this study, the LSTM model was used to predict previously computed simulation results. In the future, to achieve real-time prediction, we aim to perform prediction in parallel with ongoing numerical simulations and observation data, enabling the estimation of various trafficability indicators and behavior of the wheel in real time. In this study, the LSTM model was used to predict previously computed simulation results. In the future, to achieve real-time prediction, we aim to perform prediction in parallel with ongoing numerical simulations and observation data, enabling the estimation of various trafficability indicators and behavior of the wheel in real time. It should be noted that a measurement noise has great impact on the actual real-time data. LSTM predicts the behavior of the next step based on precursory phenomena, thus, when extending it to actual real-time data, filtering of the mobility data (Torque, velocity, etc.) becomes an important issue. Furthermore, when extending the present approach to real-time data, it will be necessary to be able to predict the overall behavior of rovers, which is our current challenge.

6. References

Applied Intuition, 2025. Off-road autonomy stack. https://www.appliedintuition.com/off-road-autonomy-stack (accessed 2025-05-13).

Arvidson, R.E., Iagnemma, K.D., Maimone, M., Fraeman, A.A., Zhou, F., Heverly, M.C., Bellutta, P., Rubin, D., Stein, N.T., Grotzinger, J.P., Vasavada, A.R., 2017. Mars science laboratory curiosity rover megaripple crossings up to Sol 710 in gale crater. J. Field Robotics 34, 495–518.

Chung, J., Gulcehre, C., Cho, K., Bengio, Y., 2014. Empirical evaluation of gated recurrent neural networks on sequence modeling. NIPS 2014 Workshop on Deep Learning.

Fediukov, V., Huhne, J., Dietrich, F., Buse, F., 2024. Uncertainty quantification for wheeled locomotion machine learning predictions on soft soil, in: Proceedings of the 21st International and 12th Asia-Pacific Regional Conference of the ISTVS, Yokohama, Japan, pp. 301–309.

Gonzalez, J., Yu, W., 2018. Non-linear system modeling using LSTM neural networks. IFAC-Pap. 51, 485–489.

Haibo G., Junlong G., Liang D., Nan L., Zhen L., Guangjun L., Zongquan D., 2013. Longitudinal skid model for wheels of planetary exploration rovers based on terramechanics, Journal of Terramechanics, Vol. 50, 327-343.

Hochreiter, S., Schmidhuber, J., 1997. Long short-term memory. Neural Comput. 9, 1735–1780.

ISECG Technology Working Group, 2019. Global Exploration Road Map Critical Technology Needs.

Ishigami, G., Miwa, A., Nagatani, K., Yoshida, K., 2007. Terramechanics-based model for steering maneuver of planetary exploration rovers on loose soil. J. Field Robotics 24, 233–250.

Ishigami, G., Nagatani, K., Yoshida, K., 2009. Slope traversal controls for planetary exploration rover on sandy terrain. J. Field Robotics 26, 264–286.

Junlong G., Liang D., Haibo G., Tianyou G, Zongquan D., Zhen L., 2017 Longitudinal skid model for wheels of planetary rovers based on improved wheel sinkage considering soil bulldozing effect, Journal of Terramechanics, Vol. 74, 45-56.

Kawakatsu, Y., Kuramoto, K., Usui, T., Sugahara, H., Ootake, H., Yasumitsu, R., Yoshikawa, K., Mary, S., Grebenstein, M., Sawada, H., Imada, T., Shimada, T., Ogawa, K., Otsuki, M., Baba, M., Fujita, K., Zacny, K., van Dyne, D., Satoh, Y., Tokaji, A., 2023. Preliminary design of Martian Moons eXploration (MMX). Acta Astronaut. 202, 715–728.

Kerr, R.A., 2009. Mars rover trapped in sand, but what can end a mission? Science 324, 998.

Kita, K., Arai, T., Tsuchiya, K., Kon, S., Katayama, M., Ozaki, S., 2024. Terramechanics analysis of climbing behavior of wheel against rigid obstacle on soft terrain. Discover Mechanical Engineering 3, 16.

Liao, Y., Tang, H., Li, R., Ran, L., Xie, L., 2023. Response Prediction for Linear and Nonlinear Structures Based on Data-Driven Deep Learning.

NASA technology taxonomy. https://techport.nasa.gov/view/taxonomy (As of, 2024).

Nishimori, H., Ouchi, N., 1993. Formation of ripple patterns and dunes by wind-blown sand. Terramechanics-Based Analysis of Autonomous Haulage Vehicle.

Shimizu, M., Ozaki, S., 2024. Traveling analysis of wheel for lunar exploration rover based on extended terramechanics model: Examination of similarity law of gravity, in: Proceedings of the 21st International and 12th Asia-Pacific Regional Conference of the ISTVS, Yokohama, Japan, pp. 220–226.

Staudemeyer, R.C., Morris, E.R., 2019. Understanding LSTM – a tutorial into Long Short-Term Memory Recurrent Neural Networks.

Small Lunar Rovers – NASA Technical Reports Server. https://ntrs.nasa.gov/ (accessed 2025-05-13).

Suzuki, H., Kawakami, H., Kobayashi, T., Ozaki, S., 2023. Extended terramechanics model for machine–soil interaction: Representation of change in the ground shape and property via cellular automata.

Tang, Z., Shi, Y., Wang, D., Feng, Y., Zhang, S., 2017. Memory visualization for gated recurrent neural networks in speech recognition. 2017 IEEE International Conference on Acoustics, Speech and Signal Processing (ICASSP), pp. 2736–2740.

Watanabe, Y., Nakano, S., Suzuki, H., Ozaki, S., 2023. Numerical analysis of wheel locomotion on soft soil using the extended terramechanics model based on cellular automata.

Wong, J.Y., 2008. Theory of Ground Vehicles, Fourth Edition. Wiley, Hoboken.

Wong, J.Y., Reece, A.R., 1967a. Prediction of rigid wheel performance based on the analysis of soil-wheel stresses. Part I. Performance of driven rigid wheels. J. Terramech. 4, 81–98.

Wong, J.Y., Reece, A.R., 1967b. Prediction of rigid wheel performance based on the analysis of soil-wheel stresses. Part II. Performance of towed rigid wheels. J. Terramech. 4, 71–25.

Zhou, F., Arvidson, R.E., Bennett, K., Trease, B., Lindemann, R., Bellutta, P., Iagnemma, K., Senatore, C., 2014. Simulations of mars rover traverses. J. Field Robotics 31, 141–160.

8969 | Survey of LiDAR effects for autonomous driving simulations

John Kaniarz [a,*], Brian Brady [b,*], Laura Walizer [c], Phillip Price [c], Mike Parker [b]

[a] *U.S. Army DEVCOM Ground Vehicle System Center, Detroit Arsenal, MI*

[b] *U.S. Army Engineer Research and Development Center (ERDC) Cold Regions Research and Engineering Laboratory, Hanover, NH*

[c] *U.S. Army ERDC Geotechnical and Structures Laboratory, Vicksburg, MS*

* *Corresponding authors: john.g.kaniarz.civ@army.mil, brian.r.brady@erdc.dren.mil*

ABSTRACT

A wide variety of software packages are available for modelling Light Detection and Ranging (LiDAR) sensors in the context of autonomous driving simulations. This report details ten effects found in point clouds generated by time-of-flight LiDAR units that can be modelled to increase realism in LiDAR simulations. Examples are provided from real-world data collected in the field, synthetic results generated in simulation, and published literature. Simulation users must carefully consider performance versus fidelity tradeoffs when selecting a suitable LiDAR simulation package for use.

Keywords
LiDAR
Simulation
Autonomous
Fidelity

1. Introduction

In a study of over 2 million vehicle collisions, the National Highway Transportation Safety Administration found that 94 percent of these accidents were due to driver error (Singh, 2018). Mistakes like these frequently result in fatalities. In 2022, 42,514 people were killed in motor vehicle crashes along roadways in the United States (Overview of Motor Vehicle Traffic Crashes in 2022). Autonomous Ground Vehicles (AGV) may be capable of reducing these fatalities by removing human decision making from the chain of events leading to an accident. Unfortunately, automobile manufacturers have not yet been able to produce a vehicle with "full driving automation" (On-Road Automated Driving Committee, 2021).

To reach this milestone developers must ensure that a vehicle can respond intelligently to a variety of high-risk edge cases. For example, how would the AGV react to oncoming traffic suddenly swerving into the lane ahead? Can it detect pedestrians and wildlife crossing into the path of travel? How well do the vehicle's perception systems function during adverse weather such as rain, snow, or fog?

It is difficult, time consuming, and expensive to replicate these edge cases in physical testing due to the 'curse of rarity' (Feng et al., 2023). A study by the RAND Corporation indicates that millions, if not billions, of road miles must be logged to collect statistically significant evidence proving that AGV algorithms meet safety requirements (Kalra and Paddock, 2016). As such, AGV designers are frequently turning to simulation to accelerate development. Researchers at Waymo utilized simulation to log over 20 billion virtual road miles (Waymo Self-Driving Cars).

There are many different AGV simulation packages available to log these virtual miles. Li et al. 2024 compiled a list of 50 different simulators developed between 1992 and 2022. These programs cover a broad spectrum of use cases including vehicle dynamics, traffic control, and, of interest to the present study, sensor performance.

Autonomous vehicles rely on a variety of sensors to perceive their surroundings and detect obstacles. This paper focuses on one modality commonly used by the automotive industry: Light Detection and Ranging (LiDAR). Major manufacturers, such as GM, Hyundai, and BMW have all incorporated LiDAR into their AGV perception toolkits (Zhao et al., 2019).

Many of these manufacturers rely on time-of-flight (ToF) LiDAR units. A ToF LiDAR unit fires a laser into the environment and records the time for the pulse to return. Distance between the transmitter and target is calculated from the measured time of flight and the known speed of light. By firing thousands of pulses across its field of view, a LiDAR system creates a three-dimensional "point cloud" of its surroundings. Royo and Ballesta-Garcia, 2019 provide details on the different types of LiDAR systems available and their operating principles.

A robust autonomous driving simulation must generate physically plausible data streams from the sensors used to perceive the virtual environment. Commercial entities such as Matlab (Simulation 3D Lidar), Ansys (Patterson, 2021), and NVIDIA (Lidar Validation: NVIDIA DRIVE Sim) all offer LiDAR simulation packages for sale. Within academia, Mississippi State University (Hudson et al., 2018) and MIT (Amini et al., 2021) both developed software to simulate LiDAR performance. Open-source projects such as CARLA (CARLA Simulator), Gazebo (Gazebo Velodyne Tutorial), and AirSim (Jansen et al., 2022) also have this capability. Within the government sector, the U.S. Army Corps of Engineers' Virtual Autonomous Navigation Environment (VANE)

https://doi.org/10.56884/O5B2MXY7

can simulate a variety of LiDAR makes and models (Carrillo et al., 2020).

The overall goal of this report is not to compare performance between the many different LiDAR simulation options available, but rather to highlight effects that a high-resolution LiDAR simulation should be capable of producing within the context of an autonomous driving simulation. The following sections describe ten of these effects. The data used to describe these phenomena was compiled from several sources. First, experiments were run using a real-world Velodyne HDL-32E to scan calibrated target panels and other objects commonly found in urban environments. The resulting point cloud files were analyzed with Velodyne's analysis software, VeloView (VeloView: The Velodyne Lidar Viewer). Other effects were demonstrated entirely within VANE simulations. Finally, several effects are described with results from published literature.

2. Calibrated Reflectivity

Two key data points captured by a ToF LiDAR system are distance and reflectivity. When calibrated, a reflectivity score in the range of 0-100 represents the energy returned from an idealized 0%-100% diffuse reflector at that range. Values above 100 may be returned by specular and retroreflective surfaces. All other factors being equal, targets with high reflectivity will bounce more energy back to the LiDAR's detector than a target with low reflectivity. This helps determines how "bright" or "dark" points will be displayed in the point cloud output. The brightness of a point is related to the amount of power collected by the LiDAR receiver, P_r, as calculated by the LiDAR Equation (Goodin et al., 2019; Rasshofer et al., 2011):

$$P_r = \frac{C_a \rho}{\pi x^2} \exp(-2\alpha x) \tag{1}$$

C_a is the LiDAR system constant, and ρ is the backscatter coefficient of the target. Distance from the sensor to the target is modeled by x. α represents the extinction coefficient of the medium through which the laser travels. Assuming a perfectly transmissive medium, the exponential term goes to unity. Laser angle of incidence on the target is accounted for by multiplying the right-hand side of Eq. 1 by the cosine of the angle θ between the laser's flight path and the target's outward normal. This reduces Eq. 1 to the following:

$$P_r = \frac{C_a \rho}{\pi x^2} \cos\theta \tag{2}$$

Equation 2 shows that P_r is strongly dependent on the distance to the target. Rather than directly use P_r to set the "brightness" of a point, a LiDAR implementing a calibrated reflectivity model like the Velodyne VLP-16 (VLP-16 User Manual) or Ouster OS1 (Ouster Firmware User Manual) uses proprietary firmware to report reflectivity values that are independent of the range to the target. These sensors accomplish this by scaling P_r according to distance.

Scaling like this can be implemented in a LiDAR simulation by defining P_n, the amount of power nominally expected at distance x from a perpendicular laser strike on a 100% reflective target:

$$P_n = \frac{C_a}{\pi x^2} \tag{3}$$

Calibrated reflectivity is determined by taking the ratio P_r/P_n which eliminates the influence of distance. Results from a LiDAR simulation demonstrating this effect are shown in Fig. 1.

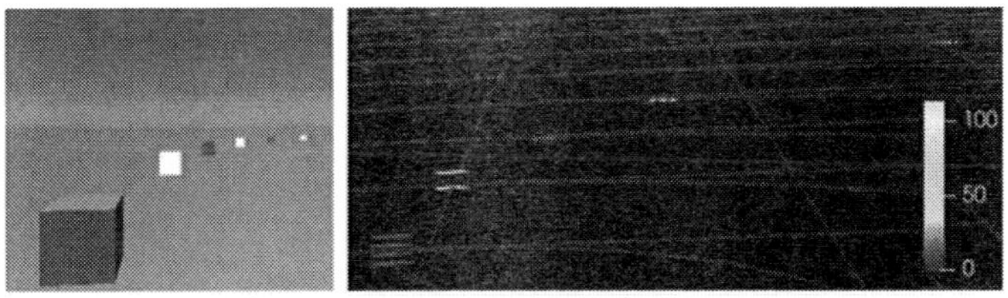

Fig. 1. Two images of a scene used for a sensor simulation. Left: RGB camera image of 3 dark (10% reflectance) cuboids and 3 light (100% reflectance) cuboids at varying distances. Right: point cloud of the cuboids captured by a LiDAR simulator with the third dark target not visible. The color scale corresponds to reflectivity. Note that reflectivity is the same across each set of targets, despite being placed at different ranges and thus reflecting different amounts of power to the receiver.

This scene contains 6 cuboid targets arranged at 10-meter intervals from the LiDAR. Targets alternate between dark (10% reflectance) and bright (100% reflectance). Each cuboid is oriented so that an outward normal on one face points directly at the LiDAR transmitter. This ensures the laser strikes the center of each target with a 90-degree angle of incidence. The point cloud shows that the center of all three bright panels have the same reflectivity, even though they are at different ranges and are reflecting different amounts of power back to the receiver. There is a noticeable reflectivity reduction on the edge of each panel, which will be discussed in the following section.

As with the bright panels, the centers of the two visible dark panels are at the same reflectivity. The most distant dark panel visible to the RGB camera is not visible to the LiDAR, even though its calibrated reflectivity is identical to that of the nearest dark panel. This phenomenon occurs because the amount of power reflecting from this target back to the receiver is not enough to exceed the sensor's minimum detectable power threshold, P_{min}:

$$P_{min} = \frac{0.9 C_a}{\pi x_{max}^2} \tag{4}$$

This is derived from Eq. 2 by assuming a perpendicular strike on a 90% reflective target at the sensor's published maximum detection range, x_{max} (Goodin et al., 2019). A target will appear in the point cloud with calibrated reflectivity P_r/P_n provided that P_r is greater than P_{min}.

The maximum detection distance of an automotive LiDAR unit typically falls between a range of roughly 20 to 300 meters. The Ouster OS1 can detect an 80% Lambertian target up to 170 meters away (OS-1 Datasheet). Velodyne models like the HDL-32 (HDL-32E User's Manual and Programming Guide) and Puck (Velodyne LiDAR Puck) are rated out to 100 meters. Under ideal conditions (no atmospheric attenuation or external radiation present) the maximum distance at which a given LiDAR can make a detection depends primarily on two factors: the reflectivity of its target and the angle of incidence at which its laser strikes. Figure 2 contains screenshots demonstrating how these effects can manifest in simulation.

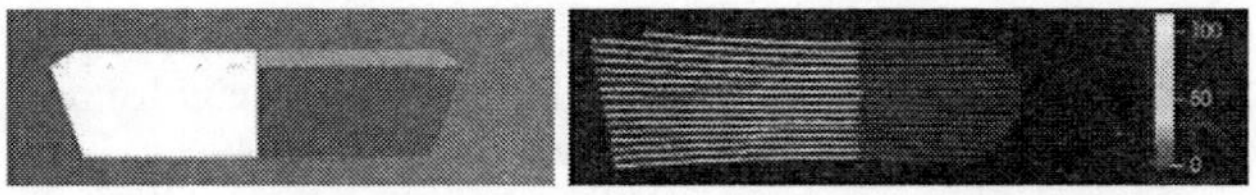

Fig. 2. Two images of a scene used for a sensor simulation. Left: RGB camera image of a composite wall constructed of two materials with differing Lambertian reflectivities. Right: point cloud of the same wall captured by a LiDAR simulator placed 30 meters perpendicular from the wall's centerline. The color scale corresponds to calibrated reflectivity. The reported reflectivity decreases as the beam moves off-center of the target. At the right extremity of the darker panel, the power returned to the sensor does not exceed the detection threshold.

The LiDAR can detect the entire width of the bright panel. Close to the centerline, the laser strikes the wall with an almost perpendicular angle of incidence. In this region the reflectivity measured by the LiDAR is nearly 100%, as shown by the bright yellow points in the cloud. As the beam traverses left across the bright panel, its angle of incidence decreases. In accordance with Eq. 2, this reduces the amount of energy that reflects from the bright surface back to the sensor, resulting in a lower output reflectivity (i.e. green points).

Only a fraction of the dark panel is visible in the point cloud. On the right side of the centerline, the beam reaches a critical angle of incidence where P_r falls below P_{min}. When this occurs, the wall disappears.

3. Reflectivity effects due to beam width and divergence

The laser departs the sensor with a finite width, not as a point source. The frustum diverges as a function of distance, which causes the beam "spot" to grow. The frustum geometry is hardware-specific and can vary between different makes and models of sensors. For example, Velodyne LiDAR models have a rectangular beam pattern composed of three parallel bands of light due to utilizing three stacked laser diodes per beam (VLP-16 User Manual), as shown in Fig. 3.

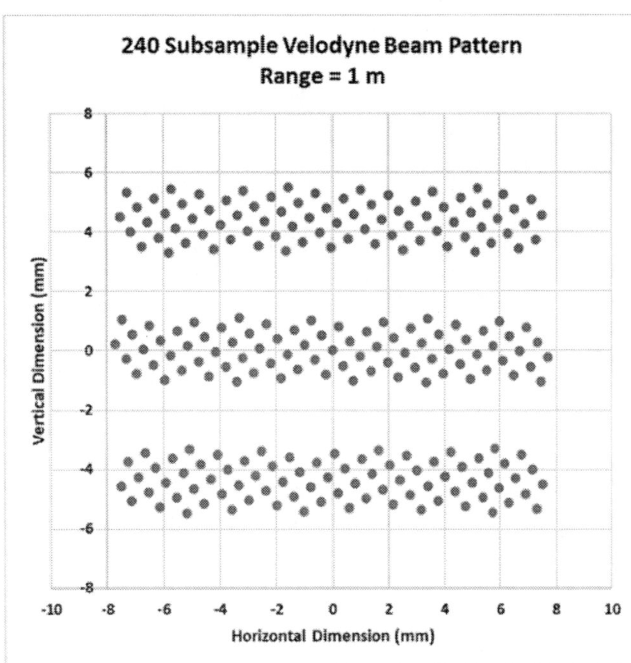

Fig. 3. Horizontal and vertical dimensions of a 240-subsample Velodyne beam pattern projected on a wall at a range of 1 meter.

In simulation, beam patterns like this can be modelled by tracing multiple subsample rays through the frustum's volume. In the case of the Velodyne pattern shown above, this was accomplished using three parallel rows of 80 ray subsamples.

Converting these subsamples to individual returns is conceptually simple. Convolve the samples (modeled as Dirac delta functions) with a filter kernel and find the peaks. But this technique is computationally expensive and impractical for simulation at interactive speeds. Instead, simulations may implement a fast approximation like the range gate algorithm described in Carrillo et al., 2020. After the signal is reconstructed, its peaks are converted to calibrated reflectivities using the equations described in Section 2.

There are some subtle effects that naturally fall out from this model. In cases where the beam pattern partially hits a target, the subsamples that miss contribute a reflectivity value of 0 to the calculation. This causes the overall calibrated reflectivity measured by the beam to be less than the target's nominal reflectivity. This effect is visible on the vertical edges of the target panels displayed in Fig.1, which are magnified below in Fig 4.

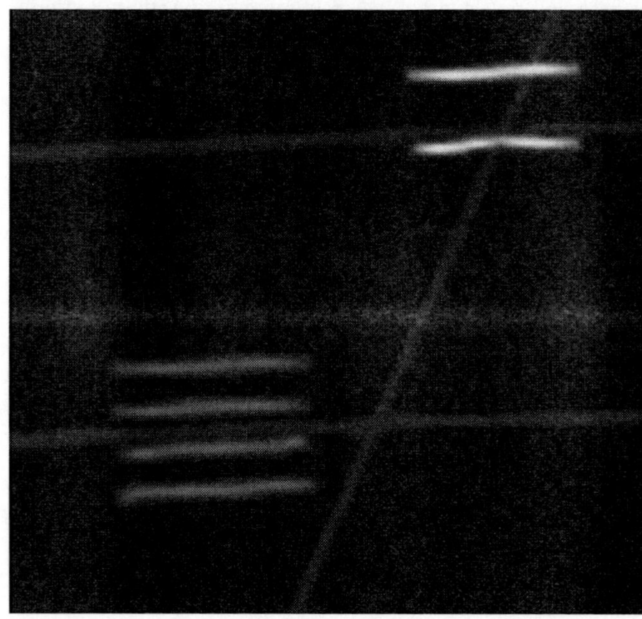

Fig. 4. Close-up of the two nearest panels depicted in Fig. 1. Points along the edges of the targets have a lower reflectivity due to a beam pattern partial hit.

The width of this edge effect is directly proportional to the divergence of the laser beam. Figure 5 depicts two panels located at identical distances from the LiDAR transmitter. The beam pattern used to measure the top panel has a divergence three times that of the laser used to measure the bottom panel. This ratio corresponds to the widths of the zone of reduced reflectivity along the panels' left and right edges. Both scans utilize a beam pattern containing the same number of sub-samples.

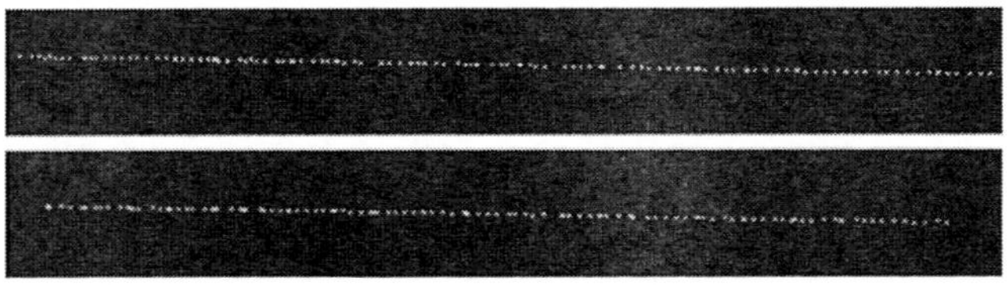

Fig. 5. LiDAR scan of two identical panels located the same distance from the transmitter. Top: scan with laser divergence of 6 milliradians. Bottom: scan with divergence of 2 milliradians. Note how a large beam divergence results in a large region of reduced reflectivity along the edge of the panel.

Although both panels in Fig. 5 have identical physical dimensions, the panel width measured in the top scan is slightly wider than that measured by the bottom scan. This phenomenon occurs because the single "point" reported by the LiDAR for each detection corresponds to the location of the central ray in the beam frustum. The "point" returned by the LiDAR on the far left and right edges of the panel contains only a few sub-samples that physically intersect the target. Most of the subsamples making up these "points" (including the central most sample) pass through empty space on the side of the target. The distance between the central ray of the beam pattern and its left/right extremities causes the measured width of the panel to increase with increasing beam divergence.

A related effect arises when the beam pattern transitions between "bright" and "dark" zones on the same target, like the area near the midpoint of the composite wall shown in Fig 2. The following figure zooms in on one laser swath as it passes through the midpoint region. Four different scans are depicted. Each scan was made with identical beam divergences, but at varying LiDAR head rotation rates.

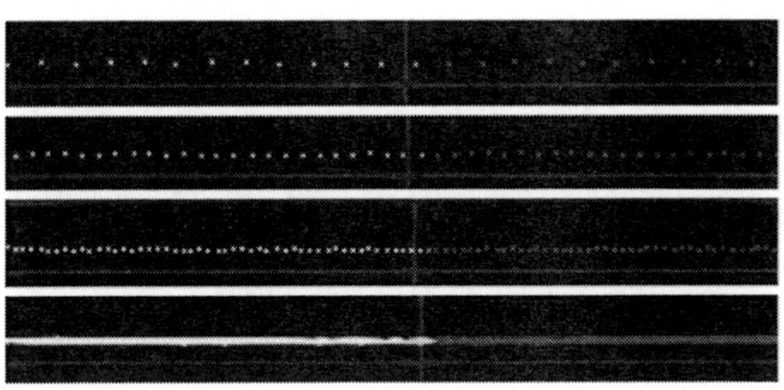

Fig. 6. Close-up of one laser swath crossing between the bright and dark panels shown on Fig. 2. Rotational scanning speeds from top to bottom: 600, 300, 15, and 6 RPM, respectively. Reflectivity averaging is visible as the beam spot transitions between the bright and dark panels.

When beam pattern's subsamples simultaneously intersect both the yellow 100% region and blue 10% region, the output reflectivity displayed in the point cloud averages out to a green 55%. The width of the green transition region is proportional to width of the beam pattern when it reaches the panel. The distance between each point depends on the frequency at which the laser is fired and the speed at which the LiDAR head is rotating. Each of the scans in Fig. 6 were made using identical firing frequencies. The distance between points is entirely due to the rotation rate of the sensor head. As the rotation rate decreases the density of points in the transition zone increases. At very slow rotational velocities the true width of the beam pattern is revealed.

4. Return Modes

Since each laser pulse is a finite width frustum, it is possible that a single beam could intersect several targets at different ranges. Some LiDAR units can harness this effect and output multiple returns for each shot fired. Velodyne models have three selectable modes available for how multiple returns are processed: "Last," "Strongest," and "Dual." When operating in "Last" mode, only the most distant reflection is populated to the point cloud. In "Strongest" mode, only the reflection with the largest signal strength is populated. In "Dual" mode, both the "Last" and "Strongest" reflections are output (VLP-16 User Manual). Ouster LiDAR units have "Single" and "Dual" modes. In each mode, returns can be sorted by distance (near to far, or far to near), or by signal strength (strongest to weakest) (Ouster Firmware User Manual). Another approach, provide by the Luminar Iris, offers only one mode of up to six returns per pulse in distance order irrespective of intensity (Iris Product Information Guide). Accurate LiDAR simulations should provide users with configuration options that mirror those available on their real-world counterparts.

Unfortunately, the user manuals for automotive LiDAR units do not always contain enough information to reverse-engineer exactly how these configuration parameters should operate. In the case of the OS1, the firmware manual states that the "strongest to weakest" configuration option "prioritizes the points based on their signal strength, with the strongest signals coming first. This can be useful in applications where identifying the most prominent or reflective objects is crucial" (Ouster Firmware User Manual). This sentence is ambiguous since signal strength as measured by P_r (or raw photon counts in the case of a physical OS1) is different from the calibrated reflectivity output by the sensor. The OS1 determines reflectivity by scaling the number of measured "sensor signal photons" by distance to the target (Ouster Firmware User Manual).

The definition of "signal strength" plays a crucial role in determining which points are selected for output. Consider a scenario in which the OS1 is set to operate in "single" return mode with points sorted according to the "strongest to weakest" option. Now assume a relatively wide beam hits two targets with the same shot: a "dark" target at close range and a "bright" target at long range, as shown below in Fig. 7.

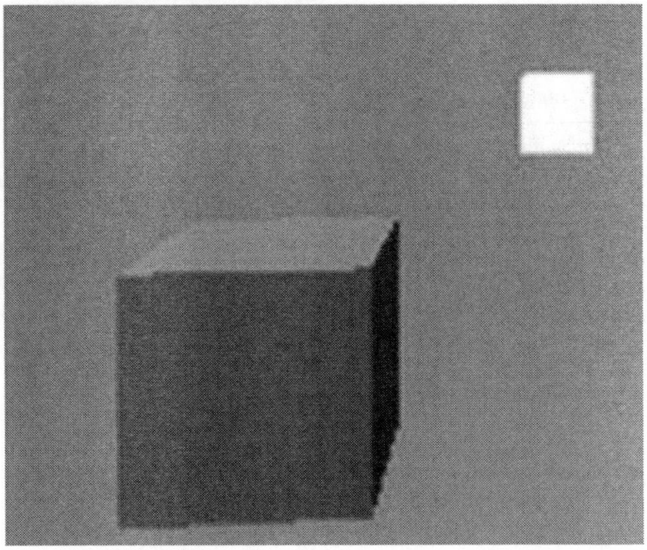

Fig. 7. Hypothetical scenario in which a laser hits two targets in the same pulse. The close target has a low Lambertian reflectance, and the far target has a high Lambertian reflectance. In a single return mode based on signal strength, will the point output by the sensor correspond to the dark surface or the bright surface?

The OS1 should always measure a higher calibrated reflectivity on the "bright" target than the "dark" target. But depending on the distance between the panels, it is possible the "dark" panel reflects more signal photons. In this case, if "signal strength" is defined by photon counts, the "dark" panel will be the single return populated to the point cloud. If "signal strength" is defined by calibrated reflectivity, the "bright" panel will be the only return. Physical testing is required to verify which interpretation Ouster has implemented.

5. Near Field Effects

Publicly available user manuals may not contain much information about how a LiDAR processes detections occurring near the receiver. If the transmitter and receiver are not coaxially located, a "dead zone" may develop around the sensor. In this region, there is no overlap between the transmitter and receiver's fields of view, and detections are not possible (Hahner et al., 2021). Fig. 8 depicts this geometry.

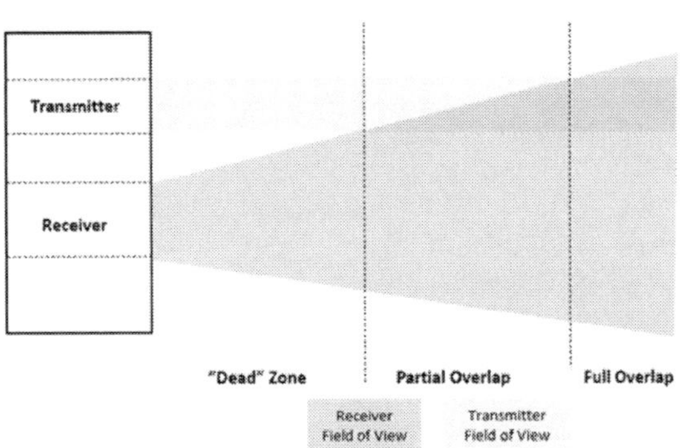

Fig. 8. "Dead" zone where receiver field of view does not overlap with transmitter field of view. Figure adapted from Hahner et al. 2021.

Typically, this "dead" zone is located within a few meters of the sensor. In the case of the SICK TIM781S, it only extends 5 cm from the transmitter (TIM781S Datasheet). Ouster OS1 units running firmware prior to version 3.1 are "subject to a minimum range limitation dictated by the hardware" (Ouster Firmware User Manual). The latest Ouster firmware updates allow the user to adjust the OS1's minimum detection distance between 0 and 50 centimeters. The Velodyne HDL-32E manual states that the sensor's minimum detection distance is 1 meter and returns closer than this should be ignored (HDL-32E User's Manual and Programming Guide).

Figure 9 displays LiDAR near-field data collected with a real-world HDL-32E. An 80% reflective target panel was positioned 0.65 meters from the transmitter. The LiDAR was activated, and the panel was pulled directly away from the sensor. At the closest ranges, the LiDAR a measured a calibrated reflectance of 100% over most of the panel's surface. As distance increased beyond one meter, the calibrated reflectivity reported by the sensor dropped to 76% almost uniformly across the panel.

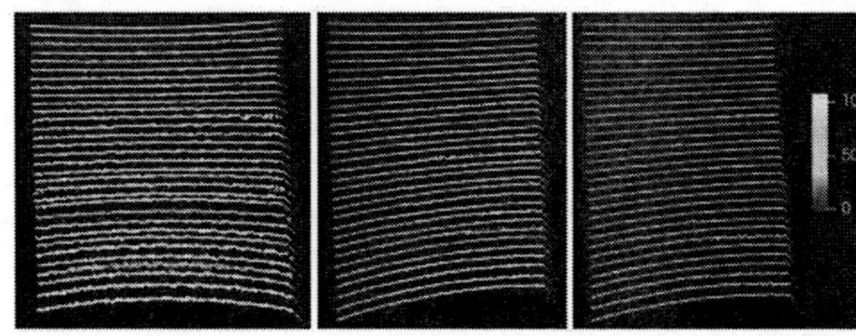

Fig. 9. Field data collected with a real-world HDL-32E of a target panel with an 80% Lambertian reflectivity at three different distances. From left to right the distances are 0.65, 0.85, and 1.04 meters respectively. The target appears brighter when located in the partial overlap/dead zone of the LiDAR.

Figure 9 provides evidence that the HDL-32E modifies its calibrated reflectivity algorithm at ranges of less than one meter. Patent documents for the Velodyne HDL-64 suggest this LiDAR unit behaves in a similar manner and has a built-in feature to ignore early reflections (Hall, 2011).

These near-field effects become particularly important during adverse weather simulations. If the sensor removes near-field detections from the point cloud, this could result in the formation of a torus pattern of "soft" returns from precipitation particles (Hahner et al. 2021). Figure 10 demonstrates this effect in a virtual scene filled with falling snow.

Fig. 10. Torus pattern of "soft" returns from snowflakes centered around the LiDAR transmitter.

6. Split Wall Effect

When the energy of the beam is spread across two surfaces, a LiDAR will return the brighter surface or both when in a multiple return mode (VLP-16 User Manual, Ouster Firmware User Manual). For this to work correctly, the surfaces must be distinguishable by the LiDAR. Surfaces closer than this distinction threshold will make the discontinuity indistinguishable from surface roughness. Our testing of the VLP-16 shows that this is not a hard cutoff. The error rate is highest with less than a meter of separation, gradually decreasing as the distance increases from 1 meter to 2 meters, after which no errors are observed. When the LiDAR fails to distinguish the two surfaces it will return a single point floating in space between the surfaces with a distance biased by the fraction of laser energy reflected by each surface. This phenomenon is sometimes referred to as "mixed pixels" or "ghost points" (Carrillo et al. 2020). An example of this effect is shown below in Fig. 11. This point cloud depicts two rectangular target panels with false returns appearing between them.

Fig. 11. Mixed pixels between two target panels.

The distinction threshold as measured in experimentation is a combination of a base distinction threshold defined by hardware and numerous geometry smoothing effects discussed in the subsequent sections of this report. Two contributors that we have identified in Velodynes are its band-pass filter and the pulse shape of the laser, both of which "blur" the energy over time signal. This effect naturally occurs when performing a full signal reconstruction as discussed in Section 3, but fast approximation algorithms might not reproduce this effect.

7. Corner rounding

In a LiDAR point cloud, a sharp corner will appear rounded. Imagine a LiDAR with a thick square beam and no divergence scanning a corner at an infinite sampling rate. In cross-section, this is equivalent to a convolution of a rectangle (the beam) and a triangle (the corner) which is effectively a low-pass filter of the geometry. Compared with the split wall effect, this smoothing occurs across space, while the split wall has smoothing over time. This effect is most prominent when the corner is viewed at a 45° angle which balances the energy on each side. At sharper angles Lambertian reflectance will dim one side more than the other which biases it toward the ideal result. As this effect is caused by beam width, the radius will increase with range due to beam divergence increasing the width. However, the direct effects of divergence are negligible at any given distance.

Figure 12 shows an example of this phenomenon with real-world data captured with a Velodyne HDL-32. This effect is predominantly created by modeling the beam as a frustum and does not require any artifact-specific modelling.

Fig. 12. HDL-32 corner rounding. (a) experiment set up; (b) point cloud output].

8. Long Hallway Curvature

In the paper "Lidar Measurement Bias Estimation via Return Waveform Modelling in a Context of 3D Mapping," Leconte et al., 2019 describe both the phenomenon of hallway walls appearing curved in LiDAR point cloud data and a means of correction. In summary, the divergence of the beam at low angles of incidence compounded by Lambertian reflectance causes the energy peak to arrive earlier than would be expected by an idealized intersection of the centerline of the beam. As the beam moves down the hallway, the angle of incidence becomes more acute thus increasing the distortion with range. This effect can naturally occur in simulations that model the beam as a frustum and account for the reflectivity of surface materials.

9. Electronics Error

LiDAR data sheets typically report the overall range accuracy of the sensor. Here is a list of phenomena that we've identified that contribute to the variance in distance returned by the LiDAR.

- Laser pulse timing – both variance in the analog circuitry releasing the energy and quantization error in the clock requesting the pulse.
- Detector clock quantization.
- Head angle measurement error – measuring a spot slightly to the side.

The combined effect of these errors is visible in analysis of experimental data. In a test case, a 10% reflective target was set up 10 meters from a Velodyne HDL-32E LiDAR unit. LiDAR returns from the panel were isolated from the raw data, and the distance between the reported location of the returns and the actual surface of the panel was computed. Distributions of these distance errors were then compiled and fit to a Gaussian curve. An example of one such distribution and corresponding curve fit is shown in Fig. 13.

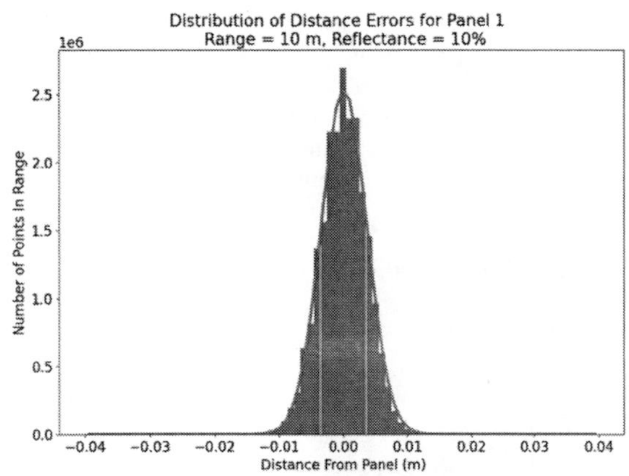

Fig. 13. Distribution of distance errors found for returns from a 10% reflectance panel set 10 m from the LiDAR unit.

On average, the confidence interval (twice the standard deviation) was found to be 1 centimeter for this experiment, which is in line with the manufacturer's specifications. This demonstrates that intrinsic LiDAR errors can be detected with analysis and analytically described in a way that allows for incorporation into LiDAR simulations.

10. Specular Lucky Shot

When a mirror is placed in the path of the beam and an object is in the reflection, the distance returned will be *d(LiDAR, mirror) + d(mirror, object)*. When this distance is later converted to a point in space the decoder will not know about the mirror and place the point behind the mirror at the measured distance along centerline of the beam (Henley et al. 2023).

While plate glass mirrors are uncommon on the road, specular surfaces are common on vehicles with glossy paint or chromed parts. Thankfully, for practical purposes, these surfaces are often curved which spreads the beams energy enough to weaken the reflected returns beyond detection. However, the retroreflective surfaces of roadway signage and work zone safety devices can create a reflected return more intense than the return from the primary surface thus creating false returns behind the specular surface. Generally speaking, retroreflective surfaces or mirror-like glass are more commonly found in cities than in nature. Therefore, modelling this effect is of higher importance in urban settings and of lesser importance in off-road simulations.

11. Single Photon Avalanche Diode (SPAD) Error

Solid state LiDAR units, such as Ouster models, use SPADs which measure the arrival time of single photons at picosecond accuracy. They also introduce two forms of error. The first is their high sensitivity results in false detections. Dark counts are false detections due to thermal noise, and after-pulsing is a false detection due to trapped charge in the silicon. The second occurs when the avalanche circuitry is recharging. Any photons arriving during recharge period are missed. This creates non-linearity in the energy response; as the total energy gets brighter, a greater percentage is lost (Erfanian et al., 2024).

12. Conclusion

This study highlights ten effects generated by real-world LiDAR units that a high-fidelity simulation should be capable of reproducing. LiDAR simulation users should consider what level of fidelity is required for their desired application. For example, an off-road driving simulation may not need to capture the effects of long hallway curvature or corner rounding, but these effects may be critical for simulations set inside a logistics warehouse.

Consideration must also be given to how an autonomous navigation algorithm is programmed to interpret LiDAR measurements. Consider a sensor with multiple return modes, such as the Luminar Iris discussed in Section 4. If the Iris' associated AGV is only programmed to read the closest measurement generated with each pulse, then there is no need to simulate the remaining five farthest returns. Likewise, some algorithms may only process the distance to each measurement and ignore reflectivity effects entirely.

In many cases, autonomous vehicles make decisions based on simultaneous input from multiple sensor modalities. Consider an example in which the AGV algorithm uses LiDAR data for long-range obstacle detection while relying on inputs from other sensors for close ranges. This could eliminate the need for modelling the near-field LiDAR effects discussed in Section 5.

Many of the effects discussed in this paper require the laser pulse to be modelled as a finite-width frustrum composed of multiple ray subsamples. As subsample count increases, so does the computational workload required for raytracing and signal processing. This frequently results in a performance versus fidelity tradeoff. If the intent of the simulation is to generate synthetic LiDAR data to feed a real-time AGV algorithm, hardware performance may limit the maximum number of subsamples and the type of effects which can be produced. If real-time performance is not required (perhaps in the case of synthetic training data generation), restrictions on subsample counts may not be necessary.

Due to the wide variety of use cases for autonomous vehicles, it is difficult to rank the effects discussed in this paper in order of relative importance. Engineering judgement must be used to compare the virtual sensor in use to the scene in which it operates, the navigation algorithm to which it is a connected, and the performance limitations of the associated simulation hardware. It is also challenging to define numeric limits on how accurately each effect should be reproduced in simulation. In some cases, it may be possible to find this data in the sensor's operating manual (refer to the HDL-32E electronics error discussed in Section 9). In most cases, accuracy limits are specific to each make and model of sensor and must be determined from empirical data. Users must balance these factors to select software capable of delivering an appropriate level of realism.

13. Declaration of Competing Interest

The authors declare that they have no known competing financial interests or personal relationships that could have appeared to influence the work reported in this paper.

14. References

Amini, A., Wang, T.-H., Gilitschenski, I., Schwarting, W., Liu, Z., Han, S., Karaman, S., Rus, D., 2021. VISTA 2.0: An Open, Data-Driven Simulator for Multimodal Sensing and Policy Learning for Autonomous Vehicles. arXiv. https://doi.org/10.48550/ARXIV.2111.12083.

"CARLA Simulator." Accessed October 22, 2024. http://carla.org.

Carrillo, J., Goodin, C., Fernandez, J., 2020. Sensor and Environment Physics in the Virtual Autonomous Navigation Environment (VANE). Engineer Research and Development Center (U.S.), August 27. https://doi.org/10.21079/11681/37968.

Erfanian, A., Rahmanpour, M., Khaje, M., Afifi, A., Fahimifar., M., Reduction of Afterpulse and Dark Count Effects on SPAD Detectors Using Processing Methods. Results in Optics 15: 100617. https://doi.org/10.1016/j.rio.2024.100617.

Feng, S., Sun, H., Yan, X., Zhu, H., Zou, Z., Shen, S., Liu, H.X., 2023. Dense Reinforcement Learning for Safety Validation of Autonomous Vehicles. Nature 615(7953): 620–27. https://doi.org/10.1038/s41586-023-05732-2.

Gazebo Velodyne Tutorial, n.d. Accessed October 22, 2024. https://classic.gazebosim.org/tutorials?tut=guided_i1.

Goodin, C., Carruth, D., Doude, M., Hudson, C., 2019. Predicting the Influence of Rain on LIDAR in ADAS. Electronics 8(1): 89. https://doi.org/10.3390/electronics8010089.

Hahner, M., Sakaridis, C., Dai, D., Van Gool, L., 2021. Fog Simulation on Real LiDAR Point Clouds for 3D Object Detection in Adverse Weather. arXiv. https://doi.org/10.48550/ARXIV.2108.05249.

Hall, D., 2011. United States Patent No. US 7,969,558 B2, June 28. https://patentimages.storage.googleapis.com/3c/dd/91/8fd123ec4300f7/US7969558.pdf.

"HDL-32E User's Manual and Programming Guide. 2011. Velodyne LiDAR Inc., April. https://archive.org/details/manualzilla-id-6960639.

Henley, C., Somasundaram, S., Hollmann, J., Raskar, R., 2023. Detection and Mapping of Specular Surfaces Using Multibounce LiDAR Returns. Optics Express 31(4): 6370. https://doi.org/10.1364/OE.479900.

Hudson, C., Goodin, C, Doude, M., Carruth, D.W., 2018. Analysis of Dual LIDAR Placement for Off-Road Autonomy Using MAVS. 2018 World Symposium on Digital Intelligence for Systems and Machines (DISA), 137–42. Kosice: IEEE. https://doi.org/10.1109/DISA.2018.8490620.

IRIS Product Information Guide, 2021. Luminar Technologies, Inc. https://www.luminartech.com/contact

Jansen, W., Huebel, N., Steckel, J., 2022. Physical LiDAR Simulation in Real-Time Engine. arXiv, December 19. http://arxiv.org/abs/2208.10295.

Kalra, N. Paddock, S., 2016. Driving to Safety: How Many Miles of Driving Would It Take to Demonstrate Autonomous Vehicle Reliability? Research Report. RAND Corporation, April 12. https://www.rand.org/pubs/research_reports/RR1478.html.

Laconte, J., Deschênes, S.-P., Labussière, M., Pomerleau, F., 2019. Lidar Measurement Bias Estimation via Return Waveform Modelling in a Context of 3D Mapping. 2019 International Conference on Robotics and Automation (ICRA), 8100–8106. https://doi.org/10.1109/ICRA.2019.8793671.

Li, Y., Yuan, W., Zhang, S., Yan, W., Shen, Q., Wang, C., Yang, M., 2024. Choose Your Simulator Wisely: A Review on Open-Source Simulators for Autonomous Driving. IEEE Transactions on Intelligent Vehicles 9(5): 4861–76. https://doi.org/10.1109/TIV.2024.3374044.

Lidar Validation: NVIDIA DRIVE Sim, n.d. Accessed October 22, 2024. https://images.nvidia.com/aem-dam/en-zz/Solutions/self-driving-cars/NVIDIA-DRIVE-Sim-Lidar-Validation-whitepaper.pdf.

On-Road Automated Driving Committee, 2021. Taxonomy and Definitions for Terms Related to Driving Automation Systems for On-Road Motor Vehicles. SAE International. https://doi.org/10.4271/J3016_202104.

OS1 Datasheet, 2024. Ouster Inc., June. https://data.ouster.io/downloads/datasheets/datasheet-rev7-v3p0-os1.pdf.

Ouster Firmware User Manual, 2024. Ouster Inc., July 29. https://data.ouster.io/downloads/software-user-manual/firmware-user-manual-v3.1.0.pdf.

Overview of Motor Vehicle Traffic Crashes in 2022, 2024. Research Note. Traffic Safety Facts. National Highway Traffic Safety Administration, June. https://crashstats.nhtsa.dot.gov/Api/Public/ViewPublication/813560.

Patterson, S., 2021. Get High-Fidelity Optical Edge Case Analyses with Lidar Simulation, June 10. https://www.ansys.com/blog/get-high-fidelity-optical-edge-case-analyses.

Rasshofer, R.H., Spies, M., Spies, H., 2011. Influences of Weather Phenomena on Automotive Laser Radar Systems." Advances in Radio Science 9, 49–60. https://doi.org/10.5194/ars-9-49-2011.

Royo, S., Ballesta-Garcia, M., 2019. An Overview of Lidar Imaging Systems for Autonomous Vehicles." Applied Sciences 9(19): 4093. https://doi.org/10.3390/app9194093.

Simulation 3D Lidar, n.d. Accessed October 22, 2024. https://www.mathworks.com/help/driving/ref/simulation3dlidar.html.

Singh, S., 2018. Critical Reasons for Crashes Investigated in the National Motor Vehicle Crash Causation Survey. Statistical Summary. Traffic Safety Facts. National Highway Traffic Safety Administration. https://crashstats.nhtsa.dot.gov/Api/Public/Publication/812506.

TIM781S Datasheet, 2024. SICK Sensor Intelligence, August 29. https://cdn.sick.com/media/pdf/9/49/149/dataSheet_TIM781S-2174104_1096363_en.pdf.

Velodyne LiDAR Puck, 2015. Velodyne Acoustics, Inc. https://navysbir.com/n23_2/N232-080-VLP16.pdf.

VeloView: The Velodyne Lidar Viewer, n.d. Accessed October 22, 2024. https://www.paraview.org/veloview/.

VLP-16 User Manual." Velodyne Lidar, March 7, 2022. https://data.ouster.io/downloads/velodyne/user-manual/vlp-16-user-manual-revf.pdf.

"Waymo Self-Driving Cars, n.d. Accessed April 22, 2025. https://waymo.com/waymo-driver/.

Zhao, F., Jiang, H., Liu, Z, 2019. Recent Development of Automotive LiDAR Technology, Industry and Trends. Eleventh International Conference on Digital Image Processing (ICDIP 2019), edited by Xudong Jiang and Jenq-Neng Hwang, 178. Guangzhou, China: SPIE. https://doi.org/10.1117/12.2540277.

9165 | Simulation of the relationship between the decreased bearing capacity while imparting vibration to ground and vibration frequency in the discrete element method

Tomohiro Watanabe [a,*], Dai Watanabe [b], Kojiro Iizuka [b]

[a] *Department of Engineering, Mechanical Engineering Program, Niigata University, Niigata City, Niigata Prefecture, Japan*

[b] *Department of Science and Engineering, Machinery and Control Systems, Shibaura Institute of Technology, Saitama City, Saitama Prefecture, Japan*

* *Corresponding author: watanabe@eng.niigata-u.ac.jp*

ABSTRACT

The demand for exploring celestial bodies is increasing, and many exploration missions have been conducted and planned. In recent years, moving underground has been focused on as a method of exploring robots' movement. Exploration rovers have difficulty moving underground because the drag force from the ground restricts their movement, hindering underground exploration. A previous study developed a mechanism that moves underground by imparting vibration to the ground and decreasing the drag force. The experimental results demonstrate the superiority of the proposed method, as the movement distance achieved with vibration is considerably greater than without vibration. The optimal vibration should be selected to improve the proposed mechanism's underground mobility. When imparting vibration, the change of the bearing capacity, which is the resistance from the ground, depends on lift-off acceleration. Lift-off acceleration is the minimum vibration acceleration that causes the flow of the ground particles. The bearing capacity increases when the rod's movement is under lift-off acceleration. Conversely, it increases when the rod's movement is over lift-off acceleration. If the simulation reproduces the change of the bearing capacity when imparting vibration and lift-off acceleration, the optimal vibration can be selected. This study conducted the experiment and the DEM simulation in which the rod is dragged while vibrating on the ground. These methods measured the bearing capacity that the rod received from the ground. In the experiment and simulation, the vibration frequency and the dragging speed of the rod were changed. As the vibration frequency increases, the acceleration of the rod's movement caused by the vibration eventually exceeds the lift-off acceleration. The tendency to change the bearing capacity depends on the dragging speed of the rod. The reproducibility of DEM simulations was evaluated by comparing the relationship between the bearing capacity and these parameters in each method.

Keywords
Loose soil
Bearing capacity
Vibration
Discrete element method

1. Introduction

In recent years, moving underground has been focused on as a method of exploring robots' movement in extraterrestrial bodies. There are two main reasons for conducting the underground exploration. The first is that the underground sample is valuable from a scientific point of view because these samples remain in good condition and are unaffected by space weathering. The fresh samples are significant for elucidating the origin of extraterrestrial bodies. The second advantage is the protection of the rover from severe temperature changes. The Moon has maximum and minimum temperatures of 121 °C and −133 °C, respectively. The electronic devices on the rovers can easily break down in these extreme temperature environments. Alternatively, the ground absorbs heat and maintains a constant internal temperature. Therefore, rovers can protect their electronic devices when underground. From these advantages, rovers moving underground have been developed. Yuan et al. developed a mole-type robot for exploring the Moon underground. This robot consists of a drill and rudders that move forward and change direction. This study provides important information for the exploration mission to obtain the underground samples (Yuan et al., 2023). However, exploration rovers face difficulty moving underground due to earth pressure, which resists motion and obstructs progress, limiting effective underground exploration. To address this issue, a mechanism has been developed that facilitates smoother underground movement by utilizing ground condition changes induced by vibration. Our previous research validated this approach. Specifically, it was observed that applying vibration increases ground density and hardens the soil (Watanabe et al., 2023). On the other hand, soil particles behave more fluid-like, softening the ground due to their mobility while imparting vibration (Watanabe et al., 2020). The proposed mechanism incorporates vibrators to apply vibrations to the ground, reducing drag force and enabling forward motion underground. Experimental results demonstrated that rovers travel a greater distance underground when vibration is applied, indicating the effectiveness of the proposed method.

The effect of vibration on altering ground conditions plays a crucial role in the proposed underground locomotion mechanism. Our research group has previously examined this phenomenon (Watanabe et al., 2024). To enhance the proposed method and refine the design for underground mobility, it is essential to simulate this behavior. Simulations can help identify effective strategies before conducting physical experiments. In this study, the discrete element method (DEM) was employed to simulate changes in the ground's bearing capacity under vibration. Vibration induces complex alterations in the ground, affecting compaction, shear behavior, and friction. DEM is well-suited for capturing these dynamics, as it models the motion of individual particles. The bearing capacity was assessed by dragging a rod across the ground in both experiments and simulations, and the outcomes were compared. This study investigates explicitly how vibration-induced changes in bearing capacity correlate with the rod's drag speed.

The structure of the rest of this paper is as follows. Section 2 reviews prior research related to underground locomotion mechanisms and the reduction of bearing capacity through vibration. Section 3 describes the experimental procedures, simulation methods, and setup used in this study. Section 4 presents the results and discusses the comparison between experimental findings and simulation data. Lastly, Section 5 provides a summary of the study.

https://doi.org/10.56884/JUY2WUYX

2. Previous study

This section describes the phenomenon where the bearing capacity of the ground changes when vibration is applied. In a previous study, a rod was dragged across the ground while vibration was imparted (Watanabe et al., 2023). The results showed that the bearing capacity under vibration was lower than without it. This reduction is attributed to the fluid-like behavior of soil particles induced by vibration. Baumgartner et al. investigated particle motion under vibration to understand the sandfish skink's movement through desert sand (Baumgartner et al., 2008), a lizard species known for burrowing. Their findings indicated that high-frequency vibrations cause localized decompaction, making the sand behave like a fluid around the vibration source. Conversely, in another study conducted by our group, the bearing capacity was found to increase when the rod was dragged after vibration had been applied to the ground (Watanabe et al., 2023). This increase is due to ground compaction caused by vibration, which rearranges particles and enhances soil density. As a result, the friction between particles increases, leading to greater bearing capacity.

Building on these insights, we previously proposed a robot that utilizes both ground-softening and ground-hardening effects of vibration for underground movement (Watanabe et al., 2024). The robot's movement relies on two key phenomena: (1) while vibration is applied, the ground becomes softer, reducing drag force at the front; and (2) after vibration, the ground compacts and hardens, increasing drag force at the rear. This differential in drag force enables forward motion underground. To test this concept, we designed and constructed a testbed system, shown in Fig. 1. The testbed consists of three vibration units—front, center, and rear—connected via linear actuators. Each unit contains an internal vibration motor. An experiment was conducted to assess the testbed's mobility underground. The operation process, illustrated in Fig. 2, involved sequential movement of each vibration unit from front to rear while applying vibration to reduce drag. Results showed that the testbed achieved greater underground movement when vibration was applied compared to when it was not.

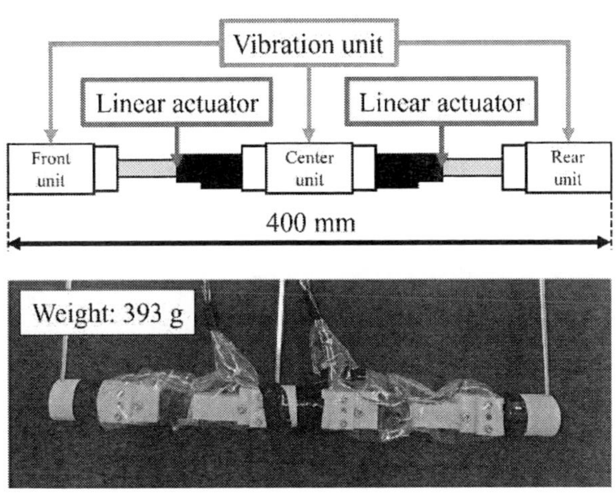

Fig. 1. Overview and construction of testbeds. This figure is reprinted from the study by Watanabe et al. (Watanabe et al., 2024).

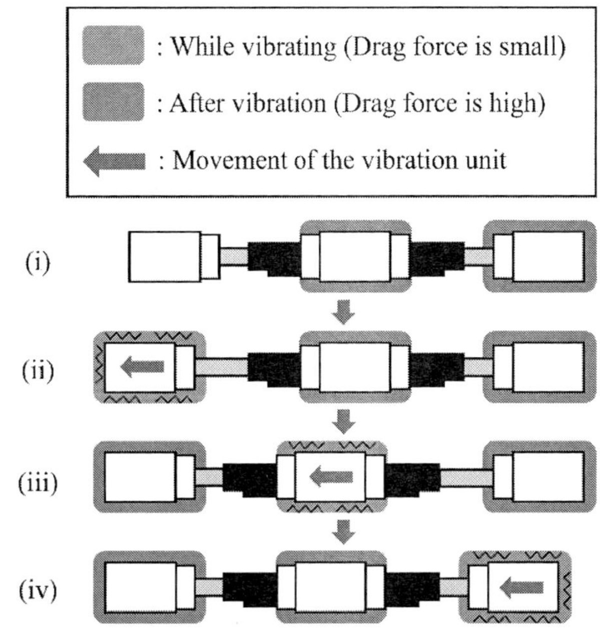

Fig. 2. Sequence of operations for moving underground This figure is reprinted from the study by Watanabe et al. (Watanabe et al., 2024).

3. Method

3.1. Experiment

In this study, the bearing capacity was measured by dragging the rod on the ground. This experiment was conducted to compare the results with those of the discrete element method (DEM) simulation. In our experiment, the vibration frequency was varied as the experimental parameter. This experimental method was designed based on related studies that measured the force caused by the ground for understanding terramechanics on extraterrestrial bodies (Yeomans et al., 2013).

Figure 3 shows the experimental setup. It consists of a soil tank, a rod, and a force sensor. Figure 4 shows the flow of the experiment. First, the ground was mixed and flattened. Next, the rod was sunk into the ground. The sinkage of the ground is 50 mm. Finally, the rod was dragged while vibrating. At this time, the bearing capacity value was measured. The dragging speed of the rod was set at 2.22 mm/s and 8.87 mm/s. The dragging distance of the rod was 40 mm. The size of the soil tank was a length, width, and height of 309, 439, and 300 mm, respectively. Fig. 3 (b) shows the size of the rod. Its size and length were a quadrangular prism and 200 mm. The bottom shape of the rod was square, and the side length was 20 mm. The kind of soil, which consisted of the ground, was Silica No. 5. Silica No. 5 is commonly used in studies about terramechanics in extraterrestrial bodies (Mori and Ishigami (2015); Shirai and Ishigami (2015)). The experimental setup was equipped with a vibration generator (Wave Maker 05) to generate vibration. The vibration frequency was changed to 10 Hz, 20 Hz, and 30 Hz. The vibration amplitude was set to 0.25 mm. The direction in which the rod vibrated was the same as the direction in which the rod was dragged. Each experimental condition was tested in five trials. Table 1 presents the experimental conditions.

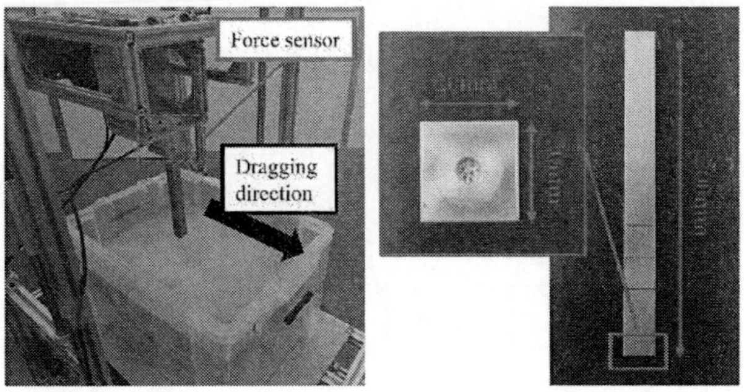

(a) Experimental environment (b) Overview of a rod

Fig. 3. Experimental setup for measuring bearing capacity.

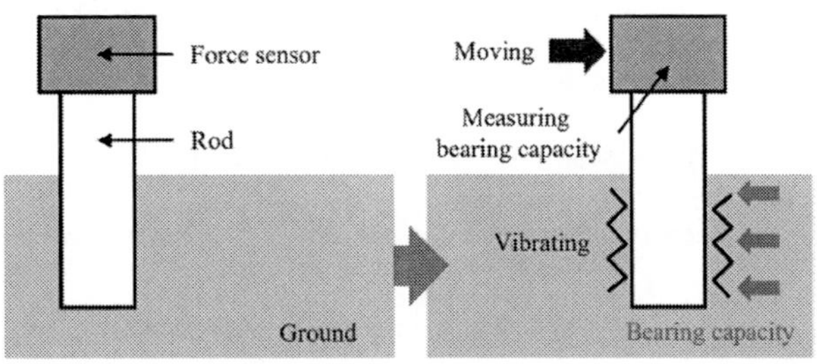

Fig. 4. Experimental flow for measuring bearing capacity. This figure is reprinted from the study by Watanabe et al. (Watanabe and Iizuka., 2025)].

Table 1
Conditions for measuring bearing capacity in experiment.

Item	Conditions (value)
Number of trials	5 trials
Sinkage of a rod [mm]	50
Dragging speed [mm/s]	2.22, 8.87
Dragging time [s]	5
Kind of sand	Silica No.5
Vibration generator	Wave Maker 05
Vibration amplitude [mm]	0.25
Vibration frequency [Hz]	10, 20, 30
Force sensor	PFS055YA251U6

3.2. Simulation

A simulation, which was the same as the experiment described in the previous section, was conducted to compare with the experimental result.

Figure 5 shows the simulation flow. First, the ground was mixed and flattened. Next, the rod was sunk into the ground. The sinkage of the ground is 50 mm. Finally, the rod was dragged while vibrating. At this time, the bearing capacity value was measured. The dragging speed of the rod was set at 2.22 mm/s and 8.87 mm/s. The dragging distance of the rod was 40 mm. The rod size was the same as the experimental one as shown in Fig. 3. The shape of the vibration wave was created by evenly obtaining 8 points from a sine wave waveform and connecting these points to mimic a sine wave (Fig. 6). The conditions of the simulation are presented in Table 2.

In the DEM model, contact forces, which include elastic repulsion and friction, are expressed. The movement of the particles is simulated by considering that these forces are applied to the particles. Fig. 7 shows the DEM model used in this study. In the normal direction, the particles are connected by a spring and dashpot in parallel to each particle (Fig. 7 (a)). In the tangential direction, friction is expressed by a friction slider (Fig. 7 (b)). In a previous study, we obtained these parameter values by conducting the experiment and used them in this study (Watanabe et al., 2025). Table 3 presents the parameters for DEM simulation.

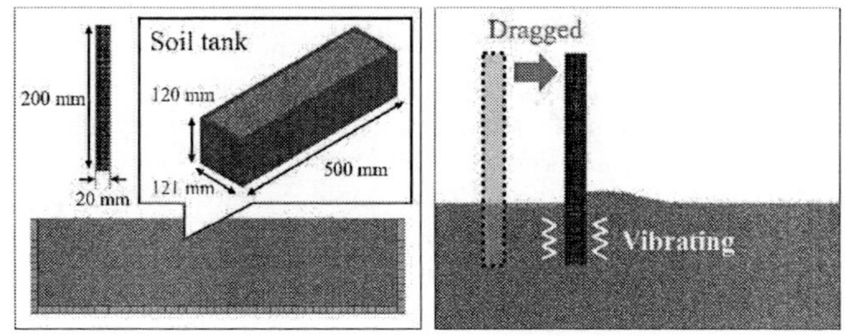

(a) Experimental condition (b) Experimental flow

Fig. 5. Experimental setup for measuring bearing capacity.

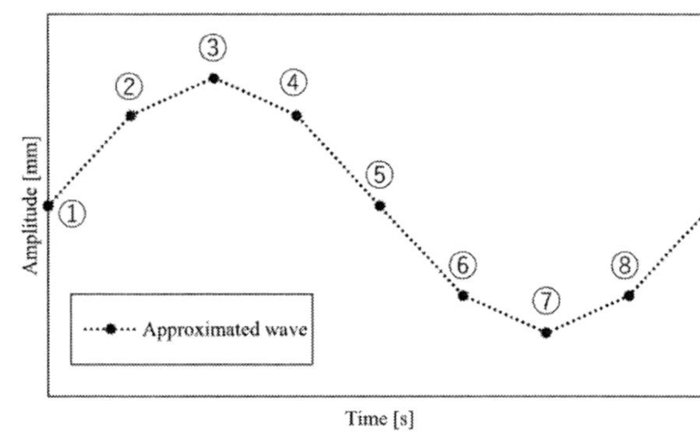

Fig. 6. Vibration waveform in DEM simulation.

Table 2
Conditions for measuring bearing capacity in simulation.

Item	Conditions (value)
Sinkage of a rod [mm]	50
Dragging speed [mm/s]	2.22, 8.87
Dragging time [s]	5
Vibration amplitude [mm]	0.25
Vibration frequency [Hz]	10, 20, 30
Software	LS-DYNA

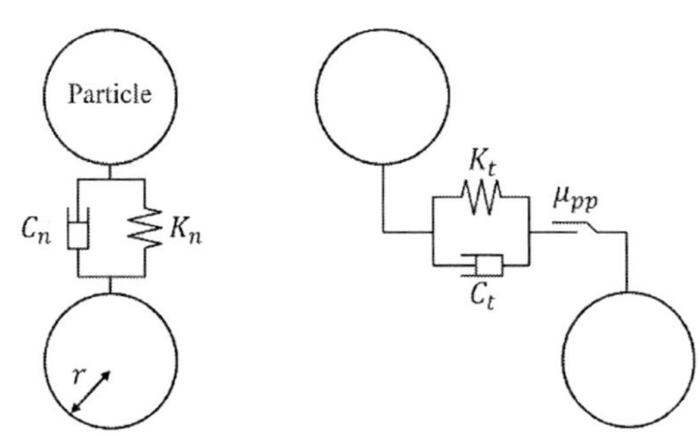

(a) Normal direction (b) Tangent direction

Fig. 7. Simulation mode.

Table 3
DEM simulation parameter.

Item	Conditions (value)
Element radius r [m]	0.002
Element density ρ [kg/m^3]	2650
Mass of the element m [kg]	$m = \frac{4}{3}\pi r^3$
Normal spring constant K_n [N/m]	300
Normal damping coefficient C_n [N/m]	$C_n = 2\sqrt{mK_n}$
Tangential spring constant K_t [N/m]	$K_t = 0.25K_n$
Tangential damping coefficient C_t [Ns/m]	$C_t = 2\sqrt{mK_t}$
Angle of repose φ_{pp} [°]	40.2
Critical slope angle of the rod φ_{pr} [°]	29.0
Friction coefficient for particle elements μ_{pp} [-]	$\mu_{pp} = \tan\varphi_{pp}$
Friction coefficient for particle-Rod μ_{pr} [-]	$\mu_{pr} = \tan\varphi_{pr}$
Coefficient of rolling friction μ_r [m]	$\mu_r = 0.02r$

4. Result and discussion

4.1. Experimental result

Figs. 8 and 9 show the variation in bearing capacity relative to shear displacement. In Fig. 8, the results are presented for both cases with and without vibration, at frequencies of 10 Hz, 20 Hz, and 30 Hz. These results are the average value of 5 trials. In all experimental results, the bearing capacity increases initially and converges finally. When the rod was dragged at 2.22 mm/s, the bearing capacity with vibration (frequency of 10 Hz) was larger than that without vibration. The values of the bearing capacity without and with vibration (frequency of 30 Hz) are almost the same. In contrast, the bearing capacity with vibration (frequency of 20 Hz) was smaller than that without vibration. When the rod was dragged at 8.87 mm/s, all bearing capacities with vibration are larger than those without vibration.

Figs. 10 and 11 compare the peak and converged values of the bearing capacity in each experimental condition. In Figs. 10 and 11, the standard errors are calculated from the five experimental trials. The converged values were calculated by averaging the bearing capacity when the dragged distance is 35 to 40 mm. The tendency between the peak and converged values of the bearing capacity and vibration frequency is almost the same. The bearing capacity increases when the vibration frequency is under 10 Hz. The reason is that vibration with a frequency of 10 Hz compacts the ground and cannot cause the ground particles to flow because the vibration acceleration is insufficient to cause them to flow. On the other hand, the bearing capacity decreases when the vibration frequency exceeds 10 Hz. The reason is that vibration, with a frequency over 10 Hz, causes the ground particles to flow, and the ground becomes soft. This tendency is confirmed at each speed at which the rod is dragged. The bearing capacity when the rod is dragged at 2.22 mm/ s is the same or larger than that when the rod is dragged at 8.87 mm/ s when the vibration frequency is under 10 Hz. The bearing capacity when the rod is dragged at 8.87 mm/ s is larger than that when the rod is dragged at 2.22 mm/ s when the vibration frequency exceeds 10 Hz.

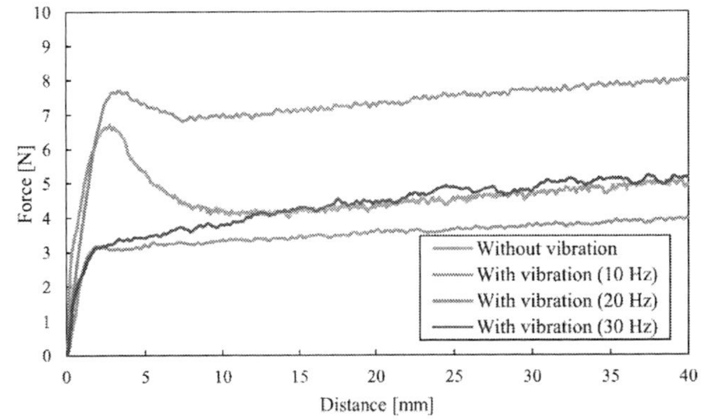

Fig. 8. Bearing capacity versus shear displacement in the experiment when the rod was dragged at 2.22 mm/s.

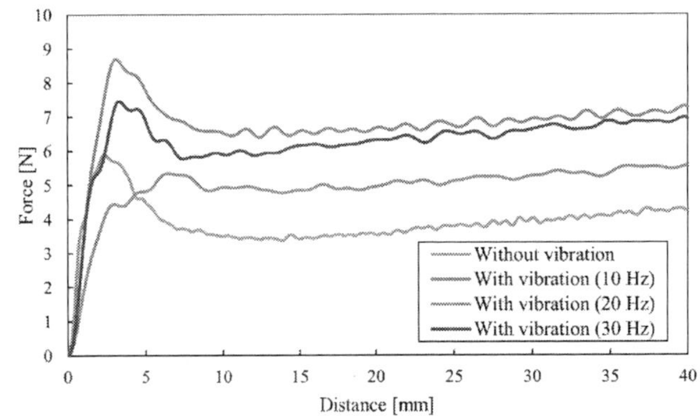

Fig. 9. Bearing capacity versus shear displacement in the experiment when the rod was dragged at 8.87 mm/s.

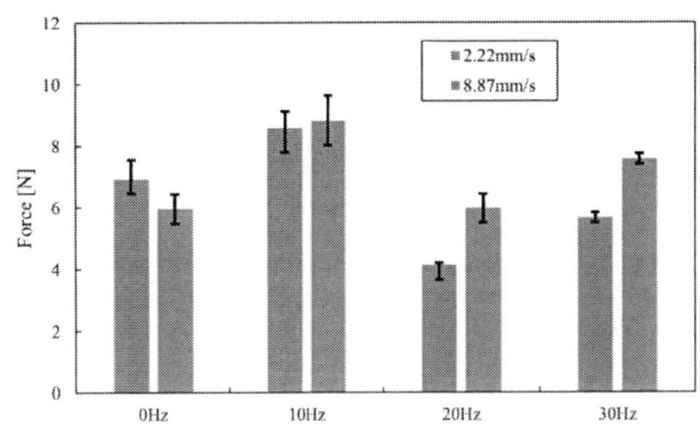

Fig. 10. Comparison of the peak value of bearing capacity in each vibration (Experiment).

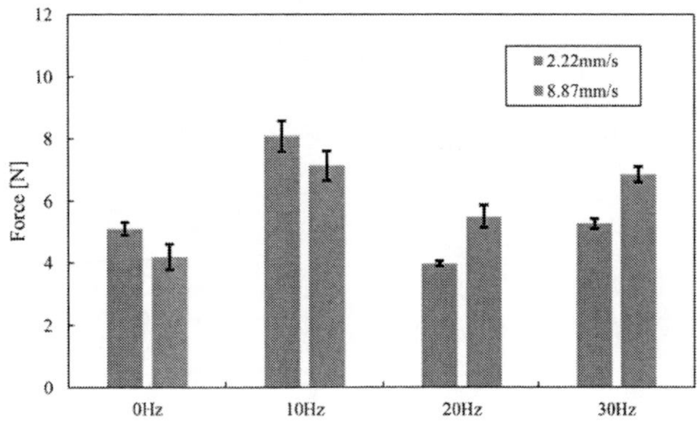

Fig. 11. Comparison of the average value of bearing capacity in each vibration (Experiment).

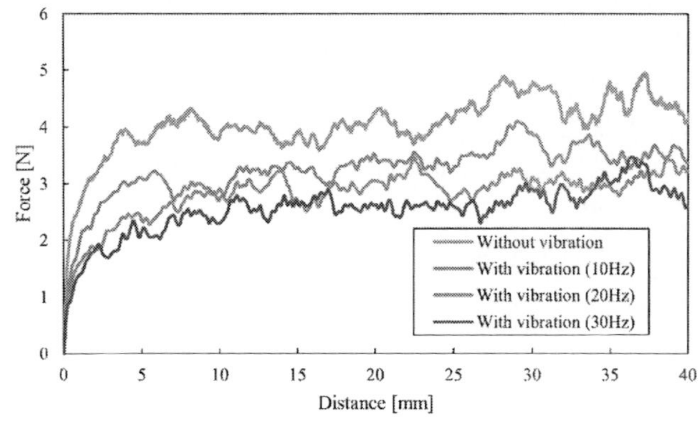

Fig. 12. Bearing capacity versus shear displacement in the simulation when the rod was dragged at 2.22 mm/s.

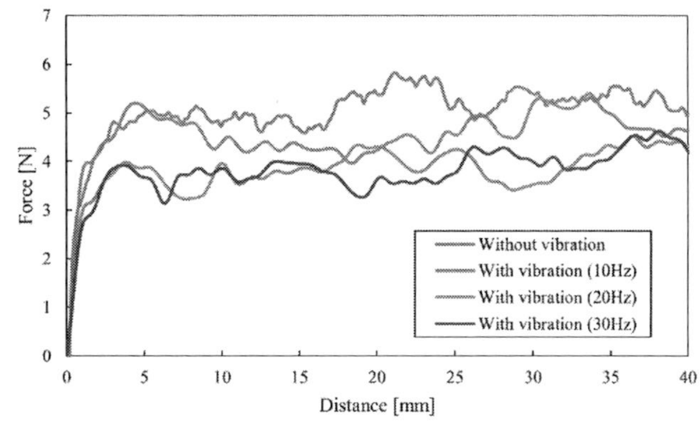

Fig. 13. Bearing capacity versus shear displacement in the simulation when the rod was dragged at 8.87 mm/s.

4.2. Simulation result

Figs. 12 and 13 show the variation in bearing capacity relative to shear displacement. In Figs. 12 and 13, there are the results without and with vibration, whose frequencies were 10 Hz, 20 Hz and 30 Hz. In the simulation, the bearing capacity initially increases and then converges. When the rod was dragged at 2.22 mm/s, the bearing capacity without vibration is larger than that with vibration. When the rod was dragged at 8.87 mm/s, the value of the bearing capacity in each vibration condition was close.

Figs. 14 and 15 compare the peak and converged values of the bearing capacity in each condition. The converged values were calculated by averaging the bearing capacity over the dragged distance range of 35 to 40 mm. The tendency between the peak and converged values of the bearing capacity and vibration frequency is the same. The higher the vibration frequency, the smaller these values of the bearing capacity. In all vibration conditions, the bearing capacity when the rod is dragged at 8.87 mm/ s is larger than that when the rod is dragged at 2.22 mm/ s. In the DEM simulation, the peak point does not exist when the vibration frequency is 10 Hz compared with the experimental result. It means the DEM simulation could not express the compaction of the ground by vibration. The reason is that the shape of the ground particle is a sphere and simple, and the connected area hardly increases by imparting vibration. The bearing capacity does not increase because the friction between the ground particles does not increase due to the lack of an increase in the connected area of ground particles. In future studies, we will reset the shape of the particle and the DEM parameters which is related to the friction to reproduce the experimental result. It is confirmed that the bearing capacity decreased by imparting vibration in the DEM simulation. This point is common to the experimental results qualitatively. We consider that the flow movement of the ground particles when imparting vibration is reproduced in the DEM simulation.

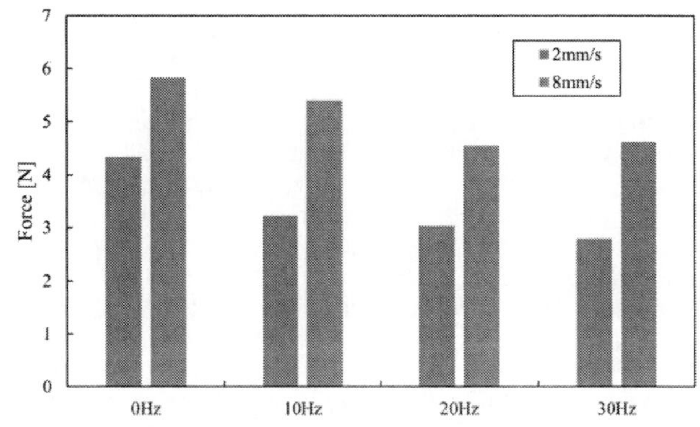

Fig. 14. Comparison of the peak value of bearing capacity in each vibration (simulation).

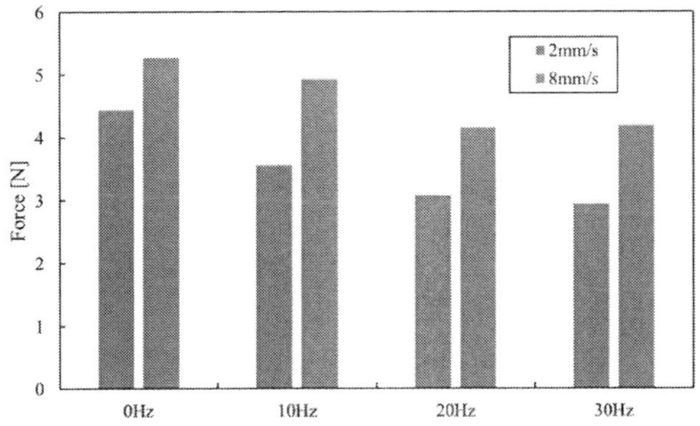

Fig. 15. Comparison of the average value of bearing capacity in each vibration (simulation).

5. Conclusion

This study conducted the experiment and the DEM simulation in which the rod is dragged while vibrating on the ground. These methods measured the bearing capacity that the rod received from the ground. In the experiment and simulation, the vibration frequency and the speed of the dragged rod were changed. The experimental results showed that the bearing capacity changed in an increasing and decreasing tendency depending on the vibration frequency. On the other hand, in the simulation result, the higher the vibration frequency, the smaller the values of the bearing capacity. This point differs between the experimental and simulated results. The reason is that the bearing capacity when using vibration is not increased because the friction force is small. In future studies, we will reset the shape of the particle and the DEM parameters that are related to the friction to reproduce the experimental result.

Acknowledgements

This study was supported by JSPS KAKENHI (Grant Number JP 21J11944).

Figure 4 has been reprinted from T. Watanabe, D. Watanabe, K. Iizuka, "Verification of Decreasing Bearing Capacity While Imparting Vibration to Ground in Discrete Element Method Simulation", Walking Robots into Real World, pp.133–143. with permission from Springer.

6. Declaration of competing interest

The authors declare that they have no competing interests.

7. References

Baumgartner W., Fidle F., Weth A., Habbecke M., Jakob P., Butenweg C., Bohme W., 2008, Investigating the Locomotion of the Sandfish in Desert Sand Using NMR-Imaging, PLoS One, 10.

Mori, D., Ishigami, G., 2015. Excavation model of soil sampling device based on particle image velocimetry. J. Terramech. 62, 19-29.

Shirai, T., Ishigami, G., 2015. Development of in-wheel sensor system for accurate measurement of wheel terrain interaction characteristics. J. Terramech. 62, 51-61. https://doi.org/10.1016/j.jterra.2015.09.001.

Watanabe, T., Iizuka, K., 2020. Experimental Investigation of Changing Traction Force Caused by a Propagation of Vibration for Planetary Exploration Rovers with Legs, in: Online, Japan, Workshop on JAXA Astrodynamics and Flight Mechanics

Watanabe, T., Iizuka, K., 2023. Experimental investigation of relationship between bearing capacity and vibration parameters for planetary exploration legged rovers, Int. ROBOMECH J., 10.

Watanabe T., Kobayashi K., Hiramoto K., Iizuka K., 2024, Method for Underground Motion Using Vibration-Induced Ground Resistance Changes for Planetary Exploration, Aerospace, 11.

Watanabe, T., Iizuka, K., 2025. Verification of Decreasing Bearing Capacity While Imparting Vibration to Ground in Discrete Element Method Simulation, Walking Robots into Real World, 133–143.

Watanabe T., Higashiyama R., Watanabe D., Iizuka K., 2025. Analyzing bearing capacity changes due to vibration in discrete element method simulations. J. Terramech. 118.

Yeomans B., Saaj C. M., and Winnendael M. V., 2013. Walking planetary rovers – experimental analysis and modelling of leg thrust in loose granular soils, J. Ter., 50, 107-120

Yuan Z., Mu R., Zhao H., Wang K., 2023, Predictive Model of a Mole-Type Burrowing Robot for Lunar Subsurface Exploration, Aerospace, 10.

9563 | Physical testing and digital twin simulations for prediction of ride quality of light utility vehicle

Eric Karpman *, Wei Huang, Nikita Yovchev, Patrick Kehoe

National Research Council Canada, Ottawa, Canada

** Corresponding author: eric.karpman@nrc-cnrc.gc.ca*

ABSTRACT

Ride quality is an important metric for assessing the comfort and safety of vehicle passengers as a vehicle traverses rough off-road terrain. The Absorbed Power metric, which sets a power threshold below which crew members can effectively perform their tasks, is a common approach for assessing ride quality. The roughness of the terrain itself is characterized by the root mean square (RMS) value of the terrain profile as per TOP-1-1-014 guidelines. Using an accelerometer to measure the acceleration experienced by the passengers, a series of tests can be conducted in which the vehicle traverses a given course at constant speeds, increasing the speed with each traversal. By calculating the absorbed power at each speed and selecting a maximum acceptable absorbed power value, a maximum speed can be determined for a given course. Because it is not always possible to take a vehicle into the field to test it on such a course, vibration testing using a four-post shaker to emulate the terrain profile can be used as an alternate test to assess the ride quality. In this work, we perform vibration testing on a light utility vehicle in an effort to assess ride quality over an asymmetric 3cm RMS terrain. Additional physical testing – including a tilt test to determine centre of gravity height – were performed to tune a digital twin of the vehicle. The same suite of tests was performed on the digital twin in a virtual environment and compared to the physical test results. Both a virtual traversal of a simulated 3cm RMS terrain and a simulated for-post shaker test representing that terrain were performed on the digital twin to compare the vibration response results.

Keywords
Ride Quality
Absorbed Power
Digital Twin
Multibody Dynamics Simulation
Physical Testing
Off-road mobility

1. Introduction

Many metrics are important when designing or predicting the suitability of off-road vehicles for a given task. One such metric is the ride quality. The US Army's Test Operations Procedure (TOP-1-1-014) relating to this topic lists several different methods for evaluating ride quality, including (US Army Development Test Command, 2007):

- Frequency Weighted RMS Acceleration (FWRMSA)
- Vibration Dose Value (VDV)
- Absorbed Power (AP)

Out of these methods, AP and FWRMSA are not sensitive to the length of exposure, while VDV is. In this work, we will focus on the AP metric, which sets a power threshold – typically 6W – below which a crew member can effectively perform their tasks. (US Army Development Test Command, 2007). The typical procedure for testing a vehicle's ride quality involves travelling over a terrain of known roughness with an accelerometer placed on the driver's (or other passenger of interests') seat. Based on the acceleration recorded by the seat pad accelerometers, the AP experienced by a passenger at that location on the vehicle for the test can be computed. Repeating the test at different speeds allows for the generation of an AP vs speed curve. An acceptable AP threshold is then set, and from this a maximum speed for the vehicle on that terrain which meets ride quality standards can be determined.

The terrain itself is characterized by its root mean square (RMS) value – meaning the root mean square of the surface amplitude. Thus, by repeating the above procedure on multiple terrain profiles, a maximum speed can be determined as a function of surface roughness.

However, it is not always possible to take vehicles out to do field tests. Test campaigns are costly and inevitably come with some level of risk to the driver. It is also simply not practical to repeat this test procedure for every new vehicle of interest. The four-post shaker test is a laboratory alternative to road testing that uses servo hydraulic actuators with flat wheel pans installed. The vibration of the posts can be controlled in a way that is meant to reproduce the vibration that would be experienced during road testing (Gräbe, Kat, van Staden, & Els, 2020) (Kat & Els, 2024) (Kat, Gräbe, van Staden, Botha, & Els, 2024). One method of achieving this is by running on-road tests, measuring wheel spindle acceleration, and then controlling the posts to accurately recreate this acceleration at the spindles. In this work the posts are controlled in a way that directly mimics the road surface profile.

A digital twin, a virtual representation of the real vehicle, was created using multibody dynamics (MBD) simulation software. Some additional physical testing of the vehicle was performed to tune the model. Once the model was sufficiently tuned to represent the physical vehicle, the four-post shaker mechanism was created in the simulation environment and the same shaker test performed on the digital twin. In addition to the four-post shaker test, a simulation of the digital twin performing road tests on terrain with the same profile simulated by the four-post shaker was performed to study the validity of vibration testing as a substitute for road testing. In the results of this work, we will compare the results of the physical shaker test with the simulation results of both the shaker and road tests using AP as the metric for comparison.

https://doi.org/10.56884/6SKEESJC

2. Absorbed Power Theory

The AP metric that TOP-1-1-04 suggests is based on using the acceleration of a body to determine the rate at which it is absorbing energy. This is done by integrating the product of the power spectral density of the acceleration and a transfer function (US Army Development Test Command, 2007). The computation includes a weighting factor which is a function of the frequency and which filters out higher frequency vibrations. The result of the integration is a scalar value for absorbed power at a given point on the vehicle. When studying the absorbed power in multiple axes, the absorbed power values in each axis can be summed. The absorbed power can be calculated at different positions of interest – the driver's seat for example. It is generally agreed that 6W in the vertical direction is the threshold below which crew members can effectively perform tasks and it is therefore often a requirement for military vehicles (US Army Development Test Command, 2007).

As per TOP-1-1-014 guidelines, the AP in a given axis is calculated using the following equations:

$$P = \sum_{i=1}^{n}(C_i)A_i^2 \tag{1}$$

$$C_i = K_1K_0(F_1F_4 - F_2F_3)/(F_3^2 + W_i^2F_4^2) \tag{2}$$

where P is the absorbed power (in Watts), A_i is the root mean squared acceleration (in ft/s^2) in the ith spectral band, W_i is the frequency, and C_i is the frequency weighting for the ith spectral band, as determined by the coefficients K_0, K_1, F_1, F_2, F_3, F_4 which vary for each axis and are listed in Table 1, which is taken from the appendix of TOP-1-1-014 (US Army Development Test Command, 2007). Note that there was an error in the F_1 term in the y-axis in TOP-1-1-014 which has been corrected in Table 1.

Table 1
Coefficients for Absorbed Power Computation

	X	Y	X
K_0	4.3532	4.353	4.3537
K_1	1.356	1.356	1.356
F_1	1.0	$0.24052124\text{x}10^{-3}\,W_i^4$ - $0.066974483\,W_i^2 + 1$	$-0.10245\text{x}10^{-9}\,W_i^6$ $+0.17583\text{x}10^{-5}\,W_i^4$ - $0.44601\text{x}10^{-2}\,W_i^2$ + 1
F_2	0.219106	$0.57384538\text{x}10^{-5}\,W_i^4$ - $0.50170413\text{x}10^{-2}\,W_i^2$ + 0.33092592	$0.12882\text{x}10^{-7}\,W_i^4 -$ $0.93394\text{x}10^{-4}\,W_i^2$ + 0.10543
F_3	$-0.0185309\,W_i^2$ + 1	$-0.14979958\text{x}10^{-5}\,W_i^6$ + $0.0010088882\,W_i^4 -$ $0.10108617\,W_i^2 + 1$	$-0.45416\text{x}10^{-9}\,W_i^6$ $+ 0.37667\text{x}10^{-5}\,W_i^4$ $- 0.56104\text{x}10^{-2}\,W_i^2$ + 1
F_4	$-0.00061893\,W_i$ + 0.219106	$-0.1713749\text{x}10^{-7}\,W_i^6$ + $0.53137351\text{x}10^{-4}\,W_i^4 -$ $0.011096507W_i^2$ + 0.33092592	$-0.21179\text{x}10^{-11}\,W_i^6$ $+ 0.51728\text{x}10^{-7}\,W_i^4$ $- 0.17947\text{x}10^{-3}\,W_i^2$ + 0.10543

The following centre frequencies (in Hz) were used to filter the acceleration data into discrete spectral bands:

$$f = [0.4, 0.5, 0.63, 0.8, 1, 1.25, 1.6, 2, 2.5, 3.15, 4, 5, 6.3, 8, 10, 12.5, 16, 20, 25, 31.5, 40, 50, 63, 80] \tag{3}$$

3. Physical Testing

3.1. Parameter Measurement

The test vehicle can be seen in Fig. 1. A digital twin of the vehicle was obtained and tuned based on manufacturer specifications. However, certain parameters required additional testing to accurately tune the digital twin.

Fig. 1. Vehicle Mounted on Four-Post Shaker

Due to the additional masses added to the vehicle to represent passengers and payloads in the cargo hold, the CG location of the loaded vehicle needed to be determined. The lateral and longitudinal positions of the test vehicle's centre of gravity (CG) were determined by placing each wheel of the vehicle on a scale. The distribution of the overall vehicle load was used to determine the lateral and longitudinal CG location. The vertical position of the CG was determined by performing a tilt test. In this test, the vehicle is mounted on a tilting platform. One side of the vehicle is lifted until the vehicle begins to tip over. At this tipping point, the vehicle's CG is directly above the lower wheel's contact point. Based on the tilt angle and the previously determined lateral position of the CG, the height can be determined using the following geometric relation:

$$h = \frac{w/_2 \mp \Delta y_{CG}}{\tan(\theta)} \tag{4}$$

where w is the track width, Δy_{CG} is the lateral CG offset towards the roadside (RS) (+ for RS tilt, - for curbside (CS) tilt), and θ is the tilt table angle at which the wheels on the lifted side lose contact. More accurate results may be possible if the suspension is blocked in the static squat position on level ground.

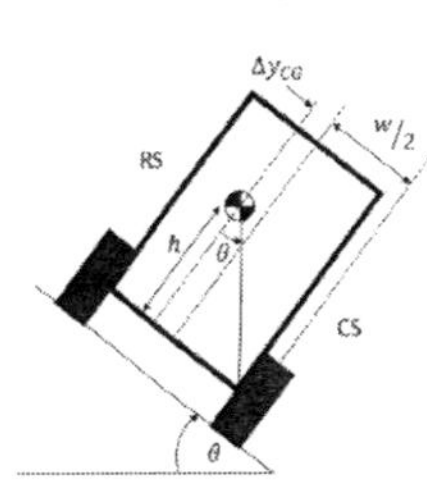

Fig. 2. Roadside Lift - CG Location Geometry

Table 2
Vehicle Specifications for Tilt Test

Gross vehicle weight rating	1686 kg (3717 lbs)
Curb weight	1088.6 kg (2400 lbs)
Payload (Cargo Box)	216.4 kg (477 lbs)
Payload (Occupants)	381 kg (840 lbs)
Tire pressure (front)	82.74 kPa (12 psi)
Tire pressure (rear)	137.9 kPa (20 psi)
Wheelbase	2.71 m
Track Width	1.34 m

3.2. Ride Quality Testing

3.2.1. Surface Profile Generation

Terrain profiles are described by their root mean square (RMS) roughness as defined by the standards laid out in ISO-8608 (International Organization for Standardization, 2016). In accordance with ISO-8608, terrains are classified according to their angular spatial frequencies, G_d, with classifications ranging from A to H in increasing roughness, as listed in Table 3.

Table 3
ISO-8608 Road Roughness Classification

Road Class	**Classified roughness $G_d(\Omega_0) \times 10^{-6}\ m^3$ where $\Omega_0 = 1$ rad/s**		
	Geometric Mean	**Lower Limit**	**Upper Limit**
A (Very Good)	1	-	8
B (Good)	4	2	8
C (Average)	16	8	32
D (Poor)	64	32	128
E (Very Poor)	256	128	512
F	1024	512	2048
G	4096	2048	8192
H	16384	8192	-

A method for generating random terrain profiles of a desired roughness was followed to create a profile for a terrain with an RMS value of approximately 3cm (Bylsma, 1999). In addition to guaranteeing the desired roughness, this method is also designed to remove trends in the random terrain profile that are unrelated to ride quality, such as variations in grade (Rosca & Bradley, 2021). The procedure that was followed to generate the surface profile is summarized below. The power spectral density (PSD) of the surface is defined by:

$$G_d(\Omega) = G_d(\Omega_0)\left(\frac{\Omega}{\Omega_0}\right)^{-2} \tag{5}$$

where Ω is the spacial frequency band, which is uniformly distributed between the lower and upper limits. TOP 1-1-014 states that "Test courses used for ride quality should have a wave number spectrum (power spectral density function of the instantaneous terrain vertical profile in the spatial domain) with a slope of approximately -2 in the log-log domain" (US Army Development Test Command, 2007).

$$\Omega = [\Omega_L : \Delta\Omega : \Omega_U] \tag{6}$$

where $\Delta\Omega$ is given by:

$$\Delta\Omega = \frac{\Omega_U - \Omega_L}{N} \tag{7}$$

where N is the number of samples required. TOP 1-1-014 recommends using $\Omega_L = 2\pi/14.75$ and $\Omega_U = 2\pi/0.15$ for vehicle speeds up to 40 km/h. The number of samples required is given by:

$$N = \frac{L}{\Delta x} \tag{8}$$

where L is the length of the track and Δx is the distance between successive coordinates of the surface profile. In this work, a track length of 510m was used with the profile defined at intervals of $\Delta x = 0.05m$. The $G_d(\Omega_0)$ for road class E in Table 3 was used to generate the surface profile – resulting in an RMS value of approximately 3cm.

Taking the inverse discrete Fourier transform of the PSD function to get the profile as a function of distance, the surface height at each point is given by:

$$z(r) = \sum_{k=1}^{N} \sqrt{\frac{2\pi}{N\,\Delta x} G_d(\Omega_k)} \times \mathrm{Re}\left(e^{i\left(\theta_k + \frac{2\pi k r}{N}\right)}\right) \text{ for } r = 1, \ldots, N \tag{9}$$

where θ_k is a set of random phase angles uniformly distributed between 0 and 2π.

3.2.2. Four-Post Shaker Test

The road surface profile combined with the vehicle speed being simulated were used to generate desired displacement time history profiles for each shaker post. The rear post on each side follows the same profile as the front post with a time delay that is based on the wheel base dimension and the simulated vehicle speed.

A tandem elliptical cam filter was applied to the road surface profiles to more accurately represent the vertical displacements resulting from the wheel-road interaction. This filter is meant to adjust the profile based on the geometry of the wheel, and takes into account the tire diameter and the static deflection of the tire under gross vehicle weight (GVW). Envelope filters are recommended when the road profile has wavelengths smaller than two to three times the contact patch of the tire.

When verifying the acceleration and velocity requirements of the actuator pans by differentiating the displacement profile with respect to time, it was determined that the higher vehicle speeds would not be achievable on the four-post shaker. The hypothesis was that the envelope filter alone was not sufficient as the double cams are modeled as rigid ellipses. With the wheel assumed to be a rigid body, a bandpass filter was applied to remove surface profile content below 0.25 Hz and above 15 Hz to mimic the effects of tire deformation during the wheel-road interaction. By limiting the higher frequencies to 15 Hz, the estimated actuator pan velocities and accelerations dropped into a range that was achievable. The assumptions made here require further investigation to validate this process. Currently, there are concerns that the 15 Hz cutoff is too aggressive and would impact the wheel-hop frequency of most vehicles that typically lies in the 10-20 Hz range. The basis for this investigation would be a frequency-domain comparison of real over-the-road versus four-post shaker generated vibration at four locations on a vehicle: spindles, driver's seat, front passenger's seat, and vehicle body inertial measurement unit (IMU). Methods of tuning and adjusting synthesized drive files would be investigated to match the vehicle response on a tire-coupled road simulator to that resulting from traversing comparable over the road surfaces.

Table 4
Vehicle Specifications for Shaker Test

Gross vehicle weight rating	1921 kg (4235 lbs)
Curb weight	1088.6 kg (2400 lbs)
Payload (Cargo Box)	288 kg (635 lbs)
Payload (Occupants)	544.3 (1200 lbs)
Tire pressure (front)	82.74 kPa (12 psi)
Tire pressure (rear)	137.9 kPa (20 psi)
Wheelbase	2.71 m
Track Width	1.34 m

4. Digital Twinning

4.1. Vehicle

A digital twin of the vehicle, the four-post shaker, and a full-scale RMS course were created using the Vortex Studio MBD simulation software to generate the results presented in this work. In order to tune the parameters of the digital twin, a tilt test was also simulated in order to validate the inertial properties of the vehicle model.

The model of the vehicle includes an accurate representation of the vehicle drivetrain with accurate transmission ratios and engine torque table. The double wishbone front suspension and trailing arm rear suspension are also represented.

Fig. 3. Digital Twin of Vehicle with Added Passenger Masses

The digital twin was also loaded with weighted dummies on the driver and three passenger seats, as well as with a payload in the cargo hold, as shown in Fig. 3. The weight of these added masses was set to match their weights in the respective physical testing. The lateral and longitudinal location of the CG was also set to match the results of physical testing. The vertical position of the chassis CG was adjusted such that the results of the simulated and experimental tilt test result showed a good match for the vertical CG position, as illustrated in Fig. 5. After tuning the digital twin, the average of the tilt angles for both CS and RS tilts yielded a CG height of 1.02 m for the digital twin and 1.03m for the physical testing.

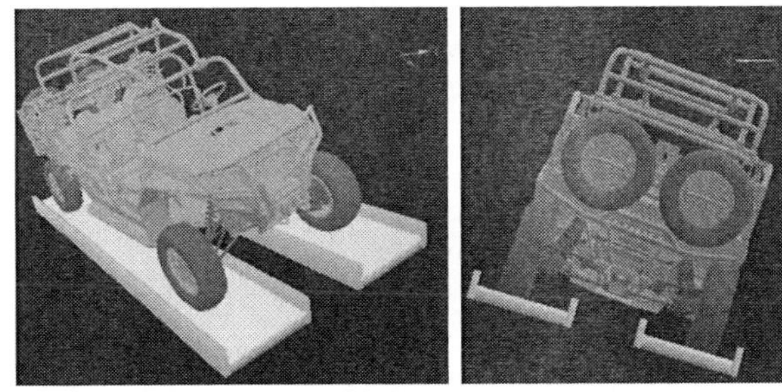

Fig. 4. Digital Twin of Tilt Table Test

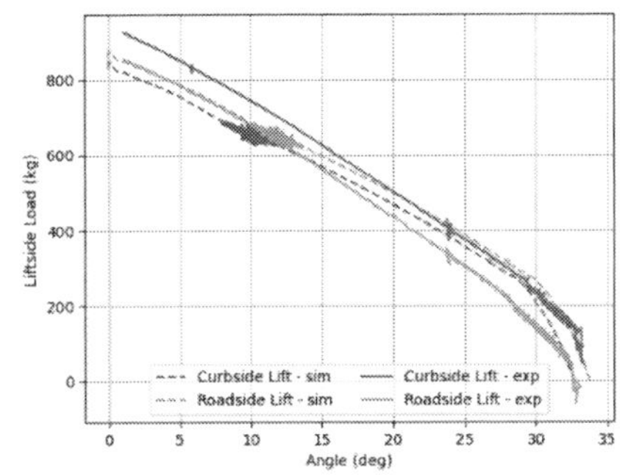

Fig. 5. Simulation vs Experimental Tilt Test Results

4.2. Four-Post Shaker

After validation of the digital twin through analysis of static and tilt test results, the digital twin could be used to simulate the four-post shaker test with some degree of confidence. In order to compare results of the physical ride quality tests that were performed on the MRZR vehicle, a digital twin of the four-post shaker mechanism was also created.

Each post consisted of a rigid body with three collision geometries that can interact with the vehicle model. These collision geometries are the cylindrical wheel pan and two side walls that prevent the vehicle from sliding too far in the lateral direction. A Coulomb tire model is used to compute the interaction forces between the wheels and wheel pans.

A prismatic constraint is added between each wheel pan and the inertial frame to control the vertical displacement of each wheel pan. Prismatic constraints in Vortex can be controlled by setting desired constraint velocities and maximum/minimum constraint forces. The limits for the constraint forces were set based on the specifications of the physical actuators, and the desired velocity was adjusted at each simulation time step by a post-controller Python script to achieve the desired displacement time history profile. Based on the desired position of the wheel pan at the next time step and the length of the time step, the controller sets the appropriate desired velocities for each wheel pan, as illustrated in Fig. 6.

Fig. 6. Digital Twin of Vehicle Mounted on Four-Post Shaker

Much like the physical four-post shaker, the controller can generate displacement time histories for the front left and front right posts based on a terrain profile and vehicle speed input. The same displacement profiles are applied to each rear wheel with a time delay computed based the vehicle speed. Thus, four displacement time histories are generated, one for each post. Alternately, the time history for each post can be directly input into the controller without the need to convert the terrain profile and vehicle speed to displacement time history profiles. This is useful for direct comparison to physical tests carried out on the four-post shaker because it ensures that the digital twin follows the true displacement time history of the physical shaker posts rather than their desired displacement time history.

4.3. RMS Terrain

The purpose of the four-post shaker test is to emulate the behaviour of a vehicle driving over a terrain of a given profile without the need for field tests. It accomplishes this by controlling the time history of the wheel pan displacement in a way that mimics what the vehicle would experience while traveling over the desired terrain at a given speed.

While we did not do field tests to confirm the validity of the four-post shaker testing approach, it is far less costly to set up this comparison in simulation. This was achieved by generating a mesh to represent the terrain which has the profile that the four-post shaker is meant to mimic and simulating the vehicle traveling over the terrain profile at a set speed. The MBD software handles the contact between the wheels and the mesh by computing a contact normal direction based on the intersection of the wheel and terrain collision geometries and incorporating this information into the tire model's wheel-terrain interaction force computation. Because this simulation requires the vehicle to drive over uneven terrain, a cruise controller and direction controller are incorporated into the model to ensure that the vehicle maintains a speed as close as possible to the desired velocity and to ensure that it stays on the course. The absorbed power of the full terrain traversal was then compared to the simulation of the vehicle on the four-post shaker. The vehicle travelling over the terrain mesh in the simulation environment is illustrated in Fig. 7.

Fig. 7. Digital Twin of Vehicle on 0.5 km-Long 3cm RMS Course - Side View with Contact Normals Visualized (Top) and Parametric View (Bottom)

5. Results

5.1. Real vs Simulated Four-Post Shaker Ride Quality Test

The results presented in this section come from the data collected during physical and virtual testing of a light utility vehicle travelling over a terrain with an RMS value of 3cm. The terrain profile is also asymmetric – meaning that the left and right wheels are travelling over different profiles with the same RMS value. The asymmetry is intended to induce vibration in the lateral direction so that the importance of vibrations in axes other than the typically studied z-axis can be observed. The acceleration in both physical and virtual testing is recorded at the location of the driver's seat. The acceleration profiles were then used to compute the absorbed power value. The absorbed power in each axis is plotted as a function of vehicle traversal speed for both the physical and virtual tests in Fig. 8 and Fig. 9, respectively. Looking at the 6W threshold in the z-axis, there is reasonable agreement between the physical and virtual testing, with the former finding that the vehicle reaches the 6W threshold in the vertical axis at a speed of 6.68 km/h while the latter crosses it at 4.26 km/h (36% error).

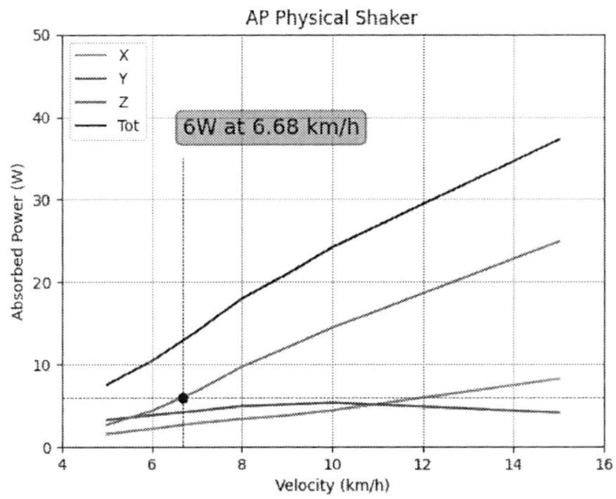

Fig. 8. Four-Post Shaker Results for Physical Testing

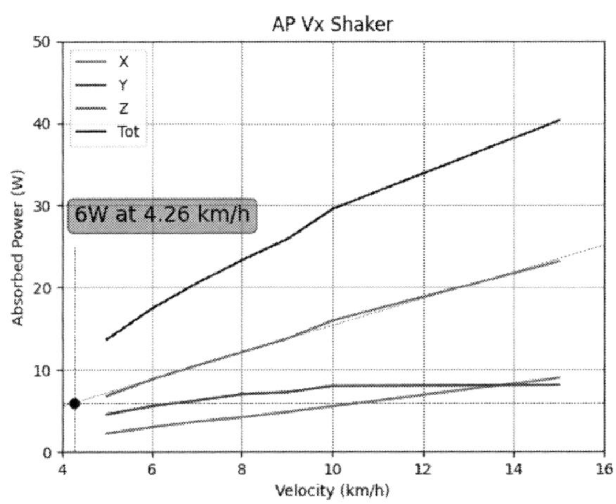

Fig. 9. Four-Post Shaker Results for Digital Twin

It is also noteworthy that the results for both the physical and virtual four-post shaker test show that the vibrations in the x- and y-axes also result in a non-negligible amount of absorbed power. While the current 6W standard only applies to vibration in the z-axis, these results could suggest that it may be beneficial to develop a standard which also considers vibration in other axes. However, that is not the focus of this paper.

5.2. Simulated Four-Post Shaker vs Simulated Full Terrain Traversal

The absorbed power values in the z-axis for each speed that was tested for the physical shaker, virtual shaker, and virtual RMS terrain traversal are summarized in Table 5 with the error relative to the virtual shaker test calculated. Fig. 10 shows the absorbed power results for the simulated vehicle traversing a fully simulated RMS terrain. The comparison between these results and the four-post shaker test would suggest that the accelerations experienced during the shaker tests could lead to an overestimation of the actual absorbed power for the course being tested. In the results obtained, the vehicle being tested only crosses the 6W threshold in the z-axis at 10.66 km/h – a value much higher than those determined during the physical and simulated shaker tests.

These results suggest that the difference in the way the vibration is applied to the vehicle in a four-post shaker test compared to a road test does have a significant effect on the resulting ride quality. This is likely a result of the difference between the contact geometry of a stationary wheel with a flat wheel pan and that of a rolling wheel with an uneven surface. The most significant difference being that the four-post shaker contact normal is always purely vertical, which is rarely the case for the contact between the wheel and the road. This result supports the findings of other works such as Kat and Els' work (Kat & Els, 2024), who also observed higher vibration magnitudes in four-post shaker test results compared to the equivalent terrain traversal and who also suggested the differing contact geometry as a potential cause.

Table 5
Z-axis Absorbed Power Results and Error Compared to Physical Testing

Speed	**Virtual Shaker**	**Physical Shaker**		**Virtual Traversal**	
(km/k)	AP (W)	AP (W)	|% Error|	AP (W)	|% Error|
5	6.82	2.69	60.56	2.77	59.38
6	8.84	4.37	50.57	3.38	61.76
7	10.53	6.76	35.80	3.99	62.11
8	12.11	9.69	19.98	4.58	62.18
9	13.76	12.02	12.65	5.18	62.35
10	15.95	14.42	9.59	5.57	65.08
15	23.19	24.90	7.37	8.82	61.97
		Average =	28.07	*Average =*	62.12

However, the authors do not consider this result to be conclusive proof that shaker tests produce much higher absorbed power predictions than their equivalent terrain traversal tests. It is possible that the assumptions made in MBD model relating to the method of detecting contacts for uneven terrain with high-frequency oscillations is responsible for the magnitude of discrepancies between the simulations of the four-post shaker test and full-scale terrain traversal. In the MBD software, an equivalent contact plane is defined at each time step based on the intersection between the wheel and terrain collision geometries. Field tests of the full terrain traversal would be required to definitively determine whether the difference in absorbed power for the simulated shaker test and full terrain traversal can be expected in reality or are a result of artificial phenomenon caused by the modelling approach used for handling uneven terrain. It is likely there are elements of both factors at play as the difference in the contact geometry will also be present in reality.

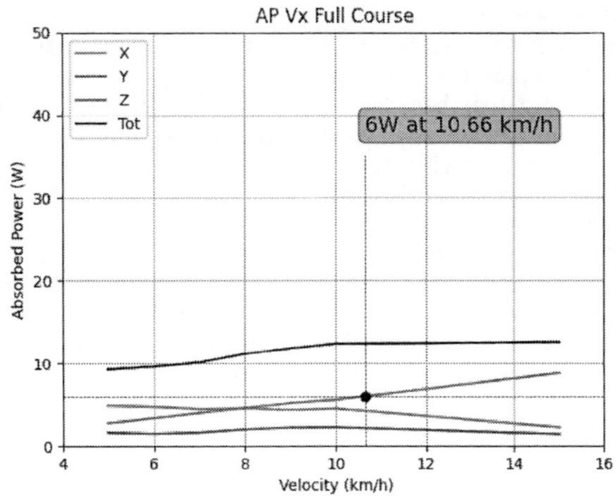

Fig. 10. Full-scale RMS Course Results for Digital Twin

6. Conclusions and future work

In this work we have detailed a method for assessing ride quality of a vehicle using a four-post shaker instead of field testing. Additionally, we have created a digital twin of the vehicle and simulated the four-post shaker test to validate the digital model. Finally, we simulated the vehicle performing a traversal of an uneven terrain course with the same profile that the four-post shaker was programmed to emulate. This final test was an effort to validate the concept of using the four-post shaker tests as a substitute for field testing. The comparison between the physical and virtual shaker testing showed good agreement and serves as a measure of validation for the quality of the digital twin and supports the use of digital twins to accurately simulate four-post shaker ride quality tests. The comparison between the virtual shaker and full-scale rough terrain course simulation was less conclusive. The absorbed power vs speed plots for the full course showed consistently lower absorbed power when compared to the equivalent test on the virtual shaker, though it is not clear whether this discrepancy is representative of reality, a result of the modelling method used by the MBD software for handling contact with uneven terrain or some combination of the two. In future work we hope to investigate the absorbed power comparison between the four-post shaker and the equivalent terrain traversal further through both simulation and field tests.

7. Declaration of competing interest

The authors declare that they have no known competing financial interests or personal relationships that could have appeared to influence the work reported in this paper.

8. References

Bylsma, W. W. (1999). Computer Generated Root-Mean-Square (RMS) Roughness Courses for Ride Quality Evaluation. Warren, Michigan: U.S. Army Tank-Automotive Research, Development, and Engineering Center (TARDEC).

Gräbe, R. P., Kat, C.-J., van Staden, P. J., & Els, P. S. (2020). Difference thresholds for a vehicle on a 4-poster test rig. Applied Ergonomics, 87, 103115.

International Organization for Standardization, 2016. ISO 8608:2016 Mechanical vibration — Road surface profiles — Reporting of measured data.

Kat, C.-J., Els, S.P., 2024. Ride comfort comparison between 4-poster and full vehicle driving simulations using difference thresholds. Journal of Terramech, 116, 101000.

Kat, C.-J., Gräbe, R., van Staden, P., Botha, T., Els, S.P., 2024. Difference thresholds for primary and secondary ride of a vehicle on a 4-poster test rig. Ergonomics, 67(11), 1702–1714.

Rosca, P., Bradley, S., 2021. Computation of Absorbed Power Using Co-Simulation TRUCKSIM-SIMULINK. Romanian Journal of Physics, 66.

US Army Development Test Command, 2007. Test Operations Procedure (TOP) 1-1-014 Ride Dynamics.

9668 | Innovating mobility: a student competition in wheel and track design

Devin Chen [a,b,*], Chaitanya Shekhar Sonalkar [c], Riku Kikuta [a], Andries Peenze [d], Varsha S Swamy [c], J. Ethan Salmon [a], Bohumir Jelinek [a], George L. Mason [a], P. Schalk Els [d], Corina Sandu [c]

[a] *Center for Advanced Vehicular Systems (CAVS), Mississippi State University, Mississippi, United States*
[b] *Department of Computer Science and Engineering, Mississippi State University, Mississippi, United States*
[c] *Department of Mechanical Engineering, Virginia Tech, Blacksburg, Virginia, United States*
[d] *Department of Mechanical and Aeronautical Engineering, University of Pretoria, Pretoria, South Africa*

** Corresponding author: dc2368@msstate.edu*

ABSTRACT

In this paper, we propose a new initiative for an ISTVS Wheel/Track Design Student Competition, aimed at engaging students in terramechanics and providing hands-on experience in off-road wheel/track design and testing. The competition will challenge student teams to design and fabricate a wheel or track system for a small unmanned ground vehicle (UGV), which will then undergo a series of tractive performance and mobility tests in selected soil types and conditions. By minimizing costs and focusing on practical, accessible design and testing methods, the competition ensures that students from various backgrounds and disciplines at any educational institution can participate. To support accessibility, our goal is to initially maintain a low barrier to entry by utilizing free software and affordable materials whenever possible. For example, teams can use freely available CAD software, 3D printing or other rapid prototyping techniques.

The competition details will vary from year to year. Initially, we propose a competition with two components: a laboratory-style terramechanics test rig with a single wheel soil test bin to test wheel/track design, and a small UGV similar to a radio-controlled (RC) car to test wheel/track design in different mobility/terrain tests. Performance tests may include measurement of drawbar pull, sinkage, slip, gross traction, and slope climbing characterization in alignment with ISTVS standards (He et al., 2020).

A single-wheeled terramechanics test rig will assess the wheel/track performance in controlled environments, supplemented by the small vehicle-based mobility tests under various soil conditions in the field. Each competition will include design presentations and performance tests evaluated using structured scoring criteria. The performance evaluation will be conducted using a test matrix that examines both structural integrity and performance. Through this competition, students will gain valuable experience, develop innovative solutions, and establish connections to ISTVS, fostering the next generation of terramechanics engineers.

Keywords
Terramechanics
Off-road mobility
Hands-on/Experiential Learning
ISTVS student competitions
Rapid Prototyping

1. Introduction

With the increasing number of graduates from engineering programs across the world, there is a need to augment traditional classroom theory with hands-on, or experiential learning. Experiential learning is defined as "a teaching philosophy in which educators purposely engage with learners in direct experience and focused reflection in order to increase knowledge, develop skills, clarify values, and develop people's capacity to contribute to their communities" (Tembrevilla et al., 2024). Terramechanics, a critical subfield of engineering, focuses on the interaction between vehicles and various types of terrain—making it highly relevant to applications involving wheeled or tracked systems.

To promote hands-on learning in this field, Mississippi State University, Virginia Tech, and University of Pretoria student chapters of the International Society for Terrain Vehicle Systems (ISTVS)—a professional network of researchers and engineers specializing in terramechanics—proposes a student competition designed to cultivate essential engineering skills. The goal is for interested universities/organizations to form an ISTVS student chapter, where members are then registered for the ISTVS Student Competition. This is to encourage interest in the field of terramechanics while increasing the number of members in ISTVS.

The competition will emphasize teamwork, engineering design, and technical communication while introducing participants to key concepts in characterizing off-road tractive performance for wheeled and tracked vehicles. For the first competition, each student team will be tasked with designing and prototyping a unique wheel or track system, culminating in the live testing and evaluation at the 56th Conference of the International Society of Terrain-Vehicle Systems scheduled in the last quarter of 2026 at the University of Pretoria, South Africa.

This paper outlines a proposal for the first of its kind ISTVS student competition: what is possible for the first competition in 2026, ideas for next competitions, and an invitation to discuss and receive feedback from interested external parties. In this paper the methodology of designing, testing, and evaluation of the wheel or track system is outlined, with flexibility to adjust requirements based on feedback. Requirements are outlined for the student teams to document their design process and include the design report as part of the competition, which can be judged through a review process as well as a design presentation hosted at the ISTVS conferences. In addition to the design requirements expected from the student teams, information is presented for the proposed test platforms that can be used to assess the performance of the student designs.

Two test platforms are proposed, consisting of a laboratory-style single-wheel soil bin test platform to ensure repeatable experiments as well as a small UGV or similar RC-like vehicle to assess dynamic maneuvers and tests. The laboratory-style single-wheel test apparatus will be used to test the wheel performance in a controlled and repeatable manner. The soil properties will be established first, followed by measurements of drawbar pull, gross traction, and sinkage of the wheel at varying forward and angular speeds. The new wheels will then be installed on a radio-controlled vehicle to test basic mobility metrics such as acceleration, deceleration, time to cover a specific distance, or a distance travelled in a specified time, aiming to correlate single-wheel performance with overall vehicle capabilities. Finally, a proposal is

https://doi.org/10.56884/IW5OTL1Q

made for scoring the various aspects of the student competition, including the design document, presentation, and experimental results, to crown a winner of the student competition.

2. Methodology

Along with the call for papers, universities will receive notification of the upcoming ISTVS Student Competition for the 2026 ISTVS conference, including a competition rulebook, deadlines, and schedules. This will be advertised by various means, including digital flyers on social media and mass email distribution. On the flyer or email will be a link to an online form platform where interested students can sign up for an interest meeting to get more information from competition organizers. During this information session, interested groups can ask any questions they may have regarding the competition. At the end of this information session, a sign-up link will be sent out to the interested parties. Teams can then register through the link, where they will be prompted to input their university/organization name and team member information before the selected registration deadline. At the conclusion of the registration period, those teams will receive further instructions via email.

2.1. Competition Organization

The student team captain will be the primary point of contact for all organization correspondence with support and involvement from the faculty advisors. They will both be required to be listed at registration. Upon confirmation of registration once the period of registration closes, the advisor and team captain will receive an email with details such as how to get started as well as receiving notification that they will be receiving a competition kit shortly with the proper materials to get their team started. Teams must design and deliver a 3D printable file in a prescribed format (e.g., .stl) containing their proposed wheel/track design to be submitted to the ISTVS competition organizers for rapid prototyping and testing.

The schedule for the competition will be provided by the student competition executive board in conjunction with that year's hosting university. The competition organizers will then print their file and test it during the conference time frame with the proposed testing apparatus discussed in sections 3.2 and 3.3. Judge selection for the ISTVS Student Competition will be made by a committee formed by the ISTVS board of directors. These judges will be chosen based on factors such as availability to help with the competition organization and judging, as well as having relevant topic area expertise. Competition judges will be responsible for helping facilitate the proceedings of the ISTVS Student Competition at the competition site. These responsibilities will include scoring students based on the grading rubric, providing feedback to students, as well as coordinating between competition organizers and competing student teams. The duties of the judges will change from year to year, depending on the metrics being tested. The debut of the ISTVS Student Competition will be held in South Africa at the University of Pretoria in 2026. A livestream will be set up for those universities unable to travel to the competition site. The competition will be held at the ISTVS conference venue.

2.2. Design Methodology

Students will design a wheel or tread track based on the small ground-vehicle platform provided as a basis for that competition year, such as an RC car. Designs will be tested with a single-wheeled testing rig, like the one shown below in Fig. 1. Each team will design their wheel or track based on the criteria being tested in a specific competition, such as maximizing drawbar pull, minimizing rolling resistance, improving traction, and provide their design in a well-established prescribed format, which can be created using software such as SolidWorks or Blender. Design parameters such as size, geometry, and orientation of parts will need to follow the specifications and reflect the parameters being tested for that competition year. These specifications may vary from year to year, depending on what hardware is chosen and other limitations.

Fig. 1. Single-wheel test platform at TMVS lab, Virginia Tech

Wheel and/or track designs submitted by students will be manufactured and prototyped with inexpensive methods such as 3D printing, as seen in Fig. 2. and Fig. 3. The type of filament or other material used to prototype each team's design will vary depending on what resources are available at the organizing university but will be the same for each competing team. Faculty advisors for the student teams are allowed to give general advice but may not design the wheel or tread pattern for the students.

Fig. 2. Examples of 3D printed wheel design, CAVS MSU

Fig. 3. Examples of 3D printed track design, CAVS MSU

3. Competition Test Cases

The competition can consist of different events, with each year having a different focus depending on the resources available and the local conference team. A detailed rulebook will be sent out to the universities every year to discuss the specifics of that event.

We propose below the different competitions and test metrics that can take place every year to engage students across universities. First, a technical design presentation is given by the teams to introduce to the audience the design methodology, leveraged technologies, modeling and simulation efforts that influenced the design decisions. After this, the design is tested using different benchmarks. We propose two methods to evaluate the design: the indoor controlled testing and the outdoor testing, depending on the resources available to the local conference team for evaluating the different teams' designs. Sensors, timers, and/or cameras will be used to measure the relevant performance parameters to provide an absolute scale on which the teams will be ranked. This section explores these ideas, proposing the different tests that can be performed, encouraging different test matrices for the dynamic evaluation and different goals for each year.

3.1. Design Presentation

The design presentation serves as the first opportunity for students to showcase their evaluation techniques and approach to the problem statement. Teams are expected to present various design alternatives, along with their evaluation criteria and the rationale behind selecting the final design. Suggested methodologies for performance prediction include but are not limited to: (a) Empirical models (e.g., tire numeric), (b) Analytical modeling (e.g., Bekker-Wong model), (c) High-fidelity simulations (e.g., FEM models), and (d) Experimental testing. A peer-review strategy is proposed for evaluating the presentations to ensure constructive feedback and knowledge exchange. Additionally, teams are encouraged to submit conference papers based on their findings, contributing to the broader research community, like the referenced article (Sinha & Gupta, 2023). Engaging in such scholarly discussions will help refine their methodologies and promote innovation in the field.

3.2. Controlled Testing

Following the design presentation and based on the competition rules and constraints from each year's conference organizers, a series of controlled, laboratory-style tests will be conducted on the submitted wheel or track systems. These test conditions will be included in the rules and will be shared with student teams well in advance to inform their design process. To promote innovation and avoid incremental design improvements, the competition parameters will vary annually. Changes may include geometry constraints, test procedures, or evaluation metrics. Accordingly, the test platform must be versatile enough to accommodate different configurations while remaining simple to install, operate, and transport between conferences. A modular design will support ease of assembly and adaptability. Soil properties and wheel performance will be measured in line with ISTVS standards (He et al., 2020). For soil characterization, unless the soil properties were already established, pressure–sinkage and/or shear strength measurements (Okello, 1991) will be performed. For wheel performance, established terramechanics performance metrics can include drawbar pull, gross traction, rolling resistance, sinkage, and their dependence on slip. The overall vehicle performance can be characterized by gradeability, startability, and obstacle traversal.

Student teams attending the conference may have the opportunity to directly interact with the controlled testing apparatus. For teams unable to attend in person, competition organizers will conduct testing on their behalf.

3.3. Single Wheel Test Platform

To assess these metrics, a standardized test setup is proposed. A scaled version of the single wheel test-rig at Virginia Tech, illustrated in Fig. 1, can be utilized to evaluate the performance of the 3D printed wheels. At its core is a small soil bin (approx. *2.0* m long, *1.0* m wide, *0.6* m deep), constructed from low-cost materials by each year's organizers. Soil selection is flexible, ideally reflecting local conditions to add variability. Careful soil preparation is proposed in between test runs to provide fair and similar initial conditions for each team. The conference organizers can measure and supply soil properties well in advance to the teams. The test platform may be fixed to the soil bin or mounted on rails or with a rack and pinion setup to traverse its length. In both cases, relative motion between the wheel/track and soil is controlled using electric motors, possibly with gearboxes and belt/pulley systems. Independent control of wheel and carriage speed enables precise slip ratio testing. Vertical load can be varied using standard weights mounted on the carriage. The total mass will be provided to teams in advance, eliminating the need for onboard load measurement. The soil bin may also be inclined or fitted with mechanisms to vary slope between tests. Careful soil preparation is critical for consistency. Obstacle traversal can be introduced using rigid or removable fixtures placed in the soil bin, defined by organizers or as part of the challenge. This modular, adaptable setup enables rigorous evaluation across a range of metrics. By varying the test focus annually, the competition can remain engaging and technically meaningful. It also offers students valuable exposure to experimental methods in terramechanics within a fun and collaborative environment.

3.4. Controlled Testing Measurement Systems

Accurate sensing and control are essential for repeatable results. A brief list of sensors and measurements required to accurately assess team designs is included. To measure the wheel or track states, encoders can be installed on the electric motors to measure motor revolutions and calculate speeds. Similar encoders can be installed on the carriage drive system to measure and infer the carriage speeds to determine the wheel slip ratios. The electric motor drive torques can be inferred via current sensors and forces measured directly using load cells between the drivetrain and frame. Laser displacement sensors can measure rut depth and sinkage, placed ahead of, next to, and behind the wheel.

3.5. Test Platform Proposed for the Inaugural ISTVS Student Competition

From this extensive list of configurations, test cases, and measurement systems, the following selection is proposed for the first competition, with flexibility for change based on recommendations and discussion from the wider ISTVS community.

- The initial design will make use of a single wheel design.
- A simple, horizontally flat soil bin with a carriage system driven with an electric motor on a rack and pinion to propel the carriage in a controlled manner along the soil bin.
- The wheel subassembly, with its own driving motor and gearbox, is attached to the carriage to control relative wheel speed.
- The wheel subassembly is mounted with linear rails to enable free movement in the vertical direction. Provision is made here to add pre-defined weights to vary the vertical load.
- Load cells are installed horizontally in between the linear rails and the carriage system to enable measurement of drawbar pull forces.
- All electric motors are instrumented with encoders, possibly torque/current sensors, and appropriate drivers to enable angular speed and torque/current measurements and control.

With this configuration, the drawbar pull, longitudinal slip ratio, power requirements, and efficiency can be measured, enabling an extensive list of testing criteria to evaluate the performance of each team's wheel design in a controlled manner. This method of evaluation, however, also has drawbacks such as limitations on the longitudinal velocity of the wheel. The test platform provides information on the performance of a single wheel; however, the performance of a wheeled vehicle cannot be estimated effectively using the single wheel test results. The test platform may also present a logistics challenge as it needs to be disassembled and transported to the next conference venue each year. Also, the teams whose facilities have the capability to run such tests may host certain elements of each year's competition until more teams can acquire the necessary equipment. As the competition evolves, the platform can be updated to accommodate additional evaluation parameters.

3.6. Tests On Small Unmanned Ground Vehicles

The competition may also test the manufactured wheel on a small UGV (e.g., RC car) to assess its behavior in dynamic maneuvers. The specific test setup will be determined based on the conference location and available resources for that year. Teams will be provided with key details, including: (a) UGV specifications (weight, dimensions, number of wheels, drivetrain type), (b) soil characteristics (e.g., cone index, engineering properties, moisture content), and (c) maneuver conditions (target speed, maximum steering angle, torque/engine power limits). Below in Fig. 4 is an example of a small UGV platform at the Center for Advanced Vehicular Systems (CAVS) that might be used as a test vehicle by competition organizers.

Fig. 4. An example small UGV or 'RC car', CAVS MSU

Given a specified soil type, the following maneuvers are proposed: (a) acceleration/braking test, (b) cornering tests, (c) hill climb test, and (d) wet soil traction test. To ensure repeatability, tests will be performed under controlled environmental conditions with consistent soil preparation. Soil properties such as moisture content, density, and strength will be measured to verify uniformity. The vehicle's onboard sensors will log real-time data, including wheel speed and lateral and longitudinal acceleration, providing objective performance metrics. Additionally, post-test analysis, including visual inspection of soil displacement and rut formation, will offer further qualitative insights into wheel effectiveness on different terrains. This method of evaluation will provide better insight into how the designed wheel performs on a vehicle. However, this method would limit the dimensions of the wheels to the size of the RC car. Also, the selected sensors should be feasible to instrument the vehicle.

3.7. Measurement Systems for Small Unmanned Vehicles

For tests on small UGVs or RC cars, the sensor suite should be accurate and reliable, as vehicle tests will be subject to more disturbances. Establish a well-characterized sand/soil pit—thoroughly measured, compacted, and calibrated—to ensure uniform driving conditions. Onboard accelerometers are sufficient for capturing vehicle acceleration, while slope climb and wet-traction performance can be assessed by recording the elapsed time for each run.

4. Assessment of Design and Scoring

Each team's design reports will be assessed by a panel of experts, ideally from the ISTVS community. It should reflect sound engineering judgement and a clear understanding of the design problem. Each panel member will independently score the report to ensure consistency and reduce bias. Ideally, this evaluation should take place before the conference so that the overall competition winner can be announced during the event.

Design presentations will be judged at the conference by a panel including ISTVS members, local organizers, and invited experts. Presentations should clearly communicate the design approach, justify key decisions, and demonstrate the team's ability to respond to questions. While in-person presentations are preferred, provisions can be made for remote participation via recordings if needed.

Scoring for the controlled testing will rely on objective performance metrics based on the competition's specific constraints and rules. These parameters will be measured using the test platforms and used to assess each design's performance.

In addition to expert reviews, peer assessment between teams is encouraged to promote collaboration, knowledge exchange, and networking. A portion of the final score may reflect these peer evaluations, contributing to a student team choice award.

Winners may be announced in each segment—design report, presentation, testing, and peer award—with an overall winner determined through a weighted combination of all segments. This assessment and scoring of the event will be done on-site after the competition by a panel of judges assembled by the committee formed by the ISTVS board of directors. While this favors teams participating in all areas, those entering select segments still have the chance for recognition. There will be 1st, 2nd, and 3rd place winners, with a special category for the Student's Choice Design Award winner. Special recognition will be given to winners of the student competition in the form of unique ISTVS-themed trophies for that competition year, certificates, useful tools or equipment, and eventually monetary prizes as funding for the student competition grows. This funding will come from external entities such as industry experts in the field of terramechanics who are interested in promoting greater student interest in the field.

The assessment and scoring framework is still under development and open to feedback. Organizers plan to draw on existing models from similar competitions, such as Formula SAE (SAE International, 2024a), Baja SAE (SAE International, 2025) and Mini Baja SAE (SAE International, 2024b), to help shape the final rulebook that will be shared with student teams. The current approach serves as an initial foundation and will be refined over time through input from participants, organizers, and the broader ISTVS community.

5. Conclusions and Future Work

This paper suggests the first of its kind ISTVS student competition series. It outlines the goal of the first competition, which is the design of a wheel or track element for an off-road vehicle. Requirements are listed for the student teams: 1) document and provide arguments for your design process, and 2) include the design report as part of a conference poster submission, which will be judged through a review process as well as a presentation hosted at the ISTVS conferences. Two test platforms for the design are proposed: 1) an indoor single-wheel test rig with soil bin, and 2) a radio-controlled (RC) car or similar small, unmanned ground vehicle (UGV) with the newly designed tractive element installed. The single-wheel test rig will measure the drawbar pull, gross traction, rolling resistance, and sinkage of the wheel at various slips in a controlled environment. A small UGV will be used to correlate the single-wheel test results with the full vehicle performance, such as acceleration, deceleration, slope climb, and time to cover a specific distance.

Multiple aspects of the student design, including the design document, presentation, and experimental results, will be used to score the competing teams and to crown a winner of the student competition. The detailed rules and schedule for the 2026 competition will be established and distributed after the 2025 ISTVS conference by the competition-hosting entity, which is the ISTVS student chapter of the University of Pretoria. The final testing and scoring of the teams will be performed at the 2026 ISTVS conference.

We actively welcome the feedback and involvement from ISTVS members, academic institutions, and industry partners that are interested in contributing to or hosting the competition. This collaborative approach will ensure the competition reflects the shared goals and insights of the broader terramechanics community.

For future work, we aim to increase participation, seek and incorporate feedback, and ask for ideas for the future competitions from the conference attendees and later from the general ISTVS community. Any ideas to expand or incorporate more areas of interest in the future, like autonomy for the UGVs, are welcome. Soliciting sponsorship from external groups will be key to growing the program in coming years. This sponsorship will only further elevate the prestige of the event and help offset costs for student teams interested in participating. These costs can include travel costs associated with travelling to the ISTVS conference as well as any design-related costs such as filament for 3D printers. Broader engagement will be key to refining the concept and ensuring the long-term success and relevance of the competition.

6. Acknowledgements

We gratefully acknowledge our peers, research teams, and academic institutions for their invaluable contributions to this paper and the successful launch of the ISTVS Student Competition. We also extend our sincere thanks to the International Society for Terrain Vehicle Systems for their support and encouragement in turning this vision into reality.

7. Declaration of Competing Interest

The authors declare that they have no known competing financial interests or personal relationships that could have appeared to influence the work reported in this paper.

8. References

Baja SAE, n.d.. https://www.bajasae.net/cdsweb/gen/DocumentResources.aspx

Formula SAE, n.d.. https://www.fsaeonline.com/cdsweb/gen/DocumentResources.aspx

He, R., Sandu, C., Mouisave, H., Shenvi, M.N., Braun, K., Kruger, R., Els, P.S., 2020. Updated Standards of the International Society for Terrain-Vehicle Systems. J. of Terramech. 91, 185-231. https://doi.org/10.1016/j.terra.2020.06.007

Okello, A., 1991. A review of soil strength measurement techniques for prediction of terrain vehicle performance. J. of Agricultural Engineering Research 50, 129-155. https://doi.org/10.1016/S0021-8634(05)80010-1

Sinha, P., Gupta, A., 2023. Design and Performance Optimization of SAE Baja ATV. https://www.researchgate.net/publication/38718491

Tembrevilla, G., Phillion, A., Zeadin, M., 2024. Experiential learning in engineering education: A systematic literature review. J. of Engineering Education 113, 195-218. https://doi.org/10.1002/jee.20575

The return of mini baja, n.d. https://www.bajasae.net/cdsweb/gen/DownloadDocument.aspx?DocumentID= d9a8c915-e910-4ab3-ba47-0f4a0322c1b8

ISTVS

Abstract-only submissions

0280 | Real-time ground segmentation and terrain topography mapping using point clouds

Jyotirmoy Mukherjee *, Giri M Kumar,
Andrea L'Afflitto *, Hannah White, Corina Sandu *

Virginia Tech, United States

** Corresponding authors: jmukherjee@vt.edu, a.lafflitto@vt.edu, csandu@vt.edu*

The abstraction of the world in a virtual space known as a map is crucial to enable motion planning. Autonomous mobility over uneven terrains requires the model of terrain topography to account for surface profile induced factors that affect the feasibility of planned trajectories. Point cloud-based environment modeling has been particularly effective due to their ability to produce depth information. However, when deployed in UGVs, they encapsulate both traversable terrain and positive obstacles which must then be distinguished. This work introduces a novel and fast point cloud segmentation algorithm called the inverse ray tracing algorithm to classify traversable and non-traversable points. The segmented point cloud is then collated together to create a comprehensive map. To this end, we present a Bayesian method using inverse sensor models that generate a 3D voxel grid of the terrain's topography and a second method using convex hulls around classified terrain points, sequentially appending points to generate the terrain topography map. The proposed results are suitable for UGVs operating in GNSS-denied environments. Our approach integrates sensor fusion from a visual-inertial sensor, wheel encoders, and an inertial measurement unit with an extended Kalman filter to maintain precise odometry. The effectiveness of the proposed algorithms has been experimentally validated on a Terramechanics rig and compared with point cloud obtained from a 3D scanner, which resulted in a chamfer distance of approximately 1 mm, demonstrating high accuracy. Additionally, the navigation stack has been deployed on the Clearpath Husky platform to generate real-time terrain topography and obstacle maps, proving the practical applicability and robustness of our proposed algorithms.

0398 | Contact dynamics slope stability

Tomas Berglund [a], Joakim Rydman [a],
Martin Servin [b,*]

[a] *Algoryx Simulation AB, Sweden*

[b] *Algoryx Simulation AB and Umeå University, Sweden*

** Corresponding author: martin.servin@umu.se*

A heavy vehicle operating near a steep slope alters the stress distribution and might cause the slope to collapse. The result can be catastrophic for the operator, the machine, and for other units in the area. Computational methods for assessing the risk for slope collapse has existed for a long time in geotechnics but are limited to static conditions without mobile equipment. We explore a contact dynamics approach to slope failure analysis. The method searches a space of potential failure surfaces for the most critical one. The soil in each potential failure surface is treated as a rigid body supported by contact forces with the interfacing (static) terrain. The body is subject to gravity and contact forces from any multibody systems, such as a vehicle, interacting with it. A Signorini-Mohr-Coulomb contact model is assumed, stating that the relative normal and tangential velocity should be zero along the interface unless the normal and tangential forces reach their limit conditions, that depends on the (internal) friction and cohesion parameters, in which case the body is prone to acceleration. The discretized equations of motion, constrained by the contact model, form a linear complementarity problem that is solved numerically. The solution reveals whether the configuration is unstable or stable, and is factor-of-safety (FOS). A search algorithm for finding the most critical failure surface, in terms of smallest FOS or highest acceleration, is developed and analysed. The method was implemented in the realtime physics engine AGX Dynamics. The tests shows that it gives reliable results, comparable with conventional LEM methods, and is computational efficient enough for being used with a realtime simulator of terrain vehicles.

0442 | Lessons learned from global UGV and autonomous systems

Patrick Shepherd [a,*], Paul Clayton [b,*]

[a] *Milrem Robotics, Finland*

[b] *Milrem Robotics, Estonia*

** Corresponding authors: patrick.shepherd@milrem.com, paul.clayton@milrem.com*

Milrem Robotics today is in operational test and use in 19 countries worldwide. Our systems have both robotic and autonomous capabilities depending on the customer and are operational in a variety of terrains from Western Europe, Middle East and Asia. Our cornerstone customer is Ukraine where Milrem has been recently awarded the largest non-USA contract for UGVs. With already over 3 years and 10,000 operational hours in support of combat operations Milrem is the largest provider of non-Ukraine UGVs in the current conflict with Russia.

0583 | Dynamic off-road route planning: integrating terrain passability and weather for optimal mobility

Wojciech Dawid *, Krzysztof Pokonieczny,
Marek Wyszyński

Military University of Technology, Poland

** Corresponding author: wojciech.dawid@wat.edu.pl*

The need to move vehicles off-road arises primarily in military operations and crisis management, where standard road networks may be insufficient or inaccessible. While vehicles typically follow existing roads, critical situations often require off-road movement to execute surprise maneuvers or reach remote locations beyond the infrastructure network. In such scenarios, effective planning is crucial and requires detailed geographical data to assess terrain conditions and vehicle mobility. To support this process, terrain passability maps are essential. These specialized maps synthesize geospatial data, classifying terrain based on how different types of vehicles can navigate it. Despite advancements in vehicle traction technology, such as improved suspension, tires, and tracks, many vehicles still struggle on muddy or waterlogged soils. The war in Ukraine underscored this issue, with reports of modern Russian tanks failing to traverse swampy, clay-rich terrain. This research presents advanced methods for dynamically determining off-road routes with varying difficulty levels while accounting for changing weather conditions. The proposed solutions not only enable real-time route adjustments based on current weather patterns but also incorporate meteorological forecasts, allowing for more strategic long-term route planning. The route determination process relies on the creation of dynamic terrain passability maps, which serve as a foundation for generating a graph that calculates optimal paths. These maps take into account the impact of precipitation on soil properties. By integrating this data, routes can be precisely adapted to both environmental conditions and terrain characteristics, significantly improving efficiency and safety in difficult operational areas.

0733 | Impact of operational environment on mobility in unmanned mode of wheeled UGVs

Marek Nowakowski *

WITPIS, Poland

* *Corresponding author: marek.nowakowski@witpis.pl*

Unmanned Ground Vehicles (UGVs) are increasingly deployed in environments where human access is limited or hazardous, such as military operations, disaster response, and remote exploration. The autonomous or semi-autonomous operation of these platforms relies heavily on advanced perception systems, which integrate various sensors, including LiDAR, radar, cameras, and inertial measurement units (IMUs). However, complex environments—especially those with dense vegetation—pose significant challenges for object detection and navigation. Occlusion, signal scattering, and distortion from foliage significantly reduce sensor effectiveness, compromising mobility and operational safety. This study analyzes the limitations of both active sensors (LiDAR, radar) and passive sensors (cameras) under adverse environmental conditions, particularly in contexts where security and reliability are critical. Special attention is given to the UGV's ability to traverse obstacles in off-road terrain, focusing on design constraints and mechanical limitations that affect mobility. Additionally, the challenges in detecting and classifying objects hidden in dense vegetation are addressed to avoid collisions and prevent vehicle damage. Object recognition based on physical properties is emphasized, especially in autonomous navigation contexts. These findings are particularly relevant to sensor systems, which provide critical data to the vehicle's navigation subsystem in unmanned mode. The study underscores the importance of optimizing sensor fusion and vehicle design to enhance reliability and mobility in challenging operational environments.

0938 | AI-based terrain passability modeling

Marek Wyszyński *, Krzysztof Pokonieczny, Wojciech Dawid

Military University of Technology, Poland

* *Corresponding author: marek.wyszynski@wat.edu.pl*

Terrain passability assessment is a crucial aspect of military operations, influencing strategic mobility and tactical decision-making. Traditional methods of evaluating terrain traversability rely on static maps and expert judgment, often lacking the precision needed for dynamic environments. This paper presents an AI-driven approach to terrain passability modeling using artificial neural networks (ANNs). In research two models ann_full and ann_fast— were developed to predict passability coefficients based on geospatial and hydrological data. The ann_full model utilizes a comprehensive dataset of terrain attributes, while the ann_fast model focuses on the most influential parameters to enhance computational efficiency. The models were trained and validated using military geospatial databases, including VMAP Level 2, applied to the Suwałki Gap region. Results indicate that the ANN-based models outperform conventional methods by providing a more granular and accurate assessment of terrain conditions. Comparative analysis of different modeling techniques, including statistical threshold methods and vegetation roughness factor (VRF) models, further underscores the advantages of machine learning approaches in terrain analysis. The proposed AI-driven methodology enhances decision support systems, providing real-time, data-driven insights for military and autonomous vehicle applications.

1056 | Optimizing vehicle navigation over terrain

Anthony Farkas *, Henry Hodges *, Cooper Madrazo *

Hodges Transportation, DBA Nevada Automotive Test Center, United States

* *Corresponding authors: afarkas@natc-ht.com, hhodgesjr@natc-ht.com, cmadrazo@natc-ht.com*

This abstract summarizes the theoretical approach and method of application used to identify available travel paths, based on the near real time assessment of the interaction between the vehicle and the deformable terrain surface. The framework of this system is based on two avenues of effort that are combined to provide high fidelity mobility corridors that the manned or unmanned vehicle can use to help ensure successful mission completion. The first avenue uses multiple layers of terrain data to determine parameters of map grid elements, including but not limited to: soil type, elevation, soil moisture, roughness, slope, and land cover. These layers are used to establish conditions the vehicle may encounter. The second avenue of effort is a 3DMDB physics-based, deformable soils vehicle model. A large array of operational conditions, are provided to the high fidelity model to produce a predicted mobility and operating speed for each. The resultant information is used to create a fast running model, which is then combined with the first avenue. By combining both avenues, the area of operation can be evaluated to determine the available terrain and the speed at which a vehicle can maneuver through a selected area. The fast running model is continuously updated with available vehicle parameters and sensor data, to assess the capability of the vehicle to continue through the operational area. The real time interaction of the vehicle with the terrain is assessed using established terramechanics, physics-based relationships, updating the prediction for the available routes, and speed which can be achieved. Using a cost based optimization model, a route based on the selectable criteria such as fastest or most efficient is then computed resulting in a recommended path.

1171 | Effect of stress distribution model in wheel traveling analysis using the extended terramechanics model

Mai Shimizu [a,*], Shingo Ozaki [b,*]

[a] *Yokohama National University, Japan*

[b] *Faculty of Engineering, Yokohama National University, Japan*

* *Corresponding authors: shimizu-mai-fx@ynu.jp, ozaki-shingo-xd@ynu.ac.jp*

The development and operation of exploration rovers is essential for human lunar exploration, which is considered an international space exploration goal. However, the lunar surface is a completely different environment from that on the Earth because it is covered with regolith. Hence, it is difficult to evaluate rover's traveling performance in advance. Under these circumstances, the extended terramechanics (xTerramechanics) model has been proposed that takes terrain surface deformation into account. Meanwhile, in a wheel traveling analysis, stress evaluation models are important in the evaluation of wheel-ground interactions such as forces, torques, and soil movements. In addition to the well-known parabolic stress evaluation model proposed by Wong and Reece, Tsubaki and Ishigami have proposed a Gaussian stress evaluation model based on direct measurement of stress distributions. This model may be able to suppress excessive shear stress behind the wheel and predict more realistic traveling performance even in assuming regolith surface. Therefore, this study systematically examined the effects of these stress distribution models by conducting a traveling analysis using xTerramechanics model. The results show that for small-diameter wheels, the Gaussian distribution model can predict practical traction performance with fewer parameters.

1536 | Automated system for determining military road and terrain passability based on terrain, meteorological and hydrological conditions

Krzysztof Pokonieczny *, Marek Wyszyński *, Wojciech Dawid *

Military University of Technology, Poland

* *Corresponding authors: krzysztof.pokonieczny@wat.edu.pl, marek.wyszynski@wat.edu.pl, wojciech.dawid@wat.edu.pl*

The article will present a system that allows automation of the process of performing terrain passability analysis. The use of automation allows this analysis to be performed quickly and in any area of interest. As part of the research, methodologies have been developed for determining the index of passability (IOP), the task of which is to calculate measurable conditions of vehicle passability in a specific area, taking into account a variety of factors. The main task of these methodologies is to integrate data from various sources (mainly geospatial data: digital maps, meteorological and soil data) and determine an easy-to-interpret coefficient that will additionally take into account the traction characteristics of military vehicles. The determined coefficients will allow the generation of easy-to-interpret passability maps. Combining such a large amount of data from different sources made the resulting maps change their content over time as the meteorological situation changes. The terramechanics experiments carried out, made it possible to take into account the ability of specific vehicles used by the armed forces to travel on roads and roadless areas under specific meteorological and hydrological conditions. Taking into account the specifics of the weather data, the system generates passability maps both taking into account the current situation and also allows for the prediction of passability conditions. In this way, an "on-call, automatically updating passability map" was created, this allows the system to provide the commander with information on the ability to move troops not only for the present moment, but also, taking into account the forecasts. This gives an entirely new capability for planning the movement of troops over roads and roadless areas.

1960 | Efficient modelling of the effect of tire deformation during contact with obstacles

Mahdi Maleki [a,*], Jozsef Kovecses [b,*]

[a] *McGill University, Canada*

[b] *McGill, Canada*

* *Corresponding authors: mahdi.maleki@mail.mcgill.ca, jozsef.kovecses@mcgill.ca*

The ability of compliant tires to deform plays a crucial role in enhancing their obstacle negotiation capability compared to rigid wheels. This work presents an efficient method for evaluating the effect of tire deformation by adjusting the contact parameters of a rigid wheel model. Instead of explicitly solving for structural deformations, the approach incorporates constraint kinetic energy at the moment of contact to account for deformation effects, ensuring computational efficiency while maintaining accuracy. The method dynamically modifies contact parameters, such as normal vectors and contact points, to reflect the altered interaction between the tire and the obstacle. This adjustment enables the introduction of a reduced model that efficiently incorporates the effects of tire deformation without explicitly modelling structural flexibility, providing a computationally efficient yet accurate approach to analyzing obstacle negotiation. The simulation results demonstrate that the proposed approach effectively replicates key deformation-induced behaviours observed in high-fidelity models while achieving real-time performance. This makes it a practical and efficient solution for vehicle dynamics studies, robotics, and autonomous vehicle applications where both computational efficiency and physical accuracy are essential.

2154 | Simulation of dynamic behavior in armored vehicles using modified off-road vehicles for civil protection applications

Jorge Lopera [a,*], Juan Pablo Angulo [b,*]

[a] *PACIFIC+, Colombia*

[b] *ANGTRAC, Colombia*

* *Corresponding authors: jlopera@pacificplus.co, info@angtrac.com*

In the realm of civil protection at countries with security concerns like Colombia, armored vehicles play a crucial role in ensuring the safety and security of persons. These vehicles undergo significant modifications, including increased mass to enhance their armor features, which in turn alters their dynamic behavior. This study explores the feasibility of using an off-road vehicle to simulate the dynamic behavior of an armored vehicle, thereby providing a cost-effective platform for the development, and testing of non-sprung components. The project utilized a modified off-road vehicle with an initial mass of 2,500 kg. By incrementally adding ballast, the vehicle's mass was increased to 3,500 kg, simulating the weight of an armored vehicle. Dynamic tests were conducted on three types of terrains (gravel, mud, and asphalt), to evaluate the vehicle's performance under different conditions. Key parameters such as suspension travel, damping characteristics, and ride comfort were measured and analyzed. Preliminary results indicate that the modified off-road vehicle successfully replicates the dynamic behavior of an armored vehicle with a reliable degree of accuracy. Components, like the shock absorber increases their load and could harmful the vehicle dynamics for this type of vehicles. With this methodology it could be possible to analyze too the energy absorption and ride comfort metrics for these types of vehicles. This research demonstrates the potential of using older off-road vehicles as a viable platform for simulating the dynamic behavior of armored vehicles. The findings provide valuable insights for the development of non-sprung components, ultimately contributing to the enhancement of armored vehicle performance in civil protection applications.

2458 | Obtaining first order tyre parameters with insitu testing

Carl M. Becker *, P. Schalk Els

Vehicle Dynamics Group, University of Pretoria, South Africa

* *Corresponding author: carl.becker@up.ac.za*

In any vehicle dynamics simulation model, the tyre model is of upmost importance as it is the only component in contact with the terrain. In many cases engineers under estimate the contribution the tyre model makes to the reliability and accuracy of the simulation results. On larger off-road vehicles this is the case due to the limited tyre data and models available to the engineers as parameterize tyres are not trivial and can be very expensive compare to the cost of the tyre. On very large tyres the infrastructure required to conduct tyre parameterization tests are not available. Vehicles used on soft terrain are also often used on hard terrain. Tyre parameters are measured on hard terrain as the tyre carcass construction, inflation pressure and vertical load on the tyre dictates the linear section of the longitudinal and lateral tyre parameters. The maximum longitudinal and lateral load generated by the tyre is dictated by the surface roughness of the terrain over which the tyre transverse. In this study we conduct in-situ test on 2 vehicles on a specific surface. These results are compared to laboratory tests conducted on the same tyres and same surface. This method enables engineers to obtain first order tyre parameters to validate simulation models with limited and non-specialized equipment.

3375 | Real-time simulation of flexible tracks for accurate vehicle-terrain interaction

Mahdi Maleki [a,*], Wing Hang Ho [a], Joe Hewlett [b], Marek Teichmann [b], Jozsef Kovecses [c]

[a] *McGill University, Canada*

[b] *CM Labs Simulations, Montreal, Quebec, Canada, Canada*

[c] *McGill, Canada*

* *Corresponding author: mahdi.maleki@mail.mcgill.ca*

Modelling flexible tracks is essential for accurately simulating the mobility of tracked vehicles, particularly in off-road environments. Traditional high-fidelity models provide detailed insights but are computationally expensive, making them unsuitable for real-time applications. This work presents an efficient approach for simulating flexible tracks by dynamically calculating track shape, tension, and contact forces while maintaining computational efficiency. The track shape is determined based on an ordinary differential equation (ODE) that relates wheel positions, track tension, wheel sinkage, and pressure distribution. Ray casting is used to estimate sinkage along the track, while track forces (including normal, shear, and lateral forces) are computed using established pressure-sinkage and shear stress-displacement relationships. The model updates track shape each time step by estimating entrance and exit angles at wheel contact points and fitting a polynomial to ensure smooth transitions. Contact forces are then distributed to the wheels, enabling accurate simulation of traction and mobility. This approach efficiently captures key deformation behaviours observed in high-fidelity models while achieving real-time performance by incorporating track tension effects and interaction forces without explicitly solving complex structural deformations. The method is well-suited for vehicle dynamics studies, robotics, and autonomous systems where fast and reliable terrain interaction modelling is required.

3985 | One AI driver for multi-mission ground operations

Bobby Bonello [a,*], Rob Brown [b,*]

[a] Private sector, United States

[b] Head of Product, United States

* Corresponding authors: b.bonello@kodiak.ai, r.brown@kodiak.ai

Military operations rarely occur in a single terrain—missions often transition from paved highways to rugged trails, unimproved roads, or entirely off-road environments. Building separate autonomy systems for each creates gaps, integration challenges, and operational risk. Kodiak takes a different approach: one AI driver that seamlessly covers all operating domains, meeting the demands of military ground operations while adhering to the law of the road, civilian traffic, safety, and rules of engagement. Kodiak is the leading provider of dual-use ground vehicle autonomy and the first to field a driverless product owned and operated by a customer. Our mission is to deliver a reliable, flexible human driver replacement for on-road and off-road operations. Since 2018, Kodiak has deployed the Kodiak Driver across commercial and DoD missions. From 2022–2024, we competed in the Army's Robotic Combat Vehicle program, integrating our system into Ford F-150s to demonstrate terrain-aware off-road navigation, secure teleoperation, and rapid adaptation to new Environments. In 2024, we integrated onto the Textron M3 Ripsaw, proving our platform-agnostic design. Our modular, vehicle-agnostic autonomy stack integrates across platforms, supports high uptime with rapid hardware replacement, and is built on a Modular Open Systems Approach (MoSA) for mission command, teleoperation, and communications. One AI driver across all domains ensures mission continuity, reduces complexity, and enables safer operations across the battle space.

4111 | Off-road autonomy using Koopman theory

Umesh Vaidya *

Clemson University, United States

* Corresponding author: uvaidya@clemson.edu

How can we make autonomous vehicles smarter in unpredictable, off-road environments? In this talk, I will share recent progress on using Koopman theory as a powerful tool for data-driven autonomy. By transforming complex, nonlinear vehicle dynamics into a linear framework, Koopman theory allows us to build linear models directly from data. This linear structure then makes it possible to design adaptive control strategies that can handle the challenges of off-road navigation with greater reliability and flexibility.

4416 | Prediction of soil rut depth and tire traction forces using Bekker and DEM-FEM techniques

Mehari Tekeste [a,*], Pius Jjagwe [b], Thomas Way [c], Kumaran Periannan [d], Raghuram Thiagarajan [d], Mohamed Abdeldayem [e]

[a] Iowa State University Ames, Iowa, United States

[b] Former Iowa State University, Ames, Iowa, United States

[c] USDA-ARS National Soil Dynamics Laboratory, Auburn, AL, United States

[d] Pratt and Miller Inc., New Hudson, MI, United States

[e] Iowa State University, Ames, Iowa, United States

* Corresponding author: mtekeste@iastate.edu

The study focuses on simulating the mobility systems of vehicles on soft terrains, which necessitates the use of physics-based soil models, tire models, and their interactions. A significant challenge lies in obtaining soil model parameters for empirical traction equations and high-fidelity DEM-FEM techniques, mainly due to the variability in soil initial conditions, such as soil bulk density. The objective is to simulate soil behaviors and traction forces related to pull and motion resistance using reduced mathematical models and DEM-FEM techniques. Bekker's soil and DEM contact soil model parameters were calibrated using soil mechanics testing data on cohesive-frictional artificial soil. Predicted soil rut depth and traction forces from the two modeling techniques were validated against soil bin data from single tire testing using a Yokohama Geolandar radial tire (235/75R15) under two vertical loads (4 kN and 6 kN) and two tire inflation pressures (179 kPa and 241 kPa), with the tire operating in towing mode on dense and loose initial conditions. DEM elasto-plastic soil model and FEM PM-tire model demonstrated better accuracy in predicting soil sinkage compared to the Bekker parameterized soil modeling technique. The Bekker approach provided relatively accurate traction force predictions for free-towing scenarios, particularly in denser soil conditions and with over-inflated tire pressure compared to correctly inflated conditions. The DEM-FEM modeling effectively captured soil behaviors, including compression and bulldozing, in a loose soil state. The comparison of Bekker's reduced physics and high-fidelity DEM-FEM approaches for predicting soil-to-tire interactions offers opportunities to accelerate the automation of mobility systems on deformable terrains.

4970 | L-band microwave radiometry for non-contact moisture measurement of field trafficability

Maik Wolleben [a], Joseph Dvorak [b], Kaleb Dempsey [c,*], Michael Peterson [c,*]

[a] *Skaha Labs, Canada*

[b] *University of Kentucky, United States*

[c] *Rainier Sensing, United States*

* *Corresponding authors: info@rainiersensing.com, mick.peterson@uky.edu*

The moisture content of soil is the single most important source of uncertainty for the trafficability of fields. Because of increasing frequency and intensity of rain events, planting or field equipment operations are more likely to occur as soon as possible after a rain events. For harvesting, pest control, planting or other field operations timeliness is critical for the effectiveness of the work and ultimately the yield produced. The trafficability necessary to complete field work is largely a factor of soil moisture from the surface down to 15 to 30 cm; however, large temporal and spatial variations complicate sensing and deploying machinery in the areas that are sufficiently dry to support fieldwork. This work focuses on high sand content soils: sandy loam, loamy sand and sand. Under controlled conditions L-band Microwave Radiometry was used to measure the moisture content in a non-contact configuration. Initial testing focused on the effect of interference from cellular, emergency and other radio sources of radio frequency interference. A second test was conducted in proximity to general aviation operations. Initial concerns regarding noise from most sources was found to be less of a problem than anticipated. However, at the frequency employed (1.4 GHz), a poor correlation with time domain reflectometry at 3 cm depth was observed, suggesting that the sensing depth of the microwave sensor is deeper than the TDR probe. Higher microwave frequencies may need to be employed to match the critical depth for sandy soils, A number of alternative applications exist for these moisture sensors which may be of value in other applications or surfaces with higher clay content or higher salinity.

6198 | Vehicle impact analysis using remote datasets

Heidi Howard *, Jaime Miranda Ramirez *, Israel Lopez Toledo *

ERDC-CERL, United States

* *Corresponding authors: heidi.r.howard@usace.army.mil, Jaime.L.MirandaRamirez@usace.army.mil, israel.j.lopeztoledo@erdc.dren.mil*

The Army currently has limited capabilities to calculate training impacts for multi-day-vehicle exercises. The ARNG has deployed a program of record for eXportable Combat Training Capability (XCTC) which utilizes a tracking system similar to those used for fleet management. The use of spatial and temporal data from these training events can then be linked to vehicle severity models developed by CERL, CRREL, and other agencies to quantify actual impacts to ground, air, and water. CERL is utilizing data from historic exercises to develop a machine learning capability that could then project out impacts to the terrain from existing and future vehicle platforms. This presentation will discuss lessons learned and accuracy of the model estimations for sinkage and disturbance.

6366 | Simulation-based optimization of a tracked vehicle suspension system

Markus Pogulis [a,*], Joel Nilsson [b], Martin Servin [c,*]

[a] *BAE Systems Hägglunds, Umeå University, Sweden*

[b] *Umeå University, Sweden*

[c] *Algoryx Simulation AB and Umeå University, Sweden*

* *Corresponding authors: markus.pogulis@umu.se, martin.servin@umu.se*

This research considers simulation-based optimization of a tracked vehicle suspension system. We focus on maximizing ride quality across a large spectrum of operational conditions rather than specific terrain/speed combinations, using surface roughness to guide our exploration of diverse scenarios. The vehicle suspension system features 14 torsion rods and dampers subject to optimization. Using multibody dynamics simulation, we evaluate ride quality over diverse terrains at varying speeds, computing whole-body acceleration and vibrational dose values (VDV). We analyze and compare alternative ways of posing and solving the problem of finding a suspension configuration that is optimal for a desired spectrum of terrains. We consider derivative-free optimization methods, including Particle Swarm Optimization (PSO) and Bayesian Optimization, and compare these to first learning and pre-building a surrogate model, prior to optimization. This surrogate strategy relates local terrain characteristics to vehicle performance metrics, such as acceleration-constrained velocity limits. Preliminary work with PSO demonstrated significant ride quality improvements, with VDV reductions of 20-60% compared to baseline configurations, but was limited to optimizing only torsion rods for specific terrain/speed combinations. Our current research extends this by simultaneously optimizing both suspension and damping parameters and implementing an aggregated approach that optimizes performance across multiple terrains and velocities in a single cycle. This work demonstrates how simulation-based optimization techniques can significantly reduce whole-body vibrations in tracked vehicles, improving both operational capabilities and crew endurance in challenging environments.

6752 | Foundational software architectures: the cornerstone of off-road autonomy and multi-mission UGV performance

Matthew Foglesong *

Parry Labs, United States

* *Corresponding author: matt.foglesong@parrylabs.com*

Delivering resilient, off-road autonomy for unmanned ground vehicles (UGVs) will not come from incremental advances in perception, planning, or mobility alone. The decisive enabler is a robust software operating environment—a foundational architecture that integrates autonomy stacks, payloads, and edge compute into a coherent and resilient system-of-systems. Today, much of the DOD's ground robotics development remains platform-centric, with limited ability to reuse or scale autonomy services across missions. Without a modular, standards-aligned software backbone, breakthroughs in autonomy cannot reliably transition to the field. Foundational software architecture must be treated as the cornerstone of UGV development. Open, containerized, and portable environments enable rapid integration of autonomy algorithms, heterogeneous payloads, and resilient communications, while ensuring certification pathways for cyber and safety. The lessons learned are informed by Parry Labs' prior work modernizing the MQ-1C Gray Eagle fleet, where software-first approaches demonstrated how containerization, abstraction, and modular integration could transform a legacy platform into a flexible, multi-mission system. Drawing on these insights and previous experience with the Ground Vehicle Systems Center and Marine Corps Warfighting Lab's previous and ongoing robotics and UGV efforts, we will outline both the risks of not adopting a software-first paradigm and the operational advantages gained when architectures are designed for reuse, adaptability, and contested environments. Ultimately, a foundational software operating environment is the enabler of resilient autonomy, accelerated capability delivery, and true multi-mission relevance for the Army's future ground vehicle fleet.

7256 | Data assimilation analysis of footpad impact behavior of lander on soft ground

Ryuta Kaneda [a,*], Masatsugu Otsuki [b], Shingo Ozaki [c,*]

[a] *Yokohama National University, Japan*

[b] *Japan Aerospace Exploration Agency, Japan*

[c] *Faculty of Engineering, Yokohama National University, Japan*

* *Corresponding authors: kaneda-ryuta-wd@ynu.jp, ozaki-shingo-xd@ynu.ac.jp*

In FY 2026, the Martian Moons eXploration (MMX) is scheduled to be launched. In the MMX mission, the spacecraft will collect a sample from Phobos, a Martian Moon, to bring back to Earth, and the safe landing of the lander is crucial to the success of the mission. However, the MMX mission and other planned landing missions on low-gravity solid bodies have difficulties in evaluating landing performance in advance using ground experiments and simulations due to uncertainties in the characteristics of the target surface layer. In this study, we examined an estimation method for the surface conditions based on the data assimilation analysis using the collision data of an advance landing probe as a precursor before landing on the main lander. Specifically, we attempted to estimate the stiffness and bulk density of the ground for a drop tower experiment using the footpad of MMX lander, by integrating the collision model based on the resistive force theory and the particle filter. In the drop tower experiments, Phobos-simulated regolith was used as the test soil under quasi-vacuum conditions (less than 100 Pa) and microgravity conditions (1/2500 G). The results of the data assimilation analysis of time series of impact reaction force and settlement under various bulk density conditions indicate that the proposed approach could estimate the initial state of the regolith and the change in state after impact to some extent. However, the accuracy of this particle filter-based estimation method depends on the performance of the collision model (physical model), and it is necessary to use the collision model that can adequately represent realistic phenomena.

7470 | Constraint-based terramechanics simulation for realtime and faster simulation

Björn Ahlman [a], Tomas Berglund [a], Mikael Lundbäck [b], Joakim Rydman [b], Martin Servin [c,*]

[a] *Algoryx Simulation AB, Sweden*

[b] *Umeå University, Sweden*

[c] *Algoryx Simulation AB and Umeå University, Sweden*

* *Corresponding author: martin.servin@umu.se*

We present models and numerical methods for realtime and faster simulation of wheeled and tracked vehicles traversing and interacting with deformable terrain. The wheel-terrain interaction is formulated as a set of kinematic constraints with constraint forces and limits that reflect the stresses on the wheel and soil failure at critical stresses. When soil failure occurs, a 3D soil displacement field is predicted and used to update the spatial distribution of soil, local packing density, and the surface heightmap. The constraint-based formulation enables stable and strong coupling between the terrain and vehicle multibody dynamics at large time-steps. Simulation tests are performed where the implemented model is compared to experiments and standard semi-empirical terramechanics models. Finally, we apply the terramechanics model on a model of a heavy forest machine traversing rough terrain and study how the motion and dynamic load forces are affected by inclusion of deformable terrain.

7475 | Validation of experimental terramechanics modelling approaches through physical testing

Eric Karpman [a,*], Jozsef Kovecses [b], Marek Teichmann [c]

[a] *McGill University, Canada*

[b] *McGill, Canada*

[c] *CM Labs Simulations, Montreal, Quebec, Canada, Canada*

* *Corresponding author: eric.karpman@mail.mcgill.ca*

The traditional semi-empirical terramechanics models have been heavily relied on over the past decades. Despite the fact that their numerous assumptions limit their applicability in certain scenarios. In particular, the real forces in transient states can results in behaviour that is not captured by the semi-empirical models. Additionally, the models are formulated with the assumption that the wheel or track is travelling over flat, level terrain. We have previously proposed two separate methods for adapting the models to address these two specific shortcomings. For the former, we proposed a hybrid modelling approach which uses trained neural networks to augment the force predicted by the semi-empirical models. Using this approach, we do not discard the valuable insight that the established models provide, but are still able to capture some dynamic and transient behaviour. For the later shortcoming, we proposed a novel variable terrain height adaptation of the semi-empirical model which computes stresses at each point along the wheel-soil interface by projecting that point in space onto the heightfield that defines the terrain. This is in contrast to other established methods for handling uneven terrain by defining least squares planes which are meant to approximate the terrain geometry, but which in practice can have the effect of smoothing out terrain roughness. In order to validate these two proposed models, an experimental test campaign was carried out in which an instrumented off-road vehicle underwent a drawbar-pull test on soft terrain, and a second test in which the same vehicle was driven over a sand course with an uneven surface profile. The results of these two sets of experiments are used to justify the modelling approaches we have proposed.

7714 | The role of the evanescent slow wave in non-contact moisture measurement of field trafficability

Zhihe Jin [a], Joseph Dvorak [b], Michael Peterson [b,*]

[a] *University of Maine, United States*

[b] *University of Kentucky, United States*

* *Corresponding author: mick.peterson@uky.edu*

The moisture content of soil is the single most important factor for trafficability of fields. Many sensing technologies are limited to surface moisture such as optical techniques or may be sensitive to variation in texture or salinity. Elastic wave methods are more likely to be sensitive to the mechanical properties of the soil and are directly related to the trafficability of the surface. Acoustic methods are also not impacted by the crowding of the electromagnetic spectrum. Previous investigators have considered the use of acoustic sensing for moisture and hardpan depth. However, the transfer of this research into applications has been limited. This work is initially focused on characterizing the underlying physics of waves in porous media at a length scale applicable to sandy soils with a narrow pore size distribution. These soils are well suited to this investigation because of potential importance of the slow wave in for moisture detection. The slow longitudinal wave has been used extensively in other areas such as geophysics and membrane characterization. The pore size in sandy soil would not lead to propagation of a slow wave at practical sensor frequencies. However, an evanescent slow wave will impact the effective acoustic impedance of the soil air interface. The energy transfer to the evanescent wave will depend on the portion of the pore structure that is filled with water instead of air. A model for reflection from a porous interface is formulated for a narrow distribution of pore sizes. Sensitivity of the model to different water content and the effect of frequency on the depth of penetration are considered. Simple experiments with a model soil are used to validate modeling, explore sensor configurations and proposes sensor field deployment.

9257 | Acoustic winter terrain classification for offroad autonomous vehicles

Anthony Fragoso *

ERDC CRREL, United States

* *Corresponding author: anthony.t.fragoso@erdc.dren.mil*

Autonomous vehicles can experience extreme changes in performance when operating over winter surfaces, and require accurate classification to transit them safely. In this work we consider acoustic classification of winter terrain, and demonstrate that a simple and efficient frequency-space analysis exposed to a small convolutional neural network, rather than recurrent architectures or temporally-varying spectrogram inputs, is sufficient to provide near-perfect classification of deep snow, hardpacked surfaces and ice. Using a dual-microphone configuration, we also show that acoustic classification performance is due to a combination of vehicle noises and vehicle-terrain interaction noises, and that engine sounds can serve as a particularly powerful classification cue for offroad environments.

9360 | Optical velocity and slip estimation for off-road UGV control

Glenn Guthrie *, Andries Peenze, P. Schalk Els *

Vehicle Dynamics Group, University of Pretoria, South Africa

* *Corresponding authors: guthrie.glenn@up.ac.za, Schalk.Els@up.ac.za*

In vehicle dynamics, the measurement of accurate vehicle velocity (magnitude and direction) using low-cost sensors is a significant challenge, yet this measurement is extremely valuable as an input for vehicle control systems. Traditional low-cost methods for measuring vehicle velocity often rely on dual antenna GPS algorithms that have reduced accuracy at low speed and GPS reception is often challenging in varied environments. This work applies an optical technique to measure the longitudinal and lateral velocities required to calculate slip using a downward facing depth camera (Intel Realsense) to capture images and a point cloud of the terrain surface. The images and point cloud are processed by a feature tracking algorithm and the vehicle velocity is calculated based on the relative motion of the ground and vehicle body. The technique is applied to a four-wheeled, skid steer, off-road, unmanned ground vehicle (UGV) that frequently encounters low-traction conditions while operating on slippery and uneven terrain. This suspensionless UGV has electric servo motors driving each wheel independently, allowing both measurement and control of wheel speed and torque. The calculated vehicle velocity is combined with wheel speeds from the motors to calculate individual longitudinal wheel slips. The measurement technique's performance is evaluated by using the vehicle velocity and longitudinal wheel slip in a control algorithm that is designed to maintain a desired vehicle attitude even when wheel slip occurs. The slip measurement technique and control algorithm are evaluated by driving the vehicle on low friction and split friction surfaces while recording the desired attitude and actual vehicle attitude.

ISTVS

Author index

Name	Affiliation	Submissions
Abdeldayem, Mohamed	Iowa State University, Ames, Iowa, United States	4416 | Prediction of soil rut depth and tire traction forces using Bekker and DEM-FEM techniques *(abstract only)*
Ahlman, Björn	Algoryx Simulation AB, Sweden	7470 | Constraint-based terramechanics simulation for realtime and faster simulation *(abstract only)*
Alger, Peter	Keweenaw Research Center, United States	6540 | KRC bevameter development and mobility data collection
Alger, Russell	Keweenaw Research Center, United States	0051 | Ground vehicle detection and environment characterization derived from a distributed acoustic sensing system deployed on an Arctic ice road 6540 | KRC bevameter development and mobility data collection
Ålstig, Sandra	Algoryx Simulation AB, Sweden	7222 | A simulation framework for autonomous lunar construction work
Angulo, Juan Pablo	ANGTRAC, Colombia	2154 | Simulation of dynamic behavior in armored vehicles using modified off-road vehicles for civil protection applications *(abstract only)*
Azuma, Ryota	Keio University, Japan	2908 | Drawbar pull experiment and analysis of grousered-track unit on sandy terrain
Becker, Carl M.	Vehicle Dynamics Group, University of Pretoria, South Africa	0270 | Performance assessment of an off-road vehicle across different terrains 2458 | Obtaining first order tyre parameters with insitu testing *(abstract only)* 6905 | Investigating data collection and processing strategies for UAV-based terrain models in vehicle dynamics simulation
Berglund, Tomas	Algoryx Simulation AB, Sweden	0398 | Contact dynamics slope stability *(abstract only)* 7470 | Constraint-based terramechanics simulation for realtime and faster simulation *(abstract only)*
Bhagat, Abhishek	U.S. Army Engineer Research and Development Center Cold Regions Research and Engineering Laboratory, United States	0051 | Ground vehicle detection and environment characterization derived from a distributed acoustic sensing system deployed on an Arctic ice road
Bodie, Mark	Army Corp of Engineers, United States	5658 | Winter route planning algorithm 8342 | Effect of snow subduction on vehicle mobility
Bonello, Bobby	private sector, United States	3985 | One AI driver for multi-mission ground operations *(abstract only)*
Boyle, Sara	U.S. Army Ground Vehicle Systems Center, United States	5493 | Effects of electronic control systems on autonomous vehicle performance in granular terrain

Name	Affiliation	Submissions
Brady, Brian	U.S. Army Engineer Research and Development Center Cold Regions Research and Engineering Laboratory, United States	3559 | Modeling light detection and ranging (LiDAR) response in snowy environments 8969 | Survey of LiDAR effects for autonomous driving simulations
Brown, Rob	Head of Product, United States	3985 | One AI driver for multi-mission ground operations *(abstract only)*
Card, Angela	CAVS, Mississippi State University, United States	5493 | Effects of electronic control systems on autonomous vehicle performance in granular terrain
Chen, Devin	Center for Advanced Vehicular Systems, Mississippi State University, United States	9668 | Innovating mobility: a student competition in wheel and track design
Clayton, Paul	Milrem Robotics, Estonia	0442 | Lessons learned from global UGV and autonomous systems *(abstract only)*
Cole, Michael	U.S. Army Ground Vehicle Systems Center, United States	4342 | Cornering performance of rigid wheel in granular media using coarse-scale DEM models 5493 | Effects of electronic control systems on autonomous vehicle performance in granular terrain
Cox, Devon	Vehicle Dynamics Group, University of Pretoria, South Africa	4335 | Extended brush model for use with gravelly terrain
Crous, Petrus Arnoldus	Department. of Civil Engineering, University of Pretoria, South Africa	8140 | Numerical modeling of plate sinkage and in situ bevameter shear testing using FEM and the density-dependent norsand critical state constitutive soil model
Dawid, Wojciech	Military University of Technology, Poland	0583 | Dynamic off-road route planning: integrating terrain passability and weather for optimal mobility *(abstract only)* 0938 | AI-based terrain passability modeling *(abstract only)* 1536 | Automated system for determining military road and terrain passability based on terrain, meteorological and hydrological conditions *(abstract only)*
Dempsey, Kaleb	Rainier Sensing, United States	4970 | L-band microwave radiometry for non-contact moisture measurement of field trafficability *(abstract only)*
Dickerson, Aidan	Mississippi State University, United States	4342 | Cornering performance of rigid wheel in granular media using coarse-scale DEM models 5493 | Effects of electronic control systems on autonomous vehicle performance in granular terrain

Name	Affiliation	Submissions
Doran, Adrian	U.S. Army Engineer Research and Development Center Cold Regions Research and Engineering Laboratory, United States	0051 | Ground vehicle detection and environment characterization derived from a distributed acoustic sensing system deployed on an Arctic ice road
Driggs-Campbell, Katherine	University of Illinois at Urbana-Champaign, United States	4558 | In-situ soil-property estimation and Bayesian mapping with a simulated compact track loader
Dvorak, Joseph	University of Kentucky, United States	4970 | L-band microwave radiometry for non-contact moisture measurement of field trafficability *(abstract only)*
		7714 | The role of the evanescent slow wave in non-contact moisture measurement of field trafficability *(abstract only)*
Elder, Bruce	ERDC CRREL, United States	4755 | Relationship between seasonal terrain properties and vehicle rut depths
		7265 | Evaluation of a new friction measurement device on Arctic terrain
Els, P. Schalk	Vehicle Dynamics Group, University of Pretoria, South Africa	0270 | Performance assessment of an off-road vehicle across different terrains
		0954 | Enhancing combined slip manoeuvres in high centre-of-mass off-road vehicles using a semi-active suspension system
		2458 | Obtaining first order tyre parameters with insitu testing *(abstract only)*
		4335 | Extended brush model for use with gravelly terrain
		6905 | Investigating data collection and processing strategies for UAV-based terrain models in vehicle dynamics simulation
		7559 | Dynamic contact patch estimation from 3D contact patch measurements inside a moving tyre under combined slip conditions
		9360 | Optical velocity and slip estimation for off-road UGV control *(abstract only)*
		9668 | Innovating mobility: a student competition in wheel and track design
El-Sayegh, Zeinab	Ontario Tech University, Canada	4579 | Analysis of passenger tire-salted pavement interaction: a traction perspective
Endo, Haruki	Yokohama National University, Japan	8964 | Application of LSTM networks to prediction of a wheel traveling phenomena
Eto, Ryosuke	National Defense Academy of Japan, Japan	2824 | Local path planning on rough terrain for unmanned ground vehicle by reinforcement learning considering search reward
Eylander, John B.	ERDC-CHL, United States	8352 | Assessing the strength of river sediments for vehicle fording

Name	Affiliation	Submissions
Fairley, Joshua	U.S. Army Engineer Research and Development Center, United States	1868 | Modeling and simulation of autonomous vehicular systems at ERDC - III Verification and validation of M&S for autonomous vehicles 6001 | Modeling and simulation of autonomous vehicular systems at ERDC - II Soft soil model support in vehicle dynamics 6775 | Modeling and simulation of autonomous vehicular systems at ERDC - I Software integration laboratory: development and applications
Farkas, Anthony	Hodges Transportation, DBA Nevada Automotive Test Center, United States	1056 | Optimizing vehicle navigation over terrain *(abstract only)*
Fenton, Erin	Ontario Tech University, Canada	4579 | Analysis of passenger tire-salted pavement interaction: a traction perspective
Figueroa-Santos, Miriam	U.S. Army Ground Vehicle Systems Center, United States	5493 | Effects of electronic control systems on autonomous vehicle performance in granular terrain
Foglesong, Matthew	Parry Labs, United States	6752 | Foundational software architectures: the cornerstone of off-road autonomy and multi-mission UGV performance *(abstract only)*
Fragoso, Anthony	ERDC CRREL, United States	9257 | Acoustic winter terrain classification for offroad autonomous vehicles *(abstract only)*
Frankenstein, Susan	Engineer Research and Development Center Cold Regions Research and Engineering Laboratory, United States	2264 | Mobility testing on frost susceptible soils at field capacity 4755 | Relationship between seasonal terrain properties and vehicle rut depths
Fujioka, Natsu	JAXA, Japan	7125 | A data-driven parameter identification method of characterization of wheel-regolith interaction for lunar surface vehicles
Fujisawa, Hiroto	Suwa University of Science, Japan	4368 | Improving the expanding/contracting speed of a push-rolling rover on loose surfaces with steep slopes
Fujiwara, Daisuke	Suwa University of Science, Japan	4368 | Improving the expanding/contracting speed of a push-rolling rover on loose surfaces with steep slopes
Galvagno, Enrico	Department of Mechanical and Aeronautical Engineering, Politecnico di Torino, Torino, Italy	0270 | Performance assessment of an off-road vehicle across different terrains

Name	Affiliation	Submissions
Gates, Burhman	U.S. Army Engineer Research and Development Center, United States	1868 | Modeling and simulation of autonomous vehicular systems at ERDC - III Verification and validation of M&S for autonomous vehicles 6001 | Modeling and simulation of autonomous vehicular systems at ERDC - II Soft soil model support in vehicle dynamics 6775 | Modeling and simulation of autonomous vehicular systems at ERDC - I Software integration laboratory: development and applications
Gibson, C. Michael	CAVS, Mississippi State University, United States	5493 | Effects of electronic control systems on autonomous vehicle performance in granular terrain
Gorsich, David	U.S. Army DEVCOM Ground Vehicle Systems Center, United States	0375 | A general model to approximate the slippage curve for applications from analysis of the tire and soil contributions to longitudinal slippage
Guthrie, Glenn	Vehicle Dynamics Group, University of Pretoria, South Africa	6905 | Investigating data collection and processing strategies for UAV-based terrain models in vehicle dynamics simulation 9360 | Optical velocity and slip estimation for off-road UGV control *(abstract only)*
Halliday, Don	Halliday Technologies, United States	7265 | Evaluation of a new friction measurement device on Arctic terrain
Hamersma, Herman	Vehicle Dynamics Group, University of Pretoria, South Africa	6905 | Investigating data collection and processing strategies for UAV-based terrain models in vehicle dynamics simulation
Hannis, Tyler	CAVS, Mississippi State University, United States	5493 | Effects of electronic control systems on autonomous vehicle performance in granular terrain
Hasenbichler, Gernot	AVL List GmbH, Austria	2469 | TERRA-V toolchain: virtual validation and optimization of autonomous vehicles in unpaved terrain
Hassler, Daniel	AVL List GmbH, Austria	2469 | TERRA-V toolchain: virtual validation and optimization of autonomous vehicles in unpaved terrain
He, Yinglong	Centre for Automotive Engineering, School of Engineering, University of Surrey, United Kingdom	4217 | A computationally efficient plug-in for high-fidelity off-road vehicle dynamics studies on deformable terrains
Henley, Greg	CAVS, Mississippi State University, United States	5493 | Effects of electronic control systems on autonomous vehicle performance in granular terrain
Hewlett, Joe	CM Labs Simulations, Montreal, Quebec, Canada, Canada	3375 | Real-time simulation of flexible tracks for accurate vehicle-terrain interaction *(abstract only)*
Ho, Wing Hang	McGill University, Canada	3375 | Real-time simulation of flexible tracks for accurate vehicle-terrain interaction *(abstract only)*

Name	Affiliation	Submissions
Hodges, Henry	Hodges Transportation, DBA Nevada Automotive Test Center, United States	1056 | Optimizing vehicle navigation over terrain *(abstract only)*
Holden, Garrett	U.S. Army Engineer Research and Development Center, United States	1868 | Modeling and simulation of autonomous vehicular systems at ERDC - III Verification and validation of M&S for autonomous vehicles 6001 | Modeling and simulation of autonomous vehicular systems at ERDC - II Soft soil model support in vehicle dynamics 6775 | Modeling and simulation of autonomous vehicular systems at ERDC - I Software integration laboratory: development and applications
Horie, Kosuke	Komatsu Ltd., Japan	2908 | Drawbar pull experiment and analysis of grousered-track unit on sandy terrain
Hoshino, Tasuku	Suwa University of Science, Japan	4368 | Improving the expanding/contracting speed of a push-rolling rover on loose surfaces with steep slopes
Howard, Heidi	ERDC-CERL, United States	6198 | Vehicle impact analysis using remote datasets *(abstract only)*
Huang, Wei	National Research Council Canada, Canada	0051 | Ground vehicle detection and environment characterization derived from a distributed acoustic sensing system deployed on an Arctic ice road 9563 | Physical testing and digital twin simulations for prediction of ride quality of light utility vehicle
Hurrell, James	Tohoku University, Japan	6650 | Reliability of gravity offload for lunar rover wheel performance
Iizuka, Kojiro	Shibaura Institute of Technology, Japan	4368 | Improving the expanding/contracting speed of a push-rolling rover on loose surfaces with steep slopes 9165 | Simulation of the relationship between the decreased bearing capacity while imparting vibration to ground and vibration frequency in the discrete element method
Ishigami, Genya	Keio University, Japan	2908 | Drawbar pull experiment and analysis of grousered-track unit on sandy terrain
Jelinek, Bohumir	CAVS, Mississippi State University, United States	4342 | Cornering performance of rigid wheel in granular media using coarse-scale DEM models 5493 | Effects of electronic control systems on autonomous vehicle performance in granular terrain 9668 | Innovating mobility: a student competition in wheel and track design

Name	Affiliation	Submissions
Jin, Zhihe	University of Maine, United States	7714 | The role of the evanescent slow wave in non-contact moisture measurement of field trafficability *(abstract only)*
Jjagwe, Pius	Former Iowa State University, Ames, Iowa, United States	4416 | Prediction of soil rut depth and tire traction forces using Bekker and DEM-FEM techniques *(abstract only)*
Kamprath, Nathan	GVSC, United States	7265 | Evaluation of a new friction measurement device on Arctic terrain
Kaneda, Ryuta	Yokohama National University, Japan	7256 | Data assimilation analysis of footpad impact behavior of lander on soft ground *(abstract only)*
Kaniarz, John	U.S. Army DEVCOM Ground Vehicle System Center, United States	8969 | Survey of LiDAR effects for autonomous driving simulations
Kapareliotis, Christos	Vehicle Dynamics Group, University of Pretoria, South Africa	0954 | Enhancing combined slip manoeuvres in high centre-of-mass off-road vehicles using a semi-active suspension system
Karpman, Eric	McGill University, Canada	7475 | Validation of experimental terramechanics modelling approaches through physical testing *(abstract only)*
		9563 | Physical testing and digital twin simulations for prediction of ride quality of light utility vehicle
Kawamura, Hiroaki	Toyota Motor Corporation, Japan	7125 | A data-driven parameter identification method of characterization of wheel-regolith interaction for lunar surface vehicles
Kehoe, Patrick	National Research Council Canada, Canada	9563 | Physical testing and digital twin simulations for prediction of ride quality of light utility vehicle
Khosravian, Amir	Centre for Automotive Engineering, School of Engineering, University of Surrey, United Kingdom	4217 | A computationally efficient plug-in for high-fidelity off-road vehicle dynamics studies on deformable terrains
Kikuta, Riku	Center for Advanced Vehicular Systems, Mississippi State University, United States	9668 | Innovating mobility: a student competition in wheel and track design
Kimball, Matthew	U.S. Army Engineer Research and Development Center Cold Regions Research and Engineering Laboratory, United States	0051 | Ground vehicle detection and environment characterization derived from a distributed acoustic sensing system deployed on an Arctic ice road
Kiss, Peter	Hungarian University of Agriculture and Life Sciences, Hungary	6965 | Correlation between optical reflectance and load-bearing capacity of soils

Name	Affiliation	Submissions
Kovecses, Jozsef	McGill, Canada	1960 | Efficient modelling of the effect of tire deformation during contact with obstacles *(abstract only)* 3375 | Real-time simulation of flexible tracks for accurate vehicle-terrain interaction *(abstract only)* 7475 | Validation of experimental terramechanics modelling approaches through physical testing *(abstract only)*
Kruger, Ray	Department of Mechanical Engineering, Iowa State University, United States	8140 | Numerical modeling of plate sinkage and in situ bevameter shear testing using FEM and the density-dependent norsand critical state constitutive soil model
Kumar, Giri M	Virginia Tech, United States	0280 | Real-time ground segmentation and terrain topography mapping using point clouds *(abstract only)*
L'Afflitto, Andrea	Virginia Tech, United States	0280 | Real-time ground segmentation and terrain topography mapping using point clouds *(abstract only)*
Letcher, Theodore W.	ERDC-CRREL, United States	8352 | Assessing the strength of river sediments for vehicle fording
Linde, Mattias	Algoryx Simulation AB, Sweden	7222 | A simulation framework for autonomous lunar construction work
Lindmark, Daniel	Algoryx Simulation AB, Sweden	7222 | A simulation framework for autonomous lunar construction work
Lopera, Jorge	PACIFIC+, Colombia	2154 | Simulation of dynamic behavior in armored vehicles using modified off-road vehicles for civil protection applications *(abstract only)*
Lopez Toledo, Israel	ERDC-CERL, United States	6198 | Vehicle impact analysis using remote datasets *(abstract only)*
Lundbäck, Mikael	Umeå University, Sweden	7470 | Constraint-based terramechanics simulation for realtime and faster simulation *(abstract only)*
Madrazo, Cooper	Hodges Transportation, DBA Nevada Automotive Test Center, United States	1056 | Optimizing vehicle navigation over terrain *(abstract only)*
Maeda, Tatsuya	Komatsu Ltd., Japan	2908 | Drawbar pull experiment and analysis of grousered-track unit on sandy terrain

Name	Affiliation	Submissions
Maleki, Mahdi	McGill University, Canada	1960 | Efficient modelling of the effect of tire deformation during contact with obstacles *(abstract only)* 3375 | Real-time simulation of flexible tracks for accurate vehicle-terrain interaction *(abstract only)*
Mange, Jeremy	U.S. Army Ground Vehicle Systems Center, United States	5493 | Effects of electronic control systems on autonomous vehicle performance in granular terrain
Mason, George L.	CAVS, Mississippi State University, United States	4342 | Cornering performance of rigid wheel in granular media using coarse-scale DEM models 5493 | Effects of electronic control systems on autonomous vehicle performance in granular terrain 9668 | Innovating mobility: a student competition in wheel and track design
Matsuura, Tadashi	Toyota Motor Corporation, Japan	7125 | A data-driven parameter identification method of characterization of wheel-regolith interaction for lunar surface vehicles
Meurer, Stephen	ATEC- ARTC, United States	7265 | Evaluation of a new friction measurement device on Arctic terrain
Meyer, Aaron	U.S. Army Engineer Research and Development Center Cold Regions Research and Engineering Laboratory, United States	0051 | Ground vehicle detection and environment characterization derived from a distributed acoustic sensing system deployed on an Arctic ice road 3559 | Modeling light detection and ranging (LiDAR) response in snowy environments
Miranda Ramirez, Jaime	ERDC-CERL, United States	6198 | Vehicle impact analysis using remote datasets *(abstract only)*
Miyai, Shinichiro	Komatsu Ltd., Japan	2908 | Drawbar pull experiment and analysis of grousered-track unit on sandy terrain
Moradi, Lee	Worcester Polytechnic Institute, United States	0375 | A general model to approximate the slippage curve for applications from analysis of the tire and soil contributions to longitudinal slippage
Mukherjee, Jyotirmoy	Virginia Tech, United States	0280 | Real-time ground segmentation and terrain topography mapping using point clouds *(abstract only)*
Muryokoji, Shunsuke	Yokohama National University, Japan	4122 | DEM analysis of effect of interparticle forces on sand-outflow behavior from hopper under low-gravity environment
Nguyen, Han	U.S. Army Engineer Research and Development Center Cold Regions Research and Engineering Laboratory, United States	3559 | Modeling light detection and ranging (LiDAR) response in snowy environments

Name	Affiliation	Submissions
Nilsson, Joel	Umeå University, Sweden	6366 | Simulation-based optimization of a tracked vehicle suspension system *(abstract only)*
Noda, Shinichiro	Toyota Motor Corporation, Japan	7125 | A data-driven parameter identification method of characterization of wheel-regolith interaction for lunar surface vehicles
Nowakowski, Marek	WITPIS, Poland	0733 | Impact of operational environment on mobility in unmanned mode of wheeled UGVs *(abstract only)*
Otsuki, Masatsugu	Japan Aerospace Exploration Agency, Japan	7256 | Data assimilation analysis of footpad impact behavior of lander on soft ground *(abstract only)*
Ozaki, Shingo	Faculty of Engineering, Yokohama National University, Japan	1171 | Effect of stress distribution model in wheel traveling analysis using the extended terramechanics model *(abstract only)*
		4122 | DEM analysis of effect of interparticle forces on sand-outflow behavior from hopper under low-gravity environment
		7256 | Data assimilation analysis of footpad impact behavior of lander on soft ground *(abstract only)*
		8964 | Application of LSTM networks to prediction of a wheel traveling phenomena
Paldan, Jesse	Worcester Polytechnic Institute, United States	0375 | A general model to approximate the slippage curve for applications from analysis of the tire and soil contributions to longitudinal slippage
Parker, Michael	Engineer Research and Development Center Cold Regions Research and Engineering Laboratory, United States	2264 | Mobility testing on frost susceptible soils at field capacity
		3559 | Modeling light detection and ranging (LiDAR) response in snowy environments
		5658 | Winter route planning algorithm
		8342 | Effect of snow subduction on vehicle mobility
		8969 | Survey of LiDAR effects for autonomous driving simulations
Parker, Owen	ERDC-CRREL, United States	7265 | Evaluation of a new friction measurement device on Arctic terrain
Patel, Amal	Vehicle Dynamics Group, University of Pretoria, South Africa	7559 | Dynamic contact patch estimation from 3D contact patch measurements inside a moving tyre under combined slip conditions

Name	Affiliation	Submissions
Peenze, Andries	Vehicle Dynamics Group, University of Pretoria, South Africa	0270 | Performance assessment of an off-road vehicle across different terrains 0954 | Enhancing combined slip manoeuvres in high centre-of-mass off-road vehicles using a semi-active suspension system 9360 | Optical velocity and slip estimation for off-road UGV control *(abstract only)* 9668 | Innovating mobility: a student competition in wheel and track design
Periannan, Kumaran	Pratt and Miller Inc., New Hudson, MI, United States	4416 | Prediction of soil rut depth and tire traction forces using Bekker and DEM-FEM techniques *(abstract only)*
Peterson, Michael	University of Kentucky, United States	4970 | L-band microwave radiometry for non-contact moisture measurement of field trafficability *(abstract only)* 7714 | The role of the evanescent slow wave in non-contact moisture measurement of field trafficability *(abstract only)*
Picucci, Jennifer	Geotechnical and Structures Laboratory, U.S. Army Corps of Engineers Research and Development Center, Vicksburg, MS, United States	0051 | Ground vehicle detection and environment characterization derived from a distributed acoustic sensing system deployed on an Arctic ice road
Pillinger, György	Hungarian University of Agriculture and Life Sciences, Hungary	6965 | Correlation between optical reflectance and load-bearing capacity of soils
Pogulis, Markus	BAE Systems Hägglunds, Umeå University, Sweden	6366 | Simulation-based optimization of a tracked vehicle suspension system *(abstract only)*
Pokonieczny, Krzysztof	Military University of Technology, Poland	0583 | Dynamic off-road route planning: integrating terrain passability and weather for optimal mobility *(abstract only)* 0938 | AI-based terrain passability modeling *(abstract only)* 1536 | Automated system for determining military road and terrain passability based on terrain, meteorological and hydrological conditions *(abstract only)*
Price, Phillip	U.S. Army Engineer Research and Development Center Geotechnical and Structures Laboratory, United States	8969 | Survey of LiDAR effects for autonomous driving simulations
Priddy, Jody D.	U.S. Army Engineer Research and Development Center, United States	4342 | Cornering performance of rigid wheel in granular media using coarse-scale DEM models 5493 | Effects of electronic control systems on autonomous vehicle performance in granular terrain

Name	Affiliation	Submissions
Quinn, Meghan	U.S. Army Engineer Research and Development Center Cold Regions Research and Engineering Laboratory, United States	0051 | Ground vehicle detection and environment characterization derived from a distributed acoustic sensing system deployed on an Arctic ice road
Rydman, Joakim	Algoryx Simulation AB, Sweden	0398 | Contact dynamics slope stability *(abstract only)* 7470 | Constraint-based terramechanics simulation for realtime and faster simulation *(abstract only)*
Salmon, John Ethan	CAVS, Mississippi State University, United States	5493 | Effects of electronic control systems on autonomous vehicle performance in granular terrain 9668 | Innovating mobility: a student competition in wheel and track design
Sandu, Corina	Virginia Tech, United States	0280 | Real-time ground segmentation and terrain topography mapping using point clouds *(abstract only)* 4217 | A computationally efficient plug-in for high-fidelity off-road vehicle dynamics studies on deformable terrains 9668 | Innovating mobility: a student competition in wheel and track design
Servin, Martin	Algoryx Simulation AB and Umeå University, Sweden	0398 | Contact dynamics slope stability *(abstract only)* 6366 | Simulation-based optimization of a tracked vehicle suspension system *(abstract only)* 7222 | A simulation framework for autonomous lunar construction work 7470 | Constraint-based terramechanics simulation for realtime and faster simulation *(abstract only)*
Shepherd, Patrick	Milrem Robotics, Finland	0442 | Lessons learned from global UGV and autonomous systems *(abstract only)*
Shimizu, Mai	Yokohama National University, Japan	1171 | Effect of stress distribution model in wheel traveling analysis using the extended terramechanics model *(abstract only)*
Shintani, Kohei	Toyota Motor Corporation, Japan	7125 | A data-driven parameter identification method of characterization of wheel-regolith interaction for lunar surface vehicles
Shoop, Sally	ERDC-CRREL, United States	7265 | Evaluation of a new friction measurement device on Arctic terrain 8352 | Assessing the strength of river sediments for vehicle fording
Skoff, Gerhard	AVL List GmbH, Austria	2469 | TERRA-V toolchain: virtual validation and optimization of autonomous vehicles in unpaved terrain

Name	Affiliation	Submissions
Skorupa, Thomas	U.S. Army Ground Vehicle Systems Center, United States	4342 | Cornering performance of rigid wheel in granular media using coarse-scale DEM models 5493 | Effects of electronic control systems on autonomous vehicle performance in granular terrain
Sonalkar, Chaitanya Shekhar	TMVS Laboratory, Virginia Tech, United States	9668 | Innovating mobility: a student competition in wheel and track design
Song, Peilin	GSL/ERDC, United States	1868 | Modeling and simulation of autonomous vehicular systems at ERDC - III Verification and validation of M&S for autonomous vehicles 6001 | Modeling and simulation of autonomous vehicular systems at ERDC - II Soft soil model support in vehicle dynamics 6775 | Modeling and simulation of autonomous vehicular systems at ERDC - I Software integration laboratory: development and applications
Soylemezoglu, Ahmet	US. Army Construction Engineering Research Laboratory, United States	4558 | In-situ soil-property estimation and Bayesian mapping with a simulated compact track loader
Staebell, Kathleen A.	ERDC-Coastal and Hydraulics Laboratory, United States	8352 | Assessing the strength of river sediments for vehicle fording
Sugimura, Shunsuke	Komatsu Ltd., Japan	2908 | Drawbar pull experiment and analysis of grousered-track unit on sandy terrain
Swamy, Varsha S	TMVS Laboratory, Virginia Tech, United States	9668 | Innovating mobility: a student competition in wheel and track design
Syed, Zamir	Singularity Solutions, United States	6010 | Validation of mechanistic DEM calibration
Taguchi, Fushiki	National Defense Academy of Japan, Japan	2824 | Local path planning on rough terrain for unmanned ground vehicle by reinforcement learning considering search reward
Takehana, Keisuke	Tohoku University, Japan	6650 | Reliability of gravity offload for lunar rover wheel performance
Tanishima, Nobutaka	JAXA, Japan	7125 | A data-driven parameter identification method of characterization of wheel-regolith interaction for lunar surface vehicles
Tavernini, Davide	Centre for Automotive Engineering, School of Engineering, University of Surrey, United Kingdom	4217 | A computationally efficient plug-in for high-fidelity off-road vehicle dynamics studies on deformable terrains
Teichmann, Marek	CM Labs Simulations, Montreal, Quebec, Canada, Canada	3375 | Real-time simulation of flexible tracks for accurate vehicle-terrain interaction *(abstract only)* 7475 | Validation of experimental terramechanics modelling approaches through physical testing *(abstract only)*

Name	Affiliation	Submissions
Tekeste, Mehari	Iowa State University Ames, Iowa, United States	4416 | Prediction of soil rut depth and tire traction forces using Bekker and DEM-FEM techniques *(abstract only)* 8140 | Numerical modeling of plate sinkage and in situ bevameter shear testing using FEM and the density-dependent norsand critical state constitutive soil model
Thiagarajan, Raghuram	Pratt and Miller Inc., New Hudson, MI, United States	4416 | Prediction of soil rut depth and tire traction forces using Bekker and DEM-FEM techniques *(abstract only)*
Tõnutare, Tõnu	Chair of Soil Science, Estonian University of Life Sciences, Estonia	4417 | The status of soil physical, mechanical, and chemical properties two years after deforestation in a military training area
Towne, Brent	Mobility System Branch, Engineer Research and Development Center, Vicksburg, MS, United States	6001 | Modeling and simulation of autonomous vehicular systems at ERDC - II Soft soil model support in vehicle dynamics 6775 | Modeling and simulation of autonomous vehicular systems at ERDC - I Software integration laboratory: development and applications
Trinidad Gonzalez, Yuderka	Department. of Civil Engineering, Iowa State University, United States	8140 | Numerical modeling of plate sinkage and in situ bevameter shear testing using FEM and the density-dependent norsand critical state constitutive soil model
Uno, Kentaro	Tohoku University, Japan	6650 | Reliability of gravity offload for lunar rover wheel performance
Vaidya, Umesh	Clemson University, United States	4111 | Off-road autonomy using Koopman theory *(abstract only)*
van Aswegen, Christian	Vehicle Dynamics Group, University of Pretoria, South Africa	6905 | Investigating data collection and processing strategies for UAV-based terrain models in vehicle dynamics simulation
Vantsevich, Vladimir	Worcester Polytechnic Institute, United States	0375 | A general model to approximate the slippage curve for applications from analysis of the tire and soil contributions to longitudinal slippage
Vecherin, Sergey	U.S. Army Engineer Research and Development Center Cold Regions Research and Engineering Laboratory, United States	3559 | Modeling light detection and ranging (LiDAR) response in snowy environments
Vennik, Kersti	Estonian Military Academy, Estonia	4417 | The status of soil physical, mechanical, and chemical properties two years after deforestation in a military training area
Wagner, W. Jacob	ERDC-CERL, United States	4558 | In-situ soil-property estimation and Bayesian mapping with a simulated compact track loader

Name	Affiliation	Submissions
Walizer, Laura	U.S. Army Engineer Research and Development Center Geotechnical and Structures Laboratory, United States	8969 | Survey of LiDAR effects for autonomous driving simulations
Watanabe, Dai	Shibaura Institute of Technology, Japan	9165 | Simulation of the relationship between the decreased bearing capacity while imparting vibration to ground and vibration frequency in the discrete element method
Watanabe, Tomohiro	Niigata University, Japan	9165 | Simulation of the relationship between the decreased bearing capacity while imparting vibration to ground and vibration frequency in the discrete element method
Way, Thomas	USDA-ARS National Soil Dynamics Laboratory, Auburn, AL, United States	4416 | Prediction of soil rut depth and tire traction forces using Bekker and DEM-FEM techniques *(abstract only)*
Wheeler, Allan	Engineer Research and Development Center Cold Regions Research and Engineering Laboratory, United States	2264 | Mobility testing on frost susceptible soils at field capacity 4755 | Relationship between seasonal terrain properties and vehicle rut depths
White, Hannah	Virginia Tech, United States	0280 | Real-time ground segmentation and terrain topography mapping using point clouds *(abstract only)*
Whitson, Jordan	U.S. Army DEVCOM Ground Vehicle Systems Center, United States	0375 | A general model to approximate the slippage curve for applications from analysis of the tire and soil contributions to longitudinal slippage
Wilson, Keith	U.S. Army Engineer Research and Development Center Cold Regions Research and Engineering Laboratory, United States	0051 | Ground vehicle detection and environment characterization derived from a distributed acoustic sensing system deployed on an Arctic ice road
Witte, Clifford	Engineer Research and Development Center Cold Regions Research and Engineering Laboratory, United States	2264 | Mobility testing on frost susceptible soils at field capacity
Wolleben, Maik	Skaha Labs, Canada	4970 | L-band microwave radiometry for non-contact moisture measurement of field trafficability *(abstract only)*
Wyszyński, Marek	Military University of Technology, Poland	0583 | Dynamic off-road route planning: integrating terrain passability and weather for optimal mobility *(abstract only)* 0938 | AI-based terrain passability modeling *(abstract only)* 1536 | Automated system for determining military road and terrain passability based on terrain, meteorological and hydrological conditions *(abstract only)*
Xing, Kai	Yokohama National University, China	8964 | Application of LSTM networks to prediction of a wheel traveling phenomena

Name	Affiliation	Submissions
Yamakawa, Junya	National Defense Academy of Japan, Japan	2824 | Local path planning on rough terrain for unmanned ground vehicle by reinforcement learning considering search reward
Yarmohammadisatri, Sadegh	Department of Mechanical Engineering, Virginia Tech, United States	4217 | A computationally efficient plug-in for high-fidelity off-road vehicle dynamics studies on deformable terrains
Yoshida, Kazuya	Tohoku University, Japan	6650 | Reliability of gravity offload for lunar rover wheel performance
Yovchev, Nikita	National Research Council Canada, Canada	9563 | Physical testing and digital twin simulations for prediction of ride quality of light utility vehicle
Zerbato, Luca	Department of Mechanical and Aeronautical Engineering, Politecnico di Torino, Torino, Italy	0270 | Performance assessment of an off-road vehicle across different terrains

International Society for Terrain-Vehicle Systems
www.istvs.org

ISBN 979-8-3313-3088-0

CURRAN ASSOCIATES INC.
proceedings.com

7th Asia Pacific International Institute of Welding International Congress 2013

Recent Development in Welding and Joining Technologies

Singapore
8 – 10 July 2013